Evolution and Palaeobiology of Pterosaurs

Special Publication reviewing procedures

The Society makes every effort to ensure that the scientific and production quality of its books matches that of its journals. Since 1997, all book proposals have been refereed by specialist reviewers as well as by the Society's Books Editorial Committee. If the referees identify weaknesses in the proposal, these must be addressed before the proposal is accepted.

Once the book is accepted, the Society has a team of Book Editors (listed above) who ensure that the volume editors follow strict guidelines on refereeing and quality control. We insist that individual papers can only be accepted after satisfactory review by two independent referees. The questions on the review forms are similar to those for *Journal of the Geological Society.* The referees' forms and comments must be available to the Society's Book Editors on request.

Although many of the books result from meetings, the editors are expected to commission papers that were not presented at the meeting to ensure that the book provides a balanced coverage of the subject. Being accepted for presentation at the meeting does not guarantee inclusion in the book.

Geological Society Special Publications are included in the ISI Index of Scientific Book Contents, but they do not have an impact factor, the latter being applicable only to journals.

More information about submitting a proposal and producing a Special Publication can be found on the Society's web site: www.geolsoc.org.uk.

It is recommended that reference to all or part of this book should be made in one of the following ways:

BUFFETAUT, E. & MAZIN, J-M. (eds) 2003. *Evolution and Palaeobiology of Pterosaurs.* Geological Society, London, Special Publications, **217**.

FREY, E., MARTILL, D.M. & BUCHY, M-C. 2003. A new crested ornithocheirid from the Lower Cretaceceous of northeastern Brazil and the unusual death of an unusual pterosaur. *In:* BUFFETAUT, E. & MAZIN, J-M. (eds) *Evolution and Palaeobiology of Pterosaurs*. Geological Society, London, Special Publications, **217**, 55–63.

GEOLOGICAL SOCIETY SPECIAL PUBLICATION NO. 217

Evolution and Palaeobiology of Pterosaurs

EDITED BY

E. BUFFETAUT

Centre National de la Recherche Scientifique, Paris, France

J-M. MAZIN

Centre National de la Recherche Scientifique, Poitiers, France

2003

Published by

The Geological Society

London

THE GEOLOGICAL SOCIETY

The Geological Society of London (GSL) was founded in 1807. It is the oldest national geological society in the world and the largest in Europe. It was incorporated under Royal Charter in 1825 and is Registered Charity 210161.

The Society is the UK national learned and professional society for geology with a worldwide Fellowship (FGS) of 9000. The Society has the power to confer Chartered status on suitably qualified Fellows, and about 2000 of the Fellowship carry the title (CGeol). Chartered Geologists may also obtain the equivalent European title, European Geologist (EurGeol). One fifth of the Society's fellowship resides outside the UK. To find out more about the Society, log on to www.geolsoc.org.uk.

The Geological Society Publishing House (Bath, UK) produces the Society's international journals and books, and acts as European distributor for selected publications of the American Association of Petroleum Geologists (AAPG), the American Geological Institute (AGI), the Indonesian Petroleum Association (IPA), the Geological Society of America (GSA), the Society for Sedimentary Geology (SEPM) and the Geologists' Association (GA). Joint marketing agreements ensure that GSL Fellows may purchase these societies' publications at a discount. The Society's online bookshop (accessible from www.geolsoc.org.uk) offers secure book purchasing with your credit or debit card.

To find out about joining the Society and benefiting from substantial discounts on publications of GSL and other societies worldwide, consult www.geolsoc.org.uk, or contact the Fellowship Department at: The Geological Society, Burlington House, Piccadilly, London W1J 0BG: Tel. +44 (0)20 7434 9944; Fax +44 (0)20 7439 8975; E-mail: enquiries@geolsoc.org.uk.

For information about the Society's meetings, consult *Events* on www.geolsoc.org.uk. To find out more about the Society's Corporate Affiliates Scheme, write to enquiries@geolsoc.org.uk.

Published by The Geological Society from:
The Geological Society Publishing House
Unit 7, Brassmill Enterprise Centre
Brassmill Lane
Bath BA1 3JN, UK
(*Orders*: Tel. +44 (0)1225 445046
Fax +44 (0)1225 442836)

Online bookshop: http://bookshop.geolsoc.org.uk

British Library Cataloguing in Publication Data
A catalogue record for this book is available from the British Library.

ISBN 1-86239-143-2

Typeset by Servis Filmsetting Limited, Manchester, UK

Printed by Cromwell Press, Trowbridge, UK.

Distributors

USA
AAPG Bookstore
PO Box 979
Tulsa
OK 74101-0979
USA
Orders: Tel. +1 918 584-2555
Fax +1 918 560-2652
E-mail bookstore@aapg.org

India
Affiliated East-West Press PVT Ltd
G-1/16 Ansari Road, Daryaganj,
New Delhi 110 002
India
Orders: Tel. +91 11 327-9113
Fax +91 11 326-0538
E-mail affiliat@nda.vsnl.net.in

Japan
Kanda Book Trading Company
Cityhouse Tama 204
Tsurumaki 1-3-10
Tama-shi, Tokyo 206-0034
Japan
Orders: Tel. +81 (0)423 57-7650
Fax +81 (0)423 57-7651
E-mail geokanda@ma.kcom.ne.jp

Contents

Evolution and palaeobiology of pterosaurs

ERIC BUFFETAUT[1] & JEAN-MICHEL MAZIN[2]

[1]*Centre National de la Recherche Scientifique, 16 cour du Liégat, 75013 Paris, France*
[2]*Centre National de la Recherche Scientifique, UMR 6046, Laboratoire de Géobiologie, Biochronologie et Paléontologie humaine, Université de Poitiers, 40 avenue du Recteur Pineau, 86022 Poitiers, France*

The first scientific description of a pterosaur was published in 1784 by Cosimo Alessandro Collini, a former secretary of Voltaire and at that time curator of the natural history cabinet of Karl Theodor, Elector of Palatinate and Bavaria. The specimen came from one of the main sources of such fossils, the Late Jurassic lithographic limestones of northern Bavaria, and Collini, after much deliberation, interpreted it as the skeleton of an unknown marine creature (Collini 1784).

In 1801, Georges Cuvier, on the basis of Collini's description and figure, identified the mysterious animal as a flying reptile (Cuvier 1801), for which he later coined the name 'Ptero-Dactyle' (Cuvier 1809). Cuvier's basically correct interpretation of the 'winged finger' marked the beginning of the study of pterosaurs as an extinct group of flying reptiles.

In the two centuries which have elapsed since those first efforts to understand what have been considered bizarre fossils, the study of pterosaurs has developed enormously. Some of the basic questions about them have long been solved: pterosaurs were neither birds, nor bats, as was suggested by various authors of the early nineteenth century, but a peculiar group of vertebrates which acquired the ability to fly in an original way, using a membrane attached to a single finger of the hand. From the few fossils from the Bavarian lithographic limestones known to Cuvier and his contemporaries, the number of pterosaur specimens has increased enormously, starting with the Early Jurassic specimens from Lyme Regis found by Mary Anning in the 1820s and first described by Buckland (1829), to the present day, when more than 60 genera have been found all over the world (see the review by Wellnhofer 1991). It has now become obvious that pterosaurs, although built on a fairly uniform basic type, showed considerable diversity in terms of size and adaptations. However, despite considerable advances in our knowledge of pterosaurs, many questions and problems remain. The aim of this volume is to bring together papers which attempt to shed some light on various aspects of pterosaurs as fossil organisms, with special emphasis on their evolution and palaeobiology.

A first and important aspect is that the fossil record of pterosaurs is far from being completely known. No fossil record can be known entirely, of course, but that of the pterosaurs is still conspicuously incomplete, because it is strongly influenced by the existence of *Konservat-Lagerstätten*, fossil localities with exceptional preservation, which have led to the preservation of the fragile, hollow-boned skeletons of these flying reptiles. The Late Triassic bituminous limestones of northern Italy, the Liassic bituminous shales of southern Germany, the Late Jurassic lithographic limestones of Bavaria, the Early Cretaceous nodules of Brazil, and the fine-grained Late Cretaceous chalk of the central United States are well-known examples of formations which have yielded a wealth of well-preserved pterosaur specimens. In rocks formed under more usual conditions, pterosaur specimens tend to be scanty and fragmentary. As a result, the evolutionary history of pterosaurs is still full of gaps, or time intervals, during which the group is poorly represented, separating periods during which good material was preserved under more or less exceptional taphonomical conditions. Things, however, are changing rather fast, as new specimens are being found both in newly discovered *Konservat-Lagerstätten*, such as the Early Cretaceous Yixian Formation of northeastern China, and in other formations, in which pterosaur fossils may be more fragmentary but are nonetheless important. Some of the papers in this volume are thus descriptions of new pterosaur fossils from various parts of the world and from various stages of the Mesozoic: the Late Triassic of Austria (**Wellnhofer**); the Late Jurassic of the western United States (**Carpenter *et al.***); the Early Cretaceous of Brazil (**Frey *et al.***) and Venezuela (**Kellner & Moody**); and the Late Cretaceous of Morocco (**Pereda-Suberbiola *et al.***) and Romania (**Buffetaut *et al.***).

One of the main problems about pterosaurs is their origin and early evolutionary history. Triassic pterosaurs in particular have been known only for the last 30 years, and yet these early forms, although already fully fledged pterosaurs, are of obvious importance for our understanding of the beginnings of the group. Both a report of a new find from Austria (**Wellnhofer**) and a review of Triassic pterosaurs (**Dalla Vecchia**) address this question in the present volume. More generally, it is only recently that the evolutionary history of pterosaurs has begun to be

From: BUFFETAUT, E. & MAZIN, J-M. (eds) 2003. *Evolution and Palaeobiology of Pterosaurs.* Geological Society, London, Special Publications, **217**, 1–3. 0305–8719/03/$15

investigated using the modern concepts of phylogenetic systematics. Two papers in this volume (**Kellner; Unwin**) propose alternative comprehensive phylogenies of the Pterosauria, which will undoubtedly serve as a basis for further discussions.

Besides their phylogeny, an enduring problem has been pterosaur biology. Because they have no real equivalent in the present living world, the mode of life of pterosaurs has been the subject of much speculation ever since it was recognized that they were flying animals. A detailed analysis of various aspects of their skeletal anatomy is a prerequisite to an understanding of the way in which they functioned, as illustrated by a study of the morphological evolution of their pectoral girdle (**Bennett**), obviously a fundamental part of the anatomy of any flying vertebrate. As discussed, much of what we know about pterosaurs depends on preservation, and even taxa which have been known for a long time can yield remarkable new information, particularly when good specimens are prepared using modern techniques, as exemplified by the description of axial pneumaticity in *Rhamphorhynchus* (**Bonde & Christiansen**), a taxon first described by Hermann von Meyer in 1847. Careful and painstaking preparation of exquisitely preserved specimens has also contributed immensely to our knowledge of the soft parts of pterosaurs, which are of obvious importance for our understanding of the biology and biomechanics of animals in which the flying apparatus consisted of a wing membrane, which in most instances has not been preserved. As described in one of the papers (**Frey *et al.***), anatomical details as delicate as blood vessels have sometimes been preserved and shed unexpected light on various aspects of pterosaur biology.

Ever since Cuvier realized that pterosaurs were winged reptiles, their locomotion, both in the air and on the ground, has been the subject of much controversy. The flight of pterosaurs can be investigated mainly on the basis of their skeletal anatomy, but comparisons with man-made flying machines can lead to interesting conclusions about the existence of various types of flight adaptations in this group of extinct vertebrates (**Frey *et al.***). Locomotion on the ground is a different matter, and totally divergent interpretations have been put forward on purely morphological grounds, with some authors supporting a bipedal gait, while others favoured a quadrupedal stance. The matter has largely been solved by the discovery and study of pterosaur footprints and trackways, in many parts of the world, which provide direct evidence as to how these animals moved when on the ground. A new discovery of pterosaur footprints from the Late Cretaceous of Mexico is described here (**Rodriguez-de la Rosa**), and a detailed analysis based on the remarkable trackways from the Late Jurassic of Crayssac (southwestern France) clearly illustrates the quadrupedal locomotion of pterodactyloid pterosaurs (**Mazin *et al.***). Pterosaurs were, however, not only able to fly and walk; they could also swim, as shown by ichnological evidence from the Late Jurassic of North America (**Lockley & Wright**), which also provides clues as to their feeding behaviour. Pterosaur trackways have been the subject of much controversy and their parataxonomy has become considerably entangled, hence the need for a critical review advocating drastic simplification (**Billon-Bruyat & Mazin**).

A further way to explore the palaeobiology of pterosaurs is the study of their bone histology. Interestingly, this approach was pioneered as early as the mid-nineteenth century by the British researchers James Bowerbank (1848) and John Quekett (1849*a*, *b*). Some of Quekett's thin sections have survived until the present day (despite the bombing of the Royal College of Surgeons, where they were kept, during the Second World War), and they are redescribed and reinterpreted here (**Steel**). Pterosaur fossils from the Brazilian *Konservat-Lagerstätten* are excellent material for histological investigations, as illustrated by a study on differential growth rates based on such specimens (**Sayão**).

Much indeed can be learned from pterosaur fossils, and the description of a new ornithocheirid taxon from Brazil also includes an interesting piece of forensic palaeontology that provides convincing evidence as to the cause of death of what is now the type specimen (**Frey *et al.***).

Although they are not very common fossils, pterosaurs were an important group of vertebrates during the Mesozoic, and their unusual and interesting adaptations are attracting the attention of a growing number of palaeontologists. The aim of the present volume is to give an idea of the diverse topics addressed by researchers working on these fascinating animals and to encourage further research and discussion.

References

BOWERBANK, J. S. 1848. Microscopical observations on the structure of the bones of *Pterodactylus giganteus* and other fossil animals. *Quarterly Journal of the Geological Society, London*, **13**, 2–10.

BUCKLAND, W. 1829. On the discovery of a new species of pterodactyle in the Lias at Lyme Regis. *Transactions of the Geological Society, London*, **3**, 217–222.

COLLINI, C. 1784. Sur quelques zoolithes du Cabinet d'Histoire Naturelle de S.A.S.E. Palatine et de Bavière, à Mannheim. *Acta Academiae Theodoro-Palatinae, Mannheim, Pars Physica*, **5**, 58–103.

CUVIER, G. 1801. Extrait d'un ouvrage sur les espèces de quadrupèdes dont on a trouvé les ossemens dans l'intérieur de la terre. *Journal de Physique, de Chimie et d'Histoire Naturelle*, **52**, 253–267.

CUVIER, G. 1809. Mémoire sur le squelette fossile d'un reptile volant des environs d'Aichstedt, que quelques naturalistes ont pris pour un oiseau, et dont nous formons un genre de Sauriens, sous le nom de Ptero-Dactyle. *Annales du Muséum national d'Histoire Naturelle, Paris*, **13**, 424–437.

MEYER, H. VON. 1847. Homoeosaurus maximiliani *und* Rhamphorhynchus (Pterodactylus) longicaudus, *zwei fossile Reptilien aus dem Kalkschiefer von Solenhofen*. Schmerber, Frankfurt.

QUEKETT, J. T. 1849a. On the intimate structure of bone, as composing the skeleton in the four great classes of animals, viz., mammals, birds, reptiles and fishes, with some remarks on the great value of the knowledge of such structure in determining the affinities of minute fragments of organic remains. *Transactions of the Microscopical Society, London*, **2**, 46–58.

QUEKETT, J. T. 1849b. Additional observations on the intimate structure of bone. *Transactions of the Microscopical Society, London*, **2**, 59–64.

WELLNHOFER, P. 1991. *The Illustrated Encyclopedia of Pterosaurs*. Salamander, London, 192 pp.

A Late Triassic pterosaur from the Northern Calcareous Alps (Tyrol, Austria)

PETER WELLNHOFER

Bayerische Staatssammlung für Paläontologie und Geologie, Richard-Wagner-Strasse 10, 80333 München, Germany (e-mail: p.wellnhofer@lrz.uni-muenchen.de)

Abstract: Disarticulated skeletal remains of an eudimorphodontid pterosaur from the Late Triassic of the Karwendel Mountains in Tyrol, Austria, are described and figured. It is the second record of Triassic pterosaurs from the Northern Calcareous Alps, after previous discoveries in the Southern Calcareous Alps of northern Italy. The fossil material – isolated jaws, bones and fragmentary skeletal parts of one individual – was collected from the Norian Seefeld Beds (also called 'Bitumenmergel' or 'Seefelder Fischschiefer') near Seefeld in Tyrol. The fossiliferous formation of largely bituminous and calcareous layers originated on top of an extended marine carbonate platform, an equivalent to the nearly contemporaneous limestones and dolomites in the Southern Calcareous Alps, as exposed in Lombardy and Friuli in Italy, that has produced pterosaur fossils since 1973. Based on the biostratigraphic significance of conodont index fossils, the Seefeld Beds can be dated as Late Norian, most likely Sevatian. Sufficiently well-preserved skeletal elements include a jugal, isolated teeth, both mandibular rami with a dentition of uni-, tri- and pentacuspid teeth, cervical, dorsal and caudal vertebrae, ribs, sternum, scapulocoracoids, humerus, first wing phalanx, pelvis and tibia/fibula. There are morphological characters that support a subadult stage of the individual with an estimated wing span of 70–80 cm. The dentition is comparable to *Eudimorphodon ranzii* Zambelli from the Norian Calcare di Zorzino of Cene near Bergamo, northern Italy. However, some skeletal proportions and osteological features are distinctive from this taxon, as well as from a second species, *Eudimorphodon rosenfeldi* Dalla Vecchia from the Norian Dolomia di Forni of Friuli, northern Italy. The meaning of these differences in the Seefeld specimen, in particular the relatively long tibia and short wing phalanx 1, is discussed. The caudal zygapophyses and haemal arches are not elongated into the rod-like bony extensions significant for other known rhamphorhynchoid pterosaurs. It appears, however, that all specimens of *Eudimorphodon* lack elongated caudal zygapophyses. This might be evaluated as a primitive trait for these basal pterosaurs. The typical dentition permits the assignment of the Seefeld pterosaur to the genus *Eudimorphodon*. The fragmentary state of the skeleton and somewhat different skeletal proportions allow the author only to refer the specimen to *Eudimorphodon* cf. *ranzii* Zambelli 1973.

Fossils of Late Triassic pterosaurs are still the oldest known evidence of these archosauromorph flying reptiles of the Mesozoic. Prior to 1973 no pterosaur older than the Early Liassic *Dimorphodon* from the Dorset coast in the southern United Kingdom was known. Since then, several fossil specimens from Late Triassic deposits have been discovered, leading to two important conclusions about early pterosaurian evolution. Firstly, pterosaurs had already reached a worldwide distribution in Late Triassic times, and, secondly, they appear to represent separate phylogenetic lineages at the very beginning of their fossil documentation, indicating a long evolutionary history prior to the Late Triassic.

Our knowledge of Triassic pterosaurs is based on quite a few specimens from several localities of the southern Alps (Bergamasc Pre-Alps, Lombardy and Friuli) in northern Italy (Zambelli 1973; Wild 1978, 1984, 1994; Padian 1981; Dalla Vecchia 1995, 1998, 2000, 2001; Dalla Vecchia *et al.* 1989; Renesto 1993), of Luxembourg (Hahn *et al.* 1984; Cuny *et al.* 1995, 1997), France (Cuny 1993, 1995; Godefroit 1997; Godefroit & Cuny 1997), Switzerland (Peyer 1956; Clemens 1980), the United Kingdom (Fraser & Unwin 1990), Texas, United States (Murry 1986; Hunt & Lucas 1993; Lucas & Hunt 1994), and eastern Greenland (Jenkins *et al.* 1994, 2001). Recently, a new pterosaur taxon has been described from the same deposits in which the new specimen presented in this paper was found (Dalla Vecchia *et al.* 2002).

The fossil material of these oldest known pterosaurs, comprising isolated teeth, partial and complete skeletons, already shows a high taxonomic diversity. Originally, they were assigned to three separate families, including four genera and six species. These are: (1) Eudimorphodontidae Wellnhofer 1978 (Campylognathoididae *sensu* Unwin 1992, 1995; Unwin *et al.* 2000), with *Eudimorphodon ranzii* Zambelli 1973 and *Eudimorphodon rosenfeldi* Dalla Vecchia 1995 from the Mid-Norian of northern Italy and *Eudimorphodon cromptonellus* Jenkins *et al.* 2001 from the Late Triassic of eastern Greenland; (2) Dimorphodontidae Seeley 1870, with *Peteinosaurus zambellii* Wild 1978; and (3) Rhamphorhynchidae Seeley 1870 (Rhamphorhynchoidea family

From: BUFFETAUT, E. & MAZIN, J-M. (eds) 2003. *Evolution and Palaeobiology of Pterosaurs*. Geological Society, London, Special Publications, **217**, 5–22. 0305–8719/03/$15

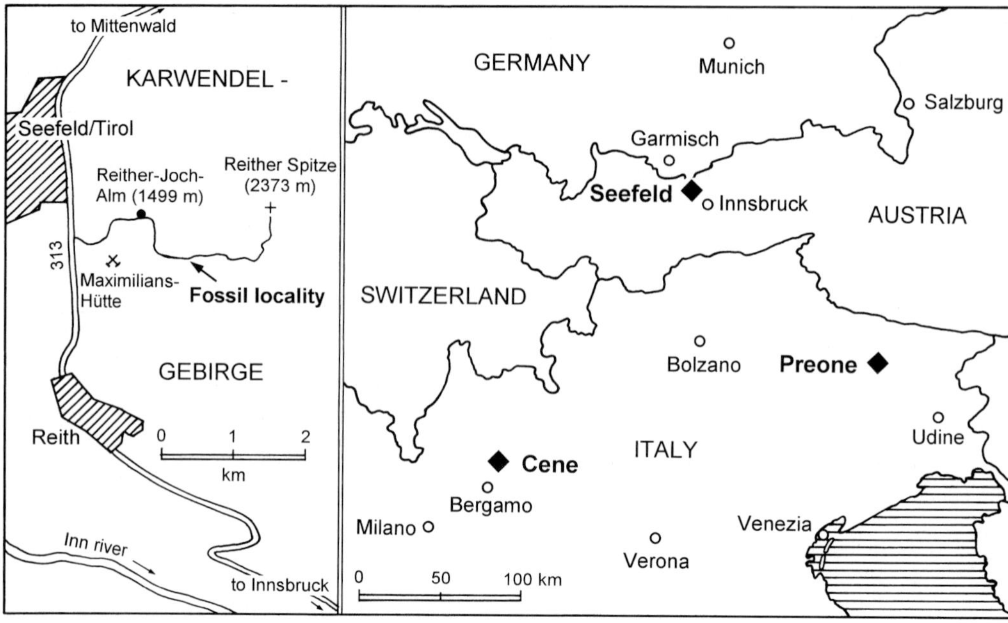

Fig. 1. (Left) Topographic map with the fossil locality of *Eudimorphodon* cf. *ranzii* near Seefeld/Tyrol, Austria, in the western Karwendel-Gebirge, Northern Calcareous Alps. (Right) Fossil locations of Alpine Late Triassic pterosaurs in the vicinity of Cene, near Bergamo, and Preone, Province of Udine, Friuli, in the Southern Calcareous Alps, and of Seefeld in Tyrol, Austria, in the Northern Calcareous Alps.

incertae sedis sensu Dalla Vecchia 1998), with *Preondactylus buffarinii* Wild 1994 from the Mid-Norian of northern Italy and *Austriadactylus cristatus* Dalla Vecchia *et al.* 2002, from the Norian of Tyrol, Austria.

The new pterosaur remains described here are disarticulated bones and skeletal parts, mostly fragments, of one individual. They were discovered in the Norian Seefelder Schichten (also called 'Bitumenmergel' or 'Fischschiefer') near Seefeld in Tyrol, Austria, by Bernd Lammerer, Geology Department of Munich University, in June 1994. Subsequent collecting produced some more skeletal material from the same individual. Professor Lammerer kindly donated his fossil finds to the Bavarian State Collection of Palaeontology and Geology in Munich, where the material is deposited under the catalogue number BSP 1994 I 51.

Institute abbreviations: BMNH, Natural History Museum, London, UK; BSP, Bayerische Staatssammlung für Paläontologie und Geologie, München, Germany (formerly Bayerische Staatssammlung für Paläontologie und historische Geologie); CM, Carnegie Museum of Natural History, Pittsburgh, USA; MCSNB, Museo Civico di Storia Naturale, Bergamo, Italy; MGUH, Geological Museum, University of Copenhagen, Denmark; MFSN, Museo Friulano di Storia Naturale, Udine, Italy, MPUM, Museo Palaeontologia Università di Milano, Italy; SMNS, Staatliches Museum für Naturkunde, Stuttgart, Germany.

Locality

The first fragments of the new pterosaur specimen were discovered by chance lying on the mountain trail to the Reither Spitze (2373 m), one of the higher peaks of the western Karwendel-Gebirge in the Northern Calcareous Alps, southeast of Seefeld in Tyrol, Austria. The locality is situated above the Reither Joch-Alm (1499 m) at an elevation of about 1600 m (Fig. 1). There, at a steep road cut, a profile of several metres of bedded, slightly bituminous, dark gray limestones was exposed. The first find was an isolated block of limestone, 8 cm thick, with a few fragments of jaw bones and some postcranial elements exposed on the surface. During subsequent visits Professor Lammerer and the author were successful in tracing the particular bed from which the isolated block had originated, and found the rest of the skeleton scattered over the surface of four more blocks, in addition to more bones and impressions contained in a few thin fragments of the overlying slab. The site is located in the centre of a former oil-shale mining area which had been operated for hundreds of years in order to produce oil, the so-called 'Steinöl'. The content of bitumen of the bituminous layers varies between 5 and 45%. The Seefeld Beds have a thickness of about 250–400 m. According to Brandner & Poleschinski (1986) they could be called an oil source rock. In the literature the terms 'Asphaltschiefer', 'Ölschiefer' and 'Fischschiefer' have also been used, the last term because of the occurrence, in certain facies, of semionotid and pholidophorid fishes (Kner 1867). From these fish fossils the brand name 'Ichthyol' (fish oil) is derived; this oil was a black, bituminous, medicinal ointment extracted from the rocks (Jung 1992). The Seefeld mining operations ceased in 1964.

Geology and stratigraphy

According to Brandner & Poleschinski (1986) the sedimentation of the Seefeld Beds started in Late Triassic times, in the Norian, in graben-like depressions on an extended marine carbonate platform. It is characterized by the deposition of alternating sequences of bituminous and calcareous layers, contemporary with the late Hauptdolomit facies elsewhere. Obviously the Seefeld Basin was in a marginal position with restricted water circulation and anoxic conditions within the bottom zone.

The problems of the biostratigraphic subdivision of the Alpine-Mediterranean Early Triassic was discussed by Krystyn (1974). He suggested, however, the elimination of the Rhaetian, leaving the Norian as the uppermost stage of the Triassic, subdivided from bottom to top into the Lacian, Alaunian and Sevatian. Based on the pelagic hydrozoan *Heterastridium conglobatum*, Brandner & Poleschinski (1986) concluded a Mid-Norian (Alaunian) age for the Seefeld Beds. Using conodont index fossils H. W. Kozur (pers. comm.) kindly provided the following stratigraphic information:

> The Seefeld Beds contain a monospecific fauna of *Mockina slovakensis* (Kozur). The holotype of this species is from the transition Hallstatt Limestone-Zlambach Beds of the Silica Nappe in the Slovak Karst, where it occurs in the uppermost Sevatian. In the Lagonegro Basin and in western Sicily *M. slovakensis* is common in the *Parvigondolella andrusovi* zone and *Misikella hernsteini* zone of the upper Sevatian, but rarely present also in the *Mockina bidentata* zone of the lower and middle Sevatian. According to unpublished data by Krystyn *M. slovakensis* occurs in the uppermost Alaunian *Halorites macer* zone of Timor. But this ammonoid fauna may already represent the basal Sevatian, because it contains, at least in its upper part, the first *M. bidentata*.
>
> In the Northern Calcareous Alps (Seefeld Beds), in Hungary (Rezi Dolomite of Kesthely Mountains and Feketehegy Limestone of the Pilis Mountains) and in Turkey (metamorphic crystalline limestone of the Izmir–Ankara Belt), *M. slovakensis* occurs mainly in shallow restricted basins as a monospecific fauna. Mostly, it is the only stratigraphic important fossil in these beds, but in the Feketehegy Limestone it occurs immediately above the upper faunal horizon of Cserepes Valley, from where Oravecz (1961) published *Rhabdoceras suessi*, a Sevatian–Rhaetian ammonoid species. Thus, in the western Tethys all dated occurrences both from open sea and restricted basin environments occur within the Sevatian, but according to the unpublished data of Krystyn the first appearance of the species in Timor is in the uppermost Alaunian. This indicates a Sevatian age for the Seefeld Beds which may start during the uppermost Alaunian.

The stratigraphic age of the Seefeld *Eudimorphodon* specimen would thus be Late Norian, most likely Sevatian, and may be somewhat younger compared to most of the Italian occurrences of that genus. In the Southern Calcareous Alps, in northern Italy, *Eudimorphodon* is known from three different formations: (1) the Calcare di Zorzino in the vicinity of Bergamo, which is dated as Mid-Norian, Late Alaunian (Jadoul *et al.* 1994); (2) the Argilliti di Riva di Solto in Lombardy, which is Late Norian, Early Sevatian (Wild 1994; Dalla Vecchia 2001, 2003); and (3) the Dolomia di Forni of Friuli of Mid-Norian, which is Late Alaunian (Dalla Vecchia 1991, 1995; Roghi *et al.* 1995).

The Seefeld semionotid fish *Paralepidotus ornatus* already described from the Seefeld Beds by Agassiz (1833–43), is known also from the Calcare di Zorzino near Bergamo. According to Tintori (1996) this species occurs also in Mid- and Late Norian deposits of northern and southern Italy and Austria.

Systematic palaeontology

Order Pterosauria Kaup 1834
Family Eudimorphodontidae Wellnhofer 1978
Genus *Eudimorphodon* Zambelli 1973
Eudimorphodon cf. *ranzii* Zambelli 1973

Material

This consists of fragments of the skeleton of one individual (Bayerische Staatssammlung für Paläontologie und Geologie, Munich, BSP 1994 I 51).

Several isolated skeletal fragments, black bones and impressions are distributed irregularly on the

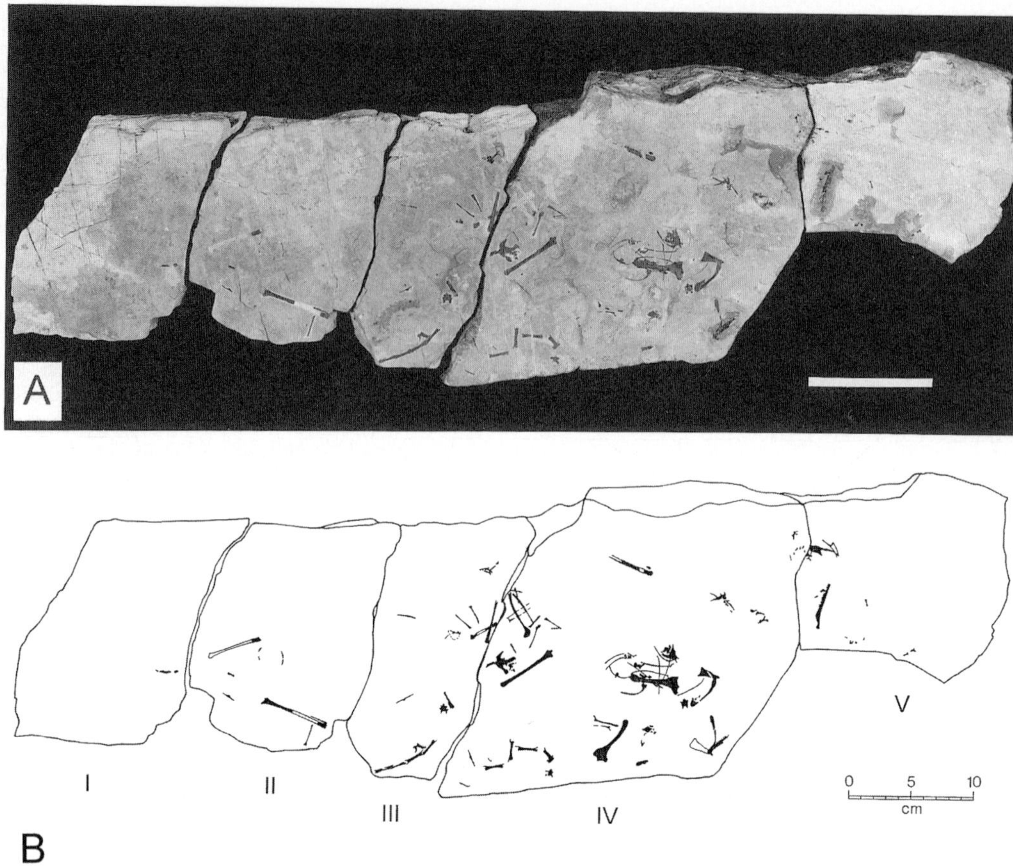

Fig. 2. Fossiliferous blocks of Late Norian Seefeld oil shale with disarticulated skeleton of *Eudimorphodon* cf. *ranzii*. (**a**) Specimen BSP 1994 I 51. Scale bar 10 cm. (**b**) Distribution of skeletal elements of the Seefeld specimen of *E.* cf. *ranzii*, on blocks I–V. BSP 1994 I 51.

surface of a light grey, marly limestone, 8 cm thick and about 60×20 cm in area, which is broken into five blocks (numbered I–V, Fig. 2). Furthermore, some isolated bones and impressions of bones are preserved on thin limestone fragments from the overlying bed. All elements are from one individual that was obviously in a progressive state of decay and whose skeleton had already fallen apart prior to its embedding in the sediment. The material includes isolated skull elements, both mandibular rami (incomplete) with teeth, a few isolated teeth, cervical, dorsal and caudal vertebrae, haemapophyses, ribs, gastralia, sternum, both scapulocoracoids, humeri, manual claws, a first wing phalanx, one half of the pelvis, ?femur, tibia/fibula, ?metatarsals, pedal phalanges and many bone fragments of uncertain identity.

Description

Skull. Several fragments of skull bones are distributed on the surfaces of blocks IV and V, but they are only partly preserved, as impressions. Most of the cranial elements are too fragmentary for reliable identification.

Jugal (Fig. 3a). On block V, next to a broken bone mass, possibly a maxilla, there is a slender, tetraradiate bone, 14 mm in length, that could be identified as a jugal. When compared with specimens of *Eudimorphodon* from Cene near Bergamo, this bone is clearly distinctive. The forked caudal processes (the processus postorbitalis and quadratojugalis) include only a small angle, indicating a narrow infratemporal fenestra, similar to the juvenile Milano specimen MPUM 6009 of *Eudimorphodon ranzii* (Wild 1978, p. 216). The orbital margin is only slightly curved, which would indicate a large orbit, usually a criterion for juvenility (Wellnhofer 1970, p. 89).

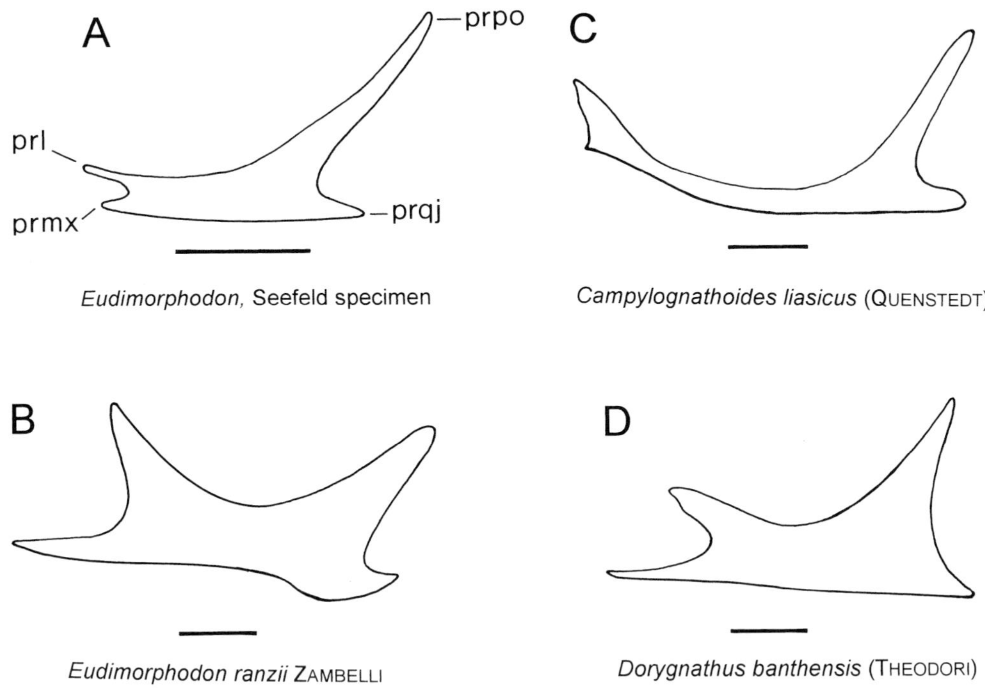

Fig. 3. Comparison of jugals. (**a**) *Eudimorphodon* cf. *ranzii*, Seefeld specimen, BSP 1994 I 51. (**b**) *E. ranzii*, holotype, Norian, Cene, MCSNB 2888 (after Wild 1978). (**c**) *Campylognathoides liasicus*, Late Liassic, Holzmaden, CM 11424 (after Wellnhofer 1974). (**d**) *Dorygnathus banthensis*, Late Liassic, Holzmaden, Museum Hauff, Holzmaden (after Wild 1978). prl, processus lacrimalis; prmx, processus maxillaris; prpo, processus postorbitalis; prqj, processus quadratojugalis. Scale bars 5 mm.

Isolated teeth. On block V, near some cranial bone fragments, an isolated unicuspid tooth is completely preserved, including its root (Fig. 4b). In size (length 4 mm) and shape – it shows a hook-like posteriorly recurved tip – it is very similar to anterior premaxillary teeth of *E. ranzii* (Wild 1978, p. 192 & Fig. 8). Its enamel surface is sculpted by longitudinal ridges, and between crown and root a shallow 'waist' is developed. The enamel–dentine boundary has a sigmoidal curvature.

Also on block V another isolated unicuspid tooth is partly preserved, showing some similarity to the large mid-maxillary teeth of *E. ranzii* (Fig. 4a). Its original length was about 3.5 mm. The root is split longitudinally in half, exposing the pulpa canal. At the tip an oval wear facet indicates a hard component in the diet, such as fish scales, as was suggested by Wild (1978) for *E. ranzii*. The tooth could be described as 'pseudo-unicuspid', because it has several small accessory cuspules slightly splaying from the central axis. This, however, is distinct from the large mid-maxillary teeth of *E. ranzii* where only two lateral cuspules are developed. Alternatively, this tooth could also be assigned to the lower jaw, the more so because it is lying next to

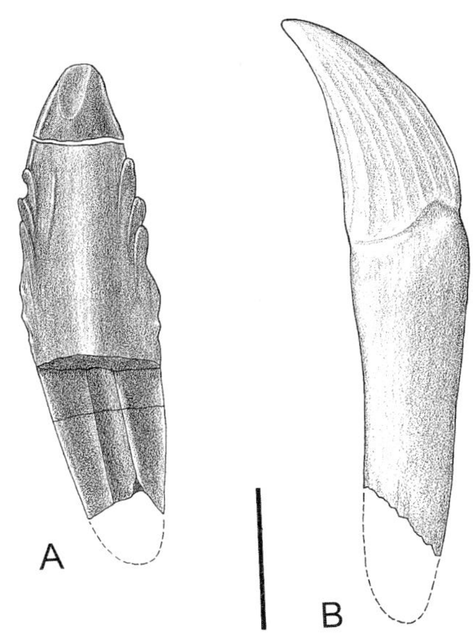

Fig. 4. *Eudimorphodon* cf. *ranzii*, Seefeld specimen, BSP 1994 I 51, isolated teeth on block V. (**a**) Pseudo-unicuspid (?mandibular) tooth. (**b**) ?Rostral premaxillary tooth (fang). Scale bar 1 mm.

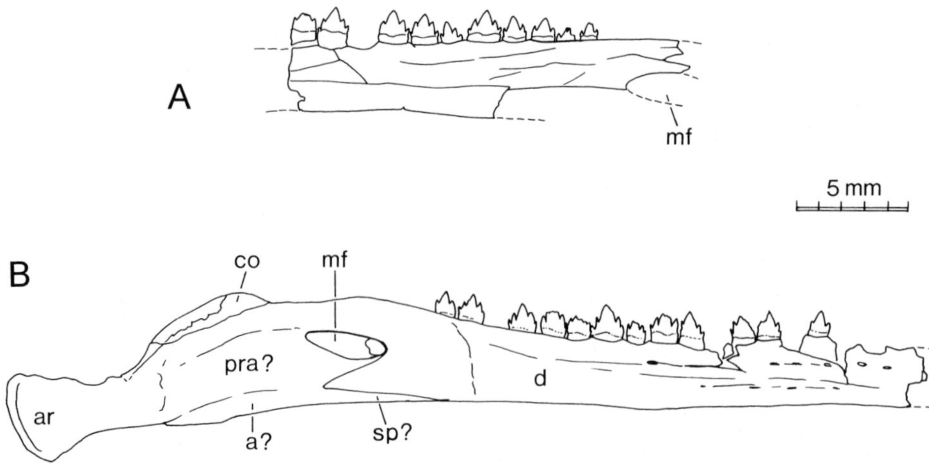

Fig. 5. *Eudimorphodon* cf. *ranzii*, Seefeld specimen, BSP 1994 I 51. (**a**) Right mandible as preserved on block IV in lingual view. (**b**) Left mandible as preserved on block V in lingual view. a, angular; ar, articular; co, coronoid; d, dentary; mf, fenestra meckeli (fossa adductoria); pra, prearticular; sp, splenial.

the left mandibular ramus, as if it had fallen out of that jaw post mortem. However, in the lower jaw of *E. ranzii* only the two rostral-most teeth are large and unicuspid, functioning as fangs. This part of the lower jaw is not preserved in the Seefeld specimen, and therefore this alternative tooth position can not be confirmed. However, in the juvenile Milano specimen, MPUM 6009, the third mandibular tooth is also large and unicuspid with lateral accessory cuspules. According to Wild (1978, p. 218) tooth morphology and tooth count may change during ontogeny. Since additional osteological criteria indicate immaturity for the Seefeld specimen, this tooth could also represent the third mandibular tooth of the left jaw.

Mandible. On blocks IV and V both mandibular rami are preserved and lie at a distance of 15 cm from each other. From the left ramus (on block V) the rostral end including the symphysis is missing (Figs 5b, 6b & 7).The distance from the caudal margin of the retroarticular process to the break is 40 mm. The jaw is exposed from its lingual surface, showing the fenestra mandibularis, which is framed dorsally by the surangular, rostrally by the dentary and ventrally by the prearticular. A pronounced elevation on top of the surangular and in front of the articular can be identified as a coronoid. Here, the mandible is much deeper than that restored by Wild (1978, fig. 4) in the holotype of *E. ranzii*. However, a deep mandible at this point was described in *E. rosenfeldi* by Dalla Vecchia (1995), and it appears to be deep in the juvenile Milano specimen MPUM 6009 of *E. ranzii* as well. The sutures between particular elements, especially between angular and splenial at the ventral margin of the jaw, can hardly be identified.

Dentition. Twelve teeth are still in place in their alveoli. The distance between the posterior-most tooth and the caudal margin of the retroarticular process is 19 mm. In the holotype of *E. ranzii* this distance is 25 mm (Wild 1978). The distance between the posteriormost tooth and the coronoid process is much greater in the Seefeld specimen than in the holotype of *E. ranzii*. Dalla Vecchia (1995) mentioned a long ‘diastema’ here in *E. rosenfeldi*. The rostral-most tooth still preserved in the left mandible is tricuspid, followed by pentacuspid teeth. Some of the alveoli are empty. So, in front of the first tooth, two teeth have fallen out, as well as one tooth behind. There follow two pentacuspid teeth and again a gap. In front of the last two teeth one tooth is

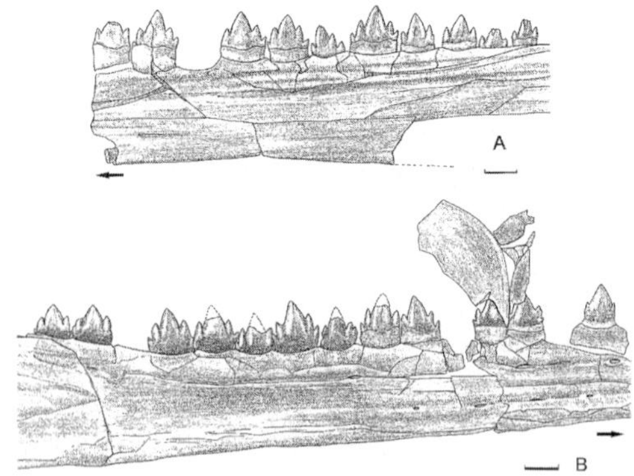

Fig. 6. *Eudimorphodon* cf. *ranzii*, Seefeld specimen, BSP 1994 I 51. Mandibular dentition as in Figure 5. (**a**) Right mandible as preserved on block IV. (**b**) Left mandible as preserved on block V. Arrows point to rostral end. Scale bar 2 mm.

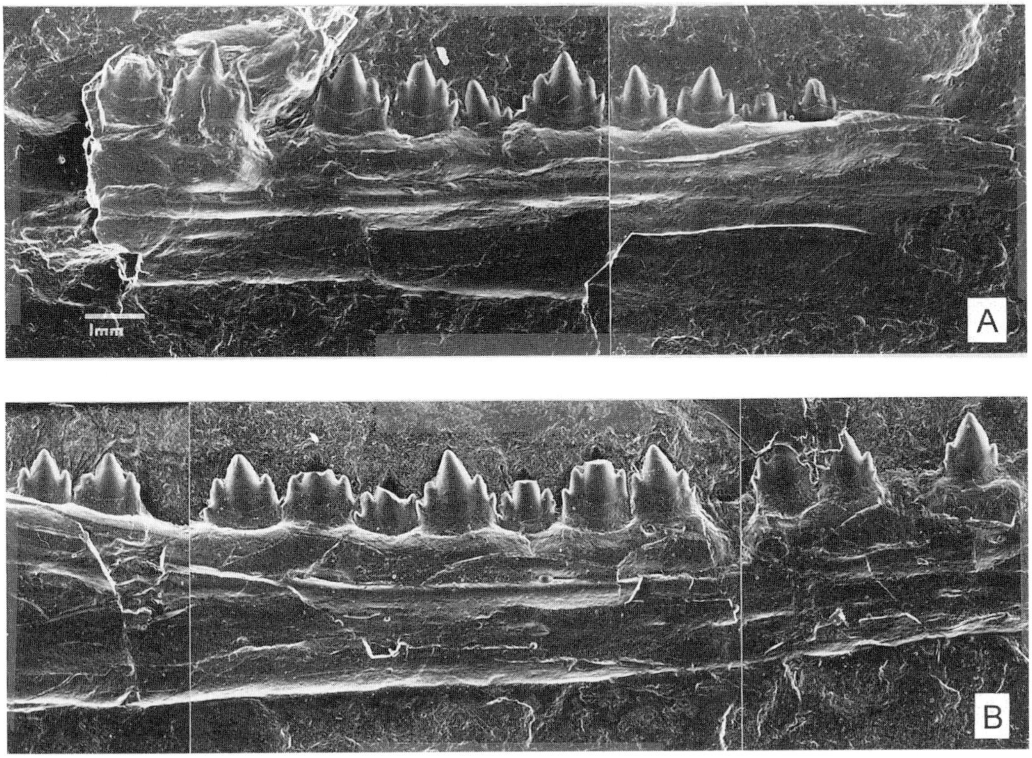

Fig. 7. *Eudimorphodon* cf. *ranzii*, Seefeld specimen, BSP 1994 I 51. Mandibles as in Figure 6. Casts in scanning electron miscroscope (SEM) photographs. (**a**) Right mandible. (**b**) Left mandible. (SEM photographs by R. Liebreich, Munich).

missing too. The last 13, perhaps 14, mandibular teeth in this left ramus are pentacuspid.

In a longitudinal groove in the upper third of the lingual surface of the dentary, small, elongated foramina for nerves and blood vessels penetrate the bone. Up to the front end of the jaw 18 tooth positions, including empty alveoli, are preserved. The total number of mandibular teeth in the holotype specimen of *E. ranzii* is 28, respectively 26 per ramus (Wild 1978, fig. 4). In these figures the anterior large, unicuspid fangs are included. Given the same number of mandibular teeth as in the holotype specimen of *E. ranzii*, tooth positions 11–28, or 9–26 respectively, would be preserved in the Seefeld specimen. If this is the case, the anterior 8–10 mandibular teeth would be missing. However, the lower jaw of the Seefeld specimen was shorter than that of the holotype of *E. ranzii*, in which it is 74.5 mm in length. If the lower jaw of the Seefeld specimen is restored proportionately, its total length could be calculated as having been 45–55 mm. This would make the Seefeld specimen about 10–25% smaller than the holotype specimen of *E. ranzii*, suggesting an immature or subadult individual.

In the juvenile Milano specimen MPUM 6009 of *E. ranzii* the mandible is 34 mm in length. According to Wild (1978) 17 teeth are present in this specimen; rostrally to caudally, these include two large unicuspid fangs, one pseudo-unicuspid, three tricuspid, one pentacuspid, one tricuspid, one tri- or pentacuspid and eight pentacuspid teeth. In size, the Seefeld specimen is somewhere between the adult holotype and the juvenile Milano specimen MPUM 6009 of *E. ranzii*. Accordingly, the number of mandibular teeth in that specimen can be estimated at about 22 teeth per mandible ramus.

Of the right mandible (on block IV) only the middle section, also in lingual view, is preserved (Figs 5a, 6a & 7). Here, not only the symphysial part but also the caudal half of the ramus is broken away. In addition, in the posterior portion, it is covered by a scapulocoracoid. What is present is just part of the dentary, bearing the last 10 tricuspid and pentacuspid teeth still in their alveoli. In front of the eighth tooth, counted from the caudal-most one, there is a gap. Including two more teeth rostrally, 11 teeth are represented. This is probably about half of the total tooth count. In contrast to the left mandible, the last four teeth in the right mandible are not pentacuspid but tricuspid. In both rostral teeth of this series a

Fig. 8. *Eudimorphodon* cf. *ranzii*, Seefeld specimen, BSP 1994 I 51. Cast of pentacuspid tooth of the right mandible in last but fifth tooth position, in lingual view. (SEM photograph by R. Liebreich, Munich.)

fourth minute accessory cuspule is developed. All other teeth preserved are pentacuspid (Fig. 8).

These differences in tooth morphology between left and right side, and between the Seefeld specimen, the Milano specimen MPUM 6009 and the holotype specimen of *E. ranzii*, are evidence not only of ontogenetic variation, but also of individual variation in general. None of the mandibular teeth displays enamel striations. The crowns are completely smooth, partly with a somewhat rugose surface. The development of enamel striations in *Eudimorphodon* is growth dependent (Wild 1978, p. 218). Vice versa, the smooth enamel surface of the multicuspid teeth in the Seefeld specimen would again indicate an immature or subadult stage. In their general morphology, however, the mandibular teeth of the Seefeld specimen agree well with the corresponding dentition of the holotype of *E.ranzii*.

Postcranial skeleton

Cervicals. On blocks III and IV three vertebrae can be identified as cervicals (Fig. 9a). Two are exposed from the dorsal surface and should have originated from the mid-cervical section. The vertebra lying on block III is about 8 mm wide and 8 mm long. The neural crest is depressed into the neural canal, which is collapsed. Strong prezygapophyses extend beyond the cranial extremity of the centrum by more than a quarter of its length. In comparison with *E. ranzii* it can be concluded that it is either the third or fourth cervical vertebra (Wild 1978, p. 202). As in all pterosaurs, the cervicals are strongly procoelous. Two more cervicals are preserved on block IV, one adjacent to the right scapulocoracoid, the other 4 cm away. The first one is about 7 mm wide and 8.5 mm long, and shows its ventral surface. Caudally the centrum terminates in a projecting, ball-like articular surface. Ventrolateral fenestrae in the centrum are interpreted as pneumatic foramina. This vertebra can also be attributed to the mid-cervical section. The second vertebra on block IV is obliquely compressed. Here too, a strong, posterior, ball-like articular process is developed as well as a low median neural crest extending over the entire length of the centrum. In size it corresponds to both cervicals as described above.

Dorsals. Only two vertebrae on block IV can reliably be identified as dorsals. One is preserved in dorsoventral position showing its dorsal surface. As a result of compression only fragments of the neural spine can be recognized. The transverse processes are oriented laterocaudally. Its length is about 6 mm; its breadth across the transverse processes is 8.5 mm. The other dorsal was imbedded craniocaudally and is still in natural articulation with its ribs. Compression also permits a somewhat oblique aspect. The transverse processes are broad and obtuse. The ribs attached to this dorsal are only 13 mm in length, suggesting a more caudal position, in contrast to other, much longer (up to 30 mm) isolated dorsal ribs on this block. A fragment of another dorsal lies next to the left humerus on block IV and is still in articular association with a double-headed rib. Its length, measured across the chord, is 23 mm, indicating a more cranial position within the dorsal vertebral column.

Several isolated dorsal ribs and rib fragments are scattered across the surface of block IV. They are double-headed and of different sizes ranging from 16 to 30 mm. Some show slight expansions at their free ends, presumably facets for the contact with cartilaginous sternal ribs.

Caudals (Figs 9b & 10). The usual number of caudals typical for long-tailed pterosaurs ('Rhamphorhynchoidea') is 30–40 segments (Wellnhofer 1975). In the Seefeld specimen only 10 caudals are preserved, suggesting that 20–30 vertebrae are missing from the long vertebral tail. Six caudals are disarticulated from one another, but four middle caudals are still associated in natural articulation. The slender, much elongated vertebrae vary in length between 12.5 and 17 mm. The pairs of both prezygapophyses and postzygapophyses extend only a little beyond the centrum. They are bifurcated and bent slightly upwards. Originally they had the function of bracing the individual vertebrae to a certain degree. However, extremely elongated processes of the caudal zygapophyses, as developed in the typical rhamphorhynchoids (so-called 'ossified tendons', which stiffened the long tail) are not present in the Seefeld specimen. There is no evidence whatsoever of these elongated zygapophysial

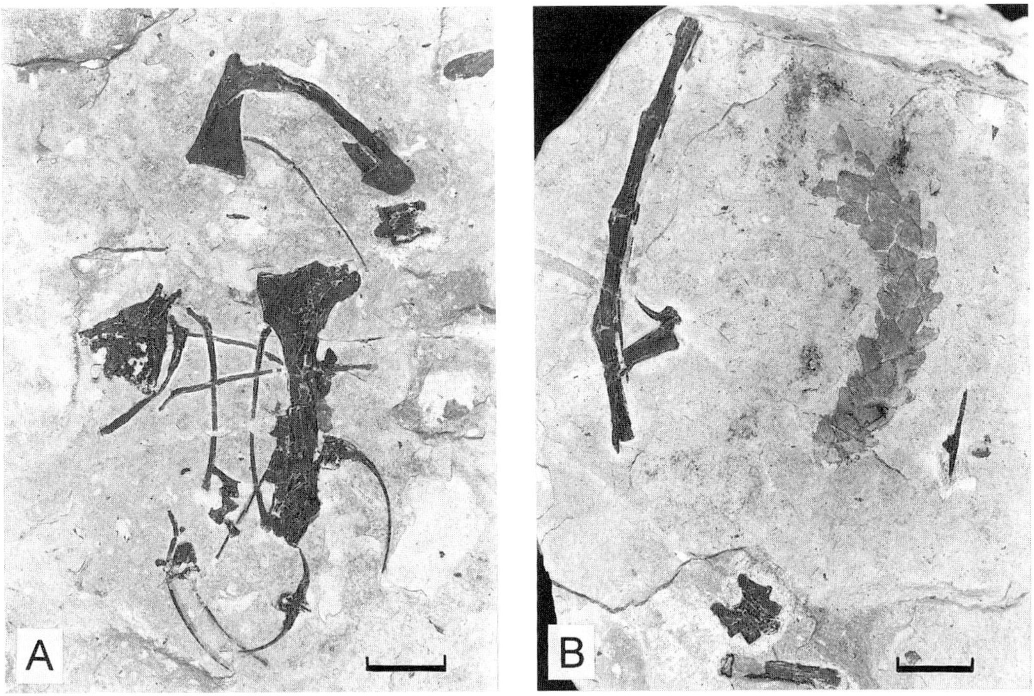

Fig. 9. *Eudimorphodon* cf. *ranzii*, Seefeld specimen, BSP 1994 I 51, postcranial elements. (**a**) Section of block IV showing right scapulocoracoid, left humerus, sternum, dorsal ribs and vertebrae. (**b**) Section of block III showing a mid-caudal series of four associated vertebrae, isolated caudals, a pedal claw, a mid-cervical vertebra and a haemapophysis (chevron) at right to the imprint of the conifer *Pagiophyllum*. Scale bar 1 cm.

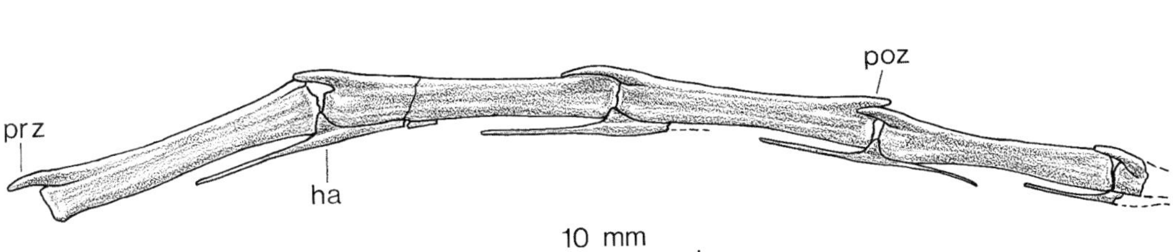

Fig. 10. *Eudimorphodon* cf. *ranzii*, Seefeld specimen, BSP 1994 I 51. Four mid-caudal vertebrae in natural association in left lateral aspect as preserved on block III. ha, haemapophysis (chevron); poz, postzygapophysis; prz, prezygapophysis.

processes, suggesting that these structures were not developed primarily. Comparison with *E. ranzii* is limited, since the tail is missing in the holotype specimen as well as in specimen MCSNB 2887 (Wild 1978), and also in *E. rosenfeldi* (Dalla Vecchia 1995). Specimens MCSNB 8950 (Wild 1994) and the juvenile Milano specimen MPUM 6009 of *E. ranzii* have some caudal vertebrae preserved, but these lack extensions of the zygapophyses, which are short. Only in specimen MCSNB 3496, originally assigned to *E. ranzii* by Wild (1978), was the presence of elongated 'ossified tendons' reported. However, according to Dalla Vecchia (2001, 2002), this specimen is not a *Eudimorphodon* but belongs to the dimorphodontid *Peteinosaurus*. *Eudimorphodon cromptonellus*, specimen MGUH VP 3393, probably a juvenile individual from the Late Triassic of eastern Greenland, also lacks elongated caudal zygapophyses (Jenkins *et al.* 2001). Therefore, this character of the Seefeld specimen, i.e. the lack of elongated caudal zygapophyses, is consistent with the caudal architecture of all specimens of *Eudimorphodon* known so far. Since the caudal zygapophyses are also short in *Austriadactylus*, this character can be interpreted as an archaic, plesiomorphic feature of these basal-most pterosaurs, as

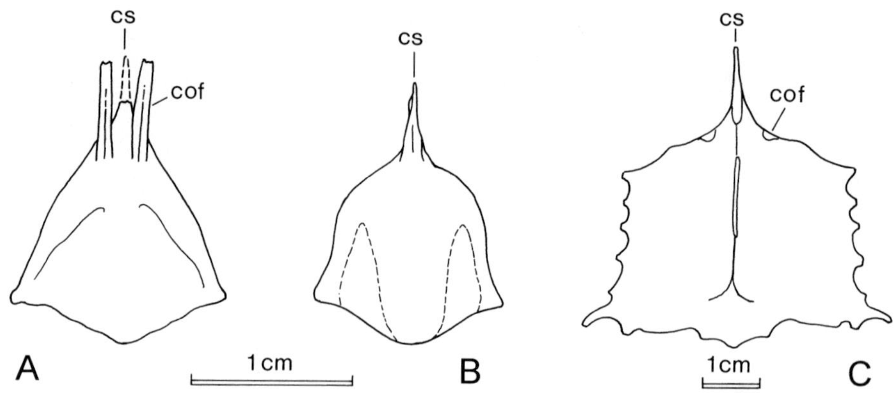

Fig. 11. Comparison of sterna. (**a**) *Eudimorphodon* cf. *ranzii*, Seefeld specimen, BSP 1994 I 51, block IV. (**b**) *Eudimorphodon ranzii*, immature specimen MCSNB 8950 (after Wild 1994). (**c**) *Eudimorphodon ranzii*, adult individual, holotype MCSNB 2888 (after Wild 1978). cof, articular facet for coracoid; cs, cristospina.

has also been suggested by Dalla Vecchia *et al.* (2002). In contrast, in both the contemporary pterosaurian taxa *Peteinosaurus zambellii* and *Preondactylus buffarinii* – which might be congeneric according to Dalla Vecchia (1998) – elongated caudal zygapophyses are present.

In the Seefeld specimen, the haemapophyses (chevrons), however, have developed rod-like anterior and posterior processes. They overlap ventrally in the middle of the centrum between them. In comparison with Late Jurassic examples of *Rhamphorhynchus*, the extensions of the haemal arches in the Seefeld specimen are much shorter (Wellnhofer 1975, p. 16).

Sternum (Figs 9a & 11a). A fragment of a sternum is preserved on block IV. Its entire shape, however, can be restored from the negative impression on the counterpart slab.The sternal plate is nearly triangular. The cranial extremity of the cristospina is not preserved. In its general shape it agrees well with the sternum of the juvenile specimen of *E. ranzii* described by Wild (1994) and is distinct from the transverse-rectangular shape of the adult individuals of that species (Wild 1978, 1994; Renesto 1993). Paired structures attached to the anterior margin of the sternal plate, interpreted by Wild (1994) and Peters (2000) as 'clavicles', are not present in the Seefeld specimen. Symmetrical to the cristospina two narrow processes of the sternal plate project cranially, obviously for articulation with the coracoids. If this is correct, the coracoids would have been oriented subparallel to the sagittal axis. Such a construction of the sternal articulation of the shoulder girdle is not known in other specimens of *Eudimorphodon* or any other pterosaur so far. The sternum seems not to be ossified completely. There are no processi for the attachment of the sternal ribs.

Shoulder girdle (Figs 9a & 12). Scapula and coracoid are co-ossified into one boomerang-shaped element. If other features might indicate a rather immature individual, the fusion of the shoulder girdle, which is growth dependent, would be in contrast to this assumption. Both scapula and coracoid enclose an angle of less than 90°, a value considerably smaller than in the holotype of *E. ranzii* (Wild 1978, fig. 15), but comparable to the scapulocoracoid of specimen MCSNB 2887, as figured by Wild (1978, fig. 16). Both scapulocoracoids are lying isolated on block IV and show their medial surface. The glenoid fossae are imbedded in the matrix.

The distal part of the left coracoid is preserved as imprint only. In the right scapulocoracoid the coracoidal part has been twisted into the bedding plane post mortem, thus presenting its caudomedial aspect. The coracoid expands distally, terminating in a gently curved ventral surface which articulated with the sternum, by the already mentioned narrow processes. Craniodorsally a strong processus acrocoracoideus is developed. Whether it served as a 'canal' for the tendon of the supracoracoideus muscle has been questioned by Bennett (2001), however. This muscle, despite its ventral position, functions as an elevator of the humerus in modern birds during the upstroke. The coracoid is relatively short, being 18 mm in length. In this and in the scapulocoracoidal angle, it is more comparable to the scapulocoracoid in specimen MCSNB 2887 than to that in the holotype of *E. ranzii* (Wild 1978).

The scapula is narrow and sabre-like in shape. It is longer than the coracoid and has a thin, spatula-like, rounded distal end, not a pointed tip as described by Wild (1978, p. 207) in *E. ranzii*. There is no suture visible between coracoid and scapula. The length of the scapula can therefore be given only approximately as 30 mm.

The distal expansion of the coracoid, a strongly developed acrocoracoid, the rounded distal end of the scapula, and the great difference in the lengths of

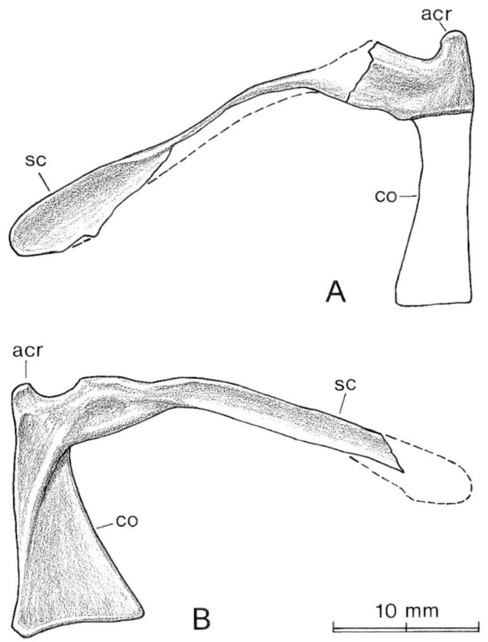

Fig. 12. *Eudimorphodon* cf. *ranzii*, Seefeld specimen, BSP 1994 I 51. (**a**) Left scapulocoracoid as preserved, partly as impression, on block IV in medial view. (**b**) Right scapulocoracoid as preserved on block IV in medial view. acr, processus acrocoracoideus; co, coracoid; sc, scapula.

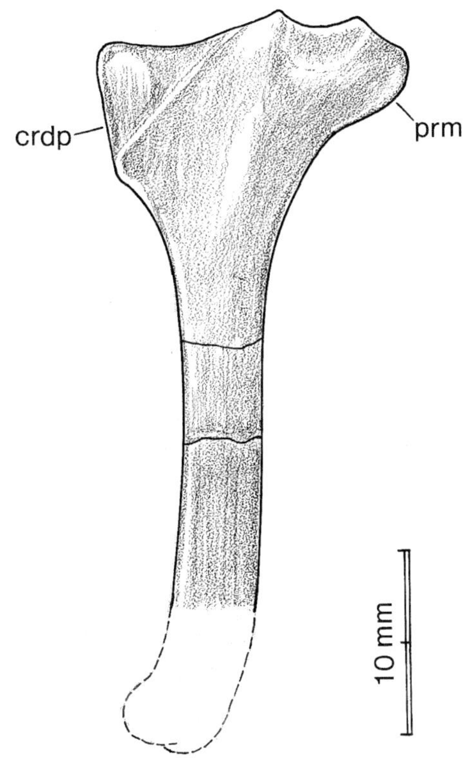

Fig. 13. *Eudimorphodon* cf. *ranzii*, Seefeld specimen, BSP 1994 I 51. Left humerus in lateral view as preserved on block IV. Distal end as impression only. crdp, crista deltopectoralis; prm, processus medialis.

both scapula and coracoid are clearly distinct from the condition in *E. ranzii*. However, they are more similar to the scapulocoracoid in the dimorphodontid *Peteinosaurus zambellii*, specimen MCSNB 3359, from the Zorzino limestone of Cene near Bergamo (Wild 1978, p. 227, fig. 34). A relatively short and wide coracoid is also present in *Eudimorphodon rosenfeldi* from the Dolomia di Forni of the Province of Udine (Dalla Vecchia 1995, p. 60).

Forelimb. Only two bones of the wing skeleton, humerus and wing phalanx I, are sufficiently well preserved for detailed analyses and measurements. Fragments of other long bones are probably those of radius and ulna, and of some distal wing phalanges.

Humerus (Fig. 9a & 13). Both humeri are represented on block IV. The humerus is preserved only as an imprint showing the outline of the bone in negative. The humerus, 40 mm in length, is a relatively slender bone, with a mid-shaft diameter of 4.5 mm and slightly curved laterally. Proximally, a large lateral process is developed, the processus deltopectoralis, for the origin of a strong depressor musculature. The processus medialis is less pronounced and set off distally. The distal articular surface with its trochlea appears to be twisted by 90° with respect to the proximal expansion. In shape, the humerus of the Seefeld specimen is not as robust as in the type specimen of *E. ranzii*, but is similar to the juvenile Milano specimen (Wild 1978, fig. 29). The ratio of maximum width to maximum length of the humerus is about 2.5. It is 2.5 in the juvenile Milano specimen MPUM 6009 (Wild 1978), 2.6 in the immature specimen MCSNB 8950 (Wild 1994), but 2.1 in the adult holotype specimen of *E. ranzii* (Wild 1978). Dalla Vecchia (1995, p. 60) found the humeral shaft in *E. rosenfeldi* proportionately longer and more slender than in *E. ranzii* and also recognized a delicate, rectangular deltopectoral crest, similar to the Seefeld specimen.

Wing phalanges. Only the first phalanx of the right wing digit is completely preserved on block IV (Fig. 15a). It is a strong, straight bone, 52.9 mm in length, with a mid-shaft diameter of 2.9 mm. The proximal articular surface has a maximum width of 7.5 mm. The distal end is 4.5 mm in diameter. The phalanx is exposed from its ventral side, showing proximally the strongly concave fossa ventralis of the twin articulation with the ginglymoid distal articular condyles of the fourth (wing) metacarpal. The fossa ventralis expands largely into the proximal

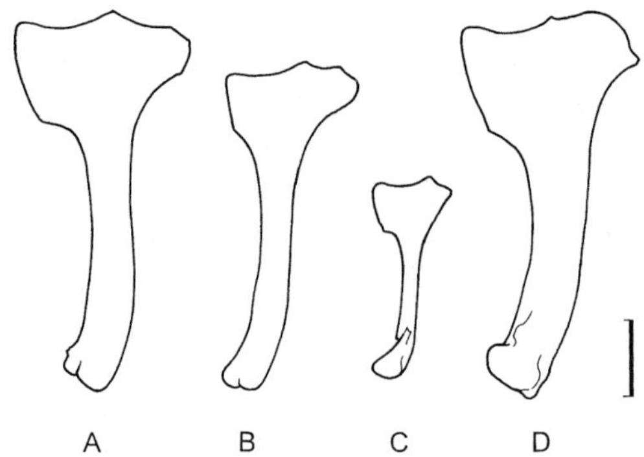

Fig. 14. Comparison of left humeri in lateral view. (**a**) *Campylognathoides liasicus*, CM 11424 (after Wellnhofer 1974). (**b**) *Eudimorphodon* cf. *ranzii*, Seefeld specimen, BSP 1994 I 51. (**c**) *Eudimorphodon ranzii*, immature specimen, MCSNB 8950 (after Wild 1994). (**d**) *Eudimorphodon ranzii*, holotype, MCSNB 2888 (after Wild 1978). Scale bar 1 cm.

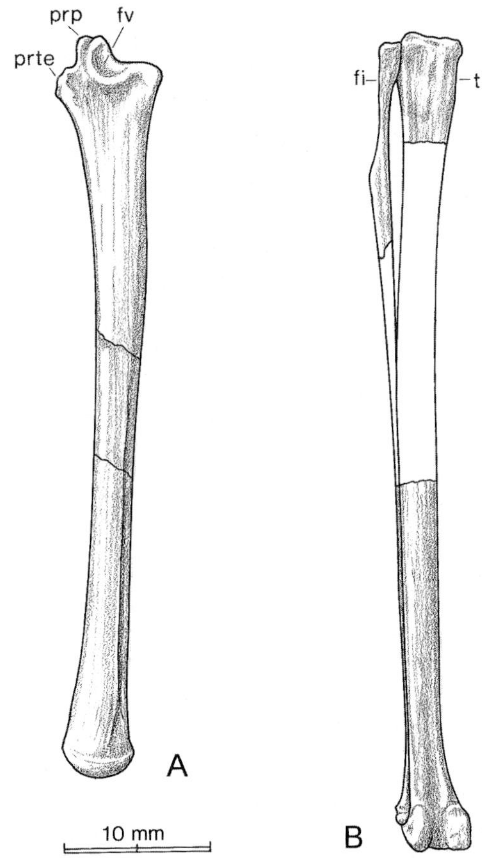

Fig. 15. *Eudimorphodon* cf. *ranzii*, Seefeld specimen, BSP 1994 I 51. (**a**) Right wing phalanx 1 as preserved on block IV in ventral view. (**b**) Right tibia and fibula as preserved on block II in anterior view. fi, fibula; fv, fossa ventralis of proximal articular surface; prp, proximal ('olecranon-like') process; prte, process of extensor tendon of wing digit; ti, tibia.

olecranon-like process typical of pterosaurs. Distally to this process a blunt projection on the cranial side served for the attachment of the extensor tendon of the wing finger. Caudally a longitudinal narrow groove with sharp edges for the attachment of the wing membrane is developed, as in most rhamphorhynchoids. The distal articular surface is well rounded, indicating a certain degree of articulation with the second phalanx of the flight digit.

Fragments of two slender long bones on block IV are probably distal phalanges of the wing finger. One of them displays its proximal articular end with a slightly concave surface. At this end it is 2.8 mm in width. Its mid-shaft diameter is 1.3 mm. It is probably the third wing phalanx, whereas the other bone, with a shaft diameter of less than 1 mm, is presumably the distal (fourth) wing phalanx.

Pelvic girdle (Fig. 16). The left pelvis is preserved on block IV in lateral view. Because of the incomplete and fragmentary condition single pelvic elements cannot be separated. Both postacetabular and preacetabular processes of the ilium are considerably long. The cranial end of the preacetabular process, however, is missing. The dorsal margin of the ilium above the acetabulum is marked by impression only. In low-angle illumination a trace of the acetabulum can be recognized which, due to compression, appears rather flat. The ischium is expanded caudally in agreement with the condition represented in specimen MCSNB 3496 (Wild 1978, p. 213, fig. 19). As in this specimen, assigned to *Peteinosaurus* by Dalla Vecchia (2001, 2003), a short, caudally oriented process at the dorsocaudal rim of the ischium is developed. Ventrally, pubis and ischium are represented by numerous bone fragments only. Consequently, the ventral margin of the ischiopubic plate can be recognized only faintly but, in general, it appears to be much expanded ventrally. A bone fragment at the ventral margin, if belonging to the ischium, indicated a ventral projection, not known in any of the other *Eudimorphodon* specimens. There is no evidence of an obturator foramen.

Hindlimb (Fig. 15b). The only bones that can be assigned to the hindlimb are a tibia (length 57.7 mm) and a fibula (length 52 mm), preserved in natural association on block II. If they do indeed show their cranial side, they represent the right lower leg. Both articular ends are preserved. A middle section of the shafts of both tibia and fibula is represented by impressions only. Although no synostosis between tibia and fibula is apparent, they form a common proximal articular surface (width 6 mm). Unlike the condition in adult individuals of *E. ranzii*, tibia and fibula are separate and not co-ossified. The narrowest diameter of the tibia shaft is 2.5 mm, 8 mm proxi-

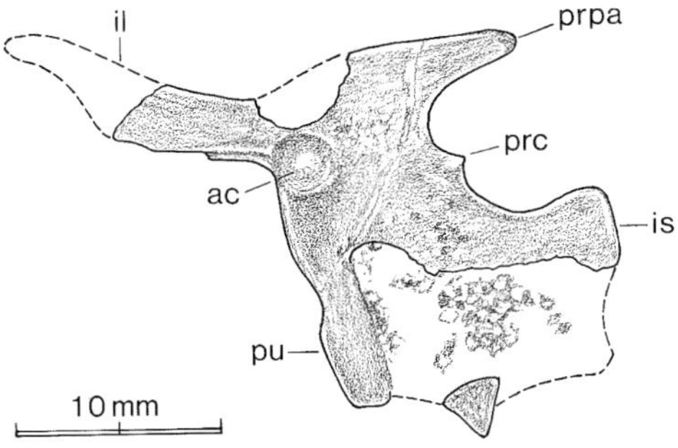

Fig. 16. *Eudimorphodon* cf. *ranzii*, Seefeld specimen, BSP 1994 I 51, pelvis as preserved on block IV, partly as impressions, in left lateral view. ac, acetabulum; il, ilium; is, ischium; prc, processus caudalis of ischium; prpa, processus postacetabularis of ilium; pu, pubis.

mally of the distal end. Distally the tibia is expanded up to a width of 5 mm where an articular trochlea bearing two condyles is developed. The fibula is separated from the tibia proximally, for about one third of its length, leaving a spatium interosseum between the two bones, about 1 mm wide. Distally the fibula is closely attached to the tibia laterally as a splint-like, slender bone. At its distal end the fibula is slightly expanded, presumably for contact with the calcaneum.

In *E. ranzii* the fibula is only a little more than half the length of the tibia, narrowing distally into a splint and fused with the tibia shaft (Wild 1978, p. 215). In *E. rosenfeldi* the fibula does not reach the distal part of the tibia either (Dalla Vecchia 1995). The Seefeld *Eudimorphodon* specimen seems to be more primitive in this character. Whether this could also indicate a juvenile condition must be questioned, however. In the immature specimen MCSNB 8950 of *E. ranzii*, the fibula, although not co-ossified with the tibia, is reduced distally at two-thirds of the tibia length. In this respect the Seefeld specimen agrees with *Campylognathoides liasicus* from the Late Liassic of Holzmaden, in which the fibula is developed in its entire length (Wellnhofer 1974, p. 21).

In association with these observations it should be emphasized that several other characters in the cranial and postcranial skeleton of both *Campylognathoides* and *Eudimorphodon* suggest a closer relationship between these two taxa. Wellnhofer (1978) and Wild (1978) have indicated that both genera showed similarities, and they concluded that *Campylognathoides* could be derived phylogenetically from the Late Triassic *Eudimorphodon*.

Discussion and taxonomic assignment

The dentition comprising uni-, tri- and pentacuspid teeth is diagnostic for the Late Triassic genus *Eudimorphodon* and is not known in any other pterosaur. There is no doubt that the Seefeld specimen belongs to this taxon. Direct comparisons with known species of *Eudimorphodon* is limited, however, because of the fragmentary condition and the incompleteness of its skeleton. Two species of *Eudimorphodon* have been described from the Norian of northern Italy: *E. ranzii* Zambelli 1973 and *E. rosenfeldi* Dalla Vecchia 1995. Wild (1978), in his first diagnosis and in his emended diagnosis (1994) of *E. ranzii*, listed the following characters that are relevant for comparison with the Seefeld specimen: up to 28 uni-, tri- and pentacuspid teeth in the mandible; lower jaw with coronoid; tetraradiate jugal; fibula a little longer than half of the tibia. Other characters are present only in adult individuals: a rectangular sternum, co-ossified scapulocoracoid, massive humerus and fused tibia/fibula. In juveniles of *E. ranzii* these characters are different due to incomplete ossification. So, the sternum is triangular, the scapulocoracoid is not co-ossified, the humerus is more slender and tibia/fibula are separate.

In the Seefeld specimen the number of mandibular teeth was probably less than 28, as mentioned above, because it was a smaller individual. It is in agreement with the diagnosis for *E. ranzii* in having a coronoid and a tetraradiate, but a more slender jugal. This slenderness can be interpreted as an immature condition, as can the triangular sternum, more slender humerus and the lack of synostosis of tibia and fibula. In contrast, however, the scapulocoracoid is firmly co-ossified in the Seefeld specimen.

Unfortunately, only a preliminary note on *E. rosenfeldi* from the Mid-Norian Dolomia di Forni of Udine Province has been published (Dalla Vecchia 1995). A detailed description is still in preparation (Dalla Vecchia, pers. comm.). Therefore, comparisons have to rely on descriptions rather than detailed illustrations. The relevant diagnostic characters of this species, which is about the size of the Seefeld specimen, are as follows: posterior part of the mandible at the coronoid process rather deep; very long diastema between posterior-most tooth and coronoid process; shaft of the coracoid short and wide; humeral shaft proportionately longer and more slender than in *E. ranzii*; deltopectoral crest narrow and rectangular in profile; tibia very long; fibula does not reach the distal part of tibia and is not fused to it proximally.

In the Seefeld specimen, the mandible at the coronoid is rather deep too. However, in the holotype of *E. ranzii*, this part of the lower jaw is obscured by cranial bones, and the restoration given by Wild (1978, fig. 4) is tentative in this respect. In the juvenile Milano specimen MPUM 6009 of *E. ranzii* the lower jaw at the coronoid is also rather deep. The shaft of the coracoid is also short and wide in the

Seefeld specimen, but it appears to be similar in the immature specimen MCSNB 2887 of *E. ranzii* (Wild 1978, fig. 16). A slender shaft of the humerus could be interpreted as a juvenile character, as present in the Milano specimen MPUM 6009, and the deltopectoral crest is rectangular in *E. ranzii* as well. The relatively long tibia of the Seefeld specimen remains the only character comparable with *E. rosenfeldi*. Comparisons of postcranial proportions are discussed below.

It is not possible here to analyse the specific characters given for *E. rosenfeldi*, nor can a revision of the taxonomic assignment of the individual specimens of *E. ranzii* be conducted. In addition, a possible and very probable effect of ontogenetic variability and sexual dimorphism on osteological morphology and skeletal proportions must be considered. However, the sample available is so small (less than ten specimens of the genus *Eudimorphodon*) that in practice it could not be decided whether the differences observed fall within the range of individual variability of the known species of *Eudimorphodon*. Since this is a problem of the palaeontological species concept in general, these uncertainties simply have to be accepted.

Five specimens from the Norian of Bergamo, Italy, have been assigned to *E. ranzii* by Wild (1978, 1994). Three of them are considered to be juveniles and immatures. However, one specimen, MCSNB 3496, has been shown by Dalla Vecchia (2001, 2003) and S. Renesto (pers. comm.) to belong to *Peteinosaurus* rather than to *Eudimorphodon*. Some peculiarities in the Seefeld specimen suggest also an immature or rather subadult stage, especially the smooth surface of the multicuspid teeth. The increase of enamel striation on the tooth crowns of *E. ranzii* was recognized as also age dependent by Wild (1978, p. 218). However, Dalla Vecchia (1995, p. 60) has shown that the pentacuspid teeth of *E. rosenfeldi* present smooth tooth crowns also, suggesting that this was more likely to be a specific character. However, both specimens are about the same size and a subadult condition in *E. rosenfeldi* should be considered. Furthermore, if the isolated pseudo-unicuspid tooth of the Seefeld specimen were to be a third mandibular tooth, it would be in accordance with the juvenile Milano specimen MPUM 6009 of *E. ranzii*, although not with the adult holotype specimen of that species. The triangular sternum of the Seefeld specimen shows more similarities to the juvenile specimen MCSNB 8950 of *E. ranzii* described by Wild (1994), and the lack of 'clavicles' might indicate that they were not yet ossified at the time of death of the individual.

The lack of elongated zygapophyses ('ossified tendons') in the caudals of the Seefeld specimen has to be taken as the original condition and has to be considered a primitive character that was retained in these basal pterosaurs. On the other hand, the normal condition in more advanced, long-tailed ('rhamphorhynchoid') pterosaurs is the development of greatly elongated caudal zygapophyses (and haemapophyses) which functioned as tail stiffeners (Wellnhofer 1975, 1978, 1991). In this respect the Late Liassic *Campylognathoides* with fully developed 'ossified tendons' (Wellnhofer 1974) has to be considered as more advanced than *Eudimorphodon*. The opinion of Unwin (1992, 1995), that both taxa were united by relatively advanced characters and constituted a distinct family, the Campylognathoididae, is therefore not supported. It is preferred here to maintain the assignment of *Eudimorphodon* to a family of its own: the Eudimorphodontidae (Wellnhofer 1978).

The same problem of interpretation emerges with regard to the morphology of the humerus. Compared to the holotype specimen of *E. ranzii*, the humerus is more slender, with a longer, narrower shaft in the Seefeld specimen. In addition to the high ratio (2.5) of maximum proximal width to maximum length, this morphology could be growth related and could again indicate a subadult stage for the Seefeld specimen. On the other hand, however, this high ratio was taken to be diagnostic for *E. rosenfeldi* by Dalla Vecchia (1995).

Generally, the degree of ossification is considered to be a growth-related criterion for individual age, independent of size. With regard to the Seefeld *Eudimorphodon* specimen there is no reason to conclude an early ontogenetic stage. Most of the bones appear to be completely ossified, as in the lower jaw, the vertebrae, the scapulocoracoid and the pelvis. Also the articular ends of the long bones are well ossified. There are two exceptions, however: the fibula, which is not fused to the tibia and is separated throughout its entire length, in contrast to the condition found in *E. ranzii* and *E. rosenfeldi*, and the sternum which, in its triangular shape, appears to be incompletely ossified.

Considering the relatively large size of the Seefeld specimen, the possibility that it was a very young individual can be excluded. Compared to the presumably adult holotype specimen of *E. ranzii*, and based on the lengths of the humeri, the Seefeld specimen was about 15% smaller. Using the same parameters it was 35% larger than the juvenile Milano specimen MPUM 6009. Assuming a similar adult size to *E. ranzii*, and considering the immature traits in the proportions of the humerus, in the incompletely ossified sternum and in the unfused tibia/fibula, the Seefeld specimen could be regarded as a subadult individual of *Eudimorphodon*. By comparison with the holotype of *E. ranzii*, its wing span can be estimated as about 70–80 cm, depending on which elements of the wing skeleton the estimations are based.

Table 1. *Lengths of humerus, wing phalanx 1 and tibia, and their proportions to each other in* Eudimorphodon *cf.* ranzii, *Seefeld specimen, BSP 1994 I 51, in comparison to a variety of Triassic and Liassic pterosaurs*

Specimen	Humerus (hu)	Wing phalanx 1 (wph1)	Tibia (ti)	hu/ti	wph1/ti	ti/hu	wph1/hu
Eudimorphodon cf. *ranzii* Seefeld specimen, BSP 1994 I 51	40	52.9	57.7	0.69	0.92	1.44	1.32
Eudimorphodon ranzii Holotype, MCSNB 2888 (Wild 1978)	47	—	—	—	—	—	—
Eudimorphodon ranzii MCSNB 2887 (Wild 1978)	28	39.5	28.5	0.98	1.39	1.02	1.41
Eudimorphodon ranzii MCSNB 8950 (Wild 1994)	26	33	25.5	1.02	1.29	0.98	1.27
Eudimorphodon ranzii MPUM 6009 (Wild 1978)	26	38.5	—	—	—	0.96	1.48
Eudimorphodon rosenfeldi MFSN 1797 (Dalla Vecchia 1995)	40.5	64	54.2	0.75	1.18	1.34	1.58
Eudimorphodon cromptonellus MGUH VP 3393 (Jenkins *et al.* 2001)	18.15	18*	20.5*	0.89*	0.88*	1.13*	0.99*
Peteinosaurus zambellii MCSNB 3359 (Wild 1978)	38.5	43	49	0.79	0.88	1.27	1.12
Preondactylus buffarinii MFSN 1770 (Dalla Vecchia 1998)	32	35.5	44	0.73	0.81	1.36	1.11
Dimorphodon macronyx BMNH 41212–13 (Wellnhofer 1978)	90	108	126	0.71	0.86	1.40	1.20
Campylognathoides liasicus CM 11424 (Wellnhofer 1974)	50.3	93.3	47.4	1.06	1.97	0.94	1.85
Dorygnathus banthensis SMNS 18969 (Wild 1978)	57	78.5	66	0.86	1.19	1.16	1.38

Lengths of wing phalanx 1 and tibia, originally estimated by Wild (1978) in specimen MCSNB 2888 (holotype) and in the Milano specimen MPUM 6009, are omitted. They are considered doubtful estimates and are not used here.
† Measurements in mm, * estimated.

Finally, the proportions in the postcranial skeleton need to be compared. The possibilities are limited, however, since reliable measurements are available only for humerus, wing phalanx 1 and tibia (Table 1). Here especially, the relatively long tibia (ti) in proportion to the humerus (hu) is remarkable. The ti/hu ratio in the Seefeld specimen is 1.44, which comes close to *E. rosenfeldi* (1.34). In *E. ranzii* this ratio ranges between 0.98 in the juvenile specimen MCSNB 8950 and 1.02 in the immature specimen MCSNB 2887. Neither in the juvenile Milano specimen MPUM 6009 nor in the adult holotype specimen of *E. ranzii* is the tibia preserved completely, and the measurements given by Wild (1978) are only estimates and cannot be used. In only two out of four specimens of *E. ranzii* is the tibia preserved in its entire length. In the holotype of *E. ranzii* only the proximal part of the tibia is preserved, and Wild (1978) has estimated its entire length to have been 50 mm, resulting in a ti/hu ratio of 1.06. This index, however, is 1.44 in the Seefeld specimen and 1.34 in *E. rosenfeldi*. Both individuals are intermediate in size between the juveniles and the adult of *E. ranzii*. This suggests that the length of the tibia in the adult holotype of *E. ranzii* was underestimated, and might have been actually 65–70 mm. If this was the case, the Seefeld specimen, as well as *E. rosenfeldi*, would easily fit into a single growth series of one species, *E. ranzii*, with the implication that the tibia during ontogeny increases in length in proportion to the humerus (and to general body size), indicating a positive allometric growth of that element. Similar relative lengths of the tibia can be found in the other two Triassic pterosaur taxa, *Peteinosaurus zambellii* (ti/hu 1.27) and *Preondactylus buffarinii* (ti/hu 1.36). For this and other reasons, these two taxa were considered more closely related to each other by Dalla Vecchia (1998).

In contrast to *E. ranzii* and *E. rosenfeldi* the Seefeld specimen is characterized by a relatively short wing phalanx 1 (wph1). It is shorter than the tibia, according to a wph1/ti ratio of 0.92. However, the first wing phalanx is longer than the tibia in the two immature specimens, MCSNB 8950 and MCSNB 2887, of *E. ranzii*, representing a wph1/ti ratio of 1.29 and 1.39 respectively. In the holotype specimen of *E. ranzii* the first wing phalanx is only preserved in its proximal part. So again, the length given by Wild (1978) as 80 mm is an estimate and cannot be used for comparison. In *E. rosenfeldi* the first wing phalanx is also longer than the tibia, the wph1/ti ratio being 1.18. Still shorter than in the Seefeld specimen is wing phalanx 1 in *Peteinosaurus zambellii* (wph1/ti 0.88) and *Preondactylus buffarinii* (wph1/ti 0.81), but also in *Dimorphodon macronyx* from the Early Liassic of the United Kingdom (wph1/ti 0.86). The latter has also similar ratios of ti/hu (1.40) and wph1/hu (1.20). *Campylognathoides liasicus* and *Dorygnathus banthensis* from the Late Liassic of Holzmaden, however, are quite different in this respect (Table 1).

Conclusions

Based on the significant dentition, the pterosaur from the Late Norian of Seefeld in Tyrol, Austria, can be assigned to the genus *Eudimorphodon*. There are features that support a subadult stage for this individual. Most skeletal morphological features and proportions available are in accordance with especially immature specimens that have been assigned to *E. ranzii*, with the exception of an unusually short first wing phalanx. This, and the incomplete and fragmentary state of the fossil material make the assignment to this taxon uncertain, however. Nonetheless, it is preferred to use open nomenclature and to refer the Seefeld pterosaur to *Eudimorphodon* cf. *ranzii* Zambelli 1973.

I am indebted to the discoverer of the pterosaur material of *Eudimorphodon* cf. *ranzii*, B. Lammerer, Munich, who brought the fossil to my attention and who donated his find to the Palaeontological State Collection in Munich. He also guided me to the fossil locality and helped in collecting further fossil remains of the specimen. H. Kozur, Budapest, kindly provided a contribution to the stratigraphy of the Seefeld Beds and gave permission to publish it in this paper. I profited greatly from discussions with R. Wild, Stuttgart; F. Dalla Vecchia, Monfalcone, Italy; W. A. Clemens, Berkeley, USA; and E. Ott, M. Kirchner and H. Wierer, Munich. F. A. Jenkins Jr, Cambridge, USA, kindly provided a galley proof of a paper (2001) on a new Triassic pterosaur from Greenland. The reviewers of the manuscript, G. Cuny, Bristol, and S. Renesto, Milano, provided valuable suggestions and information that considerably improved the paper. Lastly, but not least, I benefited greatly from the technical skills of R. Liebreich, Munich, who carried out the difficult fossil preparations, prepared the casts and took the SEM photographs. F. Höck and G. Bergmeier, Munich, took the photographs. Without the help and support of these persons this study would not have been possible.

References

AGASSIZ, L. 1833–1843. *Recherches sur les Poissons Fossiles*. 5 vols & supplement. Petitpierre, Neuchâtel et Soleure.

BENNETT, S. C. 2001. The evolution of the pterosaurian pectoral girdle. *Strata, Série 1*, **11**, 15–17.

BRANDNER, R. & POLESCHINSKI, W. 1986. Stratigraphie und Tektonik am Kalkalpensüdrand zwischen Zirl und Seefeld in Tirol (Exkursion D am 3. April 1986). *Jahresberichte und Mitteilungen des Oberrheinischen Geologischen Vereines, Neue Folge*, **68**, 67–92.

CLEMENS, W. A. 1980. Rhaeto-Liassic Mammals from Switzerland and West Germany. *Zitteliana*, **5**, 51–92.

CUNY, G. 1993. Discovery of mammals in the Upper Triassic of the Jura (France). *In*: LUCAS, S. G. &. MORALES, M. (eds) *The Nonmarine Triassic*. New Mexico Museum of Natural History and Science Bulletin, **3**, 95–99.

CUNY, G. 1995. French vertebrate faunas and the Triassic–Jurassic boundary. *Palaeogeography, Palaeoclimatology and Palaeoecology*, **119**, 343–358.

CUNY, G., GODEFROIT, P. & MARTIN, M. 1995. Micro-restes de Vertébrés dans le Trias supérieur du Rinckenbierg (Medernach, G. D. Luxembourg). *Neues Jahrbuch für Geologie und Paläontologie, Abhandlungen*, **196** (1), 45–67.

CUNY, G., DELSATE, D., GODEFROIT, P. & ROCHE, M. 1997. Syren: A new site from the Upper Triassic of Luxemburg. *Abstracts of the 2nd European Workshop of Vertebrate Palaeontology, Espéraza-Quillan, 7–10 May 1997*.

DALLA VECCHIA, F. M. 1991. Note sulla stratigrafia, sedimentologia e palaeontologia della Dolomia di Forni (Triassico superiore) nella valle del Rio Seazza (Preone, Friuli-Venezia Giulia). *Gortania*, **12**, 7–30.

DALLA VECCHIA, F. M. 1995. A new pterosaur (Reptilia, Pterosauria) from the Norian (Late Triassic) of Friuli (northeastern Italy). Preliminary note. *Gortania*, **16**, 59–66.

DALLA VECCHIA, F. M. 1998. New observations on the osteology and taxonomic status of *Preondactylus buffarinii* Wild, 1984 (Reptilia, Pterosuria). *Bollettino della Società Paleontologica Italiana*, **36** (3), 355–366.

DALLA VECCHIA, F. M. 2000. A wing phalanx of a large basal pterosaur (Diapsida, Pterosauria) from the Norian (Late Triassic) of NE Italy. *Bollettino della Società Paleontologica Italiana*, **39** (2), 229–234.

DALLA VECCHIA, F. M. 2001. Triassic pterosaurs: unravelling the puzzle. *In*: *Two Hundred Years of Pterosaurs. A Symposium on the Anatomy, Evolution, Palaeobiology and Environments of Flying Reptiles, Toulouse, France, September 5–8, 2001*. *Strata, Série 1*, **11**, 33–35 [Abstract].

DALLA VECCHIA, F. M. 2003. New morphological observations on Triassic pterosaurs. *In*: Buffetaut, E. & Mazin, J-M. (eds) *Evolution and Palaeobiology of*

Pterosaurs. Geological Society, London, Special Publications, **217**, 23–43

DALLA VECCHIA, F. M., MUSCIO, G. & WILD, R. 1989. Pterosaur remains in a gastric pellet from the Upper Triassic (Norian) of Rio Seazza valley (Udine, Italy). *Gortania*, **10**, 121–132.

DALLA VECCHIA, F. M., WILD, R., HOPF, H. & REITNER, J. (2002). A crested rhamphorhynchoid pterosaur from the Late Triassic of Austria. *Journal of Vertebrate Paleontology*, **22**, 196–199.

FRASER, N. C. & UNWIN, D. M. 1990. Pterosaur remains from the Upper Triassic of Britain. *Neues Jahrbuch für Geologie und Paläontologie, Monatshafte*, **5**, 272–282.

GODEFROIT, P. 1997. Reptilian, therapsid and mammalian teeth from the Upper Triassic of Varangéville (northeastern France). *Bulletin de l'Institut Royal des Sciences Naturelles de Belgique, Science de la Terre*, **67**, 83–102.

GODEFROIT, P. & CUNY, G. 1997. Archosauriform teeth from the Upper Triassic of Saint-Nicolas-de-Port (Northeastern France). *Palaeovertebrata*, **26**, 1–34.

HAHN, G., LEPAGE, J. C. & WOUTERS, G. 1984. Cynodontierzähne aus der Obertrias von Medernach, Großherzogtum Luxemburg. *Bulletin de la Societé Belge de la Géologie*, **93**, 357–373.

HUNT, A. P. & LUCAS, S. G. 1993. Triassic vertebrate paleontology and biochronology of New Mexico. *New Mexico Museum of Natural History Science Bulletin*, **2**, 49–60.

JADOUL, F., MASETTI, D., CIRILLI, S., BERRA, F., CLAPS, M. & FRISIA, S. 1994. Norian-Rhaetian stratigraphy and paleogeographic evolution of the Lombardy Basin (Bergamasc Alps). *15th IAS Regional Meeting, April 1994, Ischia, Italy, Field Excursions, Excursion B1*, 5–38.

JENKINS, F. A. JR, SHUBIN, N. H. *et al.* 1994. Late Triassic continental vertebrates and depositional environments of the Fleming Fjord Formation, Jameson Land, east Greenland. *Meddelelser om Grønland, Geoscience*, **32**, 1–25.

JENKINS, F. A. JR, SHUBIN, N. H., GATESY, S. M. & PADIAN, K. (2001). A diminutive pterosaur (Pterosauria: Eudimorphodontidae) from the Greenlandic Triassic. *Bulletin of the Museum of Comparative Zoology, Harvard*, **156** (1), 151–170.

JUNG, W. 1992. Öl aus dem Karwendel. *Katalog der Mineralientage München 1992*, 100–103.

KNER, R. 1867. Nachtrag zur fossilen Fauna der Asphaltschiefer von Seefeld in Tirol. *Sitzungsberichte der k. k. Akademie der Wissenschaften Wien*, **1** (56), 898–913.

KRYSTYN, L. 1974. Probleme der biostratigraphischen Gliederung der Alpin-Mediterranen Obertrias. *In*: ZAPFE, H. (ed.) *Die Stratigraphie der Alpin-Mediterranen Trias, Symposium Wien 1973*. Schriftenreihe der Erdwissenschaftlichen Kommission der Österreichischen Akademie der Wissenschaften, **2**, 137–144.

LUCAS, S. G. & HUNT, A. P. 1994. The chronology and paleobiogeography of mammalian origins. *In*: FRASER, N. C. & SUES, H.-D. (eds) *In the Shadow of the Dinosaurs*. Cambridge University Press, Cambridge, 335–351.

MURRY, P. A. 1986. Vertebrate paleontology of the Dockum Group, western Texas and eastern New Mexico. *In*: PADIAN, K. (ed.), *The Beginnings of the Age of Dinosaurs*. Cambridge University Press, Cambridge, 109–137.

ORAVECZ, J. 1961. A Gerecse- és Buda-Pilisi-Hegység közötti rögterület triász képzödményei (Die Triasbildungen des Schollengebietes zwischen den Gerecse- und Buda-Piliser Gebirgen). *Földtany Közlöny*, **91**, 173–185.

PADIAN, K. 1981. Note of a new specimen of pterosaur (Reptilia: Pterosauria) from the Norian (Upper Triassic) of Endenna, Italy. *Rivista del Museo Civico di Scienze Naturali 'E. Caffi', Bergamo*, **2**, 119–127.

PETERS, D. 2000. A reexamination of four prolacertiforms with implications for pterosaur phylogenies. *Rivista Italiana di Paleontologia e Stratigrafia*, **106** (3), 293–336.

PEYER, B. 1956. Über Zähne von Haramyiden, von Triconodonten und von wahrscheinlichh synapsiden Reptilien aus dem Rhät von Hallau, Kt. Schaffhausen, Schweiz. *Schweizerische Paläontologische Abhandlungen*, **72**, 1–72.

RENESTO, S. 1993. An isolated sternum of *Eudimorphodon* (Reptilia, Pterosauria) from the Norian (Late Triassic) of the Bergamo Prealps (Lombardy, northern Italy). *Rivista Italiana di Paleontologia e Stratigrafia*, **99**, 415–422.

ROGHI, G., MIETTO, P. & DALLA VECCHIA, F. M. 1995. Contribution to the conodont biostratigraphy of the Dolomia di Forni (Upper Triassic, Carnia, NE Italy). *Memorie di Scienze Geologiche*, **48**, 125–133.

SEELEY, H. G. 1870. *The Ornithosauria: An elementary study of the bones of pterodactyles, made from fossil remains found in the Cambridge Greensand, and arranged in the Woodwardian Museum of the University of Cambrige*. Deighton, Bell & Co., Cambridge.

TINTORI, A. 1996. *Paralepidotus ornatus* (Agassiz 1833–1843): A semionotid from the Norian (Late Triassic) of Europe. *In*: ARRATIA, G. & VIOHL, G. (eds) *Mesozoic Fishes – Systematics and Paleoecology*. Dr. Friedrich Pfeil, Munich, 167–179.

UNWIN, D. M. 1992. The phylogeny of the Pterosauria. *Journal of Vertebrate Paleontology*, **12**(3) (supplement), 57A.

UNWIN, D. M. 1995. Preliminary results of a phylogenetic analysis of the Pterosauria (Diapsida: Archosauria). *Sixth Symposium on Mesozoic Terrestrial Ecosystems and Biota, Beijing, Short Papers*, 69–72.

UNWIN, D. M., LÜ, J. & BAKHURINA, N. N. 2000. On the systematic and stratigraphic significance of pterosaurs from the Lower Cretaceous Yixian Formation (Jehol Group) of Liaoning, China. *Mitteilungen aus dem Museum für Naturkunde, Berlin, Geowissenschaftliche Reihe*, **3**, 181–206.

WELLNHOFER, P. 1970. Die Pterodactyloidea (Pterosauria) der Oberjura-Plattenkalke Süddeutschlands. *Bayerischen Akademie der Wissenschaften, Mathematisch-Naturwissenschaftliche Klasse, Abhandlungen, Neue Folge*, **141**, 1–133.

WELLNHOFER, P. 1974. *Campylognathoides liasicus* (Quenstadt), an Upper Liassic pterosaur from

Holzmaden – The Pittsburgh specimen. *Annals of the Carnegie Museum, Pittsburgh*, **45**, 5–34.

WELLNHOFER, P. 1975. Die Rhamphorhynchoidea (Pterosauria) der Oberjura-Plattenkalke Süddeutschlands. Teil I: Allgemeine Skelettmorphologie. *Palaeontographica, A*, **148** (1–3), 1–33.

WELLNHOFER, P. 1978. Pterosauria. *In*: WELLNHOFER, P. (ed.) *Handbuch der Paläoherpetologie.* Gustav Fischer, Stuttgart, vol. 19, 82 pp.

WELLNHOFER, P. 1991. *The Illustrated Encyclopedia of Pterosaurs*. Salamander, London, 192 pp.

WILD, R. 1978. Die Flugsaurier (Reptilia, Pterosauria) aus der Oberen Trias von Cene bei Bergamo, Italien. *Bolletino della Società Paleontologica Italiana*, **17** (2), 176–256.

WILD, R. 1984. A new pterosaur (Reptilia, Pterosauria) from the Upper Triassic (Norian) of Friuli, Italy. *Gortania,* **5**, 45–62.

WILD, R. 1994. A juvenile specimen of *Eudimorphodon ranzii* Zambelli (Reptilia, Pterosauria) from the Upper Triassic (Norian) of Bergamo. *Rivista del Museo Civico di Scienze Naturali 'E. Caffi', Bergamo*, **16**, 91–115.

ZAMBELLI, R. 1973. *Eudimorphodon ranzii* gen. nov., sp. nov., uno pterosaurio Triassico. *Rendiconti dell Istituto Lombardo di Scienze e Lettere* (B), **107**, 27–32.

New morphological observations on Triassic pterosaurs

FABIO M. DALLA VECCHIA

Museo Paleontologico Cittadino of Monfalcone, Via Valentinis 134, I–34074 Monfalcone (Gorizia), Italy (e-mail: fabdalla@tin.it)

Abstract: Since 1973, about 20 specimens of Triassic pterosaurs have been found in northern Italy, Austria and Greenland, belonging to *Eudimorphodon*, *Peteinosaurus*, *Preondactylus* and *Austriadactylus*. Their age is middle to late Norian and *Eudimorphodon* is the most common genus. The restudy of the specimens shows that *Peteinosaurus* presents trimorphodonty in the lower jaw and a fibula unreduced in length, distally expanded and fused to the tibiotarsus above the lateral condyle. Specimen MCSNB 3359 does not show diagnostic features of *Peteinosaurus* and is referred to it with doubt, whereas MCSNB 3496 is not *Eudimorphodon* but *Peteinosaurus*. *Preondactylus*, *Peteinosaurus*, ?*Peteinosaurus* and *Dimorphodon* could form a monophyletic group.

The tarsus of Triassic pterosaurs consists of two proximal tarsals, which fuse to the tibia during ontogeny, forming a tibiotarsus, and two distal tarsals. The larger of the two proximal tarsals was probably the calcaneum.The lateral condyle of the tibiotarsus is larger or more well formed than the medial one. The shape of the distal tarsals is similar to that of the distal tarsals in *Dimorphodon*. The metatarsals did not spread and the foot was ectaxonic; metacarpal length increases from metacarpal I to IV. This suggests that footprints of Triassic pterosaurs were different from *Pteraichnus*-like footprints.

Some features are unique to Triassic pterosaurs. *Eudimorphodon* and *Austriadactylus* have a multicuspid dentition, *Peteinosaurus* has cuspules in the distal teeth of the lower jaw and *Preondactylus* has serrated maxillary teeth. This could be a convergent feature or symplesiomorphic for their clade. *Eudimorphodon*, *Preondactylus* and *Austriadactylus* have very large maxillary teeth below the ascending process of the maxilla. *Eudimorphodon*, *Austriadactylus* and possibly *Preondactylus* do not have the bundles of elongated caudal zygapophyseal and haemapophyseal processes which are present in *Peteinosaurus* and in the Jurassic long-tailed pterosaurs.

Triassic pterosaurs have been known only since 1973, when the holotype of *Eudimorphodon ranzii* was found in a quarry near Cene, Bergamo Province, in the Lombardy region of northern Italy (Zambelli 1973). Since that discovery other specimens have been collected at Cene (Wild 1978) and some other Late Triassic sites in Lombardy (Wild 1984, 1994; Renesto 1993). In 1982 a pterosaur specimen (the holotype of *Preondactylus buffarinii*) was discovered in the Late Triassic rocks of Seazza Creek valley near Preone, in the Udine Province of Friuli region in northern Italy (Wild 1984). After that discovery many other specimens, only partly described, have been collected near Preone and nearby localities (Dalla Vecchia 1994, 1995, 1998, 2000; Dalla Vecchia *et al.* 1989). A pterosaur was reported in 1993 from the Late Triassic of Eastern Greenland (Jenkins *et al.* 1993; 2001) and recently pterosaur skeletons have been reported from the Late Triassic of Tyrol, Austria (Wellnhofer 2001; Dalla Vecchia *et al.* 2002).

In addition to the discovery of more or less complete skeletons, some fragmentary remains and isolated teeth from Late Triassic sites have been identified or reinterpreted as pterosaurian (see Dalla Vecchia 1994 for a review).

All the more or less complete specimens, excluding the Greenland one, are fossilized in laminated carbonate rock, crushed and preserved on slabs.

This paper reports some observations on Triassic pterosaur specimens, some previously described (Wild 1978, 1984, 1994; Dalla Vecchia 1995, 1998) and some new. The focus is on aspects of Triassic pterosaur osteology that have been misunderstood or not considered for their possible taxonomical relevance (e.g. dentition, metacarpus, tibia, structure of tarsus and foot, caudal segment of the vertebral column and osteological features of immaturity).

Institution abbreviations: BSP, Bayerische Staatssammlung für Paläontologie und historische Geologie, Munich; MCSNB, Museo Civico di Scienze Naturali, Bergamo, Italy; MFSN, Museo Friulano di Storia Naturale, Udine, Italy; MGUH, Geological Museum, University of Copenhagen, Denmark; MPUM, Dipartimento of Scienze della Terra, University of Milano, Italy; SMNS, Staatliches Museum für Naturkunde, Stuttgart, Germany.

Terminology

Concerning stratigraphy, the Norian stage is considered to comprise Lacian, Alaunian and Sevatian sub-

From: BUFFETAUT, E. & MAZIN, J-M. (eds) 2003. *Evolution and Palaeobiology of Pterosaurs.* Geological Society, London, Special Publications, **217**, 23–44. 0305–8719/03/$15 © The Geological Society of London 2003.

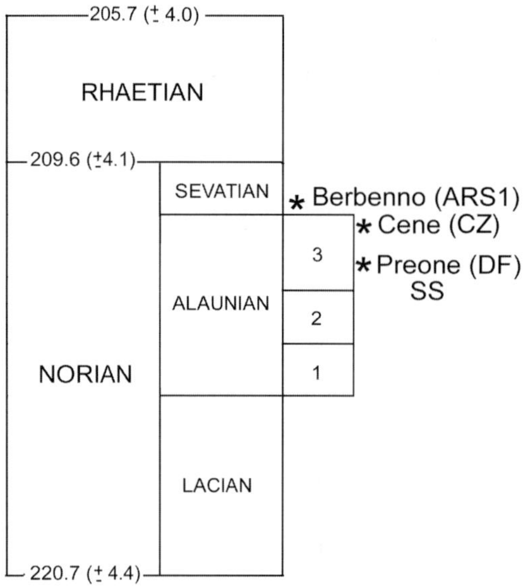

Fig. 1. Chronostratigraphical position of the lithostratigraphical formations and main sites with skeletal remains of Triassic pterosaurs. Chronostratigraphical scale based on Gradstein *et al.* (1995), modified. ARS1, lower part of the Argilliti di Riva di Solto; CZ,Calcare di Zorzino; DF, Dolomia di Forni; SS, Seefelder Schichten.

stages, corresponding respectively to the early, middle to late Norian (Fig. 1). The Sevatian corresponds to the *Quinquepunctatus* Ammonite Biochronozone, according to Gradstein *et al.* (1995). The Alaunian is divided into three ammonite zones; e.g. Alaunian 3 corresponds to the *Halorites macer* Zone of the Alpine-Tethyan region.

The terminology used for the teeth and dentition in general is that of Edmund (1969). In particular, 'mesial' and 'distal' are used instead of 'anterior' and 'posterior' to indicate the relative position of a tooth in the tooth row. The terms 'cusps' and 'cuspules' are used for topographically separate elevations on teeth. A serrate tooth is considered to be a multicuspid tooth with a high number of small cusps along the cutting edge of the crown.

The term 'basal pterosaurs' rather than 'rhamphorhynchoids' (i.e. all the genera of the suborder Rhamphorhynchoidea of the Linnnean taxonomy) is used to indicate all pterosaurs not included in the clade Pterodactyloidea of the phylogenetical taxonomy (i.e. all the genera of the suborder Pterodactyloidea of the Linnnean taxonomy).

Most of the caudal section of the vertebral column in Jurassic long-tailed pterosaurs has extremely elongated and thin pre- and postzygapophyseal processes and ventral haemapophyseal processes. The dorsal processes overlap each other, as do the ventral ones, forming dorsal and ventral bundles of filiform bones that stiffen the tail (see Wellnhofer 1991, p. 51); for brevity, these are called 'bundles' in the text. The firmness produced by the 'bundles' is such that, in many disarticulated skeletons of long-tailed Jurassic pterosaurs, the tail acts as a single compact skeletal element, and isolated mid-tail vertebrae are very rarely if ever found (e.g. Kremmling 1912, pl. 6; Broili 1939, pl. I; Wellnhofer 1975b, pls 13, fig. 1 & 15, figs 1 & 2; Wellnhofer 1975c, pl. 4 [30], fig. 3; Wellnhofer 1991, pp. 73 [lower], 76 & 82 [lower]).

Material

Eudimorphodon Zambelli 1973

This is the most common and widespread Triassic genus. It occurs in Lombardy (Calcare di Zorzino [Zorzino Limestone Formation] and Argilliti di Riva di Solto [Riva di Solto Shale Formation]), Friuli (Dolomia di Forni [Forni Dolostone Formation]), Tyrol (Seefelder Schichten [Seefeld Beds]) (Wellnhofer 2001) and eastern Greenland (Fleming Fjord Formation; Jenkins *et al.* 1993). Teeth attributed to *Eudimorphodon* have been reported from the Late Triassic of the southwestern United States (Chatterjee 1986; Murry 1986), Switzerland (Clemens 1980), Luxembourg (Cuny *et al.* 1995) and France (Godefroit 1997; Godefroit & Cuny 1997). However, teeth described by Chatterjee (1986) as *Eudimorphodon* are most probably cynodont teeth (S. Chatterjee, pers. comm.).

Three species (*E. ranzii* Zambelli 1973, *E. rosenfeldi* Dalla Vecchia 1995 and E. *cromptonellus* Jenkins *et al.* 2001) have been named. The holotype of *E. ranzii* (MCSNB 2888) is a skeleton of a relatively large individual without most of the tail, hind limbs and wing fingers (Wild, 1978, pls 1 & 2). MPUM 6009 (known also as the Milano specimen) is a decidedly smaller (Table 1), nearly complete individual, but is poorly preserved; part of the skeleton is represented only by an impression of the bones (Wild 1978, pls 4 & 5). MCSNB 8950 is an articulated skeleton of a small individual without the skull, lower jaw and most of the neck and tail (Wild 1994, figs 1–4) and is the only pterosaur from the Argilliti di Riva di Solto. MCSNB 2887 preserves part of a disarticulated skeleton without skull elements (see Wild 1978, pls 6b & 8). MCSNB 3496 is an originally articulated but only partly preserved skeleton (Wild 1978, pls 6a & 7); it belongs to an individual smaller than MCSNB 2888 and larger than MPUM 6009. MCSNB 3496 has been attributed to *E. ranzii* (Wild 1978, p. 183), but none of the characters used for this attribution is actually diagnostic of *Eudimorphodon*, whereas its features of *Peteinosaurus*

Table 1. *Measurements (mm) of long bones of Triassic pterosaurs*

	h	u	mcIV	wph1	wph2	wph3	wph4	fe	ti
Eudimorphodon									
MCSNB 2888	47	65	29	80*	—	—	—	41	50*
MCSNB 2887	28	38	—	40	36.5	—	—	21.2	28.5
MPUM 6009	26.3	36	10.5	37.5	33*	36.2	34*	18.5	25*
MCSNB 8950	26	33.5	9	34	35.3	36.2	32	19.6	25
MFSN 1797	42	55	21	14.5	58.2	63.2	51.5	37	54.2
Peteinosaurus									
MCSNB 2886	—	—	16.5	42.5	41*	—	34*	—	51.5
?*Peteinosaurus*									
MCSNB 3359	39	49.5	17	42.5	42.5	46.5	35	37	48
Preondactylus									
MFSN 1770	32	42	14.25	35.5	39	39	28	32.5	44

* measurements estimated or approximate.
fe. femur; h, humerus; mcIV, wing metacarpal; ti, tibia; u, ulna; wph1–4, wing phalanges 1–4.

are clear (see below). MCSNB 3345 is a single, isolated maxillary tooth (Wild 1978, fig. 9) and MPUM 7039 is an isolated pterosaurian sternum attributed to *E. ranzii* by Renesto (1993). *E. rosenfeldi* is represented by the holotype MFSN 1797, a nearly complete and articulated skeleton from the Friuli region lacking part of the skull and lower jaw, most of the pelvic girdle and the sacral and caudal segment of the vertebral column (Dalla Vecchia 1994, 1995). Two other *Eudimorphodon* specimens, MFSN 1922 (Dalla Vecchia 1994) and MFSN 21545 (pers. obs.) come from the same area as MFSN 1797. *Eudimorphodon cromptonellus* is based on a specimen from Greenland (MGUH VP 3393; Jenkins *et al.* 2001). A specimen from Austria (BSP 1994 I 51) has been referred to *E.* cf. *ranzii* (Wellnhofer, 2001).

Peteinosaurus Wild 1978

This monospecific genus (*P. zambellii* Wild 1978) was erected on the basis of two specimens from the Calcare di Zorzino of Cene, Lombardy. MCSNB 2886, the holotype, is a disarticulated and very incomplete skeleton (Wild 1978, pls 11 & 12). The preserved bones are mainly a mandibular ramus without both extremities, the posterior part of another jaw ramus, a few disarticulated skull bones, a tibia + fibula, wing phalanges 1 and wing metacarpals, an anterior portion of the ischiopubic plate and a probable sternal plate. MCSNB 3359 (Wild 1978, pls 13 & 14) is a pterosaur skeleton lacking most of the cervical segment of the vertebral column and the whole skull and lower jaw. It has been referred to *Peteinosaurus zambellii* (Wild 1978), but this identification is ambiguous (see below). MCSNB 3496 is considered here as a *Peteinosaurus* specimen (see below).

Preondactylus Wild 1984 (monospecific)

Preondactylus buffarinii Wild 1984 is based on the holotype (MFSN 1770 from the Dolomia di Forni of Friuli), a nearly complete skeleton represented mostly by the impression of the bones (Wild 1984, figs 1–3; Dalla Vecchia 1998, fig. 1). Another specimen (MFSN 25161) is still under study by the author, and a second, formerly kept in a private collection, has been cited by Dalla Vecchia (1994). Both come from the Dolomia di Forni.

Austriadactylus Dalla Vecchia *et al.* 2002 (monospecific)

Austriadactylus cristatus Dalla Vecchia *et al.* 2002 is represented by a nearly complete, but not well-preserved, articulated skeleton (SMNS 56342) found in the Seefelder Schichten of Tyrol (Dalla Vecchia *et al.* 2002).

Indeterminate material

MCSNB 4562 from Zogno, Lombardy, is a partial wing finger (wing phalanx 2 incomplete and 3 and 4 complete) of a large pterosaur identified as cf. *Preondactylus buffarinii* because of the ratio wph2/wph3 (Wild 1984), but it has been considered as Pterosauria indet. by Dalla Vecchia (1994, 1998). The bones preserved in a gastric pellet of a predator (MFSN 1891), derived from the Dolomia di Forni of Friuli, have been attributed to cf. *Preondactylus buffarinii* (Dalla Vecchia *et al.* 1989), but such a specific taxonomical determination now appears doubtful. In fact, the specimen does not show any osteological features to suggest that it is *Preondactylus* rather than *Eudimorphodon* or *Peteinosaurus,* and the ratios of long bone lengths, all based on estimated measurements, are similar to those of MCSNB 3359. Other taxonomically indeterminate remains from the Dolomia di Forni include a single large wing phalanx 4 (MFSN 19836) (Dalla Vecchia 2000) and a partial segment of a caudal vertebral column with two wing phalanges 4 (MFSN 19864; Dalla Vecchia 2001). Fraser & Unwin (1990) described, as pterosaurian wing metacarpals, two small bones from the fissure fillings of Gloucestershire in the United Kingdom, which are probably Norian in age.

In conclusion, there are about 20 remains of Triassic pterosaurs (excluding teeth from USA, Switzerland, Luxembourg and France). However, the incomplete and often 'non-overlapping' preservation (e.g. a taxon represented by a lower jaw without a skull, another represented mainly by the skull and with poorly preserved lower jaw) sometimes makes comparison impossible. Understanding the relationships among the specimens of Triassic pterosaurs is like unravelling a puzzle, each new specimen or specimen revision being a step towards the resolution.

Dating

Only the dating of levels containing specimens determined at the generic level will be considered here. Localities with *Eudimorphodon* teeth only are not considered.

Most Triassic pterosaurs (*Eudimorphodon*, *Peteinosaurus*, *Preondactylus* and *Austriadactylus*) come from rocks of marine origin in northern Italy (Lombardy and Friuli) and Austria (Tyrol). The pterosaur-bearing rocks represent deposition in small, tectonically controlled anoxic basins (Dalla Vecchia 1991; Jadoul *et al.* 1992; Hagen Hopf, Diplomarbeit und Diplomakartierung dissertation, University of Göttingen, 1997) in a wide carbonate platform situated at the northern margin of the western Tethyan corner (Gaetani *et al.* 2000).

Most of the pterosaurs from Lombardy were collected from a bed 15 cm thick (A. Paganoni, pers. comm.) in the uppermost part of the Calcare di Zorzino (CZ) in a quarry near Cene that is 20–30 m below the boundary with the overlying Argilliti di Riva di Solto (ARS) (Wild 1978). Specimens from this horizon include the holotype of *Eudimorphodon ranzii* (MCSNB 2888), MCSNB 2887, MPUM 6009 and MCSNB 3496, and the two specimens attributed by Wild (1978) to *Peteinosaurus zambellii* (MCSNB 2886 and MCSNB 3359). Two other fragmentary specimens (MPUM 7039 and MCSNB 4562) have been collected from the Endenna/Zogno site of the Bergamasc Pre-Alps in the uppermost part of the CZ just at the boundary with ARS (A. Tintori, pers. comm.). A single specimen (MCSNB 8950) comes from the lower part of the ARS (ARS1) at Ponte Giurino/Berbenno (Bergamo). The CZ is a sequence of dark grey to black, well-bedded limestone with a maximum thickness of 300 m (Jadoul 1986). It is the lateral equivalent of part of the Dolomia Principale Formation (DP – the Hauptdolomit Formation of German authors). The ARS is a shale-limestone sequence that was deposited above the CZ and locally directly above the DP. The ARS1 is represented by 20–180 m of black shales and marls. The uppermost CZ is dated as latest middle Norian (latest Alaunian) by palynomorphs, whereas the ARS1 at Ponte Giurino is earliest late Norian (earliest Sevatian) (Jadoul *et al.* 1994). The beginning of deposition of the ARS corresponds to a widespread and recognizable event in Europe marked by a dilution of the waters in the marine basins (possibly due to a climatic change to more humid conditions) and by the supply of terrigenous sediment from northern domains (Stefani *et al.* 1992; Jadoul *et al.* 1994; Cirilli 1995). According to Jadoul *et al.* (1992) climate change is reflected terrestrially by a change in the palynomorph assemblage and by a transition in marine settings from evaporites and dolostone to limestone, clay and shales.

The site of Cene also contains *Aetosaurus ferratus*, which permits a correlation with the Lower Stubensandstein of southern Germany (Wild 1989) and the Ørsted Dal Member of the Fleming Fjord Formation of Greenland (Jenkins *et al.* 1993).

The specimens from Friuli come from different sites of the Dolomia di Forni (DF) in a range of about 20 km (see Dalla Vecchia 1991, 1994, 2000; Roghi *et al.* 1995). The DF is a sequence of well-bedded, dark grey to black or brown, bituminous dolostones with chert, which is 700–850 m thick in the section of Seazza Creek near Preone. The middle–lower part of DF (*sensu* Dalla Vecchia 1991) in the sections of the Seazza Creek and Forchiar Creek, where MFSN 1770, MFSN 19864, MFSN 1891, MFSN 21545, a still undescribed *Preondactylus* specimen and MFSN 1797 have been found respectively, is locally abundant in the conodont *Epigondolella slovakensis*. Based on the size and morphological variation of this taxon in the samples from different positions of the stratigraphical column, and compared with the trends observed in *Epigondolella* species in the Norian of British Columbia (Orchard 1991), Roghi *et al.* (1995) preliminarily dated the level from which the holotype of *Preondactylus buffarinii* (MFSN 1770) was recovered as Early Alaunian 3 (*Epigondolella serrulata* Zone of Orchard 1991). The level that yielded the holotype of *Eudimorpohodon rosenfeldi* (MFSN 1797) is slightly older, but is still in the lower part of Alaunian 3. The lower part of the DF in the Seazza Creek section is dated to Alaunian 2 (Roghi *et al.* 1995). In samples from the Rovadia Creek (where MFSN 19836 has been found) and other western sections of the DF, *Epigondolella slovakensis* or small *E. slovakensis* and *E. postera* are preserved together, and the dating is also Alaunian 2–3 (Carulli *et al.* 1998; G. Roghi pers. comm.). The Calcare di Chiampomano (CC) is found above the DF and is still a basinal unit, but limestone instead of dolostone predominates; a palynomorph assemblage at its basal part suggests more humid conditions (see the palynomorph list in Carulli *et al.* 1998, 2000). CC corresponds to the 'dilution' event characterizing the ARS1 of Lombardy (Dalla Vecchia 1996; Carulli *et al.* 1998). DF is the lateral equivalent of part of the DP.

A monospecific conodont association consisting of *Epigondolella slovakensis* has also been found in the Rezi Dolomite Formation (RD) of the Keszthely Mountains of western Hungary (Budai & Kovacs 1986). RD is a sequence of 150 m of light brownish to grey, well-bedded bituminous dolostone, and dolomitic marls with chert and intercalations of marly shales. It is underlain by the Hauptdolomit (DP) and overlain by the marly-shaly Kössen Formation (KF) (Budai & Kovacs 1986). The comparison of the conodonts to those found in an ammonite-controlled section of Timor suggests an Alaunian 3 dating (Krystyn's unpublished data in Budai & Kovacs 1986, p. 185). The stratigraphical position of the RD is considered the same as the Plattenkalk of the Northern Calcareous Alps (Budai & Kovacs 1986).

Specimens from Tyrol (*Austriadactylus* and *Eudimorphodon*) are from the Seefelder Schichten (SS): 250–400 m of brown to black, well-bedded dolostone somewhat similar to the DF (see Brandner & Poleschinski 1986; pers. obs.). The SS overlying the Hauptdolomit (DP) is the lateral equivalent of the Plattenkalk and is covered by the KF, of which the lower part is Sevatian (see Plöchinger 1980; Brandner & Poleschinski 1986). The lower part of KF corresponds to the same events of 'dilution' and terrigenous input which led to the deposition of the ARS1 in Lombardy (Riva *et al.* 1986). The SS are considered to be Late Alaunian–Early Sevatian by Brandner & Poleschinski (1986, fig. 1). However, this formation contains the same conodonts as DF (*Epigondolella slovakensis* and *E. postera*) plus *E. carinata* (D. A. Donofrio, pers. comm. to G. Roghi 1993), and the pelagic hydrozoan *H. conglobatum* in its upper part (Brandner & Poleschinski 1986). *H. conglobatum* appears in the *Halorites macer* Zone (Alaunian 3) in a section of Timor (Krystyn & Wiedmann 1986; Brandner & Poleschinski 1986). The conodont taxa reported from the marginal facies of the SS by Poleschinski (1986) (*Epigondolella postera*, *E. abneptis abneptis*, *E. bidentata*) range from Late Lacian to Sevatian according to Kovacs *et al.* (1989). In particular, *E. abneptis abneptis*, not associated with *E. abneptis spalatus*, ranges from the Late Lacian (Lacian 3) to the Alaunian 2, whereas *E. bidentata* is a Sevatian taxon.

It should be recognized that Kozur (1989) considers *Epigondolella slovakensis* to be a typical Sevatian species and his interpretation would change the relative stratigraphical position of DF and SS with respect to CZ and ARS1. Kozur's dating scheme has not been incorporated into Figure 1 because his conclusions are not in accord with the observations of most other authors or with the sequence of events suggested by the lithostratigraphy.

The Ørsted Dal Member of the Fleming Fjord Formation of Greenland (where *Eudimorphodon cromptonellus* was found) has been considered to be middle Norian by Jenkins *et al.* (1993) based on its vertebrate fauna, but the age of the pterosaur-bearing bed appears to be Late Norian according to Clemmensen *et al.* (1998, fig. 3).

In summary, within the limits of the biostratigraphical resolution and interpretation of the microfossil ranges (e.g. Kozur's range of *Epigondolella slovakensis*), the Norian levels that have yielded pterosaur remains have a rather similar age (Fig. 1). *Eudimorphodon*, *Peteinosaurus*, *Preondactylus* and *Austriadactylus* were living during the Mid- to Late Norian in the same geographical region of the world and in a similar carbonate platform setting.

Revised characters of Triassic pterosaurs

Restudy of Triassic pterosaur specimens has permitted revision of some aspects of previous descriptions and interpretations.

Eudimorphodon

This genus will be the object of a separate monograph and thus it is only briefly considered here. The metacarpals of MCSNB 2888 show an increasing length from I to IV (Wild 1978, pl. 2). This condition is observed also in MPUM 6009, MCSNB 8950 and MFSN 1797, the holotype of *E. rosenfeldi* (Fig. 2b). In the latter specimen the shafts of the metacarpals are distally slightly curved dorsally and the distal condyle of each metacarpal is asymmetrically expanded dorsally to favour a dorsal flexion of the first phalanx. The distal curvature is observed also in some metacarpals of MCSNB 2888. A ventral extension of digit III appears impossible in MFSN 1797 because of the shape of the distal condyle of the metacarpal III and the obstacle represented by a metacarpal IV which is longer than metacarpal III (see Fig. 2b).

The fibula tapers distally and is reduced to a splint in MFSN 1797, ending just below the tibial midshaft. The same is observed in MCSNB 2887 and MCSNB 8950 from Lombardy. Thus, the fibula of *Eudimorphodon* specimens from northern Italy is

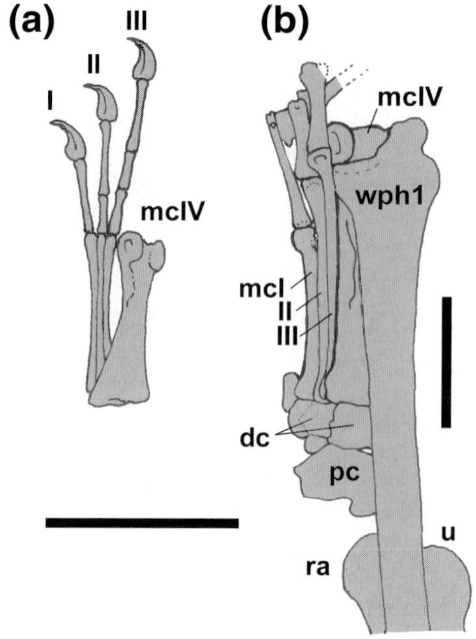

Fig. 2. Manus of pterosaurs. (**a**) Right manus of *Rhamphorhynchus longicaudus*, BSP 1889. (**b**) Right manus of *Eudimorphodon rosenfeldi*, MFSN 1797. dc, distal carpals; mcI–IV, metacarpals I–IV; pc, proximal carpal; ra,radius; u, ulna; wph1, wing phalanx 1. Scale bars 10 mm.

markedly shorter than the tibia and ends in a distal point. The tarsus is preserved and well exposed in MCSNB 8950, an immature individual with two proximal tarsals unfused to the tibia and two free distal tarsals (*contra* Wild 1994). One of the proximal right tarsals is much larger than the other and appears to be pulley-like (Wild 1994; Fig. 3a) in probable ventral view. The proximal tarsals in MFSN 1797 are fused to the tibia to form a bicondylar tibiotarsus with a lateral condyle more well formed and rounded than the medial one (Fig. 3b, c). Also in MCSNB 2887 the lateral condyle projects more anteriorly than the flat medial one. The metatarsals in MFSN 1797 are closely appressed, all parallel to each other, and have different lengths, the second being the longest. The same condition is observed in MCSNB 8950.

Only the first three caudal vertebrae and part of vertebra 4 exposed in ventral view are preserved in the holotype of *E. ranzii*. The long segment of the tail of MPUM 6009 is very poorly preserved. The postzygapophyses and prezygapophyses of the only two mid-caudal vertebrae preserved as bone are short and there is clearly no 'bundle' dorsally. When filiform bone structures are visible, they are on the ventral side, belong to the haemapophyses and do not form a 'bundle'. Two long mid-caudal vertebrae and a more proximal one (probably vertebra 4 or 5) are found isolated in MCNSB 2887, suggesting that the rigid 'bundles' were not developed. The 'ossified tendons', identified by Wild (1978, pl. 8) as being isolated from the corresponding vertebrae in a separated bundle, cannot be identified as caudal pre- and postzygapophyseal processes and are probably rib shafts. Only vertebrae 1 to 5 and part of vertebra 6 are preserved in MCSNB 8950 (Wild 1994, fig. 2). Vertebra 5 is already a rather elongated element (its centrum is about 3.5 times the length of a dorsal centrum), but clearly it is not bordered by 'bundles'. Elongated caudal vertebrae are always bordered by the 'bundles' in Jurassic long-tailed pterosaurs and also in MCSNB 3359 (see below). Finally, the caudal 'bundles' are absent in a still undescribed *Eudimorphodon* specimen (BSP 1994 I 51) from the Seefelder Schichten of Austria (Wellnhofer 2001). All of this evidence suggests the absence of elongated caudal pre- and postzygapophyseal processes and of the 'bundles' in *Eudimorphodon*.

Size-independent features of immaturity in pterosaurs have been discussed by Wellnhofer (1975a, 1991) and, above all, by Bennett (1993, 1995, 1996b). Only MCSNB 8950 and the holotype of *Eudimorphodon cromptonellus* (Jenkins *et al.* 2001) show clear evidence of non-fusion of skeletal elements (Wild 1994), whereas the other *Eudimorphodon* specimens, all noticeably smaller than the holotype of *E. ranzii*, show little evidence of this (e.g. the scapula is fused to the coracoid in MPUM 6009, MCSNB 2887 and MFSN 1797 and the proximal tarsals are fused to the tibia in MFSN 1797 and MCSNB 2887). However, the bone texture is grainy or 'orange-peel-like' in some zones of the bones of the small individuals, but this is also observed in some skeletal elements (sternum, sternal ribs, pelvis, articular surfaces of the left pteroid, left radius and right coracoid) of the large MCSNB 2888.

Peteinosaurus

MCSNB 2886, the holotype, was originally preserved on three slabs according to Wild (1978, pl. 12). The main slab A contains most of the preserved elements of the very disarticulated sleleton. The smaller slab B has the posterior part of a mandibular ramus, a few disarticulated skull bones and a probable sternal plate. The margins of slab B do not fit with those of slab A. The drawing of the bones on slab B is reversed in plate 12 of Wild (1978, compare pl. 12 to fig. 31 and pl. 15b); thus the slab B is a part of the counterslab. However, in this case, the drawing of Plate 12 in Wild (1978) shows the wrong side of the lower jaw, because it appears as a left ramus whereas Wild (1978) considers it to be a right one. Slab C is indicated by Wild (1978, p. 220) as the

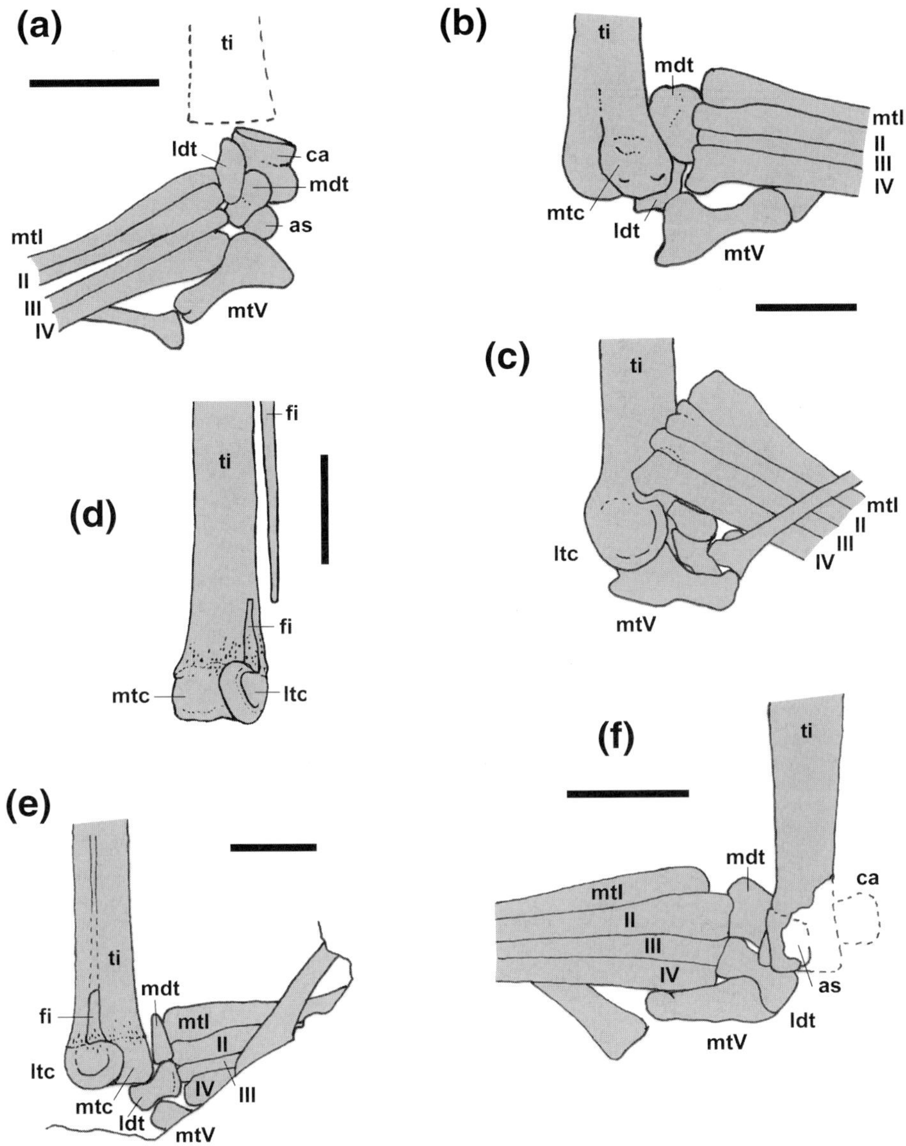

Fig. 3. The ankle of Triassic pterosaurs. (**a**) *Eudimorphodon ?ranzii*, MCSNB 8950, right hind limb. (**b**) & (**c**) *E. rosenfeldi*, MFSN 1797, left (b) and right (c) hind limb. (**d**) *Peteinosaurus zambellii* (MCSNB 2886), left tibiotarsus and fibula. (**e**) *Peteinosaurus zambellii* (MCSNB 3496), right hind limb. (**f**) *?Peteinosaurus* (MCSNB 3359), right hind limb. as, astragalus; ca, calcaneum; fi, fibula; ldt, lateral distal tarsal; ltc, lateral tibiotarsal condyle; mdt, medial distal tarsal; mtI–V, metatarsals I–V; mtc, medial tibiotarsal condyle; ti, tibia. Scale bar 5 mm.

counterslab of slab A, with the print of the anterior half of the right lower jaw ramus. Slab C cannot be found anymore at the MCSNB and probably was lost after 1978. Thus, for the anterior part of the lower jaw of *Peteinosaurus zambellii*, the only possible reference is Wild (1978).

According to the diagnosis of *Peteinosaurus* (Wild 1978, p. 219) the 'lower and upper jaw have monocuspid teeth, slightly curved distally, with mesial and distal sharp cutting edges'. In fact, these are characters of the teeth of the middle portion of the lower jaw. Nothing is preserved of the upper jaw. The dentition is defined as 'subthecodont' in the diagnosis by Wild (1978, p. 219). It can be seen in the segment of the mandibular ramus preserved in slab A that the teeth are set in crater-like alveoli placed along the margin of the lower jaw exposed to the observer ('labial' of Wild 1978). The other margin ('lingual') is lower than the crater-like structures, but between each crater-like alveolus the 'labial' margin is depressed and decidedly lower than the 'lingual' margin (see Wild 1978, fig. 31a). This arrangement changes in the posterior part of this preserved portion of the lower jaw, where the 'labial' margin, which is uniformly higher than the 'lingual' margin, obscures the 'lingual' margin. The partial ramus preserved on slab B has the margin exposed to the observer ('lingual' by Wild 1978) and is decidedly lower than the other ('labial'). The overhanging portion of the 'labial' margin tapers mesially. Here teeth seem to lean on the 'labial' wall (see Wild 1978, fig. 31b), but the bone is strongly crushed and the teeth are poorly preserved. The distal teeth differ from the preceding ones in having small cuspules (at least two–three per margin) along the mesial and distal cutting edges. Furthermore, the distal teeth are comparatively mesiodistally wider than the preceding ones and are not recurved backwards. Thus, a characteristic trimorphodonty seems to be present in the lower jaw of *Peteinosaurus*.

In the author's opinion, the possibility of the different heights of the two mandibular margins being caused or emphasized by the strong crushing and slight deformation of the mandible cannot be excluded. A further specimen is needed to disprove this possibility, but until it is discovered the author accepts this feature as valid.

The metacarpals show a slight increase in length from I to III. Distally the left tibia has a rounded, well-developed and well-formed lateral condyle and a decidedly unformed and anteriorly flat medial condyle (Figs 3d & 4a). A suture separates the condylar part from the tibial shaft and an 'orange-peel-like' texture is visible all around it. The fibula reaches the distal part of the tibia and is fused to it anterolaterally, just above the well-developed condyle. The fibula is slightly expanded distally and its distal ventral surface is fused to the upper surface of the lateral condyle, i.e. there is no free distal condyle on the fibula (Figs 3d & 4a). Its shaft is broken just above the fused distal part and is slightly translated laterally, giving the erroneous impression that the fibula is unfused and points distally.

As already noted by Dalla Vecchia (1998), MCSNB 2886 shows some evidence of osteological immaturity. Elements of the posterior part of the lower jaw and of the skull are unfused, the sternal plate is very small and probably mostly not ossified, the ischiopubic plates are not fused to each other at the ischial symphysis and to the corresponding ilium because there is an isolated plate, and a suture is identifiable between ischium and pubis. The distal part of the tibia just above the condyles is incompletely ossified ('orange-peel-like' texture) and a suture is still present between the tibial shaft and the fused proximal tarsals; evidence of incomplete ossification is found also in the distal end of the wing metacarpal and wing phalanx 1.

The anterior portion of the ischiopubic plate of MCSNB 3496 has the same shape as the anterior part of the ischiopubic plate of MCSNB 2886 (including the suture between pubis and ischium). The tibia has been misinterpreted as being the left (Wild 1978) because a splint of the crushed tibial shaft was misidentified as the fibula. In fact, the left tibia is the bone (still in connection with the left pes) identified as the wing metacarpal, and the right tibia is Wild's left tibia. In the latter, the development of the condyles is the same as in the left tibia in MCSNB 2886 and the fibula is long and slightly expanded at its distal end; the fibula is fused to the tibial shaft and to the lateral condyle just above the latter (Figs 3e & 4b). A segment of the thin fibular shaft is still preserved on the tibial shaft (Fig. 3e). Two distal tarsals are present: a wedge-shaped medial element in anterior view and a lateral one in dorsal view (Figs 3e & 4b).

The elongated mid-tail vertebrae of MCSNB 3496 have 'bundles'. The most complete centrum of the two articulated vertebrae preserved as bone, probably vertebra 6 or 7, has four to five filiform pre/postzygapophyseal processes in the partially preserved dorsal 'bundle'. Up to ten haemapophyseal processes are visible in the ventral 'bundle'. MCSNB 3496 also shows evidence of osteological immaturity. The ischiopubic plates are not fused to each other at the ischial symphysis or to the ilia; the ilia are not co-ossified to sacral ribs; there is a suture between the ischium and pubis; the ventral margins of the ischiopubic plates are scarcely ossified ('orange-peel-like' texture); the 'orange-peel-like' texture is found on the surface of the distal part of the right tibia just above the condyles and also on the proximal part of the corresponding metatarsals; and a suture is still visible between the proximal tarsals and the tibia.

The right metacarpus of specimen MCSNB 3359, the paratype of *Peteinosaurus zambelli* designated by Wild (1978), has a metacarpal III that is slightly longer than metacarpal II and a metacarpal I that is slightly shorter than metacarpal II.

The right fibula seems to taper and ends well before reaching the distal end of tibia; the left fibula is certainly longer, but also seems to taper before the tibial end. Because the posterior side of the crus is the only part exposed in both cases, the possibility exists that the distal portion of the fibulae cannot be seen because it is twisted anteriorly. However, this is conjectural; the fibulae actually appear shorter than in MCSNB 2886. The right tarsus is poorly preserved and the left one is covered by other bones, and so Wild's (1978, fig. 41b) reconstruction

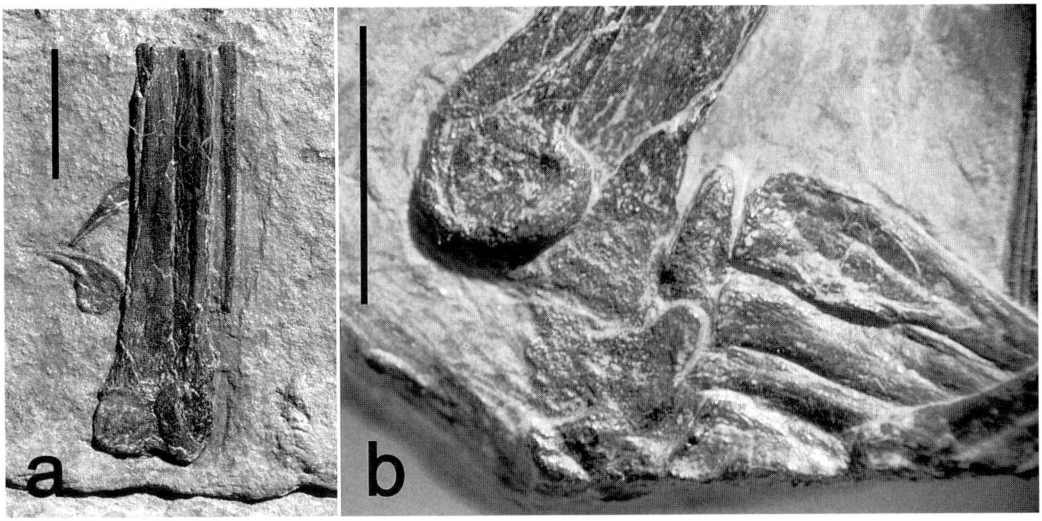

Fig. 4. Tibiotarsus of *Peteinosaurus*. (**a**) MCSNB 2886, holotype, left tibiotarsus. (**b**) MCSNB 3496, right tibiotarsus and pes. Note the distal part of the fibula, the suture between proximal tarsals and tibia with an 'orange-peel-like' texture all around and the asymmetrical development of the condyles. Scale bar 5 mm.

requires considerable interpretation. It can be interpreted in a substantially different manner (Fig. 3f). The tibia and pes are exposed in posterior (ventral) view. The tibia does not end in condyles (see also the left tibia). It appears to be anteroposteriorly flat and covers a partially preserved rounded bone identified as the astragalus by Wild, who also somewhat exaggerates its size. On the lateral side of the distal portion of the tibia there is the print of a squared or rounded bone (not reported by Wild 1978). It could be the print of the unfused calcaneum, but its actual size cannot be determined because the area is obscured by the covering glue. The bones identified as a separate calcaneum and a lateral distal tarsal by Wild (1978) are actually a single element divided by a fracture, the lateral distal tarsal in dorsoventral view. That fracture also extends into metatarsal V. In figure 41 of Wild (1978) the clear cut of the tarsal segment of the fracture is reported, but the segment of metatarsal V is omitted. The other element is quadrangular and is the medial distal tarsal in dorsoventral view. The metatarsals are closely appressed in both feet and are all the same length, unlike the condition in *Eudimorphodon* and most other basal pterosaurs.

The long segment of the caudal vertebral column of MCSNB 3359 is the best preserved and most complete among Triassic pterosaurs. The elongated prezygapophyseal processes appear as far anteriorly as the vertebra 3 (Wild 1978, pl. 14) and haemapophyses form the 'bundle' beginning at vertebra 5. At least four filiform processes are visible ventrally in vertebra 6, whereas five zygapophyseal processes are found dorsally. Ventral to vertebrae 8 and 14 there are six and five processes respectively, dorsally eight and seven to eight.

MCSNB 3359 also shows characters of immaturity, some of which have already been reported by Dalla Vecchia (1998). The ilia are not fused to the ischiopubic plates, the ilia and sacral ribs are not co-osified and the sacral vertebrae are not fused to each other. The ischiopubic plates and the process for the extensor tendon of wing phalanx 1 have a grainy aspect, indicating incomplete ossification, there is a neurocentral suture in some dorsal vertebrae, the tibia and fibula are probably not co-ossified proximally and the proximal tarsals are not fused to the tibia.

Preondactylus

Preondactylus buffarinii was first described by Wild (1984) and additional observations on the morphology and taxonomy of this species were made by Dalla Vecchia (1998). Recently, during the preparation of a new specimen from Seazza Creek, the author realized that the teeth of the upper jaw have several cuspules along each cutting edge, i.e. they are serrated (Fig. 5). This feature could not be appreciated in the holotype because the teeth are only preserved as impressions, and it was not visible in the new specimen before preparation because the serrate margins were covered by matrix. Thus, *Preondactylus* is now known to have had a multicuspid maxillary dentition. Also, the bone identified as the postorbital by Wild (1984), and considered as such also by Dalla Vecchia (1998), is more probably the jugal. Thus, all the statements in Dalla Vecchia

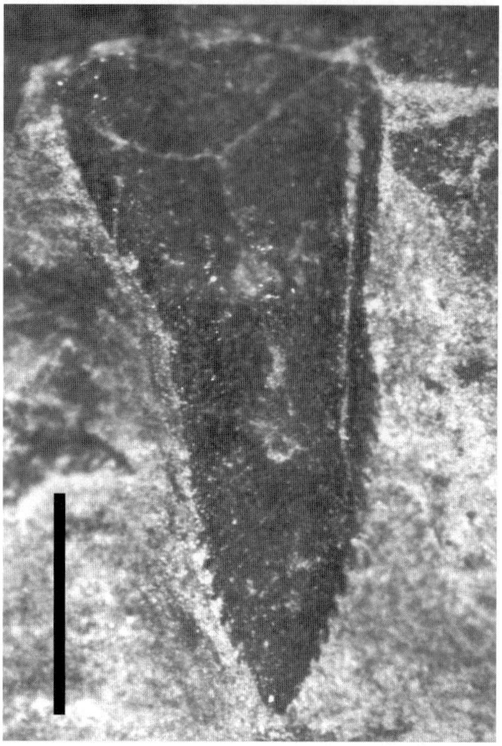

Fig. 5. The large tooth just below the ascending process of the maxilla in a specimen of *Preondactylus* (MFSN 25161). Note the cuspules along the cutting margins. Scale bar 1 mm.

(1998) based on that bone are questionable. The fibula of MFSN 1770 apparently ends well before the distal end of the tibia and a square condyle is visible medially at the distal termination of the left tibia. However, the poor preservation of the specimen and the ambiguity in the identification of the small elements, such as the tarsals, based only on impressions, suggest that better-preserved material is needed before making determinations about the tarsus and fibula of *Preondactylus*.

The caudal segment of the vertebral column of the holotype is bent at vertebra 4, and vertebrae 5 and 6 (which are elongated elements, the centrum of vertebra 6 being more than three times longer than a mid-dorsal centrum) do not show any evidence of long postzygapophyseal processes (Wild 1984, fig. 3). Ventral to the following caudal vertebrae 7–11, no evidence of a 'bundle' made of the haemapophyseal processes can be found. The 'bundles' are well developed between vertebrae 5 and 6 in MCSNB 3359 and, as seen above, elongated caudal vertebrae are always included between the 'bundles' in Jurassic long-tailed pterosaurs. This suggests the possible absence of the 'bundles' in MFSN 1770.

Some possible evidence of osteological immaturity can be identified in MFSN 1770: the bones of the skull are unfused, the mandibular rami are unfused at the symphysis, the scapula and coracoid are probably unfused, and the fibula is probably also not completely fused to the tibia proximally.

Austriadactylus

This taxon has large serrate maxillary teeth and small mandibular teeth with 4–6 cuspules on each cutting edge (Dalla Vecchia *et al.* 2002). The tail lacks the 'bundles'.

Indeterminate material

MFSN 19864, a segment of 24 vertebrae from a rather long caudal vertebral column, shows no traces of the development of 'bundles' (Dalla Vecchia 2001).

Taxonomical remarks

Eudimorphodon

This genus can be diagnosed mainly on the basis of its peculiar multicuspid dentition, with small, close-set tricuspid and pentacuspid teeth of similar size in both upper and lower jaws (see Wild 1978). Some Lombardian specimens with no teeth preserved (MCSNB 2887, MCSNB 8950) are attributed to the genus based on the concomitant presence of a square deltopectoral crest in the humerus (shared with *Campylognathoides*), a first wing phalanx that is just slightly longer than the ulna (except in immature individuals; see Tables 2 & 3), and the absence of the caudal 'bundles'. The Lombardian specimens attributed to *E. ranzii* show marked differences from each other and it is difficult to state whether more taxa are represented or the species has a high degree of variability. The holotype of *E. ranzii* is noticeably larger than all other *Eudimorphodon* specimens and could just be a very old or giant individual. It is possible that basal pterosaurs had indeterminate growth (see Bennett 1995) and, like many living reptiles, continued to grow all through life (see Andrews 1982). Thus older adults would be larger than younger adults. A greater age is suggested also by the worn and broken dentition of MCSNB 2888, which could be explained by the slowed and irregular replacement rhythm that occurs with increasing age (Edmund 1969), rather than by a particular dentition-damaging diet. The larger size of MCSNB 2888 could also be explained by the very high intraspecific variability of size in adult individuals observed in reptiles (Andrews 1982) and possibly also present in pterosaurs (Unwin 2001). Furthermore, the scarcity and ambiguity of independent osteological features of immaturity found in all other specimens excluding MCSNB 8950 and MGUH VP 3393, do not support their identification as juveniles. *E. rosenfeldi* differs from all *Eudimorphodon* specimens from Lombardy in having comparatively longer hindlimbs (Table 2).

Peteinosaurus

MCSNB 2886 (the holotype), MCSNB 3496 and MCSNB 3359 all present features of osteological immaturity and probably do not represent adult individuals.

The diagnostic features of *Peteinosaurus zambellii* found in the holotype are related to the lower dentition and fibula. The dentition is trimorphodont. The tip of the mandible bears a couple of moderately long, narrow and recurved teeth. They are followed by small, monocuspid teeth slightly recurved backwards, higher than long, with mesial and distal sharp cutting margins, set in crater-like alveoli. The set of the most distal teeth corresponds with the labial side of the mandible being higher than the lingual side. They are no smaller than the preceding teeth and are triangular and longer than high; they are not recurved backwards and bear at least two or three small cuspules along each cutting margin. The fibula is unreduced in length and slightly expanded distally and it is fused to the upper part of the lateral tibiotarsal condyle without a distal condyle.

The condition of the fibula in *Peteinosaurus* is most probably a primitive character. *Eudimorphodon ranzii* (Wild 1978, 1994), *E. rosenfeldi* (Dalla Vecchia 1994), *Dimorphodon macronyx* (Owen 1870; Padian 1983), *Dorygnathus* (Arthaber 1919; *contra* Wellnhofer 1991, p. 56; Padian & Wild 1992) and *Rhamphorhynchus* (Wellnhofer 1975a, 1978) have fibulae that do not reach the tarsus but taper and end well before it. However, a fibula 'as long as the tibia' and forming 'its own distal condyle' is present in the Jurassic *Campylognathoides liasicus* (Wellnhofer 1974, p. 21) and a fibula as long as the tibia and expanded distally, with proximal tarsals fused to the tibia to form the condyles, is found in *C. zitteli* (Plieninger 1895). A fibula unreduced in length and with its own distal condyle is also present in the Austrian specimen of *Eudimorphodon* (Wellnhofer 2001, 2003).

Thus, the unreduced fibular length is shared by pterosaurs usually considered rather phylogenetically distant, such as *Peteinosaurus* and *Campylognathoides*, and both states of the character are present in *Eudimorphodon*. However, the fibula of both *Campylognathoides liasicus* and the *Eudimorphodon* specimen from Austria differ from that of *Peteinosaurus zambellii* in retaining a distal condyle.

MCSNB 3496 is important because it shows the complete ischiopubic plate of *Peteinosaurus*, its tarsus formed by two distal elements and a mid-tail segment of the vertebral column with the 'bundles'. However, this specimen can no longer be taken as typical for the pelvic girdle of *E. ranzii*, nor for its tarsus and tail, as has been done to date. Thus, the shape of the ischiopubic plate of *E. ranzii* is still unknown. The ischiopubic plate of *Peteinosaurus* has an outline rather different from that of the ischiopubic plate of *Dimorphodon* (Owen 1870, pl. 19, fig. 2; Arthaber 1919, fig. 3; Unwin 1988), *Dorygnathus* (Wellnhofer 1978, fig. 14) and *Rhamphorhynchus* (Wellnhofer 1975a, fig. 10a, d, g).

Characters diagnostic of *Peteinosaurus zambellii* cannot be seen in MCSNB 3359 because it lacks the lower jaw, and the distal part of the fibula is, at best, not visible. The only important long bones preserved in both MCSNB 3359 and 2886 are the tibia (ti) and first wing phalanx (wph1). The ratio wph1/ti is actually similar in both specimens, but it is also the same in *Preondactylus* and *Dimorphodon* (Table 2). Actually, the two specimens were originally grouped in the same taxon just because they both differ from *Eudimorphodon*. MCSNB 3359 differs from MCSNB 2886 in the shape of the pteroid (cf. Wild 1978, p. 14, fig. 26 & pl. 12). An important feature, common to *Peteinosaurus* (MCSNB 3496) and MCSNB 3359, that seems to be absent in other Triassic pterosaurs is the presence of the 'bundles' in the tail; this, however, is synapomorphical of all Jurassic long-tailed pterosaurs. Thus, there is uncertainty about the actual taxonomic position of this specimen and the author considers it here as ?*Peteinosaurus*.

Ratios of long bone lengths reported for *Peteinosaurus* are all obtained from MCSNB 3359, except for wph1/ti length ratio and those with the wing metacarpal length.Wph1/ti length ratio of MCSNB 3359 and MCSNB 2886, and also those with the wing metacarpal length, are very similar to those of *Preondactylus* and *Dimorphodon* (see Table 2). Thus, on the sole basis of those ratios, MCSNB 3359 cannot be distinct from *Preondactylus* or *Dimorphodon*.

Because of the uncertain attribution of MCSNB 3359, most ratios of long bone lengths previously reported for *Peteinosaurus* should be considered as unknown and cannot be used in the diagnosis of the taxon. For example, a wing phalanx 1 shorter than the forearm, a diagnostic feature of *Peteinosaurus* according to Wild (1978), is a character of MCSNB 3359 unknown in the holotype; anyway, it is also found in *Preondactylus*, *Dimorphodon* and even in *Dorygnathus*, *Sordes* and *Scaphognathus* (Tables 2 & 3).

Table 2. *Ratios of long bone lengths in Triassic pterosaurs and* Dimorphodon macronyx

	Eudimorphodon cromptonellus MGUH VP 3393	*Eudimorphodon ?ranzii* MCSNB 8950	*Eudimorphodon ?ranzii* MPUM 6009	*Eudimorphodon ?ranzii* MCSNB 2887	*Eudimorphodon ranzii* MCSNB 2888	*Eudimorphodon rosenfeldi* MFSN 1797	*Preondactylus buffarinii* MFSN 1770	*Peteinosaurus zambellii* MCSNB 2886	*?Peteinosaurus* MCSNB 3359	*Dimorphodon macronyx*†
u/h	1.11	1.29	1.37	1.36	1.38	1.31	1.31	—	1.30	1.29
h/mcIV	2.16	2.89	2.50	—	1.62	1.93	2.00	—	2.29	1.95–2.36
u/mcIV	2.39	3.72	3.43	—	2.24	2.62	2.95	—	2.91	3.00
h/fe	0.92	1.33	1.42	1.32	1.14	1.13	0.98	—	1.05	1.04–1.09
h/ti	0.88*	1.04	1.05*	0.98	0.94*	0.77	0.73	—	0.81	0.71–0.76
u/fe	1.02	1.71	1.95	1.79	1.58	1.49	1.29	—	1.34	1.34–1.40
u/ti	0.98*	1.34	1.44*	1.33	1.30	1.02	0.95	—	1.03	0.97–0.98
ti/fe	1.04*	1.27	1.35*	1.34	1.22*	1.46	1.35	—	1.30	1.39–1.46
fe/mcIV	2.34	2.18	1.76	—	1.41	1.76	2.28	—	2.18	1.87–2.23
ti/mcIV	2.44*	2.78	2.38*	—	1.72*	2.58	3.09	3.12	2.82	2.74–3.15
wph1/h	0.97*	1.31	1.43	1.43*	1.70*	1.52	1.11	—	1.09	1.06–1.20
wph1/u	0.90*	1.01	1.04	1.05*	1.23*	1.16	0.85	—	0.86	0.82–0.90
wph1/mcIV	2.14*	3.78	3.57	—	2.75*	3.05	2.49	2.57	2.50	2.35–2.70
wph1/fe	0.91*	1.73	2.03	1.89*	1.95*	1.73	1.09	—	1.15	1.10–1.26
wph1/ti	0.89*	1.36	1.50*	1.40*	1.60*	1.18	0.81	0.82	0.88	0.80–0.88
wph2/wph1	1.14*	1.04	0.88	0.91*	—	0.91	1.10	0.96*	1.00	1.03–1.15
wph3/wph2	1.00*	1.02	1.10*	—	—	1.09	1.00	—	1.09	1.08–1.12
wph3/wph4	—	1.12	1.06	—	—	1.23	1.39	—	1.33	1.25
wph3/wph1	1.14*	1.06	0.96*	—	—	0.99	1.08	—	1.09	1.24–1.29

* measurements estimated or approximate.
† *D. macronyx* specimens are YPM350 and YPM9182 after Padian (1983), GSM1546 and BMNH R.1034 (holotype) after Unwin (1988), BMNH 41212 after Wellnhofer (1978).

Table 3. *Ratios of long bone lengths in Jurassic and Cretaceous basal pterosaurs*

	Dorygnathus banthensis[1]	*Campylognathoides liasicus*[2]	*Campylognathoides zitteli*[2]	*Anurognathus ammoni*[3]	*Dendrorhynchoides curvidentatus*[3]	*Sordes pilosus*[3]	*Scaphognathus crassirostris*[4]	*'Rhamphorhynchus longicaudus'*[3]	*'Rhamphorhynchus intermedius'*[3]	*'Rhamphorhynchus muensteri'*[3]	*'Rhamphorhynchus gemmingi'*[3]	*'Rhamphorhynchus longiceps'*[3]
u/h	1.36–1.72	1.18–1.24	1.17*	1.41	1.28	1.60	1.63–1.76†	1.62	1.72	1.74	1.58	1.49
h/mcIV	1.92–2.07	2.19–2.39	2.33*	2.91	2.99	2.61	1.98–2.00	1.65	1.59	1.81	2.15	1.84
u/mcIV	2.61–3.35	2.62–2.83	2.73	4.09	3.82	4.19	3.27–3.48†	2.67	2.74	3.14	3.40	2.74
h/fe	1.20–1.32	1.34–1.39	1.08*	1.18	1.46*	1.20	0.97–1.03	1.32	1.28	1.15	1.43	1.42
h/ti	0.83–1.03	1.04–1.13	0.79*	0.82	1.04	0.87	0.92	1.06	0.99	0.78	0.98	1.01
u/fe	1.70–2.12	1.60–1.68	1.26	1.67	1.87*	1.92	1.69–1.71†	2.14	2.20	2.00	2.27	2.11
u/ti	1.13–1.64	1.23–1.36	0.93	1.15	1.33	1.39	1.51+	1.72	1.70	1.36	1.54	1.50
ti/fe	1.27–1.50	1.24–1.33	1.35	1.44	1.40	1.38	1.12	1.24	1.29	1.47	1.47	1.41
fe/mcIV	1.51–1.73	1.58–1.74	2.17	2.45	2.04*	2.18	1.93–2.04	1.25	1.24	1.57	1.50	1.30
ti/mcIV	1.94–2.32	1.96–2.30	2.93	3.54	2.87	3.00	2.17	1.55	1.60	2.31	2.20	1.83
wph1/h	1.16–1.38	1.73–1.96	2.64*	1.81	1.60	1.17	1.27–1.31	2.24	2.47	3.01	2.63	2.31
wph1/u	0.73–0.89	1.46–1.59	2.26	1.29	1.25	0.73	0.74–0.77†	1.39	1.43	1.73	1.66	1.55
wph1/mcIV	2.33–2.66	4.04–4.34	6.17	5.27	4.78	3.06	2.53–2.59	3.70	3.93	5.45	5.65	4.26
wph1/fe	1.40–1.67	2.38–2.55	2.85	2.15	2.34*	1.40	1.27–1.31	2.96	3.16	3.47	3.77	3.27
wph1/ti	1.01–1.26	1.79–2.06	2.10	1.49	1.67	1.02	1.17	2.39	2.44	2.36	2.57	2.33
wph2/wph1	1.04–1.21	1.03–1.07	1.13	—	0.80	1.05	1.07–1.08	0.86	0.92	0.99	0.96	—
wph3/wph2	0.96–1.02	0.85–0.88	0.79	—	—	1.03	0.98	0.91	0.88	0.93	0.91	—
wph3/wph4	1.12–1.44	1.21–1.26	1.36	—	—	1.59	1.08	0.89	0.92	1.05	0.96	—
wph3/wph1	1.07–1.21	0.90	0.89	—	—	1.06	1.05	0.78	0.81	0.93	0.87	

* measurements estimated or approximate.

† Ratios based on radial length.

Data sources:

[1] Wild (1978)

[2] Wellnhofer (1974)

[3] Unwin *et al.* (2000).

[4] Wellnhofer (1975b). Wellnhofer reported the length of radius instead of ulna and radius is usually shorter than ulna in pterosaurs.

Concerning *Rhamphorhynchus* spp., I follow Unwin *et al.* (2000) in enclosing in quotation marks a taxonomical name where there is some doubt regarding its validity. For the different views about the validity of *Rhamphorhynchus* species see Wellnhofer (1975b) and Bennett (1995).

Preondactylus

The holotype MFSN 1770 is probably an osteologically immature individual and is similar in size to *Peteinosaurus* specimen MCSNB 2886 and ?*Peteinosaurus* MCSNB 3359. The two still undescribed specimens are not larger than the holotype, so the sample possibly represents only immature individuals. The conclusion is that *Peteinosaurus* and *Preondactylus* could be represented only by immature individuals and thus possibly do not represent the character states of the adults.

The dentition of the lower jaw in both MFSN 1770 and MCSNB 2886 comprises a couple of moderately long and recurved teeth at the tip of each mandibular ramus and numerous small teeth posteriorly (Dalla Vecchia 1998). Unfortunately, the state of preservation of MFSN 1770 does not permit determination of the actual shape of the crowns, the nature of tooth implantation, nor the condition of the lingual and labial margins of the lower jaw of MFSN 1770. The upper jaw dentition, characteristic of *Preondactylus* and fundamental in confirming or disproving that it is different from *Peteinosaurus*, is not preserved in any *Peteinosaurus* specimens. There are some differences between MFSN 1770 and MCSNB 2886: (1) the dentary forms less than half of the entire mandibular length and has a short posterior process in MFSN 1770; (2) the splenial is posterior and adjacent to the posterior process of the dentary in MFSN 1770; (3) the tip of the lower jaw is not ventrally bent in MFSN 1770 (but it is in MCSNB 2886, *fide* Wild 1978). All those differences are observed from the impression of the lower jaw of MFSN 1770, and impression has proved to be misleading with respect to the serration of the maxillary teeth. Another distinguishing feature is the relative elongation of the metacarpals (see Wild 1978, pl. 12; Dalla Vecchia 1998, fig. 2), but better-preserved specimens are needed to confirm the structure of the metacarpals in both taxa.

In contrast to MCSNB 3496 and MCSNB 3359, the caudal vertebrae in MFSN 1770 seem to lack the elongated processes of the pre- and postzygapophyses and 'bundles' are not developed. Like MCSNB 2886, MCSNB 3359 seems to differ from MFSN 1770 in the relative length of the metacarpals. In MCSNB 3359 wph2 is as long as wph1, and wph3 is slightly longer than wph2; in MFSN 1770 wph2 is as long as wph3, and wph2 is slightly longer than wph1. The differences in length between each element are actually small and only a larger sample could reveal whether these differences are taxonomically meaningful or simply reflect intraspecific individual or ontogenetical variability.

The holotype of *Preondactylus* is not the only specimen of basal pterosaur in which the femur (fe) is longer than the humerus (h), a feature reported as unique to this Triassic taxon (see Wild 1984). The holotypes of *Scaphognathus crassirostris* and *Eudimorphodon cromptonellus* have an h/fe ratio even lower than that of MFSN 1770 (Tables 2 & 3). The small difference in length between the femur and the humerus of MFSN 1770 does not mean that the femur was consistently longer than the humerus in all *Preondactylus* specimens. For example, in two specimens of *Scaphognathus crassirostris* the ratio is 0.97 and 1.03. In the author's opinion it would be more prudent to describe it as 'femur and humerus with similar lengths'.

According to the observations made above, and given the reported limits of our knowledge, the diagnosis of *Preondactylus buffarinii* by Dalla Vecchia (1998) should be emended as follows: heterodonty between lower and upper jaw; premaxillary and anterior mandibular teeth relatively narrow, elongated and recurved backwards; one to possibly three very enlarged, triangular maxillary teeth below the ascending process of the maxilla, followed posteriorly by triangular teeth decreasing regularly in size; maxillary teeth serrated; numerous small teeth in the lower jaw posterior to the first larger teeth; large, elliptical and anteroposteriorly very elongated narial opening; tip of the snout made of a short and dorsally convex premaxilla; dentary less than half the length of the complete lower jaw.

There are also some characters that link *Preondactylus*, *Peteinosaurus* (MCSNB 2886), ?*Peteinosaurus* (MCSNB 3359) and *Dimorphodon*. Numerous small teeth in the lower jaw and larger but only moderately elongated proximal teeth are present in MFSN 1770, MCSNB 2886 and *Dimorphodon*. Heterodonty (different tooth shape and size) between the maxillary and lower jaw dentition, with much larger maxillary teeth, is found in MFSN 1770 and in *Dimorphodon*, as in *Austriadactylus*; the state is unknown in *Peteinosaurus*.The shape of the humerus is similar in MFSN 1770, MCSNB 3359 and *Dimorphodon* specimens, with a relatively elongated shaft and, above all, a triangular deltopectoral crest (Dalla Vecchia 1998, fig. 6).

The ratios of the long bones (e.g. u/h, h/fe, u/fe, h/ti, u/ti, wph1/h, wph1/u, wph1/fe and wph1/ti) are similar in MFSN 1770, MCSNB 3359, MCSNB 2886 (for wph1/ti only, which is the most diagnostic ratio for the group, see Tables 2 & 3) and *Dimorphodon*. Most of these ratios differ significantly from those of the other basal pterosaurs (see Tables 2 & 3) and depend upon a relatively short wing phalanx 1 and, in general, comparatively shorter forelimb than hindlimb elements.The latter point is also clear from the wing ratio (humerus to wing phalanx 4/ femur + tibia length), which is representative of the relative elongation of the hindlimb with respect to the wing, a feature considered primitive by Wild (1978, 1984). On this basis

Preondactylus was considered more primitive than other pterosaurs. The ratio in *Preondactylus* (MFSN 1770) is 3.00. In MCSNB 3359 the ratio is 3.20 (based on measures reported in Table 1 and thus slightly different from the ratio reported in Dalla Vecchia 1998, which is based on measurements in Wild 1978), and it is 3.20 in *Dimorphodon macronyx* (GSM1546). It is worth noting that the proximal tarsals are not fused to the tibia in MCSNB 3359; thus the tibia is comparatively shorter than in MFSN 1770 and GSM1546, and as a consequence the wing ratio of MCSNB 3359 had to be slightly lower after the fusion of the tarsals. In the Lombardian *Eudimorphodon* the ratio is 4.91 for MPUM 6009 and 4.62 in MCSNB 8950 (but the proximal tarsals are unfused), whereas in *E. rosenfeldi* it is decidedly lower (3.89) because of the long hindlimbs. The gracile hindlimbs and robust forelimbs of the holotype of *E. ranzii* suggest that the ratio was also rather high in this specimen. The ratio in four specimens of *Dorygnathus banthensis*, with humeral lengths 38.8, 52, 60.5 and 68 mm, is 4.65, 4.68, 4.76 and 4.14 respectively (measurements after Wild 1978, table 7). *Campylognathoides liasicus* has ratios of 5.61 and 5.83 (the Pittsburgh and Paris specimens respectively; Wellnhofer 1974) and *C. zitteli* has 5.64 (Wellnhofer 1974). *Rhamphorhynchus* species (for growth stages, see Bennett 1995) have the highest wing ratio among basal pterosaurs, and this increases with size from 6.55 to 7.47 (according to data in Unwin *et al.* 2000). The Late Jurassic *Sordes pilosus* has a wing ratio of 3.74 (measurements after Unwin *et al.* 2000), lower than *Eudimorphodon rosenfeldi*, but higher than MCSNB 3359. Thus, *Preondactylus*, MCSNB 3359 (?*Peteinosaurus*) and *Dimorphodon* share the lowest wing ratio ($\leq$3.20) among the basal pterosaurs and are the most basal members from this point of view. More complete specimens unambiguously belonging to *Peteinosaurus* and *Austriadactylus* are needed in order to determine their state with respect to this character. *Preondactylus*, *Peteinosaurus*, MCSNB 3359 (?*Peteinosaurus*) and *Dimorphodon* may form a clade, but additional and more complete material of the Triassic taxa is needed to support this. It is possible that some shared features present in the Triassic taxa reflect ontogenetical traits rather than strict relationships. This can be proved or disproved only by comparing specimens which are unquestionably mature.

Austriadactylus

This genus from Austria is clearly distinguished from all the others on the basis of a peculiar heterodont and multicuspid dentition and the presence of a sagittal cranial crest (Dalla Vecchia *et al.* 2002). *Austriadactylus* and *Preondactylus*, like *Dimorphodon*, are united by possessing a marked heterodonty between the lower and upper jaws. On the other hand, this feature, along with a different pattern of denticulation, distinguishes its dentition from that of *Eudimorphodon*. The absence of the 'bundles' is a feature shared with *Eudimorphodon* and possibly with *Preondactylus* (Dalla Vecchia 2001).

The ankle and pes of Triassic pterosaurs

Specimens with distal condyles in the tibia have only two tarsals, whereas specimens with more than two tarsals do not bear distal condyles at the tibial end, which has a flat distal articular surface. This is observed also in many Jurassic and Cretaceous pterosaurs (e.g. Wellnhofer, 1975a, 1978; pers. obs.) and suggests that the proximal tarsals fused to the tibia during ontogeny to form a distally bicondylar tibiotarsus, a conclusion already made by several authors (e.g. Seeley 1901; Wellnhofer 1978; Padian 1983; Unwin 1988). Further supporting this view is the fact that, in MCSNB 2886 and MCSNB 3496, it is possible to see the still uncompletely ossified zone around the suture between the tibia and proximal tarsals.

The medial condyle in MCSNB 2886 and MCSNB 3496 is flat in anterior view and was possibly still made of cartilage because of immaturity. In fact, in *Eudimorphodon rosenfeldi* MFSN 1797, it is a true, ossified convex structure even if less well rounded than the lateral condyle (Fig. 3b, c). It is worth noting that, in MCSNB 2886 also, the ventral distal condyle of the wing metacarpal is practically flat. In *Eudimorphodon ranzii* MCSNB 2887 the medial condyle of the tibia is less developed than the lateral one. An asymmetry in the shape of the condyles is present also in specimens from the Liassic of the United Kingdom determined as *Dimorphodon* (Padian 1983, p. 23); thus it is a feature common among early pterosaurs. Also, the tibiotarsal lateral condyle in both MCSNB 2886 and MCSNB 3496 is pointing anteromedially rather than anteriorly only, although this feature could be caused or emphasized by crushing.

The larger, more robust and pulley-like of the two proximal tarsals has been considered to be the astragalus (e.g. Wild 1994, fig. 12). Therefore, in the case of a differential size of the condyles, the author would expect to find a more developed medial condyle because the astragalus lined up with the tibia, which is the medial element of the crus. Actually, the lateral condyle of early pterosaurs is more developed than the medial condyle, and the fibula is fused just above it (at least in *Peteinosaurus*), suggesting that the tarsal formed the lateral condyle and it aligned with the fibula. The

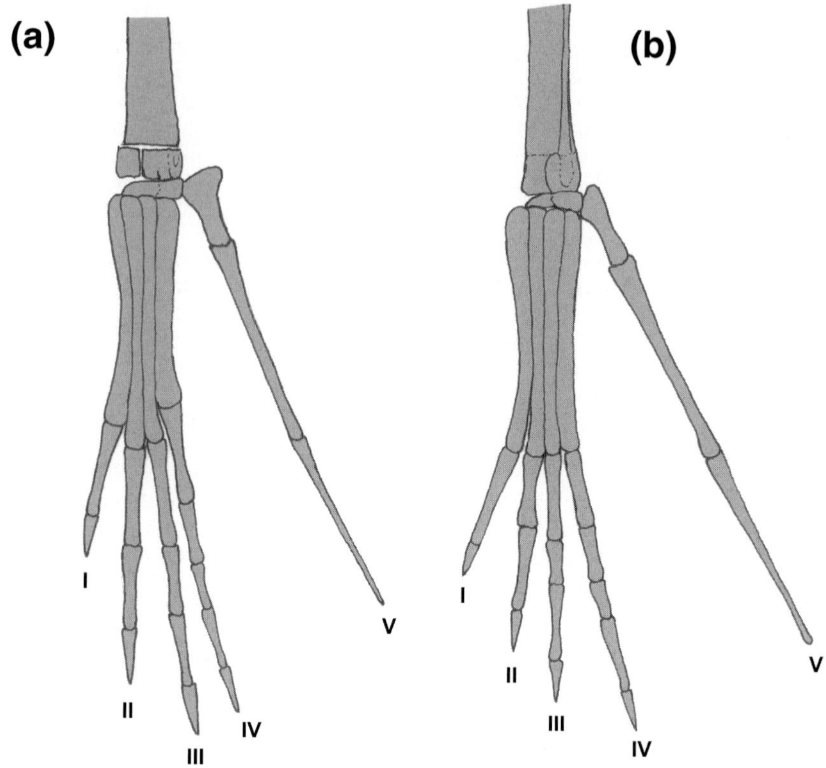

Fig. 6. Reconstruction of the ankle and pes of Triassic pterosaurs. (**a**) Immature individual with proximal tarsals unfused to tibia, based on *Eudimorphodon ?ranzii* MCSNB 8950, right foot in plantar view, proximal tarsals placed according to *Peteinosaurus* (MCSNB 3496), distal tarsals placed as in *Dimorphodon macronyx* according to Padian (1983). (**b**) Left foot in dorsal view, after the fusion of the proximal tarsals to tibia; tibiotarsus based on *Peteinosaurus* (MCSNB 2886 and MCSNB 3496), distal tarsals placed as in *Dimorphodon macronyx* according to Padian (1983), pes of MCSNB 3359 (?*Peteinosaurus*).

larger proximal tarsal in MCSNB 8950 has the shape of the lateral condyle of the tibiotarsus of more mature individuals. All this suggests that the larger proximal tarsal is the calcaneum, which forms the lateral tibiotarsal condyle co-ossifying to the tibia. Also in *Rhamphorhynchus* the lateral of the proximal tarsals is the largest (see Wellnhofer 1975a, fig. 17f). In a specimen of *Dorygnathus* (1938 I 49 BSP) the lateral condyle, exposed in lateral view, is made of a large proximal tarsal separated from the tibia by a suture; the tarsal has a semi-circular ventral outline like the larger proximal tarsal in MCSNB 8950.

A reconstruction of the ankle and foot of Triassic pterosaurs is shown in Figure 6.

The tarsus in MCSNB 3359 is actually made of two distal elements with a shape similar to that of the distal tarsals of *Dimorphodon* described by Padian (1983, figs 20 & 26), given that, in the detachment and rotation of the foot, the distal tarsals have been exposed in dorsoventral, possibly plantar, view. The structure of the tarsus of MCSNB 3359 is basically the same as that of MCSNB 3496. However, proximal tarsals of the latter are fused to the tibia and the medial distal tarsal is probably exposed in anterior view and thus appears wedge-like. The same pattern is present, as far as can be seen, in the other Triassic pterosaurs. Thus, there is nothing to show that the tarsus of Triassic pterosaurs was different from that described by Padian (1983) for *Dimorphodon*.

The reconstruction of metatarsals I–IV of Triassic pterosaurs in a spreading disposition (see Wild 1978, fig. 41b) is arbitrary. Actually, in both feet of the specimen MCSNB 3359 they are closely appressed and parallel to each other (Wild 1978, pl. 18 & fig. 41a). This 'block' (unspread) disposition is the condition of all other Triassic pterosaurs in which the foot is preserved (partly, as in MCSNB 3496 and MFSN 1770, or completely, as in MFSN 1797) and in most Early Jurassic pterosaurs (e.g. Buckland 1835, fig. 1; Owen 1870, pl. 18; Plieninger 1895, fig. 8; Arthaber 1919, figs 52 & 53; Wellnhofer 1974, fig. 10; Padian 1983, figs 2, 14 & 15). Furthermore, there are examples of disarticulated skeletons where the metatarsals drifted as a compact block (e.g. Padian

1983), showing that they were actually tightly appressed. This is not true for the Late Jurassic pterosaurs *Rhamphorhynchus* and *Pterodactylus*, in which metatarsals are often preserved as spreading, i.e. they are not parallel to each other (e.g. Owen 1870, pl. 19, fig. 5; Arthaber 1919, figs 54 & 55; Wellnhofer 1970, fig. 19 & pls 5, fig. 1, 8, figs 1 & 3, 9, fig. 1 & 10, fig. 1; Wellnhofer 1975a, pls 2, fig. 1, 5, fig. 2, 13, fig. 4, 34[20], fig. 1, 35[21], fig. 1 & 40[26], fig. 4; Frey & Martill 1998, figs 7 & 8; Frey & Tischlinger 2000, pl. 1), suggesting a different orientation of metatarsals I–IV of those later taxa with respect to those of the earlier pterosaurs and a different plantar shape. Thus the pedal print of an early pterosaur, if plantigrade, could not have the triangular outline of the *Pteraichnus*-like and *Pteraichnus*-like pedal print which matches the spreading foot of *Rhamphorhynchus* and *Pterodactylus*. Furthermore, the foot of Triassic pterosaurs was functionally ectaxonic (i.e. the longest digit is one of the outer ones) (Fig. 6) and could not leave symmetrical prints with equally long digital marks like those attributed to pterosaurs. It is clear that most, if not all, of the alleged pterosaur pedal prints reported in the literature were made by a *Pterodactylus*-like foot with spread metatarsals, non-ectaxonic foot and short digit V (as admitted by Lockley *et al.* 1995, p. 10, Bennett 1997, p. 108 and Unwin 1997, pp. 383–384) and all the speculations made on those ichnites cannot automatically be extended to earlier pterosaurs. Also, some features of the manus cannot be reconciled with the generalization of the pteraichnid footprint model for early pterosaurs. In the latter, manual digits I–III flex dorsally or dorsally and preaxially (Fig. 2b), and were clearly used for grasping. The manus of *Pterodactylus* seems to lack the distal curvature of metacarpals I–III and their asymmetrically developed distal condyles, and digit III is supposed to have had a freedom of postaxial movement not possible for *Eudimorphodon* (see Unwin 1997, fig. 1). Furthermore, whereas *Rhamphorhynchus* and *Pterodactylus* mani have metacarpals of equal length (Fig. 2a), and all digits would touch the ground at the metacarpophalangeal joint in the posture supposed by Unwin (1997) and Bennett (1997), those of the Triassic pterosaurs, as well as *Dimorphodon* (Padian 1983) and *Campylognathoides* (Wellnhofer 1974), have different lengths. This is particularly marked in *Eudimorphodon* (Fig. 2b). If the manus of *Eudimorphodon* could ever touch the ground with the manual digitigrade posture suggested by Unwin (1997, fig. 4) and Bennett (1997, fig. 3), not all digits touched the ground at the metacarpophalangeal joint and they could not leave digital prints starting from a common pad, as is the case of pteraichnid manual prints. It is probably meaningful that all ichnites attributed to quadrupedal pterosaurs with plantigrade pes and digitigrade manus are never older than Late Jurassic (Lockley *et al.* 1995; Mazin *et al.* 1995; Unwin 1997; Wright *et al.* 1997; Lockley *et al.* 1997).

Remarkable features of Triassic pterosaurs

At least three characters are found in Triassic pterosaurs that are not present in more recent pterosaurs.

(1) Teeth with accessory cusps or cuspules

Triassic pterosaurs, unlike all later pterosaurs, have accessory cusps or cuspules on many or at least some of their teeth. This feature could be a primitive feature of pterosaurs which was retained to a different degree in the earliest members and is therefore related to their taxonomical relationships and common ancestry and then lost, or just a convergent adaption of the Triassic pterosaurs to a peculiar diet.

The multicuspid dentition in general is useful in piercing and cutting hard or tough tissues in order to process relatively large prey before swallowing it. Wild (1978, p. 237) has suggested that the multicuspid dentition in the large specimen of *Eudimorphodon ranzii* (MCSNB 2888) served to pierce the hard covering of ganoid scales of pholidophorid fish because scales of this type of fish have been found inside the rib cage of MCSNB 2888. The peculiar dentition of Triassic pterosaurs could thus be an adaptation to feed on 'ganoid' fish, crustaceans, large insects, or other prey with hard exoskeletons. However, a change in pterosaur food source is not immediately evident in the Early Liassic, when pterosaurs no longer had multicuspid teeth. Fishes had a covering of ganoid scales in the Liassic of Germany and the United Kingdom, and crustaceans were present, but, of course, it is not possible to demonstrate that Jurassic pterosaurs changed their diet from one based on large and hard-covered prey to one based on softer, smaller prey. Moreover, cuspules never appeared again in pterosaur history. The cuspules in the posterior mandibular teeth of *Peteinosaurus* seem to be more a remnant than an efficient cutting device, because they are extremely reduced in size and are present only in few distal teeth. Thus, it should be considered that the presence of cusps and cuspules is a primitive feature of Triassic pterosaurs inherited from their ancestors and lost during the successive evolution of the group, and that the different patterns of denticulation are variants of a single ancestral condition. Tricuspid teeth are known among Prolacertiformes in immature specimens of *Tanystropheus* and *Macrocnemus* (Wild 1973) and in both mature and immature *Langobardisaurus* (Renesto 1994; Renesto & Dalla Vecchia 2000). Teeth with few cusps or cuspules are uncommon in Triassic archosaurs, whereas serrations are very common in carnivorous forms. In any case, the feature, if not convergent, suggests stricter relationships among Triassic pterosaurs than previously supposed. A possible evolutionary trend finally led to the disappearance of the accessory cusps or cuspules. This could have been attained by loss of the cusps or by a reduction in their number, as reported for *Tanystropheus* during ontogeny (Wild 1973). This is also supposed for *Eudimorphodon*, in which, according to Wild (1978), pentacuspid teeth were replaced by tricuspid teeth during growth. Alternatively, the size of the cusps could have been reduced, and their number possibly increased, up to their disappearance. In the first, and at present most plausible, case *Preondactylus* and *Austriadactylus* would be more primitive than *Eudimorphodon*, and vice versa in the second case. *Peteinosaurus* would be most derived in both cases because of the very reduced denticulation which occurs only in the distal teeth.

(2) Very enlarged maxillary teeth below the ascending process of the maxilla

In some Jurassic pterosaurs (e.g. *Dimorphodon* and *Campylognathoides*) the maxillary teeth below the ascending process of the maxilla are also the largest maxillary teeth, but in *Eudimorphodon ranzii* (holotype), *Preondactylus* (at least two specimens) and *Austriadactylus*, one to three of these teeth are much more developed than in Jurassic taxa. Apparently the holotype of *Eudimorphodon cromptonellus* (Jenkins *et al.* 2001) and a small specimen of *Eudimorphodon* from Lombardy (MPUM 6009) lack this feature. Wild (1978) suggested for *Eudimorphodon* that this could be a sexual character, but it could also be considered a feature of immature individuals because MPUM 6009 has been considered juvenile by Wild (1978) and the Greenland specimen is undoubtedly immature (Jenkins *et al.* 2001). However, the two *Preondactylus* specimens are probably also immature, so sexual dimorphism seems to be more plausible. Whether or not this was a sexual feature, it disappeared in Jurassic taxa and is possibly a synapomorphy of Triassic pterosaurs.

(3) Absence of elongate pre- and postzygapophyseal processes in the caudal segment of the vertebral column and haemapophyses not forming bundles of filiform processes

This is the condition in *Eudimorphodon*, *Austriadactylus* and possibly also in *Preondactylus* (Dalla Vecchia 2001). Considering the absence of very elongated pre- and postzygapophyses in the caudal vertebrae as a primitive feature, as suggested by their presence in Jurassic long-tailed pterosaurs and their absence in all supposed pterosaur relatives (Wild 1978; Sereno 1991; Bennett 1996a; Peters 2000), two pterosaurian groups are identified in the Late Triassic. One group (*Eudimorphodon*, *Austriadactylus* and possibly *Preondactylus*) is more primitive than the other (?*Peteinosaurus* MCSNB 3359 and *Peteinosaurus* MCSNB 3496) and the most primitive among long-tailed pterosaurs according to this point of view.

Triassic pterosaurs also seem to have a comparatively longer tail than that of the Jurassic long-tailed pterosaurs, with comparatively longer mid-caudal vertebrae (Wild 1978; Dalla Vecchia 2001; Dalla Vecchia *et al.* 2002).

The manus of *Eudimorphodon*, *Peteinosaurus* and MCSNB 3359 (?*Peteinosaurus*) has metacarpals that increase in length from I to IV. This condition is also found in *Dimorphodon* (Owen 1870; Padian 1983) and *Campylognathoides* (Wellnhofer 1974). Also metacarpals I and II–III of *Preondactylus* seem to be somewhat unequal in length. On the contrary, as observed above, metacarpals I–III have equal lengths in *Rhamphorhynchus* (Wellnhofer 1975a, fig. 14) and also in *Pterodactylus* (e.g. Wellnhofer 1991, figs on pp. 88–89). The difference in length is more marked in *Eudimorphodon* than in all other pterosaurs. The condition in *Eudimorphodon* is the primitive condition for reptiles (Romer 1966).

Considering a low wing ratio as primitive, *Preondactylus* and ?*Peteinosaurus* (MCSNB 3359), with *Dimorphodon*, form a group of pterosaurs that is more primitive than the other long-tailed pterosaurs.

Considering the unreduced of the fibular length as primitive, *Peteinosaurus* is, with *Campylognathoides* and the Austrian specimen of *Eudimorphodon*, more primitive than the other pterosaurs where fibular length is known. It appears improbable that this condition has been reversed, in a bone of such small utility as the fibula, in *Campylognathoides*. The Austrian specimen of *Eudimorphodon* is more primitive than the *Eudimorphodon* specimens from northern Italy, because its fibula is not reduced in length. *Campylognathoides* and the Austrian specimen of *Eudimorphodon* are more primitive than *Peteinosaurus* because they still retain a distal condyle in the fibula.

Conclusions

The record of Triassic pterosaurs is restricted to a relatively short interval of geological time. Triassic pterosaurs show features that could shed light on their taxonomic relationships and the origin of pterosaurs, but the partial preservation of the specimens and the small sample limit our understanding. Before an attempt at phylogenetic analysis of basal pterosaurs can be made, the following are needed:

(1) An improved understanding of the influence of ontogeny on the character states. The taxa we have are based only on probably immature specimens and it could be misleading to compare character states of immature individuals with those of mature individuals.
(2) An improved understanding about the range of intraspecific variability.
(3) The preparation and description of new *Preondactylus* specimens, the study of the still undescribed *Eudimorphodon* specimens and, if possible, the discovery of more complete material to fill the gap in the knowledge of characters of some important taxa, mainly *Peteinosaurus*, but also, for example, the poorly known Jurassic *Anurognathus*, *Batrachognathus* and *Sordes*.

Only the discovery of a well-preserved lower jaw of *Preondactylus* or a maxillary of *Peteinosaurus* could definitively clarify the relationships between these two taxa.

This work was made possible by a 60% MURST grant (A. Russo). I thank A. Paganoni, Museo Civico di Scienze Naturali of Bergamo, and C. Morandini and G. Muscio, Museo Friulano di Storia Naturale of Udine, for permission to study the specimens under their care and for their support during the realization of this study. Thanks also to P. Wellnhofer for information about the *Eudimorphodon* from Austria, to S. C. Bennett for comments on a first version of the manuscript and to F. A. Jenkins Jr and K. Padian for the final review. For their personal communications I am indebted to S. A. Chatterjee, F. A. Jenkins Jr., A. Paganoni, G. Roghi and A. Tintori.

References

ANDREWS, R. M. 1982. Patterns of growth in reptiles. *In*: GANS, C. & POUGH, F.H. (eds) *Biology of the Reptilia*. Academic Press, London, vol 13, 273–320.

ARTHABER, G. VON. 1919. Studien über Flugsaurier auf grund der Bearbeitung des wiener Exemplares von *Dorygnathus banthensis* Theod. sp. *Denkschriften Akademie der Wissenschaften Mathematisch-Naturwissenschaftliche Klasse*, **97**, 391–464.

BENNETT, S. C. 1993. The ontogeny of *Pteranodon* and other pterosaurs. *Paleobiology*, **19** (1), 92–106.

BENNETT, S. C. 1995. A statistical study of *Rhamphorhynchus* from the Solnhofen Limestone of Germany: year-classes of a single large species. *Journal of Paleontology*, **69**, 569–580.

BENNETT, S. C. 1996a. The phylogenetic position of the Pterosauria within the Archosauromorpha. *Zoological Journal of the Linnean Society, London*, **118**, 261–308.

BENNETT, S. C. 1996b. Year-classes of pterosaurs from the Solnhofen Limestone of Germany: taxonomic and systematic implications. *Journal of Vertebrate Paleontology*, **16**, 432–444.

BENNETT, S. C. 1997. Terrestrial locomotion of pterosaurs: a reconstruction based on *Pteraichnus* trackways. *Journal of Vertebrate Paleontology*, **17**, 104–113.

BROILI, F. VON. 1939. Ein *Dorygnathus* mit Hautresten. *Sitzungberichte der Bayerischen Akademie der Wissenschaften, Mathematisch-Naturwissenschaflitche Abteilung*, **1939**, 129–132.

BRANDNER, R. & POLESCHINSKI, W. 1986. Stratigraphie und Tektonik am Kalkalpensüdrand zwischen Zirl und Seefeld in Tirol (Excursion D am 3. April 1986). *Jahresberichte und Mitteilungen des Oberrheinischen Geologischen Vereines*, **68**, 67–92.

BUCKLAND, W. 1835. On the discovery of a new species of pterodactyle in the Lias at Lyme Regis. *Transactions of the Geological Society, London*, (**2**) **3**, 217–222.

BUDAI, T. & KOVACS, S. 1986. A Rezi Dolomit Rétegtani Helyzete a Kestzthelyi-Hegységben. *Magyar Allami Földtani Intézet Evkonive Jelentése*, **1984**, 175–191.

CARULLI, G.B., COZZI, A., LONGO SALVADOR, G., PERNANCIC, E., PODDA, F. & PONTON, M. 2000. *Note illustrative alla Carta Geologica delle Prealpi Carniche*. Edizioni del Museo Friulano di Storia Naturale, Udine, Pubblicazione no. 44, 47 pp.

CARULLI, G.B., FANTONI, R., MASETTI, D., PONTON, M., TRINCIANTI, E., TROMBETTA, G. L. & VENTURINI, S. 1998. Analisi di facies e proposta di revisione stratigrafica del Triassico superiore del Sudalpino orientale. *Atti Ticinesi di Scienze della Terra* (Serie Speciale), **7**, 159–183.

CHATTERJEE, S. 1986. The Late Triassic Dockum vertebrates: Their stratigraphic and paleobiogeographic significance. *In*: PADIAN, K. (ed.) *The Beginning of the Age of Dinosaurs*. Cambridge University Press, Cambridge, 139–150.

CIRILLI, S. 1995. Upper Triassic black shales in the western Tethys and their palaeoclimatological meaning. 3rd EPA Workshop – Black Shales Model. *Europal*, **8**, 63.

CLEMENS, W. A. 1980. Rhaeto-Liassic Mammals from Switzerland and West Germany. *Zitteliana*, **5**, 51–92.

CLEMMENSEN, L. B., KENT, D. V. & JENKINS, F. A., Jr. 1998. A Late Triassic lake system in East Greenland: Facies, depositional cycles and palaeoclimate. *Palaeogeography, Palaeoclimatology, Palaeoecology*, **140**, 135–159.

CUNY, G., GODEFROIT, P. & MARTIN, M. 1995. Micro-restes de Vertébres dans le Trias Superieur du Rinckebierg (Medernach, G-D Luxembourg). *Neues Jahrbuch für Geologie und Paläontologie, Abhandlungen*, **196**, 45–67.

DALLA VECCHIA, F. M. 1991. Note sulla stratigrafia, sedimentologia and paleontologia della dolomia of Forni (Triassico superiore) nella valle del Rio Seazza (Preone, Friuli-Venezia Giulia). *Gortania*, **12**, 7–30.

DALLA VECCHIA, F. M. 1994. *Studio sugli pterosauri Triassici con note sulla loro datazione, habitat e storia evolutiva*. PhD thesis, University of Modena.

DALLA VECCHIA, F. M. 1995. A new pterosaur (Reptilia, Pterosauria) from the Norian (Late Triassic) of Friuli (northeastern Italy). Preliminary note. *Gortania*, **16**, 59–66.

DALLA VECCHIA, F. M. 1996. Ipotesi stratigrafiche sulle unità bacinali noriche della Carnia. *Natura Nascosta*, **12**, 18–21.

DALLA VECCHIA, F. M. 1998. New observations on the osteology and taxonomic status of *Preondactylus buffarinii* Wild, 1984 (Reptilia, Pterosauria). *Bollettino della Società Paleontologica Italiana*, **36** (3), 355–366.

DALLA VECCHIA, F. M. 2000. A wing phalanx of a large basal pterosaur (Diapsida, Pterosauria) from the Norian (Late Triassic) of NE Italy. *Bollettino della Società Paleontologica Italiana*, **39** (2), 229–234.

DALLA VECCHIA, F. M. 2001. A caudal segment of a Late Triassic pterosaur (Diapsida, Pterosauria) from Northeastern Italy. *Gortania*, **23**, 5–36.

DALLA VECCHIA, F. M., MUSCIO, G. & WILD, R. 1989. Pterosaur remains in a gastric pellet from Upper Triassic (Norian) of Rio Seazza valley (Udine, Italy). *Gortania*, **10**, 121–132.

DALLA VECCHIA, F. M., WILD, R., HOPF, H. & REITNER, J. A. 2002. A crested rhamphorhynchoid pterosaur from the Late Triassic of Austria. *Journal of Vertebrate Paleontology* (Rapid Communications), **22**, 195–198.

EDMUND, A. G. 1969. Dentition. *In*: GANS, C., BELLAIRS, A. d'A. & PARSONS, N. (eds) *Biology of the Reptilia.* Academic Press, London, vol. 1, 117–200.

FRASER, N. C. & UNWIN, D. M. 1990. Pterosaur remains from the Upper Triassic of Britain. *Neues Jahrbuch für Geologie und Paläontologie, Monashefte*, **5**, 272–282.

FREY, E. & MARTILL, D. M. 1998. Soft tissue preservation in a specimen of *Pterodactylus kochi* (Wagner) from the Upper Jurassic of Germany. *Neues Jahrbuch Geologie und Paläontologie, Abhandlungen*, **210**, 421–441.

FREY, E. & TISCHLINGER, H. 2000. Weichteilanatomie der Flugsaurierfüsse und Bon der Scheitelkämme: neue Pterosaurierfunde aus den Solnhofener Schichten (Bayern) und der Crato-Formation (Brasilien). *Archaeopteryx*, **18**, 1–16.

GAETANI, M., BARRIER, E. *et al.* 2000. Map 6 – Late Norian (215–212 Ma). *In*: DERCOURT, J., GAETANI, M. *et al.* (eds) *Atlas Peri-Tethys, Palaeogeographical Maps.* CCGM/CGMW, Paris.

GODEFROIT, P. 1997. Reptilian, therapsid and mammalian teeth from the Upper Triassic of Varangéville (Northeastern France). *Bulletin de l'Institut Royale des Sciences Naturelles de Belgique*, **67**, 83–102.

GODEFROIT, P. & CUNY, G. 1997. Archosauriform teeth from the Upper Triassic of Saint-Nicolas-de-Port (northeastern France). *Palaeovertebrata*, **26**, 1–34.

GRADSTEIN, F. M., AGTERBERG, F. P., OGG, J. G., HARDENBOL, J., VAN VEEN, P., THIERRY, J. & HUANG, Z. 1995. A Triassic, Jurassic and Cretaceous time scale. *In*: BERGGREN, W. A., KENT, D. V., AUBRY, M. P. & HARDENBOL, J. (eds) *Geochronology Time Scales and Global Stratigraphic Correlation. SEPM Special Publications*, **54**, 95–126.

Hopf, H. 1997. *Fazielle und paläoökologische Untersuchungen in den Seefelder Schichten (Haupdolomit, Nor) von Seefeld, Tirol.* Diplomarbeit und Diplomakartierung dissertation, University of Göttingen, 146 pp.

JADOUL, F. 1986. Stratigrafia e paleogeografia del Norico delle Prealpi Bergamasche occidentali. *Rivista Italiana di Paleontologia e Stratigrafia*, **91**, 479–502.

JADOUL, F., BERRA, F. & FRISIA, S. 1992. Stratigraphic and paleogeographic evolution of a carbonate platform in an extensional tectonic regime: the example of the Dolomia Principale of Lombardy (Italy). *Rivista Italiana di Paleontologia e Stratigrafia*, **98** (1), 29–44.

JADOUL, F., MASETTI, D., CIRILLI, S., BERRA, F., CLAPS, S. & FRISIA, S. 1994. Norian–Rhaetian stratigraphy and paleogeographic evolution of the Lombardy Basin (Bergamasc Alps). *15th IAS Regional Meeting, April 1994, Ischia, Italy, Field Excursions, Excursion B1*, 5–38.

JENKINS, F. A. JR, SHUBIN, N. H. *et al.* 1993. A Late Triassic continental vertebrate fauna from the Fleming Fjord Formation, Jameson Land, east Greenland. *New Mexico Museum of Natural History and Science Bulletin*, **3**, 74.

JENKINS, F. A., JR, SHUBIN, N. H., GATESY, S. M. & PADIAN, K. 2001. A diminutive pterosaur (Pterosauria: Eudimorphodontidae) from the Greenlandic Triassic. *Bulletin of the Museum of Comparative Zoology, Harvard*, **156**, 151–170.

KOVACS, S., LESS, G., PIROS, D., RÉTI, Z., & RÒTH L. 1989. Triassic Formations of the Aggtelek-Rudabanya Mountains, northeastern Hungary. *Acta Geologica Ungarica*, **32**, 31–63.

KOZUR, H. 1989. Significance of events in conodont evolution for the Permian and Triassic stratigraphy. *Courier Forschunginstitut Senckenberg*, **117**, 385–408.

KREMMLING, W. 1912. Beitrage zur Kenntnis von *Rhamphorhynchus gemmingi* H. v. Meyer. Nach einem in Halle befindlichen Exemplar. *Abhandlungen der Kaiserliche Leopold Carol Deutschen Akademie der Naturforscher*, **96** (3), 349–368.

KRYSTYN, L. & WIEDMANN, J. 1986. Ein *Choristoceras*-Vorläufer (Ceratitina, Ammonoidea) aus dem Nor von Timor. *Neues Jahrbuch für Geologie und Paläontologie, Monashefte*, **8**, 449–463.

LOCKLEY, M. G., LIM, M., HUH, S-K., YANG, S-Y., CHUN, S. S. & UNWIN, D. M. 1997. First report of pterosaur tracks from Asia, Chollanam Province, Korea. *Journal of the Paleontology Society, Korea*, Special Publications, **2**, 17–32.

LOCKLEY, M. G., LOGUE, T. J., MORATALLA, J. J., HUNT, A. P., SCHULTZ, R. J. & ROBINSON, J. W. 1995. The fossil trackway *Pteraichnus* is pterosaurian, not crocodilian: implications for the global distribution of pterosaur tracks. *Ichnos*, **4**, 7–20.

MAZIN, J-M., HANTZPERGUE, P., LAFAURIE, G. & VIGNAUD, P. 1995. Des pistes des ptérosaures dans le Tithonien de Crayssac (Quercy, France). *Comptes Rendus de l'Académie des Sciences, Paris*, **321**, 417–424.

MURRY, P. A. 1986. Vertebrate paleontology of the Dockum Group, western Texas and eastern New Mexico. *In*: PADIAN, K. (ed.) *The Beginning of the Age of Dinosaurs*. Cambridge University Press, Cambridge, 109–107.

ORCHARD, M. J. 1991. Upper Triassic conodont biochronology and new index species from the Canadian Cordillera. *In*: ORCHARD, M. J. & MCCRACKEN, A. D. (eds) *Ordovician to Triassic conodont paleontology of the Canadian Cordillera*. Geological Survey of Canada Bulletin, **417**, 299–335.

OWEN, R. 1870. A Monograph of the fossil Reptilia of the Liassic Formations. III, *Monographs of the Paleontographical Society*, pp. 41–81, London.

PADIAN, K. 1983. Osteology and functional morphology of *Dimorphodon macronyx* (Buckland) (Pterosauria: Rhamphorhynchoidea) based on new material in the Yale Peabody Museum. *Postilla*, **189**, 1–43.

PADIAN, K. & WILD, R. 1992. Studies of Liassic Pterosauria, I. The holotype and referred specimens of the Liassic pterosaur *Dorygnathus banthensis* (Theodori) in the Petrefaktensammlung Banz, northern Bavaria. *Palaeontographica A*, **225**, 59–77.

PETERS, D. 2000. A redescription of four Prolacertiform genera and implications for pterosaur phylogenesis. *Rivista Italiana di Paleontologia e Stratigrafia* **106**, 293–336.

PLIENINGER, F. 1895. *Campylognathus Zitteli*. Ein neuer Flugsaurier aus dem Oberen Lias Schwabens. *Palaeontographica*, **41**, 193–222.

PLÖCHINGER, B. 1980. The Salzkammergut and its geological setting within the Northern Calcareous Alps. *2nd European Conodont Symposium, ECOS 2, Guidebook, Abstracts*, 62–65.

POLESCHINSKI, W. 1986. *Stratigraphie, fazies und sedimentologie der Seefelder Schichten im Raum Seefeld/Tirol – Ein potentielles Erdölmuttergestein aus dem ober nor der Westlichen Kalkalpen*. PhD thesis, University of Innsbruck.

RENESTO, S. 1993. An isolated sternum of *Eudimorphodon* (Reptilia, Pterosauria) from the Norian (Late Triassic) of the Bergamo Prealps (Lombardy, northern Italy). *Rivista Italiana Paleontologia e Stratigrafia*, **99**, 415–422.

RENESTO, S. 1994. A new prolacertiform reptile from the Late Triassic of northern Italy. *Rivista Italiana di Paleontologia e Stratigrafia*, **100**, 285–306.

RENESTO, S. & DALLA VECCHIA, F. M. 2000. The unusual dentition and feeding habits of the Prolacertiform reptile *Langobardisaurus* (Late Triassic, northern Italy). *Journal of Vertebrate Paleontology*, **20**, 622–627.

RIVA, A., SALVATORI, R., CAVALIERE, R., RICCHIUTO, T. & NOVELLI, L. 1986. Origin of oil in Po Basin, northern Italy. Advances in Organic Geochemistry 1985. *Organic Geochemistry*, **10**, 391–400.

ROGHI, G., MIETTO, P. & DALLA VECCHIA, F. M. 1995. Contribution to the conodont biostratigraphy of the Dolomia di Forni (Upper Triassic, Carnia, NE Italy). *Memorie di Scienze Geologiche*, **47**, 125–133.

ROMER, A. S. 1966. *Osteology of Reptiles*. University of Chicago Press, Chicago, 772 pp.

SEELEY, H. G. 1901. *Dragons of the Air. An Account of Extinct Flying Reptiles*. Methuen & Co., London, xiii + 239 pp.

SERENO, P. C. 1991. Basal archosaurs: phylogenetic relationships and functional implications. *Journal of Vertebrate Paleontology Memoirs*, **2**, 1–53.

STEFANI, M., ARDUINI, P., GARASSINO, A., PINNA, G., TERUZZI, G. & TROMBETTA, G. L. 1992. Palaeoenvironment of extraordinary fossil biotas from the Upper Triassic of Italy. *Atti della Società Italiana di Scienze Naturali e del Museo Civico di Storia Naturale di Milano*, **132**, 309–335.

UNWIN, D. M. 1988. New remains of the pterosaur *Dimorphodon* (Pterosauria: Rhamphorhynchoidea) and the terrestrial ability of early pterosaurs. *Modern Geology*, **13**, 57–68.

UNWIN, D. M. 1997. Pterosaur tracks and the terrestrial ability of pterosaurs. *Lethaia*, **29**, 373–386.

UNWIN, D. M. 2001. Variable growth rate and delayed maturation: do they explain 'giant' pterosaurs? *Journal of Vertebrate Paleontology*, **21**, [Abstract] 109A.

UNWIN, D. M., LÜ, J. & BAKHURINA, N. N. 2000. On the systematic and stratigraphic significance of pterosaurs from the Lower Cretaceous Yixian Formation (Jehol Group) of Liaoning, China. *Mitteilungen aus dem Museum für Naturkunde, Berlin, Geowissenschaftliche Reihe*, **3**, 181–206.

WELLNHOFER, P. 1970. Die Pterodactyloidea (Pterosauria) der Oberjura-Plattenkalke Süddeutschlands. *Bayerische Akademie der Wissenschaften zu München, Mathematisch-Naturwississenschaftliche Klasse, Abhandlungen*, **141**,1–33.

WELLNHOFER, P. 1974. *Campylognathoides liasicus* (Quenstedt), an Upper Liassic pterosaur from Holzmaden. The Pittsburgh Specimen. *Annals of the Carnegie Museum, Pittsburgh*, **45** (2), 5–34.

WELLNHOFER, P. 1975a. Die Rhamphorhynchoidea (Pterosauria) der Oberjura-Plattenkalke Süddeutschlands. Teil I: Allgemaine Skelettmorphologie. *Palaeontographica*, **148**, 1–33.

WELLNHOFER, P. 1975b. Die Rhamphorhynchoidea (Pterosauria) der Oberjura-Plattenkalke Süddeutschlands. Teil II: Systematische Beschereibung. *Palaeontographica*, **148**, 132–186.

WELLNHOFER, P. 1975c. Die Rhamphorhynchoidea (Pterosauria) der Oberjura-Plattenkalke Süddeutschlands. Teil III: Palökologie und Stammesgeschichte. *Palaeontographica*, **149**, 1–30.

WELLNHOFER, P. 1978. Pterosauria. *In*: WELLNHOFER, P. (ed.) *Handbuch der Paläoherpetologie*. Fischer, Stuttgart, Teil **19**, 82 pp.

WELLNHOFER, P. 1991. *The Illustrated Encyclopedia of Pterosauria*. Salamander Books, London, 192 pp.

WELLNHOFER, P. 2001. A Late Triassic pterosaur from the Northern Calcareous Alps. *In*: *Two Hundred Years of pterosaurs – A Symposium on the Anatomy, Evolution, Palaeobiology and Environments of Mesozoic Flying Reptiles, Toulouse, France, September 5–8, 2001. Strata*, Série 1, **11**, 99–100.

WELLNHOFER, P. 2003. A Late Triassic pterosaur from the Northern Calcareous Alps (Tyrol, Austria). *In*: BUFFETAUT, E. & MAZIN, J-M. (eds) *Evolution and Palaeobiology of Pterosaurs*. Geological Society, London, Special Publications, **217**, 5–22.

WILD, R. 1973. Die Triasfauna der Tessiner Kakalpen. XXIII. *Tanystropheus longobardicus* (Bassani) (Neue Ergebnisse). *Schweizerische Pälaontologische Abhandlungen*, **95**, 1–162.

WILD, R. 1978. Die Flugsaurier (Reptilia, Pterosauria) aus der Oberen Trias von Cene bei Bergamo, Italien. *Bollettino della Società Paleontologica Italiana*, **17**, 176–256.

WILD, R. 1984. A new pterosaur (Reptilia, Pterosauria) from the Upper Triassic (Norian) of Friuli, Italy. *Gortania*, **5**, 45–62.

WILD, R. 1989. *Aetosaurus* (Reptilia: Thecodontia) from the Upper Triassic (Norian) of Cene near Bergamo, Italy, with a revision of the genus. *Rivista del Museo Civico di Scienze Naturali 'E. Caffi', Bergamo*, **14**, 1–24.

WILD, R. 1994. A juvenile specimen of *Eudimorphodon ranzii* Zambelli (Reptilia, Pterosauria) from the Upper Triassic (Norian) of Bergamo. *Rivista del Museo Civico di Scienze Naturali 'E. Caffi' di Bergamo*, **16** (1993), 91–115.

WRIGHT, J. L., UNWIN, D. M., LOCKLEY, M. G. & RAINFORTH, E. C. 1997. Pterosaur tracks from the Purbeck Limestone of Dorset, England. *Proceedings of the Geologists' Association, London*, **108**, 39–48.

ZAMBELLI, R. 1973. *Eudimorphodon ranzii* gen. nov., sp. nov., uno pterosauro triassico. *Rendiconti dell Istituto Lombardo di Scienze e Lettere, (B)*, **107**, 27–32.

A new scaphognathine pterosaur from the Upper Jurassic Morrison Formation of Wyoming, USA

KENNETH CARPENTER[1], DAVID UNWIN[2], KAREN CLOWARD[3], CLIFFORD MILES[3] & CLARK MILES[3]

[1]*Department of Earth Sciences, Denver Museum of Natural History, 2001 Colorado Boulevard, Denver, Colorado, CO 80205, USA (e-mail: Kcarpenter@DMNS.org)*

[2]*Institut für Paläontologie, Museum für Naturkunde, Humboldt-Universität zu Berlin, Invalidenstrasse 43, Berlin D-10115, Germany (e-mail: h0662eka@rz.hu-berlin.de)*

[3]*Western Paleontological Laboratories, 2929 Thanksgiving Way, Lehi, Utah, UT 84043, USA (e-mail: Karen@dinosaurpoint.com)*

Abstract: A partial rostrum of a new species of scaphognathine pterosaur, distinguished by a thin median crest along its dorsal margin and a deep embayment of the dental margin, is the first identifiable cranial fragment of a pterosaur from the Upper Jurassic Morrison Formation of western North America. By contrast with pterodactyloids, cranial crests are rare in "rhamphorhynchoids" and this is the first record of such a structure. The new material provides fresh insights into the taxonomic diversity of Late Jurassic North American pterosaurs. Based on the ratio of the skull and skeleton of *Scpahognathus*, the fragment represents an individual with an estimated wing span of 2.5 m. Consequently, this is one of the largest "rhamphorhynchoids" found so far. A mandible fragment from the same quarry has closely spaced alveoli, therefore cannot be referred to the rostrum. Its large size indicates another large "rhamphorhynchoid" in the Morrison Formation.

Records of pterosaurs from the Upper Jurassic Morrison Formation (Oxfordian–Tithonian) of the western United States remain remarkably rare despite over a century of collecting. In 1878 Marsh described the first specimen from the Morrison Formation under the name *Dermodactylus montana*. Collected from Reed's Quarry 5 at Como Bluff, Wyoming, it consisted of the fragmented distal end of a wing metacarpal. Pterodactyloid in nature, this bone, together with the first records of *Pteranodon* from the Chalk of Kansas (Marsh 1871) demonstrated the existence of pterosaurs in the Western Hemisphere. Another bone collected in 1879 at Reed's Quarry 9 (mammal quarry), but not identified until 1981 by Galton, is the holotype of *Comodactylus ostromi*. Marsh (1881) later named *Laopteryx priscus*, which he identified as a Jurassic bird on the basis of a cranial fragment also from Quarry 9, but Ostrom (1986) subsequently referred the specimen to an unidentified pterosaur. Several small bones found at Dry Mesa Quarry, Colorado, were identified by Jensen & Padian (1989) as the pterodactyloid *Mesadactylus ornithosphyos*. More recently, Harris & Carpenter (1996) named *Kepodactylus grandis* on the basis of some associated bones recovered from the Small *Stegosaurus* Quarry, Cañon City, Colorado. *Kepodactylus* was initially identified as a pterodactyloid and subsequently assigned by Unwin & Heinrich (1999) to the Dsungaripteroidea because of its thick-walled bones and the shape of the humerus.

The Late Jurassic pterosaur record in North American also includes a rapidly growing track record. The first report, of a single, well-preserved trackway in the Morrison Formation of Arizona, was made by Stokes (1957) and, more recently, many new prints and tracks have been recorded from the even older Summerville Formation (Mid-Callovian–Mid-Oxfordian) of Utah (Lockley *et al.* 1995) and the Sundance Formation (Callovian–Mid-Oxfordian) of Wyoming (Logue 1994, 1997; Lockley *et al.* 1995).

The meagre body fossil record of Morrison pterosaurs is now supplemented with the first discovery of skull material. A skull fragment was collected in 1996 in the vicinity of Bone Cabin Quarry, Albany County, Wyoming, but was not recognized as pterosaurian until later (Cloward & Carpenter 1998), when it was compared with *Scaphognathus*. The skull fragment was found in a poorly consolidated, fine-grained sandstone that lies in the lower half of Dinosaur Zone 2 of Turner & Peterson (1999), which is equivalent to the upper part of the Salt Wash Member of the Colorado Plateau. The relative stratigraphic position of the other Morrison pterosaurs are shown in Figure 1. The new specimen is the lowest occurrence and hence the oldest pterosaur from the Morrison Formation.

In 1999 a fragment of a pterosaur mandible, found within a metre of the skull fragment described here, was recognized among material collected in 1996. The narrow, elongate mandibular symphysis and

From: BUFFETAUT, E. & MAZIN, J-M. (eds) 2003. *Evolution and Palaeobiology of Pterosaurs*. Geological Society, London, Special Publications, **217**, 45–54. 0305-8719/03/$15

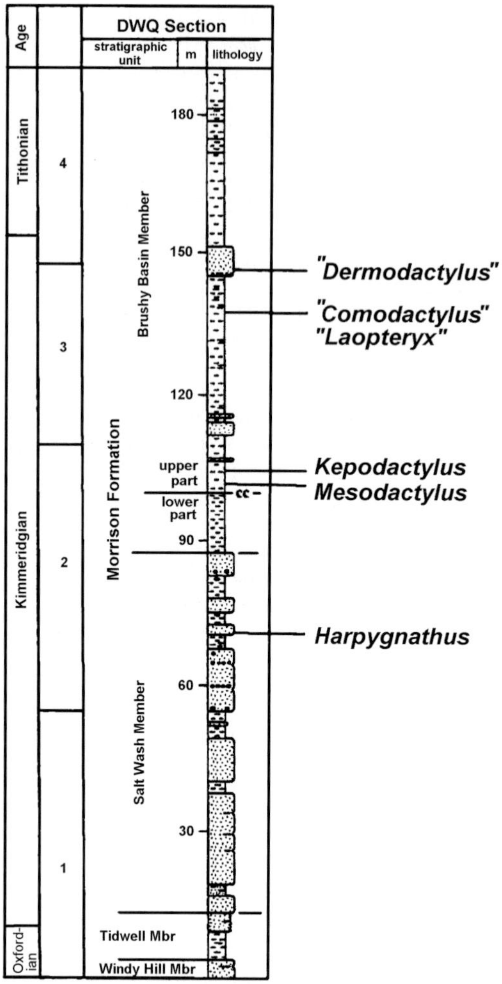

Fig. 1. Biostratigraphic position of pterosaurs against the Dinosaur National Monument (DWQ) reference section of Turner & Peterson (1999). Many of the pterosaurs of the Morrison Formation occur in unnamed strata equivalent to the named stratigraphic units shown in the column. For example, Turner & Peterson (1999) have determined that the quarry from which *Harpactognathus* came is, in strata, equivalent to the upper portion of the Salt Wash Member of the Colorado Plateau. *Nomen dubia* taxa are in quotes. cc, clay change boundary of Turner & Peterson (1999) separating the lower and upper parts of the Morrison Formation.

closely spaced, similarly sized dental alveoli do not match details of the rostral fragment described below; consequently, this specimen cannot be assigned to the same taxon. The mandibular fragment will be described elsewhere.

The term "Rhamphorhynchoidea" and its derivatives denotes a paraphyletic taxon, which is indicated by placing the name in double quotation marks. Phylogenetic relationships and pterosaur systematics follows Unwin (1995), Unwin & Lü (1997), Unwin *et al.* (2000, table 4, fig. 7) and Wellnhofer (1991).

Institute abbreviations: BMNH, Natural History Museum, London, UK; BYU, Brigham Young University, Provo, Utah, USA; NAMAL, North American Museum of Ancient Life, Lehi, Utah, USA; YPM, Yale University Peabody Museum of Natural Histroy, New Haven, Connecticut, USA.

Systemic descriptions

Family Rhamphorhynchidae Seeley 1870
Subfamily Scaphognathinae Hooley 1913
Genus *Harpactognathus* gen. nov.

Diagnosis. Thin median crest extending from tip of rostrum posteriorly above the external nares; antorbital fenestra bounded anteriorly by shallow triangular antorbital fossa; lateral surface of premaxillary and maxillary scalloped between widely spaced alveoli; ventral profile of dental margin undulating and deeply emarginated below external nares.

Type Species. *Harpactognathus gentryii*. sp. nov.

Horizon. Lower Dinosaur Zone 2 equivalent to Upper Salt Wash member of the Colorado Plateau.

Etymology. The generic name derives from Greek, *harpact* = seize or grasp, *gnathus* = jaws, in reference to the 'snatching jaws'. The specific name is given in honour of Joe Gentry, volunteer for the western Paleontological Laboratories, Lehi, Utah.

Specific diagnosis. As for genus.

Holotype. NAMAL 101, anterior portion of the rostrum.

Locality. Bone Cabin Quarry Extension (WY-79 of Turner & Peterson 1999), Albany County, Wyoming, USA.

Comments

Hooley (1913) proposed Scaphognathinae to include *Scaphognathus* and *Parapsicephalus*; *Sordes* was subsequently added to this taxon by Wellnhofer (1978). Wellnhofer listed the following characters as diagnostic for the subfamily, although many of these are vague: (1) relatively short skull; (2) steeply oriented quadrate; (3) a few, upright, widely spaced teeth in the jaw; (4) short wing finger; (5) wing phalange 1 shorter than wing phalanges 2 and 3; (6) long fifth toe; and (7) ulna longer than any of the four wing-finger phalanges. However, characters 1, 2, 4, 5, 6 and 7 are found in various basal pterosaurs, including dimorphodontids, anurognathids, campylognathids and rhamphorhynchines, and are not therefore diagnostic for Scaphognathinae.

This subfamily, nontheless, is distinguished by three characters:

(1) Only nine or less, straight (or slightly recurved), widely spaced pairs of teeth (equal to the distance of 3–4 alveoli) in the rostrum (Figure 4; Wellnhofer 1975, fig. 33; Sharov 1971). Other pterosaurs (Wellnhofer 1978, figs 2–6) generally have more teeth in the rostral dentition and the alveoli are not so widely spaced (gap equal to one alveolus) as in scaphognathines.

(2) Only six or less, widely spaced, vertically oriented pairs of teeth in the lower jaw (Wellnhofer 1975, fig. 33; Sharov 1971). All other pterosaurs have a greater number of teeth in the lower jaw and the alveoli are not so widely spaced as in scaphognathines.

(3) Phalanx two of the fifth pedal digit has a distinctive angular flexure at mid-length, such that the distal half of the phalange is 40–45° relative to the proximal half (Wellnhofer 1975, fig. 36d). In other pterosaurs that retain a second phalange in the fifth toe, this bone is straight or gently curved (Wellnhofer 1978, fig. 17).

These characters are present in *Scaphognathus*, *Sordes* and the new taxon described below (characters 2 and 3 are not yet known in the latter), but not in *Parapsicephalus*, which is remarkably similar to *Dorygnathus* and quite possibly congeneric with that taxon (Unwin unpub. data).

Description of *Harpactognathus*

The holotype of *Harpactognathus gentryii* is represented by a well-preserved, but slightly crushed fragment of the rostrum (Figs 2 & 3). The tip of the rostrum is missing, exposing a pair of alveoli near the mid-line (Fig. 2c). Posteriorly, the rostrum is sheared across the antorbital fenestra, consequently, the posterior process of the premaxillae is missing, as are the posterior terminations of the jugal and nasal processes of the maxillae and the rest of the skull. In addition, the dorsal margin of the thin median crest is also broken away dorsal of the nasal fenestra. Generally, the dental alveoli are empty (left alveoli 1–5, right 1, 5, 6). In some cases, the teeth are broken off at their base, leaving the root embedded in the dental alveolus (left 2–4); the dentition is discussed below.

The rostral region of *Haractognathus gentryii* is relatively broad and, even assuming that the holotype has suffered some dorsoventral compression, in life it would seem to have been wider than it is tall (compare Fig. 2a, b with 2d, e). This is unlike the typical condition in "rhamphorhynchoids", in which most taxa have a narrow, laterally compressed rostrum that is usually deeper than wide (Wellnhofer 1975, 1978, 1991). In *Harpactognathus*, the main body of the rostrum faces dorsally while laterally it rounds rapidly into low, somewhat convex lateral walls.

A striking feature of *H. gentryii* is the uneven outline of the dental margin of the rostrum, evident in both dorsal and lateral view (Fig. 2a, b, d). In dorsal aspect, the dental alveoli project distinctly from the jaw margin and are separated from one another by deep, rounded notches. In lateral view, the mid-region of the rostrum has a subrectangular-shaped embayment or notch below the nasal fenestra. A single alveolus is situated in this embayment.

The premaxillae taper anteriorly and may have formed a pointed tip to the rostrum, although the exact shape of this rostral apex is unknown. The premaxillae are fused along their dorsal mid-line, and this margin is drawn up into a wafer-thin, blade-like, vertical medial crest that extends posterodorsally from the anterior tip of the rostrum at least as far as the external nares. The preserved portion of the crest is of fairly even height, with a sharp dorsal margin and flat lateral surfaces that bear well-spaced, fine vertical grooves. Posteriorly, the premaxilla has a broad contact with the maxilla, although the suture is completely fused and can no longer be traced. It is likely that, as in other "rhamphorhynchoids", this suture curved forward, and downward, from the anteroventral corner of the external narial opening.

The main body of the maxilla extends anteroposteriorly and is markedly convex dorsoventrally (Fig. 2c). Posteriorly, the maxilla divides into a subhorizontal jugal process and a robust, posterodorsally directed nasal process, which has a thickened dorsal edge (best seen on the right side; it is slightly damaged on the left). In lateral view the shape of the maxilla is similar to that of *Scaphognathus* and *Sordes*. However, the maxilla in *Harpactognathus* is distinguished by the development of a small, triangular recess, part of the antorbital fossa, in the angle between the nasal and jugal processes (Fig. 2a, b).

The external narial openings are narrow, elongate, located near the dorsal mid-line of the skull and separated only by a thin process of the fused premaxillae. In lateral view the anterior half of the ventral margin of each external naris is horizontal, while, as is typical of "rhamphorhynchoids", the posterior half is angled posterodorsally. The dorsal and ventral margins of the naris converge at an acute angle anteriorly as is typical of derived "rhamphorhynchoids" such as *Dorygnathus* and *Rhamphorhynchus*, but unlike basal forms, including dimorphodontids, anurognathids and campylognathoidids.

The antorbital fenestra extends beneath the posterior margin of the narial opening and has dorsal and ventral margins that converge relatively acutely (~45°) anteriorly. This configuration resembles that of other derived "rhamphorhynchoids", such as campylognathids and rhamphorhynchids, but it is

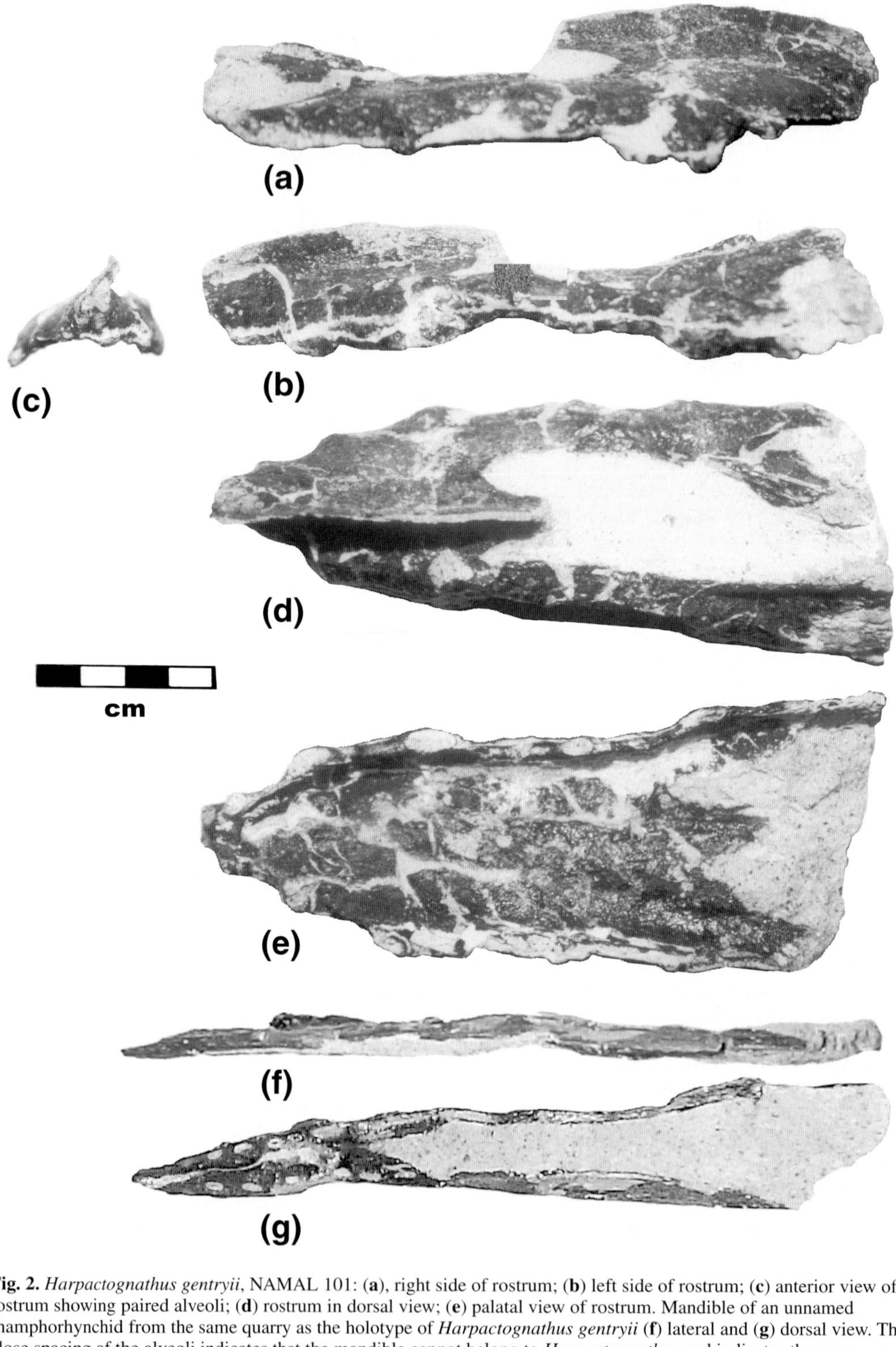

Fig. 2. *Harpactognathus gentryii*, NAMAL 101: (**a**), right side of rostrum; (**b**) left side of rostrum; (**c**) anterior view of rostrum showing paired alveoli; (**d**) rostrum in dorsal view; (**e**) palatal view of rostrum. Mandible of an unnamed rhamphorhynchid from the same quarry as the holotype of *Harpactognathus gentryii* (**f**) lateral and (**g**) dorsal view. The close spacing of the alveoli indicates that the mandible cannot belong to *Harpactognathus* and indicates the presence of yet another large rhamphorhynchid in the Morrison.

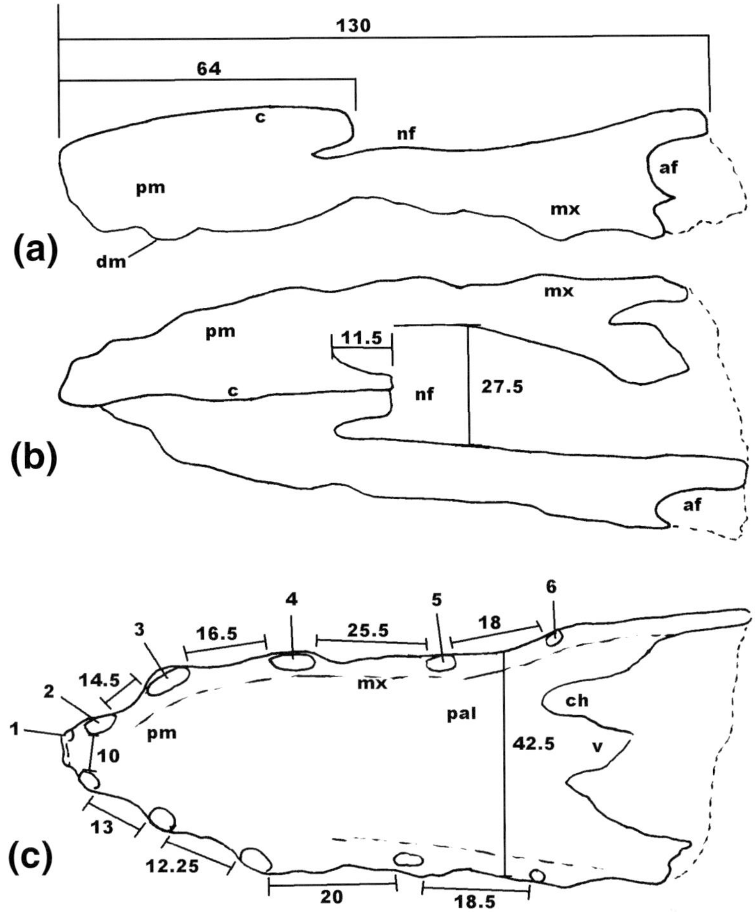

Fig. 3. Measurements of *Harpactognathus gentryii* holotype in mm: (**a**) lateral; (**b**) dorsal; (**c**) palatal. af, antorbital fenestra; af, antorbital fossa; c, median crest; ch, choana; dm, dental margin; mx, maxilla; nf, nasal fenestra; pal, palatine; pm, premaxilla; v, vomers; 1–6, dental alveoli 1–6.

unlike that in basal forms (dimorphodontids, anurognathids), where the antorbital opening is posterior to the narial opening and has ventral and dorsal margins that converge less acutely (>75°).

The floor of the palate (Fig. 2e) is deeply recessed. It consists of an almost flat, continuous sheet formed presumably by the premaxillae anteriorly, the maxillae laterally and the palatines medially, although the contacts between these elements, and hence their extent and shape, are not determinable. The tooth margins are developed into distinct ridges that project below the level of the palate and reach 10 mm in height. Anteriorly and anterolaterally in the region formed by the premaxillae, the palate curves dorsally into the dental borders, but posteriorly, in the maxilla region, the dental borders meet the palate at a right angle.

Posteriorly, the palate is perforated by the paired internal nares, or choanae (Fig. 2e). These elongate oval openings are separated medially by a slender bar of bone formed by the vomers, although the exact anterior extent of these elements is unclear. Laterally, the choanae are bounded by the narrow, flat posterior process of the palatines. The choanae terminate anteriorly opposite the sixth alveolus. This condition is similar to *Scaphognathus* (Wellnhofer 1975, fig. 34), although this depends on the exact numbering of the dental alveoli in the latter taxon (see below).

Dentition

The first six dental alveoli are preserved on each side of the rostrum (Fig. 3) and a slight swelling of the dental margin, on the left side, immediately anterior to the break, suggests the presence of a seventh alveolus. As in *Scaphognathus*, the first three pairs of alveoli on each side are presumed to be located in the premaxilla, while pairs four to six and any subsequent alveoli were borne by the maxilla. The teeth

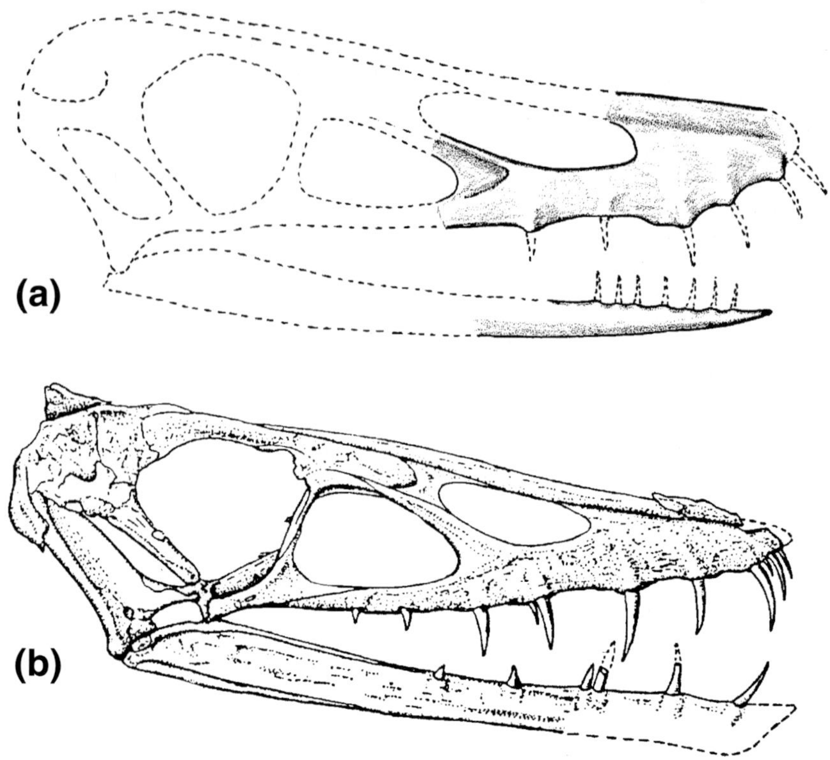

Fig. 4. Comparison of scaphognathid skulls: (**a**), *Harpactognathus gentryii* (estimated length 28–30 cm); (**b**) *Scaphognathus crassirostis* (11.5 cm), modified from Wellnhofer (1975, fig. 33). Skulls are drawn to same length.

seem to have been somewhat variable in shape: alveoli one, two and six are circular in cross-section, whereas three, four and five are an elongate oval about twice as long as wide. The latter bore relatively narrow, laterally compressed teeth as is shown by the laterally compressed base of the fourth tooth preserved on the left side.

The size of the alveoli shows a marked increase from the first to fourth pair, the latter reaching almost twice the basal length of the former. Posteriorly, however, alveolus size rapidly declines again and the sixth pair are no larger than the first pair. The gaps between the maxillary alveoli are remarkably large, being equivalent to 3–4 times the length of an alveolus; the gaps are narrower in the premaxillary part of the rostrum. Typically in 'rhamphorhynchoids' such gaps are less than a single alveolus length, although greater spacing is present in some rhamphorhynchids, most notably *Scaphognathus* (see Fig. 4b).

The first pair of teeth projected anteroventrally at about 30° to the ventral margin of the tooth row. Pairs two and three were directed downward and slightly forward, at about 75° to the ventral margin and are angled slightly outward laterally. The remaining teeth appear to have been directed almost vertical to the jaw margin (Fig. 4a).

Discussion

Systematic relationships of Harpactognathus. The remarkably thin walls of the rostrum and the smooth, unornamented texture of their external surface indicate that the holotype material of *Harpactognathus gentryii* is undoubtedly pterosaurian. The skull fragment of *H. gentryii* does not exhibit any pterodactyloid characters, such as a confluent nasoantorbital opening, but does have at least one feature that is found in basal clades of pterosaurs, collectively referred to as the paraphyletic grade group 'Rhamphorhynchoidea'. As in 'rhamphorhynchoids', there is a sloping maxillonasal bar separating the nares from the antorbital fenestra. *Harpactognathus* does share some derived 'rhamphorhynchoid' features, including an elongate, anteriorly-tapered rostrum, like those of campylognathoidids and rhamphorhynchids, but unlike the short, deep rostrum of basal forms, such as dimorphodontids and anurognathids. *Harpactognathus* also has relatively narrow, slit-like external narial openings, as found in rhamphorhynchids and some campylognathoidids, but unlike the deep rounded openings evident in dimorphodontids and anurognathids. In addition, the antorbital opening extends beneath the narial opening, as in most rhamphorhynchids and campylognathoidids, rather than being located behind the narial opening as in dimorphodontids and anurognathids.

Harpactognathus shares one derived character in common with rhamphorhynchids based on the wide spacing of the alveoli: the apparent reduction of the rostral dentition to 11, or less, pairs of teeth. Other 'rhamphorhynchoids' have more teeth in the rostrum, the only exception being anurognathids. The reduced number of teeth in this clade is presumed to have occurred independently from that in rhamphorhynchids because these clades are not thought to be closely related (Unwin 1995; Unwin *et al.* 2000).

Within 'Rhamphorhynchoidea', *Harpactognathus* shows clear similarity to scaphognathines and exhibits a distinctive apomorphy of this group: only nine, or less, relatively straight (or slightly recurved) widely spaced pairs of teeth in the rostrum. Moreover, *Harpactognathus* shares at least one unique feature in common with *Scaphognathus*: the anterior tip of the rostrum is flexed dorsally, so that, in lateral view, the profile of the palate curves dorsally to meet the dorsal margin of the skull at an obtuse angle. This is unlike the typical condition in pterosaurs, where the rostral tip is not reflexed dorsally and the ventral outline of the skull continues directly to the tip of the rostrum, a condition that is also found in the scaphognathine *Sordes*. *Harpactognathus* and *Scaphognathus* also share the derived feature (within Scaphognathinae) of having eight or less pairs of teeth in the rostrum, unlike *Sordes* which has at least nine pairs.

The general shape and proportions of the rostrum and the distribution, position and orientation of the dental alveoli of *Harpactognathus* and *Scaphognathus* are remarkably similar. *Harpactognathus* is distinguished from *Scaphognathus*, and also from *Sordes*, by the presence of a median crest on the rostrum, the development of a recess (representing the antorbital fossa) on the maxilla, the strong scalloping of the dental margin when viewed dorsally or ventrally, the presence of a deep dorsal emargination in the region of the fifth alveolus and the marked unevenness of the dental margin in lateral view. *Sordes* does not have a median crest and *Scaphognathus* has always been restored without such a structure. However, the dorsal margin of the rostrum anterior to and above the narial opening seems to bear a thin flange of bone (Wellnhofer 1975, fig. 33a; Wellnhofer 1991, p. 92) that may represent the base of a crest, but further specimens of *S. crassirostris* are needed to resolve the identity of this structure.

Discussion

Although *H. gentryii* is only poorly known, there is good evidence to show that it is a "rhamphorhynchoid" pterosaur belonging to the subfamily Scaphognathinae. This subfamily is represented by two other genera: *Scaphognathus* and *Sordes*. *Scaphognathus*, from the Upper Jurassic (Tithonian) Solnhofen Plattenkalk of Bavaria, is represented by a single species: *S. crassirostris*. *Sordes* from the Upper Jurassic (Oxfordian–Kimmeridgian) Karabastau Formation of southern Kazakhstan is also represented by a single species: *Sordes pilosus*.

Harpactognathus appears to share some derived features (dorsal flexure of the rostral tip, only eight or less pairs of teeth in the rostrum) with *Scaphognathus* that are not found in *Sordes*, suggesting that they are more closely related to each other than either is to *Sordes*. This is at least consistent with the stratigraphic distribution of these taxa in that *Sordes* is geologically somewhat older (Oxfordian–Early Kimmeridgian) than either of the other two genera (Kimmeridgian).

Size of *Harpactognathus*

Reconstruction of the original skull length, based on comparison of the rostrum of *Harpactognathus gentryii* with the corresponding region of the closely related *Scaphognathus crassirostris*, indicates a maximum length (occipital condyle to tip of rostrum) of 280–300 mm. Assuming that *H. gentryii* had a wing span that was of similar proportions to the skull, as in other scaphognathines, a possible minimum wing span of 2.5 m is indicated. (The true wing span was probably greater because this calculation does not take into account positive allometry in wing length as size increases.) *Scaphognathus crassirostris* is known from three specimens, two of which are complete and which range up to 0.95 m in wing span (Wellnhofer 1975). *Sordes pilosus* is known from eight individuals that range up to 0.75 m in wing span (Sharov 1971; Bakhurina & Unwin 1995; Unwin & Bakhurina 2000).

Most "rhamphorhynchoids" have a wing span of less than 1 m and a skull length of less than 150 mm. Among the largest "rhamphorhynchoids", large individuals of *Dimorphodon macronyx* have a skull length of just over 200 mm (Wellnhofer 1978) and a wing span of approximately 1.4 m, while the largest known individual of *Rhamphorhynchus longiceps* (BMNH 37002) has a skull length of 191.5 mm and an estimated wing span of 1.8 m (Wellnhofer 1975, p. 161). Isolated wing bones of *Rhamphocephalus*, a rhamphorhynchid from the Mid-Jurassic Stonesfield Slate of Oxfordshire, United Kingdom (Unwin 1996) including a wing phalanx II (BMNH 40126k) that is 213 mm long, representing an individual with a wing span that might have reached 2.25 m. *H. gentryii* is thus much larger than other scaphognathines (see above) and appears to be significantly larger than most other "rhamphorhynchoids".

Cranial crests in "rhamphorhynchoid" pterosaurs

Cranial crests are widely distributed in pterodactyloids and occur in all four major clades. By contrast, until recently, crests were unknown in "rhamphorhynchoids", but have now been reported in a Triassic pterosaur from Austria (Wellnhofer 2003), and are also present in some dimorphodontids (Unwin unpub. data). *Harpactognathus* represents a third "rhamphorhynchoid" with a cranial crest. The size and shape of the crest is similar to that in some ctenochasmatoids (*Ctenochasma*, *Gnathosaurus*) and dsungaripteroids (*Germanodactylus*, *Dsungaripterus*), and it may also have been continued dorsally by soft-tissue derivatives of the integument as in *Tapejara* (Martill & Frey 1998). *Harpactognathus* is unusual in that the crest is continued to the anterior tip of the rostrum rather than terminating posterior to the tip.

A variety of functions have been proposed for the cranial crests of pterosaurs including: sites for muscle attachments, a forward rudder, an airbrake, a heat exchanger and a display structure (reviewed by Bennett 1992). In a detailed study based on *Pteranodon*, Bennett (1992) presented compelling evidence that the spectacular cranial crest borne by some individuals of this pterosaur represented a sexual dimorphism and acted as a display or signalling device. A similar dimorphism involving the presence or absence of crests is found in other pterosaurs, including *Ctenochasma*, *Germanodactylus* and *Anhanguera*. This, together with the remarkable interspecific variation in crest size, shape and position, further supports the interpretation of these structures as display devices and we presume that the crest of *Harpactognathus* also served in this way.

Systematic status of Morrison pterosaurs

Apart from *Harpactognathus*, the only other certain record of a "rhamphorhynchoid" from the Morrison Formation is the holotype wing metacarpal (YPM 9150) of *Comodactylus ostromi* (Galton 1981). Except for its size, this bone is very similar to that of other "rhamphorhynchoids" (compare Galton 1981, figs 2c, p). The proportional differences of the distal condyles cited by Galton (1981) are minor, thus we follow Jensen & Padian (1989) and consider *Comodactylus ostromi* to be a *nomen dubium*. It is possible that YPM 9150 might belong to *Harpactognathus gentryii*, but more complete material is needed to demonstrate this.

The holotype and only known specimen of *Dermodactylus montanus* consists of a single, isolated, distal end of an elongate wing metacarpal (YPM 2020: Marsh 1878; Galton 1981). The elongation of the specimen indicates that it is pterodactyloid, and the shape and orientation of the distal condyles, notably their dorsal flexure, when viewed from anterior or posterior (Galton 1981, fig. 2q, s) shows that it is not ornithocheiroid. The identity of the metacarpal cannot be resolved any further, however, and in the absence of any distinctive features *Dermodactylus montanus* must be considered a *nomen dubium* (see also Jensen & Padian 1989).

The holotype and only known specimen of *Laopteryx priscus*, a fragmentary, incomplete occiput (Marsh 1881) appears to belong to a small pterodactyloid pterosaur (Ostrom 1986). The specimen is too incomplete for its affinities to be further resolved and *Laopteryx priscus* also should be considered a *nomen dubium*.

Mesadactylus ornithosphyos is based on a synsacrum (BYU 2024; Jensen & Padian 1989) and another 34 isolated bones, including vertebrae, a scapulocoracoid and various fore and hind limb elements. The synsacrum is distinctive and distinguishes *Mesadactylus* from other pterodactyloids. The affinities of this pterosaur are uncertain, but it is certainly not an ornithocheiroid. Features of the synsacrum and the humerus compare most closely with those of dsungaripteroids, but further comparative work is needed to resolve the relationships of *Mesadactylus* to other pterodactyloids.

Kepodactylus insperatus, represented by a cervical and various fore- and hindlimb elements (Harris & Carpenter 1996), was initially identified as a pterodactyloid and later tentatively assigned, on the basis of humerus morphology, to Dsungaripteroidea (Unwin & Heinrich 1999). Furthermore, as yet undescribed material of *Kepodactylus*, including fragments of a coracoid, wing metacarpal, wing phalanges, a femur, a metatarsal and ribs also exhibits distinctive dsungaripteroid characters. It seems likely therefore that *Kepodactylus* belongs within Dsungaripteroidea, although its relationship to other members of this clade is still unclear. *Kepodactylus* is the first record of dsungaripteroids in North America, and possibly also one of the earliest.

In conclusion, at least two distinct pterosaur taxa, *Harpactognathus* (a scaphognathine) and *Kepodactylus* (a dsungaripteroid), are present in the Morrison Formation of western North America. It is difficult to determine the taxonomic affinities of other pterosaur material from the Morrison Formation, but it is possible that much of what has been collected might eventually be shown to belong to *Harpactognathus*, *Kepodactylus*, or closely related forms. Certainly, at present, there is no clear evidence for pterosaur lineages other than Scaphognathinae and Dsungaripteroidea.

Harpactognathus and the palaeoecology of Late Jurassic pterosaurs

Globally, most pterosaurs have been found in marginal marine sediments. This does not mean that pterosaurs were largely confined to marginal marine environments, but is more probably a reflection that deposits suitable for preserving these animals occur more frequently in marginal marine settings. Indeed, pterosaurs have been collected from a variety of terrestrial deposits from the Cretaceous (Wellnhofer 1991). Moreover, the variety of taxa recovered and the adaptations exhibited by these pterosaurs strongly support the idea that they inhabited continental environments as well (Unwin *et al.* 2000). The situation in the Jurassic is less clear, partly because non-marine pterosaur lagerstätten deposits are rarer for this interval. Consequently, units such as the Morrison Formation are of particular significance because they provide important evidence regarding the presence of pterosaurs in continental environments in the Jurassic and their likely diversity and ecology.

The Morrison pterosaurs provide three lines of evidence to support the idea that pterosaurs inhabited interior regions of the North American continent during the Late Jurassic:

(1) First, there is a steadily accumulating number of records from different geographic localities (Marsh 1878, 1881; Galton 1981; Ostrom 1986; Jensen & Padian 1989; Harris & Carpenter 1996), some of which, such as the Dry Mesa Quarry, have yielded substantial number of bones.
(2) Pterosaur material has been found at various horizons throughout the Morrison Formation (Fig. 1).
(3) Track records (e.g. Stokes 1957) demonstrate that pterosaurs were present in these environments, and do not represent allochthonous occurrences.

We conclude therefore that pterosaurs were an integral part of the vertebrate fauna of the Morrison Formation.

It may be significance that *Harpactognathus*, the only rhamphorhynchoid to be identified with certainty from the non-marine Morrison Formation, belongs to the Scaphognathinae. The scaphognathine *Sordes pilosus* is the only rhamphorhynchid found in the lacustrine Karabastau Formation of Kazakhstan, where it is relatively common. *Scaphognathus* also occurs in the Solnhofen Limestone, but it is exceptionally rare compared with the other rhamphorhynchids, *Rhamphorhynchus*. Indeed, with the exception of the Solnhofen Plattenkalks, all other Late Jurassic localities that have yielded rhamphorhynchids have produced rhamphorhynchines. Based on this distribution, we tentatively propose the hypothesis that scaphognathines inhabited terrestrial freshwater environments, whereas rhamphorhynchines inhabited marginal marine environments. Future discoveries are need to test this hypothesis.

We thank the field crews that assisted in the excavation of Bone Cabin Quarry West where *Harpactognathus* was found.

References

BAKHURINA, N. N. & UNWIN, D. M. 1995. A survey of pterosaurs from the Jurassic and Cretaceous of the former Soviet Union and Mongolia. *Historical Biology*, **10**, 197–245.

BENNETT, S. C. 1992. Sexual dimorphism of *Pteranodon* and other pterosaurs, with comments on cranial crests. *Journal of Vertebrate Paleontology*, **12**, 422–434.

CLOWARD, K. C. & CARPENTER, K. 1998. A newly discovered pterosaur skull from the Morrison Formation of Wyoming. *Journal of Vertebrate Paleontology*, **18**, 34A–35A. [Abstract]

GALTON, P. M. 1981. A rhamphorhynchoid pterosaur from the Upper Jurassic of North America. *Journal of Paleontology*, **55**, 1117–1122.

HARRIS, J. & CARPENTER, K. 1996. A large pterodactyloid from the Morrison Formation (Late Jurassic) of Garden Park, Colorado. *Neues Jahrbuch fur Geologie und Palaontologie, Monatshefte*, **1996**, 473–484.

HOOLEY, R. W. 1913. On the skeleton of *Ornithodesmus latidens*: an ornithosaurs from the Wealden shales of Atherfield (Isle of Wight). *Quarterly Journal of the Geological Society, London,* **69**, 372–422.

JENSEN, J. A. & PADIAN, K. 1989. Small pterosaurs and dinosaurs from the Uncompahgre Fauna (Brushy Basin Member, Morrison Formation: ?Tithonian), Late Jurassic, western Colorado. *Journal of Paleontology,* **6**, 364–373.

LOCKLEY, M. G., LOGUE, T. J., MORATALLA, J. J., HUNT, A. P. SCHULTZ, R. J. & ROBINSON, J. W. 1995. The fossil trackway *Pteraichnus* is pterosaurian, not crocodilian: implications for global distribution of pterosaur tracks. *Ichnos*, **4**, 7–20.

LOGUE, T. L. 1977. Preliminary investigations of pterodactyl tracks at Alcova, Wyoming. *Earth Science Bulletin*, **10**, 29–30.

LOGUE, T. L. 1994. Alcova, Wyoming tracks of *Pteraichnus saltwashensis* made by pterosaurs. *Geological Society of America Abstracts with Program, South Central Region*, **26**, 10.

MARSH, O. C. 1871. Note on a new and gigantic species of pterodactyle. *American Journal of Science*, **1**, 472.

MARSH, O. C. 1878. New pterodactyl from the Jurassic of the Rocky Mountains. *American Journal of Science,* **3**, 233–234.

MARSH, O. C. 1881. Discovery of a fossil bird in the Jurassic of Wyoming. *American Journal of Science*, **21**, 341–342.

MARTILL, D. M. & FREY, E. 1998. A new pterosaur lagerstätte in N.E. Brazil (Crato Formation; Aptian, Lower

Cretaceous): preliminary observations. *Oryctos*, **1**, 79–85.

OSTROM, J. H. 1986. The Jurassic 'bird' *Laopteryx priscus* re-examined. *In*: FLANAGAN, K. M. & LILLEGRAVEN, J. A. (eds) *Vertebrate Phylogeny and Philosophy*. Contributions to Geology, University of Wyoming Special Papers, **2**, 11–19.

SEELEY, H. G. 1879. *The Ornithosauria: An Elementary Study of the Bones of Pterodactyles*. Deighton, Bell & Co., Cambridge, 135 pp.

SHAROV, A. G. 1971. New flying reptiles from the Mesozoic of Kazakhstan and Kyrgystan. *Palaeontological Institute, Academy of Sciences, SSSR, Transactions*, **130**, 104–113. [In Russian]

STOKES, W. L. 1957. Pterodactyl tracks from the Morrison Formation. *Journal of Paleontology*, **31**, 952–954.

TURNER, C. E. & PETERSON, F. 1999. Biostratigraphy of dinosaurs in the Upper Jurassic Morrison Formation of the Western Interior, U.S.A. *In*: GILLETTE, D. (ed.) *Vertebrate Paleontology in Utah*. Utah Geological Survey, Miscellaneous Publications, **99-1**, 77–114.

UNWIN, D. M. 1995. Preliminary results of a phylogenetic analysis of the Pterosauria (Diapsida: Archosauria). *In*: SUN, A. & WANG, Y. (eds) *Sixth Symposium on Mesozoic Terrestrial Ecosystems and Biota, Beijing*, 69–72.

UNWIN, D. M. & LÜ, J. 1977. On *Zhejianopterus* and the relationship of pterodactyloid pterosaurs. *Historical Biology*, **12**, 199–210.

UNWIN, D. M. & BAKHURINA, N. N. 2000. Pterosaurs from Russia, Middle Asia and Mongolia. *In*: BENTON, M. J., SHISHKIN, M. A., UNWIN, D. M., & KUROCHKIN, E. N. (eds) *The Age of Dinosaurs in Russia and Mongolia*. Cambridge University Press, Cambridge, 420–433.

UNWIN, D. M. & HEINRICH, W-D. 1999. On a pterosaur jaw from the Upper Jurassic of Tendaguru (Tanzania). *Mitteilungen aus dem Museum für Naturkunde, Berlin, Geowissenschaftenliche Reihe*, **2**, 121–134.

UNWIN, D. M., LÜ, J. & BAKHURINA, N. N. 2000. On the systematic and stratigraphic significance of pterosaurs from the Lower Cretaceous Yixian Formation (Jehol Group) of Liaoning, China. *Mitteilungen aus dem Museum für Naturkunde, Berlin, Geowissenschaftlichen Reihe*, **3**, 181–206.

WELLNHOFER, P. 1975. Die Rhamphorhynchoidea (Pterosauria) der Oberjura-Plattenkalke Süddeutschlands, Teil II Systematische Beschreibung. *Palaeontographica A*, **146**, 132–186.

WELLNHOFER, P. 1978. *In*: WELLNHOFER, P. (ed) *Pterosauria. Handbuch der Paläoherpetologie*, Gustav Fischer, Stuttgart, Teil 19, 87 pp.

WELLNHOFER, P. 1991. *The Illustrated Encyclopedia of Pterosaurs*. Salamander, London, 192 pp.

WELLNHOFER, P. 2003. A Late Triassic pterosaur from the Northern Calcareous Alps (Tyrol, Austria). *In*: BUFFETAUT, E. & MAZIN, J-M. (eds) *Evolution and Palaeobiology of Pterosaurs*. Geological Society, London, Special Publications, **217**, 5–22.

A new crested ornithocheirid from the Lower Cretaceous of northeastern Brazil and the unusual death of an unusual pterosaur

EBERHARD FREY[1], DAVID M. MARTILL[2] & MARIE-CÉLINE BUCHY[3]

[1]*Staatliches Museum für Naturkunde Karlsruhe, D-76133 Karlsruhe, Germany*
[2]*School of Earth and Environmental Sciences, University of Portsmouth, Portsmouth PO1 3QL, United Kingdom*
[3]*Universität Karlsruhe, Geologisches Institut, Postfach 6980, D-76128 Karlsruhe, Germany*

Abstract: An exceptionally well-preserved cranium and mandible of a new species of pterodactyloid pterosaur from the Nova Olinda Member of the Crato Formation (Aptian, Early Cretaceious) of the Araripe Basin, northeastern Brazil, is described. The new taxon is characterized by the presence of a caudally directed parietal crest similar to that seen in pteranodontids, but is referred to the Ornithocheiridae of the Ornithocheiroidea. The specimen is referred to a new genus within the Ornithocheiridae, as it lacks the diagnostic rostral crest and instead possesses this parietal crest oriented. A lanceolate leaf with frayed distal end wedged between the mandibular rami suggests the cause of death for the specimen.

The Nova Olinda Member of the Crato Formation has become a rich source of exceptionally well-preserved vertebrate remains in the last few years (Martill & Frey 1998). Although the fauna is diverse, tetrapods are rare and the vertebrate assemblage is dominated by large numbers of juvenile specimens of the gonorhynchiform fish *Dastilbe* (Davis & Martill 1999). Pterosaurs are the most abundant tetrapod group, with crocodyliforms, chelonians, squamates (Evans & Yabumoto 1998) and lissamphibians (Maisey 1991) occurring only very rarely. Birds are presently described only from isolated feathers (Martill & Filguiera 1994, Martill & Frey 1995, Martill & Davis 2001), while skeletal remains with associated feathers have been seen in a private collection (D.M.M pers. obs. 2000).

Several pterosaur taxa have been recorded from the Nova Olinda Member and include the crested tapejarid *Tapejara imperator* Campos & Kellner 1997 and a second species of *Tapejara* with a vertically oriented cranial crest (Martill & Frey 1998, Frey *et al.* 2003). Frey & Martill (1994) described *Arthurdactylus conandoylei* as a possible ornithocheirid from the Nova Olinda Member, but the holotype specimen lacks a skull and its affinities remain uncertain (Kellner & Tomida 2000). Martill & Frey (1999) also noted the presence of possible azhdarchid pterosaurs in the Nova Olinda Member on the basis of a partial wing skeleton in which the phalanges exhibit a 'T'-shaped cross-section, but again this material is too fragmentary for certain referral. Here we describe a new pterodactyloid pterosaur genus and species of based on cranial material associated with plant remains. The specimen is housed in the Staatliches Museum für Naturkunde Karlsruhe, specimen number SMNK PAL 3828.

Locality and stratigraphy

The new specimen was obtained from a commercial source. We were informed that it came from the Nova Olinda region of Ceará, northeastern Brazil and was from the stone quarries near that town. Nova Olinda is situated at the foot of the Chapada do Araripe, an extensive tableland lying at the boundaries between the states of Ceará, Pernambuco and Piaui, and is well known palaeontologically for exceptionally well-preserved Early Cretaceous faunas and floras (Martill 1993). Numerous small quarries between Nova Olinda, Santana do Cariri and Tatajuba provide a wealth of fossils which are traded globally. These fossils come from an 8–12 m-thick sequence of distinctive millimetrically laminated limestones, the Nova Olinda Member (Martill & Wilby 1993) of the Crato Formation. The matrix of the new specimen is consistent with derivation from the Nova Olinda Member and examples of the small fish *Dastilbe* on the slab help to confirm the source region and horizon.

The age of the Crato Formation has been discussed by several workers (see Martill 1993) and is considered to be Early Cretaceous Aptian on palynological grounds (Pons *et al.* 1990, 1996).

The new specimen

The specimen comprises a near complete skull and mandible on a slab of limestone typical of the Nova Olinda Member fossil Lagerstätte, and has been prepared using an air chisel and an air abrasive in the SMNK palaeontology laboratory (Fig. 1). During preparation a leaf was revealed implanted between the left and right mandibular rami (Figs 1a & 2).

From: BUFFETAUT, E. & MAZIN, J-M. (eds) 2003. *Evolution and Palaeobiology of Pterosaurs.* Geological Society, London, Special Publications, **217**, 55–63. 0305–8719/03/$15

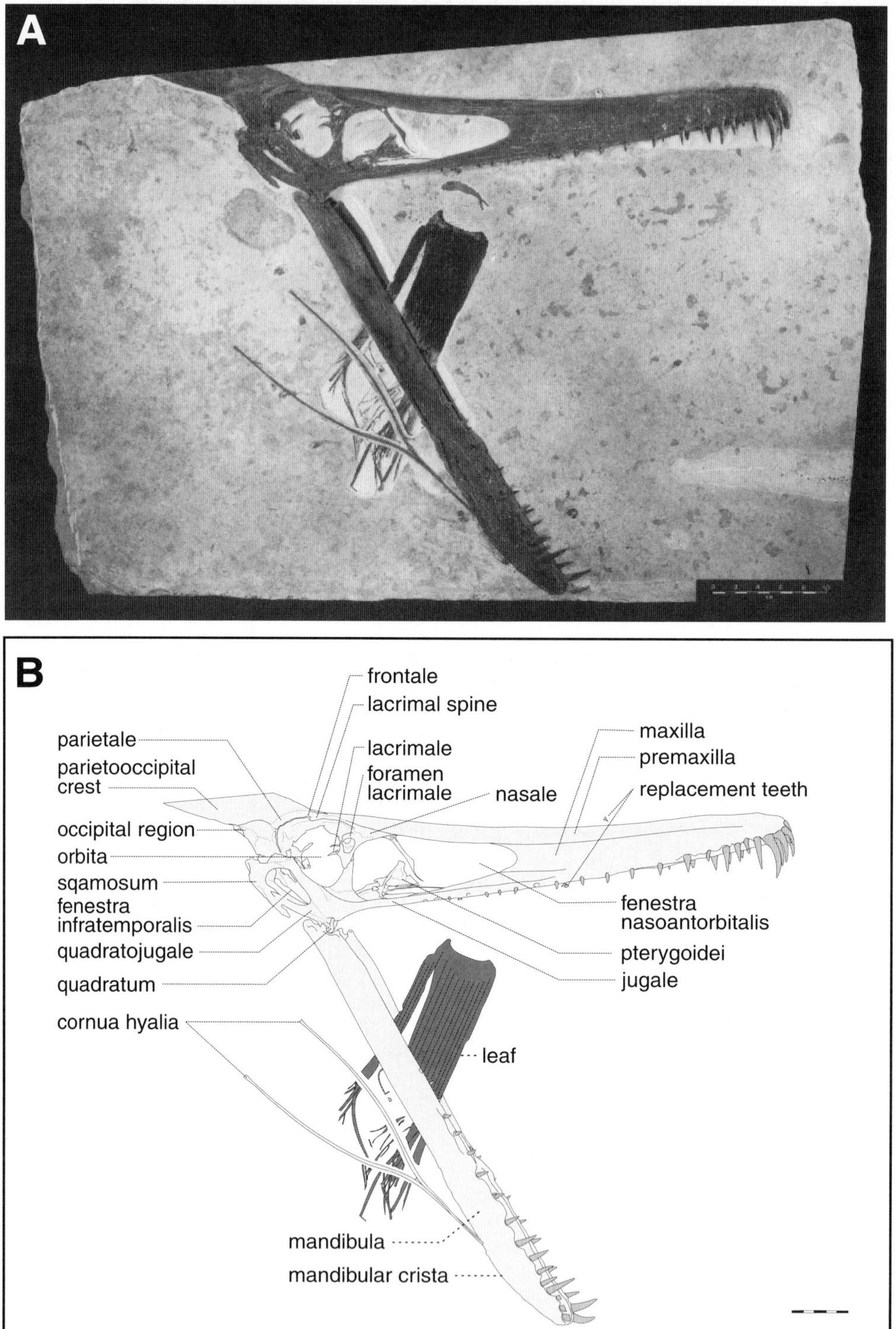

Fig. 1. *Ludodactylus sibbicki* sp. nov., holotype SMNK PAL 3828, from the Nova Olinda Member of the Crato Formation, (Early Cretaceous, Aptian) of northeastern Brazil: (**a**) photograph of the holotype; (**b**) semi-schematic line drawing of the holotype. Note the leaf which is lodged between the mandibular rami.

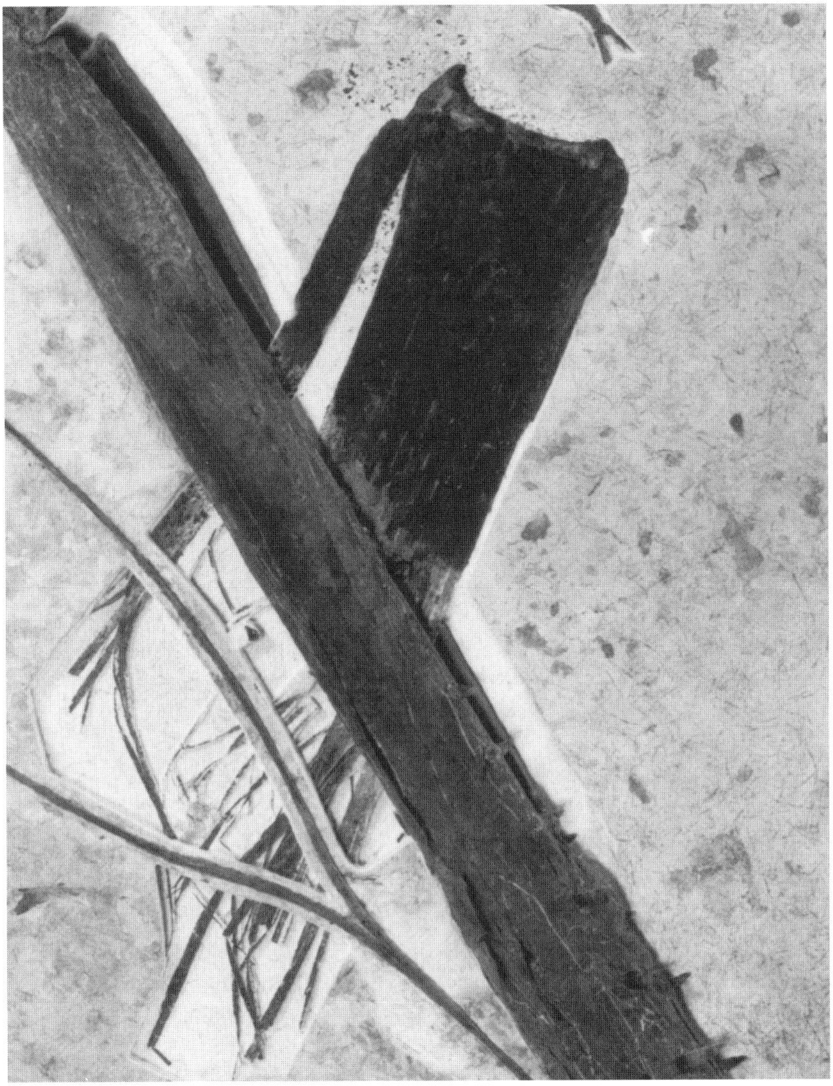

Fig. 2. *Ludodactylus sibbicki*: detail of the leaf lying between the mandibular rami. The frayed ends may be attributable to attempts by the pterosaur to remove the leaf. The small fish are young individuals of the gonorhynchiform *Dastilbe crandalli*.

The skull is well preserved. The bones are dark brown with smooth surfaces, but have been subjected to considerable fracturing due to compaction. Despite this, the teeth remain in their sockets and most of the bones are articulated. The hyoids have rotated to the right and are now seen in ventral aspect. Unusually for a toothed pterosaur there is a parieto-occipital crest which projects caudodorsally. The preserved basal part of this crest is strikingly similar to that seen in *Pteranodon ingens* (Bennett 1991, 2001). Unfortunately the distal portion of the crest is incomplete, having been cut in the stoneyard prior to preparation. Superficially, the pterosaur resembles a *Pteranodon* with teeth. Models of such pterosaurs have long been manufactured by the toy industry to make them look fierce (see etymology). Here is an animal that matches the toy industries' desires (Fig. 3).

Systematic description

Order Pterosauria Kaup 1834
Superfamily Ornithocheiroidea Seeley 1876
Family Ornithocheiridae Seeley 1870
Genus *Ludodactylus* gen. nov.

Fig. 3. A pterosaur produced by toy manufacturers: the occipital cone appears to have been based on *Pteranodon*, but the teeth are an invention by a far-sighted designer.

Diagnosis. *Ludodactylus sibbicki* is an ornithocheirid pterosaur that differs from other members of the group at present comprising the genera *Ornithocheirus*, *Coloborhynchus* and *Anhanguera* (Unwin 2001) in the following diagnostic characters: the possession of a caudally directed, laterally compressed parieto-occipital crest; presence of a dorsoventrally compressed lacrimal spine (*spina lacrimalis*) protruding caudally into the orbit; lacrimal foramen rounded triangular in outline with one corner facing ventrally; teeth of maxilla present caudally as far as mid-point of nasoantorbital fenestra; dorsal surface of the rostrum rounded with no trace of any crest; tooth row of mandible extends caudally to the rostral fourth of the nasoantorbital fenestra. The specimen is coincident with *Ornithocheirus* in the almost perpendicular orientation of the rostral-most four pairs of teeth to the long axis of the jaws, at least in lateral view which, according to Unwin (2001), is diagnostic for that genus. However, *Ludodactylus* differs from all other species within the Ornithocheiridae, including *Ornithocheirus*, in lacking a premaximillary crest (see also Unwin 2001, p. 205). The dorsal surface of the rostum of *Ludodactylus* is rounded. A blade-like occipital crest is reported from neither *Ornithocheirus* nor any other ornithocheirid pterosaur. Therefore we feel justified in erecting a new genus for specimen SMNK PAL 3828.

Type species Ludodactylus sibbicki sp. nov.

Horizon. Known only from the Nova Olinda Member of the Crato Formation; Early Cretaceous, Aptian.

Etymology. Generic name *Ludodactylus* (=play pterosaur) after Latin *ludus*=game, play, referring to the predictive toy Pterosaur in Fig. 3, a case where fantasy preceded palaeontological evidence, and *dactylus*=finger (from the Greek *dactylon*), here used in reference to pterosaur. Specific name after J. Sibbick, an artist who has recreated so many pterosaurs (see Wellnhofer 1991 for examples of his wonderful work).

Specific diangosis. As for genus.

Holotype. Staatliches Museum für Naturkunde Karlsruhe, specimen number SMNK PAL 3828: skull with associated fish and plant remains (Figs 1 & 2).

Locality. Chapada do Araripe region, Ceará, northeastern Brazil.

Description of *Ludodactylus*

The holotype of *Ludodactylus sibbicki* comprises a complete, articulated but laterally compressed skull with mandible. There are 23 tooth positions in the upper jaw and 17 in the mandible. In both jaws the third tooth is the longest. The fourth tooth pair of the rostrum is approximately the same size as the first tooth pair. The rostral-most pair of teeth of both jaws are implanted with vertical roots but their crowns are strongly curved so that the teeth interdigitate in occlusion. The fifth tooth pair is small (slightly less than half the crown height of the first and fourth tooth pairs) and there is a caudal increase in size of the tooth pairs until tooth pair seven. The eighth tooth pair is represented by an erupting tooth on the right side, but the tooth of the left side is visible and is slightly smaller than the preceding tooth (seventh tooth pair). There is then a decrease in tooth size caudally. The orbit is ovoid, with the long axis being oriented from dorsocaudally to rostroventrally. From the lacrimals a stalked, caudally directed, slightly ventrally curved and dorsoventrally compressed spine (spina lacrimalis) protrudes into the orbit. The lacrimal foramen is a rounded triangle and is large (7 mm long and 5 mm high).

The infratemporal fenestra is a rounded triangle and oriented parallel to the long axis of the orbit, with the tip directed cranioventrally. The supraorbital fenestra is crushed. The caudolateral margin of the brain case bears a sharp, caudally concave ridge that merges dorsocaudally into a blade-like crest that begins dorsal to the rostrodorsal margin of the orbit. This crest has a thickness of 1.5 mm and is formed mainly by the parietals. The exoccipitals and the supraoccipitals probably participate in the ventral part of the crest but there are no sutures to confirm this.

The mandible bears a low ventral crest that commences caudal to tooth four and extends to tooth nine. The nasoantorbital fenestra reaches rostrally to tooth position 16. The hyoids are articulated and lie adjacent to the ventral margin of the mandible. They are seen in ventral aspect and are slightly displaced rostrally so that their rostral terminus touches the ventral margin of the mandibular symphysis level with mandibular tooth nine (Fig. 1). The paired hyoids have a tuning-fork shape and diverge caudally until they are 0.25 × as wide as they are long

Systematic palaeontology

The presence of a combined nasoantorbital fenestra indicates *Ludodactylus sibbicki* is a pterodactyloid pterosaur. The teeth rule out referral to the edentulous Cretaceous pterosaur groups Pteranodontidae, Tapejaridae and Azhdarchidae. The large fang-like teeth rostrally in both the rostrum and mandible are features also seen in the Ornithocheiridae. Unfortunately the status and diagnosis of the Ornithocheiridae and other Cretaceous tooth-bearing pterosaur groups is confused (Fastnacht 2001; Unwin 2001). The Ornithocheiridae have been reviewed recently by Unwin (2001), who provides a revised diagnosis for the group which includes the following diagnostic characters: first three tooth pairs up to 10 times larger than caudal-most tooth pairs and forming a rostral rosette. The crown height in this rosette increases caudally. Fourth tooth pair smaller than first tooth pair; the caudally following teeth increase in crown height until the ninth tooth pair. Teeth decreasing in crown height caudal of ninth tooth pair. This diagnostic based on the dentition as proposed by Unwin (2001) refers only to the dentition of the upper jaw. It should also be taken into account that the tooth crowns of the holotypes of the Ornithocheiridae are scarcely preserved so that the height can only be estimated with respect to the diameter of the alveoli. Furthermore, the probably rapid tooth replacement in pterosaurs might lead to misinterpretations of the actual crown height if the tooth is not fully erupted.

Similar diagnostic characters, plus the presence of a rostral sagittal crest and a bluntly terminated rostrum were used by Campos & Kellner (1985) to define the Anhangueridae. This family was erected to accommodate a new genus and species of pterosaur, from the possibly Albian (Early Cretaceous) Romualdo Member of the Santana Formation of the Araripe Basin, which they named *Anhanguera blittersdorffi*. The holotype of *A. blittersdorffi* comprises a near-complete skull, lacking only the mandible, while a near-complete skull with mandible was also referred to the taxon. This material was preserved in three dimensions in the typical early diagenetic concretions of the Romualdo Member (Martill, 1988), which enabled unambiguous features to be identified for its diagnosis. Later Kellner (1990) and Kellner & Campos (1988) referred a number of other Romualdo Member pterosaur species to *Anhanguera*, including *Araripesaurus santanae* Wellnhofer 1985 as *Anhanguera santanae; Santanadactylus araripensis* Wellnhofer, 1985 as *Anhanguera araripensis* (Kellner 1990) and *Tropeognathus robustus* Wellnhofer 1987 as *Anhanguera robustus* (Kellner & Campos 1988). Kellner & Tomida (2000) described a fifth species of *Anhanguera* as *A. piscator*. This spectacularly preserved specimen, which includes a three-dimensional skull, also came from the Romualdo Member concretions and is characterized by a low premaxillary crest on the rostrum and mandible.

Unwin (2001) expressed some reservations regarding the status of *Anhanguera*, but provisionally accepted its validity and referred two taxa from the Lower Chalk of Kent and the Cambridge Greensand to this genus (*A. cuvieri* and *A. fittoni*). We too have reservations regarding the status of *Anhanguera* and note that the mandible of *Brasileodactylus* Kellner 1984 is almost indistinguishable from that of *Anhanguera piscator* Kellner & Tomida 2000 except in the height of the mandibular crest (see also Unwin 2001). Such a difference could easily be a result of ontogeny, sexual dimorphism or variation and should not be relied on as a diagnostic character. As *Brasileodactylus* predates *Anhanguera* it is clearly the senior synonym. Bennett (1994) has also suggested that *Anhanguera* might be a junior synonym of *Brasileodactylus*. However, the blunt termination of the rostrum caused by the dorsal deflection of the palate so clearly seen in *Anhanguera piscator* is identical to that seen in *Coloborhynchus*. Furthermore, the vertical termination in *A. piscator* bears a pair of rostroventrally strongly curved teeth as seen in *Coloborhynchus* and we here refer *A. piscator* to *Coloborhynchus*. *Coloborhynchus* also predates *Brasileodactylus*, and thus it is likely that *Brasileodactylus araripensis* should also be referred to *Coloborhynchus*. It is unclear as to whether other species referred to *Anhanguera* should also be placed within *Coloborhynchus*. The holotype of *A. santanae* lacks the rostral terminus of the skull and it is not possible to detect either a premaxillary crest or a vertically orientated termination of the palate. The situation is similar for the holotype of *A. araripensis*, though Kellner & Tomida (2000) refer a specimen to this species that possesses a prominent terminal rostral premaxillary crest. In those species of *Anhanguera* where the rostrum is preserved there is always a dorsal deflection of the premaxillae, but it is not vertical as in *Coloborhynchus robustus*. Such a slight difference may be a dimorphic or ontogenetic feature and we here tentatively suggest that other species of *Anhanguera* (*A. araripensis*, *A. blittersdorffi*, *A. cuvieri*, *A. fittoni*, *A. santanae*) be referred to *Coloborhynchus*.

Mader & Kellner (1999) described the rostral portion of a pterosaurian rostrum from the Late Cretaceous Kem Kem Formation of Morocco. They erected the genus and species *Siroccopteryx moroccensis* to accommodate this new specimen and placed it in the Anhangueridae on the basis of a slight expansion of the rostral part of the rostrum and a premaxillary crest. Such crests are present in both tooth-bearing and non-tooth-bearing pterosaurs and, as has been discussed by several authors (e.g. Bennett 1994, Carpenter *et al.* 2003), they are probably sexually dimorphic features and subject to ontogenetic change, as are the mandibular crests (see above). Their use as diagnostic characters alone should therefore be strictly avoided. The African specimen is in fact identical with a specimen of *Coloborhynchus* described by Lee (1994), as noted by Unwin (2001), and we follow Unwin (2001), in considering *Siroccopteryx* a junior synonym of *Coloborhynchus*.

The genus *Coloborhynchus* has recently been reviewed by Unwin (2001) and Fastnacht (2001), both of whom consider it to belong within the Ornithocheiridae. In his review of *Coloborhynchus*, Fastnacht (2001) drew attention to the similarity between a new pterosaurian specimen comprising the rostral portion of rostrum and mandible (SMNK PAL 2302) from the Romualdo nodules and '*Tropeognathus*' *robustus* Wellnhofer 1987 from the same horizon. Fastnacht (2001) was able to refer the isolated rostrum to the latter taxon. However, he also considered that *Tropeognathus robustus*, one of the two species referred to *Tropeognathus* should be referred instead to *Coloborhynchus*. SMNK PAL 2302 is so similar to *Coloborhynchus piscator* (see above) that they too apparently are conspecific, and we here consider *Anhanguera piscator* Kellner & Tomida 2001 to be a junior synonym of *Coloborhynchus robustus* (Wellnhofer 1987).

Unwin (2001) includes within Ornithocheiridae *Anhanguera, Criorhynchus*, and *Ornithocheirus*. *Ludodactylus sibbicki* (SMNK PAL 3828) possesses three enlarged rostral tooth pairs followed by a reduced fourth tooth pair. There is a crown height increase from tooth pair five to tooth pair seven and from then a reduction in tooth crown height caudally. This morphology allows *Ludodactylus sibbicki* to be placed in Ornithocheiridae *sensu* Unwin (2001). The only slight difference being that, in *Ornithocheiridae sensu* Unwin, the caudalwards increase in crown height proceeds as far as tooth pair nine, whereas in SMNK PAL 3828 this increase only proceeds to tooth pair seven. The lack of a premaxillary crest on the dorsally rounded rostrum excludes it from Anhangueridae *sensu* Campos & Kellner (1985), and indeed from *Anhanguera* as well as *Ornithocheirus*. The compressed nature of the specimen impendes direct comparisons with the type material of *Ornithocheirus* from the Cambridge Greensand, much of which is fragmentary.

Another difference between *Ludodactylus sibbicki* and species referred to *Ornithocheirus* is that the tooth size reduction commences caudally from tooth pair six rather than tooth pair nine. The presence of a caudodorsally directed parieto-occipital crest on specis of *Ornithocheirus*, or indeed any genus within Ornithocheiridae, has not been demonstated. Therefore the presence of a parieto-occipital crest would be another diagnostic difference and finally the lack of a premaxillary crest allows *Ludodactylus sibbicki* to be distinguished from the otherwise very similar species of *Ornithocheirus* considered by Unwin (2001) to be valid.

The only other ornithocheirid material from the Crato Formation is a complete articulated postcranium, *Arthurdactylus conandoylei* (Frey & Martill 1994). It might well be that *Ludodactylus sibbicki* could be referred to such a postcranium but this could only be clarified with new and complete finds.

Fig. 4. Chilean Pelican (*Pelecanus thagus*) in the harbour of Antofagasta, Chile, with industrial debris trapped in the throat pouch (photo Frey).

Associated plant remains

A leaf is preserved between the mandibular rami of SMNK PAL 3828 (Figs 1 & 2). The leaf is elongate and gently tapered and has a texture of coarse parallel fibres. The broad basal part of the leaf lies within the gape of the mouth and passes between the mandibles from the buccal cavity to a position ventral to the gular region. Ventral to the mandible the fibres are ragged due to physical damage and lie below the hyoids. Intact examples of such leaves occur frequently in the Nova Olinda Member, but despite their abundance they have not been formally described, though they have been figured (Martill 1993). Often these lanceolate leaves reach lengths in excess of 1 m and taper to a sharp point distally. They have a slightly concave, smooth base and resemble the leaves of recent *Cordyline*, *Yucca* or even *Agave*.

We speculate that the leaf became trapped between the left mandibular ramus and the tongue of this pterosaur and probably got stuck in a gular pouch. Such gular pouches have been reported for several pterosaur species (Wellnhofer 1991). The leaf may have been mistaken for a prey item and accidentally taken in a point-first fashion. The pterosaur would have been unable to remove the leaf as it became lodged in the flesh lateral to the tongue and attempts to remove it probably drove it deeper through the tissue of the possible gular pouch. Similar accidents are frequently seen in the Chilean Pelican (*Pelecanus thagus* Molina) which frequently collects food in industrial fish harbours. If the pelicans are unable to remove the items they die of hunger after a while. This is evidenced by abundance of pelican carcasses that float in the harbour basin of Antofagasta with their throat pouches full of rubbish (E.F. and D.M.M., pers. obs. in Antofagasta, Chile; Fig. 4). The frayed ends of the leaf ventral to the mandible of the pterosaur (Fig. 2) may have resulted from attempts by the pterosaur to dislodge the leaf by rubbing it against the ground. We have not seen frayed ends of this plant in other specimens. Being unable to close its beak fully, the pterosaur could no longer eat. The weakened, half-starved animal probably succumbed a while over the Crato lagoon, possibly suffering from a sepsis caused by the decaying leaf. Finally, it died with the leaf still painfully embedded in its throat.

Discussion

The holotype skull of *Ludodactylus sibbicki* represents a new genus and enigmatic ornithocheirid pterosaur from South America. For the first time a parieto-occipital crest has been proved for ornithocheirid pterosaur. However, Seeley (1901) did reconstruct *Ornithocheirus* with a posteriorly directed crest on the basis of an isolated element he identified as a partial supraoccipital crest (see also Unwin 2001). None of Seeley's holotypes of *Ornithocheirus* included crest bones (Unwin 2001), and this was later referred to the jaw-based pteranodontid *Ornithostoma* Seeley 1871 by Hooley (1914). Perhaps Seeley (1901) had been correct in originally referring the crest to *Ornithocheirus*. Unfortunately the specimen is too fragmentary to resolve this issue.

The Aptian age of *Ludodactylus sibbicki* is in accord with the stratigraphic range of the family ornithocheiridae at other localities, having been recorded from the Early Cretaceous Aptian to Cenomanian (Unwin 2001).

Thanks to René Kastner, Karlsruhe, for his exceptional preparation skills and Siegfried Rietschel, formerly director of SMNK, for supporting our work on pterosaurs. Thanks to D. Unwin for allowing us access to unpublished information and for helpful discussion concerning the generic status of *Ludodactylus*. D. Naish kindly read an early draft of the manuscript and provided helpful comments. Photography, with the exception of Fig. 4, was undertaken by V. Griener, Karlsruhe.

References

BENNETT, S. C. 1991. The ontogeny of *Pteranodon* and other pterosaurs. *Paleobiology*, **19**, 92–106.

BENNETT, S. C. 1994. *Taxonomy and systematics of the Late Cretaceous pterosaur* Pteranodon (Pterosauria, Pterodactyloidea). Natural History Museum, University of Kansas Occasional Papers, **169**, 1–70.

BENNETT, S. C. 2001. The osteology and functional morphology of the Late Cretaceous pterosaur *Pteranodon*. Part 1. General description of the osteology. *Palaeontographica A*, **260**, 1–112.

CAMPOS, D. A. & KELLNER, A. W. A. 1985. Un novo examplar de *Anhanguera blittersdorffi* (Reptilia, Pterosauria) da formacâo Santana, Cretáceo Inferior do Nordeste do Brasil. *Boletim de Resumos, 9th Congresso Brasileiro de Paleontologia*, **13**.

CAMPOS, D. A. & KELLNER, A. W. A. 1997. Short note on the first occurrence of Tapejaridae in the Crato Member (Aptian), Santana formation, Araripe Basin, northeast Brazil. *Anais de Academia Brasileira de Ciências*, **69**, 83–87.

CARPENTER, K., UNWIN, D. M., CLOWARD, K., MILES, C. & MILES C. 2003. A new scaphognathine pterosaur from the Upper Jurassic Morrison Formation of Wyoming, USA. *In*: BUFFETAUT, E. & MAZIN, J-M. (eds) *Evolution and Palaeobiology of Pterosaurs*. Wyoming. Geological Society of London, Special Publications, **217**, 45–54.

DAVIS, S. & MARTILL, D. M. 1999. The gonorhynchiform fish *Dastilbe* from the Lower Cretaceous of Brazil. *Palaeontology*, **42**, 715–740.

EVANS, S. & YABUMOTO, Y. 1998. A lizard from the Early Cretaceous Crato Formation, Araripe Basin, Brazil. *Neues Jahrbuch für Geologie und Paläontologie, Monatshefte*, **1998**, 349–364.

FASTNACHT, M. 2001. First record of *Coloborhynchus* (Pterosauria) from the Santana Formation (Lower Cretaceous) of the Chapada do Araripe, Brazil. *Paläontologische Zeitschrift*, **75**, 23–36.

FREY, E. & MARTILL, D. M. 1994. A new pterosaur from the Crato Formation (Lower Cretaceous, Aptian) of Brazil. *Neues Jahrbuch für Geologie und Paläontologie, Abhandlungen*, **194**, 379–412.

FREY, E., MARTILL, D. M. & BUCHY, M-C. 2003. A new species of tapejarid pterosaur with soft-tissue head crest. *In:* BUFFETAUT, E. & MAZIN, J-M. (eds) *Evolution and Palaeobiology of Pterosaurs*. Geological Society, London, Special Publications, **217**, 65–72.

HOOLEY, R. W. 1914. On the ornithosaurian genus *Ornithocheirus* with a review of the specimens from the Cambridge Greensand in the Sedgwick Museum, Cambridge. *Annals and Magazine of Natural History*, **13**, 529–557.

KAUP, J. 1834. Versuch einer Eintheilung der Saugethiere in 6 Stämme und der Amphibien in 6 ordnungen. *Isis*, **3**, 311–315.

KELLNER A. W. A. 1984. Ocorrencia de uma mandibula de Pterosauria (*Brasileodactylus araripensis* nov. gen.; nov. sp.) na formaçáo Santana, Cretáceo da Chapada do Araripe, Ceará, Brasil. *33rd Congresso Brasileiro de Geologia, Anais*, **2**, 578–590.

KELLNER A. W. A. 1990. Os répteis voadores do Cretáceo Brasileiro. *Anuario do Instituto de Geosciências, CCMN, UFRJ*, **1989**, 86–106.

KELLNER A. W. A. & CAMPOS, D. A. 1988. Sobre um novo pterosauro com crista sagital da Bacia do Araripe, Cretáceo Inferior do nordeste Brasil. *Anais da Academia Brasileira Ciências*, **60**, 459–469.

KELLNER A. W. A. & TOMIDA, Y. 2000. *Description of a New Species of Anhangueridae (Pterodactyloidea) with Comments on the Pterosaur Fauna from the* Santana Formation (Aptian–Albian), Northeastern Brazil. National Science Museum, Tokyo, Monographs, **17**, 135 pp.

LEE, Y. N. 1994. The Early Cretaceous pterodactyloid pterosaur *Coloborhynchus* from North America. *Palaeontology*, **37**, 755–763.

MADER, B. J. & KELLNER, A. W. A. 1999. A new anhanguerid pterosaur from the Cretaceous of Morocco. *Boletim do Museu Nacional, Rio de Janeiro, New Series, Geologia,* **45**, 1–11.

MAISEY, J. G. 1991. Undetermined Santana frog. *In*: MAISEY, J. G. (ed.) *Santana Fossils, An Illustrated Atlas*. Tropical Fish Hobbyist, Neptune City, New Jersey, 459 pp.

MARTILL, D. M. 1988. Preservation of fish in the Cretaceous of Brazil. *Palaeontology*, **31**, 1–18.

MARTILL, D. M. 1993. *Fossils of the Santana and Crato Formations, Brazil.* Field Guides to Fossils, Palaeontological Association, vol. 5, 159 pp.

MARTILL, D. M. & DAVIS, P. G. 2001. A feather with possible ectoparasite eggs from the Crato Formation (Lower Cretaceous, Aptian) of Brazil. *Neues Jahrbuch für Geologie und Paläontologie, Abhandlungen*, **219**, 241–259.

MARTILL, D. M. & FILGUIERA, B. J. 1994. A new feather from the Lower Cretaceous of Brazil. *Palaeontology*, **37**, 483–487.

MARTILL, D. M. & FREY, E. 1995. Colour patterning preserved in Lower Cretaceous birds and insects: the Crato Formation of N. E, Brazil. *Neues Jahrbuch für Geologie und Paläontologie, Monatshefte*, **1995**, 118–128.

MARTILL, D. M. & FREY, E. 1998. A new pterosaur Lagerstätte in N. E. Brazil (Crato Formation, Aptian, Lower Cretaceous): preliminary observations. *Oryctos*, **1**, 79–85.

MARTILL, D. M. & FREY, E. 1999. A possible azhdarchid pterosaur from the Crato Formation (Early Cretaceous, Aptian) of northeast Brazil. *Geologie en Mijnbouw*, **78**, 315–318.

MARTILL, D. M. & WILBY, P. 1993. Stratigraphy. *In*: MARTILL, D. M. (ed.) *Fossils of the Santana and Crato Formations, Brazil.* Field Guides to Fossils, Palaeontological Association, vol. 5, 159 pp.

PONS, D., BERTHOU, P.-Y. & CAMPOS, D. A. 1990. Quelques observations sur la palynologie de l'Aptien supérieur et de l'Albien du Bassin d'Araripe. *In*: CAMPOS, D. DE A., VIANA, M. S. S., BRITO, P. M. & BEURLEN, G. (eds) *Atas do Simposio Sobre a Bacia do Araripe e Bacias Interiores do Nordeste, Crato, 14–16 de Junho de 1990*, 241–252.

PONS, D., BERTHOU, P.-Y. & CAMPOS, D. A. 1996. Palynologie des unités lithostratigraphiques 'Fundão', 'Crato' et 'Ipubi' (Aptien supérieur à Albien inférieur-moyen, Bassin d'Araripe, NE du Brésil). *In*: JARDINE, S., KLASZ, L. DE & DEBANEY, J-P. (eds) *Géologie de l'Afrique et de l'Atalantique Sud (Actes Colloques Angers 1994)*. Pau, Elf Aquitaine Édition, Mémoire **16**, 383–401.

SEELEY, H. G. 1870. *The Ornithosauria: An Elementary Study of the Bones of Pterodactyles, Made from Fossil Remains Found in the Cambridge Greensand, and Arranged in the Woodwardian Museum of the University of Cambridge*. Deighton, Bell & Co., Cambridge, 137 pp.

SEELEY, H. G. 1871. Additional evidence of the structure of the head in ornithosaurs from the Cambridge Upper Greensand; being a supplement to '*The Ornithosauria*'. *Annals and Magazine of Natural History*, **37**, 20–36.

SEELEY, H. G. 1876. On the organisation of the Ornithosauria. *Zoological Journal of the Linnean Society, London*, **13**, 84–107.

SEELEY, H. G. 1901. *Dragons of the Air: An Account of Extinct Flying Reptiles*. Methuen & Co, London.

UNWIN, D. M. 2001. An overview of the pterosaur assemblage from the Cambridge Greensand (Cretaceous) of eastern England. *Mitteilungen aus dem Museum für Naturkunde, Berlin, Geowissenschaftliche Reihe,* **4**, 189–217.

WELLNHOFER, P. 1985. Neue Pterosaurier aus der Santana Formation (Apt) der Chapada do Araripe, Brasilien. *Palaeontographica A*, **187**, 105–182.

WELLNHOFER, P. 1987. New crested pterosaurs from the Lower Cretaceous of Brazil. *Mitteilungen der Bayerischen Staatsammlung für Paläontologie und Historische Geologie*, **27**, 175–186.

WELLNHOFER, P. 1991. *The Illustrated Encyclopedia of Pterosaurs*. Salamander, London, 192 pp.

A new species of tapejarid pterosaur with soft-tissue head crest

EBERHARD FREY[1], DAVID M. MARTILL[2] & MARIE-CÉLINE BUCHY[3]

[1]*Staatliches Museum für Naturkunde Karlsruhe, D-76133 Karlsruhe, Germany*
[2]*School of Earth and Environmental Sciences, University of Portsmouth, Portsmouth PO1 3QL UK*
[3]*Universität Karlsruhe, Geologisches Institut, Postfach 6980, D-76128 Karlsruhe, Germany*

Abstract: Two specimens of a tapejarid pterosaur (Pterodactyloidea, Tapejaridae) are described as representing a new species. Both specimens show evidence for soft tissues preserved in association with a sagittal bony cranial crest. Both specimens are from the Nova Olinda Member Konservat Lagerstätte of the Crato Formation of the Araripe Basin, northeastern Brazil. They represent the second tapejarid species from this formation. Comparisons are made with other crested pterosaurs and comments on the utility and aerodynamics of pterosaurian head crests are made.

The Cretaceous of South America, especially the Araripe Basin of northeastern Brazil, has yielded many unusual and well-preserved pterosaurs (Bonaparte 1970; Wellnhofer 1985, 1987, 1991a, b; Frey & Martill 1994; Martill & Frey 1998; Frey & Tischlinger 2000; Kellner & Tomida 2000; Wellnhofer & Kellner 1991). Specimens are often complete, three-dimensional and occasionally exhibit soft-tissue preservation (Kellner & Campos 1989; Frey & Tischlinger 2000; Martill & Frey 1998; Martill & Unwin 1989). In the Araripe Basin pterosaurs have been reported from two horizons: the Aptian Nova Olinda Member of the Crato Formation (Martill & Frey 1999) and the probably Albian Romualdo Member of the Santana Formation (see Kellner & Tomida 2000 for a review). There is some overlap between the faunas at the generic level, with *Tapejara* occurring in both formations (Campos & Kellner 1997) but no species have been shown to be common to both horizons. At present the Santana Formation appears to contain the most diverse pterosaur assemblage with *Anhanguera*, *Coloborhynchus*, *Criorhynchus*, *Cearadactylus*, *Santanadactylus*, *Tapejara* and *Tupuxuara*, while the Crato Formation has yielded *Ornithocheirus*, *Arthurdactylus* and *Tapejara*. Although the list for the Santana Formation is considerable, the validity of *Cearadactylus*, *Anhanguera* and *Santanadactylus* is in some doubt (Kellner & Tomida 2000, Unwin 2001). Similarly, *Arthurdactylus* lacks a cranium and may prove to be synonymous with one of the other genera.

Preservation in both formations is often exceptional, with soft-tissue preservation occurring in both assemblages (Martill & Unwin 1989; Martill & Frey 1999). Fossils from the Romualdo Member of the Santana Formation are usually enclosed in carbonate concretions that formed during early diagenesis and are frequently preserved three-dimensionally, whereas those from the Crato Formation occur in laminated limestones and are usually crushed. In both formations the pterosaur remains are associated with abundant fishes and more rarely other tetrapods, including birds, crocodyliforms, chelonians, squamates and lissamphibians in the Crato Formation (Martill 1993) and crocodyliforms, theropods and chelonians in the Santana Formation (Maisey 1991; Martill 1993).

The new specimens described here are housed in the collection of the Staatliches Museum für Naturkunde Karlsruhe. The specimen numbers are SMNK PAL 2343 and 2344.

Locality and stratigraphy

The new specimens described here (SMNK PAL 2343 and 2344; Fig. 1a–d) were both obtained by a commercial dealer from stone quarries in the region between Nova Olinda, Santana do Cariri and Tatajuba, but the exact locality cannot be determined. The matrix of the specimens is sufficiently distinctive to confirm this as the source region, with no other parts of the extensive outcrop currently being worked for ornamental stone.

Both come from the laminated limestones of the Nova Olinda Member of the Lower Cretaceous Crato Formation (Martill 1993), which is considered by Pons *et al.* (1990) to be Aptian (Early Cretaceous).

The new material

Specimen SMNK PAL 2344 comprises an almost complete cranium of a tapejarid pterosaur with associated soft tissues and is seen from its right side (Fig.

From: BUFFETAUT, E. & MAZIN, J-M. (eds) 2003. *Evolution and Palaeobiology of Pterosaurs.* Geological Society, London, Special Publications, **217**, 65–72. 0305–8719/03/$15

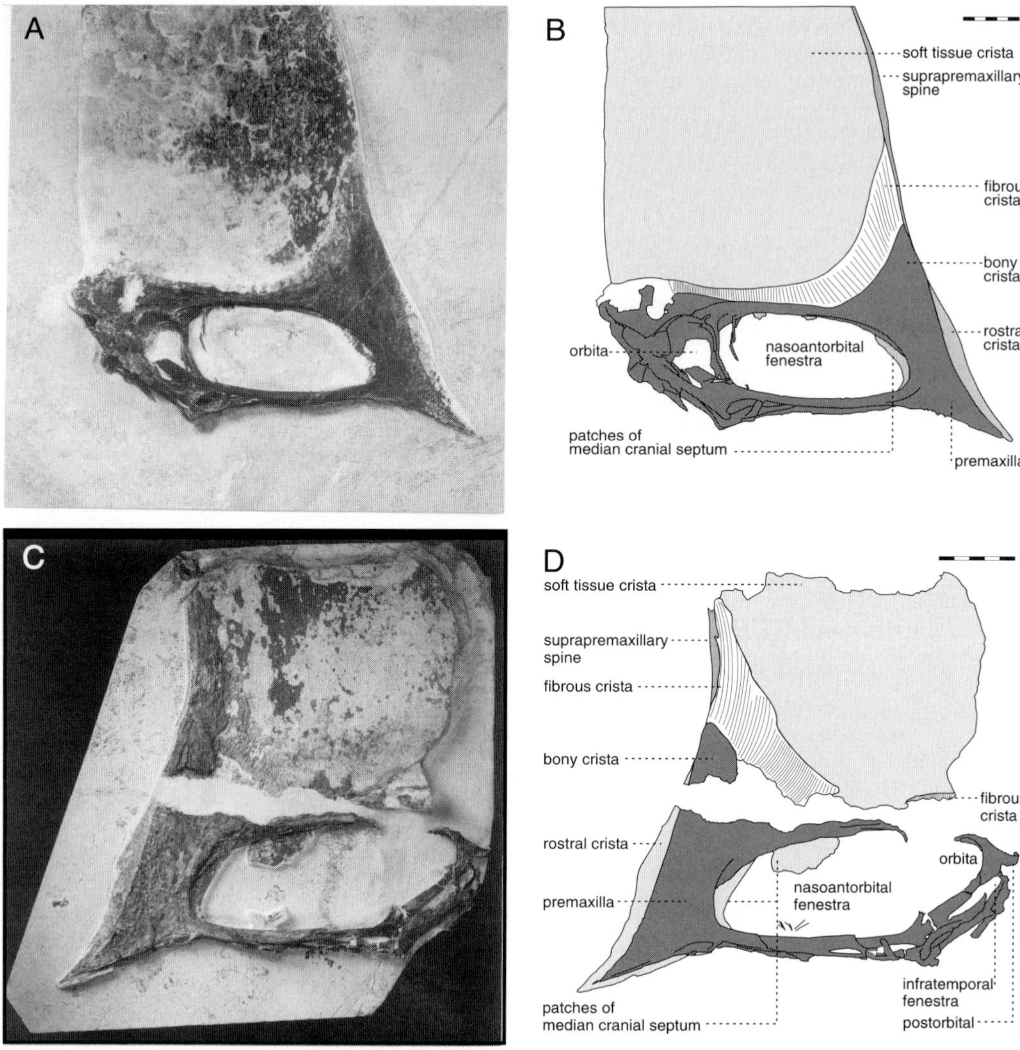

Fig 1. *Tapejara navigans*, (**a**) Holotype SMNK 2344 PAL. (**b**) Semi-schematic line drawing of the holotype; the soft-tissue crest is preserved as an internal mould. (**c**) SMNK 2343, referred specimen of *T. navigans,* (**d**) Semi-schematic line drawing of the specimen; the soft part crest is preserved as an external mould. Note the vertical trailing edge of the soft part crest and the blade-like crest rostral to the premaxilla. Scale bar 50 mm.

1a, b). It has an overall length of 375 mm (terminus of rostrum – caudal rim of occiput). The braincase, occiput and caudal palatal region are slightly crushed. The lower jaw and postcranial elements are missing. The skull lies on a rectangular slab of Nova Olinda Member limestone and has the dorsal-most part of a cranial crest cut at the margin of the slab. Specimen SMNK PAL 2343 is seen from its left side, is preserved on an irregular slab of Nova Olinda Member limestone and has been broken in several places (Fig. 1c, d). The occipital region and the dorsal half of the crest is missing and a break has destroyed the dorsal third of the nasoantorbital fenestra and rostral margin of the orbit. Both specimens arrived at the SMNK partially prepared by unknown persons. To our knowledge no counterparts were associated with the specimens.

Systematic description

Order Pterosauria Kaup 1834
Superfamily Pterodactyloidea Plieninger 1901
Family Tapejaridae Kellner 1989
Genus *Tapejara* Kellner 1989
Species *Tapejara navigans* sp. nov.

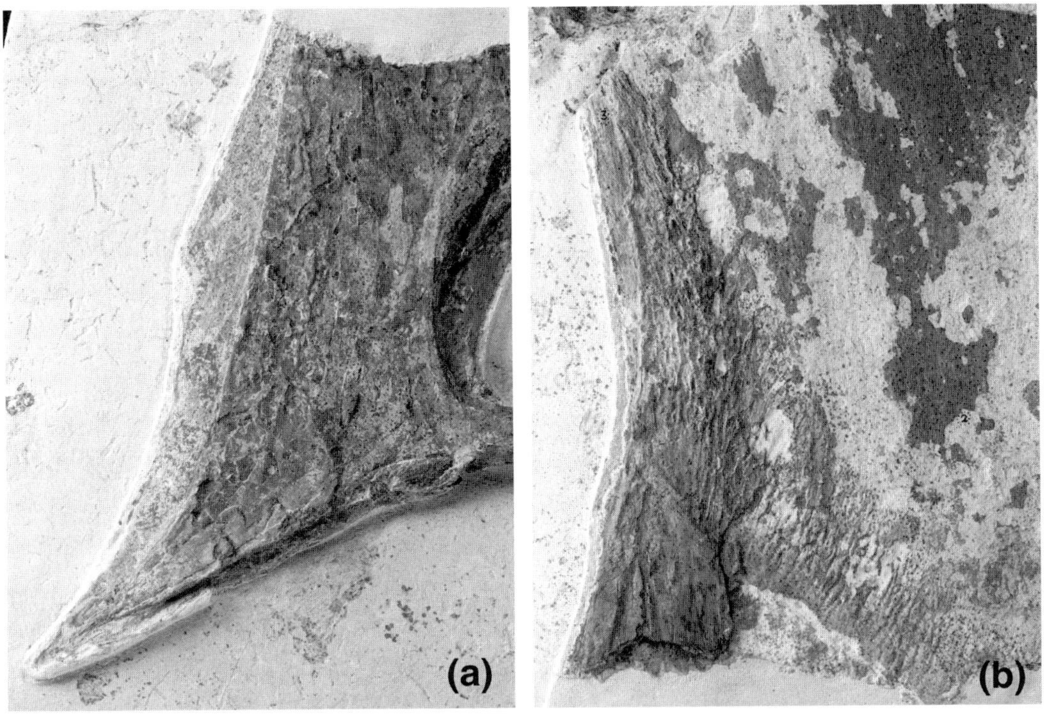

Fig. 2. *Tapejara navigans*, close-ups of the skull: (**a**) with the blade-like rostral soft tissue extension and the keratinous beak of SMNK PAL 2344; and (**b**) the transition of the bony striae into the fibres of the soft-tissue crest of SMNK PAL 2343, and also the anchoring fibres of the suprapremaxillary spine.

Diagnosis. Tapejarid pterosaur with bones of premaxillomaxillary crest striated rostrally and dorsally. Processus caudalis of jugal twice as broad as in other species of *Tapejara*. Dorsal soft tissue of cranial crest preceded by vertical spine-like suprapremaxillary ossification. Processus caudalis of premaxillae fused to cranium roof.

Horizon. Nova Olinda, Ceará, N.E. Brazil.

Etymology. Latin: *navigans* = sailing, pertaining to the shape of the cranial crest.

Holotype. SMNK PAL 2344 (Staatliches Museum für Naturkunde Karlsruhe). Cranium, with dorsal and rostral soft-tissue crests (Fig. 1a).

Referred material. Incomplete cranium with soft-tissue crest SMNK PAL 2343 (Fig. 1c).

Locality. Nova Olinda Member, Crato Formation, Aptian (Lower Cretaceous).

Description of *Tapejara navigans* sp.nov.

Despite the lateral compaction of both specimens the cranial structure of *T. navigans* can be reconstructed with confidence. The skull of *T. navigans* is characterized by a edentulous beak which is inclined ventrally at an angle of 24°. The bony cranial crest is continuous with a band of dorsally oriented ossified striae (Fig. 2b). Along the rostral margin of the premaxillae the surface of the compacta is wrinkled (Fig. 2a). The length and depth of this wrinkled surface is concordant with a goethitic area of soft tissue preserved rostrally adjacent to the bone (Figs 1a, c, 2a & 3), the distal border of which is set off sharply against the matrix. Emerging from the rostrodorsal extremity of these striae is a tapered suprapremaxillary bony spine, slightly curving caudally but cut off at the slab margin (Figs 1a, b & 2b). This spine shows a dense parallel striation and is an ossification forming the leading edge of the crest.

Caudally the dorsal margin of the bony premaxillary crest is concave and the striae gradually become shorter and less prominent. At the apex of the antorbital fenestra the crest ends in a low ridge which is approximately twice as high as the roof of the antorbital fenestra. The height of the crest increases again towards the occipital region of the cranium. A dark goethitic stain outlines a smooth, rounded caudal termination of the crest. Where bone is preserved faint striae are visible in the supraorbital part of the crest.

The most remarkable features of the new specimens are the preserved soft tissues that form a

Fig 3. Reconstruction of *T. navigans* based on both specimens (drawing by Frey).

marked extension of the bony dorsal crest, and a blade-like rostral crest along the rostral margin of the premaxilla (Figs 1 & 2a).

The soft tissue of the crest is preserved as an internal mould with mineralized replacements by goethite in the holotype SMNK PAL 2344. Goethite is most probably a product of the oxidation of an original pyrite mineralization and forms a thick layer, especially in the rostral third of the crest of SMNK PAL 2344. The goethite layer becomes less prominent caudally. The border between the dark, thick areas of soft tissue and the more buff-coloured thinner part is irregular, patchy and continues approximately vertically in the middle of the soft-tissue crest. The dark stain is in places obscured by halite pseudomorphs which are only found on the soft parts, and small light-coloured patches which might be damage-incurred during collecting. The distribution of the goethite mineralization and the three-dimensionality shows that the soft-tissue crest was shaped like a symmetrical aerofoil.

The lateral surface of the soft-tissue crest is smooth and no surface detail is apparent (e.g. scales or the 'criss-cross' pattern previously reported for pterosaur wing membrane [Martill & Unwin 1989]). The goethitic replacement has a fibrous texture in which the fibres are oriented vertically. The soft-tissue replacement extends caudally from the suprapremaxillary spine and forms a smooth, vertical border extending from the caudal margin of the brain case to the edge of the slab (Fig. 1a). The reconstruction of its original height and shape by projecting the visible margins results in a height of the cranial crest estimated at 4x or 5x greater than the occipital height of the cranium (Fig. 3).

In SMNK PAL 2343 a similar soft tissue crest with suprapremaxillary ossification is seen but the soft-tissue crest is preserved as an external mould with the vertical fibres still in place (Figs 1c, d & 2b). The dorsolateral part of the compacta of the cranial crest has broken off. This has exposed the base as well as the anchoring fibres of the suprapremaxillary spine. A rostral soft-tissue crest is also present. Additionally, this specimen shows an unossified sheath surrounding the bony tip of the premaxilla and extending along the ventral margin of the premaxilla forming a goethitic seam of 5–6 mm width (Figs 1c, d & 3). This seam is also vivible in SMNK PAL 2344 (Fig. 2a) and might represent the edge of a rhamphotheca that may have overlapped the mandible when the mouth was closed (Fig. 3).

Comparison

The general outline of the cranial skeleton is similar to that of *Tapejara wellnhoferi* Kellner 1989 but is more elongate craniocaudally with respect to its height (Fig. 5). The ratio length max./height max. of the antorbital fenestra is 2.2 in *T. navigans*, but only 1.6 in *T. wellnhoferi*. By comparison with *T. wellnhoferi*, the dorsal margin of the orbit in *T. navigans* is slightly lower than the dorsal margin of the antorbital fenestra. In a 240 mm-long cranium of *T. wellnhoferi* the difference in height is 13 mm whereas in *T. navigans* it is a maximum of 10 mm. The caudal processes of the premaxillae in *T. navigans* are fused in the roof of the antorbital fenestra that caudally arches ventrally into the anterodorsal orbital margin. *T. navigans* lacks the caudally directed occipital process, which forms a short but prominent crest projecting caudodorsally from the braincase in *Tapejara wellnhoferi*. As in *Tapejara wellnhoferi,* the compacta surface of the bony cranial crest is smooth until it terminates in a thin, irregularly broken margin.

A comparison with *Tapejara imperator* Campos & Kellner 1997 is problematic, because the preliminary description by Campos & Kellner (1997) does not provide sufficient biometric data. However, *T.*

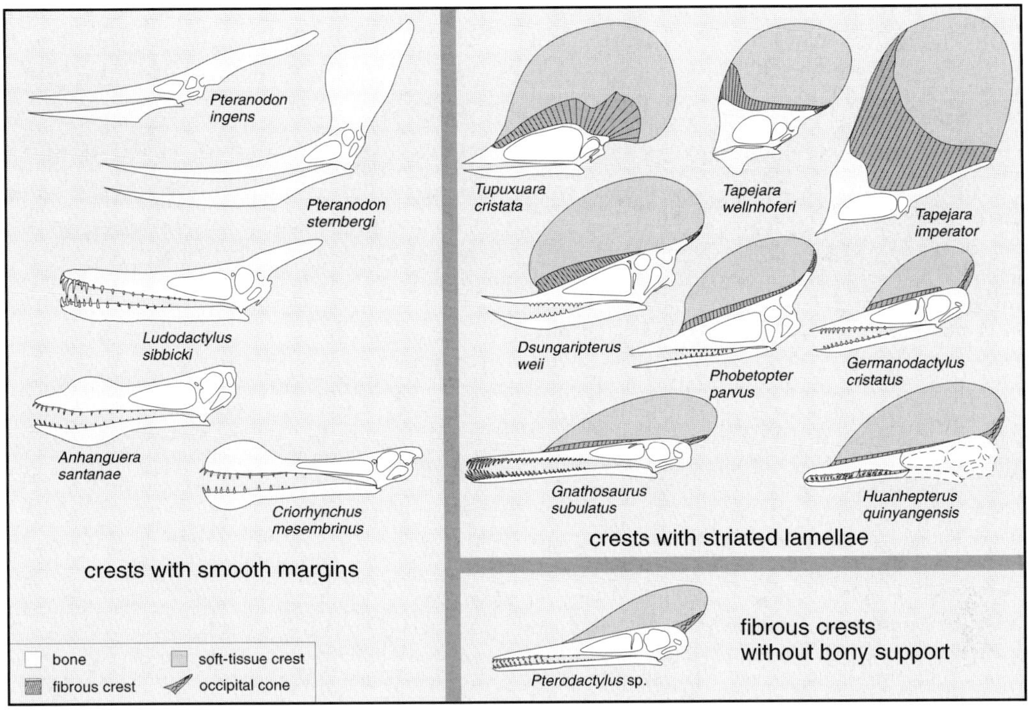

Fig. 4. Cranium constructions of pterosaurs. The supposed soft tissue crests are reconstructed according to the evidence of *Tapejara navigans*. The skull outlines are modified after Wellnhofer (1991a) and Campos & Kellner (1997).

navigans differs from *T. imperator* in the lack of an occipital spine and the inclination of the leading edge of the soft-tissue crest; in *T. navigans* it stands vertical to the long axis of the skull whereas in *T. imperator* it is inclined 15° caudally (Figs 3 & 4b). The huge dorsal extension of the striated bone lamina on the cranial crest in *T. imperator* may be a consequence of the individual age: the skull of *T. imperator* is 1.5 times as long as that *T. navigans* and *T. wellnhoferi*. The lack of an occipital spine in *T. navigans*, however, cannot be interpreted in the context of ontogeny, because the perfectly preserved caudal margin of the cranial crest in *T. navigans* stands vertical on the caudodorsal edge of the occiput. In *T. imperator*, the cranial crest is anchored continually along the dorsal margin of the occipital spine until its caudal terminus. One could speculate whether or not an adult *T. wellnhoferi* would have looked like *T. imperator*.

Crest function

Bony cranial crests have been reported for a number of pterosaur genera (Fig. 4), the most famous of which is *Pteranodon* (parietal crest; Eaton 1910), but they also occur in *Criorhynchus* (=*Tropeognathus*) (Wellnhofer 1987; Unwin 2001) and *Coloborhynchus* (rostral premaxillary and mandibular crest; Fastnacht 2001), *Tupuxuara* (cranial crest from premaxilla to occipital process, mandibular crest; Kellner & Campos 1989); *Tapejara* (cranial crest from premaxilla to occipital process, mandibular crest; Kellner & Campos 1989; Wellnhofer & Kellner 1991), *Anhanguera* (premaxillary and mandibular crest; Campos & Kellner 1985; Wellnhofer 1991b), *Ctenochasma* (premaxillary crest; Buisonié 1981), *Dsungaripterus* (premaxillary crest, occipital process; Young 1973), *Germanodactylus* (premaxillary crest; Wimann 1925), *Gnathosaurus* (premaxillary crest; Meyer 1834) and *Phobetopter* (premaxillary crest and occipital process; Bakhurina 1982), *Normannognathus* (Buffetaut *et al.* 2000), *Domeykodactylus* (premaxillary crest; Martill *et al.* 2000), *Nyctosaurus*, (occipital crest; Williston 1902), *Huanhepterus* (premaxillary crest; Dong 1982; for general review see Wellnhofer 1991a). A bony cranial crest formed at least by the premaxilla is also reported for the giant azhdarchid pterosaur *Quetzalcoatlus* (Kellner & Langston 1996).

Direct evidence for soft tissues associated with cranial crests of pterosaurs have been reported only recently (Campos & Kellner 1997; Martill & Frey

1998; Frey & Tischlinger 2000). Most recently, Frey & Tischlinger (2000) reported soft tissues associated with cranial crests for *Pterodactylus* sp. and it appears likely that most of those pterosaurs with a striated bony cranial crest, possessed a soft-tissue extension of the crest as had previously been suggested by Wellnhofer (1991b; Fig. 4). The assumption of Eaton (1910), who suggested that the parietal crest of *Pteranodon ingens* supported a membrane between the crest and the dorsal surface of the body, and was supported by Stein (1975), has not been proven.

The bony cranial crests and their possible function has been discussed widely, with the models falling into five main areas: for muscular attachment, especially the jaw musculature (Eaton 1910); for aerodynamic functions (Short 1914; Heptonstall 1971); for hydrodynamic purposes (Wellnhofer 1987); to reduce torque on the neck musculature (Mathew 1920; Whitfield & Bramwell 1971); and, rather more speculatively, for purposes of sexual display or some other behavioural activity (Bennett 1992; Stein 1975). The latter of course could be secondary options of a crest serving primarily one of the former functions. The suggestion that the caudally directed supraoccipital crest of *Pteranodon* supported an elastic membrane connected to the middle of the back, and thus functioned as a rudder (Stein 1975), was disputed because such a structure would limit movement of the head (Wellnhofer 1991a).

The possibility that cranial crests in pterosaurs were covered by a layer of keratin and may have supported even larger structures was considered by Wellnhofer (1987). He suggested that rostrally located crests on *Criorhynchus* (=*Tropeognathus*) were for hydrodynamic stabilization while fishing on the wing. If such structures served a hydrodynamic function, they would also have had an aerodynamic effect (assuming that the crest was a permanent feature and that the animal flew while in possession of a crest).

The cranial crest of *Tapejara navigans*, which had about 3x the lateral surface area of the head, must have had an aerodynamic effect. It may have acted as a stabilizer, steering device or sail, and also may have had secondary sexual or display functions. As a stabilizer such a crest would have resisted roll, but would have been sensitive to lateral winds. Each lateral movement of the head and neck would have resulted in a steering movement which must have been actively counterbalanced (Fig. 5a). For both steering and stabilization, stiffening of a short neck by bone and tendon locks would have been crucial. As a consequence the flight of *T. navigans* must have been highly unstable in pitch but extremely stable in the roll axis, but only at extremely low flight velocities.

The shape of the reconstructed cranial crest of *T. navigans* strikingly resembles that of wind-surfing sails if the suprapremaxillary ossification is regarded as a mast and the soft parts as a sail (Farke *et al.* 1994). Such a structure could have been a propulsion device, provided that the crest could produce more thrust than the entire animal could produce drag. Unlike sailing boats and sail boards, sailing *T. navigans* could not use tailwinds, only headwinds, because the head could not have turned at right-angles to the long axis of the neck. The angle of attack would have been therefore restricted to between 25° and 45° against the headwind. In this range a maximum thrust could be produced by the crest operating as a sail with an anatomically acceptable angle to the long axis of the trunk. As in sail boards, the head and crest could have been pulled back and forth, resulting in yaw control. If the animal used 'wind-surfing' while airborne, lateral drift could only have been compensated by the vertically held webbed feet and adjustment of the wings. Most probably the feet had to be held in the water to maintain course while the wings were held out to provide lift. There is one example of a recent bird providing a similar type of locomotion: storm petrels

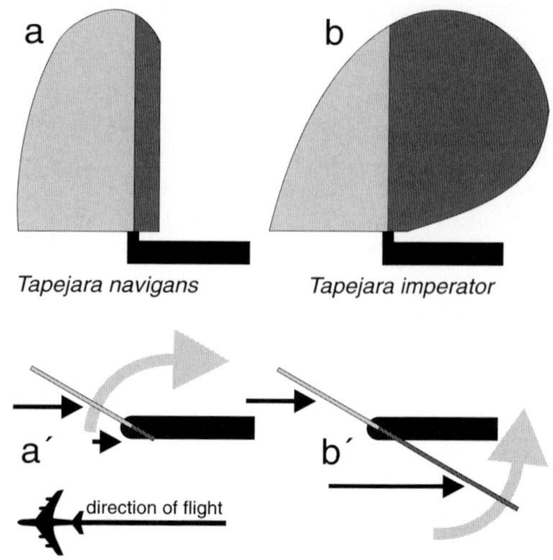

Fig. 5. Cranial crests and their aerodynamical effects; the dark areas are the areas caudal to the head articulation. (**a**) *T. navigans*: principle of construction. (**b**) *T. imperator*: principle of construction. (**a′**) *T. navigans*: the area rostral to the head articulation is about four times larger than the area caudally. The air pressure on the rostral area of the crest cannot be compensated and the result would be a turn of the construction as a whole. Such a construction is useless for steering. (**b′**) *T. imperator*: the area rostral to the head articulation is about one third of the area caudally. The air pressure on the rostral area of the crest is compensated and the result would be a readjustment of head and crest like a weather vane. Such a construction can be used for steering.

(Hydrobathidae) sail over the sea with open wings holding their feet into the water for anchoring and steering (Burton 1990, pp.114–115). It appears highly unlikely that *T. navigans* could use its head sail while swimming like a seabird. The adhesion effect of the membraneous wings on the water surface in combination with the ventral wing attachment (Frey *et al.*, 2003a) would have produced severe take-off problems. These speculations are subject to further investigation.

The construction of a cranial crest like that reconstructed for *T. imperator* would be unproblematic with respect to steering because the anchoring point of the rudder lies rostral to the largest surface (Fig. 5b). The crest thus functions like a weather vane, which orients itself automatically in the air flow and therefore works as a self-adjusting rudder.

The highly rugose and striated bone at the margins of the bony crest support of *Tapejara navigans* is also seen in the saggital crest of the tapejarid *Tupuxuara,* an unnamed Mongolian dsungaripterid (Bakhurina & Unwin 1995), and on the rostral portion of the crest of *Dsungaripterus weii*, from the Cretaceous of China (Young 1973; Dong 1988) and *Domeykodactylus* from the Cretaceous of Chile (Fig. 4; Martill *et al.* 2000; Frey *et al.*, 2003b). The Dsungaripteroidea are considered to be the 'sister-taxon' to the Azhdarchoidea, to which *Tapejara* belongs (Unwin 1995). Possession of an extended soft-tissue crest may well have been a character shared by all pterosaurs within these groups. By contrast, the cranial crests of *Criorhynchus*, *Coloborhynchus, Anhanguera* and the pteranodontids appear smooth, and perhaps only supported a keratinous covering rather than an extended soft-tissue crest. The new specimens, like those described by Campos & Kellner (1997) and Frey & Tischlinger (2000), demonstrate that at least some pterosaurs possessed large soft-tissue extensions to their bony crests and suggest that biophysical investigations on pterosaurian flight dynamics in which the effect of a crest is attempted need to be reassessed in the light of the new discoveries.

Our special thanks go to R. Kastner (Karlsruhe) who again did a wonderful job in the preparation and conservation of the specimens. We also thank D. Naish (Portsmouth), N. Bonde (Copenhagen) and M. Fastnacht (Mainz) for valuable comments on the paper. D. M. M gratefully acknowledges the support of the University of Portsmouth Palaeobiology Research Group. Photography was by V. Griener (Karlsruhe).

References

BAKHURINA, N. N. 1982. Pterodactyl from the Lower Cretaceous of Mongolia. *Palaeontological Journal*, **4**, 104–108. [Russian]

BAKHURINA, N. N. & UNWIN, D.M. 1995. A survey of pterosaurs from the Jurassic and Cretaceous of the former Soviet Union and Mongolia. *Historical Biology*, **10**, 197–245.

BENNETT, C. 1992. Sexual dimorphism of *Pteranodon* and other pterosaurs with comments on cranial crests. *Journal of Vertebrate Paleontology*, **12**, 422–434.

BONAPARTE, J. F. 1970. *Pterodaustro guinazui* gen. et sp. nov. Pterosaurio de la formación Lagarcito, provincia de San Luius, Argentina. *Acta Geologica Lilloana*, **10**, 207–226.

BUFFETAUT, E., LEPAGE, J-J. & LEPAGE, G. 2000. A new pterodactyloid pterosaur from the Kimmeridgian of the Cap de la Hève (Normandy, France). *Geological Magazine*, **135**, 719–722.

BUISONJÉ, P.H. de, 1981. Ctenochasma porocrestta nov. sp. from the Solnhofen Limestone, with some remarks on other Ctenochasmatidae. *Proceedings of the Koninklijke Nederlands Akademie van Wetenschappen, B*, **84** (4), 411–436.

BURTON, R. 1990. *Bird flight. An Illustrated Study of Birds' Aerial Mastery*. Facts On File, New York, Oxford, Sydney. 160 pp.

CAMPOS, D. A. & KELLNER, A. W. A. 1985. Panorama of the flying reptiles study in Brazil and South America. *Anais da Academia Brasileira Ciencias*, **57**, 453–466.

CAMPOS, D. A. & KELLNER, A. W. A. 1997. Short note on the first occurrence of Tapejaridae in the Crato Member (Aptian), Santana Formation, Araripe Basin, northeast Brazil. *Anais de Academia Brasileira de Ciências*, **69**, 83–87.

DONG, Z. 1982. On a new Pterosauria (*Huanhepterus quingyangensis* gen. et sp. nov.) from Ordos, China. *Vertebrata Palasiatica*, **20**, 115–121. [Chinese]

DONG, Z. 1988. *Dinosaurs of China*. British Museum (Natural History), China Ocean Press, 114 pp.

EATON, G. F. 1910. *Osteology of* Pteranodon. Yale University Press, 96 pp.

FARKE, U., MÖHLE, V. & SCHRÖDER, D. 1994. *Windsurf-Grundschein. Ich lerne surfen*. Delius Klasing Verlag, Bielefeld, 115 pp.

FASTNACHT, M. 2001. First record of *Coloborhynchus* (Pterosauria) from the Santana Formation (Lower Cretaceous) of the Chapada do Araripe, Brazil. *Paläontologische Zeitschrift*, **75**, 23–36.

FREY, E. & MARTILL, D. M. 1994. A new pterosaur from the Crato Formation (Lower Cretaceous, Aptian) of Brazil. *Neues Jahrbuch für Geologie und Paläontologie, Abhandlungen*, **194**, 379–412, Stuttgart.

FREY, E. & TISCHLINGER, H. 2000. Weichteilanatomie der Flugsaurierfusse und Bau der Scheitelkämme: Neue Pterosaurierfunde aus den Solnhofener Schichten (Bayern) und der Crato-Formation (Brasilien). *Archaeopteryx*, **18**, 1–16.

FREY, E., BUCHY, M.-C. & MARTILL, D. M. (2003a). Bottom-deckers among the Cretaceous pterosaurs, an unique design among active fliers. *In*: BUFFETAUT, E. & MAZIN, J-M (eds) *Evolution and Palaeobiology of Pterosaurs*. Geological Society, London, Special Publications, **217**, 267–274.

FREY, E., TISCHLINGER, H., BUCHY, M-C. & MARTILL, D. M. (2003b). New specimens of Pterosauria (Reptilia) with soft parts with implications for pterosaurian anatomy and locomotion. *In*: BUFFETAUT, E. & MAZIN,

J-M (eds) *Evolution and Palaeobiology of Pterosaurs*. Geological Society, London, Special Publications, **217**, 233–266.

HEPTONSTALL, W. B. 1971. An analysis of the flight of the Cretaceous pterodactyl *Pteranodon ingens* (March) [*sic*]. *Scottish Journal of Geology*, **7**, 61–78.

KAUP, J. 1834. Versuch einer Eintheilung der Saugethiere in 6 Stämme und der Amphibien in 6 Ordnungen. *Isis*, **3**, 315.

KELLNER, A. W. A. 1989. A new edentate pterosaur of the Lower Cretaceous from the Araripe Basin, northeast Brazil. *Anais Academie Brasileiro Ciências*, **61**, 439–446.

KELLNER, A. W. A. & CAMPOS, D. A. 1989. Sobre um novo pterossauro con crest sagittal da Bacia do Araripe, Cretáceo Inferior do nordeste do Brasil. *Anais Academia Brasileira Ciências*, **60**, 459–469.

KELLNER, A. W. A. & LANGSTON, W. 1996. Cranial remains of *Quetzalcoatlus* (Pterosauria, Azhdarchidae) from Late Cretaceous sediments of Big Bend National Park, Texas. *Journal of Vertebrate Paleontology*, **16**, 222–231.

KELLNER, A. W. A. & TOMIDA, Y. 2000. *Description of a new species of Anhangueridae (Pterodactyloidea) with comments on the pterosaur fauna from the Santana Formation (Aptian–Albian), northeastern Brazil*. National Science Museum, Tokyo, Monographs, **17**, 135 pp.

MAISEY, J. G. (ed.) 1991. *Santana Fossils, an Illustrated Atlas*. Tropical Fish Hobbyist, Neptune City, New Jersey, 459 pp.

MARTILL, D. M. 1993. *Fossils of the Santana and Crato Formations, Brazil*. Field Guides to Fossils, Palaeontological Association, vol. 5, 159 pp.

MARTILL, D. M. & FREY, E. 1998. A new pterosaur Lagerstätte in N.E. Brazil (Crato Formation, Aptian, Lower Cretaceous): preliminary observations. *Oryctos*, **1**, 79–85.

MARTILL, D. M. & FREY, E. 1999. A possible azhdarchid pterosaur from the Crato Formation (Early Cretaceous, Aptian) of northeast Brazil. *Geologie en Mijnbouw*, **78**, 315–318.

MARTILL, D. M. & UNWIN, D. M. 1989. Exceptionally preserved pterosaur wing membrane from the Cretaceous of Brazil. *Nature*, **340**, 138–140.

MARTILL, D. M., FREY, E., CHONG, G. & BELL, M. 2000. Reinterpretation of a Chilean pterosaur and the occurrence of Dsungeripteridae in South America. *Geological Magazine*, **137**, 19–25.

MATHEW, W. D. 1920. Flying Reptiles. *Natural History*, **20**, 73.

MEYER, H. von 1834. *Gnathosaurus subulatus* ein Saurus aus dem lithographischen Schiefer von Solenhofen. *Museum Senckenbergianum*, **1**, 3.

PLIENINGER 1901. Beiträge zur kenntnis der Flugsaurier. *Palaeontographica*, **53**, 209–213.

PONS, D., BERTHOU, P-Y. & CAMPOS, D. A. 1990. Quelques observations sur la palynologie de l'Aptien supérieur et de l'Albien du Bassin d'Araripe. *In* CAMPOS, D. A DE., VIANA, M. S. S., BRITO, P. M. & BEURLEN, G. (eds) *Atas do Simposio Sobre a Bacia do Araripe e Bacias Interiores do Nordeste, Crato, 14–16 de Junho de 1990*, 241–252.

SHORT, G. H. 1914. Wing adjustments of pterodactyls. *Aeronautical Journal*, **18**, 336–343.

STEIN, R. S. 1975. Dynamic analysis of *Pteranodon ingens*: a reptilian adaptation to flight. *Journal of Paleontology*, **49**, 534–548.

UNWIN, D. M. 1995. Preliminary results of a phylogenetic analysis of the Pterosauria (Diapsida: Archosauria). *In*: SUN, A. & WANG, Y. (eds) *Sixth Symposium on Mesozoic Terrestrial Ecosystems and Biota, Beijing*, Short Papers, 66–72.

UNWIN, D. M. 2001. An overview of the pterosaur assemblage from the Cambridge Greensand (Cretaceous) of eastern England. *Mitteilungen aus dem Museum für Naturkunde, Berlin, Geowissenschaftliche Reihe*, **4**, 189–217.

WELLNHOFER, P. 1985. Neue Pterosaurier aus der Santana-Formation (Apt) der Chapada do Araripe, Brasilien. *Palaeontographica A*, **187**, 105–182.

WELLNHOFER, P. 1987. New crestted pterosaurs from the Lower Cretaceous of Brazil. *Bayerischen Staatsammlung für Paläontologie und Historische Geologie Mitteilungen*, **27**, 175–186.

WELLNHOFER, P. 1991a. *The Illustrated Encyclopedia of Pterosaurs*. Salamander Books, London, 192 pp.

WELLNHOFER, P. 1991b. Weitere Pterosaurierfunde aus der Santana-Formation (Apt) der Chapada do Araripe, Brasilien. *Palaeontographica A*, **215**, 43–101.

WELLNHOFER, P. & KELLNER, A. W. A. 1991. The cranium of *Tapejara wellnhoferi* Kellner (Reptilia, Pterosauria) from the Lower Cretaceous Santana Formation of the Araripe Basin, Northeastern Brazil. *der Bayerischen Staatsammlung für Paläontologie und Historische Geologie Mitteilungen*, **31**, 89–106.

WHITFIELD, G. & BRAMWELL, C. 1971. Palaeoengineering: birth of a new science. *New Scientist*, **52**, 202–206.

WILLISTON, S. W. 1902. On the skull of Nyctodactylus, an Upper Cretaceous pterodactyl. *Journal of Geology*, **10**, 520–531.

WIMAN, C. 1925. Über Pterodactylus westmanni und andere Flugsaurier. *Bulletin of the Geological Insitute of the University Uppsala*, **20**, 1–38.

YOUNG, C. C. 1973. Wuerho pterosaurs. *Institute of Vertebrate Palaeontology and Palaeoanthropology, Academica Sinica, Special Publications*, **11**, 18–34. [In Chinese]

Pterosaur (Pteranodontoidea, Pterodactyloidea) scapulocoracoid from the Early Cretaceous of Venezuela

ALEXANDER W. A. KELLNER[1] & JOHN M. MOODY[2]

[1]*Setor de Paleovertebrados, Departmento de Geologia e Paleontologia, Museu Nacional/UFRJ, Quinta da Boa Vista, s/n São Cristóvão, Rio de Janeiro RJ 20940–040, Brazil (e-mail: kellner@mn.ufrj.br)*

[2]*Museo de Biologia (MBLUZ), Universidad del Zulia, Facultad de Ciencias, Apartado 526, Maracaibo 4011, Edo. Zulia, Venezuela*

Abstract: The discovery of a left scapula and coracoid (MBLUZ P-911) representing the first evidence of a pterosaur from Venezuela is reported here. The material comes from the Lower Cretaceous (Aptian) Apón Formation, in the northwestern part of the country. In MBLUZ P-911 the scapula is significantly smaller than the coracoid, a synapomorphy of the Pteranodontoidea, according to Kellner. The coracoid of the Venezuelan specimen is more elongated and gracile than those of *Istiodactylus* and *Pteranodon*, and also lacks the ventromedial coracoidal flange present in the latter. Overall MBLUZ P-911 is very similar to the scapulocoracoid of the Anhangueridae, including the presence of a longitudinal ridge on the medial surface of the coracoid and a comparatively short scapula, and is therefore tentatively referred to this taxon. This occurrence extends the pterosaur record to the northern part of the South American portion of Gondwana.

Pterosaurs constitute an important component of Mesozoic vertebrate faunas around the world. Their remains have been found in numerous deposits and over 130 nominal taxa have been described (Wellnhofer 1991b), several based on incomplete material that is difficult to diagnose (Kellner 1994). Regarding South America, most deposits with pterosaur remains are situated in Argentina (total of nine), with two in Brazil, two in Chile and one in Peru (Kellner 2001). The most important is the Aptian– Albian Romualdo Member of the Santana Formation (Araripe Basin), which leads in terms of diversity, number and preservation of specimens (e.g. Price 1971; Campos & Kellner 1985; Wellnhofer 1985; Kellner & Tomida 2000), followed by the Aptian Lagarcito Formation of San Luis (Argentina), which has a large number of mostly isolated bones apparently representing a single taxon (Bonaparte 1970; Chiappe *et al.* 2000).

In this paper we report a new occurrence of Pterosauria in South America. It consists of a left scapulocoracoid (MBLUZ P-911) from the Aptian Apón Formation, northwestern Venezuela, housed in the Museo de Biologia of the Facultad de Ciencias, La Universidad del Zulia, Maracaibo, Venezuela. A cast was made and is deposited in the collections of the Paleovertebrate Sector, Department of Geology and Paleontology, Museu Nacional/UFRJ, Rio de Janeiro, Brazil. This material, which provides the first record of these flying reptiles in Venezuela, was briefly reported (Kellner & Moody 2001) and is described here.

Geological setting and taphonomy

The discovery of the studied specimen (MBLUZ P-911) was made possible by stone-quarrying operations in the Rosarito Quarry along the Sierra de Perijá mountain front west of the town of Villa de Rosario, Zulia State, Venezuela (Fig. 1). The removal of overburden has exposed an interesting cross-section of tectonically overturned Lower Cretaceous Apón Formation (Durán *et al.* 1984). The Apón Formation consists principally of shallow-sea platform carbonates generally divided in northwestern Venezuela into three members (Piché, Machiques and Tibú), with the Machiques Member denoting the formation's middle beds. This member displays a variation in thickness in northwestern Venezuela, thinning considerably toward the north (Gonzalez de Juana *et al.* 1980; Rod & Maync 1954). The Machiques Member is considered to be Aptian based on its ammonite fauna (Renz 1982). Impure carbonate shale beds containing hard limestone lenses generally become more frequent in the Machiques Member. While the best-preserved fossils (both vertebrate and invertebrate) are found in these limestone lenses, the pterosaur bone was collected from Machiques marl beds about 2m stratigraphically above the most notable of these shale beds in the quarry (while actually below this bed in the quarry because of the overturned section). This shale is traceable across much of the northern and eastern quarry face.

Other than ammonites, the invertebrate fauna

From: BUFFETAUT, E. & MAZIN, J-M. (eds) 2003. *Evolution and Palaeobiology of Pterosaurs*. Geological Society, London, Special Publications, **217**, 73–77. 0305–8719/03/$15

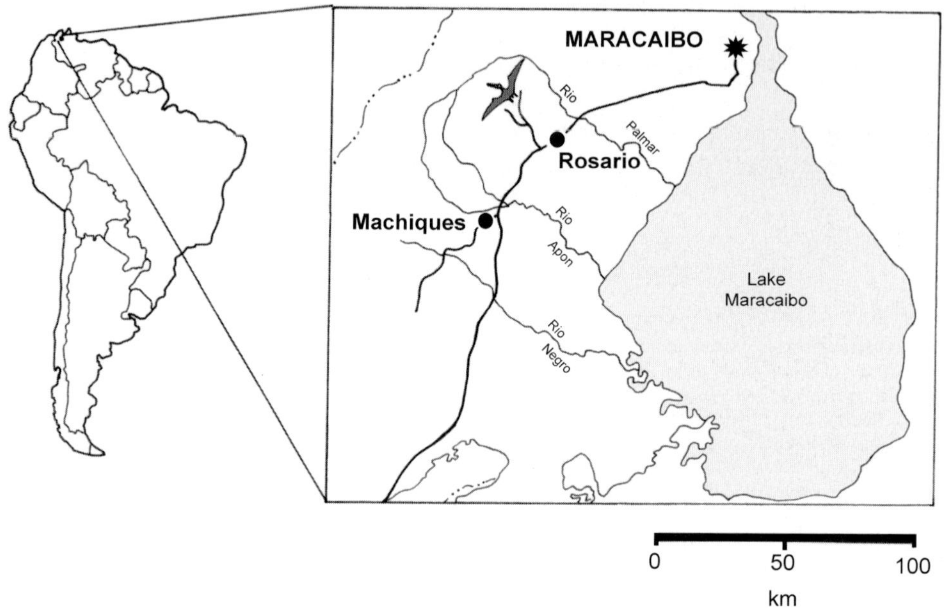

Fig. 1. Locality map showing the site where the Venezuelan pterosaur scapulocoracoid (MBLUZ P-911) was found.

includes bivalves, gastropods, an inarticulate brachiopod, an echinoid, a belemnite and a squid, some of which have not yet been properly studied. Over the past few years, the Machiques Member has become better known for its vertebrate fauna, with reports of an ichthyosaur (*Platypterygius*), a sea turtle, and various fish, including an ichthyodectid, pachyrhizodontid-like elopocephalans, the deep-bodied elopocephalan *Araripichthys*, the aspidorhynchid *Vinctifer* and a pycnodont (Moody 1993a, b; Moody & Maisey 1994; Maisey & Moody 2001). Terrestrial plant material has also been found in the quarry, suggesting that the palaeo-shoreline was very close.

The pterosaur specimen was *in situ*, found resting in a position parallel to bedding. Several fragmentary fish remains (not identifiable) were found in close association. Additional skeletal elements may be present, but it would be necessary to remove a considerable overburden from the quarry face in order to recover them. The bones have experienced some post-mortem compression but the bone surface is well preserved.

Description

The material consists mainly of the left scapula and coracoid (Figs 2–5) and was prepared mechanically. The proximal articulation of the right coracoid is also preserved and was displaced over the middle portion of the left coracoidal shaft during the fossilization process (Figs 3 & 5, cor.r). The suture between scapula and coracoid is observable only in some parts, particularly on the anterior face (the corresponding posterior region is partially damaged). Those bones are strongly connected, suggesting that they belong to an adult animal.

The scapula is about 9.5 cm long and shows a well-developed processus scapularis. The proximal articulation surface is broad and almost flat, and 'tear-drop shaped', with the apex situated dorsally. A small sharp crest is developed on the dorsal part of this bone. Overall, this element shows the same basic morphology observed in the anhanguerids, e.g. *Anhanguera piscator* (Kellner & Tomida 2000) and in *Pteranodon* (Eaton 1910; Bennett 2001).

The coracoid is the longer element (length 12.5 cm). Compared to the scapula, the coracoidal shaft is narrow and ends in a proximal and distal expansion. The proximal articulation (which articulates with the sternum) is slightly 'fork-shaped'. The lateroventral portion is not complete and the processus coracoidalis was broken off. It differs from *Pteranodon* and *Istiodactylus* (former *Ornithodesmus*; Hooley 1913; Howse *et al.* 2001) by being comparatively more elongated and gracile. Furthermore, in *Pteranodon*, the lateroventral portion of the coracoid is more developed. Compared with *Anhanguera*, other than the absence of a coracoidal process (an artefact of preservation) and the lack of a small ventral tubercle, there is no apparent difference.

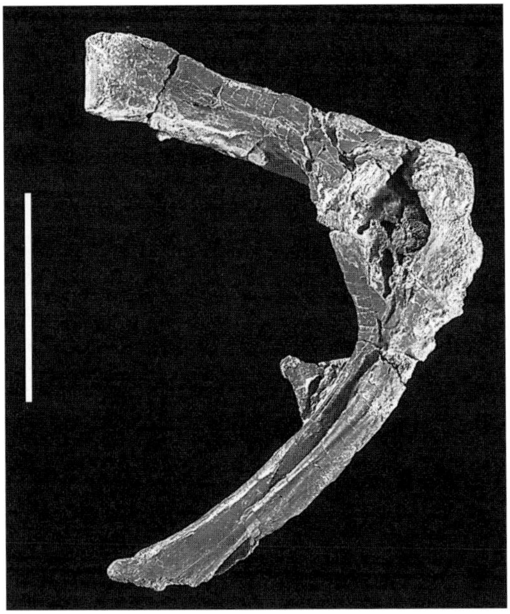

Fig. 2. Photograph of scapulocoracoid (MBLUZ P-911) in anterior view. Scale bar 50 mm.

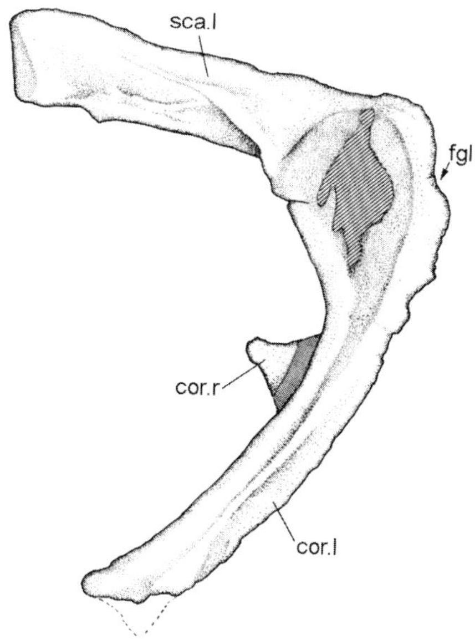

Fig. 3. Drawing of scapulocoracoid (MBLUZ p-911) in anterior view. cor, coracoid; fgl, fossa glenoidalis; sca, scapula; l, left; r, right. Scale as for Figure 2.

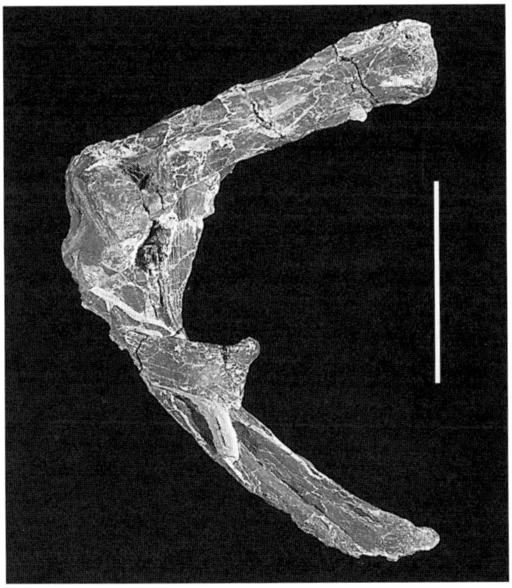

Fig. 4. Photograph of scapulocoracoid (MBLUZ P-911) in posterior view. Scale bar 50 mm.

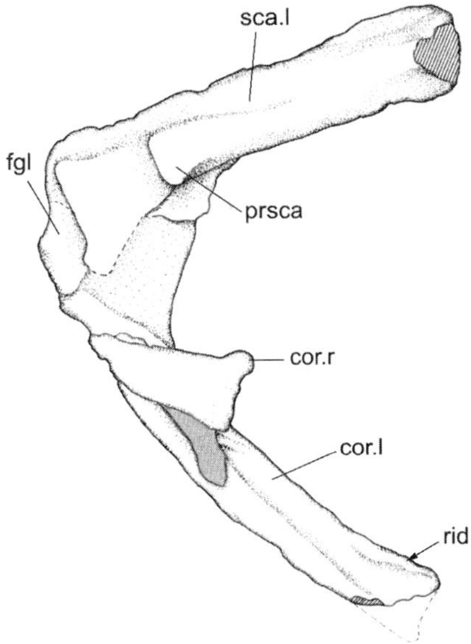

Fig. 5. Drawing of scapulocoracoid (MBLUZ P-911) in posterior view. cor, coracoid; fgl, fossa glenoidalis; prsca, processus scapularis; rid, ridge; sca, scapula; l, left; r, right. Scale as for Figure 4.

Discussion

The discovery of MBLUZ P-911 is important because it provides the first evidence of a pterosaur from Venezuela and extends the record of pterosaurs to the northern part of South America. Despite this, the systematic information that can be extracted from the specimen is very limited. The most interesting feature of MBLUZ P-911 is the proportion between the two elements, with the scapula being significantly smaller than the coracoid. In pterosaurs, these bones are usually subequal in size, with the scapula most often longer. In only one clade, Pteranodontoidea (*sensu* Kellner 1996), the reverse is observed. This clade comprises *Pteranodon*, *Istiodactylus* and the Anhangueridae, the latter having the proportionally smallest scapula (Wellnhofer 1991a; Kellner & Tomida 2000). As pointed out before, in MBLUZ the coracoid is more elongated and gracile than in *Istiodactylus* and *Pteranodon*. The latter has a small ventromedial coracoidal flange (Eaton, 1910; Bennett, 2001), absent in the Venezuelan specimen. Overall, MBLUZ P-911 is very similar to anhanguerids such as *Anhanguera santanae* (Wellnhofer 1991a) and *Anhanguera piscator* (Kellner & Tomida 2000), in having a longitudinal ridge on the medial surface of the coracoid and a proportionally short scapula, and it is tentatively referred to the Anhangueridae (Anhangueridae indet.). This occurrence further extends the record of this pterodactyloid clade to the Aptian.

The sedimentary rocks where MBLUZ P-911 was collected represent a marine environment, possibly not very far from the coast. The majority of pterosaur remains come from this kind of depositional setting (Kellner 1994). Therefore more specimens are expected to be found in the Apón Formation, which will provide more information about the pterosaurs that inhabited the former northern coast of South America.

We thank M. de Oliveira (Museu Nacional/UFRJ – fellow FAPERJ) for his skilful drawings and H. de Paula Silva (Museu Nacional/UFRJ) for preparing the specimen (MBLUZ P-911). We also wish to express our gratitude to L. Gutierrez of C.C. Faría S.A. for permitting access to Rosarito Quarry, to Ascanio Rincón and B. Moody for their help during fieldwork, and to E. Buffetaut (CNRS, Paris) and X. Pereda-Suberbiola (Universidad del País Vasco, Bilbao) for reviewing earlier versions of the manuscript. E. Buffetaut is thanked for the invitation to submit this article to this special volume.

This research was partially supported by CNPq and FAPERJ (grant to AWAK) and is a contribution to the project 'Mesozoic Archosaurs' (SID37010221003–9) which is being developed at the Museu Nacional.

References

BENNETT, S. C. 2001. Osteology and functional morphology of the Late Cretaceous pterosaur *Pteranodon*. *Palaeontographica A*, **260**, 1–153.

BONAPARTE, J. F. 1970. *Pterodaustro guiñazui* gen. et sp. nov. Pterosaurio de la Formacion Lagarcito, Provincia de San Luis, Argentina, y su significado en la geología regional (Pterodactylidae). *Acta Geológica Lilloana*, **10**, 207–226.

CAMPOS, D. A. & KELLNER, A. W. A. 1985. Panorama of the flying reptiles study in Brazil and South America. *Anais da Academia Brasileira de Ciências*, **57**, 453–466.

CHIAPPE, L.M. KELLNER, A.W.A., RIVAROLA, D., DAVILA, S. & FOX, M. 2000. Cranial morphology of *Pterodaustro guinazui* (Pterosauria: Pterodactyloidea) from the Lower Cretaceous of Argentina. *Contributions in Science*, **483**, 1–19.

DURÁN, I., KAPELLOS, C. & VAN ERVE, A. W. 1984. *Excursión a las Cuencas Sedimentarias de Machiques y Uribante (Venezuela Occidental) 25–09/02–10–1983*. Informe No. EPC-7693, Maraven, Gerencia de Exploración, Caracas, 12 pp.

EATON, G. F. 1910. *Osteology of* Pteranodon. Memoirs of the Connecticut Academy of Arts and Sciences Memoirs, **2**, 1–38.

GONZALEZ DE JUANA, C., ITURALDE DE AREZONA, J. & PICARD-CADILLAT, X. 1980. *Geología de Venezuela y sus Cuencas Petrolíferas*. Foninves, Caracas, 1030 pp.

HOOLEY, R. W. 1913. On the skeleton of *Ornithodesmus latidens*, and Ornithosaur form the Wealden shales of Atherfield (Isle of Wight). *Quarterly Journal of the Geological Society, London*, **69**, 372–422.

HOWSE, S. C. B., MILNER, A. R. & MARTILL, D. M. 2001. Pterosaurs. *In*: MARTILL, D. M. & NAISH, D. (eds) *Dinosaurs of the Isle of Wight*: Palaeontological Association, London, 324–335

KELLNER, A. W. A. 1994. Remarks on pterosaur taphonomy and paleoecology. *Acta Geologica Leopoldensia*, **39**, 175–189.

KELLNER, A. W. A. 1996. *Description of new material of Tapejaridae and Anhangueridae (Pterosauria, Pterodactyloidea) and discussion of pterosaur phylogeny*. PhD thesis, Columbia University. [Published by University Microfilms International].

KELLNER, A.W.A. 2001. A review of the pterosaur record from Gondwana. *In*: *Two Hundred Years of Pterosaurs. A Symposium on the Anatomy, Evolution, Palaeobiology and Environments of Mesozoic Flying Reptiles. Strata, serie 1*, **11**, 51–53.

KELLNER, A.W.A. & MOODY, J.M. 2001. The first occurrence of Pterosauria (Pteranodontoidea) in Venezuela. *Boletim de Resumos, 17th Congresso Brasileiro de Paleontologia, UFAC, Rio Branco, Acre*, 146.

KELLNER, A.W.A. & TOMIDA, Y. 2000. *Description of a new species of Anhangueridae (Pterodactyloidea) with comments on the pterosaur fauna from the Santana Formation (Aptian–Albian), Northeastern Brazil*. National Science Museum, Tokyo, Monographs, **17**, 135 pp.

MAISEY, J. G. & MOODY, J. M. 2001. A review of the problematic extinct teleost fish *Araripichthys*, with a description of a new species from the Lower

Cretaceous of Venezuela. *American Museum Novitates*, **3324**, 1–27.

MOODY, J. M. 1993a. First report of ichthyosaur remains from the Cretaceous of Venezuela. *Antartia*, **3**, 1–10.

MOODY, J. M. 1993b. Fósiles de reptiles cretácicos, Sierra de Perijá, Zulia, Venezuela. *In*: *VI Jornadas Científicas, La Universidad del Zulia / Facultad Experimental de Ciencias, Maracaibo*, 31.

MOODY, J. M. & MAISEY, J. G. 1994. New Cretaceous marine vertebrate assemblages from north-western Venezuela and their significance. *Journal of Vertebrate Paleontology*, **14**, 1–8.

PRICE, L. I. 1971. A presença de Pterosauria no Cretáceo Inferior da chapada do Araripe, Brasil. *Anais de Academia Brasileira de Ciências*, **43** (supplement), 451–461.

RENZ, O. 1982 *The Cretaceous Ammonites of Venezuela*. Maraven S.A., Caracas, 132 pp.

ROD, E. & MAYNC, W. 1954. Revision of Lower Cretaceous stratigraphy of Venezuela. *American Association of Petroleum Geologists Bulletin*, **38** (2), 193–283.

WELLNHOFER, P. 1985. Neue Pterosaurier aus der Santana-Formation (Apt) der Chapada do Araripe, Brasilien. *Palaeontographica*, **187**, 105–182.

WELLNHOFER, P. 1991a. Weitere Pterosaurierfunde aus der Santana-Formation (Apt) der Chapada do Araripe, Brasilien. *Palaeontographica*, **215**, 43–101.

WELLNHOFER, P. 1991b. *The Illustrated Encyclopedia of Pterosaurs*. Salamander, London, 192 pp.

A new azhdarchid pterosaur from the Late Cretaceous phosphates of Morocco

XABIER PEREDA SUBERBIOLA[1,2], NATHALIE BARDET[2], STÉPHANE JOUVE[2], MOHAMED IAROCHÈNE[3], BAÂDI BOUYA[4] & MBAREK AMAGHZAZ[4]

[1]*Universidad del País Vasco/Euskal Herriko Unibertsitatea, Facultad de Ciencias, Departamento de Estratigrafía y Paleontología, Apartado 644, 48080 Bilbao, Spain (e-mail: gpppesuf@lg.ehu.es)*

[2]*UMR 8569 du CNRS, Muséum National d'Histoire naturelle, Département Histoire de la Terre, 8 rue Buffon, 75005 Paris, France (e-mail: bardet@mnhn.fr)*

[3]*Ministère de l'Energie et des Mines, Direction de la Géologie, BP 6208, Rabat, Morocco*

[4]*Office Chérifien des Phosphates, Centre Minier de Khouribga, Khouribga, Morocco*

Abstract: A large azhdarchid pterosaur is described from the Late Maastrichtian phosphatic deposits of the Oulad Abdoun Basin, near Khouribga (central Morocco). The material consists of five closely associated cervical vertebrae of a single individual. The mid-series neck vertebrae closely resemble those of azhdarchids *Quetzalcoatlus* and *Azhdarcho* in that they are elongate, with vestigial neural spines, prezygapophysial tubercles, a pair of ventral sulci near the prezygapophyses, and without pneumatic foramina on the lateral surfaces of the centra. The Moroccan pterosaur is referred to a new genus and species of Azhdarchidae: *Phosphatodraco mauritanicus* gen. et sp.nov. It is mainly characterized by a very long cervical vertebra eight, bearing a prominent neural spine located very posteriorly. Based on comparisons with azhdarchid vertebrae, the estimated wing span of *Phosphatodraco* is close to 5 m. This discovery provides the first occurrence of Late Cretaceous azhdarchids in northern Africa. *Phosphatodraco* is one of the few azhdarchids known from a relatively complete neck and one of the latest-known pterosaurs, approximately contemporaneous with *Quetzalcoatlus*.

In recent years, pterosaur remains have been reported from several localities of Morocco, all from the Cretaceous. The first discovery was a large cervical vertebra from the Albian or Cenomanian of the Province of Ksar es Souk (southern Morocco), referred by Kellner & Mader (1996) to the Azhdarchidae. Later, Kellner & Mader (1997) described an isolated tooth from the same area, west of the Hamada du Guir, and compared it with those of anhanguerids from the Early Cretaceous of Brazil. The anhanguerid *Siroccopteryx moroccoensis* is based on an upper jaw with teeth found near Beg'aa, southwest of the town of Taouz, near the Algerian border (Mader & Kellner 1999). Moreover, Wellnhofer & Buffetaut (1999) reported jaw fragments of toothless pterosaurs and isolated teeth from the red beds of the Kem Kem region, east of Taouz, probably of the Cenomanian. These authors tentatively recognized four taxa: ?Pteranodontidae, ?Azhdarchidae, Tapejaridae (based on jaw remains) and Ornithocheiridae (based on teeth). The oldest record of pterosaurs in Morocco is from the basal Cretaceous (?Berriasian) of Anoual, eastern High Atlas Mountains, east of Talsinnt: Knoll (2000) described isolated teeth from this locality and regarded them as reminiscent of those of the Ornithocheiridae and Gnathosauridae.

During palaeontological field work in spring and summer 2000, pterosaur remains were unearthed by the Office Chérifien des Phosphates (OCP) in the Maastrichtian phosphatic deposits near the city of Khouribga. These field works have been realized as part of an active collaboration since 1997 between the OCP, the Ministère de l'Energie et des Mines and the French Centre National de la Recherche Scientifique (CNRS). The pterosaur remains described in this paper are referred to a new genus and species of azhdarchid. This is the first discovery of pterosaurs in the Late Cretaceous of Morocco and northern Africa (for a review of African pterosaurs see Dalla Vecchia *et al.* 2001; Kellner 2001).

Geological setting

The pterosaur remains were found in the eastern part of the Oulad Abdoun Phosphatic Basin, between the cities of Khouribga and Oued Zem (Fig. 1a, b). They were recovered from 'site 1' of Sidi Daoui, in the northern part of Grand Daoui, an area actively

From: BUFFETAUT, E. & MAZIN, J-M. (eds) 2003. *Evolution and Palaeobiology of Pterosaurs.* Geological Society, London, Special Publications, **217**, 79–90. 0305–8719/03/$15

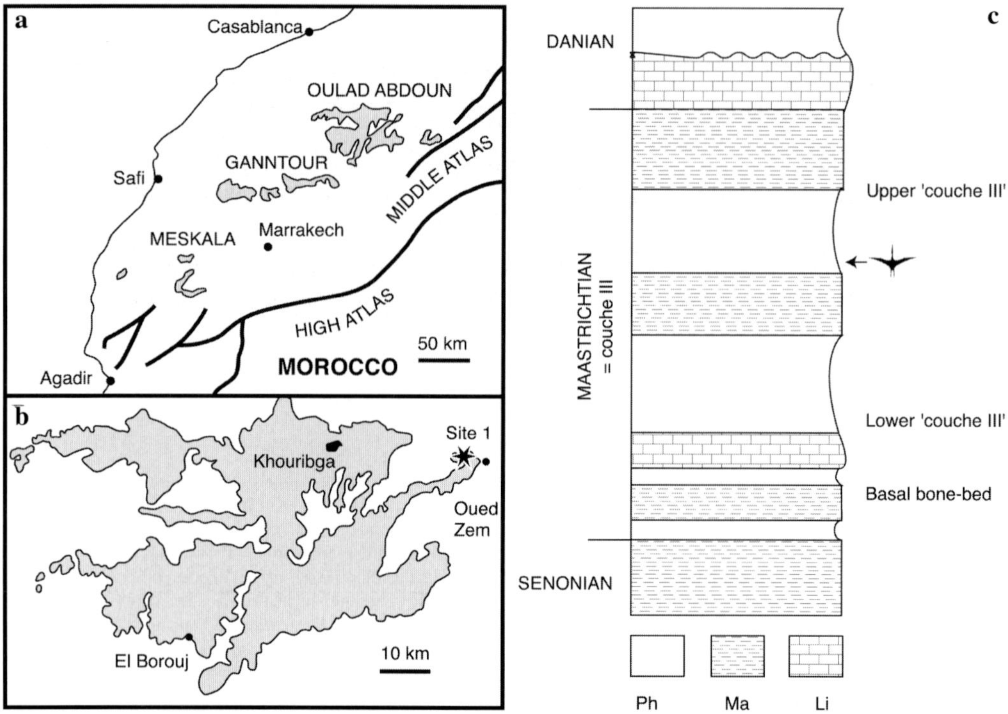

Fig. 1. Map of Morocco showing (**a**) the main phosphatic basins, (**b**) location map of the pterosaur locality in the Oulad Abdoun Basin, and (**c**) stratigraphical log showing the occurrence of the pterosaur specimen into the phosphatic series. Li, limestones; Ma, marls; Ph, phosphates.

quarried for phosphate. Stratigraphically, the pterosaur material occurs in the upper part of the phosphatic unit called 'couche III' by the miners, which is Late Maastrichtian on the basis of selachian teeth (Cappetta 1987).

The phosphatic series of the Oulad Abdoun Basin is very condensed and the Maastrichtian is only c. 2–5 m thick. The 'couche III' is composed, from bottom to top, of: thin phosphatic levels and marls; a grey limestone bone bed rich in fish remains; yellow, fine, soft phosphates (lower 'couche III'); a thick, yellow, marly level that separates the lower and upper 'couche III'; and grey, brown-striped soft phosphates overlay in by a thick level of marls (upper 'couche III'). The pterosaur comes from the lower part of the upper phosphatic unit (Fig. 1c).

The pterosaur remains are preserved in a block 98 cm long and 34 cm wide. The phosphatic matrix is grey in colour, mottled with orange. During the mechanical preparation of the specimen, other fossil remains were found associated to the pterosaur. These consist of fish vertebrae, selachian teeth (*Serratolamna serrata*, *Rhombodus binkhorsti*; determination by H. Cappetta), enchodontid teeth (*Enchodus libycus*), a pycnodontid tooth, a caudal vertebra and a few mosasaurid teeth (*Prognathodon* sp.), as well as small nodules. The pterosaur bones were found close to a partial skeleton of an indeterminate mosasaurid comprising jaw fragments and articulated caudal vertebrae. The 'site 1' of Sidi Daoui has also yielded remains of sharks and rays (Cappetta 1987; currently under study by Cappetta), actinopterygians (*Stratodus apicalis*), mosasaurids (*Platecarpus ptychodon*, *Mosasaurus beaugei*, *Halisaurus* sp. nov.), plesiosaurs (Elasmosauridae indet.) and turtles (Bothremydidae indet.). This fauna indicates a marine depositional environment (Arambourg 1952).

Systematic description

Order Pterosauria Kaup 1834
Superfamily Pterodactyloidea Plieninger 1901
Family Azhdarchidae Nesov 1984 (emend. Padian, 1986)
Genus *Phosphatodraco* gen.nov.

Diagnosis. Large azhdarchid pterodactyloid (estimated wing span 5 m) that differs in having posterior neck vertebra (cervical eight) very elongate, with a

Fig. 2. *Phosphatodraco mauritanicus* gen. et sp. nov, OCP DEK/GE 111, Late Cretaceous (Maastrichtian), Oulad Abdoun Basin, Khouribga, Morocco. Phosphatic block containing the pterosaur remains. C5–C9, cervical vertebrae; indet., indeterminate bone.

length more than 50% that of the fifth cervical, and bearing a prominent neural spine nearly as high as the centrum, squarely truncated at the top, very posteriorly located. Maximum vertebral length/anterior width between prezygapophyses ratio of mid-series cervical vertebrae approximately 4.3 (cervical five), 4.1 (cervical six).

Type species. Phosphatodraco mauritanicus sp.nov.

Horizon. Late Cretaceous (Late Maastrichtian).

Etymology. 'Phosphate' and '*draco*' (Latin), meaning 'dragon from the phosphates', and *Mauritania* (Latin), northern Africa, referring to the region where the fossil remains were found.

Specific diagnosis. As for genus.

Holotype. OCP DEK/GE 111, five cervical vertebrae and an indeterminate bone; Office Chérifien des Phosphates, Service Géologique, Khouribga Morocco (OCP field specimen number N1:1).

Locality. 'Site 1' of Sidi Daoui, northern Grand Doui, near Khouribga, central Morocco; upper 'couche III', Oulad Abdoun Phosphatic Basin.

Description of *Phosphatodraco*

The pterosaur block includes five vertebrae and an indeterminate fragment of bone (Fig. 2). All identifiable vertebrae come from the cervical region. They are disarticulated but closely associated, and therefore most probably belong to the same individual. The vertebrae are crushed and damaged. The centra are hollow and the cortical bone is approximately 1 mm thick. The external surface of the bone has been chipped off or is missing in places and there is evidence of sedimentary infilling by phosphate. Consequently, the fossil remains have not been removed from the matrix, so the following description is based on visible parts. For measurements see Table 1.

The cervical vertebrae are variable in length. The longest one (estimated length about 300 mm), is broken into two associated fragments (cervical five, Fig. 3a). The first fragment consists of the anterior third of the vertebra in ventral view (length 110 mm as preserved), the second represents the posterior two-thirds in left lateral view (approximately 190 mm long). The vertebra is badly preserved and crushed. The bone surface is crackled and some areas have collapsed. The possibility of the two fragments belonging to two vertebrae is unlikely because they lie in continuity, with no sediment between them and overlying each other in some places. The lateral expansion at the end of the anterior part is due to crushing. This kind of preservation, where fragile but well-preserved bones lie together with damaged elements of the same individual, is not unusual in the phosphatic deposits of Morocco and has been observed in other vertebrate remains from the same level. The prezygapophyses are long, horn-shaped and diverge slightly anteriorly from the lateral borders of the centrum. The cotyle area is collapsed. In lateral view, the posterior end of the vertebra shows a developed left postzygapophysis and, posterior to it, the convex articular condyle and the left postexapophysial process. Due to crushing, all these structures appear to lie in the same plane.

The best-preserved vertebra (cervical six, Fig. 3b) is shorter than the preceding vertebra, with a maximum length of about 225 mm (Table 1). It is crushed and slightly distorted; the anterior part is exposed in ventrolateral view while the posterior vertebra is seen in left lateral view. The specimen

Table 1. Phosphatodraco mauritanicus *gen. et sp. nov., cervical vertebrae measurements (mm)*

Cervical 5	
Maximum length[1]	c. 300
Length of the centrum[2]	c. 268
Maximum anterior width across the prezygapophyses	c. 70
Maximum height of the centrum	33
Cervical 6	
Maximum length[1]	c. 225
Length of the centrum[2]	196
Maximum anterior width across the prezygapophyses	c. 55
Maximum posterior height	62
Height of centrum	36
Cervical 7	
Maximum length	190*
Length of centrum[2]	158*
Maximum distance across the prezygapophyses	61
Maximum distance across the postzygapophyses	78
Minimum width of the centrum	45
Cervical 8	
Maximum length	150*
Maximum height	115
Height of the centrum	45
Height of the neural spine[3]	40
Length of the neural spine	30
Height of the neural arch[4]	56
Cervical 9	
Maximum width across the transverse processes	89*
Maximum height	75*
Maximum width of the neural arch	29
Maximum diameter of the canal neural	11
Maximum width of the condyle	41

[1] From the anterior end of prezygapophysis to end of postexapophysis posteriorly
[2] From mid-line between anterior and posterior articulations
[3] From the dorsal surface of the postzygapophysis to the top of the neural spine
[4] From the dorsal surface of the centrum to the top of the neural spine
* As preserved (broken)

preserves most of the cortical bone, but both the left pre- and postzygapophyses and the posterolateral part of the centrum, including the condyle, are eroded and in places show the internal trabecular bone. The centrum is procoelous. The horn-like prezygapophyses are nearly parallel and concave in ventral view. The right prezygapophysis has a small medial process (or prezygapophysial tubercle). There is evidence of neither accessory processes on the anterior end of the vertebra nor pneumatic foramina on the lateral surface of the centrum. The neural spine is totally absent. The anterior cotyle is slightly distorted, but it seems to be at least twice wide as high (Fig. 3f). It is ovoid, with the dorsal margin slightly concave. The ventral margin bears a prominent hypapophysis. The height of this keel diminishes rapidly towards the mid-length of the centrum. A longitudinal oval sulcus is present in the right side of the ventral surface, close to the base of the prezygapophysis. The occurrence of convergent ridges which extend posteriorly to the sulci is difficult to confirm, but is not excluded. The right lateral and ventral edges of the cotyle are broken and the bony trabeculae can be seen. The ventral surface of the centrum is almost flat. As in the preceding vertebra, the postexapophysis is well developed ventrolateral to the condyle. The postexapophyseal articulation is missing because of damage.

The posteriorly adjacent vertebra (cervical seven, Fig. 3c) is exposed in ventral view. The posterior part of the vertebra beyond the postzygapophyses is missing. As preserved, the vertebra is approximately 190 mm long (Table 1). The total length is estimated to be about the same, or slightly shorter, as that of the anteriorly adjacent vertebra. The prezygapophyses are similar in their form and disposition to those of the preceding vertebra. The cotyle is oval and dorsoventrally compressed. The ventral border is broken and thus the hypapophysis is missing. A sulcus is visible ventral to the left prezygapophysis, but there is apparently no weak ridge extending posteriorly. The centrum is slightly bulged posteriorly and becomes narrower at the mid-length. The cortical bone is thin-walled, about 1 mm thick in the middle part of the centrum. The postzygapophyses are well developed and widely divergent from the longitudinal axis of the centrum. The articular surfaces are not preserved. Between the postzygapophyses, a small triangular protuberance probably indicates the position of the roof-like dorsal margin of the neural canal. Unfortunately, the nature of the openings of the neural canal cannot be determined from the specimen.

The penultimate vertebra (cervical eight, Fig. 3d) is broken anteriorly but the centrum is very elongate (as preserved, 150 mm long). This vertebra is exposed in left lateral view. The lateral surface of the centrum is abraded and most of the centrum is preserved as an internal mould. The most striking feature is the presence of a tall neural spine, located in the posterior-most part of the vertebra. The height of the neural spine (40 mm as measured from the dorsal surface of the postzygapophysis to the top) almost reaches that of the centrum (45 mm). The anterior and posterior borders of the neural spine are vertically aligned in parallel. The top of the neural spine is squarely truncated and perpendicular to the lateral edges. The left postzygapophysis is situated at the basal part of the posterior end of the neural arch. The left postexapophyseal process is well

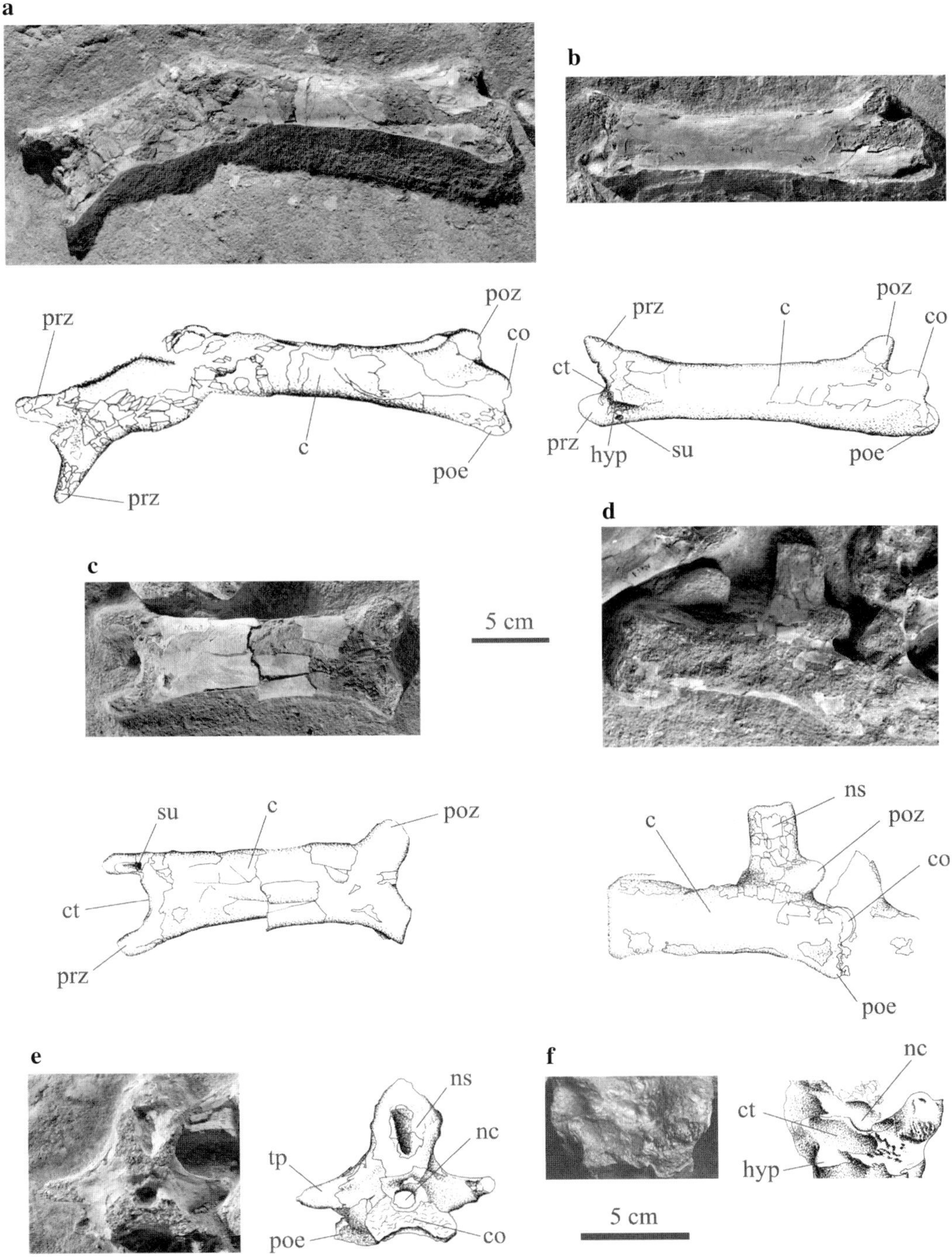

Fig. 3. *Phosphatodraco mauritanicus* gen. et sp. nov, OCP DEK/GE 111, Late Cretaceous (Maastrichtian), Morocco: (**a**) cervical five in two fragments, ventral and left lateral views; (**b**) cervical six in ventrolateral view; (**c**) cervical seven in ventral view; (**d**) cervical eight in left lateral view; (**e**) cervical nine in posterior view; (**f**) cervical six in anterior view. c, centrum; co, condyle; ct, cotyle; hyp, hypapophysis; nc, neural canal; ns, neural spine; poe, postexapophysis; poz, postzygapophysis; prz, prezygapophysis; su, sulcus; tp, transverse process.

developed posteroventrally but does not extend beyond the condyle as in the preceding vertebrae.

The last vertebrae is exposed in posterior view (cervical nine, Fig. 3e). As preserved, the vertebra is 75 mm high. The neural arch bears a massive swollen neural spine and well-developed transverse processes. The neural spine ends dorsally in a blunt process. The posterior side of the neural spine shows an oval depression bordered laterally by thick vertical edges. The postzygapophyses are missing. The transverse processes are long and slender, but are broken distally. They are projected laterally and slightly ventrally. The neural canal is small and almost circular (maximum diameter 11 mm). There are no noticeable pneumatic foramina lateral to, or just above, the neural canal. The condyle is broad, about 5 times wider than high and crescentic in cross section. The left postexapophysis is located lateral to the condyle and almost vertical.

Cervical ribs are not preserved. However, the development of the transverse processes of the last vertebra suggests that at least this vertebra probably bore ribs.

Finally, an indeterminate fragment of bone is preserved in association with the last two vertebrae (Fig. 2). The crackled texture pattern of the bone looks like that seen in the cervical vertebrae. The bone is flat and roughly crescentic in shape. As preserved, it is about 9 mm wide and 44 mm long.

Comparisons

The cervical vertebrae of *Phosphatodraco* show a disparity in length, as in pterodactyloids, with the exception of the Jurassic forms *Pterodactylus kochi* and *P. elegans* (Howse 1986; Bennett 2001). In many pterosaurs, the maximum length is reached by cervical five and then decreases posteriorly. Based on this criterion, we have assumed that the longest vertebra (Fig. 3A) from *Phosphatodraco* represents the fifth. This and the following two vertebra, regarded here as cervicals six and seven, are comparable in form to the mid-series cervical vertebrae of long-necked pterodactyloids (see Howse 1986). The last two preserved vertebrae share some features with the remainding vertebrae, such as broad ovoid cotyles and condyles, and postexapophyses. However, they differ from the remaining vertebrae in having a neural arch demarcated from the centrum and a prominent neural spine. These vertebrae are considered to be cervicals eight and nine, and could be cervicalized dorsals incorporated into the neck (see Bennett 2001). If so, *Phosphatodraco* is represented by a partial neck consisting of five vertebrae from mid-series cervical five to posterior cervical nine.

In pterosaurs, the total number of neck vertebrae varies between seven and nine (Wellnhofer 1991b). The presence of seven cervical vertebrae that lack ribs is considered to be a synapomorphy of Pterodactyloidea by Bennett (1994). This condition contrasts with that seen in *Rhamphorhynchus* and other non-pterodactyloids, which have eight cervical vertebrae and cervical ribs are at least present on cervicals three to eight (Wellnhofer 1978). Among pterodactyloids, Bennett (1994) defined an unnamed clade composed of *Gallodactylus* and Eupterodactyloidea (Nyctosauridae + Dsungaripteridae + Azhdarchidae + Pteranodontidae; Dsungaripteroidea of Young 1964; see Kellner 1996) by the presence of cervicalized sixth and seventh postaxial vertebrae. In dsungaripteroids, the two cervicalized dorsals have accessory exapophyses like the cervical vertebrae, and there is a notarium formed of fused dorsal vertebrae and ribs (Bennett 1994). The first dorsal vertebra is considered to be the one on which ribs are the first to be connected to the sternum (Wellnhofer 1991b). The monophyly of the clade Eupterodactyloidea is not supported by other phylogenetic analyses (see Unwin 1992, 1995; Peters 1997; Unwin & Lü 1997; Kellner 1995a, 1996). However, the presence of a neck composed of nine vertebrae with two cervicalized dorsals that bear exapophyseal articulations like the cervical vertebrae is common to a number of pterodactyloids, including *Pteranodon*, *Quetzalcoatlus*, *Nyctosaurus*, *Istiodactylus*, *Dsungaripterus* and *Anhanguera* (Howse 1986; Bennett 2001). Adult individuals of these genera also have a notarium, although this structural complex could have arisen convergently several times in Pterodactyloidea (see Young 1964; Unwin & Lü 1997).

The form of the mid-series cervical vertebrae of *Phosphatodraco* is reminiscent of that of azhdarchids. It shares at least two features considered by many authors (e.g. Nesov 1984, 1997; Padian 1984, 1986; Howse 1986; Wellnhofer 1991b; Padian *et al.* 1995; Kellner 1996; Company *et al.* 1999) to be diagnostic of Azhdarchidae: elongated mid-series cervical vertebrae, and low vestigial or absent neural spines. Similar cervicals occur in other long-necked pterodactyloids, such as *Huanhepterus* and *Doratorhynchus*, but these taxa probably acquired their long necks independently of azhdarchids (Bennett 1994; Unwin & Lü 1997).

Huanhepterus, from the Lower Cretaceous of China, has a long neck composed of seven vertebrae, and there is no notarium. Dong (1982) referred it to the Ctenochasmatidae on the basis of tooth form, an interpretation followed by Wellnhofer (1991b) and Unwin & Lü (1997). However, it lacks any ctenochasmatid synapomorphy, so its affinities among Pterodactyloidea remain unknown.

Recently, Howse & Milner (1995) have referred two isolated cervical vertebrae from the Purbeck of the United Kingdom, including the type of *Dorathorhychus validus* described by Seeley

(1875), to the Ctenochasmatidae: these vertebrae are elongate, have low neural spines and bear postexapophyses. *Dorathorhynchus* has usually been referred to Azhdarchidae (Howse 1986; Wellnhofer 1991b; Bennett 1994, 2001). However, the cervical vertebra has very long prezygapophyses, proportionally much longer than in *Quetzalcoatlus* and other azhdarchids, and there is a small pneumatic foramen about half way along the centrum, in contrast to azhdarchids (see Howse 1986, fig. 8). As noted by Nesov (1984), the cervical vertebra of *Doratorhynchus* 'is not tubular in the middle, with crests along the lateral sides', unlike azhdarchids. The occurrence of ovoid pneumatic foramina on the lateral sides of the cervical centra is known in many pterodactyloids, such as *Pterodactylus* (Wellnhofer 1970), *Pteranodon* (Eaton 1910; Bennett 2001), *Ornithocheirus* (Howse 1986), *Istiodactylus* (Hooley 1913; Howse *et al.* 2001), *Pterodaustro* (Bonaparte 1970), *Anhanguera* (Wellnhofer 1991a; Kellner & Tomida 2000), *Tapejara* (Kellner 1995b) and '*Santanadactylus brasilensis*' (Buisonjé 1980), but it appears to be absent in the short-necked and tall-spined *Nyctosaurus* (Williston 1903) and in all known azhdarchids. The status of *Doratorhynchus* within the Pterodactyloidea is currently unclear: it is probably not an azhdarchid and its referral to the Ctenochasmatidae by Howse & Milner (1995) is based mainly on stratigraphical ground (Tithonian–Berriasian distribution) and similarities in form and proportion with *Huanhepterus*.

With regard to other long-necked pterosaurs, it seems that the postaxial cervicals of *Ctenochasma* do not possess postexapophyses, in contrast to all large pterodactyloids (Howse 1986). Unfortunately, the cervical vertebrae of ctenochasmatids are not well known and so comparisons with those of azhdarchids are difficult.

Bennett (1994) and Unwin & Lü (1997) pointed out that the ratio between the length and width (L/W) of the mid-cervical vertebrae of azhdarchids is equal to, or greater than, 5. The total length (i.e., from the anterior end of the prezygapophysis to the posterior end of the postexapophysis) of the supposed cervical five of *Quetzalcoatlus* sp. (see Lawson 1975a, fig. 1a; Howse 1986, fig. 7; Wellnhofer 1991b, p. 144) is 5.6 times the maximum anterior width (between the prezygapophyses). Following this, Frey & Martill (1996) estimated a ratio of at least 6.9 for the type specimen of *Arambourgiania*, which they assumed to be cervical five. However, this ratio is variable according to the position of the vertebrae in the neck and is less than 5 in all other vertebrae of *Quetzalcoatlus* sp. but cervical five (Frey & Martill 1996, tab. 2). Even taking into account that many specimens are distorted by crushing and that transverse measurements are therefore increased (W. Langston Jr. pers. comm.), it is unlikely that all mid-cervicals were 5 times longer than wide along the neck. Most of the azhdarchid vertebrae are too incomplete to test this, but, when available, the ratio is much less than 5 (e.g., 3.2 in a cervical vertebra referred to cf. *Azhdarcho* by Buffetaut 1999, fig. 1b-c). In *Phosphatodraco*, the L/W ratio of the supposed cervicals five and six is 4.3 and 4.1 respectively. In terms of comparison, the L/W ratio could reach about 6.5 in the type of *Doratorhynchus* (Howse 1986, fig. 9), 2 in the paratype of '*Santanadactylus brasilensis*' (Buisonjé 1980; see Kellner & Tomida 2000), and less than 2 in *Ornithocheirus* (Owen 1859, pl. 2, figs 7–18), *Istiodactylus* (see Hooley 1913, pl. 38, fig. 2), *Nyctosaurus* (Williston 1903, pl. 44, fig. 7; Howse 1986, fig. 6), *Anhanguera* (Wellnhofer 1991a, figs 5–9), and *Pteranodon* (Bennett 2001). Further data concerning the long-necked pterodactyloids *Ctenochasma*, *Huanhepterus*, *Pterodaustro* or *Pterodactylus* are not currently available.

Kellner & Mader (1996) have suggested that the presence of two well-marked sulci leading to prominent foramina on the ventrolateral surface of the anterior part of the cervical centra, near the prezygapophyses, might represent a synapomorphy for the Azhdarchidae. In fact, such sulci, present in *Phosphatodraco*, as well as a pair of weak, longitudinal ridges which extend posteriorly, are known in most but not all azhdarchids (see discussion below). *Phosphatodraco* also has prezygapophyseal tubercles, a common azhdarchid feature (Company *et al.* 1999). Additional characters listed as possible synapomorphies of Azhdarchidae or less inclusive clades, such as the neural arch confluent with centrum, unossified neural canal or round cross-section in mid-series cervical centra (Nesov 1984, 1997; Padian 1984, 1986; Bennett 1989, 1994; Padian *et al.* 1995; Ikegami *et al.* 2000), cannot be tested in *Phosphatodraco* because of preservation of the material.

Phosphatodraco is referred to the Azhdarchidae on the basis of elongate mid-series cervical vertebrae, with vestigial or absent neural spines, prezygapophyseal tubercles, a pair of ventral sulci close to the prezygapophyses, and without oval pneumatic foramina on the lateral surfaces of the centra.

Discussion

The Azhdarchidae comprise *Azhdarcho lancicollis* Nesov, 1984 from the Turonian–Coniacian of Uzbekistan (Nesov 1984, 1997; Bakhurina & Unwin 1995; Unwin & Bakhurina, 2000), *Quetzalcoatlus northopi* Lawson 1975b and *Quetzalcoatlus* sp. from the Maastrichtian of Texas, USA (Lawson 1975a, 1975b; Langston 1981; Kellner & Langston 1996), *Arambourgiania philadelphiae* (Arambourg

1959) from the Maastrichtian of Jordan (Frey & Martill 1996; Martill *et al.* 1998), *Montanazhdarcho minor* Padian, Ricqlès & Horner 1995 from the Campanian of Montana, United States (Padian *et al.* 1995), *Zhejiangopterus linhaiensis* Cai & Wei 1994 from the Campanian of China (Cai & Wei 1994; Unwin & Lü 1997), and *Hatzegopteryx thambema* from the Maastrichtian of Romania (Buffetaut *et al.* 2002).

Padian & Smith (1992) tentatively referred the paratype cervical vertebrae of '*Santanadactylus brasilensis*' Buisonjé 1980 to the Azhdarchidae but Kellner & Tomida (2000) regarded them as Pterodactyloidea indet. These vertebrae are proportionally shorter than in azhdarchids (see above) and have a higher neural spine (Nesov 1984).

Additional azhdarchid bones have been described from the Maastrichtian of Wyoming, United States (Estes 1964), and from western Australia (Long 1998); the Campanian of Alberta, Canada (Currie & Russell 1982), Montana, United States (Padian 1984; Padian & Smith, 1992) and Israel (Lewy *et al.* 1993); the Campanian–Maastrichtian of France (Buffetaut *et al.* 1997; Buffetaut 2001), Spain (Astibia *et al.* 1990; Buffetaut 1999; Company *et al.* 1999, 2001) and Senegal (Monteillet *et al.* 1982); the Coniacian–Campanian of western Russia (Bakhurina & Unwin 1995); the Cenomanian–Turonian of Japan (Ikegami *et al.* 2000); the Late Cretaceous of New Jersey, United States (Baird, in Bennett 1989); the Albian–Cenomanian of Morocco (Kellner & Mader 1996; Wellnhofer & Buffetaut 1999); the Aptian–Albian of Texas, United States (Murry *et al.* 1991); and the Aptian of Brazil (Martill & Frey 1998; Frey & Tischlinger 2000) and Niger (Sereno *et al.* 1998). Recently, Sayão & Kellner (2001) have described azhdarchid remains from the Late Jurassic of Tanzania, extending provisionally the oldest record of this group to the Kimmeridgian–Tithonian of Africa (see also Kellner 2001).

Among azhdarchids, only *Quetzalcoatlus* and *Zhejiangopterus* are known from complete or relatively complete neck remains, while other taxa, such as *Azhdarcho*, *Arambourgiania* and *Montanazhdarcho*, are represented by more incomplete material. Unfortunately, the cervical remains from the Maastrichtian of Texas referred to as *Quetzalcoatlus* sp. (Lawson 1975a,b; Langston 1981; Howse 1986; Kellner & Langston 1996) and those of at least three different individuals of *Zhejiangopterus linhaiensis* from the Campanian of China (Cai & Wei 1994; Unwin & Lü 1997) have not yet been described in detail. The neck of *Quetzalcoatlus* has nine vertebrae, while only seven cervicals have been described in *Zhejiangopterus* (Cai & Wei 1994; Unwin & Lü 1997).

Steel *et al.* (1997) reconstructed *Arambourgiania* (formerly *Titanopteryx*) using better-known *Quetzalcoatlus* specimens (see also Frey & Martill 1996). The neck was reconstructed on the basis of cervicals three to nine of *Quetzalcoatlus* sp. (Texas Memorial Museum, Austin, United States). This assemblage comes from a population of different individuals, being of the size range within the sample of about 10% (W. Langston Jr, pers. comm.). The length formula of the postaxial neck vertebrae of *Quetzalcoatlus* is as follows: $3<4<5>6>7>8>9$ (Steel *et al.* 1997), as in *Pteranodon* (Bennett 2001) and other pterodactyloids. Differences in length between cervicals five and six or cervicals six and seven are less than 10%, like the mid-cervicals of *Phosphatodraco* (Table 1), *Zhejiangopterus* (Cai & Wei 1994) and *Pteranodon* (Bennett 2001). In contrast, the length of cervical eight of *Quetzalcoatlus* is less than 20% of cervical five, while in *Phosphatodraco* cervical eight is much longer, being at least half as long as cervical five. The cervical eight of *Phosphatodraco* resembles that of *Quetzalcoatlus* in having a neural spine squarely truncated at the top, but differs in being very posteriorly located (W. Langston Jr pers. comm.).

The centrum of the mid-cervicals of *Phosphatodraco* is less constricted at mid-length than in *Quetzalcoatlus* (Lawson 1975a; Wellnhofer 1991b) or *Azhdarcho* (Nesov 1984, pl. 7, figs. 2–5; Bakhurina & Unwin 1995, fig. 13b; Nesov 1997, pl. 14, figs 2–6), but the lateral margins are not almost straight, as in an azhdarchid cervical vertebra from Japan (Ikegami *et al.* 2000).

Arambourgiania differs from *Phosphatodraco* and other azhdarchids in that the cervical vertebrae are circular or high oval in cross-section, with condyles and cotyles higher than wide, and postexapophyses oriented ventrally rather than posteroventrally (Frey & Martill 1996; Martill *et al.* 1998). Moreover, *Arambourgiania* lacks both a pair of sulci and a pair of longitudinal ridges on the ventral surface of the anterior end of the cervical centra, unlike *Quetzalcoatlus*, *Azhdarcho* and azhdarchid specimens from Spain (Buffetaut 1999; Company *et al.* 1999). Such sulci are present in *Phosphatodraco*. No vertebra is exposed in dorsal view, so the presence of carinae or ridges lateral to the neural spine, as in *Arambourgiania* and *Azhdarcho*, cannot be attested in *Phosphatodraco*.

Among the remains of *Montanazhdarcho*, Padian *et al.* (1995) mentioned a crushed cervical vertebra which is characterized by its length, with low neural spines and neural arch confluent with centrum. In fact, these characters are common to Azhdarchidae (see above). Further comparison with *Montanazhacho* is impossible.

In summary, *Phosphatodraco mauritanicus* differs from *Quetzalcoatlus* and other azhdarchids in having a very long cervical eight which bears a prominent neural spine very posteriorly located.

Finally, comparison with other azhdarchid vertebrae suggests a wing span of about 5 m for *Phosphatodraco*. It was a large pterosaur, bigger than *Zhejiangopterus* (wing span 3.5 m; Unwin & Lü 1997), *Montanazhdarcho* (wing span 2.5 m; Padian *et al.* 1995) and most individuals of *Azhdarcho* (wing span 3–4 m, although rare remains suggest a wing span of up to 5–6 m; see Bakhurina & Unwin 1995), but smaller than *Arambourgiania* (estimated wing span 7 m; Company *et al.* in prep.) or larger individuals of *Quetzalcoatlus* (wing span about 11 m; Langston 1981; Wellnhofer 1991b). The estimated wing span of *Phosphatodraco* seems comparable to that of the small *Quetzalcoatlus* sp. (Wellnhofer 1991b) and adult individuals of an unnamed azhdarchid from Valencia, Spain (Company *et al.* 1999).

Maastrichtian pterosaur remains are known in several localities worldwide but well-dated Late Maastrichtian pterosaurs are quite rare (see Buffetaut *et al.* 1996 for a review). Only *Quetzalcoatlus northopi* from the Javelina Formation of Texas (Lawson 1975a; Langston 1981; Wellnhofer 1991b), an unnamed azhdarchid from the Sierra Perenchiza Formation of Valencia, Spain (Company *et al.* 2001), an isolated azhdarchid vertebra from the Lance Formation of Wyoming (Estes 1964), an isolated azhdarchid vertebra from the uppermost Marnes d'Auzas Formation of southern France (Buffetaut *et al.* 1997), a nyctosaurid humerus from the Gramame Formation of Brazil (Price 1953) and a possible azhdarchid ulna from the Miria Formation of western Australia (Long, 1998) are dated as Late Maastrichtian. Recently, Buffetaut *et al.* (2002) named *Hatzegopteryx*, a new azhdarchid from the Late Maastrichtian Haţeg Basin of Romania (but see López-Martínez *et al.* 2001 for a different interpretation of the age). Other specimens from Europe, Africa and New Zealand are not dated with accuracy or may be older than Late Maastrichtian (Monteillet *et al.* 1982; Wiffen & Molnar 1988; Buffetaut *et al.* 1996). Consequently, *Phosphatodraco* is one of the youngest records of pterosaurs worldwide.

Conclusions

Evidence of a large pterosaur in the Late Cretaceous (Late Maastrichtian) of the Oulad Abdoun Phosphatic Basin, near Khouribga (central Morocco) is provided by disarticulated but associated cervical vertebrae from a single individual. The partial neck consists of five vertebrae, from mid-series cervical five to posterior cervical nine. It is described as a new genus and species, *Phosphatodraco mauritanicus*, and referred to the Azhdarchidae. The mid-cervical vertebrae of *Phosphatodraco* closely resemble those of azhdarchids, especially *Quetzalcoatlus* and *Azhdarcho*, in having the following combination of features: elongated vertebrae (maximum L/W ratio$>$3, convergently developed by other pterodactyloids such as *Huanhepterus* and *Doratorhynchus*); very low or vestigial neural spines; prezygapophyseal tubercles; a pair of ventral sulci close to the prezygapophyses (absent in *Arambourgiania*); and absence of ovoid pneumatic foramina on the lateral surfaces of the centra. Additional azhdarchid characters, such as a neural arch confluent with centrum and round cross-section in mid-series cervical centra, cannot be observed in the Moroccan material because of preservation. *Phosphatodraco* is distinguished by the unusual form of the posterior cervical vertebrae: cervical eight is very long, much longer than in *Quetzalcoatlus*, as it is at least half as long as the cervical five; moreover, cervicals eight and nine bear a prominent neural spine, which is located very posteriorly on the vertebrae. Comparisons with other azhdarchid vertebrae suggest a wing span close to 5 m for *Phosphatodraco*. This is the second pterosaur genus described from Morocco (the first being the anhanguerid *Siroccopteryx moroccoensis* Mader & Kellner 1999), and the first record of azhdarchids in the uppermost Cretaceous rocks of northern Africa. Moreover, *Phosphatodraco* is one of the latest known pterosaurs.

This work has benefited from the help and collaboration ('Tripartite Convention') of the Direction de la Géologie from the Ministère de l'Energie et des Mines (Rabat) and of the Office Chérifien des Phosphates (Casablanca) from the Kingdom of Morocco. We thank especially M. Hamdi, M. Zeghnoun and all the staff of the OCP mining centre for their active support during our stay in Khouribga, and M. Sadiqui, L. Tabit and N. Aquesbi from the Ministère de l'Energie et des Mines (Rabat) for providing administrative facilities and permits. We are also grateful to W. Langston Jr for his kindness in providing valuable unpublished information on *Quetzalcoatlus* sp., and A. W. A. Kellner and E. Frey for critical revision and improvements of the manuscript. Thanks to J. M. Pacaud (MNHN, Paris) for the preparation of the material. Photographs are by D. Serrette (CNRS, MNHN, Paris). Infography of the figures is by H. Lavina (CNRS, MNHN, Paris) and by one of the authors (S. J.). This research work was supported by funds from the CNRS and the National Geographic Society (Grant 6627–99).

References

ARAMBOURG, C. 1952. Les vertébrés fossiles des gisements de phosphates (Maroc–Algérie–Tunisie). *Notes et Mémoires du Service Géologique du Maroc*, **92**, 1–372.

ARAMBOURG, C. 1959. *Titanopteryx philadelphiae* nov.

gen., nov. sp., ptérosaurien géant. *Notes et Mémoires du Moyen Orient*, **7**, 229–234.

ASTIBIA, H., BUFFETAUT, E. ET AL. 1990. The fossil vertebrates from Laño (Basque Country, Spain); new evidence on the composition and affinities of the Late Cretaceous continental faunas of Europe. *Terra Nova*, **2**, 460–466.

BAKHURINA, N. N. & UNWIN, D. M. 1995. A survey of pterosaurs from the Jurassic and Cretaceous of the former Soviet Union and Mongolia. *Historical Biology*, **10**, 197–245.

BENNETT, S. C. 1989. A pteranodontid pterosaur from the Early Cretaceous of Peru, with comments on the relationships of Cretaceous pterosaurs. *Journal of Paleontology*, **63**, 669–677.

BENNETT, S. C. 1994. *Taxonomy and systematics of the Late Cretaceous pterosaur* Pteranodon *(Pterosauria, Pterodactyloidea)*. Natural History Museum, University of Kansas, Occasional Papers, **169**, 70 pp.

BENNETT, S. C. 2001. Osteology and functional morphology of the Late Cretaceous pterosaur *Pteranodon*. *Palaeontographica A*, **260**, 1–153.

BONAPARTE, J. F. 1970. *Pterodaustro guiñazui* gen.et sp. nov., pterosaurio de la Formación Lagarcito, Provincia de San Luis, Argentina y su significado en la geología regional (Pterodactylidae). *Acta Geológica Lilloana*, **10**, 207–226.

BUFFETAUT, E. 1999. Pterosaurs from the Upper Cretaceous of Laño (Iberian Peninsula): a preliminary comparative study. *In*: ASTIBIA, H., CORRAL, J. C., MURELAGA, X., ORUE-ETXEBARRIA, X. & PEREDA-SUBERBIOLA, X. (eds) Geology and palaeontology of the Upper Cretaceous vertebrate-bearing beds of the Laño Quarry (Basque-Cantabrian Region, Iberian Peninsula). *Estudios del Museo de Ciencias Naturales de Alava*, **14** (1) (special), 289–294.

BUFFETAUT, E. 2001. An azhdarchid pterosaur from the Upper Cretaceous of Cruzy (Hérault, France). *Comptes Rendus de l'Académie des Sciences, Paris, Sciences de la Terre et des Planètes*, **333**, 357–361.

BUFFETAUT, E., CLARKE, J. B. & LE LOEUFF, J. 1996. A terminal Cretaceous pterosaur from the Corbières (southern France) and the problem of pterosaur extinction. *Bulletin de la Société Géologique de France*, **167**, 753–759.

BUFFETAUT, E., GRIGORESCU, D. & CSIKI, Z. 2002. A new giant pterosaur with a robust skull from the latest Cretaceous of Romania. *Naturwissenschaften*, **89**, 180–184.

BUFFETAUT, E., LAURENT, Y., LE LOEUFF, J. & BILOTTE, M. 1997. A terminal Cretaceous giant pterosaur from the French Pyrenees. *Geological Magazine*, **134**, 553–556.

BUISONJÉ, P. H. DE 1980. *Santanadactylus brasilensis* nov. gen., nov. sp., a long-necked, large pterosaurier from the Aptian of Brasil. *Proceedings, Koninklijke Nederlandse Akademie van Wetenschappen* (B), **83**, 145–172.

CAI, Z. & WEI, F. 1994. On a new pterosaur (*Zhejiangopterus linhaiensis* gen. et sp. nov.) from Upper Cretaceous of Linhai, Zhejiang, China. *Vertebrata Palasiatica*, **7**, 181–194. [In Chinese, English summary]

CAPPETTA, H. 1987. Mesozoic and Cenozoic Elasmobranchii, Chondrichthyes II. *In*: SCHULTZE, H-P. (ed) *Handbook of Paleoichthyology*. Gustav Fischer Verlag, Stuttgart, vol. 3B, 193 pp.

COMPANY, J., RUIZ OMEÑACA, J. I. & PEREDA SUBERBIOLA, X. 1999. A long-necked pterosaur (Pterodactyloidea, Azhdarchidae) from the Upper Cretaceous of Valencia, Spain. *Geologie en Mijnbouw*, **78**, 319–333.

COMPANY, J., UNWIN, D. M., PEREDA-SUBERBIOLA, X. & RUIZ OMEÑACA, J. I. 2001. A giant azhdarchid pterosaur from the Latest Cretaceous of Valencia, Spain – the largest flying creature ever? *Journal of Vertebrate Paleontology*, **21**(3) (supplement), 41–42A.

CURRIE, P. J. & RUSSELL, D. A. 1982. A giant pterosaur (Reptilia: Archosauria) from the Judith River (Oldman) Formation of Alberta. *Canadian Journal of Earth Sciences*, **19**, 894–897.

DALLA VECCHIA, F. M., ARDUINI, P. & KELLNER, A. W. A. 2001. The first pterosaur from the Cenomanian (Late Cretaceous) *Lagerstätten* of Lebanon. *Cretaceous Research*, **22**, 219–225.

DONG, Z. 1982. On a new Pterosauria (*Huanhepterus quingyangensis* gen. et sp. nov.) from Ordos, China. *Vertebrata Palasiatica*, **20**, 115–121 [In Chinese, with English abstract].

EATON, G. F. 1910. Osteology of *Pteranodon*. Connecticut Academy of Arts and Sciences, memoirs, **2**, 1–38.

ESTES, R. 1964. Fossil vertebrates from the Late Cretaceous Lance Formation, eastern Wyoming. *University of California Publications in Geological Sciences*, **49**, 1–180.

FREY, E. & MARTILL, D. M. 1996. A reappraisal of *Arambourgiania* (Pterosauria, Pterodactyloidea): one of the world's largest flying animals. *Neues Jahrbuch für Geologie und Paläontologie, Abhandlungen*, **199**, 221–247.

FREY, E. & TISCHLINGER, H. 2000. Weischteilanatomie der Flugsaurierfüsse und Bau der Scheitelkämme: Neue Pterosaurierfunde aus den Solnhofener Schichten (Bayern) und der Crato-Formation (Brasilien). *Archaeopteryx*, **18**, 1–16.

HOOLEY, R. W. 1913. On the skeleton of *Ornithodesmus latidens*; an ornithosaur from the Wealden Shales of Atherfield (Isle of Wight). *Quarterly Journal of the Geological Society, London*, **69**, 372–421.

HOWSE, S. C. B. 1986. On the cervical vertebrae of the Pterodactyloidea (Reptilia: Archosauria). *Zoological Journal of the Linnean Society*, London, **88**, 307–328.

HOWSE, S. C. B. & MILNER, A. R. 1995. The pterodactyloids from the Purbeck Limestone Formation of Dorset. *Bulletin of the Natural History Museum, London, Geology*, **51**, 73–88.

HOWSE, S. C. B., MILNER, A. R. & MARTILL, D. M. 2001. Pterosaurs. *In*: MARTILL, D.M. & NAISH, D. (eds) *Dinosaurs of the Isle of Wight. Field Guides to Fossils*, **10**, 324–335. The Palaeontological Association, London.

IKEGAMI, N., KELLNER, A. W. A. & TOMIDA, Y. 2000. The presence of an azhdarchid pterosaur in the Cretaceous of Japan. *Paleontological Research*, **4**, 165–170.

KAUP, 1834. Verruch einer Eintheilung der Säugethiere in 6 Stämme und der Amphibien in 6 Ordnungen. *Isis*, **3**, 1–315.

KELLNER, A. W. A. 1995a. The relationships of the Tapejaridae (Pterodactyloidea) with comments on pterosaur phylogeny. *In*: SUN, A & WANG, Y (eds) *Sixth Symposium on Mesozoic Terrestrial Ecosystems and Biota, Beijing, Short Papers*. 73–77.

KELLNER, A. W. A. 1995b. Description of a juvenile specimen of *Tapejara* (Pterodactyloidea, Tapejaridae) from Brazil. *Journal of Vertebrate Paleontology*, **15**(3) (supplement), 38–39A.

KELLNER, A. W. A. 1996. Description of new material of Tapejaridae and Anhangueridae (Pterosauria, Pterodactyloidea) and discussion of pterosaur phylogeny. PhD thesis, Columbia University. [Published by University Microfilms International].

KELLNER, A. W. A. 2001. A review of the pterosaur record from Gondwana. *Two Hundred Years of Pterosaurs. A Symposium on the Anatomy, Evolution, Palaeobiology and Environments of Mesozoic Flying Reptiles, Toulouse. Strata, Serie 1*, **11**, 51–53.

KELLNER, A. W. A. & LANGSTON, W. Jr. 1996. Cranial remains of *Quetzalcoatlus* (Pterosauria, Azhdarchidae) from Late Cretaceous sediments of Big Bend National Park, Texas. *Journal of Vertebrate Paleontology*, **16**, 222–231.

KELLNER, A. W. A. & MADER, B. J. 1996. First report of Pterosauria (Pterodactyloidea, Azhdarchidae) from Cretaceous rocks of Morocco. *Journal of Vertebrate Paleontology*, **16**(3) (supplement), 45A.

KELLNER, A. W. A. & MADER, B. J. 1997. Archosaur teeth from the Cretaceous of Morocco. *Journal of Paleontology*, **71**, 525–527.

KELLNER, A. W. A. & TOMIDA, Y. 2000. *Description of a New Species of Anhangueridae (Pterodactyloidea) with Comments on the Pterosaur Fauna from the Santana Formation (Aptian-Albian), Northeastern Brazil*. National Science Museum, Tokyo, Monographs, **17**, 1–135.

KNOLL, F. 2000. Pterosaurs from the Lower Cretaceous (?Berriasian) of Anoual, Morocco. *Annales de Paléontologie*, **86**, 157–164.

LANGSTON, W, Jr. 1981. Pterosaurs. *Scientific American*, **244**, 122–136.

LAWSON, D. A. 1975a. Pterosaur from the Latest Cretaceous of west Texas: discovery of the largest flying creature. *Science*, **187**, 947–948.

LAWSON, D. A. 1975b. Could pterosaurs fly? *Science*, **188**, 676–677.

LEWY, Z. A., MILNER, A. C. & PATTERSON, C. 1993. Remarkably preserved natural endocranial casts of pterosaur and fish from the Late Cretaceous of Israel. *Geological Survey of Israel, Current Research*, **7**, 31–35.

LONG, J. A. 1998. *Dinosaurs of Australia and New Zealand*. Harvard University Press, Cambridge, Massachusetts, 188 pp.

LÓPEZ-MARTÍNEZ, N., CANUDO, J. I. ET AL. 2001. New dinosaur sites correlated with Upper Maastrichtian pelagic deposits in the Spanish Pyrenees: implications for the dinosaur extinction in Europe. *Cretaceous Research*, **22**, 41–61.

MADER, B. J. & KELLNER, A. W. A. 1999. A new anhanguerid pterosaur from the Cretaceous of Morocco. *Boletim do Museu Nacional, Rio de Janeiro, Nova Serie, Geologia*, **45**, 1–11.

MARTILL, D. M. & FREY, E. 1998. A new pterosaur Lagerstätte in N.E. Brazil (Crato Formation; Aptian, Lower Cretaceous): preliminary observations. *Oryctos*, **1**, 79–85.

MARTILL, D. M., FREY, E., SADAQAH, R. M. & KHOURY, H. N. 1998. Discovery of the holotype of the giant pterosaur *Titanopteryx philadelphia* Arambourg, 1959 and the status of *Arambourgiania* and *Quetzalcoatlus*. *Neues Jahrbuch für Geologie und Paläontologie, Abhandlungen*, **207**, 57–76.

MONTEILLET, J., LAPPARTIENT, J. R. & TAQUET, P. 1982. Un ptérosaurien géant dans le Crétacé supérieur de Paki (Sénégal). *Comptes Rendus de l'Académie des Sciences, Paris*, **295**, 167–172.

MURRY, P. A., WINKLER, D. A. & JACOBS, L. L. 1991. An azhdarchid pterosaur humerus from the Lower Cretaceous Glen Rose Formation of Texas. *Journal of Paleontology*, **65**, 167–170.

NESOV, L. A. 1984. Upper Cretaceous pterosaurs and birds from Central Asia. *Palaeontological Journal*, **1984**(1), 38–49.

NESOV, L. A. 1997. Cretaceous non-marine vertebrates of northern Eurasia. *In*: GOLOVNEVA, L. B. & AVERIANOV, A. O. (eds), University of St. Petersburg, Institut of Earth Crust, Saint Petersburg, 1–218. [In Russian with English abstract]

OWEN, R. 1859. *Monograph of the fossil Reptilia of the Cretaceous formations. Supplement I, Pterosauria* (Pterodactylus). Palaeontographical Society, Monographs, **11**, 19 pp.

PADIAN, K. 1984. A large pterodactyloid pterosaur from the Two Medicine Formation (Campanian) of Montana. *Journal of Vertebrate Paleontology*, **4**, 516–524.

PADIAN, K. 1986. A taxonomic note on two pterodactyloid families. *Journal of Vertebrate Paleontology*, **6**, 289.

PADIAN, K. & SMITH, M. 1992. New light on Late Cretaceous pterosaur material from Montana. *Journal of Vertebrate Paleontology*, **12**, 87–92.

PADIAN, K., RICQLÈS, A. J. DE & HORNER, J. R. 1995. Bone histology determines identification of a new fossil taxon of pterosaur (Reptilia: Archosauria). *Comptes Rendus de l'Académie des Sciences, Paris, Serie IIa*, **320**, 77–84.

PETERS, D. 1997. A new phylogeny for the Pterosauria. *Journal of Vertebrate Paleontology*, **17**(3) (supplement), 69A.

PLIENINGER, F. 1901. Beiträge zur Kenntnis das Flugsaurier. *Palaeontographica*, **48**, 65–90.

PRICE, L. I. 1953. A presença de Pterosauria no Cretáceo superior do Estado da Paraíba. *Notas Preliminares e Estudos, Divisão de Geologia e Mineralogia*, **71**, 1–10.

SAYÃO, J. M. & KELLNER, A. W. A. 2001. New data on the pterosaur fauna from Tendaguru (Tanzania), Upper Jurassic, Africa. *Journal of Vertebrate Paleontology*, **21**(3) (supplement), 97A.

SEELEY, H. G. 1875. On an ornithosaurian (*Doratorhynchus validus*)from the Purbeck Limestone of Langton, near Swanage. *Quarterly Journal of the Geological Society, London*, **31**, 465–468.

SERENO, P. C., BECK, A. L., DUTHEIL, D. B., GADO, B., LARSSON, H. C. E., LYON, G. H., MARCOT, J. D., RAUHUT, O. W. M., SADLEIR, R. W., SIDOR, C. A., VARRICCHIO, D. D., WILSON, G. P. & WILSON, J. A. 1998. A long-snouted predatory dinosaur from Africa and the evolution of spinosaurids. *Science*, **282**, 1298–1302.

STEEL, L., MARTILL, D. M., KIRK, J. R. J., ANDERS, A., LOVERIDGE, R. F., FREY, E. & MARTIN, J. G. 1997. *Arambourgiania philadelphiae*: giant wings in small halls. *Geological Curator*, **6**, 305–313.

UNWIN, D. M. 1992. The phylogeny of the Pterosauria. *Journal of Vertebrate Paleontology*, **12**(3) (supplement), 57A.

UNWIN, D. M. 1995. Preliminary results of a phylogenetic analysis of the Pterosauria (Diapsida: Archosauria). *In*: SUN A. & WANG, Y. (eds) *Sixth Symposium on Mesozoic Terrestrial Ecosystems and Biota, Beijing, Short Papers*. 69–72.

UNWIN, D. M. & BAKHURINA, N. N. 2000. Pterosaurs from Russia, Middle Asia and Mongolia. *In*: BENTON, M. J., SHISHKIN, M. A., UNWIN, D. M. & KUROCHKIN, E. N. (eds) *The Age of Dinosaurs in Russia and Mongolia*. Cambridge University Press, Cambridge, 420–433.

UNWIN, D. & LÜ, J. 1997. On *Zhejiangopterus* and the relationships of pterodactyloid pterosaurs. *Historical Biology*, **12**, 199–210.

WELLNHOFER, P. 1970. Die Pterodactyloidea (Pterosauria) der Oberjura-Plattenkalke Süddeutschlands. *Bayerische Akademie der Wissenchaften, Mathematisch-Naturwissenchaften klasse, Abhandlungen*, **141**, 1–133.

WELLNHOFER, P. 1978. Pterosauria. *In*: WELLNHOFER, P. (ed) *Handbuch für Paläoherpetologie*. Gustav Fisher, Stuttgart, Teil 19, 82 pp.

WELLNHOFER, P. 1991a. Weitere pterosaurierfunde aus der Santana-Formation (Apt) der Chapada do Araripe, Brasilien. *Palaeontographica* A, **215**, 43–101.

WELLNHOFER, P. 1991b. *The Illustrated Encyclopedia of Pterosaurs*. Salamander, London, 192 pp.

WELLNHOFER, P. & BUFFETAUT, E. 1999. Pterosaur remains from the Cretaceous of Morocco. *Paläontologische Zeitschrift*, **73**, 133–142.

WIFFEN, J. & MOLNAR, R. E. 1988. First pterosaur from New Zealand. *Alcheringa*, **12**, 53–59.

WILLISTON, S. W. 1903. On the osteology of *Nyctosaurus* (*Nyctodactylus*), with notes on American pterosaurs. *Field Columbian Museum Publication, Geological Series*, **2**, 125–163.

YOUNG, C. C. 1964. On a new pterosaurian from Sinkiang, China. *Vertebrata Palasiatica*, **8**, 221–255.

Giant azhdarchid pterosaurs from the terminal Cretaceous of Transylvania (western Romania)

ERIC BUFFETAUT[1], DAN GRIGORESCU[2] & ZOLTAN CSIKI[2]

[1]*Centre National de la Recherche Scientifique, 16 cour du Liégat, 75013 Paris, France*
[2]*Laboratory of Palaeontology, Faculty of Geology & Geophysics, University of Bucharest, Boulevard N. Balcescu No. 1, 70111 Bucharest, Romania*

Abstract: Pterosaur remains from the Late Cretaceous of the Haţeg Basin of western Romania were reported by Nopcsa as early as 1899. Recent discoveries from the Late Maastrichtian Densuş-Ciula Formation include the giant azhdarchid *Hatzegopteryx thambema*, the holotype of which, consisting of skull elements and a humerus from the Vălioara locality, is described in detail. A very large femur from the same formation at Tuştea is also described. The systematic position of *Hatzegopteryx* is discussed. The wing span of *H. thambema* is estimated to be close to that of *Quetzalcoatlus northropi* (≥12 m), but its skull is especially robust and may have been remarkably long (≥ 2.5 m). The skull bones of *H. thambema* consist of a very thin outer cortex enclosing an inner meshwork of extremely thin trabeculae surrounding very numerous small alveoli, an unusual structure reminiscent of expanded polystyrene. This peculiar structure, combining strength with lightness, can probably be considered as an adaptation to flight in a very large animal, through reduction of skull weight.

Although the occurrence of pterosaurs in the Late Cretaceous non-marine deposits of Transylvania was reported as early as the end of the nineteenth century, no detailed descriptions have so far been given. The recent identification of extremely large pterosaurs in the Densuş–Ciula Formation of the Haţeg basin (Buffetaut *et al.* 2001) has led to the preliminary description of some of the remains as *Hatzegopteryx thambema* (Buffetaut *et al.* 2002), one of the largest known flying reptiles. The purpose of the present paper is to review the literature on Transylvanian pterosaurs, to describe the available specimens of giant pterosaurs from Transylvania in more detail, and to discuss some of the questions they raise.

Institutional abbreviations: FGGUB, Faculty of Geology and Geophysics of the University of Bucharest, Romania; TMM, Texas Memorial Museum, Austin, Texas, USA.

Nopcsa's Transylvanian pterosaurs

Pterosaur remains were reported from the Late Cretaceous of the Haţeg Basin of Transylvania by Franz Nopcsa as early as 1899 (Nopcsa 1899*a*, *b*). This mention was based on an identification by the Austrian palaeontologist Gustav von Arthaber, to whom the Hungarian geologist Gyula Halavats had submitted a collection of vertebrate fossils he had collected in the Haţeg Basin. In this regard, Nopcsa's German text (Nopcsa 1899*b*) is less clear than a footnote to his original Hungarian version (Nopcsa 1899*a*), which reads (translated from the Hungarian by Zoltan Csiki):

> 'Dr ARTHABER writes, besides other mentions, the following: ". . . no. 1. lower jaw of an Iguanodon-like animal . . . no. 3. Tooth fragment, very alike to those of Iguanodon Suessi, figured by Bunzel in his plate III. . . . no. 4. Three small vertebral centra . . . apparently from a small, pterosaur-like animal . . . the pterosaurs and ornithopodids went extinct at the end of the Cretaceous . . ." HALAVATS did not mention this determination in his paper "The Cretaceous of Ohaba-Ponor".'

In 1902, Nopcsa mentioned very scanty remains ('sehr dürftige Reste') of pterosaurs at Szentpéterfalva (the Hungarian name of the village known in Romanian as Sânpetru). In 1904, he further mentioned 'small characteristic fragments' that 'decidedly point to the occurrence of pterosaurs'. In 1905 (Nopcsa 1905, p. 171), he was slightly more specific and mentioned vertebral centra of an indeterminate pterosaur, as well as an isolated sacrum referred to as a 'Coeluride (?)'. There seems to be no further mention of a 'coelurid' sacrum in any of the dinosaur lists provided by Nopcsa (e.g. Nopcsa 1914, 1915, 1923), and it seems possible that this particular specimen was later reinterpreted as a pterosaurian notarium because, in 1914, Nopcsa reported that pterosaurs were represented at Szenpéterfalva by an isolated notarium and two teeth found together with unidentifiable fragments of hollow bones.

From: BUFFETAUT, E. & MAZIN, J-M. (eds) 2003. *Evolution and Palaeobiology of Pterosaurs*. Geological Society, London, Special Publications, **217**, 91–104. 0305–8719/03/$15

In 1915, Nopcsa remarked that the Haţeg pterosaurs were the youngest in the whole world (which was probably true at the time, considering their Maastrichtian age) and considered that they were reminiscent of *Ornithodesmus*, an Early Cretaceous pterosaur from the United Kingdom (now properly called *Istiodactylus*; see Howse *et al.* 2001), but gave neither a complete list nor a description of the available material.

In 1923, Nopcsa alluded again to the pterosaur material from Transylvania, remarking that it was in a poor state of preservation, but 'seemed to be related to the Ornithocheiridae, especially to *Ornithodesmus*' (Nopcsa 1923, p. 103). He added that the Romanian pterosaurs were therefore 'more primitive than the North American Pteranodontidae', which was in agreement with his general interpretation of the Haţeg vertebrate assemblage as an endemic and archaic fauna. As, again, no list of specimens or description was given, it is not clear on what kind of material the comparison with *Ornithodesmus* was based. At that time, *Ornithodesmus* was known mainly from a sacrum from the Wealden of the Isle of Wight originally described by Seeley (1887, 1901), which was later shown by Howse & Milner (1993) to belong to a maniraptoran theropod, and from a partial pterosaur skeleton, also from the Wealden of the Isle of Wight, described by Hooley (1913), which Howse *et al.* (2001) have redescribed as *Istiodactylus latidens* (see Howse *et al.* 2001 for a review of the changing interpretations of *Ornithodesmus*). As the only reference given by Nopcsa in 1923 in support of his opinion was Seeley's 1887 paper, it is possible that his interpretation was mainly based on the specimen he had probably interpreted successively as a coelurid sacrum and a pterosaurian notarium. In 1997, Jianu *et al.* erroneously asserted that Nopcsa had considered the Transylvanian pterosaur to be '*Ornithocheirus*?' in his 1926 paper on the Gosau reptiles (Nopcsa 1926*a*). In fact, this paper deals exclusively with the reptile fauna from the Campanian of Muthmannsdorf, in Austria, which does include pterosaur remains referred to *Ornithocheirus* (see Wellnhofer 1980), and makes no mention of Transylvanian material. Nopcsa's last mention of the Transylvanian pterosaurs appears to be in his *Osteologia reptilium recentium et fossilium* (1926*b*, p. 324), where he briefly repeated the attribution to *Ornithocheirus*-like ('*Ornithocheirus*-artigen') pterosaurs given in his 1923 paper.

Thereafter, Nopcsa's pterosaur material from the Haţeg Basin was long considered lost, until at least part of it was rediscovered in 1995 at the Magyar Allami Földtani Intezet in Budapest by Jianu and Weishampel (Jianu *et al.* 1997). It is hoped that a full description of this material will eventually shed some light on what kind of pterosaur specimens Nopcsa had at his disposal but never described. Jianu *et al.* (1997) also reported the recent discovery of more pterosaur material (including a notarium, a right humerus and a left femur) in the Haţeg Basin. Although they did not give a detailed description, they indicated that the notarium, with a supraneural plate, and the humerus, with a warped deltopectoral crest, could be referred to the family Pteranodontidae, but noted that the absence of a pneumatic foramen in the humerus was a plesiomorphic feature suggesting a basal position among pteranodontids.

Although very little information is available about the material reported by Nopcsa and by Jianu *et al.*, there is nothing to suggest that it belongs to unusually large pterosaurs. On the contrary, the bones we describe in the present paper are remarkable for their enormous size, which places them among the largest known pterosaurs.

Systematic palaeontology

Order Pterosauria
Family Azhdarchidae
Genus *Hatzegopteryx* Buffetaut *et al.* 2002
Species *Hatzegopteryx thambema* Buffetaut *et al.* 2002

Holotype. Fragments of a skull and an associated incomplete humerus (FGGUB R1083).

Horizon. Upper part of the middle member of the Densuş–Ciula Formation, Maastrichtian.

Locality. Vălioara, Hunedoara county, western Romania.

Referred material. Large femur (FGGUB R.1625) from the Densuş–Ciula Formation at Tuştea, western Romania.

Geological setting

The giant pterosaur bones described below come from Maastrichtian overbank deposits at two distinct localities (Fig.1) in the Haţeg Basin (Hunedoara County, western Romania).

The type of *Hatzegopteryx thambema*, consisting of associated skull elements and a humerus, was collected by one of us (D. G.) during a student field trip in 1978 near the village of Vălioara. The occurrence of Late Cretaceous vertebrate-bearing sediments in the vicinity of Vălioara (Hungarian spelling: Valiora) has been known since Nopcsa's days (Nopcsa 1905, 1914; see also Kadic 1916, 1917). The pterosaur remains found in 1978 were not part of a multi-taxon bone accumulation, but were found together in chocolate-coloured siltstones, unassociated with other bones, and undoubtedly belong to a single individual. The specimens come from the upper part of the middle member of the Densuş–Ciula Formation.

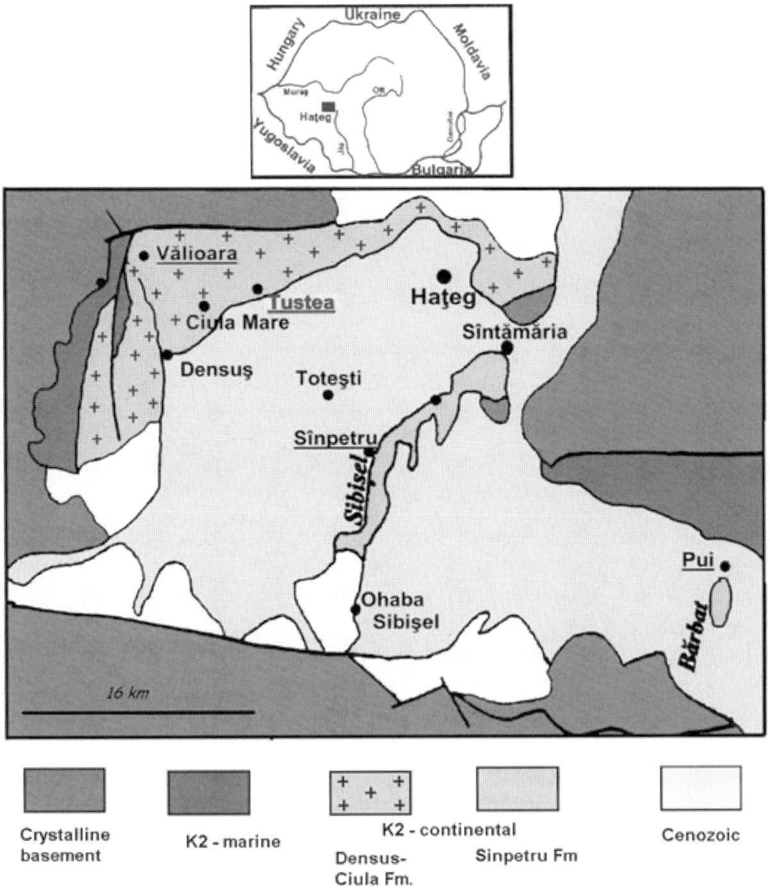

Fig. 1. Simplified geological map of the Haţeg basin of Transylvania (western Romania, see small location map), showing the distribution of Late Cretaceous (K2) formations. The main vertebrate localities are underlined. Remains of giant pterosaurs described in the present paper have been found in the Densuş–Ciula Formation at Vălioara and Tuştea, in the northwestern part of the basin. Nopcsa's pterosaur material was from Sânpetru (Szentpéterfalva).

More recently (1998), a large femur was found in red siltstones at the Tuştea locality, which has yielded an assemblage including dinosaur eggs (Grigorescu *et al.* 1990) and remains of anurans, albanerpetontids, turtles, crocodilians, dinosaurs and multituberculates (Grigorescu *et al.* 1999).

Both the Vălioara and Tuştea localities are in the Densuş–Ciula Formation (see Grigorescu 1992; Grigorescu *et al.* 1999), which was usually considered as Late Maastrichtian (Weishampel *et al.* 1991; Grigorescu 1992). This age assignment was based partly on the age of underlying marine sediments, which were supposed to be as young as Early Maastrichtian; however, recent studies suggest that they extend only up to the Late Campanian (Grigorescu & Melinte in prep.). On the other hand, according to Antonescu *et al.* (1983), the diverse palynoflora from the continental beds of the Haţeg Basin indicates that they are Late Maastrichtian, which is also supported by gastropods, so a Late Maastrichtian age can be accepted for the Densuş–Ciula Formation. However, recently doubts were raised concerning the biostratigraphic significance of the Maastrichtian *Pseudopapilopollis praesubhercynicus* assemblage as indicating the Late Maastrichtian (Antonescu pers. comm. 2001)

The holotype of *Hatzegopteryx thambema*

The best specimen of giant pterosaur currently available from the Haţeg Basin, consisting of fragments of a skull and an associated incomplete humerus (FGGUB R1083) from Vălioara (see above), was described as a new taxon, *Hatzegopteryx thambema* (Buffetaut *et al.* 2002). Interestingly, the skull remains are so large and so robustly built that they were first briefly described as belonging to a large

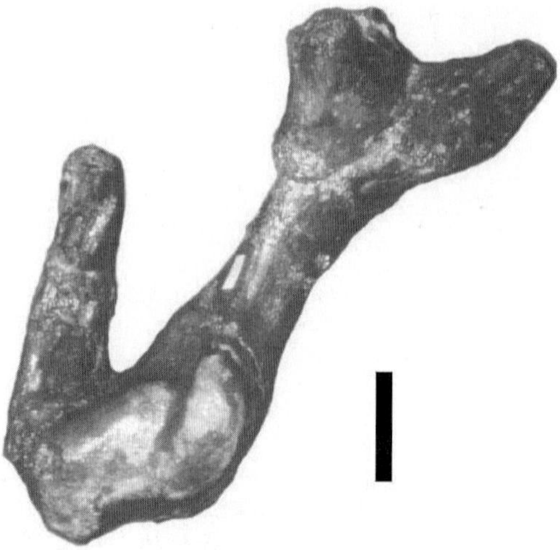

Fig. 2. Holotype of *Hatzegopteryx thambema*, FGGUB R1083, right half of posterior part of palate with suspensorium. Scale bar 50 mm.

Fig. 3. Holotype of *Hatzegopteryx thambema*, FGGUB R1083, occipital face of skull. Scale bar 50 mm.

theropod by Weishampel *et al.* (1991). However, as already noted (Buffetaut *et al.* 2001, 2002), and as shown below, they exhibit many characteristic features demonstrating their pterosaurian nature, which is fully confirmed by the characters of the typically pterosaurian-associated humerus.

The skull remains

Skull elements from Vălioara comprise the right half of the posterior part of the palate, with the right suspensorium, and the occipital region (Figs 2–4). Unfortunately, no contact can be found between those two groups of bones.

Palate and suspensorium. The available material consists of the right half of the palate at the level of the jaw articulation, comprising parts of the jugal, quadratojugal, quadrate, and pterygoid (Figs 2 & 4b). Those bones are firmly fused together, so that few sutures can be distinguished. The most striking element is the quadrate. Most of its shaft, including the region which contacted the squamosal, is missing, but the articular region for the lower jaw is well preserved; a ridge is present on the rostrodorsal surface of what is left of the shaft. The jaw articulation consists of two large oval condyles, the medial being narrower and slightly more rostral than the lateral, separated by an oblique groove, which results in a typically helical structure. A similar condition has long been known in *Pteranodon* (Plieninger 1901; Eaton 1910; Wellnhofer 1980; Bennett 2001a), and it is present in many advanced pterodactyloids (Kellner & Tomida 2000). Its functional significance has been discussed in great detail by Wellnhofer (1980), who concluded that the helical jaw joint permitted a very wide gape. Interestingly, a helical jaw joint is also present, with various modifications, in many theropods, such as *Dromaeosaurus* (Colbert & Russell 1969), *Allosaurus* (Madsen 1976), *Ceratosaurus* (Madsen & Welles 2000), *Baryonyx* (Charig & Milner 1996) and *Spinosaurus* (E.B. pers. obs. on unpublished material from Morocco). However, the general shape of the quadrate of the Vălioara specimen is quite unlike that of a theropod quadrate; in particular, there is no evidence of the anteromedial wing-like pterygoid flange seen in all theropods (see below). The condyles of the quadrate of *H. thambema* appear to be more bulbous and rounded than those of *Pteranodon*. Although in both taxa the jaw articulation is helical, there are notable differences between *H. thambema* and *Quetzalcoatlus* sp. (see below).

Laterally, the quadrate is flanked by a flat blade of bone, which appears to be formed mainly by the jugal. The limits of the quadratojugal are not very clearly visible, but it seems to have been a thin sliver of bone sandwiched between the quadrate and the jugal, and largely covered by the latter. There is no indication that the quadratojugal took part in the jaw articulation, contrary to the condition in *Pteranodon* (Bennett 2001a).

The jugal is incompletely preserved. Rostrodorsally, it forms the lower rim of a vast opening in the lateral face of the skull, which in all likelihood is the nasopreorbital opening. Below this opening, it forms a laterally compressed bar, the anterior end of which is broken. Caudodorsally, there is no evidence

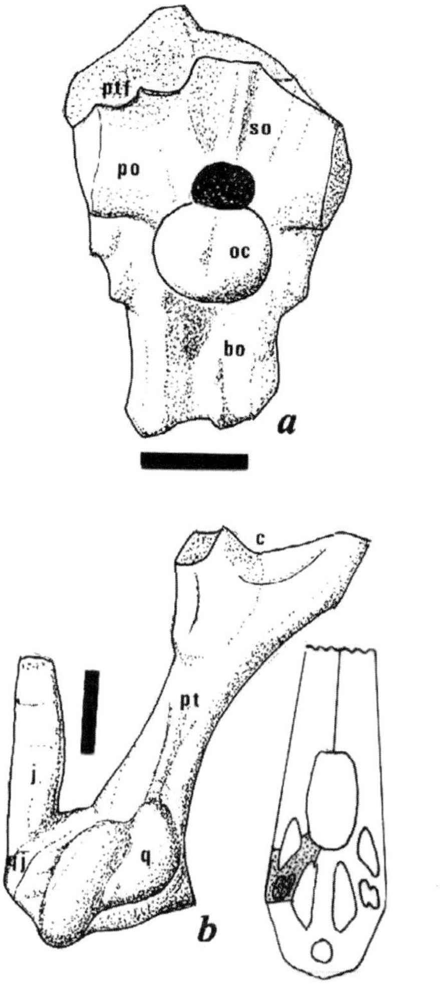

Fig. 4. Holotype of *Hatzegopteryx thambema*, skull elements: (**a**) occipital face, (**b**) right half of posterior part of palate with suspensorium (insert shows the position of this element in a pterosaur skull). (After Buffetaut *et al.* 2002). bo: basioccipital; c: choana; j: jugal; oc: occipital condyle; po: paroccipital process; pt: pterygoid; ptf: post-temporal fenestra; q: quadrate; qj: quadratojugal; so: supraoccipital. Scale bars 50 mm.

of the orbital rim, so that the exact position of the orbit (high on the skull as in *Pteranodon* or low as in *Quetzalcoatlus*) is uncertain.

Rostromedially, the pterygoid process of the quadrate is a robust bony rod which emerges rostrodorsally to the articular condyles and is directed towards the mid-line of the skull. The limit between the quadrate and the pterygoid is not clear because of strong fusion of the bones; no suture can clearly be seen, and several breaks obscure the situation.

Together with the rostrodorsal process of the quadrate, the pterygoid forms a strong bony rod (diameter up to 50 mm) which bears a well-marked ventral ridge (issuing from the medial condyle of the quadrate) and two, more or less parallel, dorsal ridges. Rostrally, the pterygoid becomes forked, with a broad medial branch and a more slender dorsolateral branch. These two branches form the posterior rim of an opening which must be the choana. The caudal edge of this opening is bevelled. The pterygoid rises rostrally relative to the level of the jaw articulation, as in the *Pteranodon* specimen described by Plieninger (1901). As a result, the posterior part of the palate must have been highly vaulted.

Occipital region. This part of the skull has undergone a slight deformation, but very little crushing (see Figs 3 & 4a). The anterior part of the specimen is a poorly preserved mass of highly cancellous bone showing no details. The occipital face, however, is fairly well preserved. The bones are strongly fused together, and sutures are not visible. Ventrally, the basioccipital forms a rectangular bony plate, which apparently merges with the basisphenoid rostrally. The surface of this plate is markedly concave and rugose, probably indicating the attachment of strong neck muscles. More dorsally, the basioccipital apparently forms most or all of the occipital condyle. The condyle is large (diameter 55 mm), hemispherical, well defined and strongly protruding caudally, without a well-marked 'neck'. The dorsal surface of the condyle is flat and merges rostrally with the floor of the foramen magnum. The foramen magnum is a circular opening. Its diameter (43 mm) is smaller than that of the occipital condyle, which seems to be unusual in pterosaurs (see, for instance, the condition in the uncrushed skull of *Tapejara wellnhoferi* described by Kellner 1996), although the same condition is observed in a *Pteranodon* specimen described by Plieninger (1901) and in *Anhanguera piscator* (Kellner & Tomida 2000). The exoccipitals form the paroccipital processes, which are incompletely preserved but appear to have been robust and wing-shaped. On the left side, which is better preserved, the dorsolateral edge of the paroccipital process forms the ventral rim of a large post-temporal fenestra. This fenestra opens rostrally into a vast supratemporal fenestra, the dorsal and medial sides of which are partially preserved. The presence of a vast post-temporal fenestra shows that this occiput cannot belong to a theropod dinosaur, as was previously suggested (Weishampel *et al.* 1991): in dinosaurs generally, the supratemporal fenestrae are reduced (Romer 1956), often being no more than a narrow slit, whereas in pterosaurs they are relatively large (see, for instance, Wellnhofer 1985; Kellner & Tomida 2000; Bennett 2001a) Lateral to the occipital condyle, there seems to be a foramen on the exoccipital, which is probably the opening for cranial nerves usually seen in this position, but this region is poorly preserved. Dorsally to the foramen magnum, the supraoccipital forms a tall plate, with a dorsoventrally and transversally concave caudal surface, bearing a distinct median ridge (for the insertion of neck muscles).

Few uncrushed pterosaur occiputs have been described, which restricts possible comparisons (e.g. *Pteranodon* occiputs from the Niobrara Formation have all suffered considerable compression, so that reconstructions must be considered as tentative). When compared with a relatively early pterosaur, such as the Liassic *Parapsicephalus purdoni* (Newton 1888), the occiput of *H. thambema* appears remarkably high, although the general morphology is similar. *Tapejara wellnhoferi*, from the Early Cretaceous Santana Formation of Brazil (Kellner 1996), also shows a lower occiput. However, comparison with other relatively well-preserved skulls from the Santana Formation reveals considerable resemblances. Comparisons with the well-described types of *Anhanguera santanae* (Wellnhofer 1985) and *Anhanguera piscator* (Kellner & Tomida 2000) are especially revealing. Although the Brazilian skulls are considerably smaller, the comparison reveals significant morphological resemblances. The tall supraoccipital plate with a well-marked median ridge is found in all three specimens. *A. piscator* shows a broad and concave basioccipital plate, as in *H. thambema*. The relative positions and proportions of the foramen magnum and the post-temporal fenestrae are also similar. The occiput of *H. thambema* can thus be considered as typically pterosaurian, and there is definitely no reason to suspect that it might belong to a theropod. On the other hand, and partly because so few comparable specimens are available, few features can be seen that could be considered as characteristic of the taxon *H. thambema*. The general robustness of the occiput is remarkable, but it may simply be a consequence of its very large size (for comparison, the diameter of the occipital condyle in the *Pteranodon* specimen described by Plieninger [1901] is 14.5 mm, versus 55 mm in the Vălioara pterosaur).

The humerus

The left humerus of the type of *H. thambema* is represented by two fragments which unfortunately cannot be fitted together because of the lack of contacts. One consists of a large part of the proximal articular head, while the other corresponds to most of the proximal half of the shaft, with the deltopectoral crest and the base of the medial process.

The articular head (Fig. 5) shows a well-rounded articular surface, which is roughly oval in proximal view. It is broken laterally, showing that in cross-section the articular surface is semi-circular. Medially, the surface curves ventrodistally in the direction of the medial process. Ventrally, the articular surface is sharply separated from the more distal part of the bone by a strongly marked, curved 'step'. Dorsally, the convex edge of the articular surface forms a well-defined lip which slightly overhangs the shaft. This articular head is clearly different from that of *Pteranodon* (Bennett 2001*a*), in which there is a greater expansion of the articular surface onto the dorsal surface of the bone, and a depressed area which is not present in the Vălioara specimen. It closely resembles the proximal articular area of the humerus of azhdarchids, such as *Bennettazhia* (Gilmore 1928; Bennett 1989), the azhdarchid from Montana described by Padian and Smith (1992) and *Quetzalcoatlus* (casts and photographs kindly made available by W. Langston Jr, J. Cunningham and E. Frey) in its dorsoventral thickening, its mediolateral extent, and the development of a ventral 'step' and a dorsal 'lip'.

Fig. 5. Holotype of *Hatzegopteryx thambema*, proximal articular head of left humerus in (**a**) proximal, (**b**) ventral and (**c**) dorsal views. Scale bar 50 mm.

The internal structure of the bone at the level of the articular head is similar to that of the skull bones (see below). A thin outer cortex, up to 1 mm in thickness, covers a mass of highly cancellous bone, con-

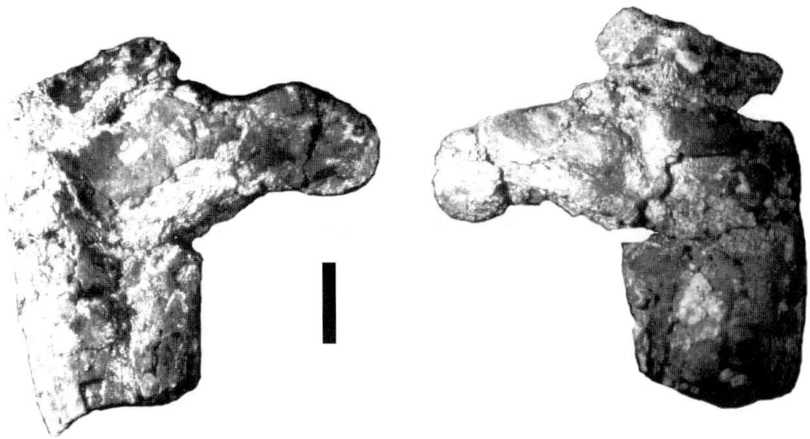

Fig. 6. Holotype of *Hatzegopteryx thambema*, part of left humerus, in (**a**) ventral and (**b**) dorsal views, showing the well-developed unwarped deltopectoral crest. Scale bar 50 mm.

sisting of numerous elongated alveoli separated by extremely thin bony walls.

The measurements of the articular head are as follows: maximum thickness 84 mm; mediolateral width, as preserved, 165 mm.

The specimen is thus much larger than the corresponding part in *Quetzalcoatlus* sp. (thickness of the proximal articular head in humerus TMM 41544–9 is 25 mm). Its dimensions are very similar to those of the humeral head in *Quetzalcoatlus northropi* (specimen TMM 41450–3), which is 81.3 mm thick, and has a mediolateral width of 190 mm (measurements provided by James Cunningham).

The larger humerus fragment consists of a 236 mm-long proximal section of the shaft (Fig. 6) with part of its processes (relatively complete deltopectoral crest, largely destroyed medial process), which has undergone very little crushing. Its surface, however, is not well preserved, the outer cortex having been destroyed in various places, thus exposing the internal bony structure. The shaft is very robust, with a roughly D-shaped cross-section, the dorsal surface of the bone being strongly convex, whereas the ventral surface is slightly concave. Only the base of the medial process is preserved, showing that it was at a marked angle to the deltopectoral crest. The deltopectoral crest is strongly developed. Its outline is slightly uncertain, because its proximal and distal edges are incompletely preserved. However, it can be seen that it was gently curved and at right angles to the longitudinal axis of the shaft. Its shape was therefore quite different from that of the 'warped' crest of pteranodontids (see Bennett 1989) and similar to the condition in azhdarchids. There is no thickening of the extremity of the crest. On its ventral surface, striations probably corresponding to a muscle insertion can be seen (muscle scars in a similar position have been described in *Pteranodon* by Bennett 2001*a*). In areas where the surface of the bone has been damaged, it can be seen that a thin cortex (2–4 mm in thickness) encloses a highly cancellous bone tissue, with densely packed alveoli separated by paper-thin trabeculae only a fraction of a millimetre in thickness. The orientation of the trabeculae parallels that of the crest.

The shaft is broken some distance beyond the distal insertion of the deltopectoral crest (diameter at the level of the break is 90 mm). The cross-section shows that the bone is hollow, with an outer cortex varying in thickness from 4 to 7 mm, which is suddenly replaced by highly cancellous bone, with large irregular spaces separated by extremely thin trabeculae; the trabeculae become less dense toward the centre of the shaft, which is almost devoid of them.

Systematic position of *Hatzegopteryx thambema*

In this respect, there is little to add to what has already been published (Buffetaut *et al.* 2001, 2002). *Hatzegopteryx thambema* can be referred to the family Azhdarchidae mainly on the basis of several characters of its humerus. Its unwarped deltopectoral crest separates it from the Pteranodontidae. However, an unwarped deltopectoral crest is found in many groups of pterosaurs and can be considered as the plesiomorphic condition. Derived characters of the azhdarchid humerus have been listed by Padian and Smith (1992); they include a great thickening of the articular head in the palmar–anconal plane, a character clearly present in *H. thambema*, in which, as mentioned above, the articular head is massive and very reminiscent of the condition in

Quetzalcoatlus. The lack of a pronounced bulbous expansion at the distal terminus of the deltopectoral crest is another derived azhdarchid character listed by Padian and Smith which is observable in *H. thambema*. According to Padian and Smith (1992, p. 90), 'the proximal margin of the deltopectoral crest that rises to meet the head of the humerus is less concave' than in primitive pterosaurs. This seems to be the case in *H. thambema*, too, but the exact condition is slightly uncertain because this area is poorly preserved. To sum up, the humerus of *H. thambema* shows great resemblances to that of previously described azhdarchids, especially *Quetzalcoatlus*, and because of this we refer the Romanian pterosaur to the family Azhdarchidae.

However, there are few distinctive features in the humerus of *H. thambema* that can be used to separate it from other azhdarchids. In this respect, the skull elements are more useful, although comparisons are made difficult by lack of information about the posterior regions of the skull in most azhdarchids: rostral fragments have been described in *Azhdarcho lancicollis* (Nesov 1984) and in the Laño azhdarchid (Buffetaut 1999), but they are useless for comparisons with the Romanian material. In *Zhejiangopterus linhaiensis*, complete skulls are known, but they are crushed flat and visible only in lateral view (Cai & Wei 1994; Unwin & Lü 1997), so that no useful comparisons can be made with the skull bones from Vălioara. The only azhdarchid skull material hitherto published that warrants significant comparisons with *H. thambema* is that of the small morph of *Quetzalcoatlus*, *Q.* sp. (Kellner & Langston 1996), which unfortunately does not include the occipital region. However, the quadrate is preserved in *Quetzalcoatlus* sp., and it differs in several respects from that of *H. thambema*. As described and figured by Kellner and Langston (1996, p. 226), 'the lateral condyle consists of anteriorly and posteriorly facing flat surfaces, set off from one another by a sharp edge instead of the rounded surface of most reptilian quadrates'. This is very different from the condition in *H. thambema*, where both condyles are smoothly rounded. Furthermore, in *Quetzalcoatlus* sp., 'the posterior joint surface is contained in a sharply excavated depression whose medial edge is formed by a dorsolateral extension of the thread of the medial condyle' (Kellner & Langston 1996, p. 226); this depression matches a corresponding process on the posterolateral edge of the glenoid fossa of the mandible. No such depression is visible in *H. thambema*.

It thus appears that the pterosaur from Vălioara can safely be assigned to the Azhdarchidae on the basis of humerus morphology, but that it is different from the only azhdarchid in which the jaw articulation is known, viz. *Quetzalcoatlus* sp. The general robustness of the skull bones may also be a distinctive feature. We therefore consider that the pterosaur material from Vălioara can be considered as the type of a distinct genus and species, to which the name *H. thambema* has been applied (Buffetaut *et al.* 2002), with the following diagnosis:

> 'A very large azhdarchid pterosaur with a robustly built and posteriorly broad skull. Helical articulation of the quadrate with the mandible massive, with smoothly rounded rather than angular condyles, and no notch posterior to the lateral condyle'.

A note on the size of *Hatzegopteryx thambema*

Estimating the size of a pterosaur (usually expressed in terms of wing span) on the basis of fragmentary remains is always fraught with difficulties. In the case of *H. thambema*, the incomplete humerus is of course the best guide to wing span estimates, while the cranial elements can be used to estimate the dimensions of the skull.

The length of the humerus can only be estimated, since the distal part is missing. Moreover, the proximal articular head cannot be fitted to the shaft, which introduces a further uncertainty. However, as noted above, the thickness of the proximal articular head is very similar in *H. thambema* and in *Quetzalcoaltus northropi*; as this part of the humerus is very similar morphologically in both forms, it seems legitimate to assume that the humerus as a whole was of roughly the same size. The available shaft portion from Vălioara is 236 mm long. It is broken distal to the deltopectoral crest, apparently somewhat proximal to mid-length. Considering that a few centimetres should be added proximally to account for the articular head, it is likely that the complete humerus was somewhat over 500 mm in length. Several slightly different lengths have been published for the humerus of *Q. northropi* (TMM 41450–3), but, according to W. Langston Jr (pers.comm.), 544 mm is the correct length. It therefore appears that the humerus of *H. thambema* was very close in size to that of *Q. northropi*. This suggests that those azhdarchid pterosaurs had similar wing spans. The wing span of *Q. northropi* has been the subject of much speculation, with early estimates up to 15.5 m or even 21 m (Lawson 1975). Subsequently, Langston's revised estimate of 11–12 m, based on the proportions of *Quetzalcoatlus* sp. (Langston 1981), has generally been followed (Wellnhofer 1991). Frey & Martill (1996), in their discussion of the wing span (estimated at 12 m) of *Arambourgiania philadelphiae*, suggest 11 m for *Q. northropi*. On the basis of humerus length, and supposing that the proportions of the wings of the

Vălioara pterosaur were similar to those of *Quetzalcoatlus*, it seems likely that *H. thambema* also had a wing span of 11–12 m.

As to the skull of *H. thambema*, its width at the level of the quadrate articulation can be estimated easily enough, since the right half of the palatal region at this level is almost completely preserved, the pterygoid part of the rim of the choana being apparently broken close to the sagittal plane of the skull. The width of the preserved part of the palate being 250 mm, the total width of the skull at the level of the jaw articulation was close to 500 mm. This is considerable by pterosaurian standards. The skull of *Pteranodon* reconstructed by Bennett (2001a, fig. 2) is only about 80 mm in width at this level, for a total length (minus the occipital crest) of about 750 mm. In the small morph of *Quetzalcoatlus*, the lower jaw at the level of the articulation is 120 mm wide, and this suggests that the skull of *Q. northropi* was about 240 mm wide at this level, this being a minimum value (W. Langston Jr pers. comm.). If this estimate is correct, the skull of *H. thambema* may have been about twice the width of that of *Q. northropi.*

Estimating the length of the skull of *H. thambema* on the basis of its width is difficult, because the width/length ratio is unknown in this species, and rather poorly known in pterosaurs in general. In fact, in many pterosaurs, skull width can only be roughly estimated because of the considerable crushing which has affected most specimens. This applies in particular to *Pteranodon* specimens from the Niobrara Formation (Williston 1892). It seems reasonable to assume that *H. thambema* was a long-snouted form, since this is the condition in other azhdarchids in which the rostrum is known (*Quetzalcoatlus* sp., *Azhdarcho lancicollis, Zhejiangopterus linhaiensis*, Laño azhdarchid). Of course, some pterosaurs, such as batrachognathids and tapejarids, had short rostra, but there is no special reason to believe that *Hatzegopteryx* is closely related to them. A further problem is that skull width is not known with any accuracy in most described azhdarchids, for the simple reason that few azhdarchid skulls are known. Several skulls of *Z. linhaiensis* are known (Cai & Wei 1994; Unwin & Lü 1996), but they are crushed flat on slabs of limestone and their width cannot be measured. A length estimate can be based on *Quetzalcoatlus* sp., in which the lower jaw is 120 mm wide at the level of the articulation with the skull (W. Langston Jr pers. comm.), for a total skull length of about 1000 mm (Kellner & Langston 1996): if the same proportions are assumed for *H. thambema*, the skull would have been more than 4 m long, which seems enormous; this may suggest undetected mediolateral compression in specimens of *Quetzalcoatlus* sp., or different skull proportions in *Quetzalcoatlus* and *Hatzegopteryx*. We have attempted other estimates based on two fairly complete non-azhdarchid pterosaur skulls which appear to have suffered little mediolateral compression, viz. the type of *Anhanguera piscator* from the Santana Formation (Kellner & Tomida 2000) and the well-preserved *Nyctosaurus* skull from the Niobrara Formation described by Williston (1902) and Huene (1914). Assuming that the skull of *H. thambema* had a width/length ratio similar to that of *Anhanguera piscator*, it may have been 2.75 m long, while the estimated skull length based on the proportions of *Nyctosaurus gracilis* is 2.9 m. All those estimates must be considered as tentative, but there seems to be no doubt that *H. thambema* had a very long skull, which probably places it among the non-marine vertebrates with the longest skulls, together with some of the large ceratopsians such as *Torosaurus* and *Triceratops*, in which the skull was up to 2.4 m long (Dodson 1996).

The size estimates based on the humerus and on the skull elements thus suggest a pterosaur with a wing span similar to that of *Q. northropi*, but with a skull which probably was broader posteriorly, and longer. It should be remembered, however, that nothing is known of the skull of *Q. northropi*, so that reconstructions have to be made by scaling up the smaller *Quetzalcoatlus* sp. In view of the numerous uncertainty factors involved in the comparison, it is difficult to demonstrate that *Hatzegopteryx* was larger than *Quetzalcoatlus*, which in any case would be of moderate scientific interest. Both were clearly extremely large pterosaurs, and only further discoveries can show whether there were significant size differences between the two taxa. If pterosaur growth was characterized by variable rates and delayed maturation, as suggested by Unwin (2001), the question of maximum size in the giant pterosaur taxa becomes to some extent irrelevant, because 'giant' individuals could occasionally appear when unusually favourable conditions occurred. However, Unwin's suggestion that 'pterosaurs retained a non-derived sauropsid growth mechanism', which was 'strikingly different' from that of neornithine birds (Unwin 2001, p. 110A), does not seem to be consistent with recent histological evidence suggesting that pterosaur growth rates were 'much more like those of birds than of typical reptiles' (Ricqlès *et al.* 2000, p. 380).

Internal structure of the bones

As mentioned above, in both the humerus fragments and the skull elements of the type of *H. thambema* the outer cortex of the bones has been damaged in many places, thus exposing their internal structure, which is all the more clearly visible because the specimens have undergone virtually no crushing. In this respect, their preservation is quite different from

Fig. 7. Close-up of bone structure in a bone fragment from the holotype of *Hatzegopteryx thambema*, showing the thin outer cortex (upper right) and the inner meshwork of bony trabeculae enclosing elongated alveoli. Scale bar 30 mm.

that of many other very large pterosaur specimens, especially those from the Niobrara Formation of the United States, which are usually crushed flat and therefore show few details of internal structures (Williston 1892; Bennett 2001a).

As already mentioned (Buffetaut *et al.* 2002), both the proximal part of the humerus and the skull bones of the type of *H. thambema* show a peculiar structure. Under a very thin outer cortex (which is usually no more than 1 mm in thickness), the bones consist of a dense meshwork of paper-thin bony trabeculae which enclose relatively small empty spaces (Fig. 7). These alveoli are very numerous, closely packed and elongate, being usually a few millimetres in width and up to more than 10 mm in length. This structure is rather reminiscent of expanded polystyrene, and probably resulted in a fairly rigid but light material.

Contrary to what one might think, relatively little is known of the internal structure of the skull bones of pterosaurs, because of the above-mentioned crushing of many specimens, which obscures details of the bony structure. Hollow skull bones have been reported in several taxa, but relatively little is known about the size and shape of the vacuities inside the bones. According to Bennett (2001a, p. 9), in *Pteranodon*, 'the majority of skull bones in the skull consist of two thin plates connected by internal reticulate reinforcing ridges' (see also Williston 1892), which seems rather different from the dense network of trabeculae enclosing small aveoli seen in *Hatzegopteryx*. Extensive pneumatization of skull bones has been described in detail in well-preserved specimens of *Tapejara* and *Anhanguera* from the Santana Formation of Brazil (Kellner 1996), but in those forms, the vacuities appear to be much larger, relative to the size of the skull, than in *H. thambema*.

The peculiar internal structure of the bones of *Hatzegopteryx* should probably be understood in terms of weight reduction in a large flying vertebrate. Pterosaurs in general were lightly built and many of their bones were pneumatized. What is surprising in the skull of *H. thambema* is its great robustness, which led to its misinterpretation as a dinosaur skull. At first sight, it may seem difficult to understand how a pterosaur with such a large and stoutly built skull could fly – unless the weight of the skull was somehow substantially reduced. We suggest that the very large size and considerable robustness of the skull of *H. thambema* were compensated for by the peculiar honeycombed structure of the skull bones, which combined strength with lightness. Although it is difficult to estimate the weight of the skull of *H. thambema*, or the weight of the whole animal, on the basis of the available material, it is very likely that the total bony mass of the skull was considerably reduced because most of the volume of its bones was filled up by vacuities rather than by the extremely thin intervening bony partitions. The relatively small size of the vacuities, by comparison with the larger pneumatic spaces described in the Santana pterosaurs, is possibly linked to the very large size of *H. thambema*: a structure consisting of a large number of densely packed small alveoli may have been more resistant than large, thin-walled bones containing a smaller number of larger vacuities. Be that as it may, it is difficult at the moment to know how widespread the kind of bony structure seen in *Hatzegopteryx* was among large pterosaurs. What is known of skull bone structure in *Pteranodon* suggests relatively large spaces rather than small alveoli (Bennett 2001a). No data are available about the internal structure of skull bones in azhdarchids other than *Hatzegopteryx*: it is apparently obscured by crushing in both *Zhejiangopterus lianhaiensis* and *Quetzalcoatlus* sp.

The femur from Tuştea

The only pterosaur specimen so far found at Tuştea (in 1998) is a large femur (FGGUB R.1625, Fig. 8), apparently from the left side, lacking both articular ends, but otherwise fairly well preserved, the shaft being almost uncrushed. The shaft is slightly bowed and subcircular to D-shaped in cross-section. It is hollow and thin-walled, the thickness of the bony wall varying from 1 to 2.5 mm, for a maximum diameter of 26 mm.

This bone shows few reliefs or muscle scars, except for a fairly well-marked oblique ridge arising in the proximal region of the shaft and extending to

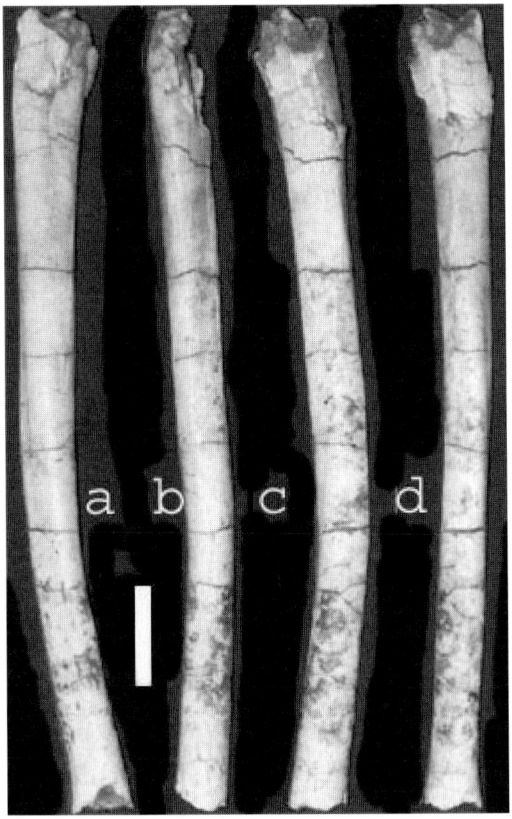

Fig. 8. Pterosaur femur from Tuştea (FGGUB R.1625) in (**a**) lateral, (**b**) posterior, (**c**) medial and (**d**) anterior views. Scale bar 50 mm.

the distal part, where it becomes fainter. This ridge begins on the medial surface of the bone and then passes onto the posterior surface. This appears to be the adductor ridge for the insertion of M. adductor femoris. Unlike the condition in *Pteranodon* (Bennett 2001a), there is no marked tuberosity corresponding to the fourth trochanter. A small bony knob close to the proximal end of the ridge may correspond to the internal trochanter.

Because it is incompletely preserved, this specimen provides relatively little systematic information, all the more so since the azhdarchid femur is poorly known, which makes comparisons difficult. The incomplete femora referred to *Azhdarcho lancicollis* by Nesov (1984, 1991,1997) show only the proximal articular end and part of the shaft, and the complete femur of *Zhejiangopterus linhaiensis* figured by Cai & Wei (1994) shows few distinguishing features. One of the notable points of the femur from Tuştea is its large size. Its total length as preserved is 365 mm; considering that the articular ends are missing, the original length of the bone must have been at least 400 mm. For comparison, the longest *Pteranodon* femora reported by Eaton (1910) and Bennett (2001a) are 270 mm and 250 mm long, respectively, and the longest *Z. linhaiensis* femur recorded by Cai & Wei (1994) is 222 mm long. The Tuştea femur thus confirms the occurrence of extremely large pterosaurs in the Densuş–Ciula Formation of Transylvania.

Conclusions

As remarked by Bennett (2001a, p. 2), until the description of *Quetzalcoatlus* by Lawson in 1975, '*Pteranodon* was the archetypal large pterosaur in both the popular and scientific literature'. Since 1975, a small number of 'giant' Late Cretaceous pterosaurs, apparently exceeding in size the largest known specimen of *Pteranodon* (which has a wing span of 7.25 m according to Bennett, 2001b), have been described or redescribed (the type of *Arambourgiania philadelphiae*, from Jordan, a cervical vertebra, had first been misidentified as a wing bone: see Arambourg 1954, 1959; Frey & Martill 1996). All those extremely large pterosaurs are from the Maastrichtian and have been referred to the Azhdarchidae Nesov, 1984 (see Nesov 1991 for a discussion). They include specimens from Texas (Lawson 1975; Langston 1981), Jordan (Martill *et al.* 1998), France (Buffetaut *et al.* 1997) and possibly Spain – the material from Tous, in the Province of Valencia, was first described by Company *et al.* (1999) as belonging to pterosaurs with a wing span of about 5.5 m, but a more recent report by Company *et al.* (2001) mentions individuals with a wing span possibly exceeding 12 m. The occurrence of *H. thambema* in the Late Maastrichtian of Transylvania confirms the wide geographical distribution of those huge pterosaurs at the end of the Cretaceous. Trying to estimate, on the basis of incomplete material, which was the largest of all these giant flying reptiles is fraught with considerable difficulties and of limited scientific interest. Beyond its very large size, *H. thambema* is interesting because it illustrates how little we know about such huge pterosaurs. Although fragmentary, the uncrushed bones from Transylvania reveal hitherto undiscovered aspects of the cranial morphology of giant azhdarchids. The idea of a pterosaur with a robustly built skull has met with some resistance, and the initial misidentification of the skull bones from Vălioara as those of a dinosaur may be an unconscious expression of the reluctance to accept such a concept. However, it is now clear that pterosaurs did not necessarily have a slenderly built skull consisting of thin bony struts and plates. Extrapolating from smaller forms, especially when they are known from crushed specimens that do not give a perfectly accurate image of three-dimensional structures, can be misleading. The discovery of *H. thambema* shows that at least some of the largest pterosaurs had a massive occiput and a palate built of robust bony rods, which contrasts with the usual image of pterosaur skulls. Moreover, the uncrushed bones of the Romanian pterosaur reveal their internal structure, which is frequently not clearly discernible in other large pterosaur skulls because of crushing. The peculiar structure of the bones of *Hatzegopteryx*, consisting of numerous small alveoli enclosed in a meshwork of paper-thin partitions, can probably be interpreted as a response to constraints imposed by flight on a very large animal. It may turn out to have been widespread in giant azhdarchid pterosaurs, when more uncrushed specimens are discovered.

We thank W. Langston Jr (Austin), J. Cunningham (Collierville), and E. Frey (Karlsruhe), for valuable information about *Quetzalcoatlus*, and F. Escuillié (Gannat) for help with casting some of the Romanian material.

References

ANTONESCU, E., LUPU, D. & LUPU, M. 1983. Corrélation palynologique du Crétacé terminal du Sud-Est des Monts Metaliferi et des dépressions de Haţeg et de Rusca Montană. *Anuarul Institutului de Geologie şi Geofizică*, **59**, 71–77.

ARAMBOURG, C. 1954. Sur la présence d'un Ptérosaurien gigantesque dans les phosphates de Jordanie. *Comptes Rendus de l'Académie des Sciences, Paris*, **238**, 133–134.

ARAMBOURG, C. 1959. *Titanopteryx philadelphiae* nov. gen., nov. sp., ptérosaurien géant. *Notes et Mémoires sur le Moyen-Orient*, **7**, 229–234.

BENNETT, S. C. 1989. A pteranodontid pterosaur from the Early Cretaceous of Peru, with comments on the relationships of Cretaceous pterosaurs. *Journal of Paleontology*, **63**, 669–677.

BENNETT, S. C. 2001a. The osteology and functional morphology of the Late Cretaceous pterosaur *Pteranodon*. Part I. General description and osteology. *Palaeontographica A*, **260**, 1–112.

BENNETT, S. C. 2001b. The osteology and functional morphology of the Late Cretaceous pterosaur *Pteranodon*. Part II. Size and functional morphology. *Palaeontographica A*, **260**, 113–153.

BUFFETAUT, E. 1999. Pterosauria from the Upper Cretaceous of Laño (Iberian Peninsula) : a preliminary comparative study. *Estudios del Museo de Ciências Naturales de Alava*, **14**, 289–294.

BUFFETAUT, E., LAURENT, Y, LE LOEUFF, J. & BILOTTE, M. 1997. A terminal Cretaceous giant pterosaur from the French Pyrenees. *Geological Magazine*, **134**, 553–556.

BUFFETAUT, E., GRIGORESCU, D. & CSIKI, Z. 2001.Giant pterosaurs from the Upper Cretaceous of Transylvania (western Romania). *Strata, Serie 1*, **11**, 26–28.

BUFFETAUT, E., GRIGORESCU, D. & CSIKI, Z. 2002. A new giant pterosaur with a robust skull from the latest Cretaceous of Romania. *Naturwissenschaften*, **89**, 180–184.

CAI, Z. & WEI, F. 1994. On a new pterosaur (*Zhejiangopterus linhaiensis* gen. et sp. nov.) from Upper Cretaceous in Linhai, Zhejiang, China. *Vertebrata Palasiatica*, **32**, 194.

CHARIG, A. J. & MILNER, A. C. 1997. *Baryonyx walkeri*, a fish-eating dinosaur from the Wealden of Surrey. *Bulletin of the Natural History Museum, Geology Series*, **53**, 11–70.

COLBERT, E. H. & RUSSELL, D. A. 1969. The small Cretaceous dinosaur *Dromaeosaurus*. *American Museum Novitates*, **2380**, 1–49.

COMPANY, J., RUIZ-OMEÑACA, J. I. & PEREDA-SUBERBIOLA, X. 1999. A long-necked pterosaur (Pterodactyloidae, Azhdarchidae) from the Upper Cretaceous of Valencia, Spain. *Geologie en Mijnbouw*, **78**, 319–333.

COMPANY, J., UNWIN, D. M., PEREDA-SUBERBIOLA, X. & RUIZ-OMEÑACA, J. I. 2001. A giant azhdarchid pterosaur from the latest Cretaceous of Valencia, Spain – the largest flying creature ever? *Journal of Vertebrate Paleontology*, **21**(3) (supplement), 41A–42A.

DODSON, P. 1996. *The Horned Dinosaurs*. Princeton University Press, Princeton.

EATON, G. F. 1910. Osteology of *Pteranodon*. Connecticut Academy of Arts and Sciences, Memoirs, **2**, 1–38.

FREY, E. & MARTILL, D. M. 1996. A reappraisal of *Arambourgiania* (Pterosauria, Pterodactyloidea): One of the world's largest flying animals. *Neues Jahrbuch für Geologie und Paläontologie, Abhandlungen*, **199**, 221–247.

GILMORE, C. W. 1928. A new pterosaurian reptile from the marine Cretaceous of Oregon. *United States National Museum, Proceedings*, **73**, 1–5.

GRIGORESCU, D. 1992. Nonmarine Cretaceous formations of Romania. *In*: MATEER, N.J. & CHEN, P. (eds) *Aspects of Nonmarine Cretaceous Geology*. China Ocean Press, Beijing, 142–164.

GRIGORESCU, D., SECLAMEN, M., NORMAN, D. B. & WEISHAMPEL, D. B. 1990. Dinosaur eggs from Romania. *Nature*, **346**, 417.

GRIGORESCU, D., VENCZEL, M., CSIKI, Z. & LIMBEREA, R. 1999. New latest Cretaceous microvertebrate fossil assemblages from the Haţeg Basin (Romania). *Geologie en Mijnbouw*, **78**, 301–314.

HOOLEY, R.W. 1913. On the skeleton of *Ornithodesmus latidens*: an ornithosaur from the Wealden Shales of Atherfield (Isle of Wight). *Quarterly Journal of the Geological Society, London*, **69**, 372–421.

HOWSE, S. & MILNER, A.R. 1993. *Ornithodesmus* – a maniraptoran theropod dinosaur from the Lower Cretaceous of the Isle of Wight, England. *Palaeontology*, **36**, 425–438.

HOWSE, S., MILNER, A.R. & MARTILL, D.M. 2001. Pterosaurs. *In*: MARTILL, D.M. & NAISH, D. (eds) *Dinosaurs of the Isle of Wight*. Palaeontological Association, London.

HUENE, F. VON 1914. Beiträge zur Kenntnis des Schädels einiger Pterosaurier. *Geologische und Paläontologische Abhandlungen*, **13**, 55–65.

JIANU, C.M., WEISHAMPEL, D.B. & STIUCĂ, E. 1997. Old and new pterosaur material from the Haţeg Basin (Late Cretaceous) of western Romania, and comments about pterosaur diversity in the Late Cretaceous of Europe. *2nd European Workshop on Vertebrate Palaeontology, Espéraza–Quillan, Abstracts*. Musée des Dinosaures, Espéraza, 1 p.

KADIC, O., 1916. Jelentés az 1915. évben végzett ásatásaimról. *A Magyar Királyi Földtani Intézet Évi Jelentése*, **1915**, 568 – 576.

KADIC, O. 1917. Bericht über meine Ausgrabungen im Jahre 1915. *Jahresbericht der Königlich. Ungarischen Geologischen Reichsanstalt*, **1915**, 610–618.

KELLNER, A. W. A. 1996. Description of the braincase of two Early Cretaceous pterosaurs (Pterodactyloidea) from Brazil. *American Museum Novitates*, **3175**, 1–34.

KELLNER, A. W. A. & LANGSTON, W. 1996. Cranial remains of *Quetzalcoatlus* (Pterosauria, Azhdarchidae) from Late Cretaceous sediments of Big Bend National Park, Texas. *Journal of Vertebrate Paleontology*, **16**, 222–231.

KELLNER, A. W. A. & TOMIDA, Y. 2000. *Description of a New Species of Anhangueridae (Pterodactyloidea) with Comments on the Pterosaur Fauna from the Santana Formation (Aptian–Albian), Northeastern Brazil.* National Science Museum, Tokyo, Monographs, **17**, 135 pp.

LANGSTON, W. 1981. Pterosaurs. *Scientific American*, **244**, 122–136.

LAWSON, D. A. 1975. Pterosaur from the latest Cretaceous of West Texas: discovery of the largest flying creature. *Science*, **187**, 947–948.

MADSEN, J. H. 1976. *Allosaurus fragilis* : a revised osteology. *Bulletin of the Utah Geological Survey*, **106**, 1–163.

MADSEN, J. H. & WELLES, S. P. 2000. *Ceratosaurus* (Dinosauria, Theropoda). A revised osteology. *Utah Geological Survey, Miscellaneous Publications*, **00–2**, 1–80.

MARTILL, D. M., FREY, E., SADAQAH, R. M. & KHOURY, H. N. (1998). Discovery of the holotype of the giant pterosaur *Titanopteryx philadelphiae* Arambourg 1959, and the status of *Arambourgiania* and *Quetzalcoatlus*. *Neues Jahrbuch für Geologie und Paläontologie, Abhandlungen*, **207**, 57–76.

NESOV, L. A. 1984. [Upper Cretaceous pterosaurs and birds from Central Asia.] *Paleontologicheskii Zhurnal*, **1984**(1), 47–57. [In Russian].

NESOV, L. A. 1991. [Giant flying reptiles of the family Azhdarchidae. I. Morphology, systematics.] *Vestnik Leningradskogo Universiteta, series* 7, **2**,14–23. [in Russian].

NESOV, L. A. 1997. *Cretaceous Nonmarine Vertebrates of Northern Eurasia.* University of Saint-Petersburg Institute of Earth Crust, Saint-Petersburg. (In Russian].

NEWTON, E. T. 1888. On the skull, brain and auditory organ of a new species of pterosaurian (*Scaphognathus Purdoni*), from the Upper Lias near Whitby, Yorkshire. *Philosophical Transactions of the Royal Society, London*, **179**, 503–536.

NOPCSA, F. 1899a. Jegyzetek Hátszeg vidékének geologiájához. *Földtani Közlöny*, **29**, 332–335.

NOPCSA, F. 1899b. Bemerkungen zur Geologie der Gegend von Hátszeg. *Földtani Közlöny*, **29**, 360–362.

NOPCSA, F. 1902. Ueber das Vorkommen der Dinosaurier bei Szentpéterfalva. *Zeitschrift der Deutschen Geologischen Gesellschaft*, **1902**, 34–39.

NOPCSA, F. 1904. Dinosaurierreste aus Siebenbürgen. III. Weitere Schädelreste von *Mochlodon*. *Denkschriften der Akademie der Wissenschaften in Wien*, **64**, 229–263.

NOPCSA, F. 1905. Zur Geologie der Gegend zwischen Gyulafehérvár, Déva, Ruszkabánya und der rumänischen Landesgrenze. *Mitteilungen aus dem Jahrbuche der Königlichen Ungarischen Geologischen Anstalt*, **14**, 93–279.

NOPCSA, F. 1914. Die Lebensbedingungen der obercretacischen Dinosaurier Siebenbürgens. *Centralblatt für Mineralogie, Geologie und Paläontologie*, **1914**, 564–574.

NOPCSA, F. 1915. Die Dinosaurier der siebenbürgischen Landesteile Ungarn. *Mitteilungen aus dem Jahrbuche der Königlichen Ungarischen Geologischen Reichsanstalt*, **23**, 1–25.

NOPCSA, F. 1923. On the geological importance of the primitive reptilian fauna in the uppermost Cretaceous of Hungary, with a description of a new tortoise (*Kallokibotion*). *Quarterly Journal of the Geological Society, London*, **72**, 100–116.

NOPCSA, F. 1926a. Die Reptilien der Gosau in neuer Beleuchtung. *Centralblatt für Mineralogie, Geologie und Paläontologie*, **1926**, 520–523.

NOPCSA, F. 1926b. Osteologia reptilium fossilium et recentium. *Fossilium Catalogus. I Animalia*, **27**, 1–391.

PADIAN, K. & SMITH, M. 1992. New light on Late Cretaceous pterosaur material from Montana. *Journal of Vertebrate Paleontology*, **12**, 87–92.

PLIENINGER, F. 1901. Beiträge zur Kenntnis der Flugsaurier. *Palaeontographica*, **48**, 65–90.

RICQLES, A. J. DE, PADIAN, K., HORNER, J. R. & FRANCILLON-VIEILLOT, H. 2000. Palaeohistology of the bones of pterosaurs (Reptilia:Archosauria): anatomy, ontogeny and biomechanical implications. *Zoological Journal of the Linnean Society*, London, **129**, 349–385.

ROMER, A. S. 1956. *Osteology of the Reptiles*. Chicago University Press, Chicago.

SEELEY, H. G. 1887. On a sacrum apparently indicating a new type of bird, *Ornithodesmus cluniculus* Seeley. *Quarterly Journal of the Geological Society, London*, **43**, 206–211.

SEELEY, H. G. 1901. *Dragons of The Air. An Account of Extinct Flying Reptiles*. Methuen & Co., London, 239 pp.

UNWIN, D. M. 2001. Variable growth rate and delayed maturation: do they explain "giant" pterosaurs? *Journal of Vertebrate Paleontology*, **21**(3) (supplement), 109A–110A.

UNWIN, D. M. & LÜ, J. 1997. On *Zhejiangopterus* and the relationships of pterodactyloid pterosaurs. *Historical Biology*, **12**, 199–210.

WEISHAMPEL, D. B., GRIGORESCU, D. & NORMAN, D. B. 1991. The dinosaurs of Transylvania. *National Geographic Research and Exploration*, **7**, 196–215.

WELLNHOFER, P. 1980. Flugsaurierreste aus der Gosau-Kreide von Muthmannsdorf (Niederösterreich) – ein Beitrag zur Kiefermechanik der Pterosaurier. *Mitteilungen der Bayerischen Staatssammlung für Paläontologie und historische Geologie*, **20**, 95–112.

WELLNHOFER, P. 1985. Neue Pterosaurier aus der Santana-Formation (Apt) der Chapada do Araripe, Brasilien. *Palaeontographica A*, **187**, 105–182.

WELLNHOFER, P. 1991. *The Illustrated Encyclopedia of Pterosaurs*. Salamander, London, 192 pp.

WILLISTON S. W. 1892. Kansas pterodactyls. *Kansas University Quarterly*, **1**, 1–13.

WILLISTON S. W. 1902. On the skull of *Nyctodactylus*, an Upper Cretaceous pterodactyl. *Journal of Geology*, **10**, 520–531.

Pterosaur phylogeny and comments on the evolutionary history of the group

ALEXANDER W. A. KELLNER

Setor de Paleovertebrados, Departamento de Geologia e Paleontologia, Museu Nacional/UFRJ, Quinta da Boa Vista, s/n São Cristóvão, Rio de Janeiro, RJ 20940–040, Brazil (e-mail: kellner@mn.ufrj.br)

Abstract: A cladistic analysis based on 39 terminal taxa and 74 characters (several multistate) using PAUP (Phylogenetic Analysis Using Parsimony) (3.1.1 for MacIntosh and 4.0b10 for Microsoft Windows) presents a new hypothesis of pterosaur inter-relationships. This study suggests that the most primitive taxon is the Anurognathidae, followed by *Sordes* and all remaining pterosaurs. *Dendrorhynchoides* is confirmed as a member of the Anurognathidae, being closely related to *Batrachognathus*. *Preondactylus* occupies a more derived position than *Sordes*, which questions its previous assignment as the most primitive pterosaur. The hypothesis of rhamphorhynchoid paraphyly is confirmed, with the Rhamphorhynchidae more closely related to the Pterodactyloidea than to more basal forms. The Pterodactyloidea shows a basal dichotomy: the Archaeopterodactyloidea and the Dsungaripteroidea. The Archaeopterodactyloidea is formed by *Pterodactylus* + *Germanodactylus* and a clade formed by Gallodactylidae + Ctenochasmatidae. The Nyctosauridae occupies the basal position within dsungaripteroids and is followed by the Pteranodontoidea and the Tapejaroidea. Pteranodontoids have *Pteranodon* at the base, followed stepwise by *Istiodactylus*, *Ornithocheirus* and the Anhangueridae. Tapejaroids are composed of the Dsungaripteridae at the base followed by the Tapejaridae and the Azhdarchidae.

Major trends within pterosaur evolutionary history are: general increase in size (wing span and body); increase of wing metacarpal and pteroid; decrease of proportional length of the second and third wing phalanx relative to the first; gradual increase of rostrum (anterior to external nares); and anterior shift of the skull–mandible articulation. Cranial crests are present in most pterodactyloids, but markedly in the Ornithocheiroidea, where all taxa show some sort of crest on the skull. The loss of teeth, previously assumed to have occurred independently in several lineages, seems to be a general trend among dsungaripteroids.

Several nodes recovered by this analysis are supported by very few characters, a result at least partially attributable to the limited available information from several taxa due to poor preservation and/or preparation.

Despite being known for over 200 years, the inter-relationship of pterosaurs has received little attention in the literature (Kellner 1995). This can be at least partially explained by the fact that most taxa are based on incomplete material (see Wellnhofer 1991b; Kellner & Tomida 2000). Furthermore, most relationships were traditionally established on 'overall' similarities, without distinction of primitive and derived features. A few studies employing the cladistic methodology to recover pterosaur phylogeny were performed. These were either based on limited taxa and incomplete specimens (Howse 1986; Bennett 1989, 1994) or cannot be tested (e.g. Unwin 1995). Most suffer from methodological problems and, except for the analysis done by Bennett (1989), in none could the published tree be recovered based on the data set provided (see Kellner 1995, 1996a for a detailed discussion of previous analyses).

This paper presents a comprehensive study of pterosaur inter-relationships. The employment of taxa for which complete material is available is optimized, but a limited number of species based on incomplete remains had to be used in order to test a particular relationship proposed in the literature (e.g. *Azhdarcho*, '*Phobetor*', *Ornithocheirus compressirostris*). The purpose of this approach is to establish a primary hypothesis of pterosaurian phylogeny, minimizing missing data and allowing the combination of cranial and postcranial information from basal and more derived taxa. The present analysis is an expansion of a previous work done by the author (Kellner 1996a,b, 1997), modified by the inclusion of more taxa (39 instead of the original 32) and more characters (74 instead the original 66). The results are further compared with previous pterosaur phylogenies and evolutionary aspects of those flying reptiles are discussed in the light of the new hypothesis.

Material and methods

Most of the data used in this analysis were obtained by direct examination of the respective specimens

From: BUFFETAUT, E. & MAZIN, J-M. (eds) 2003. *Evolution and Palaeobiology of Pterosaurs.* Geological Society, London, Special Publications, **217**, 105–137. 0305–8719/03/$15

(most holotypes). In a few cases information was obtained from the literature because the specimens were not available for direct examination when this research was being conducted (e.g. '*Phobetor*', *Dendrorhynchoides*). One problem was to define an outgroup for pterosaurs since there is no known species in the fossil record that bridges the morphological changes from basal archosaurs to this clade of flying reptiles. This could also explain the comparatively large number of autapomorphies of the taxon Pterosauria (Kellner 1996a). According to most researchers, pterosaurs are ornithodirans, closely related to dinosauromorphs (e.g. Sereno, 1991), a view supported here (Kellner 1996a; but see Peters 2000). All basal non-pterosaurian ornithodirans are very incomplete and many features observed in the present analysis could not be scored. In order to polarize the data, the basal ornithodiran *Scleromochlus* (Huene 1914; Sereno 1991), the basal theropod *Herrerasaurus* (Novas 1994; Sereno 1994; Sereno & Novas 1994) and the ornithosuchid *Ornithosuchus* (Walker 1964) were used as successive more distantly related outgroups. Although not directly included in the analysis, comparisons with the basal ornithodirans *Lagerpeton* (Romer 1971; Sereno & Arcucci 1994b) and *Marasuchus* (Sereno & Arcucci 1994a) were also made, always based on the literature.

A data matrix with 42 taxa (3 outgroups + 39 pterosaur taxa) with 74 characters (mostly multistate) was generated and analysed by the PAUP (Phylogenetic Analysis Using Parsimony) computer program versions 3.1.1 for MacIntosh (Swofford 1993) and 4.0b10 for Microsoft Windows (Swofford 2000). Characters were given equal weight and all multistate characters were treated as unordered. Due to the comparative large number of characters and taxa, the general heuristic search option was used. The consensus tree was further analysed by MacClade 3.04 (Maddison & Maddison 1992).

Institutional abbreviations: AMNH, American Museum of Natural History, New York, USA; BMNH, British Museum (Natural History), London, UK; PIN, Paleontological Institute, USSR Academy of Sciences, Moscow, Russia; SMNS, Staatliches Museum für Naturkunde Stuttgart, Stuttgart, Germany.

Results

The search executed with both PAUP versions produced the same results: 237 equally parsimonious cladograms with a length of 161 steps. In a second set of runs, the taxon *Scleromochlus* was excluded from the analysis, which considerably reduced the number of recovered trees: 80 of 161 steps (consistency index, CI = 0.8075; retention index, RI = 0.9246; rescaled consistency index, RC = 0.7466). The explanation for this is the large amount of missing data for *Scleromoclus* and the attempt with PAUP to resolve the relationships of this taxon respective to *Herrerasaurus* and *Ornithosuchus*. In both cases, the strict consensus cladogram shows the same topology and is discussed below (Fig. 1). Although not properly considered a phylogenetic tree, the consensus cladogram below illustrates the present state of knowledge regarding the relationships of the studied taxa. Each formally named taxon is defined and the temporal range is presented. The biochronology of the main pterosaurs clades and generic taxa is shown (Fig. 2) based on the consensus tree (Fig. 1) and the presently recorded temporal range. Where pertinent, the synapomorphies are discussed at each node, following the character list at the end of the paper (Appendix 1). The variation of the cranial and wing morphology of selected taxa is presented (Figs 3–6). The classification of the pterosaur taxa used in this study is provided at the end (Table 1).

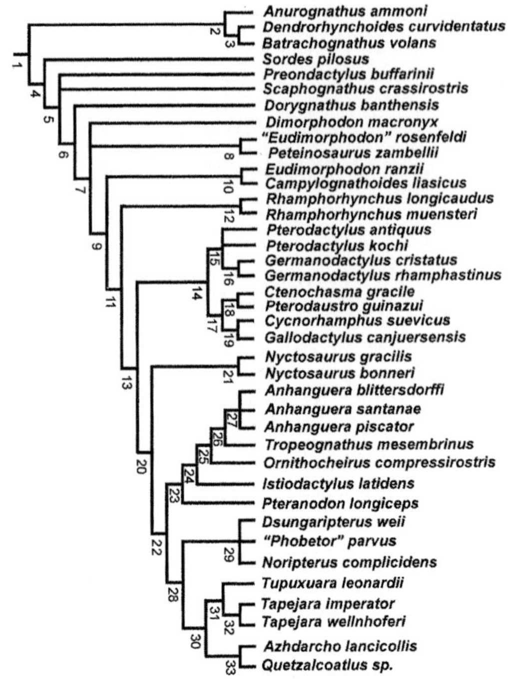

Fig. 1. Strict consensus cladogram of the 80 most parsimonious cladograms recovered in the cladistic analysis: 1, Pterosauria; 9, Novialoidea; 13, Pterodactyloidea; 14, Archaeopterodactyloidea; 20, Dsungaripteroidea. Outgroups were excluded from the figure. See text for details.

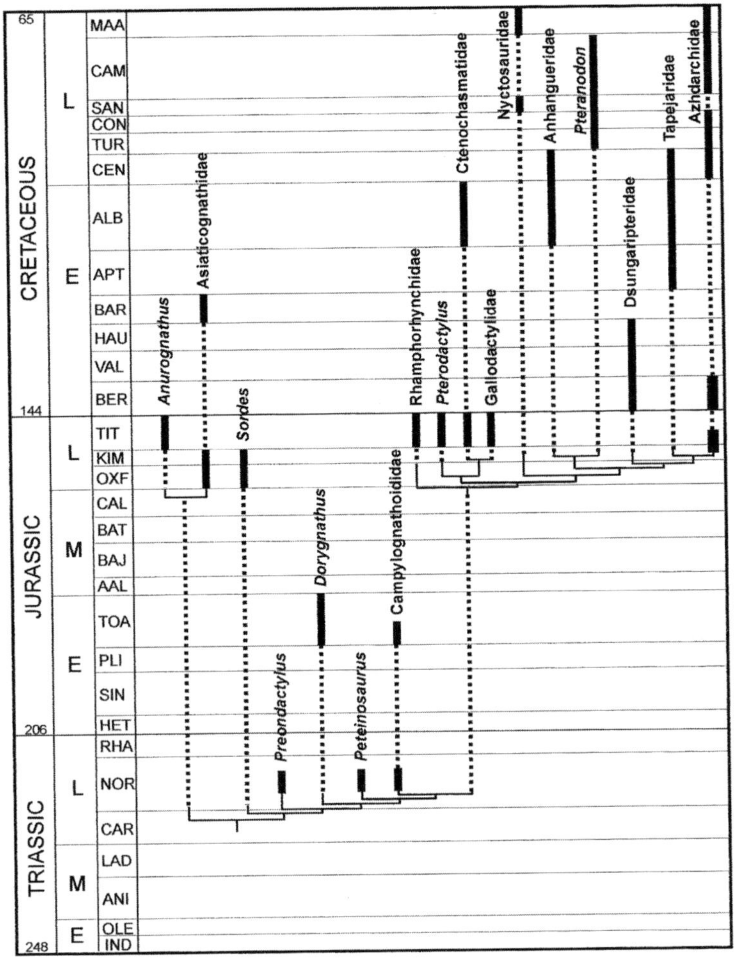

Fig. 2. Biochronology of pterosaurs based on the cladogram of Figure 1 and recorded temporal range. Time scale follows the Geological Society of America Time Scale of 1999. Dark column represents the recorded temporal range for the taxon, dotted line indicate gaps in the fossil record, and thin lines the relationships of the taxa. Due to lack of stratigraphic refinement of most pterosaur occurrences, stages where a particular taxa was recorded were filled out. AAL, Aalenian; ALB, Albian; ANI, Anisian; APT, Aptian; BAJ, Bajocian; BAR, Barremian; BAT, Bathonian; BER, Berriasian; CAL, Callovian; CAM, Campanian; CAR, Carnian; CEN, Cenomanian; CON, Coniacian; HAU, Hauterivian; HET, Hettangian; IND, Induan; KIM, Kimmeridgian; LAD, Ladinian; MAA, Maastrichtian; NOR, Norian; OLE, Olenekian; OXF, Oxfordian; PLI, Pliensbachian; RHA, Rhaetian; SAN, Santonian; SIN, Sinemurian; TIT, Tithonian; TOA, Toarcian; TUR, Turonian; VAL, Valanginian; E, Early, L, Late; M, Mid.

Tree structure and character analysis

Node 1. Pterosauria Kaup 1834

Definition. The most recent common ancestor of the Anurognathidae, Preondactylus and *Quetzalcoatlus* and all their descendants.

Recorded temporal range. Norian (Late Triassic) to Maastrichtian (Late Cretaceous). The oldest pterosaur records are *Preondactylus buffarinii* and '*Eudimorphodon*' *rosenfeldi* from the mid-Norian of Friuli, Italy (Wild 1978; Dalla Vecchia 1994). Among the youngest pterosaur records are '*Nyctosaurus*' *lamegoi* from the Gramame Formation, Brazil (Price 1953) and *Quetzalcoatlus* from the Javelina Formation, Texas (Lawson 1975a, b; Kellner & Langston 1996). A detailed discussion of pterosaur synapomorphies is presented by Kellner (1996a). Most of the unique features that unite pterosaurs have been eliminated from this analysis, unless they were part of a multistate character that changed within the group (e.g. presence of cristospine).

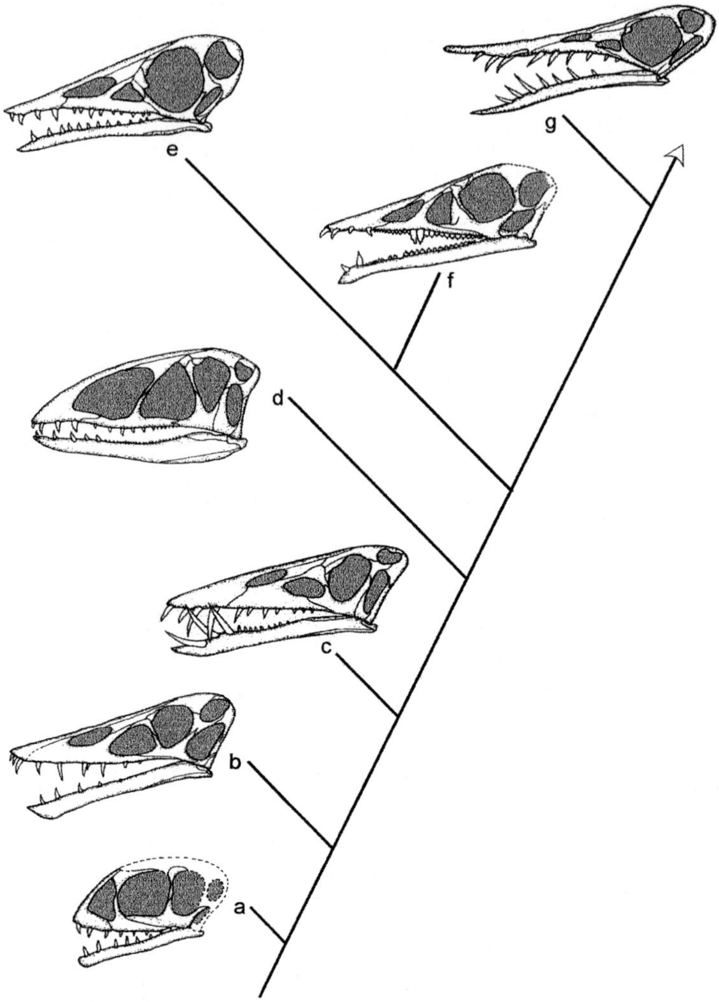

Fig. 3. Evolution of the pterosaur skull I. Basal forms, according to the cladogram of Figure 1: (**a**) *Anurognathus ammoni*; (**b**) *Scaphognathus crassirostris*; (**c**) *Dorygnathus banthensis*; (**d**) *Dimorphodon macronyx*; (**e**) *Eudimorphodon ranzii*; (**f**) *Campylognathoides liasicus*; (**g**) *Rhamphorhynchus muensteri*. (Modified from previous illustrations as follows: (a), (b), (g) from Wellnhofer 1975a; (c), (d) from Wellnhofer 1978; (e) from Wellnhofer 1974; and (f) from Wild 1978.) Drawings not to scale.

Node 2. Anurognathidae Kuhn 1937
Definition. The most recent common ancestor of *Anurognathus* and *Batrachognathus* and all their descendants. This taxon includes *Anurognathus* and the Asiaticognathidae (n. taxon).
Recorded temporal range. Oxfordian/Kimmeridgian (Late Jurassic) to Barremian (Early Cretaceous).
Synapomorphies.

(2) Upper and lower jaw comparatively broad. In all other pterosaurs, as is the general condition of many ornithodirans (and of the outgroups used in the present analysis), the skull is laterally compressed. Anurognathids differ by having a comparatively broader skull (Rjabinin 1948; Wellnhofer 1991b; Ji & Ji 1998; Unwin *et al.* 2000).

(5) Process separating external nares narrow. In all other pterosaurs, as in the outgroups, this region of the skull tends to be more massive, with the process separating the external nares broader. A narrow process separating the nares is observed in *Anurognathus* and in the asiaticognathid *Batrachognathus*. According to the published illustrations (Unwin *et al.* 2000), the skull of *Dendrorhynchoides* is very crushed and, although it is very likely that the process mentioned is

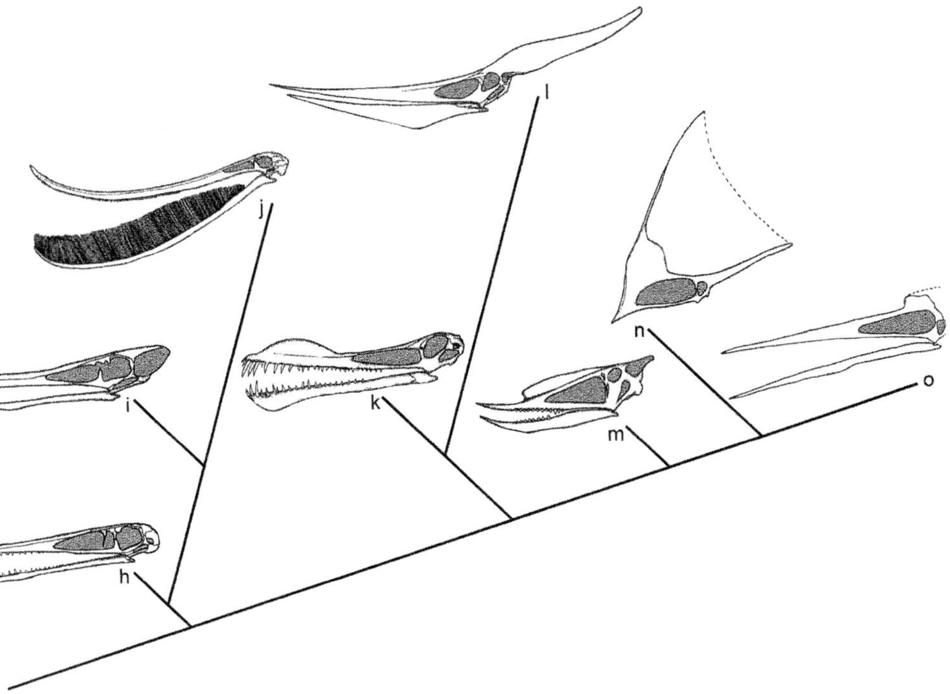

Fig. 4. Evolution of the pterosaur skull II. Pterodactyloids, according to the cladogram of Figure 1: (**h**) *Pterodactylus antiquus*; (**i**) *Cycnorhamphus suevicus*; (**j**) *Pterodaustro guinazui*; (**k**) *Anhanguera blittersdorffi*; (**l**) *Pteranodon longiceps*; (**m**) *Dsungaripterus weii*; (**n**) *Tapejara imperator*; (**o**) *Quetzalcoatlus* sp. (Modified from previous illustrations as follows: (h) from Wellnhofer 1970; (i), (m) from Wellnhofer 1978; (j) from Chiappe *et al.* 2000; (k) from Kellner & Tomida 2000; (l) from Bennett 2001; (n) from Campos & Kellner 1997; (o) from Kellner & Langston 1996.) Drawings not to scale.

very narrow, the specimen was not available for the present study in order to verify the thickness of the mentioned bone.

(39.1) Dentition formed by less than 15 peg-like teeth. Other pterosaurs also have peg-like teeth (e.g. *Pterodactylus*) but are more numerous. The teeth structure in *Sordes* is unknown. Apparently this taxon also has a reduced dentition but, based on a cast of a still undescribed specimen housed at the British Museum (cast BMNH R-10044), the teeth are elliptical, differing from the condition found in anurognathids.

Remarks. The skull of the Anurognathidae is not very well known. In the only specimen of *Anurognathus* several bones from the skull roof are missing, resulting in different interpretations of some cranial openings (cf. Döderlein 1923 and Wellnhofer 1975 a, b), while in both known specimens of *Batrachognathus* and *Dendrorhynchoides*, the skull is crushed dorsoventrally. From the known information, however, it seems that the skull in those taxa is comparatively short, possibly the shortest among pterosaurs. This feature might be another synapomorphic character of this clade, as has already been suggested several times in the literature (e.g. Wellnhofer 1991b).

Bakhurina (1988 *apud* Bakhurina & Unwin, 1995) pointed out that several bones of the palatal region of *Batrachognathus* are reduced to long, thin rod-like structures. Unfortunately the palatal region of *Anurognathus* is not known and the published illustrations of *Dendrorhynchoides* do not allow a clear analysis of those bones. Based on the general similar structure of the skull, however, it is likely that both taxa probably had a similar reduction of the palatal elements, which is therefore a further potential synapomorphic character of the Anurognathidae.

Anurognathus has a comparatively short tail (character 48.1, convergent with Pterodactyloidea – see node) that is reduced to 11 vertebrae (Döderlein 1923). It is uncertain whether some of the caudal vertebrae (preserved only as impressions) are fused, forming a ‘pygostyl’, the presence of which would be unique for this taxon, and therefore different from the short tail present in pterodactyloids. The tail in *Batrachognathus* is unknown and the tail in *Dendrorhynchoides* is debated (Ji *et al.* 1999; Unwin *et al.* 2000).

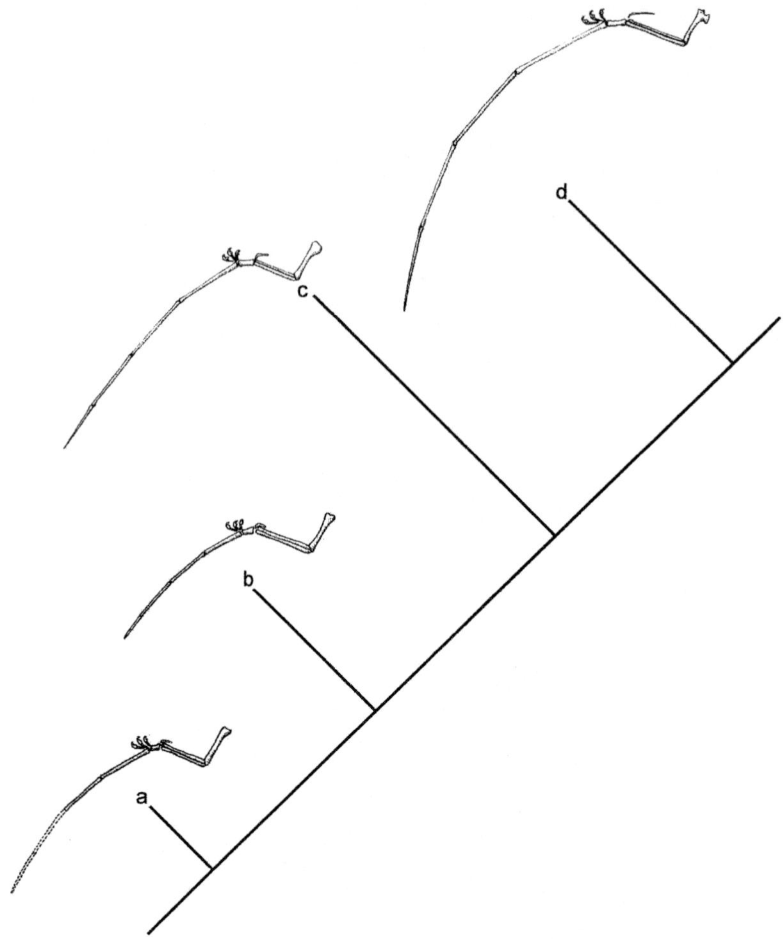

Fig. 5. Evolution of the pterosaur wing I. Basal forms, according to the cladogram of Figure 1: (**a**) *Dendrorhynchoides curvidentatus*; (**b**) *Sordes pilosus*; (**c**) *Campylognathoides liasicus*; (**d**) *Rhamphorhynchus muensteri*. In order to facilitate comparisons, the humerus in all wings was drawn to the same size. Drawings not to scale.

The monophyly of the Anurognathidae is widely accepted, although its position within pterosaurs might be disputed. Wellnhofer (1978) regarded *Anurognathus* as directly descending from *Dimorphodon*. Apparently this suggestion was based on the general shape of the skull, which in both taxa is high. There are several cranial features, however, that differ between anurognathids and *Dimorphodon*, such as the anterior extension of the skull and the position of the external naris. The skull of *Dimorphodon* is also narrower than in anurognathids, and no synapomorphy uniting these taxa was found in the present study.

Unwin (1995) listed three features that positioned the Anurognathidae closer to other pterosaurs relative to Dimorphodontidae (*Dimorphodon*): quadrate inclined forward, pteroid rod-like, and ulna larger or equal in length with tibia. According to Döderlein (1923) and Wellnhofer (1975a), no quadrate is known for *Anurognathus*. In the two and only known specimens of *Batrachognathus* (and also of *Dendrorhynchoides*, discovered only very recently, Ji & Ji 1998) the skull is crushed dorsoventrally, which hinders any observation of the original orientation and position of the quadrate. Although this bone was probably vertically or subvertically oriented, it cannot be verified that it was actually inclined forward. The pteroid in *Anurognathus* is very fragmentary (incomplete or preserved as an impression), while none was described in *Batrachognathus* (Rjabinin 1948). Nevertheless, a more flattened pteroid (and therefore non-rod-like) is present in several other basal taxa (e.g. *Eudimorphodon*), as is apparently also the case in the anurog-

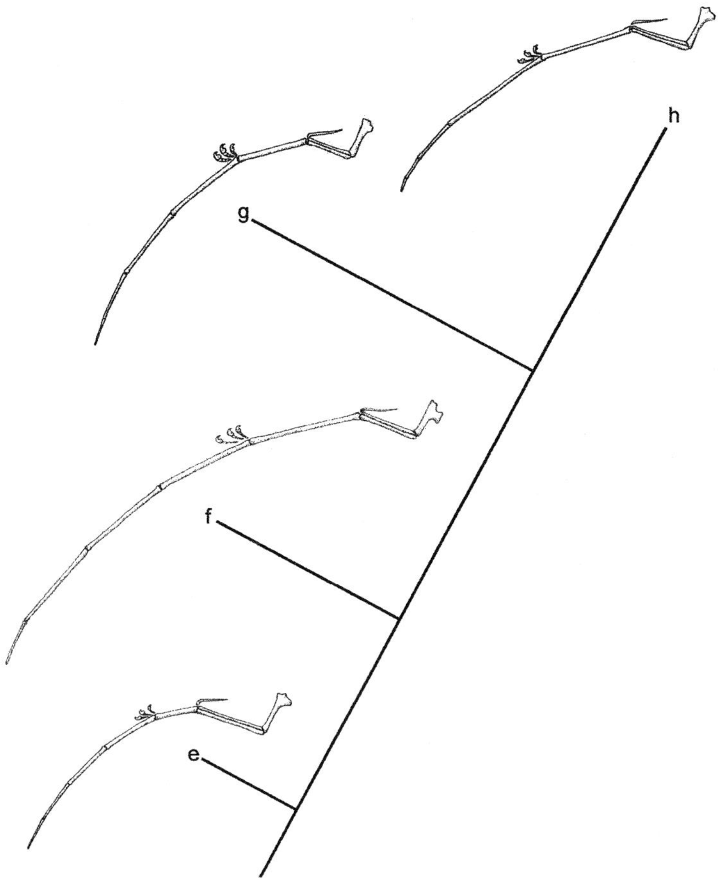

Fig. 6. Evolution of the pterosaur wing II. Pterodactyloids, according to the cladogram of Figure 1: (**e**) *Pterodactylus antiquus*; (**f**) *Nyctosaurus bonneri*; (**g**) *Pteranodon*; (**h**) *Quetzalcoatlus* sp. In order to facilitate comparisons, the humerus in all wings was drawn to the same size. Drawings not to scale.

nathid *Dendrorhynchoides* (see Unwin *et al.* 2000, fig. 2).

The ulna in *Anurognathus* is indeed larger than the tibia, while in *Dimorphodon* it is smaller (BMNH R-1034), as Unwin (1995) pointed out. Among pterosaurs, however, this feature varies considerably. The tapejaroids (Dsungaripteridae, Tapejaridae and Azhdarchidae), *Germanodactylus cristatus* and *Cycnorhamphus suevicus* have the ulna smaller than the tibia (in different proportions), similar to *Dimorphodon*. Other pterosaurs, such as *Campylognathoides liasicus*, *Anhanguera piscator* and *Rhamphorhynchus*, have the ulna longer than the tibia, while in some taxa (e.g. *Pteranodon*, *Nyctosaurus* and *Pterodactylus*) these bones are subequal in length. Therefore this feature seems to have changed considerably within the Pterosauria and its phylogenetic signal is difficult to be evaluated.

Node 3. Asiaticognathidae n. taxon

Definition. The most recent common ancestor of *Batrachognathus* and *Dendrorhynchoides* and all its descendants.

Recorded temporal range. Oxfordian–Kimmeridgian (Late Jurassic) to Barremian (Early Cretaceous).

Synapomorphies.

(55.1) Very large humerus, with proportional length of humerus relative to femur (hu/fe) larger than 1.40. As pointed out previously (Kellner 1996a), *Batrachognathus* has a very large humerus that is almost 50% longer than the femur. *Dendrorhynchoides* shares this feature (differing from the proportionally smaller humerus of *Anurognathus*, hu/fe = 1.19), which suggests that they form a monophyletic group within the Anurognathidae.

Remarks. Another potential synapomorphy of the Asiaticognathidae relative to *Anurognathus* is the number of teeth. Although the exact number of teeth is

Table 1. *Pterosaur classification (taxa used in the present analysis)*

- Pterosauria
  - Anurognathidae
    - *Anurognathus*
    - Asiaticognathidae n. taxon
      - *Batrachognathus*
      - *Dendrorhynchoides*
  - Unnamed taxon
    - *Sordes*
    - Unnamed taxon
      - *Scaphognathus*
      - *Preondactylus*
      - Unnamed taxon
        - *Dorygnathus*
        - Unnamed taxon
          - *Dimorphodon*
          - Unnamed taxon
            - *Peteinosaurus*
            - "*Eudimorphodon*" *rosenfeldi*
          - Novialoidea n. taxon
            - Campylognathoididae
              - *Eudimorphodon ranzii*
              - *Campylognathoides*
            - Unnamed taxon
              - Rhamphorhynchidae
                - *Rhamphorhynchus*
              - Pterodactyloidea
                - Archaeopterodactyloidea
                  - Unnamed taxon
                    - *Pterodactylus*
                    - *Germanodactylus*
                    - Unnamed taxon
                      - Ctenochasmatidae
                        - *Ctenochasma*
                        - *Pterodaustro*
                      - Gallodactylidae
                        - *Gallodactylus*
                        - *Cycnorhamphus*
                - Dsungaripteroidea
                  - Nyctosauridae
                    - *Nyctosaurus*
                  - Ornithocheiroidea
                    - Pteranodontoidea
                      - *Pteranodon*
                      - Unnamed taxon
                        - *Istiodactylus*
                        - Unamed taxon
                          - *Ornithocheirus*
                          - Anhangueridae
                            - *Tropeognathus*
                            - *Anhanguera*
                    - Tapejaroidea
                      - Dsungaripteridae
                        - *Dsungaripterus*
                        - "*Phobetor*"
                        - *Noripterus*
                      - Azhdarchoidea
                        - Tapejaridae
                          - *Tupuxuara*
                          - *Tapejara*
                        - Azhdarchidae
                          - *Quetzalcoatlus*
                          - *Azhdarcho*

difficult to establish in *Batrachognathus* and *Dendrorhynchoides* (in both cases less than 15) based on the available information, they have more than the eight teeth of the upper jaw reported for *Anurognathus* (Döderlein 1923, 1929; Wellnhofer 1978).

Node 4. Unnamed taxon including *Sordes*, *Preondactylus*, *Scaphognathus*, *Dorygnathus*, *Dimorphodon*, *Peteinosaurus*, '*Eudimorphodon*' *rosenfeldi* and the Novialoidea (n. taxon).
Recorded temporal range. Mid-Norian (Late Triassic) to Maastrichtian (Early Cretaceous).
Synapomorphies.

(3.1) Rostral part of skull anterior to external nares elongated (less than 60% of the skull length). In the Anurognathidae, the rostral part of the premaxillae anterior to the external nares is very short. In *Sordes*, as in all other pterosaurs, this region is laterally compressed and elongated. This condition reaches an extreme in the Ctenochasmatidae (see node 18).

(6) External naris displaced posterior to premaxillary tooth row. In most archosaurs, but also in the Anurognathidae, the external naris is situated very near to the anterior margin of the skull, above the premaxillary tooth row. In *Sordes* and all remaining pterosaurs where the skull is known, the naris is displaced posteriorly and starts behind the last premaxillary tooth.

Remarks. Two more putative synapomorphies related to the proportions of the wing phalanges (phd4) might support this clade: character 69 (state 1) – ph3d4 about the same length or larger than ph1d4, and character 70 (state 1) – ph3d4 about the same size or larger than ph2d4. The condition of both features, however, is not known in the anurognathids.

Sordes has been regarded as closely related to *Scaphognathus* but was classified in the Dimorphodontidae (Sharov 1971). Wellnhofer (1978) agreed with the close relationship of *Sordes* and *Scaphognathus* but regarded both as part of the Scaphognathinae of Hooley (1913). Among the features used by Wellnhofer (1978) to establish this relationship is the supposedly particular dentition of those taxa, formed by a few and well-spaced teeth compared to other pterosaurs. Bakhurina & Unwin (1995) also used this feature and the particular shape (like a boomerang) of the last phalanx of the pedal digit to advocate a close relationship between *Sordes* and *Scaphognathus*.

Although it is true that very few teeth form the dentition of *Sordes* and *Scaphognathus*, this feature is not unique to this taxon: *Anurognathus* too has a very reduced number of teeth. The shape of the teeth in *Sordes* is not clear, but they seem to be reduced in size (cast BMNH R-10044), contrary to the longer and stronger teeth of *Scaphognathus* (SMNS 59395). Furthermore, the 'boomerang-shaped' last phalanx of the pedal digit is present in at least another taxon (*Dorygnathus*), while the condition is unknown in *Batrachognathus* and not clear in *Anurognathus*, the latter showing this bone preserved only as a faint impression or partially covered by other elements. No unequivocal synapomorphic feature uniting *Sordes* and *Scaphognathus* could be found in the present analysis.

Node 5. Unnamed taxon including *Preondactylus*, *Scaphognathus*, *Dorygnathus*, *Dimorphodon*, *Peteinosaurus*, *'Eudimorphodon' rosenfeldi* and the Novialoidea.
Recorded temporal range. Norian (Late Triassic) to Maastrichtian (Late Cretaceous).
Synapomorphies

(54.1) Humerus less than 2.5 times but more than 1.5 times longer than the metacarpal IV (1.50< hu/mcIV<2.50). In the outgroups (*Ornithosuchus* and *Herrerasaurus*, condition in *Scleromoclus* is difficult to determine) the humerus is at least 5 or more times longer than metacarpal IV. Primitively, pterosaurs have comparatively small metacarpals, with the humerus at least 2.5 times longer. *Scaphognathus*, *Preondactylus* and all other pterosaurs increase the length of the wing metacarpal (mcIV), a trend that was continuously followed by more derived taxa, particularly pterodactyloids.

(61.1) Length of ulna between 2 and 4 times the length of metacarpal IV (4>ul/mcIV>2). The size of the wing metacarpal not only became larger relative to the humerus. *Anurognathus*, the asiaticognathid *Dendrorhynchoides* (condition in *Batrachognathus* unknown) and *Sordes* have the ulna more than 4 times longer than the metacarpals. Starting with *Scaphognathus* and *Preondactylus*, pterosaurs also continuously increased their wing metacarpals in respect to the ulna.
Remarks. Based on the present analysis, two more features fall out at this node: character 29 (state 1) – presence of an interpterygoid opening that is larger than the subtemporal fenestra; and character 53 (state 1) – the presence of a shallow and elongated cristospine. The palatal region of most cladistically primitive pterosaurs is not well known. The presence of an interpterygoid opening is an apomorphic feature of pterosaurs, although its proportional size may vary. In *Scaphognathus*, *Campylognathoides* and *Rhamphorhynchus* the interpterygoid opening is larger than the subtemporal fenestra, contrary to some pterodactyloids where this situation is reversed (i.e., interpterygoid opening smaller than subtemporal fenestra). The condition in *Sordes*, *Preondactylus* and in anurognathids, however, is unknown. The same is true for the cristospine. Its presence is a pterosaur synapomorphy, but the condition in *Preondactylus* and anurognathids is unknown. Therefore both features might define a more inclusive group.

Described by Wild (1984), *Preondactylus* has been regarded by some authors as the most primitive pterosaur, essentially based on the proportions of the hindlimbs relative to the forelimbs (e.g. Wellnhofer 1991b; Unwin 1995). A re-examination of the holotype and sole specimen of this taxon showed that the wing metacarpal and tibia were considerable shorter and the first wing phalanx considerable longer than thought, casting doubt about its basal position within the Pterosauria (Dalla Vecchia & Kellner 1995; Dalla Vecchia 1998). The analysis presented here shows *Preondactylus* to be more derived than anurognathids and *Sordes*, confirming the later hypothesis. The phylogenetic position of *Preondactylus* and *Scaphognathus* relative to all remaining pterosaurs remains to be resolved.

Node 6. Unnamed taxon including *Dorygnathus*, *Dimorphodon*, *Peteinosaurus*, *'Eudimorphodon' rosenfeldi* and the Novialoidea.
Recorded temporal range. Mid-Norian (Late Triassic) to Maastrichtian (Early Cretaceous).
Synapomorphies.

(71.1) Femur longer but less than twice the length of metacarpal IV (1.00<fe/mcIV<2.00). Within pterosaurs there is a tendency for the wing metacarpal to increase, not only in respect to bones of the forelimb (characters 54 & 61), but also in respect to elements of the hindlimb. In more primitive forms (Anurognathidae, *Sordes*, *Scaphognathus*, *Preondactylus*), the length of the femur is about twice or longer than the metacarpal IV. Starting with *Dorygnathus*, the relation fe/mcIV becomes smaller than 2.00. This trend continues in pterodactyloids, where the metacarpal reaches the same length of the femur (character 71, state 2) and in some taxa becomes up to more than twice as long (*Nyctosaurus*, *Pteranodon*: fe/mcIV<0.50). This condition is reversed in *Peteinosaurus*, which shows the primitive state (see also Dalla Vecchia 1998).
Remarks. *Dorygnathus* is further separated from all other pterosaurs by the particular shape of the deltopectoral crest of the humerus, which is distally expanded (character 58, state 3). This condition differs from the cladistically more primitive pterosaurs that have the proximal margin of the deltopectoral crest straighter and lacking an expansion on the distal part (character 58, state 1: e.g. *Scaphognathus*). It also differs from the subrectangular condition found in the Campylognathoididae (see Wild 1978). The deltopectoral crest of *Dorygnathus* shows some similarities with that found in the Rhamphorhynchidae, but the latter differs by being proportionally more developed and distally expanded in a hatchet-shaped form.

Dorygnathus was regarded by Wellnhofer (1975c, 1978) as being closely related to *Rhamphorhynchus*.

This relationship is based mainly on characters that have a wide distribution within pterosaurs (e.g. size of femur relative to humerus, small lower temporal fenestra). *Dorygnathus*, however, lacks several rhamphorhynchid features, such as the reduction of the naris and antorbital fenestra (character 7). The dentition of these taxa, which was another feature used to unite them (Wellnhofer 1978) also differs: *Dorygnathus* has strong anterior teeth that are 'fang-like' and much larger than the subsequent ones; *Rhamphorhynchus* has more delicate teeth and no accentuated heterodonty in size. Furthermore, *Rhamphorhynchus* shares several characters with the Pterodactyloidea (e.g. character 20, state 1: inclination of the quadrate) which are absent in *Dorygnathus*.

It has to be noted, however, that *Dorygnathus* shares two features with *Rhamphorhynchus*: the projection of the tip of the dentary anteriorly (character 32) and the development of a fused and comparatively long mandibular symphysis (character 30); the latter is also present in pterodactyloids. From the analysis performed in the present study, these features were developed independently by *Dorygnathus* and *Rhamphorhynchus*.

Node 7. Unnamed taxon including *Dimorphodon*, *Peteinosaurus*, '*Eudimorphodon*' *rosenfeldi* and the Novialoidea.
Recorded temporal range. Mid-Norian (Late Triassic) to Maastrichtian (Early Cretaceous).
Synapomorphies.

(62.1) Diameter of radius no more than half that of ulna. In cladistically primitive pterosaurs the diameter of radius and ulna are subequal. In *Dimorphodon* the radius is smaller, but still greater than half the diameter of the ulna, which is also the condition in *Campylognathoides* (Wellnhofer 1974; Wild 1978) and all other pterosaurs where these bones are known. Within pterodactyloids, in *Istiodactylus* and anhanguerids the diameter of the radius is reduced even more (node 24, character 62, state 2).
Remarks. The present analysis showed a trichotomy between *Dimorphodon*, a clade formed by *Peteinosaurus* + '*Eudimorphodon*' *rosenfeldi*, and the Novialoidea. *Dimorphodon* shares one feature with *Peteinosaurus*: the last phalanx of pedal digit V is elongated but straight (character 74, state 1; condition in '*Eudimorphodon*' *rosenfeldi* unknown). Furthermore, it has one unique feature that has not been found in other pterosaurs: the particular dentition showing accentuated tooth dimorphism. The dentition of *Dimorphodon* is formed by long anterior teeth (but not as long as in *Rhamphorhynchus* or fang-like as in *Dorygnathus*), followed by very small and closely spaced teeth, particularly in the lower jaw (e.g. BMNH R-41212/41213). Although several non-pterodactyloid taxa show accentuated variation in the shape and sizes of their teeth (e.g. *Eudimorphodon ranzii*), none approach the condition found in *Dimorphodon*, which is here regarded as an autapomorphy of this taxon.

Dimorphodon shows another feature that is worth mentioning: the size of the external naris, which is the largest opening of the skull in this taxon. In all other non-pterodactyloids, the orbit is the largest lateral cranial opening. The Anurognathidae also have a large external naris, but it is apparently smaller than the antorbital openings (see Wellnhofer 1975b, p. 181). All pterodactyloids differ from the condition in *Dimorphodon* by having the naris and antorbital fenestra confluent, forming the nasoantorbital fenestra.

The present analysis could not resolve the relationships between *Dimorphodon*, *Peteinosaurus* + '*Eudimorphodon*' *rosenfeldi* and the remaining pterosaurs.

Node 8. Unnamed taxon composed of *Peteinosaurus* and '*Eudimorphodon*' *rosenfeldi*.
Recorded temporal range. Mid-Norian (Late Triassic).
Synapomorphies.

(38) Presence of multicusped teeth. The only feature of the present study that unites *Peteinosaurus* and '*Eudimorphodon*' *rosenfeldi* is the presence of multicusped teeth. However, this feature is not unique to these taxa and is also present in *Eudimorphodon ranzii*; it should therefore be viewed with caution.
Remarks. The position of *Peteinosaurus* and '*Eudimorphodon*' *rosenfeldi* relative to other pterosaurs, particularly *E. ranzii*, is very difficult to establish, given the present state of knowledge. *Peteinosaurus* is known from a very incomplete skeleton (the holotype), with an incomplete lower jaw (anterior and posterior ends missing), and a second partial skeleton that lacks the skull (Wild 1978). '*Eudimorphodon*' *rosenfeldi* is known by a partial skeleton that shows the posterior end of the skull (Dalla Vecchia 1994). Although not reported in the original description, some teeth of *Peteinosaurus* are multicusped (pers. obs.). No complete mandible is known for either, so it cannot be established whether the anterior tip of the dentary is downturned, as in *Eudimorphodon ranzii* and *Campylognathoides*. To make this situation worse, no information is available regarding the deltopectoral crest of the humerus of '*Eudimorphodon*' *rosenfeldi* (the one of *Peteinosaurus* differs from *E. ranzii*, see character 58) and no complete wing is known for the holotype of *E. ranzii* (specimens associated with this taxon are being questioned, see Kellner 1996a and discussion below). A more comprehensive study of the *Eudimorphodon* complex is presently being undertaken by Fabio Dalla Vecchia (pers. comm. 2001) and will shed more light on the phylogenetic position of those Triassic pterosaur taxa.

Node 9. Novialoidea n. taxon
Definition. The most recent common ancestor of *Campylognathoides* and *Quetzalcoatlus* and all their descendants.
Recorded temporal range. Norian (Late Triassic) to Maastrichtian (Late Cretaceous).
Synapomorphies.

(69.2) Third phalanx of manual digit IV shorter than first (ph3d4 <ph1d4). Primitively in pterosaurs the third phalanx of the wing finger is about the same size or longer than the first (condition in anurognathids unknown). Starting with *Campylognathoides* (condition in *Eudimorphodon ranzii* unknown), ph3d4 becomes shorter than ph1d4.

(70.2) Third phalanx of manual digit IV shorter than second (ph3d4 <ph2d4). In primitive pterosaurs, the third phalanx of the wing finger is about the same size or longer than the second (condition in anurognathids unknown). In *Campylognathoides* (condition in *E. ranzii* unknown), ph3d4 is shorter than ph2d4, which is the case for all remaining taxa studied here where complete wings are known.

Remarks. Besides the synapomorphies presented above, there is at least one other potential synapomorphy of the Novialoidea: maxilla excluded from the internal naris (character 28). The palatal region of most non-dsungaripteroid pterosaurs is unknown. The most cladistically primitive pterosaur for which the palatal region was reconstructed is *Scaphognathus*. According to Wellnhofer (1975a), the maxilla in this taxon participates in the lateral margins of the choanae. This same condition is present in the outgroups (*Ornithosuchus* and *Herrerasaurus*) used in this analysis and is also apparently present in *Batrachognathus* (pers. obs.). In *Campylognathoides* (Wellnhofer 1974), *Rhamphorhynchus* (Wellnhofer 1975a) and all dsungaripteroids where the palatal region is known, the maxilla is excluded from the internal nares, whose lateral borders are made by the palatines. However, the palatal region of *Dimorphodon, Peteinosaurus, 'Eudimorphodon' rosenfeldi* and *Dorygnathus* (and also of the campylognathoidid *E. ranzii*), which according to the present cladistic analysis are successively more distantly related to the Campylognathoididae, is unknown.

In *Campylognathoides* and *Rhamphorhynchus* the first phalanx of manual digit IV is enlarged to about or more the length of the tibiotarsus (character 67, state 2). Primitively, the first phalanx of the wing finger in pterosaurs is less than twice the length of the tibiotarsus (ph1d4/ti<2.00), as observed in anurognathids, *Sordes* and *'E' rosenfeldi* (among others). This feature, however, varies within the remaining pterosaur taxa: in archaeopterodactyloids and tapejaroids the length of the first phalanx of the wing finger is less than twice the length of the tibiotarsus, while nyctosaurids and pteranodontoids display the same condition found in *Campylognathoides* and in *Rhamphorhynchus* (i.e. ph1d4/ti>2.00). Therefore its phylogenetic signal is difficult to interpret.

The main problem of this clade lies in the incompleteness of *Eudimorphodon ranzii*. No specimen with complete wings can be referred to this species (see discussion below). The classification of this Triassic taxon within the Novialoidea is based on its current phylogenetic position as a member of the Campylognathoididae (node 10).

Node 10. Campylognathoididae Kuhn 1967
Definition. All pterosaurs more closely related to *Campylognathoides* than to other pterosaurs. This taxon presently includes *Campylognathoides* and *Eudimorphodon ranzii*.
Recorded temporal range. Norian (Late Triassic) to Toarcian (Early Jurassic).
Synapomorphies.

(31) Anterior tip of dentary downturned.

(58.2) Deltopectoral crest of humerus subrectangular, extending down humerus shaft for at least 30% of humerus length.

Remarks. The close relationship of *Campylognathoides* and *Eudimorphodon ranzii* was first pointed out by Wild (1978) and is confirmed by the present analysis. *Eudimorphodon ranzii* shows the presence of multicusped teeth, a feature also present in *Peteinosaurus* and *'Eudimorphodon' rosenfeldi*. Furthermore this taxon has teeth on the pterygoids. The palatal region in most primitive pterosaurs is unknown, except in *Scaphognathus* and *Campylognathoides*, which lack this feature. Palatal teeth are also present in one basal theropod (*Eoraptor*, not yet fully described), while the palate of most basal ornithodirans, such as *Scleromochlus*, *Marasuchus* and *Lagerpeton*, is not appropriately known. If present in these taxa, this feature might suggest a more basal position of *Eudimorphodon ranzii* within pterosaurs than the one presented in this study.

While studying *Eudimorphodon*, Wild (1978) referred three more specimens to this taxon (MCSNB 2887, 3496 and another specimen deposited in the Università degle Studi de Milano, Italy). Later he also attributed a fourth (MCSNB 8950), found in slightly younger Triassic beds, to this taxon (Wild 1993). Although a review of *Eudimorphodon* is not the purpose of this paper, it should be noted that there is no apomorphic feature in any of these specimens that unequivocally supports their classification as *Eudimorphodon ranzii* (see Kellner 1996a for details). On the contrary, in MCSNB 2887 the scapula is fused to the coracoid, an indication that it probably represented an adult individual, despite being about 40% smaller than the type specimen of *Eudimorphodon ranzii* (MCSNB

2888). Furthermore, the scapula is considerable longer than the coracoid, contrasting with the relation of these bones in the holotype (*Eudimorphodon ranzii*: sca/cor = 1.09). A very tightly connected scapulocoracoid is also present in the Milano specimen, which further differs from the holotype of *Eudimorphodon ranzii* by having fewer teeth which are more uniform in size (e.g. lacking large fang-like teeth in the middle part of the upper dentition). The presence of multicusped teeth, the main argument for classifying the Milano specimen as *Eudimorphodon ranzii*, is also found in *Peteinosaurus* and '*Eudimorphodon*' *rosenfeldi*. Based on these observations it seems more prudent to restricted *Eudimorphodon ranzii* to the holotype.

The position of the Campylognathoididae as the sister-taxon of *Rhamphorhynchus* + Pterodactyloidea was also suggested by Unwin (1995) based on the following features:

(1) *Skull low and elongate, with a length/depth ratio of 3.0.* This particular length/depth ratio of the skull in the outgroups used in the present analysis is also about 3.0. While this ratio is apparently smaller in the Anurognathidae, in other primitive pterosaurs, such as *Sordes* (cast of PIN No. 2585/3 in the BMNH), *Scaphognathus* (cast AMNH 1692) and *Dimorphodon* (BMNH R-41212/41213) this ratio is at least 2.8. Although, primitively, pterosaurs apparently had proportionally higher skulls (e.g. Anurognathidae), the exact proportion and differences cannot be established with certainty for the time being.

(2) *Orbit larger than preorbital and nasal opening.* Although the orbit is larger than any other lateral cranial opening in Campylognathoididae (*Campylognathoides* + *Eudimorphodon ranzii*), this condition is also true for other primitive pterosaurs, e.g. *Sordes* and *Scaphognathus* (and also for the outgroups used in the present analysis). From the non-pterodactyloid taxa, only in *Dimorphodon* is the orbit smaller than the preorbital and nasal opening; the condition in Anurognathidae is unknown. A modified version of this character has been used in the present analysis (character 7: naris and antorbital fenestra reduced relative to the orbit) and diagnoses another clade (the Rhamphorhynchidae).

(3) *Premaxillae separating frontals.* In all pterosaurs where the dorsoposterior region of the skull is preserved, the premaxillae separate the frontals (e.g. Wellnhofer 1978), and the contact of those bones has been regarded as an autapomorphy of Pterosauria (e.g. Romer 1956). The condition for the Anurognathidae, however, is unknown.

Node 11. Unnamed taxon formed by Rhamphorhynchidae + Pterodactyloidea.
Recorded temporal range. Tithonian (Late Jurassic) to Maastrichtian (Late Cretaceous).
Synapomorphies

(20.1) Quadrate inclined backwards for about 120° relative to ventral margin of skull. In the outgroups used here (*Ornithosuchus*, *Herrerasaurus* and *Scleromochlus*) the quadrate is either vertical or subvertical. In most other archosaurs the condition is the same, although in some taxa the quadrate may even be inclined slightly forwards. To measure the angle between the quadrate and the ventral margin of the skull in pterosaurs is very difficult since most specimens are crushed and the bones tend to be deformed and displaced from their natural position. For the Anurognathidae this feature is not known. In *Dimorphodon* the posterior region of the skull is incomplete but apparently the quadrate has an almost subvertical position. In other primitive pterosaurs, the quadrate tends to be slightly inclined backwards, less than 120°, which is here regarded as a primitive feature within pterosaurs. In the Rhamphorhynchidae the inclination of the quadrate is around 120°, which is a trend followed by all pterodactyloids, with the Archaeopterodactyloidea taking a step further (150°, see node 14).

(21.1) Articulation between skull and mandible positioned approximately under middle portion of orbit. Primitively the articulation between the skull and the lower jaw is positioned under the posterior half of the orbit or further posteriorly. In the Rhamphorhynchidae and the Pterodactyloidea the articulation between skull and mandible is shifted anteriorly, reaching the middle portion of the orbit. In more derived pterodactyloids (the Ornithocheiroidea) this condition is taken a step further, with the skull–mandible articulation under the anterior half of the orbit (character 21, state 2).

(30) Presence of mandibular symphysis forming at least 30% of mandible length. Primitively, pterosaurs lack a mandibular symphysis; if present, it was very short (condition in *Scaphognathus* is unknown). Rhamphorhynchids and pterodactyloids have a well-developed mandibular symphysis, which might reach even larger proportions in some taxa (e.g. *Ctenochasma*, *Pterodaustro*). It must be noted, however, that *Dorygnathus*, which in this analysis is positioned in a more basal part of the cladogram, also has a well-developed mandibular symphysis (including the large anterior extension) that based on this study was acquired independently.

Remarks. Another potential synapomorphy of Rhamphorhynchidae and Pterodactyloidea is the second wing phalanx being about 30% smaller than the first (character 68, state 2). This condition, however, is also present in *Dendrorhynchoides* and '*Eudimorpodon*' *rosenfeldi*.

The position of Rhamphorhynchidae (or *Rhamphorhynchus*) as sister-taxon of the Pterodactyloidea has also been proposed in previous analyses. Howse (1986) suggested this relationship, but from his paper it is not clear which character would support this. Contrary to the more uniform length present in primitive 'rhamphorhynchoids', Howse (1986) pointed out that the cervical vertebrae in *Rhamphorhynchus* does change in length according to the position within the neck (as is the case in some pterodactyloids). However, similar variations are present in other primitive pterosaurs (e.g. *Sordes* [cast BMNH R-10044]; *Scaphognathus* [cast AMNH 1692]).

Unwin (1995) suggested a close relationship of Rhamphorhynchidae and Pterodactyloidea based on five synapomorphies, two of which have been redefined here (characters 20.1 and 30, discussed above). The three others are: reduction in the length of metatarsal IV, loss of size dimorphism in mandibular dentition and metacarpals I–III subequal.

The pes of most pterosaurs is not well known. Where preserved, metatarsal IV tends to be reduced with respect to metatarsal III. In *Rhamphorhynchus* metatarsal IV is apparently slightly more reduced then in other primitive pterosaurs but this seems to be a function of the enlargement of metatarsal III (see Wellnhofer 1978). A reduced metatarsal IV is present in the pterodactyloid *Pterodactylus* but also in *Scaphognathus* (Wellnhofer 1975b, p.178) and in *Sordes*. Until the pes of more pterosaurs are studied in detail, the phylogenetic significance of the proportional length of metatarsal IV remains uncertain.

Size dimorphism in mandibular dentition is present in a few pterosaur taxa. *Dorygnathus*, *Dimorphodon* and *Eudimorphodon ranzii* show size dimorphism in the upper and lower jaws, but a close examination shows that their dentitions are very different: *Dorygnathus* has long fang-like teeth followed by wide-spaced smaller ones; *Dimorphodon* has long but not fang-like anterior teeth followed by very small and closely spaced ones; and the teeth in *Eudimorphodon ranzii* are multicusped. Therefore the dentition of those taxa does not suggest any particular relationships among them. Furthermore, accentuated dental size dimorphism is absent in the outgroup used in the present analysis and in the Anurognathidae, *Sordes*, *Scaphognathus* and *Campylognathoides* (Wellnhofer 1974).

Another supposed synapomorphy of Rhamphorhynchidae + Pterodactyloidea presented by Unwin (1995) is the subequal length of metacarpals I–III. In all non-pterodactyloid specimens where this region is known metacarpals I–III are about the same length. Therefore this feature cannot be used as a synapomorphy of Rhamphorhynchidae + Pterodactyloidea.

Node 12. Rhamphorhynchidae Nopcsa 1928
Definition. All pterosaurs closer related to *Rhamphorhynchus muensteri* than to other pterosaurs. Based on Wellnhofer (1975 a, b, c), there are several species of what can be called the *Rhamphorhynchus* complex (not used in the present study) whose relationships has still to be determined.
Recorded temporal range. Tithonian (Late Jurassic).
Synapomorphies

(7) Naris and antorbital fenestra extremely reduced. Although the size of the naris and antorbital fenestra vary within pterosaurs, *Rhamphorhynchus* has the naris and the antorbital fenestrae very reduced. In all other pterosaurs where the skull is known, particularly the non-pterodactyloid taxa, these openings are proportionally larger.

(58.4) Deltopectoral crest of humerus expanded distally in hatchet-shaped form and positioned near caput humeri. The deltopectoral crest in pterosaurs varies significantly. The condition of this portion of the humerus in *Rhamphorhynchus* differs from that found in *Dorygnathus* by being more expanded distally in a hatchet-shaped configuration. A similar condition of the deltopectoral crest is found in *Nyctosaurus* (Kellner & Hasegawa 1993). However, the deltopectoral crest in *Rhamphorhynchus* is placed more closely to the proximal articulation of the humerus compared to nyctosaurids.

(74.2) Last phalanx of pedal digit V elongated and curved. The condition of the last pedal digit V in *Rhamphorhynchus* is different from the straight condition found in *Dimorphodon* and from the boomerang shape of this bone in several cladistically primitive taxa (e.g. *Dorygnathus*). It also differs from the reduced condition present in *Campylognathoides*. As has been pointed out before, however, the pes in most pterosaurs is not very well known. Therefore, although falling out as a synapomorphy of the Rhamphorhynchidae, more data of the condition of the fifth pedal digit in other pterosaurs is needed to confirm whether this character state is unique to this clade.

Node 13. Pterodactyloidea, composed of the Archaeopterodactyloidea (= *Pterodactylus*, *Germanodactylus*, Ctenochasmatidae, Gallodactylidae) and the Dsungaripteroidea.
Definition. The most recent common ancestor of *Pterodactylus* and *Quetzalcoatlus* and all their descendants.
Recorded temporal range. Tithonian (Late Jurassic) to Maastrichtian (Late Cretaceous).
Synapomorphies.

(8.1) Naris and antorbital fenestra confluent, shorter than 45% of skull length. This condition is present in all pterodactyloids, although some taxa have a bony bar formed by the nasal partially separating the naris and the antorbital fenestra (e.g.

Pterodactylus, *Germanodactylus*). The size of the nasoantorbital fenestra is largest in the Tapejaridae, where this opening is longer than 45% of the skull length (node 31).

(46) Absence of cervical ribs on mid-cervical vertebrae. Primitively in pterosaurs (and in several other archosaurs), all cervical vertebrae bear ribs. In pterodactyloids, cervical ribs are lost in the mid-cervical vertebrae, although in some taxa a reduced cervical rib can be found in the last two cervicals (e.g. *Pteranodon*, Bennett, 2001).

(48) Number of caudal vertebrae less than 15. All pterodactyloids where the tail is known have less than 15 caudal vertebrae. Although this number might actually be higher (very few specimens have a complete series of caudal vertebrae) the reduction of the tail (in size and number of caudal vertebrae) is an autapomorphic feature of the Pterodactyloidea. As pointed out before, *Anurognathus* also has a reduced tail (condition of *Batrachognathus* and *Dendrorhynchoides* not known) which, according to the present analysis, was achieved independently. Furthermore the structure of the caudals in pterodactyloids where the tail is known lack extensions of the pre- and postzygapophysis that form a stiffened elongated tail in all non-pterodactyloids (except probably anurognathids).

(54.2) Proportional length of humerus relative to metacarpal IV greater than 0.40 but less than 1.50 (0.40<hu/mcIV<1.50).

(58.7) Deltopectoral crest of humerus long, proximally placed, curving ventrally. Among pterodactyloids, the deltopectoral crest can display the following basic shapes: (1) proximally placed and curving ventrally; (2) hatched shaped and displaced away from the *caput humeri*; and (3) warped. The archaeopterodactyloids and the tapejaroids show condition (1), which differs from all the non-pterodactyloids previously discussed. Within the Pterodactyloidea, however, the Nyctosauridae display condition (2) and the Pteranodontoidea the condition (3). Therefore the different shapes of the deltopectoral crest of nyctosaurids and pteranodontoids have apparently been derived independently, possibly from condition (1).

(61.2) Ulna less than twice the length of metacarpal IV (ul/mcIV<2).

(66.2) First phalanx of manual digit IV less than twice the length of metacarpal IV.

(71.2) Femur about same length or less than metacarpal IV (fe/mcIV≤1.00).

(73.2) Fifth pedal digit reduced with one or no phalanx. The trend of reducing the fifth digit is not unique to pterosaurs and has occurred elsewhere, e.g. among dinosaurs. *Herrerasaurus*, one of the outgroups used in this analysis, also has only one phalanx on the pedal digit V. Within pterodactyloids where the pes is known, *Pteranodon* apparently lacks a phalanx on pedal digit V, while *Pterodactylus* and *Tapejara* still have one.

Remarks. The Pterodactyloidea is the best supported clade in the present analysis (nine synapomorphies). The monophyly of this group was never seriously questioned. The major exception was Hooley (1913), who suggested that *Istiodactylus* (previously called *Ornithodesmus* see Howse *et al.* 2001) was related to *Scaphognathus* (both classified by Hooley in the Scaphognathidae). *Istiodactylus*, however, is well nested within Pterodactyloidea, as has been suggested before (e.g. Wellnhofer 1978).

Node 14. Archaeopterodactyloidea, Kellner 1996 formed by *Pterodactylus*, *Germanodactylus*, Ctenochasmatidae and Gallodactylidae.

Definition. The most recent common ancestor of *Pterodactylus*, *Ctenochasma* and *Gallodactylus* and all their descendants.

Recorded temporal range. Tithonian (Late Jurassic) to Albian (Early Cretaceous).

Synapomorphies.

(15.1 & 2). Presence of laterally placed nasal process. A nasal process is present in several pterodactyloid taxa. In the Archaeopterodactyloidea, it is situated laterally (character 15, states 1 and 2). The shape of this process, however, varies and is either long and oriented vertically (*Pterodactylus*, *Germanodactylus*, *Pterodaustro*), or reduced (*Gallodactylus*, *Cycnorhamphus*). The sole exception is *Ctenochasma*, where this process is not present, either due to poor preservation or because it has been secondarily lost in this taxon.

(19) Posterior region of skull rounded with squamosal displaced ventrally. In *Pterodactylus*, *Germanodactylus*, *Cycnorhamphus suevicus* and the Ctenochasmatidae, the posterior region of the skull is rounded, with the squamosal shifted ventrally to a position that levels with the bottom part of the orbit (condition in *Gallodactylus canjuersensis* cannot be determined, although it is probably the same). In all other pterosaurs where this feature can be observed, the squamosal is positioned near the top level of the orbit.

(20.2) Quadrate inclined backwards for about 150° relative to ventral margin of skull. All archaeopterodactyloids have the quadrate inclined backwards relative to the margin of the skull, more than in any primitive pterosaur taxa and in dsungaripteroids. The condition in *Gallodactylus canjuersensis* is unknown. No measurements of the inclination of the quadrate could be taken in the Nyctosauridae. According to Bennett (2001), the inclination of the quadrate in *Pteranodon* varies. In the specimens that could be examined (AMNH 7515 and 5099) the angle was less than 145°. It has to be noted that all specimens of *Pteranodon* are very

compressed laterally, which can introduce errors to the measurements.

(45.1) Mid-cervical vertebrae elongated. All archaeopterodactyloids (condition in *Gallodactylus canjuersensis* unknown) have elongated mid-cervical vertebrae but not to the same degree as azhdarchids (node 33).

(47.2) Neural spines of mid-cervical vertebrae low and blade like. Pterodactylus, Cycnorhamphus suevicus and the Ctenochasmatidae have very characteristic neural spines of the cervical vertebrae, which are low and blade-like. The condition in *Germanodactylus* and *Gallodactylus canjuersensis* is unknown.

Remarks. A clade uniting *Pterodactylus, Germanodactylus*, Gallodactylidae and Ctenochasmatidae has been suggested before by Kellner (1996a, 1997). It is likely that several other pterodactyloids not discussed in the present analysis are members of this group (e.g. *Gnathosaurus*). Although very well supported in the present analysis, it should be noted that all specimens of archaeopterodactyloids are poorly prepared, despite some being found in large numbers (particularly *Pterodactylus*), hindering the detailed observation of several important anatomical features (e.g. articulation surface between quadrate and lower jaw, occipital region, features of the palate, detailed structure of the carpus).

Node 15. Unnamed taxon formed by Pterodactylus antiquus, Pterodactylus kochi and Germanodactylus.

Recorded temporal range. Tithonian (Late Jurassic).

Synapomorphies.

(15.1) Nasal process present on lateral side of skull, straight and directed ventrally, not connected with maxilla. Non-pterodactyloid pterosaurs have an extended nasal process that was connected with a maxillary process and separated the naris from the antorbital fenestra. A laterally placed nasal process, vertically oriented (but not contacting the maxilla) is present only in *Germanodactylus* and *Pterodactylus* (Wellnhofer 1970), and *Pterodaustro*. Other pterosaurs also have a nasal process which differs by being positioned medially (e.g. *Anhanguera,* see Wellnhofer 1985; Kellner & Tomida 2000) or directed anteriorly (*Gnathosaurus*, not included in the present study).

(39.2) More than 15 peg-like teeth on each side of the jaws. Peg-like teeth are present in the Anurognathidae, but they are reduced in number. In *Germanodactylus* and *Pterodactylus* the dentition is formed by more than 15 small peg-like teeth on each side of the jaws.

Remarks. Young (1964) placed *Germanodactylus* within a new family (Germanodactylidae) that he regarded as being related to *Dsungaripterus* (Dsungaripteridae), a view reiterated by Wellnhofer (1978). Unwin (1995) also regarded *Germanodactylus* as closely related to the Dsungaripteridae, based on the following characters: (1) toothless jaw tips; (2) short and broad maxillary teeth; (3) largest maxillary teeth located caudally; and (4) distal ends of paroccipital processes expanded.

In the holotype of *Germanodactylus cristatus*, only the tip of the jaws are toothless (as in some non-pterodactyloids, such as *Rhamphorhynchus*), which is not the case in *Germanodactylus rhamphastinus*, while in the dsungaripterids *Dsungaripterus* and '*Phobetor*' the whole anterior portion of the rostrum lacks teeth. The dentition in *Germanodactylus* is formed by small teeth that have basically the same shape and structure as those present in *Pterodactylus* but differ from the broad teeth of dsungaripterids. Furthermore, the largest teeth of *Germanodactylus* are not located caudally, as can be seen on the left mandibular ramus in the holotype of e.g. *G. cristatus* (*contra* Unwin 1995). *Germanodactylus* also lacks the degree of expansion observed in the paroccipital processes of *Dsungaripterus* (and '*Phobetor*'). Another difference is found in the shape and extension of the premaxillary crest, which is much shorter and lower in *Germanodactylus* (character 12).

In the present analysis no synapomorphic character was found to support a close relationship of *Pterodactylus antiquus* and *Pterodactylus kochi* relative to *Germanodactylus*.

Node 16. Unnamed taxon formed by *Germanodactylus cristatus* and *Germanodactylus rhamphastinus*.

Recorded temporal range. Tithonian (Late Jurassic).

Synapomorphies.

(12.3) Premaxillary sagittal crest low, displaced near anterior margin of nasoantorbital fenestra, reaching skull roof but not extended backwards. Premaxillary sagittal crests are present in several pterodactyloid taxa. In both species of *Germanodactylus* this crest is low, starts near the anterior margin of the nasoantorbital fenestra, and reaches the skull roof above the orbit but does not extend backwards. This cranial crest starts in a similar position in Dsungaripteridae (condition in *Noripterus* is not known), but in the latter it is much stronger and extends posteriorly above the occipital region.

Node 17. Unnamed taxon formed by Ctenochasmatidae and Gallodactylidae.

Recorded temporal range. Tithonian (Late Jurassic) to Albian (Early Cretaceous).

Synapomorphy.

(1.1) Dorsal margin of skull concave. Ctenochasmatids and gallodactylids have the dorsal margins of the skull concave instead of straight or

curved downwards. This feature is also present in *Pteranodon* (see Bennett 2001) and, according to the present hypothesis, developed independently in those taxa.

Remarks. The close relationship between ctenochasmatids and gallodactylids was previously presented by Kellner (1996a, 1997). Only one synapomorphy supports this relationship, indicating that further studies are necessary to confirm the present hypothesis.

Node 18. Ctenochasmatidae Nopcsa 1928

Definition. The most recent common ancestor of *Ctenochasma* and *Pterodaustro* and all their descendants.

Recorded temporal range: Tithonian (Late Jurassic) to Albian (Early Cretaceous).

Synapomorphies.

(3.2) Extremely elongation of the rostrum, reaching more than 60% of skull length. An elongated portion of the skull anterior to the external nares is a feature that diagnoses all non-anurognathid pterosaurs (node 4). Ctenochasmatids have taken this feature a step further and show an extreme elongation of the anterior portion of the skull.

(40) Dentition formed by over 150 long and slender teeth. This feature is unique to *Ctenochasma* and *Pterodaustro*, suggesting that they are sister-taxa. All other pterosaurs have a smaller number of teeth, even those where the teeth are long and thin (e.g. *Anhanguera*).

Remarks. The condition of the teeth in *Pterodaustro* further differs by being filiform. This is an apomorphic character for this taxon, as has been previously recognized by Bonaparte (1971) and Sanchez (1973), and was confirmed by more specimens (Chiappe *et al.* 2000)

Unwin (1995) used the inclination of the quadrate to support a close relationship between Ctenochasmatidae and Pterodactylidae (although it is not clear which taxa he included in those more inclusive taxonomic units). However, *Cycnorhamphus suevicus*, *Germanodactylus* and *Pterodaustro* all have a similar inclination of the quadrate relative to the ventral margin of the skull, especially when compared with *Pterodactylus* (see Wellnhofer 1978, p.8). Therefore this feature seems to diagnose a more-inclusive group than previously supposed. The existence of another condition, where the quadrate is even more inclined than in some of the abovementioned taxa, remains to be demonstrated.

To further support the Ctenochasmatoidea (Ctenochasmatidae + Pterodactylidae), Unwin (1995) listed the low position of the squamosal (character 19) as a synapomorphy. This feature, however, is present in *Cycnorhamphus suevicus* and in *Pterodaustro*, and applies at a higher taxonomic level (the Archaeopterodactyloidea, node 14). According to the hypothesis of pterosaur phylogeny presented here, the Ctenochasmatoidea of Unwin (1995) is paraphyletic.

Node 19. Gallodactylidae Fabre 1974

Definition. The most recent common ancestor of *Gallodactylus* and *Cycnorhamphus* and all their descendants.

Recorded temporal range. Tithonian (Late Jurassic).

Synapomorphies.

(15.2) Nasal process placed on lateral side of skull and reduced. Several pterodactyloid taxa have a nasal process. This structure is positioned laterally in all archaeopterodactyloids (absent in *Ctenochasma*) but in Gallodactylidae it is reduced.

(18.2) Parietal crest present, laterally compressed, expanded posteriorly with rounded posterior margin. Parietal crests are present in several pterodactyloid taxa, but gallodactylids are the sole members of the Archaeopterodactyloidea to have one. It is unique among pterosaurs by being comparatively extended and showing a rounded posterior margin.

(34.2) Teeth confined to anterior portion of jaws.

Remarks. In *Cycnorhamphus suevicus*, the length of the humerus + ulna is about 80% or less than the length of femur + tibia (character 56, state 1), which is the same condition found in the outgroup and in the Tapejaroidea (the condition in *Gallodactylus canjuersensis* is unknown). This differs from the condition found in all other pterosaurs, where the humerus + ulna are proportionally larger, and might differentiate gallodactylids from other archaeopterodactyloids.

When describing a new pterosaur found in Cretaceous strata of southern France, Fabre (1974, 1976) erected the clade Gallodactylidae to classify *Gallodactylus canjuersensis* and the Late Jurassic *'Pterodactylus' suevicus*, which he renamed *Gallodactylus suevicus*. Reviving the taxonomic status of the latter, Bennett (1996a) revalidated the name *Cycnorhampus*, regarding *Gallodactylus* as a junior synonym and maintained two valid species: *C. suevicus* and *C. canjuersensis*. There are, however, several differences between these species, particularly the proportions of the preserved limb elements (particularly forelimb and hindlimb). Therefore, until a more comprehensive study of both taxa is made, and in order to avoid more nomenclatural confusion (the same that occurred with *Cycnorhamphus,* see Bennett 1996a), I opt for a more conservative approach maintaining the generic status of *Gallodactylus* and confining it to the species *G. canjuersensis*.

Node 20. Dsungaripteroidea Young 1964

Definition. The most recent common ancestor of *Nyctosaurus* and *Quetzalcoatlus* and all their descendants.

Recorded temporal range. Kimmeridgian–Tithonian (Late Jurassic) to Maastrichtian (Late Cretaceous).

Synapomorphies.

(34.3) Jaws toothless. All primitive pterosaurs have teeth. The basal dsungaripteroids (Nyctosauridae, *Pteranodon*) are toothless, as are all azhdarchoids (Tapejaridae + Azhdarchidae, node 30). Some taxa nested within this clade, however, have well-developed teeth (e.g., *Anhanguera, Istiodactylus, Ornithocheirus, Dsungaripterus*). Therefore, according to the present cladistic analysis, the toothless condition is a synapomorphy of dsungaripteroids, but is reversed in the clade composed of *Istiodactylus* + *Ornithocheirus* + Anhangueridae and in the Dsungaripteridae.

(41) Notarium present. In cladistically primitive pterosaurs the dorsal vertebrae do not fuse into a notarium. This structure is present in dsungaripteroids, where it is formed by the fusion of the centra and the neural spines of at least three dorsal vertebrae. The number of vertebrae involved may vary depending on the taxon and also on the ontogenetic stage of the specimen: the older the individual, more dorsal vertebrae tend to be fused, forming the notarium (similar to the condition of the sacrum, pers. obs.). The only tapejaroids where a notarium was not found are *Tapejara* and *Anhanguera*, complete specimens of which are exclusively non-adults.

(42) Atlas and axis fused.

(43) Postexapophyses on cervical vertebrae present.

(57. 1–3) Presence of pneumatic foramen on proximal part of humerus. The humerus in non-dsungaripteroids lacks any pneumatic foramen on the humerus (proximally or distally; condition in the Anurognathidae, Gallodactylidae and Ctenochasmatidae unknown). This feature is present in all dsungaripteroid taxa, although its position may vary. In *Anhanguera* and *Noripterus* it is present on the dorsal side while in *Nyctosaurus gracilis* (Bennett 1994), *Pteranodon* (Kellner & Hasegawa 1993; Bennett 2001), *Tupuxuara, Quetzalcoatlus* sp., and *Azhdarcho* (see Nessov & Jarkov 1989, pl. II, fig. 8) it is present on the ventral side. *Tapejara wellnhoferi* differs by having two pneumatic foramina (on the dorsal and ventral side). The condition of this character is not known in *Dsungaripterus, Istiodactylus* and *Tropeognathus*.

(64.2) Pteroid more than half the length of ulna. Complete pteroids of pterodactyloid taxa are generally unknown. In *Nyctosaurus gracilis* and *Pteranodon* the pteroid is longer than half the length of the ulna, in contrast to the short condition found in cladistically more primitive taxa (e.g. *Rhamphorhynchus, Pterodactylus*). No information about the extent of this bone is known in most dsungaripteroids.

Remarks. There are three more potential synapomorphies of this clade. With the exception of *Istiodactylus* and *Tapejara wellnhoferi*, the connection between skull and lower jaw is made by a helical jaw joint (character 22), as observed in *Pteranodon, Anhanguera, Tropeognathus, Dsungaripterus, Tupuxuara* and *Quetzalcoatlus*. Where observable, all dsungaripteroids have a comparatively elongated basisphenoid (character 26; condition in Azhdarchidae unknown) and an interpterygoid opening that is smaller than the subtemporal fenestra (character 29, state 2; condition in Azhdarchidae unknown). The states of all three characters, however, are unknown in the Archaeopterodactyloidea, the sister-group of Dsungaripteroidea, and could diagnose a more inclusive group.

The Dsungaripteroidea is a well-supported clade nested within the Pterodactyloidea. The first researcher to recognize the Dsungaripteroidea was Young (1964), who based it essentially on the presence of a fused notarium (character 41). Bennett (1994, p.54) renamed this clade (Euptero-dactyloidea) since it was unclear to him what Young (1964) meant by a notarium. The present analysis suggests that Dsungaripteroidea is monophyletic and essentially groups together the same taxa that were used by both authors. Therefore the term used by Young (1964) has priority.

Node 21. Nyctosauridae (= Nyctosaurinae) Williston 1903

Definition. All pterosaurs more closely related to *Nyctosaurus gracilis* than to other pterosaurs. The Nyctosauridae is composed of *Nyctosaurus gracilis* and *Nyctosaurus bonneri*. '*Nyctosaurus*' *lamegoi* is also considered a member of this clade (Price, 1953).

Recorded temporal range. Santonian to Maastrichtian (Late Cretaceous).

Synapomorphies.

(54.3) Humerus less than 40% of metacarpal IV length (hu/mcIV<0.40). The Nyctosauridae (*Nyctosaurus gracilis* and *Nyctosaurus bonneri*) have the largest proportional wing metacarpal, which is unique among pterosaurs.

(58.5) Deltopectoral crest of humerus hatched-shaped, displaced away from caput humeri. The condition of the deltopectoral crest of the humerus is similar to the condition found in Rhamphorhynchus, differing essentially by being positioned further down the humerus shaft.

Remarks. Nyctosaurids also have metacarpals I-III not articulating with the carpus (character 65, state 2), a condition also reported in *Pteranodon* (Bennett 2001)

According to Brown (1986), one specimen (UNSM 93000) attributed to *Nyctosaurus gracilis* presents only three phalanges in the wing finger, which he regarded as unique to this taxon. Bennett

(1989, 1994) used this feature to support the monophyly of Nyctosauridae. Miller (1972a), however, reported the presence of a ph4d4 in the type specimen of *Nyctosaurus bonneri* (wrongly labelled *Nyctosaurus gracilis*; see Miller 1972a, p.16) and in the '*Nyctosaur*' specimen SMM 7968 (see Miller 1972b, p.32). Therefore it has still to be demonstrated that nyctosaurids have a reduced number of phalanges in the wing finger, a feature that would be unique among pterosaurs.

Node 22. Ornithocheiroidea Bennett 1994
Definition. The most recent common ancestor of *Anhanguera, Pteranodon, Dsungaripterus* and *Quetzalcoatlus* and all their descendants.
Recorded temporal range. Kimmeridgian–Tithonian (Late Jurassic) to Maastrichtian (Late Cretaceous).
Synapomorphies.

(18.3) Parietal crest constituting base of posterior portion of cranial crest. The presence of a parietal crest constituting the base of the cranial crest is a character shared by *Pteranodon* and some other ornithocheiroids. In Anhanguerids, however, this crest is reduced to a blunt and short extension (condition in *Istiodactylus* and the Azhdarchidae unknown). Other pterosaurs also have a parietal crest, but it differs by being expanded and by forming the major part of the cranial crest (gallodactylids).

(21.2) Articulation between skull and mandible positioned under anterior half of orbit. In the more primitive pterosaurs, as in the Archosauria in general, the articulation of the skull and mandible is positioned under the posterior half of the orbit or further backwards. In rhamphorhynchids, archaeopterodactyloids and nyctosaurids, this articulation is shifted under the middle portion of the orbit. In all ornithocheiroids the skull–mandible articulation is shifted even more, lying under the anterior half of the orbit (*Anhanguera, Tupuxuara*) or even reaching the posterior margin of the nasoantorbital fenestra (e.g. *Pteranodon*).

(44) Presence of a lateral pneumatic foramen on centrum of cervical vertebrae. Where known, the cervical vertebrae in ornithocheiroids present a lateral pneumatic foramen, except in the Azhdarchidae. According to Howse (1986, p.315), this feature is absent in *Nyctosaurus*. In *Rhamphorhynchus* there are some pneumatic foramina beneath the diaphysis for the cervical ribs, but not on the centrum (see Wellnhofer 1975b). Howse (1986) also postulated the presence of a pneumatic foramen in *Pterodactylus*, but this observation could not be confirmed in the specimens of this taxon that were examined. Therefore the presence of a lateral pneumatic foramen on the centrum of the cervical vertebra appears to be a synapomorphy of the Ornithocheiroidea that was secondarily lost in the Azhdarchidae.

Remarks. In the present analysis, two more characters have fallen at this node. The first one is the presence of a foramen pneumaticum piercing the supraoccipital (character 24), which is found in all ornithocheiroids where the occipital region of the skull is known (e.g., *Pteranodon, Anhanguera, Tupuxuara*). Without access to any nyctosaurid specimen where the occipital region was well preserved I was unable to confirm the condition in this clade. It is very likely that *Nyctosaurus* and their kin also had such a foramen and this character might diagnose a more inclusive group (perhaps at the level of Dsungaripteroidea). The second feature is the presence of a deep and short cristospine (character 53, state 2). Unfortunately the shape and extension of this typical pterosaurian structure is unknown in most taxa. Furthermore, *Pteranodon* apparently has a long and shallow cristospine (Bennett 2001).

Node 23. Pteranodontoidea Kellner 1996
Definition. The most recent common ancestor of *Anhanguera* and *Pteranodon* and all their descendants.
Recorded temporal range: Barremian (Early Cretaceous) to Campanian (Late Cretaceous).
Synapomorphies.

(47.1) Neural spines of mid-cervical vertebrae tall and spike-like.

(50) Proximal surface of scapula suboval.

(49.1) Scapula shorter than coracoid (1 > sca/cor >0.80). In *Nyctosaurus gracilis* the scapula is also shorter compared to the coracoid, but not to the same degree as in the Pteranodontoidea. A still shorter scapula relative to the coracoid (character 49, state 2) is found in *Anhanguera* (see node 27).

(58.6) Deltopectoral crest of humerus warped.

(59.1) Medial crest of humerus directed posteriorly. A developed medial (=ulnar) crest in the humerus has only been reported in ornithocheiroids. All members of the Pteranodontoidea where postcranials are available show a comparatively thin crest that is directed posteriorly (condition in *Tropeognathus, Ornithocheirus compressirostris* and *Anhanguera blittersdorffi* is unknown), and is here regarded as a synapomorphy of this group.

(60) Distal end of humerus subtriangular. In all pteranodontoids where the humerus is known, the distal end of this bone is subtriangular, differing from the oval or D-shaped condition present in all other pterosaurs. The condition in the Nyctosauridae, the sister-group of Pteranodontoidea + Tapejaroidea, is unknown.

Remarks. Pteranodon lacks two characters present in *Istiodactylus* and anhanguerids (see below) and therefore seems to be the most primitive member of the Pteranodontoidea. *Pteranodon* also has the following autapomorphies: a large cranial crest composed essentially of the frontal (character 17, state 3) and a very reduced interpterygoid opening (proportionally the smallest in all pterosaurs where the palatal region is known). *Pteranodon* is also the only taxon in which the skull–mandible articulation is shifted anteriorly under the posterior margin of the nasoantorbital fenestra (an extreme condition of character 21).

Node 24. Unnamed taxon composed of *Istiodactylus*, *Ornithocheirus compressirostris* and the Anhangueridae.
Recorded temporal range. Barremian (Early Cretaceous) to Cenomanian (Late Cretaceous).
Synapomorphies.
(51.1) Stout scapula, with constricted shaft. The scapula of *Pteranodon*, the most basal pteranodontoid, has no particular constriction on the shaft. *Anhanguera piscator*, *Anhanguera santanae* and *Istiodactylus* have a stout scapula, with a well-developed constriction on the middle portion of the shaft (condition in *Ornithocheirus compressirostris* and *Tropeognathus* unknown).
(62.2) Diameter of radius less than half that of ulna. In *Istiodactylus* and all anhanguerids where postcranial material is known, the diameter of the radius is very reduced, less than half the diameter of the ulna, a unique condition within pterosaurs.
Remarks. All members of this clade have teeth that are evenly distributed along the jaws (character 30, state 0), contrasting with *Pteranodon* and most dsungaripteroids (dsungaripterids are also an exception and have strong, short teeth, node 29). This condition is regarded as a reversal to the primitive state, present in most archosaurs and all non-dsungaripteroids. It has to be noted, however, that the dentition of those taxa is very different. In *Istiodactylus*, teeth are laterally compressed, comparatively short, triangular in shape and subequal in size (an autapomorphy of this taxon). The upper and lower teeth are intercalated during occlusion, with the labial surface of the upper teeth contacting the lingual edges of the lower teeth. The teeth in anhanguerids are thin and long, particularly the anterior ones, contrasting markedly with the ones of *Istiodactylus*.

Another autapomorphy of *Istiodactylus* was identified in this study: the low and rounded anterior part of the rostrum. Most pterosaur taxa have the rostral portion laterally compressed and pointed. The Anurognathidae constitute an exception since the complete skull is apparently broad but comparatively high. *Istiodactylus* differs by having the rostral part of the skull low, wider than high, and the anterior margin rounded.

Node 25. Unnamed taxon composed of *Ornithocheirus compressirostris* and the Anhangueridae.
Recorded temporal range. Albian (Early Cretaceous) to Cenomanian (Late Cretaceous).
Synapomorphy.
(27.1) Discrete palatal ridge, tapering anteriorly. The only feature that unites *Ornithocheirus compressirostris* with the Anhangueridae is the presence of a palatal ridge. In *Tropeognathus*, the palatal ridge is more developed than in *Anhanguera* and in *Ornithocheirus compressirostris* (character 27, state 2). One of the problems with the latter is its incompleteness. Nevertheless, *Ornithocheirus compressirostris* lacks all cranial anhanguerid synapomorphies and lies outside of the Anhangueridae (node 26).

Node 26. Anhangueridae Campos & Kellner 1985
Definition. The most recent common ancestor of *Anhanguera* and *Tropeognathus* and all their descendants.
Recorded temporal range. Albian (Early Cretaceous) to Cenomanian (Late Cretaceous).
Synapomorphies.
(12.1) Premaxillary sagittal crest confined to anterior portion of skull.
(13) Tip of premaxilla slightly expanded. *Gnathosaurus* (not included in the present study) also has the anterior tip of the premaxillae expanded, but differs from the anhanguerid condition by being 'spoon-shaped'.
(33.1) Short, blade-like dentary sagittal crest. No pterosaur where the lower jaw is known has a blade-like sagittal crest positioned on the ventral margin of the anterior part of the dentary; this feature is unique to anhanguerids. *Tapejara* also has a dentary sagittal crest, but this is very deep and massive, unlike the condition found in the Anhangueridae.
Remarks. There are at least two more potential synapomorphies of this clade: the presence of a low and blunt bony frontal and parietal (characters 17, state 1 and 18, state 1). Such crests are present in *Anhanguera* and *Tropeognathus* and differ from the condition found in *Pteranodon* (the condition in *Ornithocheirus compressirostris* and *Istiodactylus* is unknown). Therefore, although this character is very likely a synapomorphy of the Anhangueridae, it could potentially apply to a more-inclusive group.

The taxon Anhangueridae was first identified as a distinct clade of pterodactyloids by Campos & Kellner (1985). Some taxa that are here regarded as members of this clade have been previously classified in the Ornithocheiridae (*Anhanguera santanae*) and in the Criorhynchidae (*Tropeognathus*). Those two pterosaur clades are based on very fragmentary material and have become a 'waste-basket' for several specimens, particularly from the Cambridge Greensand, some of which might belong to different pterosaur groups (see discussion in Kellner 1990). In any case, the present analysis demonstrates that *Ornithocheirus compressirostris* is closely related to, but lies outside of, the Anhangueridae.

Anhanguerids were first found in the Albian Romualdo Member of the Santana Formation. Confirmed specimens are known from Cenomanian strata of Morocco (Mader & Kellner 1999) but are likely present in several other deposits (see Kellner & Tomida 2000)

Node 27. Anhanguera Campos & Kellner 1985
Recorded temporal range. Albian (Early Cretaceous) to Cenomanian (Late Cretaceous).
Synapomorphies.

(15.3) Long nasal process placed medially.

(16) Foramen perforating nasal process.

(36) Particular variation in size of anterior teeth, with 5th and 6th smaller than 4th and 7th.

(49.2) Scapula substantially shorter than coracoid (sca/cor <0.80).

(52.2) Coracoidal contact with sternum forming an oval articulation surface, with posterior expansion.

(57.2) Pneumatic foramen on proximal part of humerus situated dorsally. This feature is also present in *Noripterus*. *Tapejara wellnhoferi* has two pneumatic foramina (on the dorsal and ventral side near the proximal articulation of the humerus).

(65.1) Metacarpal III long, articulating with carpus; metacarpals I and II reduced.

Remarks. There are several additional postcranial features shared by *Anhanguera piscator* and *Anhanguera santanae*, the only members of the five species reported for this genus that show enough of the postcranial skeleton to allow comparisons. Among these is the presence of a small ventral tuberculum (Wellnhofer 1991a; Kellner & Tomida 2000) so far unique among pterosaurs. The deltopectoral crest of those taxa has the distal end of the warped deltopectoral crest expanded, with a marked distal concave surface. The main problem in correctly evaluating these features (and several listed above) is the lack of postcranial material for *Tropeognathus mesembrinus* and *Ornithocheirus compressirostris*. Therefore, some of the synapomorphies presented above might diagnose a more-inclusive group.

Node 28. Tapejaroidea Kellner 1996
Definition. The most recent common ancestor of *Dsungaripterus, Tapejara* and *Quetzalcoatlus* and all their descendants.
Recorded temporal range. Kimmeridgian–Tithonian (Late Jurassic) to Maastrichtian (Late Cretaceous).
Synapomorphies.

(17.2) Low and elongated frontal crest.

(23) Supraoccipital extending backwards. In all tapejaroids where the occipital region of the skull is known, the supraoccipital is extended backwards and forms the basis of the cranial crest (condition in the Azhdarchidae unknown). In no other pterosaur does this seem to be the case, not even in *Pteranodon*, which also has a large cranial crest that extends backwards (Bennett 2001).

(25) Expanded distal ends of paroccipital processes.

(59.2) Medial crest of humerus massive, with developed proximal ridge.

Remarks. Besides the above, in all dsungaripteroid the humerus is about 80% or less than of the femur length (character 55, state 0), which is a reversal to the primitive state. The same happens with the forearm relative to the hindlimb (character 56, state 0), with the proportions of humerus + ulna relative to the femur + tibia being about 80% or less (hu + ul/fe + ti <0.80). Tapejaroids, but also *Cycnorhamphus suevicus*, reverse to the primitive condition. Another feature that falls out at the tapejaroid node is the length of metatarsal III less than 30% of tibia length (character 72). The primitive condition among pterosaurs is to have the metatarsal III more than 30% of the tibia length; tapejaroids (condition in the Azhdarchidae unknown) have a comparatively smaller metatarsal III. This feature, however, is also found in *Germanodactylus cristatus* (Unwin *et al.* 2000), gallodactylids and *Anhanguera piscator*, and therefore its phylogenetic signal is unclear.

Other features might also be unique of this group, but could not be evaluated at the present time. Among these is the particular triangular proximal articulation surface of the wing metacarpal (present in *Tupuxuara leonardii*, *Tapejara wellnhoferi* and *Dsungaripterus*; condition in azhdarchids unknown) that differs from the more subrectangular outline of this bone in *Rhamphorhynchus* and anhanguerids (condition in *Pteranodon* and *Nyctosaurus* unknown). The dorsal articulation surface of this bone in tapejaroids is formed by a quadratic projection that fits in a deep excavation into the distal carpal series, differing particularly from anhanguerids.

Kellner and Hasegawa (1993) suggested that *Dsungaripterus* is the sister-taxon to Azhdarchidae + Tapejaridae. The present analysis supports this relationship (i.e. the Dsungaripteridae as sister-group of the Azhdarchoidea).

Node 29. Dsungaripteridae Young 1964
Definition. The most recent common ancestor of *Dsungaripterus, Noripterus, 'Phobetor'* and all their descendants.
Recorded temporal range. Berriasian–Hauterivian (Early Cretaceous).
Synapomorphies.

(9) Orbit comparatively small and positioned very high in skull.

(11) Suborbital opening present.

(12.2) Premaxillary sagittal crest high, displaced, near anterior margin of nasoantorbital fenestra, reaching skull roof above orbit, and extending backwards.

(14) Posterior ventral expansion of maxilla.

(34.1) Teeth absent from anterior portion of jaws. Although *Dsungaripterus* and '*Phobetor*' have teeth (also *Noripterus*), the rostral part of the jaws are toothless (condition of *Noripterus* is unknown since the only known portion of the jaws lacks the rostral part).

(35) Largest maxillary teeth positioned posteriorly.

(37) Teeth with broad, oval base. The main features that diagnose the Dsungaripteridae are found in the dentition. Only the members of the Dsungaripteridae have broad and short teeth, with an oval base. Several of those teeth show wear facets on the tip of the crown. Although no complete skull or mandible is known for *Noripterus*, the jaw fragment figured by Young (1973) indicates that the latter has the same dentition as '*Phobetor*' and *Dsungaripterus*.

Remarks. It should be noted that the only character supporting this node that is found in all three members of this clade is character 37. *Dsungaripterus* is known by some specimens (Young 1964, 1973) but the remains of *Noripterus* (Young 1973) and '*Phobetor*' (Bakhurina 1982; Wellnhofer 1991b) are incomplete. Therefore, all other likely synapomorphies of this clade are unknown, for either *Noripterus* or '*Phobetor*'. Only more complete material of those taxa will show if they diagnose the Dsungaripteridae or a less-inclusive group within this clade. *Dsungaripterus* further differs from all other pterosaurs by having the rostral part of the skull upwards (character 1, state 2).

Node 30. Azhdarchoidea Unwin 1995
Definition. The most recent common ancestor of *Tapejara* and *Quetzalcoatlus* and all their descendants.
Recorded temporal range. Kimmeridgian–Tithonian (Late Jurassic) to Maastrichtian (Late Cretaceous).
Synapomorphies.

(10) Orbit positioned lower than dorsal rim of nasoantorbital fenestra.

(68.3) Second phalanx of manual digit IV more than one-third smaller than first phalanx of manual digit IV (ph2d4/ph1d4 <0.70).

Remarks. A sister-group relationship between Azhdarchidae and Tapejaridae, as previously suggested (Kellner & Campos 1992; Kellner & Langston 1996) is supported in the present analysis. Kellner (1989) originally regarded the particular position of the orbit as an autapomorphy of *Tapejara*. Since then more complete cranial material of *Tupuxuara* (Kellner & Hasegawa 1993) and *Quetzalcoatlus* sp. (Kellner & Langston 1996) became available, showing that this feature (character 10) diagnose a more-inclusive group, the Azhdarchoidea. As pointed out by Kellner (1996a), *Zhejiangopterus* (not included in the present analysis) from Late Cretaceous deposits of China shows the same low position of the orbit in the skull and is also referable to the Azhdarchoidea rather than to the Nyctosauridae, as originally regarded by Cai & Wei (1994). This was latter corroborated by Unwin & Junchang (1997).

Node 31. Tapejaridae Kellner 1989
Definition. The most recent common ancestor of *Tapejara* and *Tupuxara* and all their descendants.
Recorded temporal range. Aptian (Early Cretaceous) to Cenomanian (Late Cretaceous).
Synapomorphies.

(8.2) Naris and antorbital fenestra confluent, longer than 45% of skull length.

(12.4) Premaxillary sagittal crest starting on anterior portion of skull and extending posteriorly above occipital region.

Remarks. In this analysis, *Tupuxuara leonardii* shows one synapomorphy: a strong palatal ridge that tapers anteriorly (character 27, state 3). This feature contrasts with the flat condition found in the palate of *Tapejara*.

The monophyly of the Tapejaridae was first suggested by Kellner (1989) and was accepted by subsequent workers (e.g. Unwin & Junchang 1997). A detailed description of more tapejarid specimens is being elaborated (Kellner in prep.). Besides the Aptian–Albian Santana Formation deposits, tapejarids were recently recorded in Cenomanian strata of Morocco (Wellnhofer & Buffetaut 1999).

Node 32. Tapejara Kellner 1989
Recorded temporal range. Aptian–Albian (Early Cretaceous).
Synapomorphy

(4) Rostral end of the premaxillae/maxillae downturned. The particular configuration of the rostral end (character 4) is, so far, the only unequivocal synapomorphy that unites *Tapejara wellnhoferi* and *Tapejara imperator*. The former has two apomorphies: a massive and deep bony sagittal crest (character 33, state 2; Wellnhofer & Kellner 1991) and the presence of two pneumatic foramina on the dorsal and ventral side near the proximal articulation of the humerus (character 57, state 3). Since there is no lower jaw or postcranial material known for *Tapejara imperator* (Campos & Kellner 1997) it is not known if those features are unique to *Tapejara wellnhoferi* or constitute a synapomorphy of the genus *Tapejara*.

Node 33. Azhdarchidae Nessov 1984
Definition. All pterosaurs closer related to *Quetzalcoatlus* than to any other pterosaur.
Recorded temporal range. Kimmeridgian–Tithonian (Late Jurassic) to Maastrichtian (Late Cretaceous).

Synapomorphies.

(45.2) Mid-cervical vertebrae extremely elongated.

(47.3) Neural spines of mid-cervical vertebrae extremely reduced or absent.

Remarks. Besides the synapomorphies above, the cervical vertebrae of azhdarchids lack a lateral pneumatic foramen on the centrum of the cervical vertebrae (character 44, state 0). Primitively pterosaurian cervical vertebrae lack any lateral pneumatic foramen, on the centrum. In the Ornithocheiroidea the cervical vertebrae have such a foramen with the exception of the Azhdarchidae, which apparently lost this feature secondarily.

One apomorphic feature of *Quetzalcoatlus* sp. was discovered in the present analysis: the position of the premaxillary sagittal crest that starts on the posterior half of the nasoantorbital fenestra (character 12, state 5). This taxon also shows several unique features in the postcranial skeleton, such as a well-developed and deep coracoidal flange. Since almost all azhdarchid specimens are known from cervical vertebrae only (see Ikegami *et al.* 2000 for a review) it cannot be determined whether those features are unique to this taxon or shared by other azhdarchids.

Nessov (1984) was the first author to formally recognize the monophyly of the 'long-necked' pterodactyloids. He used the name Azhdarchinae which he regarded as a subfamily of the Pteranodontidae. At the same time, Padian (1984) proposed the family Titanopterygiidae for essentially the same group. Padian (1986) subsequently raised the Azhdarchinae to family level (i.e. Azhdarchidae) and considered Titanopterygiidae as a junior synonym of the latter.

Nessov (1984) used seven characters of the cervical vertebrae and notarium to define the Azhdarchidae: (a) very large pterosaurs; (b) atlas and epistropheus (axis) fused; (c) postaxial cervical vertebrae extremely long, with (d) cross-section near the middle portion round, and (e) three air canals present on the anterior part with the medial one above the neural canal; (f) notarium present, with (g) wide anterior unpaired articulation surface.

Although it is true that all known azhdarchids have a very large wing span (character a of Nessov 1994), including the so-far largest-known flying creature (Lawson 1975a, b), other groups of pterosaurs also include very large animals, such as *Pteranodon* (Pteranodontoidea) and *Tupuxuara* (Tapejaridae). Those pterosaurs are of comparable size to at least some azhdarchids, such as *Quetzalcoatlus* sp. (see Kellner & Langston, 1996). Therefore, although there seems to be a trend to more derived pterosaurs, particularly pterodactyloids, to increase their wing span, this feature cannot be used as a synapomorphy of the Azhdarchidae.

Two other supposedly azhdarchid characters indicated by Nessov (1984) (character b, fused atlas and axis, character c, and presence of a notarium) are present in several other pterodactyloids and diagnose a more-inclusive group (the Dsungaripteroidea of the present analysis). The shape and the proportion of the anterior surface of the notarium (character g) are difficult to pinpoint since this portion is unknown in most pterosaurs. More direct comparisons of the dorsal vertebrae that compose the notarium in dsungaripteroid taxa is needed in order better to understand the variation regarding the size and breath of the anterior surface of this structure.

Regarding the long cervical vertebrae (character 45, state 2 in the present analysis) Nessov (1984) clearly recognized that this was a shared derived feature of the 'long-necked' pterodactyloids. Padian (1984) pointed out that the members of this clade have three continuous hyper-elongated cervicals, possibly from the third to the fifth (see also Howse 1986). Although this might be the case, the only azhdarchid taxa with a complete series of cervicals is *Quetzalcoaltus* sp. In the latter, both subsequent cervical vertebrae to the fifth are also hyper-elongated, comparable in size at least to the third.

Nessov (1984), pointed out that the cross-section of the middle portion of the cervical vertebrae was rounded (character d). This feature is due to the reduction of the neural spine (Padian 1984; Howse 1986), which is also recognized here as an azhdarchid synapomorphy (character 47, state 3 of the present analysis). It is worthwhile mentioning that the neural spines of the third and seventh cervical vertebrae in *Quetzalcoaltus* sp. (the only azhdarchid taxon where a complete cervical series is known) are low but well developed, and gradually reduce in size towards the fifth cervical. Any variation between the cross section of the centrum within azhdarchids, although possible, has still to be demonstrated.

The last character that Nessov (1984) listed as diagnostic for the Azhdarchidae (character e) is the presence of three 'air canals' on the anterior articulation of the cervical vertebrae. Whether there are true canals, i.e. bony tubes parallel to the neural canal, cannot be determined. This feature seems unlikely, however, since it is not present in some vertebrae attributed to Azhdarchidae where the cross-section could be examined (Currie & Russell 1982).

Another synapomorphic character reported for azhdarchids is the presence of neural canals that are only ossified at their ends (Padian 1984, 1986). So far this feature was only observed by Currie & Russell (1984) in one cervical vertebrae which is attributed to *Quetzalcoaltus* sp. Therefore, until more cross-sections of azhdarchid vertebrae are available, this character must be regarded with caution.

Several long cervical vertebrae with reduced neural spines, which were found isolated, were also referred to the Azhdarchidae in the literature (see

Howse 1986; Kellner & Mader, 1996; Ikegami *et al.* 2000; Sayão & Kellner 2001).

Discussion

The phylogenetic results obtained here differ in several points from previous proposed pterosaur phylogenies. Based only on cervical vertebrae, Howse (1986) has essentially recognized eight different groups: (1) primitive 'rhamphorhynchoids' forming a basal trichotomy (*Eudimorphodon*, *Dimorphodon* and *Scaphognathus*); (2) advanced 'rhamphorhynchoids' (*Rhamphorhynchus*); (3) *Pterodactylus kochi* and *Pterodactylus elegans* forming a trichotomy with the remaining pterosaurs; (4) *Pterodactylus micronyx*, *Germanodactylus rhamphastinus* and *Gallodactylus canjuersensis* forming a polytomy with the remaining pterosaurs; (5) *Pterodactylus antiquus*, *Pterodactylus longicollum* and *Ctenochasma gracile* forming a polytomy with the remaining pterosaurs; (6) Azhdarchidae (*Doratorhynchus*, *Azhdarcho*, *Arambourgiana* (previously known as *Titanopteryx*, see Nessov & Jarkov 1989) and *Quetzalcoatlus*; (7) *Ornithocheirus* and *Nyctosaurus*; and (8) *Pteranodon*. The present study recognised the basal position of the primitive 'rhamphorhynchoids', the isolated position of *Rhamphorhynchus* (grouping 2) closely related to the pterodactyloids (the remaining taxa studied by Howse 1986) and grouping 6 (Azhdarchidae). However, neither the paraphyly of *Pterodactylus antiquus* and *Pterodactylus kochi* (separated by Howse in two distinct groups) was recognized, nor the sister-group relationship of *Nyctosaurus* and *Pteranodon*. The present study further differs in recognizing *Germanodactylus*, *Pterodactylus* (*P. kochi* + *P. antiquus*), *Gallodactylus* and *Ctenochasma* as part of a monophyletic group (the Archaeopterodactyloidea, node 14).

Bennett (1989) recognized four pterodactyloid clades: Nyctosauridae, Dsungaripteridae, Pteranodontidae and Azhdarchidae. All these groupings (some of which were recognized by previous authors, although comprising different taxa, e.g. Williston 1903; Young 1964;) were recovered in the present analysis. However, Bennett (1989) regarded Pteranodontidae (= Pteranodontoidea) as the sister-group of Azhdarchidae with Dsungaripteridae the next related taxon. The result obtained by the present study differs, i.e. in having dsungaripterids closer related to azhdarchids relative to pteranodontoids, as has been suggested before (Kellner & Hasegawa 1993).

In a more comprehensive review of pterodactyloid relationships, Bennett (1994) recognized 13 clades (for methodological problems of this analysis see Kellner 1995, 1996a). As in his previous study (Bennett 1989), four main groups of pterodactyloid pterosaurs were recognized (Nyctosauridae, Dsungaripteridae, Pteranodontidae and Azhdarchidae). On this occasion, Bennett (1994) regarded dsungaripterids as being more closely related to azhdarchids relative to pteranodontids (he did not include tapejarids in his analysis), as suggested by Kellner & Hasegawa (1993) and also supported by the present study. The main difference is the position of *Ornithocheirus*, which Bennett (1994) regarded as closely related to the Dsungaripteridae (not substantiated by any synapomorphy) and which here is in sister-group relationship with the Anhangueridae (node 25), within the clade Pteranodontoidea (node 23). Bennett (1994) could not resolve the relationships of *Pterodactylus*, *Germanodactylus*, *Pterodaustro*, *Ctenochasma* and *Cycnorhamphus suevicus*. This study indicates that these taxa are closely related, forming the clade Archaeopterodactyloidea (node 14), with *Pterodaustro* + *Ctenochasma* forming the Ctenochasmatidae (node 18), in sister-group relationship with *Cycnorhamphus suevicus* (+ *Gallodactylus canjuersensis*, node 17), followed by *Pterodactylus* + *Germanodactylus* (node 15).

The phylogenetic hypothesis presented by Unwin (1995) lacks a proper data matrix, the species content of the clades used as terminal taxa being therefore untestable. Among the main differences between his study and the present one is the position of *Preondactylus* and anurognathids. Unwin (1995) regarded the former as the most primitive pterosaur, probably following other authors (e.g. Wild 1984). The present analysis shows that *Preondactylus* is more derived relative to anurognathids and *Sordes* (node 5). Furthermore, the basal position of anurognathids (node 2) is justified by the lack of at least two synapomorphies that unite *Sordes* and all other pterosaurs: rostral part of the skull anterior to the external nares elongated (less than 60% of skull length; character 3, state 1) and the position of the external naris, located posterior to the premaxillary tooth row (character 6). A second important difference between Unwin´s study (repeated by Unwin & Lü 1997) is the position of *Germanodactylus*, regarded by Unwin as the sister-taxon of Dsungaripteridae. The present analysis shows both species of this genus (*G. cristatus* and *G. rhamphastinus*) forming a monophyletic group (node 16) well nested in the Archaeopterodactyloidea. Bennett (1994) also considered *Germanodactylus* to be not closely related to the Dsungaripteridae.

Comparing this analysis with a previous one (Kellner 1996a) no main difference was detected. The addition of seven taxa and eight new characters (and some changes in the character states) has not affected the main typology of the consensus tree in both studies. *Dendrorhynchoides* proved to be an anurognathid, as suggested by Unwin *et al.* (2000), closely related to *Batrachognathus* (node 3). *Preondactylus* forms a trichotomy with *Scaphognathus* and the remaining pterosaurs.

Perhaps the most surprising result of the present analysis is the close relationship of *Peteinosaurus zambellii* and '*Eudimorphodon*' *rosenfeldi* (node 8). The only feature supporting this relationship is the presence of multicusped teeth (character 38), a feature also present in the campylognathoid *Eudimorphodon ranzii*. This situation is complex since several features of these Triassic taxa that support this portion of the cladogram (nodes 8–10) cannot be properly compared because of the incompleteness of the specimens involved (*Peteinosaurus* lacks the skull, '*Eudimorphodon*' *rosenfeldi* has only the posterior portion of the skull preserved, and *Eudimorphodon ranzii* lacks complete wings). Therefore this part of the cladogram is expected to change as more information about these taxa becomes available.

Within pterodactyloids (node 13), the basic structure recovered by Kellner (1996a) is maintained and the addition of taxa did not change the topology of the tree or cause any clade to collapse.

Based on the cladogram that summarizes the current knowledge of the relationships of the studied taxa (Fig. 1), there are some general remarks that can be made about pterosaur evolution. Overall there is a trend towards increased size. Starting with anurognathids (node 2), which have an estimated wing span ranging c. 40–60 cm, all early forms have a wing span of less than 1 m (e.g. *Sordes, Scaphognathus, Peteinosaurus*), with *Dimorphodon* being the main exception (c. 1.4 m, Wellnhofer 1991b). Starting at the level of the Novialoidea (node 9), pterosaurs become larger, with wing spans greater than 1 m (e.g. *Campylognathoides*) up to less than 2 m (Rhamphorhynchidae). Within Archaeopterodactyloidea, sizes of adult animals were around 2 m, with a maximum of a. 2.5 m found in a few specimens attributed to *Pterodactylus* (Wellnhofer 1991b). True large forms developed only within the Dsungaripteroidea (node 20), particularly in the Pteranodontoidea (node 23), where several specimens of *Pteranodon* and *Anhanguera* have a wing span of over 5 m. The maximum recorded sizes for pterosaurs were found in the Azhdarchidae (node 33), to which several fossils (most incomplete) of very large animals have been assigned (e.g. Frey & Martill 1996). The largest flying animal ever recorded is the azhdarchid *Quetzalcoatlus northropi* (known from a partial wing), with an estimated wing span of c. 10 m. Within dsungaripteroids, the main exception is the tapejarid *Tapejara wellnhoferi* that has an estimated wing span of between 1.5 and 2 m.

Other general trends could be detected in the relations of several bones of the wing (Figs 5 & 6). The wing metacarpal tends to increase in size, reaching an extreme in nyctosaurids (node 21; Fig. 6 f). The pteroid bone (neomorph exclusively found in pterosaurs) also increases from the very short condition in anurognathids and *Sordes*, reaching its maximum size in the Dsungaripteroidea (node 20). Where known, the sizes of the first three wing phalanges vary in the earlier forms but, starting at the Novialoidea (node 9), assume the derived condition (ph1d4>ph2d4>ph3d4).

Regarding the skulls, the main observation relates to the variation of sizes and shapes (Figs 3 & 4). Among the general trends that could be detected is the increase of the rostral end (anterior to the external nares): while anurognathids show a comparatively short rostrum, there is general trend to increase the anterior portion of the skull and lower jaw which reaches an extreme in the Ctenochasmatids (node 18). There was also a gradual anterior shift of the skull–mandible articulation, with the quadrate changing from a more vertical or subvertical position, present in the basal members to the group, to a more inclined condition (about 120°) which was reached by *Rhamphorhynchus* and all other pterosaurs (node 11). In one clade, the Archaeopterodactyloidea (node 14), this inclination reached its maximum (about 150^0).

Crests vary in shape and size, but there is a tendency for those structures to be present in more derived forms, particularly in the Ornithocheiroidea (node 22); with the possible exception of *Istiodactylus*, all taxa where the skull is known show some kind of cranial crest. The extreme condition is found in *Tapejara imperator*, where the crest makes up about five-sixths of the lateral cranial surface (Campos & Kellner 1997). Anhanguerids and *Tapejara wellnhoferi* also have a sagittal crest on the dentary (Wellnhofer & Kellner 1991), which has developed independently.

Loss of teeth seems to be a general trend among dsungaripteroids (node 20) and was observed in the Nyctosauridae (node 21), *Pteranodon* (node 23) and azhdarchoids (node 30). This is a direct result of the position of *Pteranodon* as the basal member of the Pteranodontoidea. If the result of future work changes the current phylogenetic position of this taxon, placing it in a higher position within the Pteranodontoidea, loss of teeth will have to be reinterpreted as having occurred independently in nyctosaurids, azhdarchoids and *Pteranodon*. Dsungaripterids have a particular dentition, being comparatively short with a broad oval base (in *Dsungaripterus* showing subhorizontal wear facets, pers. obs.), *Istiodactylus* shows closely spaced subequal triangular teeth, and anhanguerids (node 26) have long, thin teeth, particularly the premaxillary ones.

Another important result of the present study is the

recognition of large gaps and obligatory ghost lineages of pterosaur evolution. One of the largest is the gap between the Campylognathoididae and the Rhamphorhynchidae (a direct consequence of the phylogenetic position *Eudimorphodon ranzii*). Large gaps are also found in the basal pterosaur lineages, notably between the anurognathids, *Sordes* and *Preondactylus*. This confirms the general perception among researchers who work with pterosaurs that the fossil record is extremely incomplete for this group of flying reptiles.

The recognition of anurognathids as the most primitive pterosaurs known to date does not have an immediate impact on the discussions about the phylogenetic position of the Pterosauria (e.g. Sereno 1991; Bennett 1996b; Peters 2000). These creatures are regarded as insectivores (e.g. Döderlein 1923; Rjabinin 1948; Wellnhofer 1991b) and are probably one of the few examples of inland living species. Although admittedly speculative, based on the basal position of the Anurognathidae, it is possible that the 'proto-pterosaurs' may have been animals which had a similar-shaped skull (broad and high) and which preyed on insects.

Conclusion

There are several problems that hamper pterosaur phylogenetic studies. Perhaps the main limiting factor is the poor knowledge about pterosaur diversity in general. Only four deposits (Solnhofen Limestone, Romualdo Member of the Santana Formation, Cambridge Greensand and Niobrara Formation) account for over 90% of the specimens and about 50% of all known diversity (Kellner 1994). Because they are flying animals, the chances of them being preserved in the fossil record is rather slim (the same is true for birds and bats). This leads to exceedingly large gaps in the geological record and several ghost lineages were detected in the present study (Fig. 2). Furthermore, the fossil record of pterosaurs is strongly biased towards lagoonal, marine or marginal-marine deposits, with the inference that inland living taxa are exceedingly underrepresented (Kellner 1994).

A second set of problems is related to the incompleteness and/or poor preservation of specimens. The Cambridge Greensand material is a good example: although it has yielded hundreds of pterosaur remains, which have been used as a basis for about three dozen nominal species (Wellnhofer 1991b), the specimens consist of isolated and incomplete bones. The fragmentary nature of this material makes the several pterosaur clades based on those elements (e.g. Ornithocheiridae, Criorhynchidae) very difficult to diagnose (Kellner 1990).

Another reason that makes pterosaur phylogeny a difficult topic to address is the lack of information, too often due to poor preparation. It is surprising how limited is the amount of information available for some pterosaur species that are based on one or more nearly complete skeletons. This is particularly the case for the smaller pterosaurs (e.g. *Sordes*; *Pterodactylus*) which could provide a great deal of information if the material was to be better prepared. Consequently, several of the nodes presented in this study are based on few characters.

I have no illusion about this present hypothesis of pterosaur inter-relationships being considered a 'final word' on this subject. My aim with this paper, however, is to provide a testable hypothesis that could constitute a basis for future analyses with a gradually increasing number of taxa (and characters) that ultimately will contribute to a better understanding of the evolutionary history of this amazing and (even after more than 200 years of study) still mysterious group of flying reptiles.

I would like to express my gratitude to D. de Almeida Campos (DNPM, Rio de Janeiro) and J. G. Maisey (AMNH, New York) for encouraging me to pursue my PhD studies at the joint programme of Columbia University and the American Museum of Natural History and for several stimulating discussions during the elaboration of my PhD dissertation; M. McKenna (AMNH), M. Norell (AMNH), P. Olsen (Columbia University, New York), J. Maisey (AMNH) and W. Pitman (Columbia University) for several comments on the original text; and R. Tedford (AMNH) for supporting trips to institutions to examine several specimens first hand. I am also indebted to several colleagues and institutions for access of specimens under their care (far too many to be listed here). I profited from several discussions with colleagues, particularly D. de Almeida Campos, F. M. Dalla Vecchia, P. Wellnhofer, K. Padian, R. Wild, C. Bennett, W. Langston, D. Peters, D. Unwin, N. Bakhurina and J. M. Sayão. M. S. de Oliveira, J. M. Sayão and L. B. de Carvalho (all from the Museu Nacional/UFRJ) are acknowledged for help with the illustrations. Finally I would like to thank E. Buffetaut (CNRS, Paris) for his invitation and support to submit this article to this volume.

This project was supported by grants and fellowships from the following institutions: Conselho Nacional de Desenvolvimento Científico e Tecnológico (CNPq, Brasília), American Museum of Natural History (Frick Fund, Axelrod Fund, Theodore Roosevelt Memorial Fund), Columbia University (GSAS – Graduate Faculties Alumni Fellowships, Summer Field Grants of the Geology Department), Geological Society of America, Sigma Xi, and Fundação Carlos Chagas Filho de Amparo à Pesquisa do Estado do Rio de Janeiro (FAPERJ).

Appendix 1: Character list (per anatomical region)

Skull

1. Dorsal margin of the skull:
   0 – straight or curved downwards
   1 – concave
   2 – only rostrum curved upwards
2. Upper and lower jaw:
   0 – laterally compressed
   1 – comparatively broad
3. Rostral part of the skull anterior to the external nares:
   0 – reduced
   1 – elongated (less than 60% of skull length)
   2 – extremely elongated (more than 60% of skull length)
4. Rostral end of premaxillae/maxillae downturned:
   0 – absent
   1 – present
5. Process separating the external nares:
   0 – broad
   1 – narrow
6. Position of the external naris:
   0 – above the premaxillary tooth row
   1 – displaced posterior to the premaxillary tooth row
7. Naris and antorbital fenestra elongated and reduced relative to the orbit:
   0 – absent
   1 – present
8. Naris and antorbital fenestra:
   0 – separated
   1 – confluent, shorter than 45% of the skull length
   2 – confluent, longer than 45% of the skull length
9. Orbit comparatively small and positioned very high in the skull:
   0 – absent
   1 – present
10. Position of the orbit relative to the nasoantorbital fenestra (naris + antorbital fenestra):
   0 – same level or higher
   1 – orbit lower than the dorsal rim of the nasoantorbital fenestra
11. Suborbital opening:
   0 – absent
   1 – present
12. Premaxillary sagittal crest:
   0 – absent
   1 – confined to the anterior portion of the skull
   2 – high, displaced, near the anterior margin of the nasoantorbital fenestra, reaching the skull roof above the orbit, and extending backwards
   3 – low, displaced near the anterior margin of the nasoantorbital fenestra, reaching the skull roof above the orbit but not extending backwards
   4 – starting at the anterior portion of the skull and extended posteriorly above the occipital region
   5 – starting at the posterior half of the nasoantorbital fenestra
13. Tip of the premaxilla expanded:
   0 – absent
   1 – present
14. Posterior ventral expansion of the maxilla:
   0 – absent
   1 – present
15. Nasal process:
   0 – absent
   1 – placed on the lateral side of the skull, long, straight, and directed ventrally (not fused with maxillae)
   2 – placed laterally, reduced
   3 – placed medially, long
   4 – placed medially, reduced
16. Foramen on nasal process:
   0 – absent
   1 – present
17. Bony frontal crest:
   0 – absent
   1 – low and blunt
   2 – low and elongated
   3 – high and expanded posteriorly
18. Bony parietal crest:
   0 – absent
   1 – present, blunt
   2 – present, laterally compressed and posteriorly expanded, with a rounded posterior margin
   3 – present, constituting the base of the posterior portion of the cranial crest
19. Posterior region of the skull rounded with the squamosal displaced ventrally:
   0 – absent
   1 – present
20. Position of the quadrate relative to the ventral margin of the skull:
   0 – vertical or subvertical
   1 – inclined about 120° backwards
   2 – inclined about 150° backwards

21. Position of the articulation between skull and mandible:
    0 – under the posterior half of the orbit or further backwards
    1 – under the middle part of the orbit
    2 – under the anterior half of the orbit
22. Helical jaw joint:
    0 – absent
    1 – present
23. Supraoccipital:
    0 – does not extend backwards
    1 – extends backwards
24. Foramen pneumaticum piercing the supraoccipital:
    0 – absent
    1 – present
25. Expanded distal ends of the paroccipital processes:
    0 – absent
    1 – present
26. Basisphenoid:
    0 – short
    1 – elongated
27. Palatal ridge:
    0 – absent
    1 – discrete, tapering anteriorly
    2 – strong, tapering anteriorly
    3 – strong, confined to the posterior portion of the palate
28. Maxilla excluded from the internal naris:
    0 – absent
    1 – present
29. Opening between pterygoids and basisphenoid (interpterygoid opening):
    0 – absent or very reduced
    1 – present and larger than subtemporal fenestra
    2 – present but smaller than subtemporal fenestra
30. Mandibular symphysis:
    0 – absent or very short
    1 – present, at least 30% of mandible length
31. Anterior tip of the dentary downturned:
    0 – absent
    1 – present
32. Tip of the dentary projected anteriorly:
    0 – absent
    1 – present
33. Dentary bony sagittal crest:
    0 – absent
    1 – blade-like and short
    2 – massive and deep
34. Position and presence of teeth:
    0 – teeth present, evenly distributed along the jaws
    1 – teeth absent from the anterior portion of the jaws
    2 – teeth confined to the anterior part of the jaws
    3 – jaws toothless
35. Largest maxillary teeth positioned posteriorly:
    0 – absent
    1 – present
36. Variation in the size of the anterior teeth with the 5th and 6th smaller than the 4th and 7th:
    0 – absent
    1 – present
37. Teeth with a broad and oval base:
    0 – absent
    1 – present
38. Multicusped teeth
    0 – absent
    1 – present
39. Peg-like teeth:
    0 – absent
    1 – present, 15 or less on each side of the jaws
    2 – present, more than 15 on each side of the jaws
40. Long slender teeth:
    0 – absent or less than 150
    1 – present, more than 150

Axial skeleton

41. Notarium:
    0 – absent
    1 – present
42. Atlas and axis:
    0 – unfused
    1 – fused
43. Postexapophyses on cervical vertebrae:
    0 – absent
    1 – present
44. Lateral pneumatic foramen on the centrum of the cervical vertebrae:
    0 – absent
    1 – present
45. Mid-cervical vertebrae:
    0 – short, subequal in length
    1 – elongated
    2 – extremely elongated
46. Cervical ribs on mid-cervical vertebrae:
    0 – present
    1 – absent
47. Neural spines of the mid-cervical vertebrae:
    0 – tall, blade-like
    1 – tall, spike-like
    2 – low, blade-like
    3 – extremely reduced or absent
48. Number of caudal vertebrae:
    0 – more than 15
    1 – 15 or less

Pectoral girdle

49. Length of the scapula:
  0 – subequal or longer than coracoid
  1 – scapula shorter than coracoid (1>sca/cor>0.80)
  2 – substantially shorter than coracoid (sca/cor<0.80).
50. Proximal surface of scapula:
  0 – elongated
  1 – sub-oval
51. Shape of scapula:
  0 – elongated
  1 – stout, with constructed shaft
52. Coracoidal contact surface with sternum:
  0 – no developed articulation surface
  1 – articulation surface flattened, lacking posterior expansion
  2 – articulation surface oval, with posterior expansion
53 – Cristospine:
  0 – absent
  1 – shallow and elongated
  2 – deep and short

Forelimb

54. Proportional length of the humerus relative to the metacarpal IV (hu/mcIV):
  0 – hu/mcIV>2.50
  1 – 1.50<hu/mcIV<2.50
  2 – 0.40<hu/mcIV<1.50
  3 – hu/mcIV<0.40
55. Proportional length of the humerus relative to the femur (hu/fe):
  0 – hu/fe<0.80
  1 – 1.4>hu/fe>0.80
  2 – hu/fe>1.40
56. Proportional length of the humerus + ulna relative to the femur + tibia (hu+ul/fe+ti):
  0 – humerus plus ulna about 80% or less of femur plus tibia length (hu+ul/fe+ti<0.80)
  1 – humerus plus ulna larger than 80% of femur plus tibia length (hu+ul/fe+ti>0.80)
57. Pneumatic foramen on the proximal part of the humerus:
  0 – absent
  1 – present on ventral side
  2 – present on dorsal side
  3 – present on dorsal and ventral side
58. Deltopectoral crest of the humerus:
  0 – reduced, positioned close to the humerus shaft
  1 – enlarged, proximally placed, with almost straight proximal margin
  2 – subrectangular, extending down the humerus shaft for at least 30% of humerus length
  3 – distally expanded
  4 – enlarged, hatchet-shaped, proximally placed
  5 – enlarged, hatched-shaped, positioned further down the humerus shaft
  6 – enlarged, warped
  7 – long, proximally placed, curving ventrally
59. Medial (ulnar) crest of the humerus:
  0 – absent or reduced
  1 – present, directed posteriorly
  2 – present, massive, with a developed proximal ridge
60. Distal end of the humerus:
  0 – oval or D-shaped
  1 – subtriangular
61. Proportional length of the ulna relative to the metacarpal IV (ul/mcIV):
  0 – ulna 3.6 times longer than metacarpal IV (ul/mcIV>3.6)
  1 – length of ulna between four and two times the length of metacarpal IV (3.6>ul/mcIV>2)
  2 – ulna less than two times the length of metacarpal IV (ul/mcIV<2)
62. Diameter of radius and ulna:
  0 – subequal
  1 – diameter of the radius no more than half that of the ulna
  2 – diameter of the radius less than half that of the ulna
63. Distal syncarpals:
  0 – unfused
  1 – fused in a rectangular unit
  2 – fused in a triangular unit
64. Pteroid:
  0 – absent
  1 – present, shorter than half the length of the ulna
  2 – present, longer than half the length of the ulna
65. Metacarpals I – III:
  0 – articulating with carpus
  1 – metacarpal III articulates with carpus, metacarpals I and II reduced
  2 – not articulating with carpus

66. Proportional length of the first phalanx of manual digit IV relative to the metacarpal IV (ph1d4/mcIV):
  0 – both small and reduced
  1 – both enlarged with ph1d4 over twice the length of mcIV
  2 – both enlarged with ph1d4 less than twice the length of mcIV

67. Proportional length of the first phalanx of manual digit IV relative to the tibiotarsus (ph1d4/ti):
  0 – ph1d4 reduced
  1 – ph1d4 elongated and less than twice the length of ti (ph1d4/ti smaller than 2.00)
  2 – ph1d4 elongated about or longer than twice the length of ti (ph1d4/ti subequal/larger than 2.00)

68. Proportional length of the second phalanx of manual digit IV relative to the first phalanx of manual digit IV (ph2d4/ph1d4):
  0 – both short or absent
  1 – elongated with second phalanx about the same size or longer than first (ph2d4/ph1d4 larger than 1.00)
  2 – elongated with second phalanx up to 30% shorter than first (ph2d4/ph1d4 between 0.70 and 1.00)
  3 – elongated with second phalanx more than 30% shorter than first (ph2d4/ph1d4 smaller than 0.70)

69. Proportional length of the third phalanx of manual digit IV relative to the first phalanx of manual digit IV (ph3d4/ph1d4):
  0 – both short or absent
  1 – ph3d4 about the same length or larger than ph1d4
  2 – ph3d4 shorter than ph1d4

70. Proportional length of the third phalanx of manual digit IV relative to the second phalanx of manual digit IV (ph3d4/ph2d4):
  0 – both short or absent
  1 – ph3d4 about the same size or longer than ph2d4
  2 – ph3d4 shorter than ph2d4

Hindlimb

71. Proportional length of the femur relative to the metacarpal IV (fe/mcIV):
  0 – femur about twice or longer than metacarpal IV ($\text{fe/mcIV} > 2.00$)
  1 – femur longer but less than twice the length of metacarpal IV ($1.00 < \text{fe/mcIV} < 2.00$)
  2 – femur about the same length or shorter than metacarpal IV ($\text{fe/mcIV} < 1.00$)

72. Length of metatarsal III:
  0 – more than 30% of tibia length
  1 – less than 30% of tibia length

73. Fifth pedal digit:
  0 – with four phalanges
  1 – with 2 phalanges
  2 – with 1 or no phalanx (extremely reduced)

74. Last phalanx of pedal digit V:
  0 – reduced or absent
  1 – elongated, straight
  2 – elongated, curved
  3 – elongated, very curved (boomerang-shaped)

Appendix 2: Data matrix

Ornithosuchus longidens (Huxley 1877)
0000000000 0000000000 0000000000 0000000000
0000000000 0000000000 0000000000 0000

Herrerasaurus ischigualastensis Reig 1963
0000000000 0000000000 0000000000 0000000000
0000000000 0000000000 0000000000 0020

Scleromochlus taylori Woodward 1907
000??00000 0000000000 0????????? 0?000?0000
0???00?0?? ????00?0?? ?0?0?????? ????

Anurognathus ammoni Döderlein 1923
0100100??? ?000000??? ?????????0 0000000010
0???0??1?? ???011?100 00?1011??? 001?

Batrachognathus volans Rjabinin 1948
?100100??? ?000000??? ???????0?0 0000000010
??0?0?0??0 01??21?10? ?????????? ????

Dendrorhynchoides curvidentatus (Ji & Ji 1998)
0100?????? ?0000?0??? ?????????? 0000000010
0???00??0? ???021?1?? 00?10112?? 00??

Sordes pilosus Sharov 1971
0010010000 0000000000 0?0?0????0 00000000?0
0?0?000000 01?011?100 00?1011111 0013

Preondactylus buffarinii Wild 1984
0010010100 0000000000 0????????0 0000?00?00
0???0??0?? ???111?1?? 10?1011111 00??

Scaphognathus crassirostris (Goldfuss 1861)
0010010000 0000000000 000?00001? 0000000000
0?00000000 0111110100 10?1011111 0013

Dorygnathus banthensis (Theodori 1830)
0010010000 0000000000 000?0???11 0100000000
0???000000 01?1110300 10?1011111 1013

Dimorphodon macronyx (Buckland 1829)
0010010000 0000000000 0?0?0????0 0000000000
0?00000000 01?1110100 1111011111 1011

Peteinosaurus zambellii Wild 1978
?0???????? ?????????? ?????????? ???00?0100
0??????000 01?111?100 11?1011111 0011

"*Eudimorphodon*" *rosenfeldi* Dalla Vechia 1995
?????????? ??????00?? ???000???? ??????01??
0????????? ???111???? 1???011211 101?

Campylognathoides liasicus (Quenstedt 1858)
0010010000 0000000000 000?000110 1000000000
0000000000 0111110200 11?1012122 1010

Eudimorphodon ranzii Wild 1978
0010010000 0000000000 0?0?0????0 1000000100
0?0000?000 01?11?0200 11?10????? 1???

Rhamphorhynchus muensteri (Goldfuss 1831)
0010011000 0000000001 1000000111 0100000000
0000000000 0111110400 11?1012222 1012

Rhamphorhynchus longicaudus (Münster 1839)
0010011000 0000000001 100?000??1 0100000000
000?000000 0111110400 11?1012222 1012

Pterodactylus kochi (Wagner 1837)
0010010100 0000100012 1?0?0?0??1 0000000020
0000112100 0112110700 21?1021222 2020

Pterodactylus antiquus (Soemmerring 1812)
0010010100 0000100012 1?0?0?0??1 0000000020
0000112100 0112110700 21?1021222 2020

Germanodactylus cristatus (Wiman 1925)
0010010100 0300100012 1?0?0????1 0000000020
0???11??00 01?211?7?0 21???21222 21??

Germanodactylus rhamphastinus (Wagner 1851)
0010010100 0300100012 1?0??????? 0000000020
0???11???0 01?211?7?0 21???21??? 2???

Ctenochasma gracile Oppel 1862
1020010100 0000000012 1?0?0????1 0000000001
??00112??0 01?211?7?? 21???212?? 2020

Pterodaustro guinazui Bonaparte 1970
1020010100 0000100012 1?0?0????1 0000000001
???01121?0 01?211?7?0 2?1??21222 2020

Gallodactylus canjuersensis Fabre 1974
1010010100 00002002?? 1???0????? ?002000000
??0????00 01???????? ??????1222 ?1??

Cycnorhamphus suevicus (Quenstedt 1855)
1010010100 0000200212 1???0?0??1 0002000000
?00?112?00 01?210?700 21??021222 21??

Nyctosaurus gracilis (Marsh 1876)
0010010100 000000000? 110?010121 0003000000
1110010100 011311150? 2122222222 20??

Nyctosaurus bonneri Miller 1972
0010010100 000000000? 1?0?0????1 0003000000
1?1?010??? ???31??50? ???222222 2???

Pteranodon longiceps Marsh 1876
1010010100 0000403301 2101010121 0003000000
1111011111 0112111611 2122222222 2020

Istiodactylus latidens (Seeley 1901)
0010010??? ?00?00???? ?0????0??1 0000000000
1?1?0???11 112?1??611 ?2?????1?? ????

Ornithocheirus compressirostris (Owen 1851)
0010?1???? ?000?????? ??????1??? 0000000000
?????????? ?????????? ?????????? ????

Tropeognathus mesembrinus Wellnhofer 1987
0010010100 0110??1101 2101012121 0010000000
?????????? ?????????? ?????????? ????

Anhanguera santanae (Wellnhofer 1985)
0010010100 0110311101 2101011121 00?0010000
0111011121 12????2611 ?22?1????? ????

Anhanguera blittersdorffi Campos & Kellner 1985
0010010100 0110??1101 2101011121 0010010000
?????????? ?????????? ?????????? ????

Anhanguera piscator Kellner & Tomida 2000
0010010100 0110311101 210101?121 0010010000
0111011121 1222112611 222?1????? 2120

Dsungaripterus weii Young 1964
2010010110 1201002301 2111110121 0001101000
1111010100 012200?7?0 211?021222 2120

"*Phobetor*" *parvus* (Bakhurina 1982)
0010010110 1201002301 2?1?1????1 0001101000
?????????? ?????????? ??1??????? ????

Noripterus complicidens Young 1973
?????????? ?????????? ?????????? ??????1000
??11010??? ???2002720 211??212?? 2120

Tupuxuara leonardii Kellner & Campos 1994
0010010201 0400??2301 2111113121 0003000000
1111010?00 0122001720 211??213?? 2???

Tapejara wellnhoferi Kellner 1989
0011010201 0400??2301 2011110121 0023000000
0?11010?00 01?2003720 2112?21??? 2120

Tapejara imperator Campos & Kellner 1997
0011010201 0400??2301 2?1??????? ???3000000
?????????? ?????????? ?????????? ????

Quetzalcoatlus sp.
0010010101 050000???1 21????01?1 0003000000
1110213?00 01?2001720 211??21322 2?20

Azhdarcho lancicollis Nessov 1984
?????????? ?????????? ?????????? ???3000000
1110213??? ??????17?0 ?????????? ????

References

BAKHURINA, N. N. 1982. A pterodactyl from the Lower Cretaceous of Mongolia. *Paleontological Journal*, **4**, 105–109.

BAKHURINA, N. N. & UNWIN D. M. 1995. A survey of pterosaurs from the Jurassic and Cretaceous of the former Soviet Union and Mongolia. *Historical Biology*, **10**, 197–245.

BENNETT, S. C. 1989. A pteranodontid pterosaur from the early Cretaceous of Peru, with comments on the relationships of Cretaceous pterosaurs. *Journal of Paleontology*, **63**, 669–667.

BENNETT, S. C. 1994. *Taxonomy and systematics of the Late Cretaceous Pterosaur* Pteranodon *(Pterosauria, Pterodactyloidea).* Natural History Museum, University of Kansas, Occasional Papers, **169**, 1–70.

BENNETT, S. C. 1996a. On the taxonomic status of *Cycnorhamphus* and *Gallodactylus* (Pterosauria: Pterodactyloidea). *Journal of Paleontology*, **70**, 335–338.

BENNETT, S. C. 1996b. The phylogenetic position of the Pterosauria within the Archosauromorpha. *Zoological Journal of the Linnean Society, London*, **118**, 261–308.

BENNETT, S. C. 2001. The osteology and functional morphology of the Late Cretaceous Pterosaur *Pteranodon. Palaeontographica A*, **260**, 1–112

BONAPARTE, J. F. 1971. Descripcion del craneo y mandibulas de *Pterodaustro guiñazui* (Pterodactyloidea–Pterodaustriidae nov.), de la formacion Lagarcito, San Luis, Argentina. *Publicaciones del Museo Municipal de Ciencias Naturales, Mar del Plata*, **1**, 263–272.

Brown, G. W. 1986. Reassessment of *Nyctosaurus*: new wings for an old pterosaur. *Nebraska Academy of Science, Proceedings*, **47** [abstract].

Cai, Z. & Wei, F. 1994. On a new pterosaur (*Zhejiangopterus linhaiensis* gen. et sp. nov.) from Upper Cretaceous in Linhai, Zhejiang, China. *Vertebrata Palasiatica*, **32**, 181–194.

Campos, D. A. & Kellner A. W. A. 1985 Panorama of the flying reptiles study in Brazil and South America. *Anais da Academia Brasileira de Ciências*, **57**, 453–466.

Campos, D. A. & Kellner, A. W. A. 1997. Short note on the first occurrence of Tapejaridae in the Crato Member (Aptian), Santana Formation, Araripe Basin, Northeastern Brazil. *Anais da Academia Brasileira de Ciências*, **69**, 83–87.

Chiappe, L. M., Kellner, A. W. A., Rivarola, D., Davila, S. & Fox, M. 2000. Cranial morphology of *Pterodaustro guinazui* (Pterosauria: Pterodactyloidea) from the Lower Cretaceous of Argentina. *Contributions in Science*, **483**, 1–19.

Currie, P. J. & Russell, D. A. 1982. A giant pterosaur (Reptilia: Archosauria) from the Judith River (Oldman) Formation of Alberta. *Canadian Journal of Earth Science*, **19**, 894–897.

Dalla Vecchia, F. M. 1994. A new pterosaur (Reptilia, Pterosauria) from the Norian (Late Triassic) of Friuli (Northeastern Italy), preliminary note. *Gortania*, **16**, 59–66.

Dalla Vecchia, F. M. 1998. New observations on the osteology and taxonomic status of *Preondactylus buffarinii* Wild, 1984 (Reptilia, Pterosauria). *Bolletino della Societa Paleontologica Italiana*, **36**, 355–366.

Dalla Vecchia, F. M. & Kellner, A. W. A. 1995. Comments on the pterosaurs from the Late Triassic (Middle Norian) of Friuli; Northeastern Italy. *Journal of Vertebrate Paleontology*, **15**(3) (supplement), 25A.

Döderlein, L. 1923. *Anurognathus ammoni*, ein neuer Flugsaurier. *Sitzungberichte der Bayerischen Akademie der Wissenshaften, Mathematisch-Naturwissenschaftlitch Klasse*, 117–164.

Döderlein, L. 1929. Über *Anurognathus ammoni* Döderlein. *Sitzungberichte der Bayerischen Akademie der Wissenshaften, Mathematisch-Naturwissenschaftlitch Klasse*, 47–63.

Fabre, J. A. 1974. Un nouveau Pterodactylidae sur le gisement 'Portlandien' de Canjuers (Var): *Gallodactylus canjuersensis* nov. gen., nov. sp. *Comptes Rendus de l'Académie des Sciences*, Paris, **279**, 2014.

Fabre, J. A. 1976. Un nouveau Pterodactylidae sur le gisement Portlandien de Canjuers (Var): *Gallodactylus canjuersensis* nov. gen., nov. sp. *Annales de Paléontologie (Vertébrés)*, **62**, 35–70.

Frey, E. & Martill, D. M. 1996. A reappraisal of *Arambourgiania* (Pterosauria, Pterodactyloidea): one of the world's largest flying animals. *Neues Jahrbuch für Geologie und Paläontologie, Abhandlungen*, **199**, 221–247.

Hooley, R. W. 1913. On the skeleton of *Ornithodesmus latidens*, an Ornithosaur from the Wealden shales of Atherfield (Isle of Wight). *Quarterly Journal of the Geological Society, London*, **69**, 372–422.

Howse, S. C. B. 1986. On the cervical vertebrae of the Pterodactyloidea (Reptilia: Archosauria). *Zoological Journal of the Linnean Society, London*, **88**, 307–328.

Howse, S. C. B., Milner, A.R. & Martill, D. M. 2001. Pterosaurs. *In*: Martill, D. M. & Naish, D. (eds.), *Dinosaurs of the Isle of Wight*. Paleontological Association, London, 324–335.

Huene, F. von. 1914. Beiträge zur Geschichte der Archosaurier. *Geologische und Paläontologische Abhandlungen*. **13**, 1–53.

Ikegami, N., Kellner, A.W.A. & Tomida, Y. 2000. The presence of an azhdarchid pterosaur in the Cretaceous of Japan. *Paleontological Research*, **4**, 165–170.

Ji, S. & Ji, Q. 1998. [A new fossil pterosaur (Rhamphorhynchoidea) from Liaoning, China.] *Jiangsu Geology*, **22** (4), 199–206. [In Chinese]

Ji, S. & Ji, Q. & Padian, K. 1999. Biostratigraphy of new pterosaurs from China. *Nature*, **398**, 573–574.

Kellner, A. W. A. 1989. A new edentate pterosaur of the Lower Cretaceous from the Araripe Basin, Northeast Brazil. *Anais da Academia Brasileira de Ciências*, **61**, 439–446.

Kellner, A. W. A. 1990. Os répteis voadores do Cretáceo Brasileiro. *Anuário do Instituto de Geociências*, **1989**, 86–106.

Kellner, A. W. A. 1994. Remarks on pterosaur taphonomy and paleoecology. *Acta Geologica Leopoldensia*, **39**, 175–189.

Kellner, A.W.A. 1995. The relationships of the Tapejaridae (Pterodactyloidea) with comments on pterosaur phylogeny. *In*: Sun, A. & Wang, Y. (eds) Sixth *Symposium on Mesozoic Terrestrial Ecosystems and Biota, Beijing, Short Papers*, 73–77.

Kellner, A.W.A. 1996a. *Description of new material of Tapejaridae and Anhangueridae (Pterosauria, Pterodactyloidea) and discussion of pterosaur phylogeny*. PhD thesis, Columbia University. [Published by University Microfilms International/ ProQuest].

Kellner, A. W. A. 1996b. Pterosaur Phylogeny. *Journal of Vertebrate Paleontology*, **16**(3) (supplement),45A.

Kellner, A.W.A. 1997. On the relationships of some early pterodactyloid pterosaurs. *In*: *Boletim de Resumos, 15th Congresso Brasileiro de Paleontologia. UNESP, Rio Claro, São Paulo*, 100.

Kellner, A. W. A. & Campos, D. A. 1992. A new tapejarid from the Santana Formation (Lower Cretaceous) from the Araripe Basin, northeast Brazil. *Journal of Vertebrate Paleontology*, **12**(3) (supplement), 36A–37A.

Kellner, A. W. A. & Hasegawa, Y. 1993. Postcranial skeleton of *Tupuxuara* (Pterosauria, Pterodactyloidea, Tapejaridae) from the Lower Cretaceous of Brazil. *Journal of Vertebrate Paleontology*, **13**(3) (supplement) 44A. [Abstract]

Kellner, A. W. A & W. Langston Jr. 1996. Cranial remains of *Quetzalcoatlus* (Pterosauria, Azhdarchidae) from the Late Cretaceous sediments of Big Bend National Park, Texas. *Journal of Vertebrate Paleontology*, **16**, 222–231.

Kellner, A.W.A. & Mader, B. J. 1996. First report of Pterosauria (Pterodactyloidea, Azhdarchidae) from Cretaceous rocks of Morocco. *Journal of Vertebrate Paleontology*, **16**(3) (supplement), 45A.

Kellner, A.W.A. & Tomida, Y. 2000. *Description of a New Species of Anhangueridae (Pterodactyloidea) with Comments on the Pterosaur Fauna from the Santana Formation (Aptian–Albian), Northeastern Brazil.* National Science Museum, Tokyo, Monographs, **17**, 135 pp.

LAWSON, D. A. 1975a. Pterosaur from the latest Cretaceous of west Texas: discovery of the largest flying creature. *Science*, **187**, 947–948.

LAWSON, D. A. 1975b. Could pterosaurs fly? *Science*, **188**, 676–677.

MADDISON, W. P. & MADDISON, N. D. R. 1992. *MacClade, 3.04*. Sinauer Associates, Inc. Sunderland, Massachusetts.

MADER, B. J. & KELLNER, A.W.A. 1999. A new anhanguerid pterosaur from the Cretaceous of Morocco. *Boletim do Museu Nacional, Rio de Janeiro, Nova Serie, Geologia*, **45**, 1–11.

MILLER, H. W. 1972a. The taxonomy of the *Pteranodon* species from Kansas. *Kansas Academy of Science, Transactions*, **74**, 1–90.

MILLER, H. W. 1972b. A skull of *Pteranodon* (*Longicepia*) *longiceps* Marsh associated with wing and body bones. *Kansas Academy of Science, Transactions*, **74**, 20–33.

NESSOV, L. A. 1984. Upper Cretaceous pterosaurs and birds from central Asia. *Paleontological Journal*, **1984** (1), 38–49.

NESSOV, L. A. & JARKOV, A. A. 1989. New Cretaceous–Paleogene birds of the USSR and some remarks on the origin and evolution of the Class Aves. **197**, 78–97. [In Russian]

NOVAS, F. E. 1994. New information on the systematics and postcranial skeleton of *Herrerasaurus ischigualastensis* (Theropoda: Herrerasauridae) from Ischigualasto Formation (Upper Triassic) of Argentina. *Journal of Vertebrate Paleontology*, **13**, 400–423.

PADIAN, K. 1984. A large pterodactyloid pterosaur from the Two Medicine Formation (Campanian) of Montana. *Journal of Vertebrate Paleontology*, **4**, 516–524.

PADIAN, K. 1986. A taxonomic note on two pterodactyloid families. *Journal of Vertebrate Paleontology*, **6**(3), 289.

PETERS, D. 2000. A reexamination of four prolacertiforms with implications for pterosaur phylogenesis. *Rivista Italiana di Paleontologia e Stratigrafia*, **106** (3), 293–336.

PRICE, L. I. 1953. A presença de Pterosauria no Cretáceo Superior do Estado da Paraíba. *Notas Preliminares e Estudos, Divisão de Geologia e Mineralogia, DNPM*, **71**, 1–10.

RJABININ, A. N. 1948. [Remarks on a flying reptile form the Jurassic of Karatau.] Trudy Akad. Nauk, Paläont. Inst., **15**, 86–93. [In Russian]

ROMER, A. S. 1956. *Osteology of the Reptiles*. University of Chicago Press, Chicago, 772 pp.

ROMER, A. S. 1971. The Chañares (Argentina) Triassic reptile fauna. X. Two new but incompletely known long-limbed pseudosuchians. *Breviora*, **378**, 1–10.

SAYÃO, J. M. & KELLNER, A. W. A. 2001. New data on the pterosaur fauna from Tendaguru (Tanzania), Upper Jurassic, Africa. *Journal of Vertebrate Paleontology*, **21**(3) (supplement), 97A.

SANCHEZ, T. M. 1973. Redescripcion del craneo y mandibulas de *Pterodaustro guiñazui* Bonaparte (Pterodactyloidea, Pterodaustriidae). *Ameghiniana*, **10**, 313–325.

SERENO, P.C. 1991. Basal archosaurs: phylogenetic relationships and functional implications. *Journal of Vertebrate Paleontology*, **11**, 1–53.

SERENO, P. C. 1994. The pectoral girdle and forelimb of the basal theropod *Herrerasaurus ischigualastensis*. *Journal of Vertebrate Paleontology*, **13**, 425–450.

SERENO, P. C. & ARCUCCI A. B. 1994a. Dinosaurian precursors from the Middle Triassic of Argentina: *Lagerpeton chanarensis*. *Journal of Vertebrate Paleontology*, **13**, 385–399.

SERENO, P. C. & ARCUCCI A. B. 1994b. Dinosaurian precursors from the Middle Triassic of Argentina: *Marasuchus lilloensis*, gen. nov. *Journal of Vertebrate Paleontology*, **14**, 53–73.

SERENO, P.C. & NOVAS F. E. 1994. The skull and neck of the basal theropod *Herrerasaurus ischigualastensis*. *Journal of Vertebrate Paleontology*, **13**, 451–476.

SHAROV, A. G. 1971. [New flying reptiles from the Mesozoic of Kazakhstan and Kirghizia]. Akad. Nauk SSSR Trudy Paläont. Inst., 130, 104–113. [In Russian]

SWOFFORD, D. L. 1993. *PAUP: Phylogenetic Analysis Using Parsimony, Version 3.1.1.* Smithsonian Institution, Washington D.C.

SWOFFORD, D. L. 2000. *PAUP: Phylogenetic Analysis Using Parsimony, Version 4.0B10 (for Microsoft Windows)*. Sinauer Associates, Inc. Sunderland, Massachusetts.

UNWIN, D. M. 1995. Preliminary results of a phylogenetic analysis of the Pterosauria (Diapsida:Archosauria). *In*: SUN, A. & WANG, Y. (eds) *Sixth Symposium on Mesozoic Terrestrial Ecosystems and Biota, Beijing, Short Papers*. 69–72.

UNWIN, D. M. & LÜ, J. 1997. On Zhejiangopterus and the relationships of pterodactyloid pterosaurs. *Historical Biology*, **12**, 199–210.

UNWIN, D. M. LÜ, J. & BAKHURINA, N. N. 2000. On the systematic and stratigraphic significance of pterosaurs from the Lower Cretaceous Yixian Formation (Jehol Group) of Liaoning, China. *Mitteilungen aus dem Museum für Naturkunde, Berlin, Geowissenschofflich*, Reihe, **3**, 181–206.

WALKER, A. D. 1964. Triassic reptiles from the Elgin area: *Ornithosuchus* and the origin of carnosaurs. Philosophical Transactions of the Royal Society, London (B), **248**, 53–134.

WELLNHOFER, P. 1970. Die Pterodactyloidea (Pterosauria) der Oberjura-Plattenkalke Süddeutschlands. *Bayerischen Akademie der Wissenschaften, Mathematisch-Naturwissenschaftliche Klasse, Ahandlungen, Neue Folge*, **141**, 1–133.

WELLNHOFER, P. 1974. *Campylognathoides liasicus* (Quenstedt), an Upper Liassic Pterosaur from Holzmaden. The Pittsburgh specimen. *Annals of the Carnegie Museum*, Pittsburgh, **45**(2), 5–34.

WELLNHOFER, P. 1975a. Die Rhamphorhynchoidea (Pterosauria) der Oberjura-Plattenkalke Süddeutschlands. I: Allgemeine Skelettmorphologie. *Paläontographica A*, **148**, 1–33.

WELLNHOFER, P. 1975b. Die Rhamphorhynchoidea (Pterosauria) der Oberjura-Plattenkalke Süddeutschlands. II: Systematische Beschreibung. *Palaeontographica A*, **148**, 132–186.

WELLNHOFER, P. 1975c. Die Rhamphorhynchoidea (Pterosauria) der Oberjura-Plattenkalke Süddeutschlands. III: Palökologie und Stammesgeschichte. *Palaeontographica A*, **149**, 1–30.

WELLNHOFER, P. 1978. Pterosauria. *In*: WELLNHOFER, P. (ed.) *Handbuch der Palaeoherpetologie*. Stuttgart: Gustav Fischer, Stuttgart, Teil 19, 82 pp.

WELLNHOFER, P. 1985. Neue Pterosaurier aus der Santana-Formation (Apt) der Chapada do Araripe, Brasilien. *Palaeontographica*, **187**, 105–182.

WELLNHOFER, P. 1991a. Weitere Pterosaurierfunde aus der Santana-Formation (Apt) der Chapada do Araripe, Brasilien. *Palaeontographica A*, **215**, 43–101.

WELLNHOFER, P. 1991b. *The Illustrated Encyclopedia of Pterosaurs*. Salamander, London, 192 p.

WELLNHOFER, P. & BUFFETAUT, E. 1999. Pterosaur remains from the Cretaceous of Morocco. *Paläontologische Zeitschrift*, **73**, 133–142.

WELLNHOFER, P. & KELLNER 1991. A. W. A. The skull of *Tapejara wellnhoferi* Kellner (Reptilia, Pterosauria) from the Lower Cretaceous Santana Formation of the Araripe Basin, Northeastern Brazil. *Mitteilungen der Bayerischen Staatsmmlung für Paläontologie und Historische Geologie*, **31**, 89–106.

WILD, R. 1978. Die Flugsaurier (Reptilia, Pterosauria) aus der Oberen Trias von Cene bei Bergamo, Italien. *Bolletino della Società Paleontologica Italiana*, **17**(2), 176–256.

WILD, R. 1984. A new pterosaur (Reptilia, Pterosauria) from the Upper Triassic (Norian) of Friuli, Italy. *Gortania*, **5**, 45–62.

WILD, R. 1993. A juvenile specimen of *Eudimorphodon ranzii* Zambelli (Reptilia, Pterosauria) from the Upper Triassic (Norian) of Bergamo. *Rivista del Museo Civico di Scienze Naturali*. 'E. Caffi', Bergamo, **16**, 95–120.

WILLISTON, S. W. 1903. On the osteology of *Nyctosaurus* (*Nyctodactylus*), with notes on American pterosaurs. *Field Columbian Museum Publications, Geological Series*, **2**, 125–163.

YOUNG, C. C. 1964. On a new pterosaurian from Sinkiang, China. *Vertebrata Palasiatica*, **8**, 221–256.

YOUNG, C. C. 1973. [Pterosaurian fauna from Wnerho, Sinkiang. *In*: *Reports of Paleontological Expeditions to Sinkiang II*.] Institute of Vertebrate Palaeontology and Palaeoanthropology, Academica Sinica, Memoirs, 11, 18–35. [In Chinese]

On the phylogeny and evolutionary history of pterosaurs

DAVID M. UNWIN

Institut für Paläontologie, Museum für Naturkunde, Zentralinstitut der Humboldt-Universität zu Berlin, D-10115 Berlin, Germany (e-mail: david.unwin@rz.hu-berlin.de)

Abstract: Previous cladistic studies of pterosaur relationships suffer from restricted numbers of taxa and characters, incomplete data sets and absence of information on characters, tree structure and the robustness of trees. Parsimony analysis of a new character data set (60 characters, 20 terminal taxa, 93.75% complete) yielded six trees. In the strict consensus tree *Preondactylus* is the most basal taxon followed, stepwise, by the Dimorphodontidae and the Anurognathidae. Beyond this basal group, more derived pterosaurs (Campylognathoididae (Rhamphorhynchidae + Pterodactyloidea)) share a suite of characters principally associated with elongation of the rostrum. The Pterodactyloidea consists of four major clades. The Ornithocheiroidea is the most basal taxon consisting, stepwise, of *Istiodactylus*, the Ornithocheiridae, *Nyctosaurus* and the Pteranodontidae. The remaining taxa, Ctenochasmatoidea, Dsungaripteroidea and Azhdarchoidea, are weakly united in a clade of non-ornithocheiroid pterodactyloids, but their inter-relationships are difficult to resolve. *Cycnorhamphus* is the basal-most ctenochasmatoid, while the remaining taxa (*Pterodactylus*, Lonchodectidae, Ctenochasmatidae) form an unresolved trichotomy. The Dsungaripteroidea (*Germanodactylus* + Dsungaripteridae) is strongly supported by features of the skull and dentition. The Azhdarchoidea (*Tapejara* [*Tupuxuara* + Azhdarchidae]) is united by cranial characters such as elevation of the antorbital region, and relative shortening of the wing finger. The pattern of pterosaur evolution suggested by the results of this analysis is broadly similar to traditional ideas, but has greater resolution, more complexity and reveals several previously unrecognized 'events'.

Phylogenetic analysis has had a profound impact on our views of the relationships of extinct vertebrates, both with regard to taxa within principal clades and of these clades to each other (e.g. Benton 1997; Carroll 1997). Moreover, combining the results of such studies with data on the stratigraphic distribution of terminal taxa has often resulted in major changes to our understanding of the history of such clades. Several prominent groups, such as Mammalia and Theropoda, have been subject to intense study, but other taxa, including pterosaurs, an important clade of Mesozoic flying reptiles, have been relatively neglected. The few preliminary cladograms that have been published (Howse 1986; Bennett 1989, 1994; Unwin 1992, 1995a; Kellner 1996a; Peters 1997; Unwin & Lü 1997; Viscardi *et al.* 1999; Unwin *et al.* 2000) suggest a substantially different pattern of evolution from the traditional reconstruction, best epitomized by the *Handbuch* (Wellnhofer 1978), but these new ideas have yet to be explained in detail and, so far, have had little impact on general understanding of pterosaurs.

The current situation is not unusual in that pterosaur systematics has long been a 'poor cousin' to other more contentious aspects of these animals, such as their anatomy, physiology, and flight ability. Early accounts, for example by Meyer (1859) and Seeley (1870) were rather vague and it was not until 1901, more than 100 years after pterosaurs were first recognized, that the major division into "Rhamphorhynchoidea" and Pterodactyloidea was formally proposed by Plieninger. The early twentieth century saw a burst of activity, with various taxonomies proposed by Williston (1903), Hooley (1913), Arthaber (1919), Nopcsa (1928) and, most importantly, Plieninger (1930). While these studies were quite detailed, there was relatively little discussion of the characters underlying the various taxonomic arrangements, there were no attempts to depict relationships and accounts of the evolutionary history of the group were highly generalized.

The first detailed illustration of pterosaur relationships was published by Young (1964; Fig. 1a) and was followed by a similar reconstruction by Kuhn in 1967 (Fig. 1b). Subsequently, Wellnhofer (1975a) presented a "rhamphorhynchoid" phylogeny, and later combined this with information on pterodactyloids in a tree that has come to represent the traditional view of pterosaur relationships (Wellnhofer 1978; Fig. 2). Wild (1978, fig. 47) also depicted and discussed the general relationships of Late Triassic and Early Jurassic pterosaurs. In all these studies, taxa were aligned in ancestor-descendant lineages on the basis of overall similarity: thus *Anurognathus*, a tall-skulled form with large fenestrae and a steeply oriented quadrate was presumed to have descended from *Dimorphodon*, a similar, tall-skulled form, also with large cranial fenestrae and a subvertical quadrate (Wellnhofer 1975b, 1978).

Howse (1986) published the first phylogenetic study to make use of cladistic techniques (Fig. 3a). This work, based on a subset of pterodactyloids, with *Rhamphorhynchus* as an outgroup, used only characters of cervical vertebrae, but was able to

From: BUFFETAUT, E. & MAZIN, J-M. (eds) 2003. *Evolution and Palaeobiology of Pterosaurs.* Geological Society, London, Special Publications, **217**, 139–190. 0305-8719/03/$15

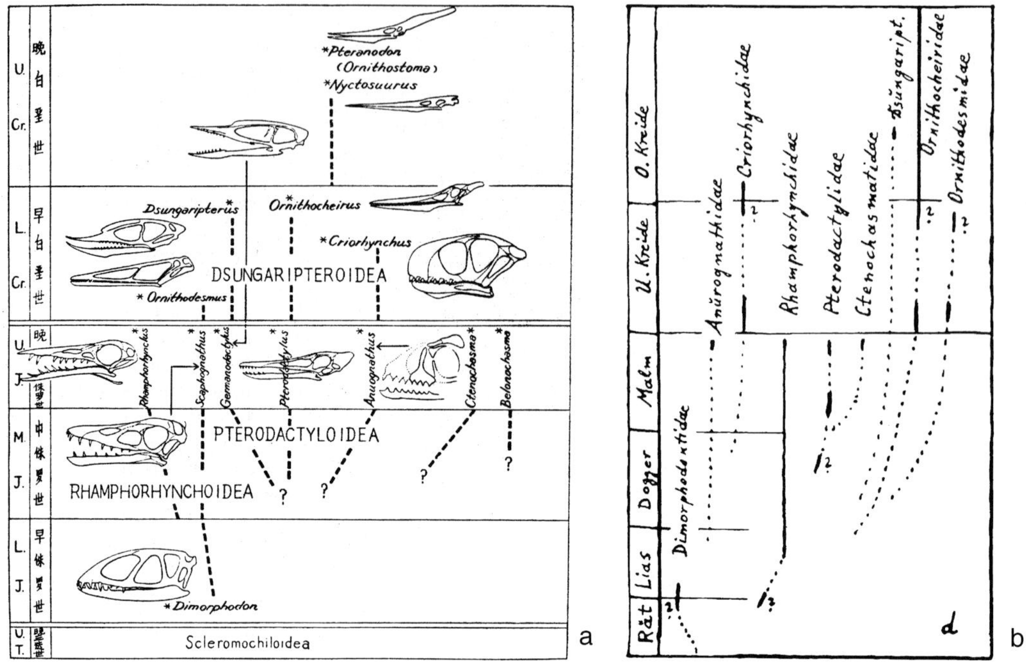

Fig. 1. Traditional evolutionary trees for pterosaurs proposed by (**a**) Young (1964) and (**b**) Kuhn (1967).

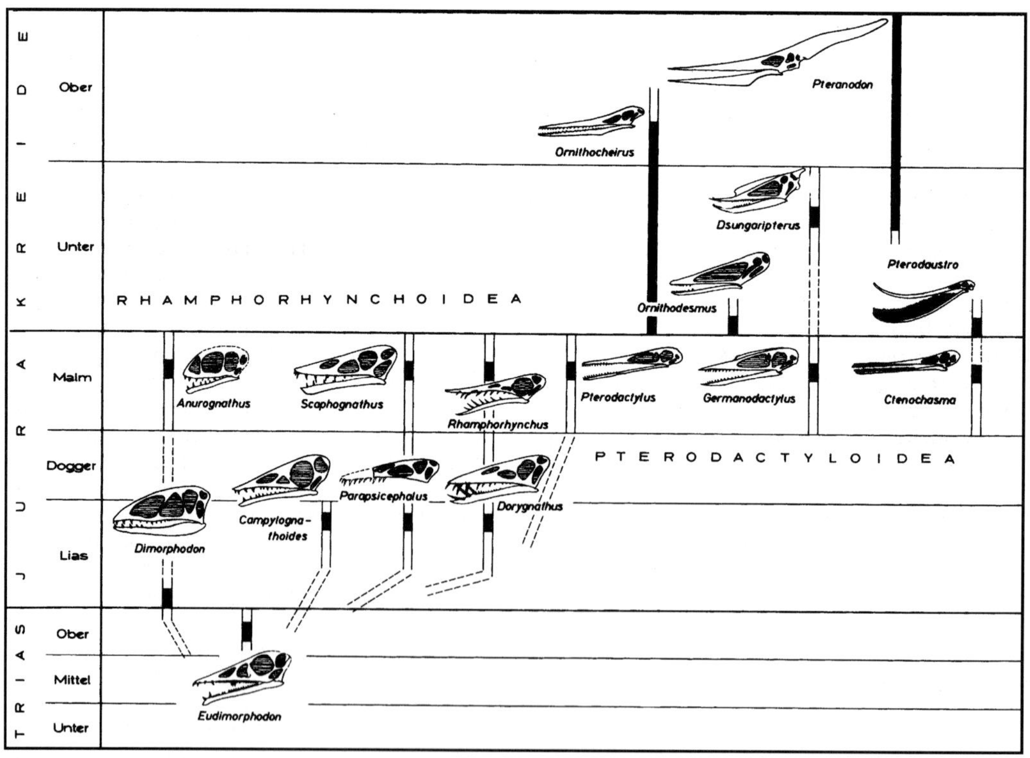

Fig. 2. Pterosaur evolutionary tree reconstructed by Wellnhofer (1978).

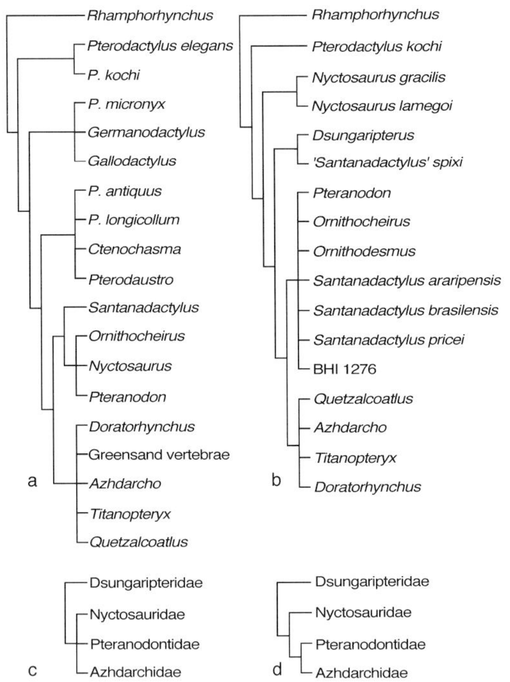

Fig. 3. Cladistic analyses of pterosaur relationships presented by (**a**) Howse (1986), and (**b**)–(**d**) Bennett (1989).

discover groups of long-necked and short-necked forms. Re-analysis of Howse's study by Kellner (1995), using Phylogenetic Analysis Using Parsimony (PAUP) and MacClade 3.03, resulted in little taxonomic resolution and also showed that the trees illustrated did not represent the most parsimonious solution (see also Kellner & Langston 1996).

Bennett (1989) presented the first analysis to utilize PAUP, including 19 taxa (*Rhamphorhynchus* and 18 pterodactyloids) and 14 characters, all drawn from postcranial anatomy and using *Rhamphorhynchus* and *Pterodactylus kochi* as outgroups (Fig. 3b–d). Notably, two major clades, *Pteranodontidae* (*Pteranodon*, *Istiodactylus* [= †*Ornithodesmus*] and various ornithocheirids) and Azhdarchidae (*Quetzalcoatlus*, *Azhdarcho*, *Arambourgiania* [= †*Titanopteryx*] and *Gnathosaurus* [= †*Doratorhynchus*]), were recognized. As Bennett observed, however (1989, p. 675), the topology was supported by relatively few characters some of which, such as the appearance of a notarium, may be size related. Moreover, the data set consisted of almost 50% missing data (Kellner 1995), and much of the topology was dependent on interpretation of the polarity of a composite character based on the distal syncarpal. Kellner (1995) and Unwin (1995a) argued that the character state considered by Bennett as derived probably represented the primitive condition, because it is widely distributed in pterodactyloids and even occurs in the outgroups. Importantly, reversing the coding for this character collapses all but one of the groups (Nyctosauridae) in Bennett's original cladogram (Kellner 1995; Wilkinson 1995; Kellner & Langston 1996).

Later, Bennett (1994) published a more substantial study involving 27 taxa (*Rhamphorhynchus* and 26 genera or species of pterodactyloids) and 37 characters. The single cladogram (Fig. 4a), reconstructed using MacClade (3.01), is generally similar to the principal cladogram previously published by Bennett (1989, fig. 3.1). A series of predominantly Jurassic pterodactyloids form progressively closer sister-taxa to three families of Cretaceous pterosaurs: *Pteranodontidae*, Dsungaripteridae and Azhdarchidae. The content of these families is similar to that in the 1989 cladogram (cf. Figs 3b & 4a) although, notably, *Ornithocheirus* lies outside *Pteranodontidae* and separate from other ornithocheirids, such as *Anhanguera*. Also of note is the pairing of Dsungaripteridae with Azhdarchidae, an arrangement not evident in any of Bennett's earlier cladograms (1989, fig. 3). Although more comprehensive than the 1989 study, Bennett's 1994 analysis suffered from the same difficulties: a large amount of missing data and problems with the polarity of some characters, especially those pertaining to the distal syncarpal (Kellner 1995). Analysis of the data set using PAUP supported only a single group, Nyctosauridae (Kellner 1995; Kellner & Langston 1996).

Utilizing data from a taxonomic review of Cretaceous pterodactyloids, Unwin (1991) carried out a cladistic analysis of pterosaurs based on 18 taxa and 103 characters. The final cladogram was constructed by hand and a brief account appeared in 1992 (Fig. 4b). This was the first cladistic study to incorporate a broad range of taxa, including "rhamphorhynchoids" and pterodactyloids. The study confirmed earlier suggestions that "Rhamphorhynchoidea" was paraphyletic, consisting of taxa successively more distant to Pterodactyloidea, and recovered a number of pterodactyloid clades (Dsungaripteroidea, Ornithocheiroidea and Azhdarchidae) that were similar in content to those identified by Bennett (1989, 1994). The principal distinction was in the recognition of a single clade of long-necked forms, the Azhdarchoidea, containing Jurassic pterosaurs such as *Pterodactylus* and *Ctenochasma*, as well as azhdarchids.

In a later analysis of pterosaur interrelationships Unwin (1995a) published a tree based on 60+ characters and 40+ taxa (Fig. 4c). This tree was one of four most parsimonious trees (MPTs) resulting from a PAUP (version 3.1.1) analysis of the data set, although, because of time constraints, details of the

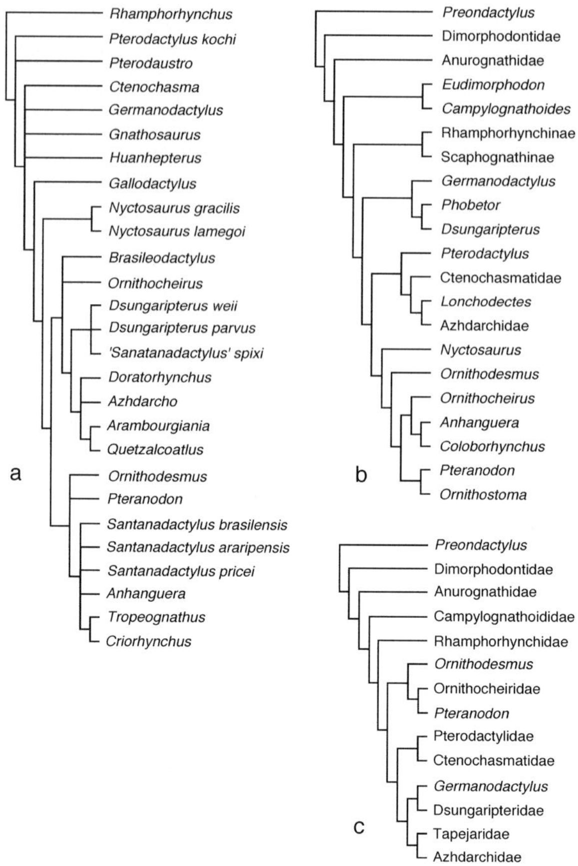

Fig. 4. Cladistic analyses of pterosaur relationships presented by (**a**) Bennett (1994), (**b**) and (**c**) Unwin (1992, 1995a).

statistical analysis were not included in the published material. The favoured tree (Fig. 4c), discussed in some detail by Unwin & Lü (1997), exhibits a similar topology to that previously described by Unwin (1992), except for some changes in the relationships of the major pterodactyloid clades, with the Ornithocheiroidea at the base of Pterodactyloidea, and long-necked pterosaurs split into two distinct groups: azhdarchids with tapejarids and dsungaripteroids as successively more distant sister taxa, and ctenochasmatoids as a sister-taxon to Dsungaripteroidea + Azhdarchoidea.

A second, comprehensive analysis of pterosaur inter-relationships, based on a data set consisting of 32 taxa and 66 characters and analysed using PAUP (version 3.1.1) was briefly outlined by Kellner (1996a; Fig. 5a), but the study upon which this was based (Kellner 1996b) has not yet been published. The structure of the tree is broadly similar to that presented by Unwin (1995a; cf. Figs 4c & 5a), but there are some differences in the arrangement of basal clades of "rhamphorhynchoids". Notably, *Dimorphodon* occupies a relatively derived position. Differences are also evident in the arrangement of the main pterodactyloid clades: the Ctenochasmatidae and related forms occupy a basal position while the *Pteranodontoidea* [=Ornithocheiroidea *sensu* Unwin 1995a] is the sister-taxon to Dsungaripteridea + Azhdarchoidea, although the content of these clades is almost identical to that proposed by Unwin (1995a).

Recently, Peters (1997) presented the results of a study of pterosaur relationships in which *Sharovipteryx* and other prolacertiforms were used as the outgroups. It is not possible to reconstruct a tree from the data presented, but details suggest some similarity to previous trees, at least regarding "rhamphorhynchoids", with, e.g. *Preondactylus*, dimorphodontids, eudimorphodontids and rhamphorhynchids as successively more derived groups. The arrangement and content of clades within Pterodactyloidea appears to be radically different from that in previous studies, e.g. in the unification of dsungaripterids, tapejarids, nyctosaurids and pteranodontids in a single clade, but

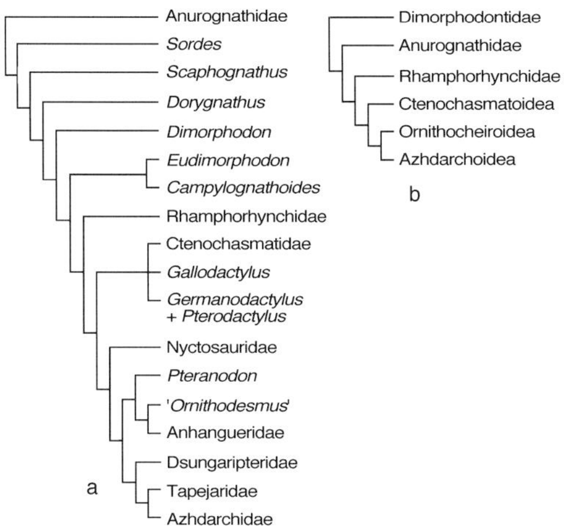

Fig. 5. Cladistic analyses of pterosaur relationships presented by (**a**) Kellner (1996a) and (**b**) Viscardi *et al.* (1999).

further assessment of these ideas is not possible at present.

Results of another recent study, based on data culled from published analyses, have been briefly outlined by Viscardi *et al.* (1999). A tree reconstructed from this account (Fig. 5b), exhibits a similar topology to that published by Kellner (1996a). According to Viscardi *et al.* (1999) the composition of the clades differs markedly from previous analyses, but details have not yet been published.

In summary, most cladistic studies published over the last 16 years share some basic topology in common (Unwin & Lü 1997; Unwin *et al.* 2000). "Rhamphorhynchoidea" is paraphyletic and composed of basal forms, e.g. dimorphodontids and anurognathids, and derived taxa such as *Rhamphorhynchus* and its relatives. Pterodactyloidea is monophyletic and most taxa can be assigned to just three or four principal clades (e.g. Dsungaripteridae, Ornithocheiroidea and Azhdarchoidea) dominated by Cretaceous forms. The content, relationships and even names of these clades are highly variable, however, and there are strongly divergent opinions regarding the relationships of Jurassic pterodactyloids. While some studies locate these taxa within the major pterodactyloid clades (e.g. *Germanodactylus* as a sister-taxon to Dsungaripteridae, forming the Dsungaripteroidea [e.g. Unwin 1995a; Fig. 4c]), in other cases the Jurassic pterodactyloids form a cluster of basal taxa lying outside the principal clades (Bennett 1994; Fig. 4a). Moreover, the basic phylogenetic topology outlined above has been directly challenged by Peters (1997) and Viscardi *et al.* (1999).

Published analyses also suffer from other problems. At least eight different cladograms have been presented, but only two of these were accompanied by data matrices and both are substantially incomplete. In addition, there has been relatively little discussion of tree structure, the definition, variation and distribution of characters or the robustness of trees.

This paper presents the results of a new, comprehensive study of the inter-relationships of pterosaurs using cladistic techniques. The aims of this study were to: (1) diagnose terminal taxa using apomorphies and describe their content and stratigraphic range; (2) define, describe and discuss phylogenetically significant characters and their distributions; (3) analyse the relationship of pterosaurs using a data matrix compiled as far as possible from direct study of fossil material; and (4) assess the robustness of the resulting trees and compare them with previous studies.

Institutional abbreviations: AMNH, American Museum of Natural History, New York, USA; BMNH, Natural History Museum, London, UK; BSP, Bayerische Staatssammlung für Paläontologie und Geologie, München, Germany; CAMSM, Sedgwick Museum, Cambridge, UK; CAMMZ, Museum of Zoology, University of Cambridge, Cambridge, UK; DMNH, Denver Museum of Natural History, Denver, USA; FMNH, Field Museum of Natural History, Chicago, USA; GSM, Museum of the Geological Survey, Keyworth, UK; IMCF, Iwaki Coal and Fossil Museum, Iwaki, Japan; IVPP, Institute for Vertebrate Palaeontology and Palaeoanthropology, Beijing, China; JM, Jura Museum, Eichstätt, Germany; MANCH, Manchester Museum, University of Manchester, Manchester, UK; MB, Museum Für Naturkunde, Berlin, Germany; MBH, Museum Bergèr, Harthof bei Eichstätt, Germany; MGUV Museo del Departamento de Geologiá, Universidad de Valencia, Valencia, Spain; MNHN, Museum National d'Histoire Naturelle de Paris, France; MPV, Museo Paleontológico Municipal de Valencia, Valencia, Spain; MSA, Museum am Solenhofer Aktienverein, Maxberg bei Solnhofen, Germany; MT, Institut und Museum für Geologie und Paläontologie der Universität Tübingen, Tübingen, Germany; NAMAL, North American Museum of Ancient Life, Lehi, Utah, USA; NHMW, Naturhistorisches Museum, Wien, Austria; NMING, National Museum of Ireland, Dublin, Republic of Ireland; NSM, National Science Museum, Tokyo, Japan; OUM, Oxford University Museum, Oxford, UK; PIN, Palaeontological Institute, Russian Academy of Sciences, Moscow, Russia; PTH, Philosophisch-Theologische Hochschule, Eichstätt, Germany; SM, Natur-Museum und Forschungsinstitut Senckenberg, Frankfurt, Germany; SMNK,

Staatliches Museum für Naturkunde Karlsruhe, Germany; SMNS, Staatliches Museum für Naturkunde Stuttgart, Germany; TMM, Texas Memorial Museum, Austin, Texas, USA; YORM, Yorkshire Museum, York, UK; ZMNH, Zhejiang Museum of Natural History, Hangzhou, China.

Symbol conventions: Citation of a name in single quotation marks indicates that the validity of this taxon has yet to be clearly established, except in the case of '*Phobetor*', which is a valid taxon awaiting a replacement name. Citation in double quotation marks indicates that, here, this taxon is considered to be paraphyletic. Citation within asterisks indicates an alternative name for what is treated here as a valid monophyletic taxon (e.g. Ornithocheiroidea here = *Pteranodontidae* *sensu* Bennett 1989, 1994). Invalid names are prefixed by the symbol †.

Materials and methods

Terminal taxa

The most recent general account of pterosaurs (Wellnhofer 1991a) recognized approximately 100 species and a little over 40 genera. Several new taxa have been described in the last decade (Dalla Vecchia 1993, 1995; Cai & Wei 1994; Frey & Martill 1994; Kellner & Campos 1994; Lee 1994; Padian *et al.* 1995; Howse & Milner 1995; Harris & Carpenter 1996; Campos & Kellner 1997; Ji & Ji 1997, 1998; Buffetaut *et al.* 1998; Clark *et al.* 1998; Mader & Kellner 1999; Unwin & Heinrich 1999; Martill *et al.* 2000; Wang & Lü 2001; Dalla Vecchia *et al.* 2002; Jenkins *et al.* 2001; Wang *et al.* 2002; Carpenter *et al.* 2003; Frey *et al.* 2003), adding to this list. On the debit side, however, taxonomic reviews (Bennett 1994, 1995, 1996a; Howse & Milner 1995; Unwin & Heinrich 1999; Unwin 1991, 2001, 2002; Fastnacht 2001; Carpenter *et al.* 2003) have dismissed significant numbers of invalid taxa.

On the basis of published accounts and examination of fossil material, 93 valid species and a further 8 putatively valid species of pterosaur are recognised here (Table 1 & Appendix 1). Some of these species may eventually disappear into synonymy if they are found to be sexual dimorphs (e.g. Bennett 1992, 2002; Unwin 2001) or part of an ontogenetic sequence (e.g. Wellnhofer 1970; Bennett 1993, 1995, 1996a; Wild 1994; Unwin 1995b) of a previously erected species, but this is unlikely to have a significant impact on the character distributions described below. In this analysis character data was collected for 84 of the 93 species cited in Table 1, of which 65 were studied on the basis of fossil material or casts, while the remainder were assessed from the

Table 1. *Classification, based on the phylogenetic relationships shown in Figure 7a, of all valid and potentially valid species of pterosaur*

- Pterosauria
  - *Preondactylus*
    - *P. buffarinii*
  - Macronychoptera
    - Dimorphodontidae
      - *Dimorphodon*
        - *D. macronyx*
        - *D. weintraubi*
      - *Peteinosaurus*
        - *P. zambellii*
    - Caelidracones
      - Anurognathidae
        - *Anurognathus*
          - *A. ammoni*
        - *Batrachognathus*
          - *B. volans*
        - *Dendrorhynchoides*
          - *D. curvidentatus*
        - *Jeholopterus*
          - *J. ningchengensis*
      - Lonchognatha
        - Campylognathoididae
          - *Austriadactylus*
            - *A. cristatus*
          - *Campylognathoides*
            - *C. liasicus*
            - *C. zitteli*
          - *Eudimorphodon*
            - *E. cromptonellus*
            - *E. ranzii*
            - *E. rosenfeldi*
        - Breviquartossa
          - Rhamphorhynchidae
            - Rhamphorhynchinae
              - *Angustinaripterus*
                - *A. longicephalus*
              - *Dorygnathus*
                - *D. banthensis*
                - *D. mistelgauensis*
                - *D. purdoni*
              - *Nesodactylus*
                - *N. hesperius*
              - *Rhamphocephalus*
                - *R. bucklandi*
              - *Rhamphorhynchus*
                - *R. 'longiceps'*
                - *R. muensteri*
            - Scaphognathinae
              - *Scaphognathus*
                - *S. crassirostris*
              - *Sordes*
                - *S. pilosus*
          - Pterodactyloidea
            - Ornithocheiroidea
              - *Istiodactylus*
                - *I. latidens*
              - Euornithocheira
                - Ornithocheiridae
                  - *Anhanguera*

Table 1 ***(continued)***

A. blittersdorffi
A. cuvieri
A. fittoni
A. santanae
Arthurdactylus
A. conandoylei
Brasileodactylus
B. araripensis
Coloborhynchus
C. capito
C. clavirostris
C. moroccensis
C. robustus
C. sedgwickii
C. wadleighi
Haopterus
H. gracilis
Ornithocheirus
O. mesembrinus
O. simus
Pteranodontia
Pteranodontidae
Ornithostoma
O. sedgwicki
Pteranodon
P. longiceps
P. sternbergi
Nyctosaurus
N. gracilis
N. lamegoi
Lophocratia
Ctenochasmatoidea
Cycnorhamphus
C. canjuersensis
C. suevicus
Euctenochasmia
Pterodactylus
P. antiquus
P. kochi
'P'. micronyx
Lonchodectidae
Lonchodectes
L. compressirostris
L. giganteus
L. machaerorhynchus
L. microdon
L. platysomus
?L. sagittirostris
Ctenochasmatidae
Ctenochasmatinae
Ctenochasma
C. gracile
C. porocristata
C. roemeri
?Eosipterus
E. yangi
Pterodaustro
P. guinazui
Gnathosaurinae
Cearadactylus
C. atrox

Table 1 ***(continued)***

Gnathosaurus
G. macrurus
G. subulatus
Huanhepterus
H. quingyangensis
Plataleorhynchus
P. streptophorodon
'Pterodactylus' longicollum
Dsungaripteroidea
Germanodactylus
G. rhamphastinus
G. cristatus
Herbstosaurus
H. pigmaeus
Kepodactylus
K. insperatus
Normannognathus
N. wellnhoferi
Tendaguripterus
T. recki
Dsungaripteridae
Domeykodactylus
D. ceciliae
Dsungaripterus
D. weii
Noripterus
N. complicidens
'Phobetor'
'P'. parvus
Azhdarchoidea
Tapejaridae
Tapejara
T. imperator
T. wellnhoferi
Neoazhdarchia
Tupuxuara
T. longicristatus
T. leonardii
Azhdarchidae
Arambourgiania
A. philadelphiae
Azhdarcho
A. lancicollis
Montanazhdarcho
M. minor
Quetzalcoatlus
Q. northropi
Q. sp.
Zhejiangopterus
Z. linhaiensis

The following taxa are of uncertain validity and in many cases their position within the classification is unclear: '*Araripesaurus castilhoi*' Price 1971; '*Mesadactylus ornithosphyos*' Jensen & Padian 1989; '*Puntanipterus globosus*' Bonaparte & Sanchez 1975; '*Santanadactylus araripensis*' Wellnhofer 1985; '*Santanadactylus brasilensis*' Buisonjé 1980; '*Santanadactylus pricei*' Wellnhofer 1985; and '*Santanadactylus spixi*' Wellnhofer 1985.

Table 2. *Morphometric data for terminal taxa used in this study (see Appendix 1)*

	Forelimb/ Hindlimb	Humerus/ Femur	Ulna/ Tibia	Humerus/ Wing metacarpal	Source of data
Preondactylus	2.48	0.98	0.95	0.45	Wild 1984a; Dalla Vecchia 1998
Dimorphodontidae	2.58–2.67	1.04–1.06	0.96–0.98	0.22–0.44	Owen 1870; Wild 1978; Padian 1983; Unwin 1988b
Anurognathidae	>2.64	1.19–1.46	1.15–1.33	0.33–0.34	Döderlein 1923; Riabinin 1948; Wellnhofer 1975b; Ji & Ji 1998; Unwin *et al.* 2000
Campylognathoididae	4.45–8.72	1.08–1.39	1.00–1.67	0.38–0.62	Plieninger 1895; Wellnhofer 1974; Wild 1978; Dalla Vecchia 1995
Scaphognathinae	3.19–3.40	1.00–1.21	1.36–1.58	0.44–0.50	Wellnhofer 1975b, author
Rhamphorhynchinae	3.40–5.49	1.03–1.48	1.25–1.81	0.39–0.68	Arthaber 1919; Wiman 1923; Salée 1928; Wild 1971; Wellnhofer 1975b, author
Istiodactylus	-	1.1	-	-	Hooley 1913
Ornithocheiridae	3.88 (est.)	1.09	1.36	1	Wellnhofer 1985, 1991b; Frey & Martill 1994; Kellner & Tomida 2000; Wang & Lü 2001, author
Nyctosaurus	4.84	1.06–1.11	1.04–1.18	2.53–2.71	Williston 1903; Brown 1986
Pteranodontidae	4.61	1.05–1.15	1.00–1.06	2.09–2.54	Bennett 2001
Cycnorhamphus	2.99	0.85	0.71–0.75	1.65–1.68	Wellnhofer 1970; Fabre 1976, author
Pterodactylus	2.66–3.25	0.82–1.28	0.80–1.17	0.80–1.79	Wellnhofer 1970; Tischlinger 1994; Frey & Martill 1998; Frey & Tischlinger 2001
Ctenochasmatidae	3.22	0.79–1.33	0.70–1.14	0.91–2.10	Wellnhofer 1970, author
Germanodactylus	2.85–2.90	0.96–0.99	0.86–1.09	1.09–1.18	Wellnhofer 1970, Unwin 1988a, author
Dsungaripteridae	>2.67	0.91	0.62	1.82	Young 1964, 1973; author
Tapejara	2.61	0.78	0.74	1.43	author
Tupuxuara	>2.50	0.78	0.71	1.54	author
Azhdarchidae	2.58	0.62	0.88	2.45	Lawson 1975a; Langston 1981; Cai & Wei 1994, author

published literature. Ideally, phylogenetic analyses should be conducted at the most exclusive level possible (e.g. Bininda-Edmonds *et al.* 1998), i.e. the species, in order to reduce *a priori* assumptions of monophyly at higher levels. This is problematic for pterosaurs, however, because many species are incompletely known (Wellnhofer 1978, 1991a), resulting in data sets with high levels of missing entries (e.g. Bennett 1994, p. 70). This, in turn, usually results in large numbers of MPTs. Neither approach to this problem, i.e. exclusion of poorly represented taxa or production of consensus trees, is entirely satisfactory (e.g. Smith 1994).

One partial solution, adopted here, is to group taxa into more inclusive clades. This enables poorly known taxa to contribute phylogenetic information and also improves the completeness of matrices. Consequently, this analysis focused on 19 terminal taxa consisting of 8 genera, 2 subfamilies and 9 families (Appendix 1). Each suprageneric taxon and named ingroup taxon (see below) was formally defined in a phylogenetic sense (Queiroz & Gauthier 1992; Gauthier & Queiroz 2001) and their content specified. Most of these terminal taxa are well established, uncontroversial and can be diagnosed by a series of autapomorphies (Appendix 1), but several (e.g. *Preondactylus*, Scaphognathinae, *Tapejara* and *Tupuxuara*) are only weakly supported and two (*Germanodactylus* and *Pterodactylus*) lack any apparent apomorphies and are treated here as metataxa *sensu* Gauthier (1986). Each of the terminal taxa exhibits a unique combination of character states for the 60 characters coded (Table 3), consequently all were included in the parsimony analysis (Wilkinson & Benton 1995).

Characters

A preliminary database of characters was compiled from published cladistic studies (see above) and in some cases from the older literature. These characters were critically assessed by examination of a wide range of fossil material (see Appendix 1 for details of specimens studied) which also led to the discovery of many additional phylogenetically useful characters. Ultimately, the database contained more than 120 characters, including some based on morphometric data (Table 2), but more than half of these were excluded before statistical analysis for the following reasons. Firstly, many published characters were relatively poorly defined and taxa could not be unambiguously scored for character states. In some cases it was possible to modify the character definition so that scoring was possible, but in other cases this was impossible and the character was rejected. Secondly, the derived state for several binary characters was found to be unique to a single terminal taxon included in this study. These uninformative characters were also excluded. Thirdly, a small subset of characters were only discovered at a late stage of this study and could only be certainly scored for a few taxa. These characters were also excluded, but are mentioned in the text at relevant points.

Finally, 60 osteological characters (Appendix 2), 30 based on cranial and 30 on postcranial anatomy, were selected for this study, primarily because their character states were clear-cut and could be unambiguously coded, and also because they had been assessed on the basis of fossil material for 18 of the 19 terminal taxa (only *Preondactylus* was not examined directly). Additional data was compiled from the literature (see Appendix 1 for citations).

Methods

The taxon-character matrix (Table 3) was analysed using the cladistics package PAUP 3.1.1 (Swofford, 1993), with the 'branch and bound' search option and addition sequence 'furthest'. Runs were executed using both 'Acctran' and 'Deltran' settings. Multiple-state characters were always treated as unordered. Characters that exhibited more than one state for a particular terminal taxon (shown as 0/1 in the matrix) were treated as polymorphic. Character states that could not be coded because of excessive morphological transformation are denoted by 'x' (Table 3). Revised versions of the data set in which 'x' was treated as a discrete character state '3', or as uncertain '?', were analysed separately to assess the significance of this recoding. Continuously variable characters (Table 2) were broken into discrete states on the basis of breaks between distribution peaks.

Rooting of trees was dealt with in two ways. Initially, an outgroup was left unspecified and trees were rooted both by rooting the tree at an internal node with a basal polytomy and by making the ingroup monophyletic. Subsequently, the analyses were rerun using a specified outgroup. Selection of a suitable outgroup is difficult because the relationship of pterosaurs to other diapsids is controversial (Unwin 1999, Brochu 2002). Recent studies (Padian 1984b, 1997; Gauthier & Padian 1985; Gauthier 1986; Benton 1990, 1999; Sereno & Arcucci 1990; Sereno & Novas 1990; Sereno 1991, 1996) have tended to locate pterosaurs within Ornithodira (*sensu* Gauthier 1986), although there is no consensus as to their exact position within this group. Bennett (1996b) has shown, however, that if the hindlimb characters of pterosaurs and dinosauromorphs are not treated as homologous, character analysis supports a position near the base of archosauriforms, an idea that was previously suggested by Benton (1982, 1984, 1985) and, more recently, by Atanassov (2001). Peters (1997, 2000) has argued that there is some evidence to support a relationship between pterosaurs and various prolacertiforms, including *Sharovipteryx* and *Cosesaurus*, essentially a modern version of a precladistic hypothesis proposed by Wild (1978, 1984b).

At present it is not clear which, if any, of these hypotheses is correct. Happily, however, with regard to the polarization of characters used for establishing ingroup relationships of pterosaurs, this is largely irrelevant, because pterosaur skeletal anatomy is so derived that in almost all cases the plesiomorphic condition is common to each of the three outgroups used in this study: basal ornithodirans, basal archosauriforms and prolacertiforms.

Support for clades was calculated using bootstrapping techniques. In each case 1000 replicates were made, using the 'branch and bound' search option and the addition sequence 'furthest'. A decay analysis was performed by rerunning the initial data set and increasing the maximum length of permitted trees by a single step each time.

Results

Statistical analyses

Initial runs, which did not include an outgroup, treated 'x' as uncertain ('?') and rooted the tree internally, yielded six MPTs, each 111 steps in length. The first of these trees, together with standard descriptive data, is shown in Fig. 6a. The other five trees were identical except that the Lonchodectidae was variously paired with *Pterodactylus* or the Ctenochasmatidae, or in one case the three formed a trichotomy, and in three of the trees Dsungaripteroidea paired with Ctenochasmatoidea rather than with Azhdarchoidea. The same results were achieved using either the Acctran or Deltran setting. Treating 'x' as a unique character state ('3') resulted in trees 10 steps longer, but otherwise the results were the same.

Including an outgroup (Fig. 6b) resulted in six trees that were identical to those resulting from the 'unrooted' runs, but with a length of 112 steps, slightly different statistical parameters, and resolution of the trichotomy between *Preondactylus*, the Dimorphodontidae and all other pterosaurs. A strict consensus tree of these six trees is shown in Fig. 7a. and a summary of character state changes at each node is given in Table 4. Results of the bootstrap analysis are shown in Fig. 6c.

Table 3. *Distribution of character states (0, 1, 2) among the terminal taxa used in this analysis*

					5					10					15					20					25					30					35					40					45					50					55					60
Outgroup	0	0	0	0	0	0	0	0	0	0	0	0	0	0	0	0	0	0	0	0	0	0	0	0	0	0	0	0	0	0	0	0	0	0	0	0	0	0	0	0	?	0	0	0	0	0	0	0	0	0	0	0	0	0	0	0	0	0	0	0
Preondactylus	0	0	0	0	0	0	0	?	0	0	0	0	?	0	0	0	0	0	0	0	0	0	0	0	?	0	0	0	0	0	0	0	0	?	?	0	0	0	?	?	?	0	0	0	?	0	0	0	0	0	0	0	0	?	0	0	0	?	0	?
Dimorphodontidae	1	1	1	1	1	0	0	0	0	0	0	0	0	0	0	0	0	0	0	0	0	0	0	0	?	0	0	0	0	0	0	0	0	0	0	0	0	0	0	0	?	0	0	0	0	0	0	0	0	?	0	0	0	0	0	0	0	1	0	0
Anurognathidae	1	1	1	1	1	?	1	1	0	0	0	0	0	0	0	0	0	0	0	0	0	0/1	0	0	0	1	1	0	0	0	0	0	0	0	?	0	0	0	0	0	1	0	0	0	0	0	0	0	0	0	0	0	0	?	0	0	0	1	0	0
Campylognathoididae	1	1	1	1	1	1	1	0/1	1	1	1	1	1	1	0	0	0	0	0	0	0	0	0	0	0	0	0	0	0	0	0	0	0	0	0	0	0	0	0	0	1	0	0	0	0	0	0	0	0	0	0	0	0	0	0	0	0	1	0	0
Scaphognathinae	1	1	1	1	0	1	1	1	1	1	1	1	1	1	1	1	1	0	1	1	1	1	1	0	0	0	0	0	0	0	0	0	0	0	0	0	0	0	0	0	0	0	0	0	0	0	0	0	0	0	0	0	0	0	0	0	0	1	0	0
Rhamphorhynchinae	1	1	1	1	1	1	1	1	1	1	1	1	1	1	1	1	1	1	1	0/1	1	1	1	0	0	0	0	0/1	0	0	0	0	0	0	0	0	0	0	0	0	0	0	0	0	0	0	0	0	0	0	0	0	0	0	0	0	0	1	0	0
Istiodactylus	1	1	?	1	1	1	?	1	1	1	x	x	x	x	1	1	1	0	1	?	?	0	0	1	?	1	?	1	1	?	1	1	1	1	1	1	?	1	0	0	0	0	0	0	0	0	0	0	0	0	0	0	0	0	0	0	0	1	0	0
Nyctosaurus	1	1	x	1	1	1	1	1	1	1	x	x	x	x	1	1	1	1	1	x	1	1	0	1	1	1	1	1	1	1	1	1	2	0	?	1	0	1	1	1	1	1	1	1	1	1	0	0	0	0	0	0	0	0	0	0	x	1	0	0
Ornithocheiridae	1	1	1	1	1	1	1	1	1	1	x	x	0	x	1	1	1	1	1	1	1	0	0	1	1	1	1	1	1	1	1	1	1	1	1	1	1	1	1	1	1	0	0	0	0	0	0	0	0	0	0	0	0	0	0	0	0	1	0	0
Pteranodontidae	1	1	1	1	1	1	0/1	1	1	1	x	x	x	x	1	1	1	1	1	1	1	0	0	1	1	1	1	1	1	1	1	1	1	0	1	1	2	1	1	1	1	1	1	1	1	1	0	0	0	1	0	0	0	0	0	0	x	1	0	0
Lonchodectidae	1	?	?	?	?	?	?	?	1	?	?	?	?	x	?	?	1	0	1	?	?	0	0	?	?	1	?	?	?	?	?	?	0	0	0	?	?	0	?	?	?	?	0	0	1	?	1	0	?	?	?	1	1	?	0	0	0	1	0	?
Pterodactylus	1	1	1	1	0/1	1	0/1	1	1	1	x	x	0	x	1	1	1	1	1	1	1	0	0	1	1	1	1	1	1	1	0	0	0	0	0	0	0	0	0	0	?	0	0	0	0	0	1	0	1	1	1	1	1	0	0	0	0	1	0	0
Cycnorhamphus	1	1	0	1	0	1	0	1	1	1	x	x	x	x	1	1	1	1	1	1	1	1	0	1	?	1	1	1	1	1	0	0	0	0	0	0	0	0	0	0	?	0	0	0	?	0	1	0	1	1	1	0	0	0	0	0	0	1	0	0
Ctenochasmatidae	1	1	0	1	0/1	1	0/1	1	1	1	x	x	x	x	1	1	1	1	1	1	1	0	0	1	1	1	1	1	1	1	0	0	0	?	?	0	0	0	0	0	?	0	0	0	?	0	1	1	1	1	1	1	1	0	0	0	0	?	0	0
Germanodactylus	1	1	0	1	0	1	0/1	1	1	1	x	x	0	x	1	1	1	0	1	1	1	0	0	1	?	1	1	1	1	1	0	0	0	?	0	0	0	0	0	0	?	0	0	0	?	0	1	1	0	0	0	0	0	1	1	0/1	0/1	0	1	0
Dsungaripteridae	1	1	0	1	0	1	0	1	1	1	x	x	x	x	1	1	1	1	1	1	1	0	0	1	1	1	1	1	1	1	1	0	0	0	0	0	?	0	0	2	0	0	0	0	0	0	1	1	0	0	0	0	0	1	1	1	1	0	1	0
Tapejara	1	1	1	1	0	1	0	1	2	1	x	x	x	x	1	1	1	1	1	1	1	1	0	1	1	1	?	1	1	1	0	0	0	0	0	0	1	0	0	0	1	0	1	0	1	0	1	1	0	0	0	0	0	0	0	0	x	1	0	1
Tupuxuara	1	1	1	1	0	1	0	1	2	1	x	x	x	x	1	1	1	1	1	1	1	1	0	1	1	1	1	1	1	1	1	0	0	0	0	0	?	0	0	0	0	0	1	0	1	0	1	1	0	0	0	0	0	1	0	0	x	1	0	1
Azhdarchidae	1	1	1	1	0	1	0	1	2	1	x	x	x	x	1	1	1	1	1	1	1	1	0	1	?	1	1	1	1	1	1	0	0	0	0	0	2	0	0	0	0	0	1	0	1	1	1	0/1	0	1	0/1	1	1	0	0	0	x	1	0	1

?, uncertainty because of incomplete preservation; 0/1, polymorphism; x, highly transformed morphologies that cannot be assigned to a particular character state.

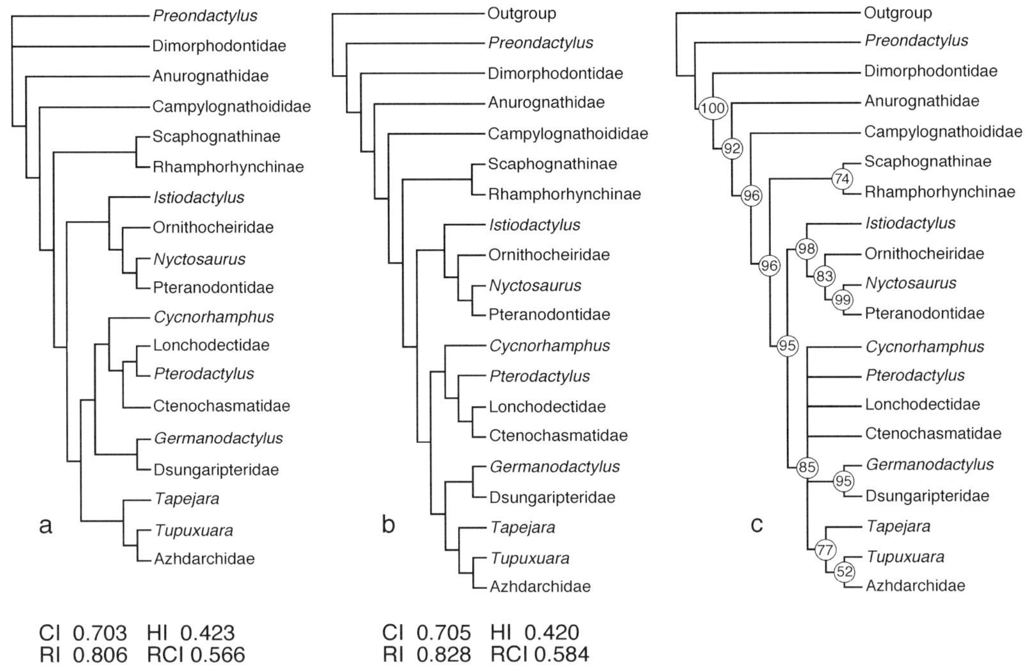

Fig. 6. Cladograms of pterosaur relationships based on a PAUP analysis of the character data matrix shown in Table 3. (**a**) One of six trees resulting from an unrooted analysis. (**b**) The same analysis, but including an outgroup. (**c**) Bootstrap values for the same analysis. Nodes occurring in less than 50% of all trees collapsed to a polytomy. CI, consistency index; HI, homoplasy index; RCI, rescaled consistency index; RI, Retention index.

Tree structure and character analysis

The "Rhamphorhynchoidea" is paraphyletic and consists of a series of clades that are successively closer to Pterodactyloidea (Fig. 7a). Characters used by Kuhn (1967) and Wellnhofer (1978) to define the "Rhamphorhynchoidea" are plesiomorphic for pterosaurs and do not support a monophyletic group.

Clade 1. Pterosauria Kaup 1834 (converted clade name).
Definition. Preondactylus buffarinii, *Quetzalcoatlus northropi*, their most recent common ancestor, and all its descendants (Peters 2000).
Content. Preondactylus, Dimorphodontidae, Anurognathidae, Campylognathoididae, Rhamphorhynchidae and Pterodactyloidea
Remarks. This clade, the ingroup for this analysis, is generally accepted to be monophyletic and is supported by numerous clear-cut characters, many related to the forelimb (see Romer 1956; Wellnhofer 1978; Padian 1984b; Sereno 1991; Bennett 1994, 1996b; Peters 2000). Surprisingly, Kuhn (1967) argued that pterosaurs were polyphyletic, an idea that was rejected by later workers (Wellnhofer 1978).

Clade 2. Macronychoptera (new clade name).
Etymology: Greek, *macro* = long or large, *onykh* = claw, *pteron* = wing, in reference to the relatively large size of the claws borne by manus digits I–III of pterosaurs in this clade.
Definition. Dimorphodon macronyx, *Quetzalcoatlus northropi*, their most recent common ancestor, and all its descendants.
Content. Dimorphodontidae, Anurognathidae, Campylognathoididae, Rhamphorhynchidae and Pterodactyloidea.
Synapomorphies

(1) The dentary forms more than 75% of the length of the mandible. (See Unwin 1995a.) In the outgroups and *Preondactylus* (Fig. 8a) the dentary forms approximately half, or less, of the total length of the mandible (Wild 1984a; Dalla Vecchia 1998). In all other pterosaurs (Figs 8b, c & 9–14) the dentary occupies at least 90% of the length of the mandible, except perhaps in *Dsungaripterus* (Bennett 2001), although this development must be homoplastic.

(2) Coracoid at least 66% the length of the scapula. In the outgroups, the coracoid is a rounded, plate-like structure. *Preondactylus* has an elongated coracoid that is approximately two-thirds the length

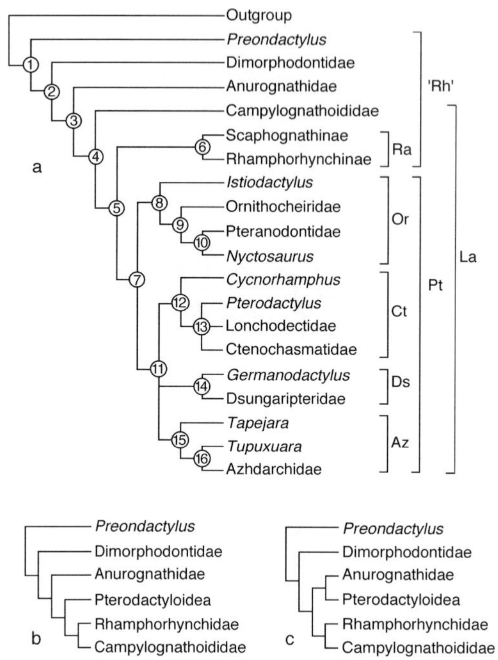

Fig. 7. Strict consensus tree of six trees resulting from PAUP analysis of the character data matrix shown in Table 3 and including an outgroup (**a**) Apomorphies supporting numbered clades are discussed in the text. Definitions and diagnoses of terminal taxa are given in Appendix 1. Alternative phylogenies depict Pterodactyloidea as (**b**) a sister-group to Campylognathoididae + Rhamphorhynchidae, or (**c**) as a sister-group to Anurognathidae. Az, Azhdarchoidea; Ct, Ctenochasmatoidea; Ds, Dsungaripteroidea; La, Lonchognatha; Or, Ornithocheiroidea; Pt, Pterodactyloidea; Ra, Rhamphorhynchidae; "Rh", "Rhamphorhynchoidea".

Table 4. *Character (Ch) state changes between the nodes shown in Figure 7a*

Branch	*Ch.*	*Change*
node 0 -> node 1	58	0=>1
node 1 -> node 2	1	0=>1
	2	0=>1
	3	0=>1
	4	0=>1
	5	0=>1
node 2 -> node 3	6	0=>1
	7	0=>1
	8	0=>1
node 3 -> node 4	9	0=>1
	10	0=>1
	11	0=>1
	12	0=>1
	13	0=>1
	14	0=>1
node 4 -> node 5	15	0=>1
	16	0=>1
	17	0=>1
	18	0=>1
	19	0=>1
	20	0=>1
	21	0=>1
	41	1=>0
node 5 -> node 6	22	0=>1
	23	0=>1
node 5 -> node 7	13	1=>0
	24	0=>1
	25	0=>1
	26	0=>1
	27	0=>1
	28	0=>1
	29	0=>1
	30	0=>1
node 7 -> node 8	31	0=>1
	32	0=>1
	33	0=>1
	34	0=>1
	35	0=>1
	36	0=>1
	37	0=>1
	38	0=>1
node 8 -> node 9	39	0=>1
	40	0=>1
	41	0=>1
node 9 -> node 10	34	1=>0
	42	0=>1
	43	0=>1
	44	0=>1
	45	0=>1
	46	0=>1
node 7 -> node 11	3	0=>1
	5	0=>1
	7	0=>1
	47	0=>1
	48	0=>1
node11 -> node 12	49	0=>1
	50	0=>1
	51	0=>1
node 12 -> node 13	52	0=>1
	53	0=>1
node 11 -> node 14	54	0=>1
	55	0=>1
	56	0=>1
	57	0=>1
	58	1=>0
	59	0=>1
node 11 -> node 15	3	0=>1
	9	1=>2
	22	0=>1
	37	0=>1
	43	0=>1
	45	0=>1
	60	0=>1
node 15 -> node 16	31	0=>1
	37	0=>1

of the scapula (Wild 1984a; Fig. 16a), although it is still relatively broad and robust compared to the condition in derived pterosaurs (Fig. 16d, e). Other basal pterosaurs such as dimorphodontids (Fig. 16b), also have a relatively short, broad coracoid (Wild 1978, p. 240), but in this case it is slightly more than 70% the length of the scapula, while in anurognathids it reaches 75% the length of the scapula (Unwin *et al.* 2000, fig. 2). In all other pterosaurs the scapula is more slender and rod-like (Fig. 16c-e) and usually approaches, or in some groups even exceeds, the length of the scapula (see character 32).

(3) Manual phalanges relatively robust compared to pedal phalanges. In the outgroups, manual and pedal phalanges are usually of similar dimensions and robustness, or the manual phalanges are relatively small. *Preondactylus* also has manual and pedal phalanges, and unguals, of similar dimensions (Fig. 20a), whereas many pterosaurs, including all "rhamphorhynchoids" other than *Preondactylus*, have manual phalanges and unguals that are much more robust than the corresponding pedal elements (Fig 20b, c). This size disparity is retained in some basal ctenochasmatoids, such as *Pterodactylus* (Fig. 20d), where it is more pronounced in adults than in juveniles, but more derived ctenochasmatoids, such as *Pterodaustro* (Wellnhofer 1991a, p. 131) and dsungaripteroids (e.g. *Germanodactylus*), appear to have manual and pedal phalanges of similar sizes. Clearly, this must represent a character reversal since both these taxa are firmly nested within the Pterodactyloidea.

(4) Forelimb length more than 2.5 times the length of the hindlimb. (See Unwin 1995a; Unwin *et al.* 2000.) Outgroup taxa either have forelimbs that are similar in length to the hindlimbs or, in some cases (e.g. prolacertiforms, *Scleromochlus*), the forelimb is much shorter than the hindlimb (femur + tibia + metatarsal III). In *Preondactylus* the forelimb is already highly elongated, as in other pterosaurs, but is slightly less than 2.5 times the length of the hindlimb (Table 2). In all other pterosaurs sufficiently complete for this index to be calculated the forelimb length is at least, and often far greater than, 2.5 times the length of the hindlimb (see Unwin *et al.* 2000, p. 186). Other basal "rhamphorhynchoids", such as dimorphodontids and anurognathids, also have relatively short forelimbs compared to the hindlimbs, but in more derived clades (Campylognathoididae and Rhamphorhynchidae) they have relatively elongate forelimbs. Remarkably, basal ctenochasmatoids, such as *Pterodactylus*, and azhdarchoids also have relatively short forelimbs (Table 2), although in the latter case this may, in part, be attributable to elongation of the hindlimbs.

(5) Humerus longer than the femur. (See Unwin 1995a.) In the outgroups, the humerus is always shorter than the femur. This is also the case in

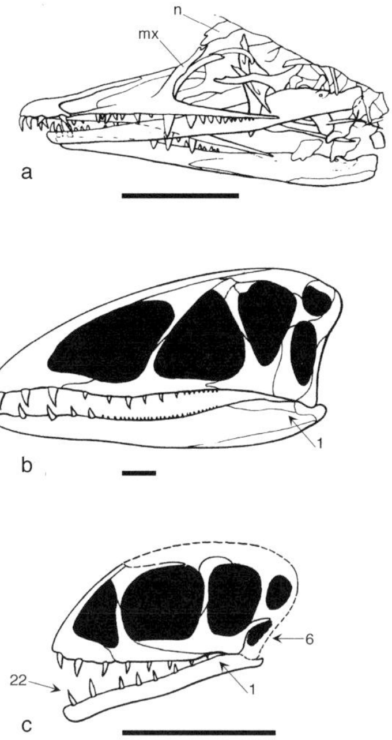

Fig. 8. Skulls of basal pterosaurs drawn to a uniform size and shown in left lateral view. (**a**) *Preondactylus buffarinii* (after Wild 1984a, reversed), (**b**) *Dimorphodon macronyx* (after Wellnhofer 1978), (**c**), *Anurognathus ammoni* (after Wellnhofer 1978). Numbered arrows (here and in Figs 9–20) indicate derived states of characters described in the text. mx, maxilla; n, nasal. Scale bar 20 mm.

Preondactylus (Table 2), but not in any other "rhamphorhynchoid" clades, where the humerus is either the same length or longer than the femur. One exception is the single known specimen of *Eudimorphodon cromptonellus* (Jenkins *et al.* 2001), but in this case the relative shortness of the humerus (92% the length of the femur) may be related to the immaturity of this individual. All known members of the Ornithocheiroidea also have a humerus/femur index greater than 1, whereas, with a few exceptions (some *Pterodactylus*, derived ctenochasmatids), the humerus is shorter than the femur in other, non-ornithocheiroid pterodactyloids. It seems likely that the primary derived state for pterosaurs (humerus longer than femur) is plesiomorphic for pterodactyloids, because it is present in basal pterodactyloids (ornithocheiroids) and is also universally present in those "rhamphorhynchoid" clades likely to be sister-taxa to Pterodactyloidea (Rhamphorhynchidae, or Campylognathoididae + Rhamphorhynchidae, see below).

Remarks. Preondactylus exhibits the plesiomorphic state for the 48 characters that can be scored for this taxon (Table 3) and thus, in agreement with the results of other studies (Unwin 1992, 1995a, Peters

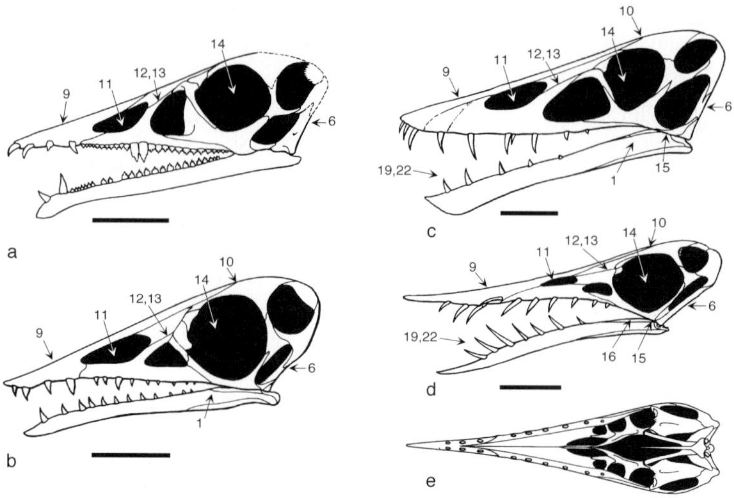

Fig. 9. Skulls of derived "rhamphorhynchoids" drawn to a uniform size and shown in left lateral view (a)–(d) and ventral view (e). (**a**) *Eudimorphodon ranzii*, (**b**) *Campylognathoides liasicus*, (**c**) *Scaphognathus crassirostris*, (**d**) and (**e**) *Rhamphorhynchus muensteri*. (All redrawn from Wellnhofer 1978.) Scale bar 20 mm.

1997; Dalla Vecchia 1998), is identified in this analysis as the most basal pterosaur and the sister-taxon to all other ingroup members. In the first description of *Preondactylus*, Wild (1984a) argued, on the basis of morphometric data and characters of the dentition, that this pterosaur was an early member of Rhamphorhynchidae, a proposal that, if correct, would place it in clade 6 of this analysis (Fig. 7a). Comparison with other taxa shows, however, that the dental and morphometric characters cited are found in other pterosaurs (e.g. *Dimorphodon*) and do not support a sister-group relationship for *Preondactylus* and the Rhamphorhynchidae. Dalla Vecchia (1998) corrected some of the meristic data and noted similarities with the dimorphodontid *Peteinosaurus* (Appendix 1), including the shape of the lower jaw and humerus and relative lengths of some of the limb bones. These characters, and a new interpretation of the skull as relatively short and deep (see Appendix 1), further emphasise the basal position of *Preondactylus*, but do not necessarily indicate that it is a dimorphodontid because they may be generally plesiomorphic for pterosaurs. In any case, the subequal size of the phalanges and unguals in the hands and feet of *Preondactylus* contrast sharply with the size disparity evident in these structures in *Peteinosaurus* (cf. Figs 20a and 20b), indicating that these taxa are not congeneric.

Clade 3. Caelidracones (new clade name)
Etymology. Latin, *caelum* = the air or the sky, *draco* = dragon, in reference to Seeley's (1901) characterization of pterosaurs as 'dragons of the air'.
Definition. Anurognathus ammoni, *Quetzalcoatlus northropi*, their most recent common ancestor, and all its descendants.
Content. Anurognathidae, Campylognathoididae, Rhamphorhynchidae and Pterodactyloidea.
Synapomorphies

(6) Quadrate inclined anteriorly. (See Unwin 1995a.) In the outgroups the quadrate is either vertically inclined or the ventral end that articulates with the mandible is located posterior to the dorsal end. In dimorphodontids (Fig. 8b), and apparently also in *Preondactylus* (Wild 1984a; Wellnhofer 1991a; Dalla Vecchia 1998), the quadrate is vertical, the ventral end located directly below the dorsal end when the skull is oriented with its ventral margin horizontal. The single described specimen of *Anurognathus* has been reconstructed with a quadrate in which the ventral end lies somewhat anterior to the dorsal end (Fig. 8c), but this cannot be confirmed at present because this and all other anurognathid skulls (Ryabinin 1948; Ji & Ji 1997; Unwin *et al.* 2000; Wang *et al.* 2002) are strongly compressed and preserved in dorsoventral orientation. It is evident, however, that in all other pterosaurs (Campylognathoididae + Rhamphorhynchidae + Pterodactyloidea) the quadrate slopes forward (Figs 9–13).

(7) Ulna longer than the tibia. (See Unwin 1995a; Unwin *et al.* 2000.) The majority of outgroup taxa have an ulna that is shorter than the tibia, as do the basal groups *Preondactylus* and dimorphodontids (Table 2). More derived "rhamphorhynchoids" and all known ornithocheiroids exhibit what is interpreted here, using outgroup comparison, as the

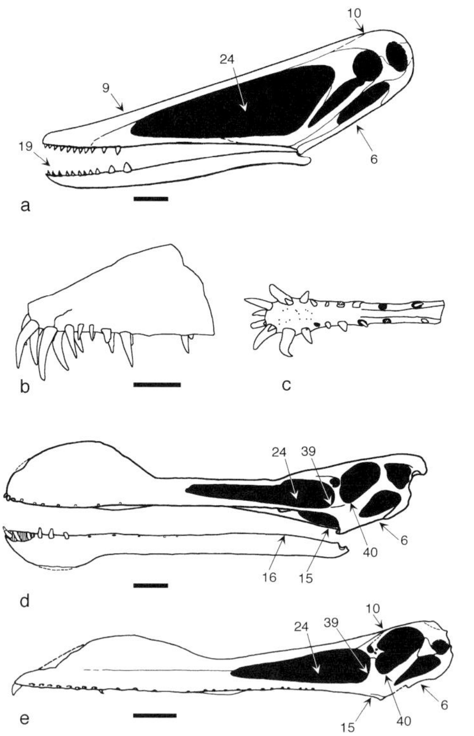

Fig. 10. Skulls of toothed ornithocheiroid pterosaurs drawn to a uniform size and shown in left lateral view (a), (b), (d) and (e), and ventral view (c). (**a**) *Istiodactylus latidens* (after Wellnhofer 1987), (**b**) and (**c**) *Coloborhynchus robustus* (after Fastnacht 2001), (**d**) *Ornithocheirus mesembrinus* (after Wellnhofer 1978), (**e**) *Anhanguera blittersdorffi* (after Kellner & Tomida 2000). Scale bar 50 mm.

derived condition with an ulna that is longer than the tibia, but in most non-ornithocheiroid pterodactyloids (with the exception of some specimens of *Pterodactylus* and ctenochasmatids) the tibia is longer than the ulna (see Unwin *et al.* 2000, p. 186). The latter, reflecting elongation of the hindlimb rather than shortening of the forelimb, is most parsimoniously interpreted as a secondarily derived condition. The alternative optimization, that non-ornithocheiroid pterodactyloids retain the primitive condition while the derived condition arose twice (in Anurognathidae + Campylognathoididae + Rhamphorhynchidae and in Ornithocheiroidea), requires location of Pterodactyloidea among the basal clades of Pterosauria and is less parsimonious (see below).

(8) Fibula less than 80% the length of the tibia. In outgroup taxa the fibula contacts the calcaneum. In basal pterosaurs the fibula is reduced to a slender, splint-like bone, but certainly contacted the calcaneum in dimorphodontids (BMNH 41347b, BMNH 43051), contrary to published illustrations and in *Campylognathoides* (Fig. 19j). In other clades, including the Anurognathidae (e.g. Unwin et al. 2000, fig. 2), *Eudimorphodon* (Wild 1978), Rhamphorhynchidae and Pterodactyloidea, the distal end of the fibula tapered to a fine point that fused with the shaft of the tibia (Fig. 19k). It is more parsimonious to presume that the plesiomorphic condition evident in *Campylognathoides* represents a reversal, because the alternative requires the derived condition to have arisen three times, rather than once.

Remarks. Another synapomorphy that potentially supports this clade is the elongate preacetabular process of the ilium. In *Preondactylus* (Wild 1984a, fig. 3) and in dimorphodontids (Owen 1870; Wild 1978), the preacetabular process of the ilium is shorter or of similar length to the postacetabular process, whereas in anurognathids (e.g. Wang *et al.* 2002, fig. 1) and more derived pterosaurs (Wellnhofer 1978, figs 14 & 15) it is considerably longer.

Wellnhofer (1975a, p. 22; Wellnofer 1978, Fig. 2) suggested a close relationship between *Anurognathus* and *Dimorphodon*, but characters cited in support of this idea (short, high skull, upright quadrate, large antorbital fenestra) also occur in *Preondactylus* and some of the outgroups, and are probably plesiomorphic for pterosaurs. A close relationship between Anurognathidae and †Criorhynchidae (Kuhn 1967; Fig. 1b) or direct descent of †*Criorhynchus* from *Anurognathus* (Young 1964; Fig. 1a) must be rejected because †*Criorhynchus* and Criorhynchidae are invalid taxa and the short, deep skull, first proposed by Arthaber (1919), is incorrect (Wellnhofer 1987; Unwin 2001).

Clade 4. Lonchognatha (new clade name)

Etymology: Greek, *logkhos* = spear, *gnathos* = jaw, in reference to the relatively elongate, spear-shaped jaws of pterosaurs in this clade.

Definition. Eudimorphodon ranzii, Rhamphorhynchus muensteri, their most recent common ancestor, and all its descendants.

Content. Campylognathoididae, Rhamphorhynchidae and Pterodactyloidea.

Synapomorphies

(9) Rostrum low with straight or concave dorsal outline. (See Unwin 1995a; Unwin *et al.* 2000.) Basal pterosaurs and outgroup taxa have short deep rostra, with a high-arched, convex outline in lateral aspect (Figs 8b, c). Campylognathoidids, rhamphorhynchids and pterodactyloids have low elongate rostra, with either a straight or concave dorsal profile (Figs 9–13). *Preondactylus* has also been reconstructed with a low elongate rostrum (Wellnhofer 1991a; Dalla Vecchia 1998), but this is questionable because the elongate subvertical nasal process of the maxilla (Fig. 8a) shows that the snout was probably

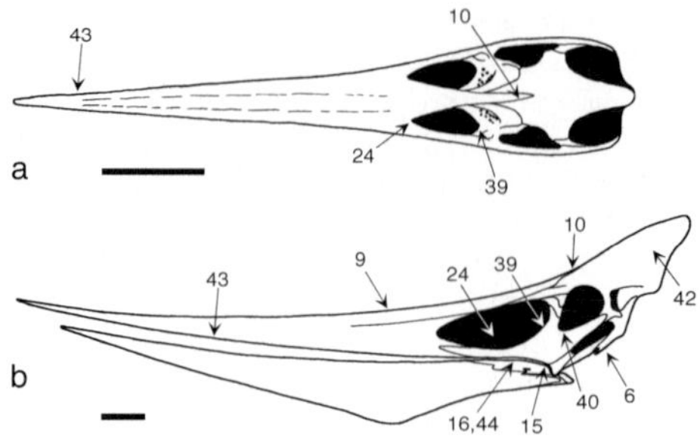

Fig. 11. Skulls of edentulous ornithocheiroid pterosaurs drawn to a uniform size and shown in dorsal view (a) and left lateral view (b). (**a**) *Nyctosaurus gracilis* (after Williston 1902), (**b**) *Pteranodon longiceps* (after Bennett 2001). Scale bar 50 mm.

as deep as that of *Dimorphodon* at this point (see Appendix 1).

Some azhdarchoids such as *Tapejara* also have relatively short deep rostra (Fig. 13c), but this probably represents a secondarily derived condition within Pterosauria, especially since the presence of an elongate rostral tip in *Tapejara* appears to indicate the existence of the primary apomorphic state in forms ancestral to Azhdarchoidea.

(10) Posterior process of premaxillae interfingers between frontals. (See Unwin 1995a.) In outgroup taxa the posterior process of the premaxillae contacts the nasals but does not extend between them (e.g. Sereno 1991; Benton 1999), whereas in pterosaurs the premaxillae extend further caudally, separating or overlapping the nasals and contacting the frontals. In basal forms, such as dimorphodontids (Fig. 8b) and anurognathids (Fig. 8c), the premaxillae terminate at the contact with the frontals, but in campylognathoidids (Fig 9a, b), rhamphorhynchids (Fig 9c, d) and pterodactyloids (Figs 10–13) the premaxillae extend even further posteriorly, projecting between the mid-line contact of the frontals. Presumably, this arrangement enabled the skull to resist larger bending loads at the premaxilla-nasal-frontal contacts than in basal pterosaurs, a construction probably related to elongation of the snout.

(11) External narial opening low and elongate. Basal pterosaurs have a narial opening that in lateral aspect is as high (dorso-ventrally) or higher than it is long (antero-posteriorly) (Fig 8a-c). In more derived "rhamphorhynchoids" (Fig. 9), the narial opening is more elongate than it is high, a morphology that is, presumably, correlated with elongation of the rostrum. Most outgroup taxa exhibit a condition wherein the narial opening is more elongate than it is high, thus it could be argued that the short, deep narial opening of basal pterosaurs represents an apomorphic condition and that the coding should be reversed. This has no effect when the outgroup is excluded, however, and no significant impact on tree structure when an outgroup is incorporated into the analysis.

(12) Nasal process of maxilla inclined backwards. In the outgroups and in basal pterosaurs (Fig 8a-c) the nasal process of the maxilla is oriented vertically. By contrast, in campylognathoidids and rhamphorhynchids, this process slants posterodorsally, its reorientation presumably at least partly associated with elongation of the rostrum. In pterodactyloids the nasal process of the maxilla is absent, although its original location can be roughly estimated from the orientation of the maxilla process of the nasal, which is present in some Jurassic taxa (Figs 12b & 13a).

(13) Broad maxilla–nasal contact. The nasal of basal pterodactyloids has a relatively elongate process that extends downwards to make a narrow contact with the nasal process of the maxilla and thus form the bony buttress separating the narial and antorbital fenestrae (Fig 8a–c). In campylognathoidids and rhamphorhynchids the maxillary process of the nasal is shortened and the nasal process of the maxilla has a broad contact with the nasal, often extending along a considerable portion of the basal margin of this bone (Fig 9a–d). In pterodactyloids the bony buttress separating the narial and antorbital fenestrae is lost, but the long, narrow maxillary process of the nasal present in some Jurassic taxa (Figs 12b & 13a) suggests that pterodactyloid ancestors had a narrow contact between the nasal and maxilla as in basal pterosaurs.

(14) Orbit larger than antorbital fenestra. (See Unwin 1995a.) The antorbital fenestra is larger than the orbit in basal pterosaurs (Fig. 8). By contrast, in campylognathoidids and rhamphorhynchids, the orbit is the larger of the two openings (Fig 9a–d). The orbit is proportionately a little larger in these taxa than in basal pterosaurs, but the main processes contributing to this morphological change are the infilling of the antorbital fenestra as a result of the expansion of the bones surrounding this opening and the reduction in the height of the rostrum anterior to the orbit (Arthaber 1919). The condition in pterodactyloids is not directly comparable because of the confluency of the nasal and antorbital fenestrae, but interestingly, where the antorbital fenestra is still discernible in some Jurassic forms (Fig 12b & 13a), it is relatively large.

Remarks. As noted by Arthaber (1919, p. 412), who commented specifically on *Dimorphodon*, basal pterosaurs (*Preondactylus*, dimorphodontids and anurognathids) have relatively short deep skulls that retain a general construction similar to that seen in the outgroups. Lonchognathans are characterized by fundamental changes in cranial morphology, principally related to elongation of the rostrum: possibly an adaptation for enhancing the ability of these

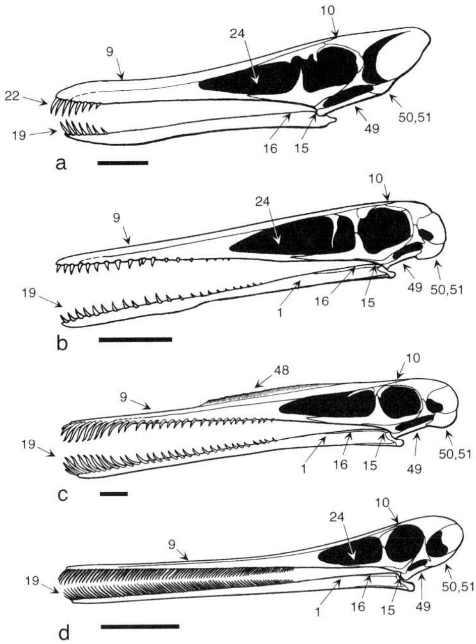

Fig. 12. Skulls of ctenochasmatoid pterosaurs drawn to a uniform size and shown in left lateral view. (**a**) *Cycnorhamphus suevicus*, (**b**) *Pterodactylus antiquus*, (**c**) *Gnathosaurus subulatus*, (**d**) *Ctenochasma gracile*. (All redrawn from Wellnhofer 1978.) Scale bar 20 mm.

pterosaurs to capture prey on or just beneath the water surface during flight. The character polarities adopted here are further supported by ontogenetic patterns described for pterodactyloids such as *Pterodactylus kochi* (Wellnhofer 1970). In this species the rostrum is relatively short and deep in early ontogenetic stages, but becomes long and low in later stages (Wellnhofer 1970, fig. 18).

Clade 5. Breviquartossa (new clade name)
Etymology. Latin, *brevis* = short, *quartus* = fourth, *ossum* = bone, in reference to the relatively short fourth metatarsal of pterosaurs in this clade.
Definition. Rhamphorhynchus muensteri, *Quetzalcoatlus northropi*, their most recent common ancestor, and all its descendants.
Content. Rhamphorhynchidae and Pterodactyloidea.
Synapomorphies

(15) Ventral margin of skull curved downwards caudally. (See Unwin 1995a.) The ventral profile of the skull essentially follows a straight line in the outgroups, basal pterosaurs (Fig. 8) and campylognathoidids (Fig. 9a, b). By contrast, in rhamphorhynchids and pterodactyloids the caudal region of the skull posterior to the antorbital fenestra is curved downwards such that the ventral articular end of the quadrate lies well below the level of the ventral margin of the rostrum (Figs 9c, d & 10–13).

(16) Loss of 'coronoid' eminence on caudal end of mandible. The dorsal margin of the mandible of outgroup taxa, basal pterosaurs (Figs 8a–c) and campylognathoidids (Figs 9a, b & 14a) bears a low 'coronoid' eminence just anterior to the articular region. In rhamphorhynchids (Figs 9a, b & 14c) and pterodactyloids (Figs 10–13) this eminence is lost and the dorsal margin of each mandible is flat anterior to the articular region.

(17) Development of bony mandibular symphysis. (See Unwin 1995a). The mandibles of outgroup taxa, basal pterosaurs (Figs 8a & 14b) and campylognathoidids (Fig. 14a) contact at their anterior tip and were probably held together by fibrous connective tissue, but do not appear to have been fused. In adult rhamphorhynchids and pterodactyloids, the anterior tips of the mandibles were co-ossified and often formed a symphysis (Fig. 14c–f).

(18) Mandibular symphysis forms more than 30% the length of the mandible. (See Unwin 1995a.) As mentioned above, the mandibles of basal pterosaurs and campylognathoidids are unfused, whereas in rhamphorhynchids and pterodactyloids they are fused to form a symphysis. Moreover, in rhamphorhynchines (Fig. 14d) and most pterodactyloids (Figs 14e, f), the symphysis forms more than 30% of the total length of the lower jaw (see also Bennett 2001). Optimization of this character is difficult. It may be apomorphic for clade 5 (Fig. 7a), reversing to the plesiomorphic condition in scaphognathines, *Istiodactylus*, lonchodectids and *Germanodactylus*. Alternatively, the apomorphic condition may have arisen homoplastically in rhamphorhynchines, derived ornithocheiroids, ctenochasmatoids, dsungaripterids and azhdarchoids. The former, more parsimonious, proposal is accepted here.

(19) Loss of heterodonty in the mandibular dentition. (See Unwin 1995a.) Typically, in the outgroup taxa, and in basal pterosaurs and campylognathoidids, the rostral end of the mandibular dentition begins with two, large, fang-like teeth (Figs 8a–c & 9a, b). The remaining mandibular teeth are much smaller and usually of similar size. In rhamphorhynchids and pterodactyloids this heterodonty is lost, the first two teeth in each mandible usually being of similar size to the remaining teeth (Figs 9c, d, 10, 12 & 13). Heterodonty reappears in ornithocheirids (Fig. 10b–d) and in some gnathosaurines (Dong 1982; Leonardi & Borgomanero 1985; Unwin 2000), but the relative enlargement of the teeth at the anterior end of the mandible always extends beyond the first two positions and thus is probably not homologous with the condition in basal pterosaurs.

(20) Metacarpals I, II and III of equivalent length. Metacarpals I-III of outgroup taxa, basal pterosaurs

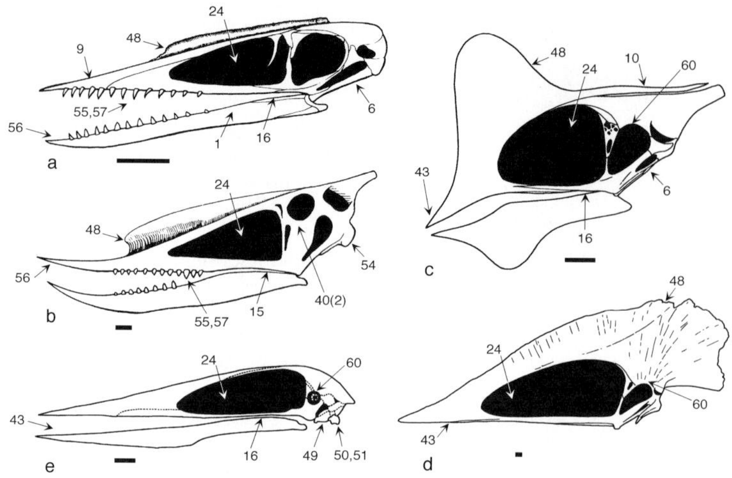

Fig. 13. Skulls of dsungaripteroid and azhdarchoid pterosaurs drawn to a uniform size and shown in left lateral view. (**a**) *Germanodactylus cristatus* (after Wellnhofer 1978), (**b**) *Dsungaripterus weii* (after Wellnhofer 1978), (**c**) *Tapejara wellnhoferi* (after Wellnhofer & Kellner 1991), (**d**) *Tupuxuara longicristatus* (IMCF 1052), (**e**) *Zhejiangopterus linhaiensis* (after Unwin & Lü 1997). Scale bar 20 mm.

(Figs 20a, b) and campylognathoidids (Fig. 18a) are of unequal length: most noticeably, metacarpal I is always distinctly shorter than II and III. By contrast, in most rhamphorhynchids and pterodactyloids metacarpals I–III are of equivalent length and their distal terminations are level with each other (Figs 18o, p & 20d). The rhamphorhynchine *Dorygnathus* exhibits the plesiomorphic condition (e.g. Arthaber 1919, fig. 22), by contrast to other rhamphorhynchids, which have metacarpals I–III of the same length. Consequently, it is most parsimonious to assume that the condition in *Dorygnathus* represents a reversal to the plesiomorphic state.

(21) Short metatarsal IV. (See Unwin 1995a; Unwin *et al.* 2000.) Metatarsals I–IV of outgroup taxa, basal pterosaurs (Figs 20a, b) and campylognathoidids are of the same, or similar, length, their distal ends terminating level with each other (Unwin *et al.* 2000, p. 192). Metatarsal IV is markedly shorter than I–III in rhamphorhynchids and pterodactyloids (Figs 20c, d).

Remarks. This and some previous studies (Unwin 1992, 1995a; Kellner 1996a; Viscardi *et al.* 1999) have presented evidence in support of a relationship between Rhamphorhynchidae and Pterodactyloidea. Another possibility is that Rhamphorhynchidae is more closely related to Campylognathoididae than to Pterodactyloidea and that this large clade forms a sister-taxon to Pterodactyloidea (Fig. 7b). This hypothesis is supported by character states that are shared by Rhamphorhynchidae + Campylognathoididae and the retention by pterodactyloids of character states otherwise found only in basal pterosaurs.

Thus, Rhamphorhynchidae + Campylognathoididae is supported by derived states for characters 11–14, while basal pterodactyloids (or at least some of them) exhibit the plesiomorphic condition for characters 3, 5 and 7 (Table 3) and a condition that is only slightly derived compared to the plesiomorphic state for character 4 (Table 2). Moreover, stem-group pterodactyloids are likely to have retained the plesiomorphic state for characters 11–14.

Optimizing character states to support the relationships shown in Fig. 7b is less parsimonious than the arrangement shown in Fig. 7a, but this idea merits further study, as does another intriguing possibility, that Pterodactyloidea might share a closer relationship with Anurognathidae than with any other "rhamphorhynchoid" clade. Wellnhofer (1975b, p. 183) noted similarities between the pelvis of *Anurognathus* and pterodactyloids, and two characters utilized in this analysis (character 26, reduction of cervical ribs; character 27, reduction in the length of the caudal series) also support this relationship. If pterodactyloids are assumed to have the plesiomorphic condition for characters 11–14, then a sister-group relationship between Pterodactyloidea and Campylognathoididae + Rhamphorhynchidae has only the same level of support (two characters: 9 and 10) as the pairing of Anurognathidae + Pterodactyloidea.

Clade 6: Rhamphorhynchidae Seeley 1870 (converted clade name)

Definition. Sordes pilosus, *Rhamphorhynchus muensteri*, their most recent common ancestor, and all its descendants.

Content. Rhamphorhynchinae and Scaphognathinae.

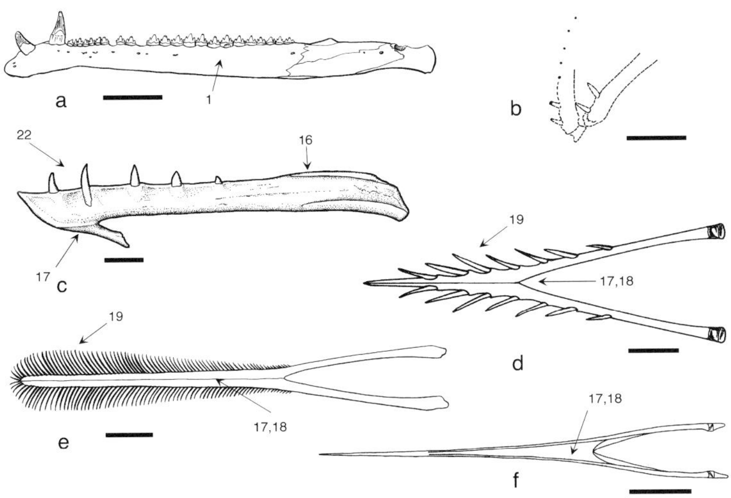

Fig. 14. Pterosaur lower jaws drawn to a uniform size and shown in left lateral view (**a**)–(**c**) and dorsal view (**d**)-(**f**). (**a**) *Eudimorphodon ranzii* (after Wild 1978), (**b**) *Anurognathus ammoni* (after Wellnhofer 1975b), (**c**) *Scaphognathus crassirostris* (after Wellnhofer 1975b), (**d**) *Rhamphorhynchus muensteri* (after Wellnhofer 1975c), (**e**) *Ctenochasma gracile* (after Wellnhofer 1978), (**f**) *Pteranodon longiceps* (after Bennett 2001). Scale bar 10 mm, except for (f) 100 mm.

Synapomorphies

(22) Less than 11 pairs of teeth in the rostrum. All outgroup taxa and most pterosaurs have at least, and often far more than, 11 pairs of teeth in the rostrum (Figs 8a, b, 9a, b, 10, 12 & 13). Rhamphorhynchids have, at most, 9 pairs of rostral teeth (Fig. 9c–e) and in many taxa this number is further reduced. *Anurognathus* (Fig. 8c) also shows a strong reduction in the numbers of rostral teeth with only 8 pairs, but the stratigraphically older form, *Batrachognathus*, has at least 11 pairs of teeth (Ryabinin 1948); thus the condition in *Anurognathus* may be homoplastic with respect to rhamphorhynchids. Several pterodactyloids also exhibit this apomorphy: *Cycnorhamphus* (Fig. 12a) has only 8 pairs of teeth and *Nyctosaurus* (Fig. 11a), pteranodontids (Fig. 11b) and azhdarchoids (Figs 13c–e) are edentulous. This suggests an alternative possibility, that this character is apomorphic for clade 5 (Fig. 7a), although this would require at least three reversals to the primitive condition: in ornithocheiroids, *Pterodactylus* + Lonchodectidae + Ctenochasmatidae, and dsungaripteroids. This is equally parsimonious to the optimization adopted here, and the only grounds for preferring the latter is that reduction in tooth numbers is far more common within clades (e.g. found in campylognathoidids, rhamphorhynchines, ornithocheiroids and dsungaripteroids) than an increase in number (e.g. ctenochasmatids).

(23) Deltopectoral crest tongue-shaped, with necked base. Outgroups and all other clades of pterosaur have a deltopectoral crest that is broad-based and either tapers towards its distal, free margin, or remains of similar depth (Fig. 17a, b, d, g–k). Rhamphorhynchids are distinguished by a deltopectoral crest that is tongue-shaped, with a constricted base and an expanded distal portion (Fig. 17c). This condition is most clearly developed in *Rhamphorhynchus* (e.g. Wellnhofer 1975a; Fig. 17c) but is also evident in *Dorygnathus* (Arthaber 1919) and in *Sordes* (Sharov 1971; Bakhurina 1986, p.33; Ivakhnenko & Korabelnikov 1987, fig. 262). *Nyctosaurus* also has a deltopectoral crest with a constricted base (Williston 1903; Fig. 17i), but the 'axe-head'-shaped distal expansion and the displacement of the deltopectoral crest further down the shaft clearly distinguishes the nyctosaurid humerus from that of rhamphorhynchids.

Remarks. Comparison with previous systematic studies shows that opinions regarding the content of this clade are variable. Wellnhofer (1978) included *Campylognathoides* together with scaphognathines and rhamphorhynchines in Rhamphorhynchidae, but the former taxon shares characters in common with *Eudimorphodon* (Wild 1978; Appendix 1), which are not found in other pterosaurs, and lacks apomorphies of Rhamphorhynchidae + Pterodactyloidea, or Rhamphorhynchidae.

Sordes, a member of Scaphognathinae (Unwin & Bakhurina 2000; Appendix 1), is considered by Kellner (1996a; Fig. 5a) to be a basal pterosaur, although no character data was cited in support of this, and by Peters (2001) to be a member of the Dimorphodontidae. Among the 12 characters listed by Peters, one (character 9) supports a derived position for *Sordes* with regard to other "rhamphorhynchoids", one (character 12) is autapomorphic for *Sordes* (Sharov 1971), five (characters 2, 3, 8, 10 and 11) are not consistent with or determinable on the basis of available fossil evidence, four (characters 1, 4, 5 and 7) are not confined to *Sordes* and dimorphodontids and are probably plesiomorphic for pterosaurs, and one (character 6) cannot be unambiguously determined for other pterosaurs. Apart from exhibiting all apomorphies of Scaphognathinae (Appendix 1), *Sordes* also shows the derived state for all characters (1–23) so far described in this analysis, strongly contradicting the possibility of a basal position within Pterosauria.

Clade 7: Pterodactyloidea Plieninger 1901 (converted clade name)
Definition. Pteranodon longiceps, Quetzalcoatlus northropi, their most recent common ancestor, and all its descendants.
Content. Ornithocheiroidea, Ctenochasmatoidea, Dsungaripteroidea and Azhdarchoidea.
Synapomorphies

(24) Narial and antorbital fenestrae confluent. (See Unwin 1995a). In the outgroups and in all "rhamphorhynchoids" (Figs 8 & 9) the narial and antorbital fenestrae are separated by a bony bar consisting of the nasal process of the maxilla and the maxilla process of the nasal. In all pterodactyloids this bony bar is absent (Figs 10–13), although in some Jurassic forms (*Pterodactylus*, *Germanodactylus*) a slender elongate process, presumably homologous with the maxilla process of the nasal of "rhamphorhynchoids"' is retained.

(25) Basipterygoids united to form a median bar of bone. (See Unwin 1995a.) In the outgroups and in "rhamphorhynchoids" the basipterygoids remain separate (Fig. 9e). In pterodactyloids, however, the basipterygoids fuse to form an elongate median bar of bone (sometimes labelled the basisphenoid, e.g. Wellnhofer 1978) which runs forwards from the basioccipital region to contact the quadrates (Wellnhofer 1978, fig. 6). Occasionally, juvenile pterodactyloids (e.g. '*Pterodactylus*' *micronyx*, Wellnhofer 1970, fig. 8a) exhibit the plesiomorphic condition, with separate basipterygoids.

(26) Reduction of cervical ribs. (Howse 1986; Bennett 1994; Unwin 1995a.) Elongate cervical ribs are present in outgroup taxa and in most "rhamphorhynchoids" (Fig. 15b). Cervical ribs are widely thought to be absent in pterodactyloids (Kuhn 1967; Wellnhofer 78; Howse 1986; Bennett 1994), and indeed there is no direct evidence for them in most taxa (Figs 15c–g). Remnants of cervical ribs are found, however, in some ornithocheirids, where they take the form of a small bony arch spanning the gap between the prezygapophysial process and the centrum. Generally, these structures are completely fused to the vertebra and their limits cannot be traced, but they are well seen in the cervical vertebrae of *Coloborhynchus robustus* (NSM-PV 19892). Their presence can also be inferred in some azhdarchids, which retain a narrow passage between the prezygapophysial process and the centrum, and whose outer wall is probably formed from a remnant of the cervical rib, now completely fused to the vertebra. Until recently, the condition in anurognathids, where cervical ribs seem to be absent, was difficult to determine because of the poor preservation of the neck region (Ryabinin 1948; Wellnhofer 1975b). New, well-preserved specimens from the Yixian Formation (Ji & Ji 1998, Unwin *et al.* 2000), one with a complete neck (Wang *et al.* 2002), also appear to lack cervical ribs, a synapomorphy that the Anurognathidae share with the Pterodactyloidea (Table 3).

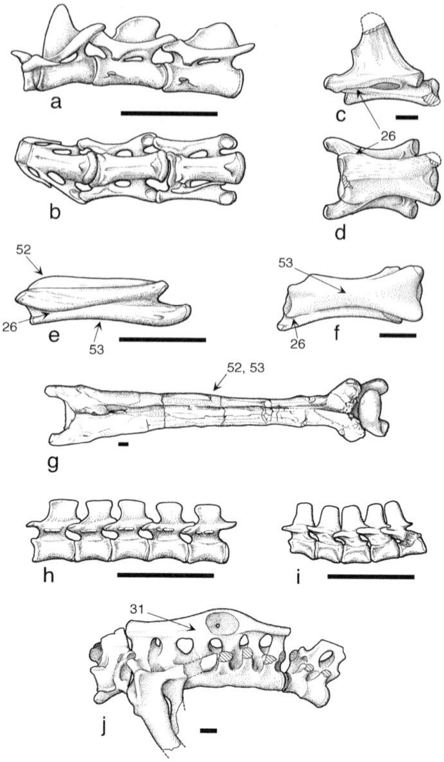

Fig. 15. Cervical and dorsal vertebrae of pterosaurs drawn to a uniform size. Cervical vertebrae shown in left lateral view (a), (c), (e), ventral view (b), (d), (f) and dorsal view (g). Dorsal vertebra shown in left lateral view. (**a**), (**b**) and (**h**) *Rhamphorhynchus muensteri* (after Wellnhofer 1975c), (**c**) and (**d**) *Anhanguera santanae* (after Wellnhofer 1991c), (**e**) and (**i**) *Pterodactylus antiquus* (after Wellnhofer 1970), (**f**) *Lonchodectes* sp. (BMNH R2287c), (**g**) *Quetzalcoatlus* sp. (after Howse 1986), (**j**) *Pteranodon longiceps* (after Bennett 2001). Scale bar 10 mm.

(27) Caudal vertebral series shorter than dorsal

series. (see Unwin 1995a; Unwin *et al.* 2000.) The complete caudal series in outgroup taxa and in all "rhamphorhynchoids" except anurognathids easily exceeds in length the complete dorsal series (Bennett 1994). In pterodactyloids the caudal series is sharply reduced to fewer than 15 caudals which usually form a short stubby tail, although in some taxa (e.g. *Dsungaripterus* and *Pteranodon*) the distal caudals are occasionally elongate (Bennett 1987). Despite this variation, the caudal series in pterodactyloids is shorter than the dorsal series in all known taxa. Anurognathids also appear to have a sharply reduced caudal series that is remarkably similar to the tail of pterodactyloids. Ji & Ji (1998) and Ji *et al.* (1999) have argued that a late surviving anurognathid, *Dendrorhynchoides*, from the Lower Cretaceous Yixian Formation has a long tail. More recently, however, Unwin *et al.* (2000) suggested that the tail may be a fake and argued that *Dendrorhynchoides* has a short tail. This interpretation has been supported by Wang *et al.* (2002), who described another anurognathid, *Jeholopterus*, from the Yixian Formation, that also clearly lacks a long tail.

(28) Pteroid long and slender. (See Unwin 1995a; Unwin *et al.* 2000.) The pteroid, a rod-like bone that articulates with the medial carpal, is a true bone (Unwin *et al.* 1996) and apparently unique to pterosaurs. In "rhamphorhynchoids" it is a relatively short stubby structure, its total length no greater than 5 times its basal width (Fig. 18a). By contrast, in pterodactyloids the pteroid is highly elongate and always at least 7 times longer than its basal width (Fig. 18b, see also Unwin *et al.* 2000, p. 186) and in many cases proportionately much longer, as for example in *Cycnorhamphus* (Plieninger 1907). The only exception to this pattern occurs in *Rhamphorhynchus*, where the pteroid is relatively elongate in some specimens of *R. muensteri* (Wellnhofer 1975a, fig. 12).

(29) Wing metacarpal (IV) at least 80% the length of the humerus. (See Unwin *et al.* 2000.) In all "rhamphorhynchoids", as in outgroup taxa, the wing metacarpal is relatively short and reaches, at most, only 68% the length of the humerus (Table 2). In all pterodactyloids the wing metacarpal is at least 80% the length of the humerus and often far exceeds the length of the latter (see also Unwin *et al.* 2000).

(30) Pes digit V with a single phalanx or entirely absent. (See Bennett 1994; Unwin 1995a; Unwin *et al.* 2000.) The fifth toe is present in most outgroup taxa, although not all (e.g. *Scleromochlus*, Benton 1999), and in all "rhamphorhynchoids" where it consists of two elongate phalanges, but no ungual (Fig. 20b). A single, highly reduced phalanx is retained in some Jurassic pterodactyloids (Fig. 20d), but the fifth toe appears to be entirely absent in other taxa (Wellnhofer 1978, 1991a).

Remarks. In addition to the synapomorphies listed above, this highly distinctive clade of 'short-tailed' pterosaurs, recognized as early as the mid-nineteenth century, is supported by numerous other apomorphies including elongation of the skull, forward inclination of the quadrate, and shift of the glenoid from wholly on the scapula to a position between the scapula and coracoid (see also Kuhn 1967; Wellnhofer 1978; Howse 1986; Bennett 1994).

Clade 8. Ornithocheiroidea Seeley 1891 (converted clade name)

Definition. Istiodactylus latidens, *Pteranodon longiceps*, their most recent common ancestor, and all its descendants.

Content. Istiodactylus, Ornithocheiridae, Pteranodontidae and *Nyctosaurus*.

Synapomorphies

(31) Development of a notarium. (See Bennett 1989, 1994; Unwin & Lü 1997.) The dorsal vertebrae of outgroup taxa, all known "rhamphorhynchoids" and some pterodactyloids remain separate and unfused (Fig. 15h, i). In ornithocheiroids early dorsal vertebrae fuse to form a notarium, that in some, but not all cases (e.g. *Nyctosaurus* Bennett 2001) articulates with the distal end of the scapula via an articular facet on a supraneural plate formed from the fused neural processes of the dorsal vertebrae (Fig. 15j).

A similar structure has been reported in *Dsungaripterus* (Young 1964, 1973) and '*Phobetor*' (Bakhurina pers comm. 2001) and also in *Tupuxuara* (IMCF 1052; Kellner & Hasegawa 1993) and azhdarchids (Nesov 1984, Buffetaut 1999). According to Bennett (1989, 1994) all these taxa and ornithocheiroids form a single clade diagnosed by the presence of a notarium. By contrast, the results of this analysis suggest that a notarium developed homoplastically in the Ornithocheiroidea, Dsungaripteroidea and Azhdarchoidea because it is absent from basal taxa within these clades (*Germanodactylus*, *Tapejara*) and, by inference, from the common ancestor of the Ornithocheiroidea and other pterodactyloids.

Non-homology of the notarium across Pterodactyloidea, as originally suggested by Wellnhofer (1978, p. 53) is supported by several observations. The degree of fusion and number of vertebrae involved is highly variable (Buffetaut 1999, p. 291; Bennett 2001, p. 50). The ornithocheiroid scapula is oriented almost perpendicular to the notarium and occupies a subhorizontal position, whereas in *Tupuxuara*, for example, (IMCF 1052), the scapula is obliquely oriented at 45° to the notarium, and slopes both downwards and forwards to its contact with the coracoid. This is a completely different arrangement from that found in ornithocheiroids, but

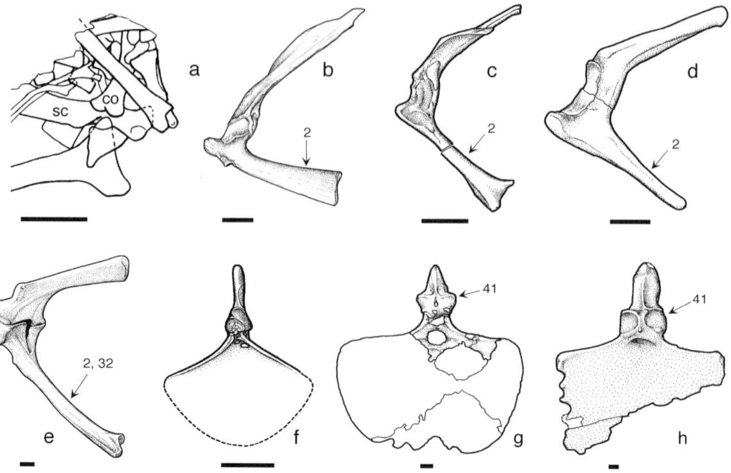

Fig. 16. Pterosaur scapulocoracoids, in left lateral view, and sterna, in dorsal view, drawn to a uniform size. (**a**) *Preondactylus buffarinii* (after Wild 1984b), (**b**) *Peteinosaurus zambellii* (after Wild 1978), (**c**) *Campylognathoides liasicus* (after Wellnhofer 1973), (**d**) *Cycnorhamphus suevicus* (after Wellnhofer 1978), (**e**) *Anhanguera santanae* (after Wellnhofer 1991c), (**f**) *Rhamphorhynchus muensteri* (after Wellnhofer 1975c), (**g**) *Coloborhynchus robustus* (after Kellner & Tomida 2000), (**h**) *Pteranodon longiceps* (after Bennett 2001). Scale bar 10 mm. co, coracoid; sc, scapula.

is remarkably similar to the situation in *Pterodactylus kochi* (e.g. Broili 1938), except that the latter taxon has no notarium. Finally, in so far as can be determined from fossil material, all medium to large and giant pterosaurs have a notarium, but they have never been reported in small pterosaurs, suggesting that the principal control on the appearance of this structure is size rather than phylogeny.

(32) Coracoid longer than scapula. (See Bennett 1994; Unwin 1995a; Unwin & Lü 1997.) The coracoid is shorter than the scapula in "rhamphorhynchoids" and outgroup taxa (see character 2, Fig. 16a–d). By contrast, in ornithocheiroids the coracoid is longer than the scapula (Fig. 16e), principally as a result of shortening of the scapula as it was reoriented to contact the notarium (see above), rather than through elongation of the coracoid.

(33) Humerus with warped deltopectoral crest. (See Padian 1984a, 1986; Bennett 1989, 1994; Unwin 1995a; Unwin & Lü 1997.) In outgroup taxa and all non-ornithocheiroids the deltopectoral crest extends only a short way down the shaft and is a relatively simple, flat, or slightly curved, flange-like structure (Fig. 17a–d, i–k). As first mentioned by Hooley (1913, p. 406) and discussed in some detail by Padian (1984a, 1986) and Bennett (1989, 1994, 2001), ornithocheiroids have a highly distinctive deltopectoral crest that has a long base and bears a terminal expansion that is twisted obliquely to the humeral shaft (Fig. 17g). The only exception is *Nyctosaurus*, but this pterosaur has a highly unusual 'axe-head'-shaped deltopectoral crest that is unlike that found in any other taxon (Bennett 1994; Fig. 17i) and is thus coded here as a distinct character state (Table 3). Moreover, in this analysis, *Nyctosaurus* forms a sister-group to Pteranodontidae (Figs 6 & 7) and is thus nested deep within Ornithocheiroidea, suggesting that the unusual morphology of the *Nyctosaurus* humerus was derived from the ornithocheiroid condition.

(34) Pneumatopore piercing the anconal surface of the proximal part of the humerus. (See Bennett 1989, 1994, 2001; Unwin 1995a; Unwin & Lü 1997.) In non-ornithocheiroid pterosaurs and outgroup taxa, approximately at the point at which the medial crest meets the shaft, the anconal surface of the proximal end of the humerus does not bear any evidence of a pneumatic opening (Wellnhofer 1978; Fig. 17j), although in some pterodactyloids (e.g. *Lonchodectes*) there is a small foramen, possibly for transmission of vascular or nerve structures, at this point. At the same position some ornithocheiroids have a prominent pneumatic opening (Bennett 1989; Fig. 17h), although this appears to be absent in pteranodontids (Bennett 2001, fig. 69) and *Nyctosaurus* (Williston 1903).

(35) Distal end of humerus has a triangular outline. (See Bennett 1989, 1994; Unwin 1995a; Unwin & Lü 1997.) In non-ornithocheiroid pterosaurs the distal end of the humerus has a subrectangular or D-shaped outline (Fig. 17e). By contrast, in ornithocheiroids the distal end has a triangular outline (Wellnhofer 1985; Kellner & Tomida 2000, Fig. 17f).

(36) Ornithocheiroid carpus. (See Unwin 1995a; Unwin & Lü 1997.) Pterosaurs have a highly unusual wrist construction, consisting of a proximal and distal syncarpal, a medial carpal and a pteroid,

that is unlike that of any other known tetrapod, living or extinct (Wellnhofer 1978, 1985, 1991a, b).

The typical condition, at least for pterodactyloids, is exemplified by wrist material originally described by Wellnhofer (1985) under the name of '*Santanadactylus spixi*', assigned by Bennett (1989, 1994) to the Dsungaripteridae, but probably referable to *Tupuxuara* in that it is remarkably similar to well-preserved remains of this pterosaur (IMCF 1052; Kellner & Hasegawa 1993; Kellner 1996a). In this case, as in other pterodactyloids, including *Pterodactylus*, *Cycnorhamphus*, *Germanodactylus*, '*Phobetor*' (Bakhurina 1982, fig. 1), *Dsungaripterus*, *Azhdarcho* and *Quetzalcoatlus*, the proximal syncarpal has a rectangular outline in proximal aspect (Fig. 18c), with a steeply oriented radial facet, and paired facets for the distal syncarpal that occupy almost the entire distal face of the proximal syncarpal (Fig. 18d). The distal syncarpal is also rectangular in proximal and distal view (Figs 18f, g), with a large, inverted triangular facet for the wing metacarpal on the distal face, bounded posterodorsally by a much smaller, subrectangular facet, which also made contact with the wing metacarpal. In dorsal view (Fig. 18h) the distal syncarpal is relatively short (anteroposteriorly) and blocky, with a short, rounded buttress bearing an articular facet for the medial carpal. *Rhamphorhynchus* (Wellnhofer 1975a, fig. 12) and *Dimorphodon* (Unwin 1988b, fig. 4) also have relatively short, blocky, rectangular syncarpals.

Ornithocheiroids exhibit a derived condition (see also Bennett 2001, p. 90), in which the proximal syncarpal is relatively elongate (anteroposteriorly), and has a proximal face with a pentagonal outline (Fig. 18i), a subhorizontal radial facet and a deep notch in the ventral margin. This syncarpal also has a large posterodistal process (Fig. 18k) that 'hugs' the posterior surface of the distal syncarpal while the distal aspect has a large non-articular area anterior to the facets for the distal syncarpal (Fig. 20j). The latter element is also elongate anteroposteriorly (Fig. 18l–n) and has a distal aspect with a rounded outline (Fig. 18m), and rounded, subequally sized facets for the wing metacarpal. In addition, the buttress bearing the facet for the medial carpal is relatively elongate and narrow and deflected distally (Fig. 18n).

(37) Reduction of proximal ends of metacarpals I–III. (See Bennett 1994.) In the outgroup taxa, "rhamphorhynchoids" and many pterodactyloids, metacarpals I–III each articulate with (in pterosaurs) the distal syncarpal (Fig. 18a, b). In those ornithocheiroids in which this region is preserved only a single metacarpal still articulates with the distal syncarpal (Fig. 18o) or, as in the case of *Pteranodon* and *Nyctosaurus*, this contact is also lost (Bennett 2001). In *Pteranodon*, metacarpals I–III are splinted to the distal end of the wing-metacarpal (Fig. 18p), while in *Nyctosaurus* these elements are apparently lost altogether (Bennett 2000). A similar, but presumably homoplastic pattern is also evident in azhdarchoids. In *Tapejara* only a single claw digit metacarpal contacts the distal syncarpal, while in azhdarchids even this contact is lost (Bennett 2001, p. 90).

(38) Femur with stout neck and steeply directed caput. (See Unwin 1995a; Unwin & Lü 1997.) The femur of derived "rhamphorhynchoids" and non-ornithocheiroid pterodactyloids has a constricted neck and a caput that is directed inwards at approximately 135° to the long axis of the shaft (Fig. 19b–e), interpreted here as the plesiomorphic state for pterodactyloids. Ornithocheiroids are distinguished by a derived condition in which the neck is relatively robust and the caput is directed upward at 160° to the shaft (Fig. 19f, g). The situation in basal pterosaurs is uncertain, but campylognathoidids also have femora with a steeply directed caput and stout collum (Wellnhofer 1978, fig. 16; Fig. 19a). Further fossil material and character analysis is needed to determine if, as it seems, this is a case of homoplasy.

Remarks. Ornithocheiroids have a remarkably diverse cranial anatomy (Figs 10 & 11), but the unusual morphology of the shoulder girdle and forelimbs and strong reduction in the size of the hindlimbs clearly distinguishes members of this group from all other pterodactyloids. Further putative synapomorphies, in addition to those listed above, include: sagittal cranial crests that have a smooth, rounded free margin (Bennett 1994); a fused atlas-axis (Howse 1986), although this also occurs in *Dsungaripterus* and *Azhdarcho* (Bennett 1989, 1994); radius less than half the diameter of the ulna (Bennett 1994; Kellner & Tomida 2000); the distinctive morphology of the proximal end of wing phalanx 1 (Wellnhofer 1991c, fig. 34; Bennett 2001, fig. 90); and strong medial rotation of the distal end of the femur (Bennett 1989; Unwin & Lü 1997).

Bennett (1994) proposed that the presence of a median ridge on the palate and corresponding median groove on the mandible characterized his *Pteranodontidae* (similar in content to Ornithocheiroidea). These structures are present in ornithocheirids (Wellnhofer 1985, 1991c), but not in the Pteranodontidae (Bennett 2001; Unwin 2001), in the sense that it is defined here, nor in *Nyctosaurus* (Bennett 2001), and the condition in *Istiodactylus* is unknown. Similar structures are also present in *Dsungaripterus* (Bennett 2001), Lonchodectidae (Unwin 2001), *Tupuxuara* (Bennett 2001), variably in *Azhdarcho* (Nesov 1984), and in *Gnathosaurus macrurus*, although according to Howse & Milner (1995) in the latter they are narrower and sharper than in ornithocheirids. At present the distribution of this character and its variations are not sufficiently well understood to be phylogenetically useful.

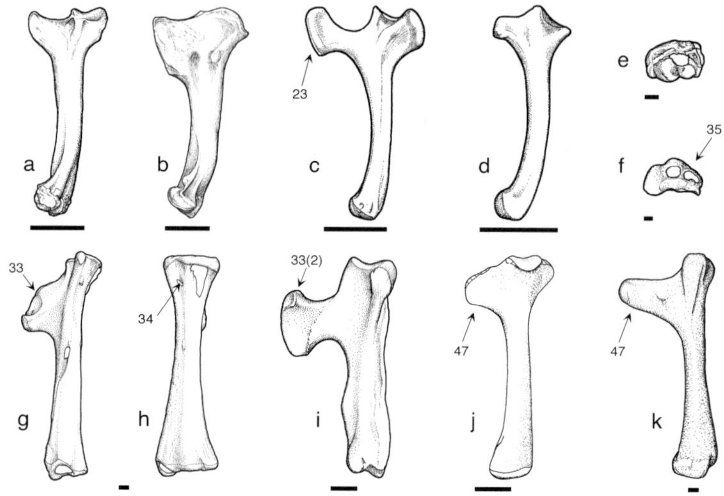

Fig. 17. Pterosaur humeri drawn to a uniform size and shown in anconal view (a), (b) and (j), palmar view (c), (d), (g), (i) and (k), posterior view (h) and distal view (e) and (f). (**a**) *Peteinosaurus zambellii* (after Wild 1978), (**b**) *Eudimorphodon ranzii* (after Wild 1978), (**c**) *Rhamphorhynchus muensteri* (after Wellnhofer 1975c), (**d**) *Pterodactylus kochi* (after Wellnhofer 1978), (**e**) ?*Tupuxuara* sp. (after Wellnhofer 1985), (**f**)–(**h**) *Coloborhynchus robustus* (after Kellner & Tomida 2000), (**i**) *Nyctosaurus gracilis* (after Williston 1902), (**j**) *Lonchodectes* sp. (CAMSM B54.081), (**k**) *Tupuxuara longicristatus* (IMCF 1052). Scale bar 10 mm.

Several authors (Padian 1984a, 1986; Bennett 1989, 1994; Unwin 1992, 1995a; Kellner 1996a; Unwin & Lü 1997), including some precladistic workers (e.g. Seeley 1891; Williston 1903; Hooley 1913; Arthaber 1922; Plieninger 1930), have recognized this clade, although its name, content and internal relationships vary from author to author. Here, for the first time, it is formally defined and diagnosed in a phylogenetic sense, using the name Ornithocheiroidea. This is the earliest available suprafamilial term for the clade and its conception here corresponds reasonably well to previous usage.

Clade 9. Euornithocheira (new clade name)
Etymology. Latin, *eu* = true, *ornithocheira*, from the root *Ornithocheirus*, one of the key members of this clade.
Definition. Ornithocheirus mesembrinus, *Pteranodon longiceps*, their most recent common ancestor, and all its descendants.
Content. Ornithocheiridae, Pteranodontidae and *Nyctosaurus*.
Synapomorphies

(39) Concave posterior margin of nasoantorbital fenestra. (See Unwin 1995a; Unwin & Lü 1997). In those outgroup taxa that have an antorbital fenestra, "rhamphorhynchoids" (Figs 8 & 9), all non-ornithocheiroid pterodactyloids (Figs 12 & 13) and *Istiodactylus* (Fig. 10a) the antorbital (or in the case of pterodactyloids the nasoantorbital) fenestra has a straight posterior margin that meets the ventral margin at (or near) a right angle. In ornithocheirids (Figs 10d, e), *Nyctosaurus* (Fig. 11a) and pteranodontids (Fig. 11b) this margin has a distinctive, rounded, concave outline, resulting largely from a partial infilling of the angle between the lacrimal and maxilla processes of the jugal. Unwin & Lü (1997) incorrectly cited the derived condition of this character as a synapomorphy of Ornithocheiroidea, but it is not present in *Istiodactylus* (Hooley 1913).

(40) Basal region of orbit infilled. (See Unwin 1995a; Unwin & Lü 1997). In most outgroup taxa, all "rhamphorhynchoids" and most non-ornithocheiroid pterodactyloids the orbit extends virtually to the ventral margin of the skull and the portion of the jugal that borders the base of the orbit is relatively slender (Figs 8, 9, 12 & 13). The orbit is at least partially infilled in *Istiodactylus*, but retains a narrow slit separating the postorbital and lacrimal processes of the jugal (Fig. 10a). By contrast, in ornithocheirids (Wellnhofer 1985, 1987; Fig. 10d, e), *Nyctosaurus* (Fig. 11a) and *Pteranodon* (Eaton 1910; Bennett 2001; Fig. 11b) the angle between the postorbital and lacrimal processes of the jugal is filled with a thin sheet of bone, restricting the orbit to a posterodorsal position within the cheek region.

Infilling of the base of the orbit also occurs in some dsungaripteroids (Fig. 13b), but the process, which involves the development of a bony bar extending from the jugal to the lacrimal (Young 1973; Ivakhnenko & Korabelnikov 1987, fig. 264), seems to be different from that observed in ornithocheiroids and is probably not homologous. The results of this analysis, indicating a sister-group relationship between dsungaripterids and *Germanodactylus*, support this interpretation because there is no evidence of orbital infilling in the latter taxon (Fig. 13a). Consequently, the state in dsungaripterids is coded differently from that in ornithocheiroids (Table 3).

(41) Coracoid facets on sternum lateral to each other. (See Bennett 1994). In pterosaurs the sternal cristospine bears coracoid facets that exhibit two morphologies. In rhamphorhynchids and some pterodactyloids (Bennett 2001) one facet is located anterior to the other (Fig. 16f), considered here to be the plesiomorphic condition, at least within Pterodactyloidea. By contrast, in ornithocheiroids other than *Istiodactylus*, and in *Tapejara*, the facets are located side by side (Figs 16g, h) and treated here as the derived state. The reverse coding is only marginally less parsimonious, however, and this character needs further exploration.

Remarks. In addition to the apomorphies cited above, this clade is also supported by the derived state for character 18 (see above) in that *Istiodactylus* has a relatively short mandibular symphysis (Hooley 1913), whereas in all euornithocheirans it is more than 30% of the total length of the lower jaw. Other putative synapomorphies of the Euornithocheira include extension of the maxilla process of the jugal to the anterior end of the nasoantorbital fenestra, near exclusion of the squamosal from the post-temporal fenestra, and a spiral mandibular articulation.

An alternative, less parsimonious arrangement, in which *Istiodactylus* is paired with the Ornithocheiridae, is supported by a single character: radius less than half the diameter of the ulna (Bennett 1994). In the topology presented here (Fig. 7a) this character is most parsimoniously optimized as apomorphic for Ornithocheiroidea, and subsequently reversed in Pteranodontidae + *Nyctosaurus*.

Clade 10. Pteranodontia Marsh 1876 (converted clade name)
Definition. Nyctosaurus gracilis, Pteranodon longiceps, their most recent common ancestor, and all its descendants.
Content. Pteranodontidae and *Nyctosaurus*.
Synapomorphies

(42) Tall, narrow frontal crest. (See Bennett 1994). Outgroup taxa and "rhamphorhynchoids" generally lack cranial crests (Figs 8 & 9), although such structures are present in at least one basal pterosaur (Dalla Vecchia 2001; Dalla Vecchia *et al.* 2002). By contrast, many pterodactyloids have cranial crests of some type, although their morphology and location is varied (Figs 10–13). Non-ornithocheiroid pterodactyloids have large cranial crests that arise from a considerable portion of the dorsal margin of the skull (Figs 12c & 13), usually involving the premaxillae and possibly also the frontals, and that are continued distally in cartilaginous tissues, though evidence of this is only rarely preserved (Campos & Kellner 1997; Martill & Frey 1998). Within Ornithocheiroidea, ornithocheirids have relatively small cranial crests that are usually located on the rostrum and the mandibular symphysis (Fig. 10b, d, e), although a small posterodorsally projecting frontoparietal crest occasionally occurs on the posterodorsal apex of the skull (Fig. 10d). The clade *Nyctosaurus* + Pteranodontidae is characterized by the presence of large, usually highly elongate frontal or frontoparietal crests that arise from the dorsal margin of the skull (Bennett 2000, 2001; Fig. 11b). Some individuals have relatively small crests, or in some cases are crestless (Eaton 1910; Miller 1972; Bennett 2001), but this variation is most convincingly interpreted as a sexual dimorphism (Bennett 1992).

(43) Dentition absent. (See Bennett 1994; Unwin 1995a; Unwin & Lü 1997). Teeth are present in all outgroup taxa, all known "rhamphorhynchoids" (Figs 8 & 9), and most pterodactyloids (Figs 10, 12 & 13). Within Ornithocheiroidea, *Nyctosaurus* and Pteranodontidae are united by the complete absence of teeth in these taxa (Fig. 11). Edentulousness is also an apomorphy of Azhdarchoidea (see below), but as this taxon does not share a close relationship with *Nyctosaurus* or pteranodontids either in this or in other cladistic analyses (e.g. Howse 1986; Bennett 1989, 1994; Unwin 1992, 1995a; Unwin & Lü 1996; Kellner 1996a; Figs 3–5), toothlessness must have evolved at least twice in pterosaurs (Bennett 1994).

(44) Mandibular rami elevated well above level of lower jaw symphysis. Viewed in lateral aspect the lower jaw of outgroup taxa, "rhamphorhynchoids" (Figs 8 & 9) non-ornithocheiroids (Figs 12 & 13), *Istiodactylus* (Fig. 10a) and ornithocheirids (Fig. 10d) is generally straight with the posterior portion usually at the same level as the anterior symphysial or tooth-bearing section. *Nyctosaurus* and pteranodontids share an unusual condition in which the postsymphysial rami curve upwards, well above the level of the symphysis (Fig. 11). This is related to a second feature of the lower jaw that also distinguishes this clade from other ornithocheiroids: pronounced deepening of the symphysis posteriorly, such that, at the caudal termination of the symphysis, the jaw is much deeper than elsewhere. The lower jaws of other pterosaurs are of even depth or achieve the greatest depth in the region of the jaw articulation (Figs 8–10, 12 & 13), the only exception being *Zhejiangopterus* (Fig. 13e), which exhibits a condition similar to that in *Nyctosaurus* and *Pteranodon*, although this must be homoplastic since this azhdarchid is not thought to share a close relationship with these pterosaurs (Figs 3–7).

(45) Pneumatic opening in palmar surface of proximal part of humerus. Outgroup taxa and "rhamphorhynchoids" appear to lack any pneumatization of the humerus (Fig. 17a–c) or other appendicular elements, but this does occur in some pterodactyloids (see character 34). *Nyctosaurus* and pteranodontids (Bennett 1989, 2001, p. 76) are united by the presence of a pneumatic opening that pierces the palmar surface of the humerus just proximal to the base of the deltopectoral crest. This pneumatopore appears to be absent in ornithocheirids, *Istiodactylus* and most other pterodactyloids, but is present in lonchodectids and the Azhdarchoidea (Nesov & Yarkov 1989, pl. 2, fig. 8).

(46) Hyper-elongation of wing metacarpal. (See Bennett 1994; Unwin & Lü 1997). The fourth (wing) metacarpal of outgroup taxa and "rhamphorhynchoids" is always shorter than the humerus (Table 2) and in most pterodactyloids it is the same or up to twice the length of the humerus. *Nyctosaurus* and pteranodontids, by contrast to other ornithocheiroids, have hyper-elongate wing metacarpals (Bennett 2001) that are more than twice the length of the humerus. This character state is rare in pterosaurs and otherwise only occurs in a single specimen of *Pterodactylus longicollum* (Wellnhofer 1970, exemplar 55) and in azhdarchoids (Table 2; Bennett 1994).

Remarks. Traditionally, *Nyctosaurus* has been allied with *Pteranodon* and the latters close relative *Ornithostoma* (e.g. Wellnhofer 1978) and this relationship is well supported here. Among characters previously discussed, this relationship is also supported by the complete loss of contact between metacarpals I–III and the distal syncarpal (character 37; see also Bennett 1994, 2001, p. 90), and putatively by the disappearance of the pneumatic opening that pierces the anconal surface of the proximal part of the humerus (reversal of character 34; see also Bennett 1994). Additional derived characters that potentially unite *Nyctosaurus* and Pteranodontidae include a greater degree of infilling of the base of the orbit than in other ornithocheiroids; the presence of marginal ridges on the jaws (Bennett 1994, 2001); beaks long and slender with premaxillae and dentaries tapering to points (Bennett 1994, 2001); mandibular symphysis approximately two-thirds the length of the mandible (Bennett 1994, 2001); and the absence of a palatal ridge and mandibular groove (though see comments above).

Several cladistic studies indicate alternative relationships for *Nyctosaurus*, either as a sister-taxon to all other ornithocheiroids (Unwin 1992, 1995a; Unwin & Lü 1997), or to a larger clade consisting of dsungaripterids, azhdarchids and ornithocheiroids (Bennett 1989, 1994; Kellner 1996a). Characters of the humerus and notarium, structures in which nyctosaurids differ somewhat from other ornithocheiroids, are relatively important in these analyses, but their impact is diluted by the inclusion of greater numbers of characters in this study. A detailed description of this pterosaur, currently being prepared by Chris Bennett (pers. comm.) may throw some more light on this problem.

Clade 11. Lophocratia (new clade name)
Etymology. Greek, *lophos* = crest, *kratos*, = head. The name refers to the prominent cranial crest borne by many members of this clade.
Definition. Pterodaustro guinazui, *Quetzalcoatlus northropi*, their most recent common ancestor, and all its descendants.
Content. Ctenochasmatoidea + Dsungaripteroidea + Azhdarchoidea
Synapomorphies

(47) Humerus with elongate, rectangular deltopectoral crest. The deltopectoral crest of outgroup taxa and "rhamphorhynchoids" is generally flange-like, and does not project far from the shaft of the humerus (Fig. 17). Dimorphodontids (Fig. 17a) and rhamphorhynchids (Fig. 17c) do not conform to this basic pattern, but in the former the deltopectoral crest is triangular in outline, while in the latter it is distinctly necked and has a rounded, tongue-like distal termination. Lophocratians have a distinctive deltopectoral crest morphology, consisting of an elongate process (measured from the base to the free margin) that has a rectangular outline with a 'squared-off' free margin (Fig. 17d, j, k). The same morphology occurs consistently throughout non-ornithocheiroid pterodactyloids but there is some variation in the length of the crest and its proximity to the proximal end of the humerus.

(48) Extensive sagittal cranial crest. (See Unwin & Lü 1997). Outgroup taxa, "rhamphorhynchoids" and ornithocheiroids either lack cranial crests (Figs 8, 9 & 10a) or, if present, they are relatively small, with smooth, rounded free margins and restricted to the rostrum, symphysial part of the mandible or the apex of the skull (Figs 10b, d, e & 11b). Ctenochasmatoids, dsungaripteroids and azhdarchoids are united by the presence, in most taxa, of an extensive sagittal cranial crest (Figs 12c & 13a–d) that extends from anterior to the nasoantorbital fenestra as far as the apex of the skull, or beyond, and is continued dorsally by stiffened integumentary structures (Campos & Kellner 1997; Martill & Frey 1998).

Despite its widespread distribution within Lophocratia, this character is problematic because it appears to be absent in some taxa, including: *Pterodactylus* (Wellnhofer 1970); some species of *Ctenochasma* (Wellnhofer 1970) and the Lonchodectidae (Unwin 2001); *Pterodaustro* (Chiappe *et al.* 2000); and *Zhejiangopterus* (Cai & Wei 1994; Unwin & Lü 1997). In other cases, such as *Cycnorhamphus* (Fabre 1976) and some crested

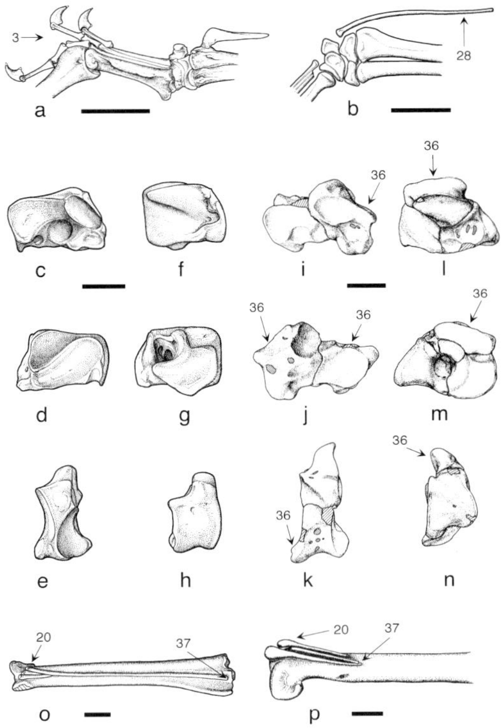

Fig. 18. The carpus and metacarpus of various pterosaurs drawn to a uniform size. (**a**) Left carpus and metacarpus of *Eudimorphodon ranzii* in posterodorsal view (after Wild 1978), (**b**) left carpus of *Pterodactylus kochi* in dorsal view (after Wellnhofer 1968), (**c**)–(**e**) proximal syncarpal and (**f**)–(**h**) distal syncarpal of ?*Tupuxuara* sp. (after Wellnhofer 1985), (**i**)–(**k**) proximal syncarpal and (**l**)–(**n**) distal syncarpal of *Coloborhynchus robustus* (after Kellner & Tomida 2000). Top row of carpals, proximal view; middle row, distal view; bottom row, dorsal view. (**o**) right metacarpus of '*Anhanguera pricei*' in anterior view (after Wellnhofer 1991c) and (**p**) left metacarpus of *Pteranodon longiceps* in dorsal view (after Bennett 2001). Scale bar 20 mm.

species of the Lonchodectidae (Unwin 2001), the crest has a different morphology or location from that of typical lophocratians (Fig. 12). The absences may be related to sexual dimorphism (Bennett 1992) or to the immaturity of specimens upon which these taxa are based (Bennett 2002), while the unusual crests of some species could be interpreted as further elaboration of the derived condition. Consequently, the plesiomorphic coding assigned to several lophocratians (Table 3) may be inappropriate.

Remarks. Lophocratia is also supported by a series of character reversals including: manual and pedal phalanges of similar dimensions and robustness (reversal of character 3), although azhdarchoids show the derived state for this character; humerus shorter than the femur (reversal of character 5, Table 2), rather than longer as in ornithocheiroids and most "rhamphorhynchoids"; and ulna shorter than the tibia (reversal of character 7, Table 2).

The relationships of the three main clades within Lophocratia (Ctenochasmatoidea, Dsungaripteroidea, Azhdarchoidea) are difficult to resolve. The pairing of Dsungaripteroidea + Ctenochasmatoidea (Fig. 6a) is supported by only a single character state change (reversal of character 3), while in some trees dsungaripteroids and azhdarchoids (Fig. 6b) are united by the derived condition for character 48 (treated as plesiomorphic for most ctenochasmatoids). It could be argued that characters shared by the Dsungaripteridae and the Azhdarchoidea (Kellner 1996a), or the Azhdarchidae (Bennett 1994; characters 11, 13 and 21–24) further support this relationship, but most are widely distributed within lophocratians, while the two remaining apomorphic states (deep supracoracoid flange and a deltopectoral crest that is long and curves ventrally) are absent in basal dsungaripteroids (*Germanodactylus*) and azhdarchoids (*Tapejara*). A basal trichotomy (Fig. 7a) would seem to be the most pragmatic assessment of relationships within Lophocratia at present.

Clade 12. Ctenochasmatoidea Unwin 1995a (converted clade name)

Definition. Cycnorhamphus suevicus, *Pterodaustro guinazui*, their most recent common ancestor, and all its descendants.

Content. Cycnorhamphus, *Pterodactylus*, Lonchodectidae and Ctenochasmatidae.

Synapomorphies

(49) Quadrate oriented in subhorizontal position. (See Unwin 1995a; Unwin & Lü 1997; Kellner cited in Chiappe *et al.* 2000.) The quadrate of outgroup taxa, "rhamphorhynchoids" and most non-ctenochasmatoid pterodactyloids is steeply or vertically oriented with regard to the ventral margin of the skull (see character 6). The quadrate of ctenochasmatoids lies in a subhorizontal position, at an angle of between 160° and 170° to the ventral margin of the skull (Fig. 12). This reorientation of the quadrate reflects a major expansion of the neurocranium in a posteroventral direction, which shifted the opisthotics and their contact with the dorsal end of the quadrate ventrally to a position behind and below the orbit.

Expansion of the neurocranium is a general feature of pterodactyloids, resulting in a relatively larger cranial capacity than in any "rhamphorhynchoid", but appears to have developed further in ctenochasmatoids than in other clades. The quadrate also occupies a near subhorizontal position in some azhdarchids (e.g. *Zhejiangopterus*, Fig. 13e), but this would appear to be a homoplasy. First, because the construction of the cranial region of the azhdarchid skull is different from that in ctenochasmatoids, with a short robust quadrate and with a frontal

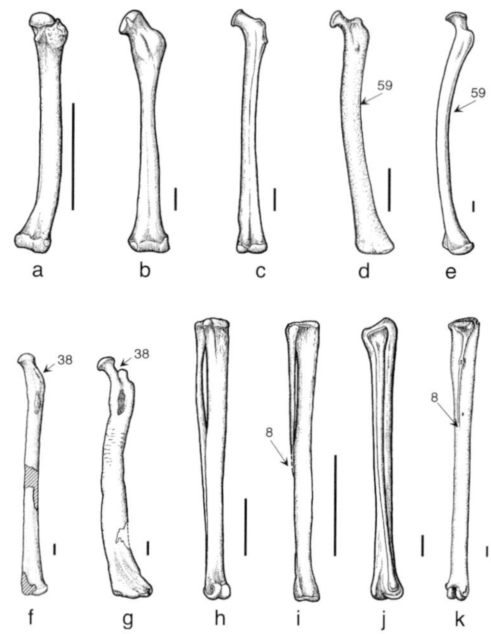

Fig. 19. Pterosaur femora and tibiotarsi drawn to a uniform size and shown in anterior view (**a**)–(**d**) and (**f**)–(**i**) and lateral view (**e**), (**j**) and (**k**). (**a**) *Eudimorphodon ranzii* (after Wild 1978), (**b**) *Dorygnathus banthensis* (after Wellnhofer 1978), (**c**) '*Pterodactylus*' *longicollum* (after Wellnhofer 1970), (**d**) *Germanodactylus cristatus* (NMING F15005), (**e**) *Dsungaripterus weii* (after Wellnhofer 1978), (**f**) *Coloborhynchus robustus* (after Kellner & Tomida 2000), (**g**) *Pteranodon longiceps* (after Bennett 2001), (**h**) *Peteinosaurus zambellii* (after Wild 1978), (**i**) *Eudimorphodon ranzii* (after Wild 1978), (**j**) *Campylognathoides zitteli* (after Plieninger 1895), (**k**) *Pteranodon longiceps* (after Bennett 2001). Scale bar 10 mm.

that extends posteroventrally to a point level with the ventral margin of the orbit (Unwin & Lü 1997). Secondly, because sister-taxa to azhdarchids (*Tapejara* and *Tupuxuara*) have more steeply oriented quadrates (Fig. 13c, d), indicating that the condition in azhdarchids is secondarily derived.

(50) Squamosal located level with or below the base of the lacrimal process of the jugal. (See Unwin 1995a; Unwin & Lü 1997; Chiappe et al. 2000). In non-ctenochasmatoid pterosaurs the squamosal occupies a position above or opposite the lacrimal process of the jugal (Figs 8–11 & 13). In ctenochasmatoids expansion of the brain case (see character 49) resulted in a marked shift in the position of the squamosal so that, in this clade, it is located at a point below a horizontal line extending posteriorly from the base of the lacrimal process of the jugal. This condition also occurs in some azhdarchids (e.g. *Zhejiangopterus*, Fig. 13e), but is interpreted as homoplastic for the same reasons given for character 49).

(51) Occiput faces ventrally. (See Unwin 1995a; Unwin & Lü 1997.) In outgroup taxa and "rhamphorhynchoids" the occiput faces posteriorly (Figs 8 & 9) and the cervical vertebrae continue in the line of the long axis of the skull. In pterodactyloids the occiput is reoriented to face posteroventrally, as a result of enlargement of the neurocranium (see above), and the neck forms an obtuse angle with the long axis of the skull (Figs 10 & 13). Ctenochasmatoids are distinguished by an occiput that faces ventrally (Fig. 12a–d) such that the neck is perpendicular (or near-perpendicular) to the long axis of the skull, reflecting the greater degree of enlargement of the neurocranium in this clade (see above). The occiput also faces ventrally in some azhdarchids (e.g. *Zhejiangopterus*, Fig. 13e), but this is interpreted as a homoplasy for the same reasons given as for character 49.

Remarks. This diverse clade comprises various specialized filter-feeding pterosaurs, such as *Ctenochasma* and *Pterodaustro*, together with a number of relatively unspecialized basal forms. The clade is mainly distinguished by characters of the skull that seem to be related to major changes in the shape and size of the brain case. Another potential synapomorphy, present in all ctenochasmatoids except *Pterodactylus kochi* and *P. antiquus*, is a spatulate condition of the jaws, wherein the anterior tip of the rostrum and mandible are dorsoventrally compressed and wider (transversely) than they are deep (dorsoventrally).

Several authors (Kuhn 1967; Wellnhofer 1970; Fabre 1974) have suggested a close relationship between the "Pterodactylidae" (containing *Pterodactylus* and *Cycnorhamphus*) and the Ctenochasmatidae. More recently, Kellner (1996a; see also Chiappe *et al.* 2000) proposed the name *Archaeopterodactyloidea* for a clade that is practically the same in content and thus synonymous with the Ctenochasmatoidea (Unwin 1995a).

Clade 13. Euctenochasmatia (new clade name)
Etymology. Latin, *eu* = true, *ctenochasmatia* from the root *Ctenochasma*, a key member of this clade.
Definition. *Pterodactylus kochi*, *Pterodaustro guinazui*, their most recent common ancestor, and all its descendants.
Content. *Pterodactylus*, Lonchodectidae and Ctenochasmatidae.
Synapomorphies

(52) Neural arch of mid-series cervicals depressed and with low neural spine. (See Howse 1986; Bennett 1989, 1994, Unwin 1995a; Unwin & Lü 1997). The neural arch of mid-series cervicals of outgroup taxa (excluding some prolacertiforms),

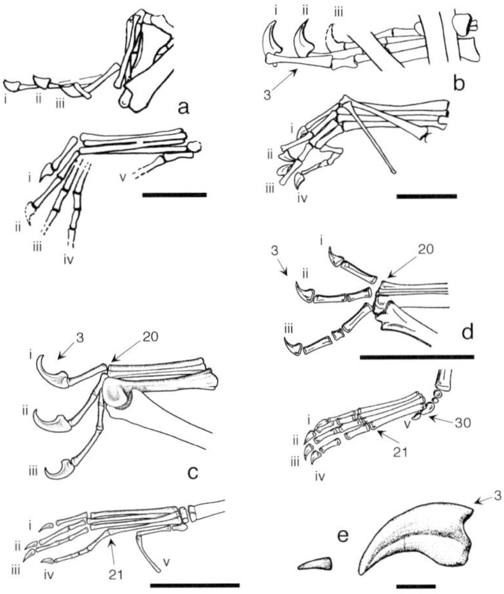

Fig. 20. The left manus and pes of various pterosaurs drawn to a uniform size. (**a**) *Preondactylus buffarinii* (after Wild 1984a, reversed), (**b**) *Peteinosaurus zambellii* (after Wild 1978), (**c**) *Rhamphorhynchus 'longicaudus'* (after Wellnhofer 1975c), (**d**) *Pterodactylus kochi* (after Wellnhofer 1970, reversed), (**e**) *Coloborhynchus robustus* (after Kellner & Tomida 2000). Scale bar 10 mm.

"rhamphorhynchoids" and most pterodactyloids including *Cycnorhamphus* (Plieninger 1907, pl. 18) is relatively tall and bears a high neural spine (Fig. 15a, c). In adult individuals of *Pterodactylus* (Fig. 15e), lonchodectids (Owen 1861) and ctenochasmatids (e.g. Broili 1924; Bonaparte 1970; Dong 1982) these vertebrae have a neural arch that is depressed down onto the centrum, and a low, rectangular neural spine (Howse 1986). Significantly, early juvenile individuals of *Pterodactylus kochi* (Wellnhofer 1970, figs 5 & 6) and *'P.' micronyx* (Wellnhofer 1970, fig. 9) exhibit the primitive state for this character, with relatively short vertebrae and high neural spines. By contrast, more mature individuals of *P. kochi* (Wellnhofer 1970 pl. 5, fig. 1; Wellnhofer 1987, fig. 1) and *'P.' micronyx* (Wellnhofer 1970, pl. 6, fig. 1) exhibit the derived condition (see also Bennett 1996a). The same pattern is evident in *Ctenochasma gracile* and putative juveniles of this species currently assigned to *'Pterodactylus elegans'* (Bennett 1996a).

A morphology similar to the derived condition described above is present in azhdarchids (Fig. 15g); consequently in some previous analyses (Howse 1986; Bennett 1989, 1994; Unwin 1992) azhdarchids and various ctenochasmatoids were grouped together on the basis of this and other characters of the cervical vertebrae. There are some important differences, however, between these two groups. Azhdarchids have a neural arch that is entirely confluent with the vertebral centrum, forming a single tubular structure (Martill *et al.* 1998, figs 5–7), whereas in derived ctenochasmatoids the neural arch, though depressed, remains distinct from the vertebral centrum (Fig. 15e). In addition, a low neural spine is retained throughout the cervical series in the latter group, whereas in the former the neural spine is absent on the fifth cervical at least (Howse 1986). The results of this analysis (Fig. 6 & 7a) also indicate that ctenochasmatoids and azhdarchids do not share a close relationship, and both have sister-groups in which the neural arch and neural spine show the plesiomorphic condition.

(53) Elongate mid-series cervicals. (See Howse 1986; Bennett 1989, 1994, Unwin 1995a; Unwin & Lü 1997.) The length of the centrum of mid-series cervicals of outgroup taxa (excluding some long-necked prolacertiforms) "rhamphorhynchoids" and most pterodactyloids, reaches, at most, 3 times the minimum width of the centrum (Fig. 15b, d). In adult individuals of *Pterodactylus*, lonchodectids (Fig. 15f) and ctenochasmatids, centrum length is at least 4 times the width and this index may reach 8 or more in some long-necked derived forms, such as *Pterodaustro* (Wellnhofer 1991a). As for character 52, juveniles of *Pterodactylus kochi*, *'P.' micronyx*, and *Ctenochasma gracile* exhibit the plesiomorphic condition with relatively short cervicals.

Similar and, in some cases, even greater degrees of relative elongation of the mid-series cervicals were achieved in azhdarchids (e.g. *Quetzalcoatlus* [Fig. 15g], *Zhejiangopterus* and *Arambourgiania*), but this would appear to be a homoplastic development, for the same reasons given for character 52.

Remarks. Euctenochasmatia is also partially supported by two character reversals (Table 3). Thus the humerus is shorter than the femur (reversal of character 5) and the ulna is shorter than the tibia (reversal of character 7) in some species of *Pterodactylus* and Ctenochasmatidae (Table 2). Kellner (1996a; see also Chiappe *et al.* 2000) suggested that *Cycnorhamphus* and Ctenochasmatidae were united by the markedly concave outline of the dorsal margin of the skull. The phylogenetic utility of this state, not uniformly present in all ctenochasmatids (e.g. absent in *Gnathosaurus*), but present in some non-ctenochasmatoids (*Pteranodon*), is unclear and needs further study.

This study further confirms the supposed paraphyly of "Pterodactylidae" (Unwin 1995a; Bennett 1996c; Unwin & Lu 1997), in that a pairing of *Cycnorhamphus* and *Pterodactylus*, traditionally the principal members of this family (e.g. Wellnhofer 1978) did not occur in any of the MPTs found here. Encouragingly, however, there is some support for the location within Ctenochasmatoidea of the Lonchodectidae, a relatively poorly known family of Cretaceous pterodactyloids, whose relationships within Pterodactyloidea have been uncertain (Unwin 2001).

Clade 14. Dsungaripteroidea Young 1964 (converted clade name)

Definition. Germanodactylus cristatus, Dsungaripterus weii, their most recent common ancestor, and all its descendants.

Content. Germanodactylus, the Dsungaripteridae and variously poorly known dsungaripteroids including *Herbstosaurus* (Casamiquela 1975; Unwin 1996), *Kepodactylus* (Harris & Carpenter 1996; Unwin & Heinrich 1999), *Normannognathus* (Buffetaut *et al.* 1998) and *Tendaguripterus (*Unwin & Heinrich 1999).

Synapomorphies

(54) Distal ends of paroccipital processes expanded. (See Unwin 1995a; Unwin & Lü 1997.) The occipital region of outgroup taxa, and all non-dsungaripteroid pterosaurs has a rounded outline in lateral view (Figs 8–12 & 13c, e). The distal ends of the paroccipital processes of dsungaripteroids are strongly expanded, resulting in a distinct protuberance in the lateral profile of the occiput (Bakhurina 1986, p. 32; Fig. 13a, b). The reconstruction of *Germanodactylus cristatus* by Wellnhofer (1970, fig. 12), reproduced in Fig. 13a, is based on the holotype specimen preserved on the main slab (BSP 1892 IV 1) and shows an occiput profile with a rounded outline. The rear part of the skull is better preserved on the counterslab (NMING F15005) and shows the strongly expanded paroccipital process and associated protusion in the lateral profile of the occiput. A similar development occurs in *Tupuxuara* (IMCF 1052; Fig. 13d), but this would appear to be a homoplasy since *Tapejara,* which is closer to dsungaripteroids than *Tupuxuara* (Fig. 7a), exhibits the non-derived condition.

(55) Dsungaripteroid teeth. (See Unwin 1995a; Unwin & Lü 1997.) Most dentate pterosaurs have simple, sharp-pointed, relatively elongate teeth that taper from the base to the tip, are lightly compressed labiolingually and gently recurved (Figs 8–10 & 12). Dsungaripteroid teeth are relatively short, with a broad, parallel-sided columnar base and a rather obtusely pointed tip (Fig. 13a, b). At least part of the dentition contains teeth of this type in *Germanodactylus* (Wellnhofer 1970, figs 12 & 13) while in dsungaripterids the entire dentition has this morphology, with teeth at the caudal end of the tooth row being particularly squat and obtuse in *Dsungaripterus* (Young 1973) and '*Phobetor*' (Bakhurina 1986, p. 32).

(56) Jaw tips toothless, but followed by a tooth row. (See Unwin 1995a; Unwin & Lü 1997.) In all

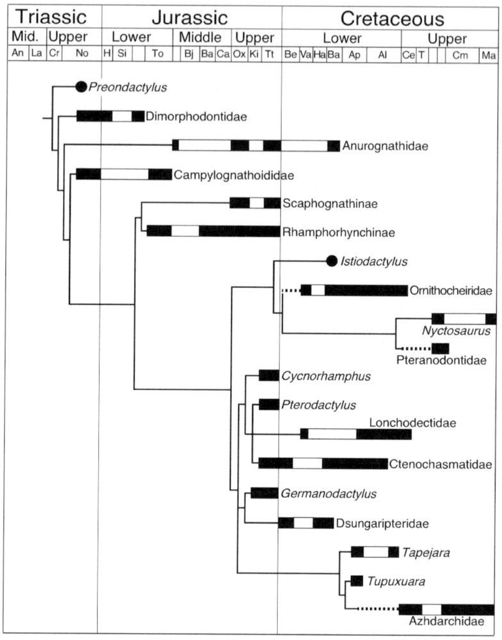

Fig. 21. Pterosaur evolutionary tree constructed from the strict concensus tree shown in Figure 7a and known stratigraphic ranges, indicated by solid shading, of the principal clades (Appendix 1). Possible range extensions based on as yet unverified records are shown by a dashed line. Al, Albian; An, Anisian; Ap, Aptian; Ba, Barremian; Bj, Bajocian; Bt, Bathonian; Be, Berriasian; Bt, Bathonian; Ca, Callovian; Cr, Carnian; Ce, Cenomanian; Cm, Campanian; H, Hettangian; Ha, Hauterivian, Ki, Kimmeridgian; La, Ladinian; Ma, Maastrichtian; No, Norian; Ox, Oxfordian; Si, Sinemurian; T, Turonian; To, Toarcian; Tt, Tithonian; Va, Valanginian.

other dentate pterosaurs the teeth extend to the tips of the jaws (Figs 8–10 & 12). In dsungaripteroids, however, the jaw tips are toothless (Young 1964; Wellnhofer 1970, fig. 12; Bakhurina 1986, p. 32; Figs 13a, b). The one exception is *Germanodactylus rhamphastinus* (Wellnhofer 1970, fig. 13) which exhibits the plesiomorphic condition, with teeth extending to the tips of the jaws. It is equally parsimonious to treat this as a reversal or as retention of the primitive condition.

(57) Largest teeth occur in the caudal half of the dentition. (See Unwin 1995a; Unwin & Lü 1997.) In non-dsungaripteroid pterosaurs teeth with the largest basal diameter occur either towards the mid-point of the tooth row, as in many basal forms such as *Preondactylus* (Fig. 8a) and *Eudimorphodon* (Fig. 9a), some basal pterodactyloids including *Pterodactylus kochi* (Wellnhofer 1968), and some ornithocheirids (Fig. 10d, e), or towards the front of the tooth row, as for example in rhamphorhynchids

(Fig. 9c–e), *Cycnorhamphus* (Fig. 12a) ctenochasmatids (Fig. 12c, d), and some ornithocheirids (Fig. 10b–e). In dsungaripteroids the teeth with the largest basal diameter occur in the caudal half of the tooth row. This pattern is less pronounced in *Germanodactylus* (Fig. 13a), but is clear in dsungaripterids, where the largest teeth occur at the caudal end of the tooth row (Young 1964, 1973; Bakhurina 1986; Ivakhnenko & Korabelnikov 1987; Fig. 13b).

(58) Limb bones with relatively thick bone walls. A distinctive feature of pterosaur limb bones that distinguishes them from outgroup taxa is the remarkably thin cortex, which rarely exceeds 2 mm in thickness. The limb bones of dsungaripteroids are similar to those of outgroup taxa in that they have relatively thick walls and only a narrow central lumen. This condition has been observed in *Germanodactylus* (Unwin 1988b), '*Phobetor*' (Bakhurina pers. comm. 2001), *Dsungaripterus* (Bennett 1989) and in other material assigned to dsungaripteroids, e.g. *Tendaguripterus* from Tendaguru, Tanzania (Galton 1980; Unwin & Heinrich 1999). Dsungaripteroids are deeply nested within Pterodactyloidea (Figs 6 & 7a); therefore, there can be no doubt, that the evolution of relatively thick bone walls represents a secondarily derived state in dsungaripteroids (Table 3), and not retention of the plesiomorphic condition.

(59) Strongly bowed femur. The femur of outgroup taxa tends to be either relatively straight or gently sigmoid. In non-dsungaripteroids it is almost always straight in anterior view (Fig. 19a–c, f, g), but in some non-ornithocheiroid pterodactyloids it may exhibit a gentle forward curvature in lateral view. Dsungaripteroid femora are unusual in that they are markedly curved in two planes: in anterior view there is a distinctive inward bowing of the shaft (Fig. 19d), while in lateral view there is a pronounced forward bowing, most clearly developed in *Dsungaripterus* (Fig. 19e).

Remarks. Young (1964) first proposed that the Late Jurassic pterosaur *Germanodactylus* was related to *Dsungaripterus* from the Early Cretaceous of China. This idea was supported by Wellnhofer (1968, 1970, 1978), and was considerably strengthened by the discovery of a dsungaripterid-like form in the Early Cretaceous of western Mongolia, '*Phobetor*', that has an intermediate morphology between *Germanodactylus* and *Dsungaripterus* (Bakhurina 1982, 1986, 1993). Apart from preliminary studies by Unwin (1992, 1995a) other cladistic analyses have so far failed to find evidence to support this hypothesised relationship, although several putative apomorphies have already been pointed out by Young (1964) and Bakhurina (1993).

One additional putative synapomorphy of the Dsungaripteroidea, found in *Germanodactylus* (BSP 1892 IV 1) '*Phobetor*' (Ivakhnenko & Korabelnikov 1987, fig. 264) and *Dsungaripterus* (Young 1964, 1973), is a strong anterior bowing of the wing phalanx 1. The plesiomorphic state, a straight or slightly posteriorly bowed wing phalanx 1, is typical for pterosaurs (e.g. Wellnhofer 1978, fig. 13), although some degree of forward bowing is also evident in phalanges probably referrable to the rhamphorhynchid *Rhamphocephalus* (Owen 1874).

Clade 15. Azhdarchoidea Unwin 1992 (converted clade name)
Definition. Tapejara wellnhoferi, Quetzalcoatlus northropi, their most recent common ancestor, and all its descendants.
Content. Tapejara, Tupuxuara and Azhdarchidae.
Synapomorphies

(60) Orbit located well below level of dorsal margin of nasoantorbital fenestra. (See Kellner 1995; Unwin 1995a; Kellner & Langston 1996; Unwin & Lü 1997.) Viewed in lateral aspect, the dorsal margin of the orbit in the skull of "rhamphorhynchoids" (Figs 8 & 9) and non-azhdarchoid pterodactyloids (Figs 10–12 & 13a, b) approaches close to the dorsal boundary of the skull. By contrast, in azhdarchoids the orbit occupies a relatively low position (Fig. 13c–e) and in azhdarchids, for example, is nearer to the ventral rather than the dorsal margin of the skull. This apparent shift in position can be attributed to a dorsal elevation of the posterior rostral region and the infilling, by ventrolateral extension of the frontal, of the region corresponding to the upper half of the orbit in other pterosaurs.

Remarks. Early cladistic studies (Bennett 1989, 1994; Unwin 1992; Figs 3b & 4b) tended to link azhdarchids with various ctenochasmatoids because of the development of long, low cervical vertebrae in both these groups. Kellner and Hasegawa (1993), Kellner (1995, 1996a) and Unwin (1995a) pointed out shared derived features of the skull that are present in both tapejarids and azhdarchids and which suggest that elongate vertebrae had evolved independently in the Ctenochasmatoidea and the Azhdarchoidea. Even though tapejarids were not described until later, Howse (1986) foresaw this phylogenetic arrangement in one of his trees (Fig. 3a), grouping various species of *Pterodactylus* with ctenochasmatids (Ctenochasmatoidea) in one group, and azhdarchids plus some taxa now thought to be ctenochasmatoids in a second group. This analysis (Fig. 7a) strongly supports the idea that the acquisition of long necks by ctenochasmatoids was entirely independent of the evolution of long necks in azhdarchoids, in part because basal members of both clades have relatively short cervicals.

Azhdarchoidea is well supported by a variety of cranial and postcranial characters. Derived states, although also apomorphic elsewhere within Pterosauria, include: absence of teeth (character 43);

dorsal expansion of the caudal half of the rostrum (see character 9); development of a pneumatic opening in the palmar surface of the proximal part of the humerus (character 45); loss of contact between the distal syncarpal and metacarpals II and III (character 37); and the redevelopment of manual phalanges that are relatively large compared to the pedal phalanges. Additional putative apomorphies include the extension of the frontal anterior to the lacrimal-jugal bar (Fig. 13c–e), and a femur length more than 1.25 times that of the humerus (Table 2).

Clade 16. Neoazhdarchia (new clade name)
Etymology. Latin, *neo* = new, *azhdarchia* from the root *Azhdarcho*, a principal taxon within this clade.
Definition. Tupuxuara longicristatus, *Quetzalcoatlus northropi*, their most recent common ancestor, and all its descendants.
Content. Tupuxuara and Azhdarchidae.
Remarks. Synapomorphies of the Neoazhdarchia, a clade initially postulated by Unwin and Lü (1997), include: presence of a notarium (character 31), reported in azhdarchids (Buffetaut 1999; pers. obs.) and *Tupuxuara* (Kellner 1996a; Kellner & Hasegawa 1993), but absent in *Tapejara*; and loss of contact between metacarpals I–III and the distal syncarpal (character 37), in contrast to *Tapejara* where at least one metacarpal retains this contact. In addition, the rostrum of *Tapejara* (measured from the anterior margin of the orbit to the anterior end of the premaxillae), forms only 70% of total skull length (anterior tips of premaxillae to occipital condyle), which falls within the range occupied by "rhamphorhynchoids" (58–76%) and is certainly plesiomorphic within Pterodactyloidea. By contrast, neoazhdarchians share a derived condition in which the rostrum forms more than 88% of total skull length.

Kellner (1995, see also Kellner 1989, 1996a; Kellner & Langston 1996) united *Tapejara* and *Tupuxuara* in the Tapejaridae on the basis of two characters: a sagittal premaxillary crest extending from the tip of the snout to the occipital region and a comparatively large nasoantorbital fenestra. The cranial crest of *Tapejara* is rather differently constructed from that of *Tupuxuara* (Fig. 13c, d) and does not extend to the tip of the snout. Moreover this general type of crest is widely distributed within lophocratians (character 48). A comparatively large nasoantorbital fenestra is also present in azhdarchids (Fig. 13e); consequently this character is apomorphic for the Azhdarchoidea (see character 9). Tapejaridae is thus paraphyletic as Unwin & Lü (1997) supposed.

Discussion

Tree robusticity and alternative topologies

Phylogenetic analyses published so far contain little or no discussion of the robustness or likely reliability of the trees presented. The only exception is the brief commentary by Kellner (1995) on previously published trees, though this largely focused on Bennett's work (1989, 1994), and was confined to rerunning data sets and discussion of character states and their distribution. The absolute and relative robusticity of trees recovered in this analysis was assessed using a variety of techniques, some statistical, some purely comparative. Results of these analyses provided insights into the strength of support for principal nodes, the robusticity of the preferred tree (Fig. 7a) compared to the results of other studies and the possible impact of future fossil discoveries.

Quality of the original character data set. The cladistic data set analysed here (Table 2) is 94% complete. This is a far higher value than for other pterosaur data sets reported so far (Bennett 1989, 51%; Bennett 1994, 41%) and generally higher than that for other studies of fossil vertebrates (Wilkinson 1995). Most taxa and most characters have high levels of completeness: only characters 25 and 41 show relatively low values (70% and 60% complete respectively), and only two taxa, *Preondactylus* and the Lonchodectidae, are relatively poorly represented (80% and 58% complete respectively). Neither of the two characters are especially important: character 25 forms part of a large cluster that diagnose Pterodactyloidea; character 41 helps resolve relationships within the well-supported clade Ornithocheiroidea. Likewise, while character distributions represented by the Lonchodectidae have some influence on relationships within the Ctenochasmatoidea they appear to have little impact on general tree topology. Incomplete knowledge of *Preondactylus*, the most basal known pterosaur (Fig. 7a), is more problematic, but in that it is generally similar to dimorphodontids (Dalla Vecchia 1998), which are much better known, this is not a critical issue.

The character data set samples all parts of the skeleton, although the emphasis is on cranial anatomy (50% of all characters), which traditionally (Young 1964; Kuhn 1967; Wellnhofer 1978) has played a critical role in establishing the relationships of pterosaurs. By sharp contrast, previous data sets focus largely (Bennett 1994, 84%) or entirely (Howse 1986; Bennett 1989) on postcranial anatomy.

The character data set also contains data from almost all known species of pterosaur, more than two-thirds of which were inspected directly. The pterosaur fossil record is often claimed to be poor, in the sense that it is highly incomplete (e.g. Carroll 1987). It could be asserted, therefore, that known taxa significantly under-represent true pterosaur diversity, to the extent that several large clades may remain unknown, thereby compromising phylogenetic analyses. Two lines of evidence suggest that this may not be a significant problem. First, in morphometric studies focused on the wings (Hazlehurst & Rayner 1992) and on the hindlimbs (Elvidge & Unwin 2001), pterosaurs were found to plot in a relatively tight cluster, with no outliers and relatively small gaps between points. This suggests that, at general taxonomic levels (e.g. family or higher), known diversity may be a reasonable proxy for true diversity. Thus, new discoveries are more likely to fit within existing clades, rather than revealing major new lineages and will not usually destabilize the general set of relationships described here.

The history of discovery provides a second independent test regarding our understanding of taxonomic diversity. If major new clades remain to be discovered we might expect to find their representatives in deposits such as the Yixian Formation that sample depositional environments and time intervals from which pterosaurs are poorly known (Unwin *et al.* 2000). Taxonomic studies of Yixian pterosaurs show, however, that the four genera found so far can all be assigned to existing clades: *Dendrorhynchoides* and *Jeholopterus* are anurognathids (Unwin *et al.* 2000; Wang *et al.* 2002), *Haopterus* appears to be an ornithocheirid (Unwin 2001) and *Eosipterus* probably belongs in the Ctenochasmatidae (Unwin *et al.* 2000; Unwin 2002). This observation also applies to other recently discovered assemblages, such as the Crato and Santana Formations (e.g. Fastnacht 2001), and revisions of older assemblages (e.g. Unwin 2001) have also failed to recover evidence of distinctly new clades.

Bootstrap and decay analyses. Results of the bootstrap analysis (Fig. 6c) indicate that nearly all the "rhamphorhynchoid" clades are well supported, with values of >90%. The only exception is the clade Rhamphorhynchidae, although this occurs in almost 75% of all replicates. The situation within Pterodactyloidea is much more variable. A few clades have values of 90% or more, but most are either only moderately well supported or, as for example, in the case of Ctenochasmatoidea, fall below the 50% mark.

A decay analysis, in which the constrained length of the preferred tree was increased in incremental fashion, beginning with the minimum number of steps (112), showed a rapid decline in resolution. In trees only one step longer, Rhamphorhynchidae collapsed, as did the relationships within Azhdarchoidea. The relationships of major pterodactyloid clades collapsed to a polytomy in trees only two steps longer, although the clades themselves, and some subclades within them, remained distinct. In trees three steps longer only Ornithocheiroidea was preserved within Pterodactyloidea, although most "rhamphorhynchoid" clades remained distinct. All resolution was lost in trees only five steps longer than the MPT.

Partitioning. The degree to which trees generated from subsets of characters and taxa resemble the preferred tree resulting from the main data set can also provide some insights into the latter's robusticity. In this study, taxa were partitioned into "rhamphorhynchoid" and pterodactyloid subsets, and characters were divided into cranial and postcranial subsets. Fortuitously, the resulting subsets were of roughly similar size. Remarkably, the strict consensus tree of 558 MPTs yielded by the cranial subset was almost identical topologically to the main tree (Fig. 7a), except that the relationships of the four major clades of pterodactyloids were unresolved. Similarly, the strict consensus tree of six MPTs yielded by the postcranial subset showed much the same topology as the main tree. The only notable difference was in the pairing of *Cycnorhamphus* with Dsungaripteroidea + Azhdarchoidea. Analysis of the "rhamphorhynchoids" resulted in a single tree that was identical to the main tree irrespective of whether or not an outgroup and/or a pterodactyloid taxon was included. The pterodactyloid subset yielded varying numbers of MPTs (though never more than 12), depending on which outgroup (the outgroups described above, a basal "rhamphorhynchoid" or a derived "rhamphorhynchoid") was selected. Strict consensus trees generated from these analyses had an identical, or near identical topology to the main tree.

Stratigraphic congruence. The degree of concordance between the stratigraphic distribution of taxa and their sequence of occurrence in a cladogram offers a means of comparing different trees (Benton 1995). This study found a good correlation between the order of branching (Figs 6 & 7) and the order of appearance of clades in the stratigraphic record (Fig. 21). Only three clades (Anurognathidae, *Istiodactylus* and Lonchodectidae) appear in a different stratigraphic order from that predicted by the results of this analysis. Cladograms by Howse (1986; Fig. 3a), Viscardi *et al.* (1999; Fig. 5b) and Unwin (1992; Fig. 4b) show a similar level of concordance, except for the position of *Nyctosaurus* in the latter work. Other studies are comparably less concordant, especially with regard to *Nyctosaurus*, *Pteranodon*, *Dsungaripterus* and *Pterodaustro* (Bennett 1989, 1994; Figs 3b & 4b) or to *Sordes*, *Dimorphodon* and *Nyctosaurus* (Kellner 1996a; Fig. 5a).

The proportion of predicted stratigraphic range

(also known as 'ghost range') compared to known stratigraphic range, a measure of relative completeness (RCI) that can be assessed in various ways (Smith 1994; Benton 1995), provides another method for comparing different trees. The cladogram presented here (Figs 7a & 21), and others described by Howse (1986; Fig. 3a), Unwin (1992, 1995a; Figs 4b, c) and Viscardi *et al.* (1999; Fig. 5b) have relatively small, similarly sized RCIs and few individual lineages with ghost ranges that approach or exceed the known stratigraphic range. Trees presented by Bennett (1989, 1994; Figs 3b & 4c) and Kellner (1996a; Fig. 5a) have larger summed RCIs and a number of individual lineages (e.g. *Nyctosaurus*, *Pteranodon*, Azhdarchidae) with ghost ranges longer than the known stratigraphic range. This is particularly marked in the case of Kellner's study, wherein a series of "rhamphorhynchoids" (Anurognathidae, *Sordes*, *Scaphognathus* and *Dorygnathus* also have remarkably long ghost ranges.

Summary. Most of the analyses reported above indicate a relatively high degree of robustness for the strict consensus tree (Fig. 7a) and show that it is better supported, or more concordant with alternative lines of evidence, than other cladograms of pterosaur relationships presented so far. This is consistent with the small number of MPTs generated in the main analysis, in contrast to the large numbers of trees generated in previous studies (see Kellner 1996a) and the general concordance of the topology of the strict consensus tree with the results of other phylogenetic studies (see below). The rapid decline in resolution in the decay analysis suggests that the results should be treated with some caution, but does not indicate that they are generally unreliable.

Importantly, a number of well-supported relationships that crop up repeatedly in precladistic and cladistic studies (see below) can be identified. These include the monophyly of Pterosauria, Pterodactyloidea and Ornithocheiroidea, and a basal position for the Dimorphodontidae and similar taxa such as *Preondactylus*. In addition, other clades, including Dsungaripteroidea and *Nyctosaurus*+ Pteranodontidae, are also well supported here and, though infrequent in other cladistic studies, were widely accepted in traditional phylogenies (e.g. Young 1964; Kuhn 1967; Wellnhofer 1978). Because these six clades act as a strong constraint on tree shape, if the majority of them are supported by new analyses, much of the basic topology of the consensus tree presented here is likely to persist into the future.

Inevitably, some clades are only weakly supported (e.g. Lophocratia, Ctenochasmatoidea, Neoazhdarchia), but in many cases the collapse of these clades, or alternative configurations, are unlikely to have a major impact on the distribution of character states or overall tree shape.

Comparison with previous studies

Traditional phylogenies. The main precladistic studies (Young 1964; Kuhn 1967; Wellnhofer 1975a, 1978) used general similarities, often based on the skull, to identify various lineages within "Rhamphorhynchoidea" and Pterodactyloidea, but were unable to determine the relationships of these lineages to one another (Figs 1 & 2). Several of these lineages, notably Rhamphorhynchinae, Dsungaripteroidea (*sensu* Wellnhofer 1978; Fig. 2) and Ctenochasmatoidea (in part) are supported by this study, but many others are not. In these cases, such as *Dimorphodon* and *Anurognathus*, the general similarities of the skull reflect the retention of features that appear to be plesiomorphic for pterosaurs. In other cases, such as the supposed relationship of *Scaphognathus* and *Istiodactylus*, this was partly based on a misunderstanding of the skull anatomy in the latter taxon, which was erroneously thought by Hooley (1913) to have a maxillo-nasal bar as in "rhamphorhynchoids". Notwithstanding the above, most of the contrasts between the strict consensus tree generated by this analysis (Fig. 7a) and the traditional phylogenies can be attributed to new fossil finds in the last two decades, and the attempt here, as in other cladistic studies, to distinguish between derived and primitive characters.

Cladistic studies. Insofar as comparisons can be made, the results of this study share more in common with other cladistic studies than with earlier precladistic schemes. Howse (1986) recognized three clades that correspond fairly well with the Ctenochasmatoidea, Ornithocheiroidea and Azhdarchoidea as defined here, the main difference being the grouping of †*Doratorhynchus* (=*Gnathosaurus*) and Greensand vertebrae (=*Lonchodectes*) with the latter clade rather than the former. In Howse's study most Jurassic pterodactyloids clustered at or near the base of Pterodactyloidea, which is to be expected considering their relatively uniform postcranial anatomy and the limited extent of his analysis. Greater resolution was achieved in this study, but as the bootstrap and other analyses show, evidence in support of these relationships is relatively fragile.

Where comparable, the main tree resulting from the first analysis by Bennett (1989) is generally similar to the consensus tree presented here (cf. Figs 3a & 7a). The content of Bennett's *Pteranodontidae* and Azhdarchidae largely overlap with that of Ornithocheiroidea and Azhdarchoidea, and Dsungaripteridae is also recognized as a distinct taxon, although with a different content from Dsungaripteroidea. The placement of †*Doratorhynchus* in Azhdarchidae is perhaps inevitable in that only one other relatively basal ctenochasmatoid

(*Pterodactylus kochi*) was included in the taxon list, although interestingly the latter falls outside all other pterodactyloid clades.

The main difference is in the placement of *Nyctosaurus* near the base of Pterodactyloidea, well away from *Pteranodon* and even well outside the clade equivalent to Ornithocheiroidea. This is because, in Bennett's study, *Nyctosaurus* lacks some features of the shoulder girdle–notarium complex present in other pterosaurs (possibly related to the relatively small size of *Nyctosaurus*) and has a highly derived type of humerus wherein ornithocheiroid features have been further transformed. As shown here, many other features not included in Bennett's character analysis support a relationship within Ornithocheiroidea, as a sister-taxon to Pteranodontidae. This result is also far more consistent with the stratigraphic occurrence of *Nyctosaurus* than that of previous studies, since a basal position within Pterodactyloidea implies a remarkably long ghost range (Unwin 2001, fig. 14).

The cladogram presented in the second analysis by Bennett (1994) is topologically similar to that of the 1989 study, differs mainly in the inclusion of more taxa and is similar in many respects to the tree presented here. One significant difference, not discussed with regard to the 1989 analysis, is the exclusion of *Ornithocheirus* and *Brasileodactylus* from the *Pteranodontidae* (=Ornithocheiroidea here). The main reason for this appears to be the absence of data for these taxa in this study (only three entries for the 37 characters) and thus a lack of evidence supporting their assignment to *Pteranodontidae* (*sensu* Bennett 1989, 1994). A second apparent difference concerns the row of Late Jurassic and Early Cretaceous pterodactyloids clustered at or near the base of Pterodactyloidea. Inspection of this list shows, however, that most of these taxa (with the exception of *Germanodactylus*) belong in what was identified in the present study as Ctenochasmatoidea. This taxon was not supported in the bootstrap analysis (Fig. 6c) and the resulting cladogram is thus topologically similar to that presented by Bennett (1994, fig. 8). The lack of resolution in Bennett's 1994 tree is probably for the same reason as that proposed to account for the same phenomenon in Howse' 1986 study (see above).

The two studies published by Unwin (1992, 1995a) used earlier versions of the data set shown in Table 3; thus, not surprisingly, the resulting trees are remarkably similar to that obtained here. The 1992 tree differs in three ways:

(1) Long-necked pterodactyloids were lumped together in a single clade principally because that study did not include *Tapejara* or *Tupuxuara*, or characters of the skull that distinguish ctenochasmatoids from azhdarchoids.

(2) *Nyctosaurus* emerged as the most basal taxon within Ornithocheiroidea, principally because it lacked some ornithocheiroid characters pertaining to the humerus, several features that link Pteranodontidae and *Nyctosaurus* here were not included in the data set (e.g. hyperelongate wing metacarpal IV, shape of the mandible) and one feature (narrow, elongate, frontal crest) was not then known in *Nyctosaurus*.

(3) Dsungaripteroids formed the most basal pterodactyloid group, rather than the Ornithocheiroidea. Only a single character (pneumatization of the limb bones) supported the clade uniting Azhdarchoidea (including Ctenochasmatoidea) and Ornithocheiroidea. This character is problematic, however, in that limb-bone pneumatization is difficult to identify in many Jurassic pterodactyloids and appears to follow a different pathway in ornithocheiroids than, for example, in azhdarchids, raising doubts as to its homology. If this character is ignored, the resulting trichotomy closely reflects the topology of the consensus tree (Fig. 7a).

In the cladogram described by Kellner (1996a; Fig. 5a), the topology crownwards from the clade uniting *Dimorphodon* with derived "rhamphorhynchoids" and pterodactyloids is almost identical to that of the consensus cladogram presented here (cf. Figs 5a & 7a). The only difference is that in Kellner's tree Nyctosauridae is placed at the base of a clade equivalent to Ornithocheiroidea, and this clade forms a sister-group to Dsungaripteridae + Azhdarchoidea rather than a sister-group to all other pterodactyloids. Both these arrangements have already been discussed above.

Kellner's cladogram (Fig. 5a) is clearly distinguished from the strict consensus cladogram presented here in that Anurognathidae forms the most basal taxon and three rhamphorhynchids (*Sordes*, *Scaphognathus*, *Dorygnathus*) lie near the base of the tree and outside Rhamphorhynchidae. According to Kellner (1996a) pterosaurs other than anurognathids are united by a posteriorly displaced external narial opening. This is certainly true of Campylognathoididae and other more derived forms (Figs 9 & 13), but an anteriorly located narial opening is also present in *Preondactylus* and dimorphodontids (Fig. 8a, b); thus, while the apparent distribution of states for this character support the results of this analysis, that anurognathids occur among basal forms (Figs 6 & 7), it does not necessarily support the placement of this group as the most basal taxon. Kellner (1996a) mentions that the placement of the other taxa, including the basal position of the three rhamphorhynchids, is dependent on derived states for the proportions of postcranial bones, especially the wing elements, but does not describe or discuss these characters.

The relationships described by Viscardi *et al.* (1999) largely match those evident in the consensus tree (Fig. 7a). According to these authors Ornithocheiroidea is the sister-group to Azhdarchoidea, rather than Ctenochasmatoidea but, as already discussed, only one or two character states provide evidence regarding the relationships of the major pterodactyloid clades. Therefore alternative topologies are only slightly less parsimonious than the scheme shown in Fig. 7a, although this particular grouping did not occur in any of the six MPTs found in this study.

Pterosaur evolutionary history

At a general level, the basic pattern evident in the 'standard' pterosaur evolutionary tree (Wellnhofer 1978; Fig. 2) is similar to that seen in the evolutionary history shown in Fig. 21. There is a basal radiation in the Late Triassic–Early Jurassic, followed by a second radiation, of pterodactyloids, in the Late Jurassic. At finer scales, however, there are many differences: the new study reveals a more complex history and helps to clarify several important events.

The first, basal, radiation of pterosaurs seems to have been well established by the Late Triassic and may have begun earlier. According to the results of this study at least five distinct clades existed prior to the Triassic–Jurassic boundary, although only three of these are known from the fossil record (Appendix 1). Notably, the majority of Triassic records appear to belong to the Campylognathoididae (Wild 1978, 1984b; Dalla Vecchia 1995, 1996; Jenkins *et al.* 1999, 2001; Wellnhofer 2001, 2003), one of two lonchognathan lineages that existed contemporaneously with various clades of basal pterosaurs.

An important event occurred in the Early Jurassic with the replacement of several basal clades (*Preondactylus*, dimorphodontids, campylognathoidids) by a more derived and highly successful clade of "rhamphorhynchoids": the Rhamphorhynchidae. Rhamphorhynchids are unreported prior to the Toarcian (Appendix 1) but, apart from a few anurognathids, the numerous records of Middle and Late Jurassic "rhamphorhynchoids" all appear to belong to this clade. A striking feature of Jurassic "rhamphorhynchoids", when compared to Late Jurassic and Cretaceous pterodactyloids, is the relatively restricted degree of morphological and taxonomic diversity and their typically small size, with wing spans of no more than 2.5–3m.

The basal radiation of pterodactyloids is first documented in the Late Jurassic (Fig. 21), although this clade probably dates back to at least the Early Jurassic. Two of the four major clades (Ctenochasmatoidea and Dsungaripteroidea) are recorded in the Late Jurassic, and were already diverse and widely distributed by this time (Appendix 1). Pterosaurs seem to have achieved their highest levels of taxonomic, morphological and ecological diversity during the Early Cretaceous, when all four major pterodactyloid clades coexisted simultaneously (Unwin *et al.* 2000), although two of these clades, Ctenochasmatoidea and Dsungaripteroidea, do not appear to have survived much beyond the end of this interval.

An important, previously unrecognized event seems to have taken place in the early Late Cretaceous, resulting, principally, in a strong reduction in pterosaur diversity, with the complete disappearance of toothed forms (represented in the Cenomanian by ornithocheirids and lonchodectids) and some edentulous taxa (tapejarids and tupuxuarids). Among the two surviving lineages, pteranodontians are only known from the middle Upper Cretaceous of North America (Bennett 1994) and a single record from the Maastrichtian of South America (Price 1953); thus, counter to popular notions (e.g. Carroll 1987), *Pteranodon* was probably not the dominant or even a typical Late Cretaceous pterosaur. Recent fossil finds and revision of older discoveries (Unwin & Lü 1997; Company *et al.* 1999) show that almost all Campanian and Maastrichtian records represent azhdarchids and this group seems to have dominated the last 20 Ma of pterosaur history.

Conclusions

The reconstruction of pterosaur phylogeny using cladistic techniques is still in its infancy but already some basic patterns, evident to some extent in precladistic studies, are beginning to emerge. Pterosauria, Pterodactyloidea and Ornithocheiroidea are well supported and other relationships, such as a basal position for *Preondactylus* and dimorphodontids, and the monophyly of Dsungaripteroidea and Azhdarchoidea, are unlikely to be substantially modified. Together, these relationships impose considerable constraints on tree shape. Future studies are likely to focus on areas of greater uncertainty, aiming to clarify the relationship of clades of derived "rhamphorhynchoids" to the Pterodactyloidea, to test the monophyly of Ctenochasmatoidea and to resolve the relationships of taxa within all four major pterodactyloid lineages.

Redescriptions of older material and detailed accounts of the many new taxa named, but often only briefly described, in recent years, will be important for such studies. Utilizing species, or even specimens, as terminal taxa would reduce the number of untested hypotheses of relationships, although a preliminary attempt by the author resulted in a character data set with a high proportion of missing data, a large number of MPTs and relatively poor resolution of relationships. One partial solution to these problems might be to partition the data set, treating "rhamphorhynchoids" and pterodactyloids separately. Exploration and refinement of characters offers another promising approach. Recent studies have identified more than 100 characters and, while some of these are now well understood, most, especially those related to morphometric data, would benefit from further analysis. Doubtless, many further characters remain to be described: the pelvis, for example, exhibits considerable variation, but character states based on this have yet to be cited.

Beyond the task of improving and refining cladograms, understanding of pterosaur phylogeny could be used as a framework for investigations of other aspects of these animals. In recent years, pterosaur anatomy, functional morphology, locomotory abilities, growth and physiology have been the subject of intense scrutiny and debate (see Wellnhofer 1991a; Unwin 1999 for reviews) but with some rare exceptions (e.g. Bennett 2003), these studies have been conducted practically without reference to pterosaur phylogeny. Indeed, it is often implied, though perhaps not intentionally, that what holds true for one pterosaur holds true for them all, even though the basic construction of, e.g. "rhamphorhynchoids" and pterodactyloids, differs in many fundamental ways (Schaller 1985). In addition to the fresh insights that it might yield, integration with phylogenetic studies would help to avoid this problem.

I am grateful to the many museums and institute staff who permitted me to inspect their collections and generously loaned material. Special thanks go to A. Milner, C. Walker and S. Chapman of The Natural History Museum, London, UK; C. Forbes, S. Etchells-Butler and D. Norman, Sedgwick Museum, Cambridge, UK; T. Kemp and P. Powell, Oxford University Museum, Oxford, UK; P. Wellnhofer, Bayerische Staatssammlung für Paläontologie und Geologie, Munich, Germany; E. Frey, Staatliches Museum für Naturkunde, Karlsruhe, Germany; R. Wild, Staatliches Museum für Naturkunde, Stuttgart, Germany; the late L. Nesov, Department of Geology, St Petersburg University, Russia; W. Langston Jr, University of Texas, Austin, Texas, USA; M. Manabe and Y. Tomida, National Science Museum, Tokyo, Japan; S. Nabana, Iwaki Museum of Coal and Fossils, Iwaki, Japan; Y. Hasegawa and Y. Takakua, Gunma Museum of Natural History, Gunma, Japan; Y. Okazaki and Y. Yabumoto, Kita-Kyushu City Museum of Natural History, Kita-Kyushu, Japan; H. Taru, Kanagawa Prefectural Museum of Natural History, Kanagawa, Japan; E. Gaffney, M. Norell and A. Kellner, American Museum of Natural History, New York, USA; K. Carpenter, Denver Museum of Natural History, Denver, USA; P. Currie, Royal Tyrrell Museum of Palaeontology, Drumheller, Canada; D. Zhiming and J. Lü, Institute for Vertebrate Paleontology and Paleoanthropology, Beijing, China; and Z. Cai, Zhejiang Museum of Natural History, Hangzhou, China. The Department of Earth Sciences, Bristol University, Institut für Paläontologie, Museum für Naturkunde, Humboldt Universität, the Royal Society, NERC and the British Council are thanked for funding aspects of this work. I thank N. Bakhurina, C. Bennett, M. Benton, S. Howse, M. Fastnacht, E. Frey, D. Martill, K. Padian, D. Peters, P. Wellnhofer and M. Wilkinson for useful discussions of pterosaur phylogeny; G. Arratia and T. Pfeiffer for help with PAUP; E. Siebert, J. Mendau and D. Lazarus for assistance with the production of the figures; and R. Schoch and, in particular, B. Creisler for advice regarding the formulation of taxonomic names. I am especially grateful to M. Benton and C. Bennett for their constructive reviews and to E. Buffetaut for his editorial advice. This paper is dedicated to P. Wellnhofer, the father of modern pterosaurology.

Appendix 1 Terminal taxa used in this analysis

Content, diagnosis and where appropriate comments are given for each taxon. The heading 'Source of phylogenetic data' lists the specimens examined. Casts are indicated by 'c'. Where additional information was collected from the literature, or only the literature was used, all citations examined are listed.

Preondactylus Wild 1984a
Content. Preondactylus Wild 1984a
Diagnosis. (Modified from Dalla Vecchia [1998] to include only characters found in *Preondactylus.*) Two or three enlarged, triangular maxillary teeth between the naris and the antorbital fenestra followed distally by 10 triangular teeth decreasing regularly in size; dentary less than half the length of the complete lower jaw.
Remarks. The relatively low, elongate snout shown in recent reconstructions (Wellnhofer 1991a; Dalla Vecchia 1998) does not match with details of the skull as illustrated by Wild (1984a, figs 1 & 2). The elongate, subvertically oriented nasal process of the maxilla, angle between dorsal and maxillary processes of the premaxilla and reinterpretation, as the nasal (Fig. 8a), of the element labelled as a frontal by Wild (1984b, fig. 3) indicate that the snout region was much higher than shown by Wellnhofer (1991a) and Dalla Vecchia (1998), with large narial and antorbital fenestrae similar to the condition in dimorphodontids. Moreover, since the premaxilla and maxilla reach 80% of the length of skull (calculated from the length of the mandible), the region from the anterior edge of the orbit to the posterior margin of the occiput must have formed less than 20% of the skull length. Since the orbit must have been accommodated in this space it is likely to have been considerably smaller than the narial or antorbital fenestrae.

Recorded temporal range. Late Triassic (Norian).
Source of phylogenetic data. Preondactylus buffarinii (Wild 1984a; Dalla Vecchia *et al*. 1989; Dalla Vecchia 1998, 2001).

Dimorphodontidae Seeley 1870
Definition. Dimorphodon macronyx, Peteinosaurus zambellii, their most recent common ancestor, and all its descendants.
Content. Dimorphodon Owen 1859; *Peteinosaurus* Wild 1978; Charmouth dimorphodontid (Unwin unpub. data).
Diagnosis. (See also Wellnhofer 1978) External narial opening forms the largest skull opening; mandible with true dimorphodontid dentition: two large teeth followed by at least 30 small, close-set, lancet-shaped teeth (Owen 1870; Wild 1978; Dalla Vecchia 1998, fig. 5); deltopectoral crest of humerus subtriangular with apex directed proximally (Wild 1978, fig. 43; Dalla Vecchia 1998, fig. 6); strongly broadened proximal end of wing phalanx 1, with short extensor tendon process (Wild 1978, p. 241); phalanx 2 of pes digit V straight and equal in length to the first phalanx. Phalanx 2 is relatively long in some rhamphorhynchids (*Sordes Scaphognathus, Rhamphorhynchus*), but in these pterosaurs it is curved or bent (Sharov 1971; Wellnhofer 1975a).
Recorded temporal range. Late Triassic (Norian) to ?Mid-Jurassic (Aalenian–Bajocian).
Source of phylogenetic data. Dimorphodon macronyx (BMNH 41212, BMNH 41213, BMNH 41346, BMNH 41347b, BMNH 43051, BMNH 43974, BMNH R1034, BMNH R1035, BMNH R1596, BMNH R1598, GSM 1546; Owen 1870; Padian 1983; Unwin 1988a). *Dimorphodon weintraubi* (Clark *et al*. 1994, 1996, 1997, 1998). *Peteinosaurus zambellii* (Wild 1978; Dalla Vecchia 1998, 2001). Charmouth dimorphodontid (OUM J.53070).

Anurognathidae Nopcsa 1928
Definition. Anurognathus ammoni, Batrachognathus volans, their most recent common ancestor, and all its descendants.
Content. Anurognathus Döderlein 1923, *Batrachognathus* Ryabinin 1948, *Dendrorhynchoides* Ji *et al*. 1999, *Jeholopterus* Wan *et al*. 2002, unnamed anurognathid Bakhurina & Unwin 1995.
Diagnosis. Skull very short, high and broad (Döderlein 1923; Ryabinin 1948; Kuhn 1967; Wellnhofer 1975b, 1978), possibly kinetic and with palatal elements reduced to thin bars of bone (Bakhurina 1988; Unwin *et al*. 2000); teeth are small, peg-like, widely spaced and greatly reduced in number to only three on the premaxilla and eight or less on the maxilla (Ryabinin 1948; Unwin *et al*. 2000); combined length of the dorsal + sacral vertebrae is almost the same length as the ulna (Döderlein 1923; Unwin *et al*. 2000); tail reduced to 11 vertebrae or less (Döderlein 1923; Ryabinin 1948; Unwin *et al*. 2000); subsymmetric, angular profile of the proximal end of the humerus in dorsal view (Ryabinin 1948, fig. 1; Wellnhofer 1975b, fig. 37; Unwin *et al*. 2000, fig. 3); metacarpal IV relatively short and only 33% the length of the humerus; digit III of manus reduced to only three phalanges, not four as in other pterosaurs and archosaurs. Both Döderlein (1923) and more recently Wellnhofer (1975b) have restored the third digit of *Anurognathus* with four phalanges, but the author's observations suggest that as a result of either loss or fusion, only three phalanges were present. Furthermore, even though slightly disarticulated, the third digit of *Batrachognathus* also appears to contain only three phalanges (Unwin & Bakhurina 2000). Further putative apomorphies include: wing phalanx 1 longer than the combined length of the ulna + wing metacarpal and wing phalanx 2 longer than the ulna (Ji & Ji 1998; Unwin *et al*. 2000, table 2).
Recorded temporal range. Mid-Jurassic (?Aalenian–Bajocian) to Lower Cretaceous (Barremian).
Source of phylogenetic data. Anurognathus ammoni (BSP 1922 I 42; undescribed specimen in SMNS collections; Döderlein 1923, 1929; Petronievics 1928, Wellnhofer 1975b). *Batrachognathus volans* (PIN 52–2, PIN 2585/4a; Ryabinin 1948; Bakhurina 1988; Bakhurina & Unwin 1995; Unwin & Bakhurina 2000). *Dendrorhynchoides curvidentatus* (Ji & Ji 1998; Ji *et al*. 1999; Unwin *et al*. 2000). *Jeholopterus ningchengensis* (Wang *et al*. 2002). Undescribed Mongolian anurognathid (Bakhurina & Unwin 1995; Unwin & Bakhurina 2000).

Campylognathoididae (Campylognathoidinae) Kuhn 1967
Definition. Eudimorphodon ranzii, Campylognathoides liasicus, their most recent common ancestor, and all its descendants.
Content. Campylognathoides Strand 1928; *Eudimorphodon* Zambelli 1973.
Diagnosis. Anterior tip of the mandible downturned; supratemporal fenestra largest skull opening after the orbit (Wild 1978); humerus with a large rectangular deltopectoral crest and relatively large medial crest (Wild 1978, fig. 45); rectangular sternum with short cristospine and short rectangular processes on each posterolateral corner (Wild 1978); wing finger from 67–79% of total wing length.
Remarks. Austriadactylus (Dalla Vecchia *et al*. 2002) and *Eudimorphodon* share one derived character state – multicusped teeth (Jenkins *et al*. 2001; Wellnhofer 2001) and *Austriadactylus* also appears to have a relatively elongate wing finger, a condition otherwise found only in campylognathoidids and rhamphorhynchids. Weak as it is, the available evidence suggests that *Austriadactylus* should be

referred to the Campylognathoididae, a proposal that is consistent with the general similarity of this pterosaur to *Eudimorphodon*.
Recorded temporal range. Late Triassic (Norian) to Lower Jurassic (Toarcian).
Source of phylogenetic data. Austriadactylus cristatus (Dalla Vecchia *et al*. 2002). *Campylognathoides liasicus* (MT, MNHN, SMNS; Plieninger 1907; Wiman 1923; Wild 1971; Wellnhofer 1974). *Campylognathoides zitteli* (SMNS; Plieninger 1895, 1907; Wild 1971). *Eudimorphodon cromptonellus* (Jenkins *et al*. 1999, 2001). *Eudimorphodon ranzii* (Zambelli 1973; Wild 1978, 1994; Dalla Vecchia 2001). *Eudimorphodon rosenfeldi* (Dalla Vecchia 1995, 1996, 2001). *Eudimorphodon* sp. (Wellnhofer 2001).

Rhamphorhynchinae Nopcsa 1928
Definition. Dorygnathus banthensis, *Rhamphorhynchus muensteri*, their most recent common ancestor, and all its descendants.
Content. Angustinaripterus He *et al.* 1983; *Dorygnathus* Wagner 1860; *Nesodactylus* Colbert 1969; *Parapsicephalus* (=*Dorygnathus*) Arthaber 1919; *Rhamphocephalus* Seeley 1880; *Rhamphorhynchus* Meyer 1847.
Diagnosis. Antorbital fenestra lies behind and below the naris; elongate antorbital fenestra twice as long as it is deep; mandible tips fused into a short symphysis bearing a forward-projecting prow and a number of large, fang-like, procumbent teeth forming a fish grab; wing finger 63% or more of total wing length; rear margin of wing finger grooved (see also Wellnhofer 1975b, 1975c, 1978).
Recorded temporal range. Early Jurassic (Toarcian) to Late Jurassic (Tithonian).
Source of phylogenetic data. Angustinaripterus longicephalus (He *et al*. 1983). *Dorygnathus banthensis* (MB 1905.15, 1986.6 [c], SMNS 18969, SMNS 18880, SMNS SO164, SMNS SO702, Tübingen 1536; Arthaber 1919; Wiman 1923; Saleé 1928; Wild 1975; Padian & Wild 1992). *Dorygnathus mistelgauensis* (Wild 1971). *Dorygnathus* (=†*Parapsicephalus*) *purdoni* (GSM); Newton 1888) *Nesodactylus hesperius* (Colbert 1969). *Rhamphocephalus bucklandi* (BMNH 47991; Huxley 1859; Owen 1874; Unwin 1996). *Rhamphorhynchus* '*longiceps*' (BMNH R37002, MT unnumb.; Smith Woodward 1902; Plieninger 1907; Wellnhofer 1975a). *Rhamphorhynchus muensteri* (=*R.* †*gemmingi*, *R.* †*intermedius*, *R.* †*longicaudus*) (AMNH 1943, BMNH 37002, BMNH 37787, BMNH 42738, BMNH R2786, BSP AS VI 34, BSP AS I 771, BSP 1867 II 2, BSP 1877 X 1, BSP 1889 XI 1, BSP 1927 I 36, BSP 1929 I 69, BSP 1934 I 36, BSP 1938 I 503, BSP 1955 I 28, BSP 1960 I 470, MB 3965, MB 3966, MB 3967, MB 69/2191b, MGUH VP2304, MGUH V45/1, MMK V. 45/1, MMK 1891.470, MMK 1891.738, NSMT 1776–8, PTH 1951.14, PTH 1954.39, PTH 1966.66, PTH unnumb., SMF R4128, SMF R4158; Meyer 1859; Zittel 1882; Winkler 1883; Plieninger 1907; Broili 1927; Koh 1937; Wellnhofer 1975a).

Scaphognathinae Hooley 1913
Definition. Sordes pilosus, *Scaphognathus crassirostris*, their most recent common ancestor, and all its descendants.
Content. Scaphognathus Wagner 1861; *Sordes* Sharov 1971; Morrison Formation scaphognathine (see Carpenter *et al*. 2003).
Diagnosis. Only nine, or less, relatively straight (or slightly recurved), widely spaced pairs of teeth in the rostrum (Sharov 1971, pl. 5, fig. 2; Wellnhofer 1975b, fig. 33; Ivakhnenko & Korabelnikov 1987, fig. 262; Carpenter *et al*., 2003). Only six, or less, widely spaced, vertically oriented pairs of teeth in the lower jaw (Sharov, 1971, pl. 5, fig. 2; Wellnhofer 1975b, fig. 33). Phalanx 2 of pedal digit V has a distinctive angular flexure at mid-length, such that the distal half of the phalanx lies at 40–45° to the proximal half (Sharov 1971, fig. 2; Wellnhofer 1975b, fig. 36d). See also Wellnhofer (1975b, 1978), Bakhurina & Unwin (1995) and Carpenter *et al*. (2003).
Recorded temporal range. Late Jurassic (Oxfordian–Tithonian).
Source of phylogenetic data. Scaphognathus crassirostris (GPIB 1304 [c]; MSA 110; SMNS 59395; Wellnhofer 1975b). *Sordes pilosus* (PIN 104/73, PIN 2470/1, PIN 2585/3, PIN 2585/25, PIN 2585/36, PIN 2585/37; Sharov 1971; Bakhurina 1986; Unwin & Bakhurina 1994, 2000; Bakhurina & Unwin 1995). New genus and species of scaphognathine (NAMAL 101; Cloward & Carpenter 2001; Carpenter *et al*. 2003).

Istiodactylus Howse *et al*. 2001.
Content. Istiodactylus (= *Ornithodesmus*) Howse *et al*. 2001.
Diagnosis. (Modified from Wellnhofer 1978, Howse *et al*. 2001.) Unusually extensive nasoantorbital fenestra occupying much of the snout. Orbit continuous with long, narrow suborbital vacuity. Teeth labiolingually compressed with sharply pointed crowns, sharp anterior and posterior edges, a vertical ridge on the lingual surface and truncated triangular roots shorter than crowns. The rear-most pairs of teeth on the mandible fit into deep embayments in the lower jaw. See also Hooley (1913).
Recorded Temporal Range. Early Cretaceous (Barremian).
Source of phylogenetic data. Istiodactylus (= *Ornithodesmus*) *latidens* (BMNH 3877, BMNH 3878, BMNH R176, CAMMZ T706; Seeley 1901; Hooley 1913; Howse & Milner 1993; Howse *et al*. 2001).

Ornithocheiridae Seeley 1870
Definition. Haopterus gracilis, Ornithocheirus simus, their most recent common ancestor, and all its descendants.
Content. Anhanguera Campos & Kellner 1985; *Brasileodactylus* Kellner 1984; *Coloborhynchus* Owen 1874; *Haopterus* Wang & Lü 2001; *Ornithocheirus* Seeley 1869; new genus and species of ornithocheirid (Frey *et al.* 2003).
Diagnosis. (Modified from Bakhurina & Unwin 1995; Unwin 2001.) The first three pairs of teeth in the rostrum are relatively large, forming a terminal rosette and show a marked increase in size posteriorly. The fourth and fifth tooth pairs are much reduced in size and smaller than the first pair of teeth. Proceeding posteriorly, there is a steady increase in tooth size up to, typically, the ninth pair, which are of similar basal dimensions to the largest teeth in the terminal rosette. Further posteriorly, tooth size declines again. Consequently, the rostrum has an expanded anterior tip that accommodates the large anterior teeth, is narrowest in the region of the fourth or fifth tooth pair, and gradually widens posteriorly (e.g. Fastnacht 2001, figs 2–5; Fig. 10b, d, e). The expansion of the anterior end of the rostrum is most marked in large species and adult individuals, but may be practically absent in small species (e.g. *Haopterus*) and juveniles. The mandibular dentition and symphysis show similar morphological patterns to the rostrum.
Remarks. Various taxa from the Santana Formation ('*Araripesaurus*' and '*Santanadactylus*') and the Crato Formation (*Arthurdactylus*) of Brazil probably also belong in the Ornithocheiridae (Unwin 2001), but their taxonomic status and precise relationships to other ornithocheirids are uncertain (Bakhurina & Unwin 1995; Unwin & Lü 1997; Unwin & Bakhurina 2000; Unwin *et al.* 2000; Fastnacht 2001). Anhangueridae has a content (e.g. Kellner & Tomida 2000) and diagnosis (Campos & Kellner 1985) similar to that of Ornithocheiridae and is treated here as a junior synonym of the latter taxon (Unwin 2001).
Recorded temporal range. Early Cretaceous (Valanginian) to Late Cretaceous (Cenomanian).
Source of phylogenetic data. Anhanguera araripensis (BSP 1982 I 89; Wellnhofer 1985; Kellner & Tomida 2000). *Anhanguera blittersdorffi* (BMNH R11978; Campos & Kellner 1985; Kellner & Tomida 2000). *Anhanguera cuvieri* (BMNH 39409, CAMSM B54.431; Bowerbank 1851; Owen 1859; Unwin 2001). *Anhanguera fittoni* (CAMSM B54.423, CAMSM B54.556; Owen 1859; Unwin 2001). *Anhanguera santanae* (AMNH 22555, BSP 1982 I 90; BSP 1982 I 91, IMCF 1053; Wellnhofer 1985, 1991b). *Anhanguera* sp. (SMNK 1136 PAL; Frey & Martill 1994). *Arthurdactylus conandoylei* (SMNK 1132 PAL, Frey & Martill 1994). *Brasileodactylus araripensis* (Kellner 1984; Sayão & Kellner 2000). *Coloborhynchus capito* (BMNH R481, CAMSM B54.625; Unwin 2001). *Coloborhynchus clavirostris* (BMNH R1822; Owen 1874; Lee 1994; Fastnacht 2001, Unwin 2001). *Coloborhynchus* (= †*Siroccopteryx*) *moroccensis* (Mader & Kellner 1999; Unwin 2001). *Coloborhynchus robustus* (= †*Anhanguera piscator*, †*Tropeognathus robustus*) (BSP 1987 I 47, NSM-PV 19892, SMNK 1133 PAL, SMNK 2302 PAL; Wellnhofer 1987; Kellner & Tomida 2000, Fastnacht 2001; Unwin 2001). *Coloborhynchus sedgwickii* (CAMSM B54.422; Owen 1859; Unwin 2001). *Coloborhynchus wadleighi* (Lee 1994); *Haopterus gracilis* (Wang & Lü 2001; see also Unwin 2001). *Ornithocheirus mesembrinus* (BSP 1987 I 46; Wellnhofer 1987; Fastnacht 2001). *Ornithocheirus simus* (CAMSM B54.428, MANCH L10832; Owen 1861; Unwin 2001). *Ornithocheirus* sp. (CAMSM B54.890). Khuren-Dukh ornithocheirid (Bakhurina & Unwin 1995). New genus and species of ornithocheirid (Viohl 2000, pl. 9, fig. 3; Frey *et al.* 2003).

Pteranodontidae Marsh 1876
Definition. Ornithostoma sedgwicki, Pteranodon longiceps, their most recent common ancestor, and all its descendants.
Content. Pteranodon Marsh 1876; *Ornithostoma* Seeley 1871.
Diagnosis. (Modified from Bennett 1994.) Cranial crest formed by frontals and directed upwards and backwards from skull. Premaxillary crest with relatively straight dorsal margin, not rounded in profile. Edentulous jaws with raised marginal ridges. Premaxillae extending beyond the end of the mandible. Ceratobranchials of hyoid apparatus greatly reduced or unossified. Notarial and synsacral supraneural plates formed of ossified interspinous ligaments. Postacetabular process of ilium contacting neural spines of posterior synsacral vertebrae and fused with them in mature adults. Proximal caudal vertebrae with duplex centra. Distal caudal vertebrae reduced and co-ossified to form caudal rod. Distal end of wing-metacarpal without an elevation between condyles. See also Bennett (2001) and Unwin (2001).
Recorded temporal range. Early Cretaceous (Albian) to Late Cretaceous (Campanian).
Source of phylogenetic data. Ornithostoma sedgwicki (CAMSM B54.485, Owen 1859; Seeley 1871; Unwin 2001). *Pteranodon longiceps* (AMNH 149, AMNH 4908, AMNH 6158, FMNH PR 464, FMNH PR 468–470, FMNH PR 494, FMNH PR 676, FMNH PR 1332; Eaton 1910; Bennett 1994, 2001). *Pteranodon sternbergi* (Harksen 1966; Bennett 1994, 2001). *Pteranodon* sp. (AMNH 5840, BMNH 2959, BMNH 4006, BMNH 4008, BMNH 4534, BMNH 4538; Eaton 1910; Bennett 1994, 2001).

Nyctosaurus Marsh 1876
Content. Nyctosaurus Marsh 1876.
Diagnosis. (Based on Brown 1986; Bennett 1989, 1994, 2000.) Neural spines of mid-notarial vertebrae T-shaped in anterior view. Dorsal centra crescentic in cross-section. Humerus with hatchet-shaped deltopectoral crest. Loss of manus digits I–III. Reduction of the wing finger to just three phalanges.
Recorded temporal range. Late Cretaceous (Santonian–Maastrichtian).
Source of phylogenetic data. Nyctosaurus gracilis (FMNH 25026; Williston 1902, 1903; Miller 1972; Bennett 1989, 1994, 2000). *Nyctosaurus lamegoi* (Price 1953; Bennett 1989). *Nyctosaurus nanus* (Miller 1972; Bennett 1994).

Cycnorhamphus Seeley 1870
Content. Cycnorhamphus Seeley 1870.
Diagnosis. Reduction of the dentition to a few, slender, curved teeth on the tips of the jaws (Wellnhofer 1978). A large, deep, wing-like, parietal crest projecting backwards from the occiput (Wellnhofer 1978; Bennett 1996c). Relatively long terminal wing phalanx (Padian & Warheit 1989).
Recorded temporal range. Late Jurassic (Tithonian).
Source of phylogenetic data. Cycnorhamphus canjeursensis (MNHN CNJ-71; Fabre 1974, 1976). *Cycnorhamphus suevicus* (MT unnumb.; Quenstedt 1855; Wagner 1858; Fraas 1878, Plieninger 1907; Wellnhofer 1970, Bennett 1996c).

Pterodactylus Cuvier 1809
Content. Pterodactylus Cuvier 1809 (*partim*).
Remarks. In the last major review of this genus Wellnhofer (1968, 1970, 1978) included the following species from the Solnhofen Limestone: *P. antiquus*, *P. kochi*, *P. micronyx*, *P. elegans* and *P. longicollum*. Bennett (1996a) has suggested that '*P.*' *micronyx* may be a juvenile form of *Gnathosaurus subulatus*. This is not unreasonable but, in any case, '*P.*' *micronyx* is quite distinct from other species of *Pterodactylus* (Wellnhofer 1968, 1970; Unwin 1995b) and for the present the author prefers not to include '*P.*' *micronyx* in *Pterodactylus sensu stricto* (see also Atanassov 2000). Bennett (1996a) also proposed that specimens of †*P. elegans* represent juveniles of *Ctenochasma gracile*, an idea that is accepted here. '*Pterodactylus*' *longicollum* has at least one diagnostic character of gnathosaurines: expanded spatulate jaw tips with a rosette composed of numerous elongate teeth; however, its exact taxonomic status, a distinct species of *Gnathosaurus*, or perhaps a representative of another genus (?*Diopecephalus* Seeley 1871), has yet to be established. In any case, '*P.*' *longicollum* cannot be retained in *Pterodactylus*. Consequently, in this study, *Pterodactylus* contains just two species: *P. antiquus* and *P. kochi*. Whether these species are truly distinct from one another, and how they differ from other similar taxa, such as *Germanodactylus rhamphastinus*, requires further study (see also Atanassov 2000).

Other taxa assigned to this genus, including †*P. arningi*, †*P. brancai* and †*P. maximus*, are *nomina dubia* (Unwin & Heinrich 1999), as are †*P. manselli* and †*P. pleydelli*.
Diagnosis. None of the characters previously used to define *Pterodactylus* (e.g. Wellnhofer 1968, 1970, 1978) are truly diagnostic because they occur in other pterosaurs. A group including *P. antiquus* and *P. kochi* is distinguished by the rostral dentition: 16 or more tall-triangular teeth, the largest located in the mid-region of the tooth row, which extends from the jaw tips posteriorly to a point beneath the nasoantorbital fenestra. While this excludes other ctenochasmatoids, it also holds true for *Germanodactylus rhamphastinus*, although not *G. cristatus*.
Recorded temporal range. Late Jurassic (Tithonian).
Source of phylogenetic data. Pterodactylus antiquus (BMNH R388, BSP AS I 739, BSP 1968 I 95, Wellnhofer 1970). *Pterodactylus kochi* (AMNH 1942, BMNH 42736, BSP AS V 29, BSP AS XIX 3, BSP 1878 VI 1, BSP 1883 XVI 1, BSP 1924 V1, BSP 1929 I 18, BSP 1937 I 18, BSP 1967 I 276, MB 1876.2059, NSMT PV19893, SM R4702, SM R4074; Zittel 1882; Abel 1925; Broili 1938; Huene 1951; Wellnhofer 1970, 1987; Tischlinger 1994; Frey & Martill 1998).

Lonchodectidae Unwin *et al.* 2000
Definition. Lonchodectes giganteus, *Lonchodectes machaerorhynchus*, their most recent common ancestor, and all its descendants.
Content. Lonchodectes Hooley 1913.
Diagnosis. (Modified from Unwin 2001.) Pterosaurs with distinctive, parapet-like alveolar borders to the jaws. Each border bears small, subequal sized, subcircular, well-spaced alveoli with margins raised into a low collar, and containing teeth with constricted bases. There is a prominent, sharply ridged, median keel on the occlusal surface of the rostrum which corresponds with a deep, V-shaped median sulcus on the occlusal surface of the mandibular symphysis.
Recorded temporal range. Early Cretaceous (Valanginian) to Late Cretaceous (Cenomanian).
Source of phylogenetic data. Lonchodectes compressirostris (BMNH 39410, CAMSM B54.584; Owen 1851, pl. XXVIII, figs 8–10; Seeley 1870, p.114). *Lonchodectes giganteus* (BMNH 39412; Bowerbank 1846, pl. I, figs 1 & 2; Bowerbank 1848, pl. I, fig. 1; Owen 1851). *Lonchodectes machaerorhynchus* (CAMSM B54.885; Seeley 1870, pl. XII, figs 1 & 2). *Lonchodectes microdon* (BMNH R2268, BMNH R2269, CAMSM B54.439, CAMSM B54.486, GSM 87822; Seeley 1870, pl. XII, figs 6 &

7). *Lonchodectes platystomus* (BMNH 43074, CAMSM B54.835, YORM 1983/113F; Seeley 1870; Owen 1874, pl. I, figs 5 & 6). *Lonchodectes* sp. indet. (BMNH R2287c, BMNH R2296, BMNH R3019, BMNH R3694, BMNH 35229, CAMSM B54.074, CAMSM B54.075, CAMSM B54.081, CAMSM B54.262, CAMSM B54.536, CAMSM B55030, GSM 87850, GSM 87887, HMG V1487/2, HMG 1486/3, HMG V1486/4, HMG V1487/4, HMG V1488/3, HMG 1505/3). ?*Lonchodectes sagittirostris* (BMNH R1823; Owen 1874, pl. II, figs 1–8).

Ctenochasmatidae Nopcsa 1928
Definition. Gnathosaurus subulatus, Pterodaustro guinazui, their most recent common ancestor, and all its descendants.
Content. Cearadactylus Leonardi & Borgomanero 1985; *Ctenochasma* Meyer 1851; ?*Eosipterus* Ji & Ji 1997; *Gnathosaurus* Meyer 1834; *Huanhepterus* Dong 1982; *Plataleorhynchus* Howse & Milner 1995; '*Pterodactylus*' *longicollum* Meyer, 1854; *Pterodaustro* Bonaparte 1970.
Diagnosis. (Modified from Wellnhofer 1978; Buisonjé 1981; Howse & Milner 1995; Unwin 2002.) Rostrum anterior to nasoantorbital fenestra forms more than half the total length of the skull (tip of rostrum to occipital condyle). Anterior end of rostrum dorsoventrally compressed and rounded (Wellnhofer 1970, 1978). At least 25 teeth per side in the rostrum. Seven or more pairs of teeth in the premaxillae. Teeth project laterally from the dental border of the rostrum at least in the anterior part of the dentition (Wellnhofer 1970, 1978). Teeth in anterior part of dentition relatively elongate, with a long cylindrical crown of even width and a short tapering distal tip (Buisonjé 1981; Knoll 2000). Metatarsal III more than one-third the length of the tibia (Unwin *et al.* 2000, fig. 6a).
Recorded temporal range. Late Jurassic (Tithonian) to Early Cretaceous (Albian).
Source of phylogenetic data. Cearadactylus atrox (Leonardi & Borgomanero 1985). *Ctenochasma gracile* (=†*Pterodactylus elegans*) (BSP AS VI 30, BSP 1867 II 1, BSP 1875 XIV 501, BSP 1920 I 57, BSP 1935 I 24, PTH 1950.33, PTH unnumb.; Zittel 1882, Broili 1919, 1924, 1936; Wellnhofer 1970; Buisonjé 1981; Bennett 1996a). *Ctenochasma porocristata* (JM SoS-2179; Buisonjé 1981). *Ctenochasma roemeri* (Meyer 1851; Broili 1924; Buisonjé 1981). *Ctenochasma* sp. (Taquet 1972). *Eosipterus yangi* (Ji & Ji 1997, Ji *et al.* 1999; Unwin *et al.* 2000). *Gnathosaurus macrurus* (CAMSM J5339; Howse & Milner 1995). *Gnathosaurus subulatus* (BSP AS VII 369, PTH 1951.84; Wellnhofer 1970). *Huanhepterus quingyangensis* (IVPP V 9070; Dong 1982). *Plataleorhynchus streptophorodon* (Howse & Milner 1995). '*Pterodactylus*' *longicollum* (BMNH Wellnhofer 1970, Exemplar 55 [c]; SMNS Wellnhofer 1970, Exemplar 58; Meyer 1859; Plieninger 1907; Wellnhofer 1970). *Pterodaustro guinazui* (PVL 3860 [c], PVL 3968 [c]; Bonaparte 1970, 1971; Sanchez 1973; Chiappe *et al.* 2000; Davila & Chiappe 2000).

Germanodactylus Young 1964
Content. Germanodactylus Young 1964.
Diagnosis. True synapomorphies uniting *Germanodactylus cristatus* and *G. rhamphastinus*, have yet to be identified; thus *Germanodactylus* is treated here as a metataxon (Gauthier 1986).
Recorded temporal range. Late Jurassic (Tithonian).
Source of phylogenetic data. Germanodactylus cristatus (BSP 1892 IV 1, NMING F15005, SMNK unnumb.; Plieninger 1901; Wellnhofer 1968, 1970). *Germanodactylus rhamphastinus* (BSP AS I 745, BSP 1977 XIX 1; Wellnhofer 1970).

Dsungaripteridae Young 1964
Definition. Dsungaripterus weii, Noripterus complicidens, their most recent common ancestor, and all its descendants.
Content. Dsungaripterus Young 1964; *Noripterus* Young 1973; '*Phobetor*' Bakhurina 1986. *Domeykodactylus* Martill *et al.* 2000.
Diagnosis. Strong variation in tooth size with the largest teeth located at the caudal end of the dentition. Ventral region of the orbit partially ossified: the postorbital process of the jugal sends forward a bar of bone to the lacrimal enclosing a small suborbital slit. Presence of a short, rectangular sagittal crest projecting backwards from the posterodorsal apex of the skull. (See also Kuhn 1967; Bennett 1989, 1994; Martill *et al.* 2000.)
Recorded temporal range. Early Cretaceous.
Source of phylogenetic data. Domeykodactylus ceciliae (Martill *et al.* 2000). *Dsungaripterus weii* (IVPP V. 2776, IVPP V. 2777, IVPP 64041–1, IVPP 64041–3, IVPP 64043–4, IVPP 64043–12, IVPP 64045–1, IVPP 64045–2, IVPP 64045–3, IVPP 64045–5, IVPP 64045–9; Young 1964, 1973). *Noripterus complicidens* (IVPP 64043–3; Young 1973). '*Phobetor*' *parvus* (Bakhurina 1982, 1983, 1986, 1993; Ivakhnenko & Korabelnikov 1987; Bakhurina & Unwin 1995; Unwin & Bakhurina 2000).

Tapejara Kellner, 1989.
Content. Tapejara Kellner 1989.
Diagnosis. (Modified from Kellner 1989; Wellnhofer & Kellner 1991.) High premaxillary sagittal crest on anterior part of skull tapering to a low posterior extension closely following the midline of the skull roof. Short frontoparietal crest. Rostrum inclined downwards with concave depression in palatal view. Upper margin of symphysis

inclined downwards with concave depression dorsally.
Recorded temporal range. Early Cretaceous (Aptian) to Late Upper Cretaceous (Cenomanian).
Source of phylogenetic data. Tapejara imperator (Campos & Kellner 1997). *Tapejara wellnhoferi* (AMNH 24440, SMNK unnumb.; IMCF unnumb.; Kellner 1989; Kellner 1991; Wellnhofer & Kellner 1991). *Tapejara* sp. (SMNK 2343 PAL; Martill & Frey 1998).

Tupuxuara Kellner & Campos 1988.
Content. Tupuxuara Kellner & Campos 1988.
Diagnosis. Large sagittal cranial crest involving the premaxillae, frontals and parietals, and forming a prominent sail-like structure with an origin extending from a little posterior to the tip of the rostrum over the cranium, round onto the occipital plate. See also Kellner & Campos (1988).
Recorded temporal range. Early Cretaceous (Albian).
Source of phylogenetic data. Tupuxuara longicristatus (IMCF 1052; Kellner & Campos 1988). *Tupuxuara leonardii* (Kellner & Campos 1994).

Azhdarchidae Nesov 1984
Definition. Azhdarcho lancicollis, Quetzalcoatlus northropi, their most recent common ancestor, and all its descendants.
Content. Arambourgiania Nesov & Yarkov 1989; *Azhdarcho* Nesov 1984; *Hatzegopteryx* Buffetaut *et al.* 2002; *Montanazhdarcho* Padian *et al.* 1995; *Quetzalcoatlus* Lawson 1975a; *Zhejiangopterus* Cai & Wei 1994.
Diagnosis. Orbit reduced in size (only one-third the height of the nasoantorbital fenestra), subcircular and located entirely below the mid-height level of the nasoantorbital fenestra. The fifth cervical is at least 8 times longer than it is broad and the neural arch merges with the centrum to form a tube-like structure (Lawson 1975a; Nesov 1984; Padian 1984a, 1986; Howse 1986; Bennett 1994; Frey & Martill 1996; Martill *et al.* 1998). The neural spine is absent in the mid-section of the fifth cervical, but forms low crests anteriorly and posteriorly. Expanded coracoidal flange occupying more than half the length of the coracoid (Kellner & Langston 1996). Extremely shallow concavity of the caudal region of the distal end of wing phalanx 1 (Frey & Martill 1996). Longitudinal ridge on the ventral surface of wing phalanges 2 and 3, resulting in a T-shaped cross-section (Nesov 1991; Bennett 1994). Femur more than 1.6 times the length of the humerus (Table 2).
Remarks. Padian (1986) suggested that azhdarchids might also be distinguished by the absence of ossification of the neural canal, at least in the mid-series cervicals, but an ossified neural canal is clearly present in some forms, including *Azhdarcho* (Unwin & Bakhurina 2000) and *Arambourgiania* (Martill *et al.* 1998, fig. 5; see also Bennett 1994). Other putative diagnostic characters include: an elongate nasoantorbital fenestra (Unwin & Lü 1997); occiput that faces ventrally; a relatively short wing finger (~50% total wing length) and various features of the humerus (Padian & Smith 1992). Many of these characters also seem to be present in *Tupuxuara* and may thus be apomorphies of Neoazhdarchia.
Recorded temporal range. Late Cretaceous (?Cenomanian–Maastrichtian).
Source of phylogenetic data. Arambourgiania philadelphiae (BSP 1978/1 [c]; Arambourg 1959; Frey & Martill 1996; Martill *et al.* 1998). *Azhdarcho lancicollis* (Ch.B.I 12454, 1/11915, 3/11915, 4/11915, 5/11915, 6/11915, 7/11915, 8/11915, 9/11915, 10/11915, 17/11915, 18/11915, 41/11915; Nesov 1984, pl. 7, figs 1–11 & 13; Nesov 1986, pl. 2, fig. 1; Nesov 1991; Nesov & Yarkov 1989, pl. 2, figs 2–8; Bakhurina & Unwin 1995). cf. *Azhdarcho* (Buffetaut 1999). *Hatzegopteryx thambema* (Buffetaut et al. 2002). *Montanazhdarcho minor* (Padian *et al.* 1995). *Quetzalcoatlus* sp. (TMM 41450, TMM 41541, TMM 41544, TMM 41546, TMM 41547, TMM 41954, TMM 41961, TMM 42138, TMM 42157, TMM 42161, TMM 42180, TMM 42422–30, TMM 42462; Lawson 1975a, b; Langston 1981; Kellner & Langston 1996). *Quetzalcoatlus northropi* (TMM 4150–3; Lawson 1975a, b; Langston 1981). ?*Quetzalcoatlus* (YPM-PU 22446 [c], MOR 553, MOR 656, MOR 1071; Padian 1984, 1986; Padian & Smith 1992). Solana azhdarchid (MGUV 2271, MGUV 2194, MGUV 2195, MGUV 2239, MGUV 2271, MGUV 3207, MGUV 3209, MGUV 3210, MPV TT48, MPV TT49; Company *et al.* 1999). *Zhejiangopterus linhaiensis* (ZMNH M1323, ZMNH 1328, ZMNH 1330; Cai & Wei 1994; Unwin & Lü 1997).

Appendix 2. List of characters that vary within Pterosauria, used in this analysis

1. Dentary: less (0), or more than 75% length of lower jaw (1).
2. Coracoid: less than (0), or at least 75% length of scapula (1).
3. Unguals of manus and pes: similar in size (0), manual unguals twice the size, or more, of pedal unguals (1).
4. Forelimb length: less (0), or at least 2.5 times the length of hindlimb (femur + tibia + mtiii) (1).
5. Humerus: shorter (0), or longer than femur (1).
6. Quadrate: vertical (0), inclined anteriorly (1).
7. Ulna: shorter (0), subequal or longer than tibia (1).

8. Fibula: subequal in length (0), or less than 80% the length of the tibia (1).
9. Rostrum: high, with convex outline (0), low with straight or concave dorsal outline (1), anterior region of rostrum low, but antorbital region expanded dorsally (2).
10. Posterior extent of premaxilla: terminates at (0), or interfingers between frontals (1).
11. External nasal opening: height similar to or greater than anteroposterior length (0), low and elongate (1).
12. Nasal process of maxilla: vertical–subvertical (0), inclined backwards (1).
13. Maxilla–nasal contact: narrow (0), broad (1).
14. Orbit: smaller than (0) or larger than the antorbital opening (1).
15. Ventral margin of skull: straight (0), curved downwards caudally (1).
16. Caudal end of mandible with distinct dorsal 'coronoid' eminence: present (0), absent (1).
17. Bony mandibular symphysis: absent (0), present (1).
18. Mandibular symphysis: less than (0), or more than 30% the length of the mandible (1).
19. Two, large, fang-like mandibular teeth: present (0), absent (1).
20. Metacarpals I–III: disparate lengths (0), or the same length (1).
21. Length of metatarsal IV: subequal (0), shorter than metatarsals I–III (1).
22. Rostral dentition: more than (0) or less than 11 pairs of teeth (1).
23. Deltopectoral crest of humerus tongue-shaped, with necked base: absent (0), present (1).
24. Narial and antorbital fenestrae: separate (0), confluent (1).
25. Basipterygoids: separate (0), or united to form a median bar of bone (1).
26. Cervical ribs: present (0), strongly reduced or absent (1).
27. Combined length of caudal vertebrae: longer (0), or shorter than the dorsal series (1).
28. Pteroid: short and stubby (0), or long and slender (1).
29. Metacarpal IV: less (0), or 80%, or more, of humerus length (1).
30. Pes digit V: two phalanges (0), one or less (1).
31. Notarium: absent (0), present (1).
32. Coracoid: shorter than (0), longer than scapula (1).
33. Humerus with warped deltopectoral crest: absent (0), present (1).
34. Pneumatopore on anconal surface of humerus: absent (0), present (1).
35. Distal end of humerus: D-shaped (0), triangular (1).
36. Ornithocheiroid carpus: absent (0), present (1).
37. Contact between distal syncarpal and metacarpals I–IV: all four in contact (0), only I and IV contact syncarpal (1), only IV contacts the syncarpal (2).
38. Femur with stout neck and steeply directed caput: absent (0), present (1).
39. Posterior margin of nasoantorbital fenestra: straight (0), concave (1).
40. Basal region of orbit: open (0), infilled (1).
41. Coracoid facets on sternum: one anterior to the other (0), lateral to each other (1).
42. Tall, narrow, frontal crest: absent (0), present (1).
43. Dentition: present (0), absent (1).
44. Mandibular rami: at same level (0), or elevated well above symphysis (1).
45. Pneumatic opening in palmar surface of humerus: absent (0), present (1).
46. Metacarpal IV: less (0), or more than twice the length of the humerus (1).
47. Deltopectoral crest of humerus with elongate rectangular profile: absent (0), present (1).
48. Sagittal cranial crest continued dorsally in soft tissues and extending from anterior to nasoantorbital fenestra to apex of skull, or beyond: absent (0), present (1).
49. Quadrate: vertical or inclined (0), subhorizontal position (1).
50. Squamosal above (0), level with or below base of lacrimal process of jugal (1).
51. Occiput faces posteriorly or posteroventrally (0), ventrally (1).
52. Neural arch of cervicals: high with high neural spine (0), or low with low neural spine (1).
53. Elongate mid-series cervicals: absent (0), present (1).
54. Distal ends of paroccipital processes: unexpanded (0), expanded (1).
55. Dsungaripteroid teeth: absent (0), present (1).
56. Dentition: extends to jaw tips (0), jaw tips toothless, but followed by tooth row (1).
57. Location of largest teeth: rostral half of dentition (0), caudal half (1).
58. Appendicular bones: with thick cortex and narrow or absent lumen (0), thin cortex and wide lumen (1), secondarily thickened cortex (2).
59. Strongly bowed femur: absent (0), present (1).
60. Dorsal margin of orbit: level with dorsal margin of nasoantorbital fenestra (0) or located well below it (1).

References

ABEL, O. 1925. On a skeleton of *Pterodactylus antiquus* from the lithographic shales of Bavaria, with remains of skin and musculature. *American Museum Novitates*, **192**, 1–12.

ARAMBOURG, C. 1959. *Titanopteryx philadelphiae* nov. gen., nov. sp., ptérosaurien géant. *Notes et Mémoires sur le Moyen-Orient*, **7**, 229–234.

ARTHABER, G. VON. 1919. Studien über Flugsaurier auf Grund der Bearbeitung des Wiener Exemplares von *Dorygnathus banthensis* Theod. sp. *Denkschrift der Akademie der Wissenschaften Wien, Mathematisch-Naturwissenschaftliche Klasse*, **97**, 391–464.

ARTHABER, G. VON. 1922. Über Entwicklung, Ausbildung und Absterben der Flugsaurier. *Paläontologische Zeitschrift*, **4**, 1–47.

ATANASSOV, M. 2000. Morphometric analysis of Late Jurassic pterodactyloids from Germany and France. *Journal of Vertebrate Paleontology*, **20**(3) (supplement) 27A.

ATANASSOV, M. 2001. Two new archosauromorphs from the Late Triassic of Texas. *Journal of Vertebrate Paleontology*, **21**(3) (supplement), 30A.

BAKHURINA, N. N. 1982. [A pterodactyl from the Lower Cretaceous of Mongolia]. *Palaeontologicheskii Zhurnal* **1982**(4), 104–108. [In Russian]

BAKHURINA, N. N. 1983. [Early Cretaceous pterosaur localities in western Mongolia]. *Transactions of the Joint Soviet-Mongolian Palaeontological Expedition*, **24**, 126–129. [In Russian]

BAKHURINA, N. N. 1986. [Flying reptiles]. *Priroda*, **7**, 27–36. [In Russian]

BAKHURINA, N. N. 1988. [On the first rhamphorhynchoid from Asia: *Batrachognathus volans* Ryabinin 1948, from Upper Jurassic beds of Karatau]. *Abstracts of the Bulletin of the Moscow Society for the Study of Natural History, Geological Section*, **63** (5), 132. [In Russian]

BAKHURINA, N. N. 1993. Early Cretaceous pterosaurs from western Mongolia and the evolutionary history of the Dsungaripteroidea. *Journal of Vertebrate Paleontology*, **13**(3) (supplement) 24A.

BAKHURINA, N. N. & UNWIN, D. M. 1995. A survey of pterosaurs from the Jurassic and Cretaceous of the former Soviet Union and Mongolia. *Historical Biology*, **10**, 197–245.

BENNETT, S. C. 1987. New evidence on the tail of pterosaur *Pteranodon* (Archosauria: Pterosauria). *Fourth Symposium on Mesozoic Terrestrial Ecosystems and Biota*, Tyrrell Museum of Palaeontology, Drumheller, Short Papers, 18–23.

BENNETT, S. C. 1989. A pteranodontid pterosaur from the Early Cretaceous of Peru, with comments on the relationships of Cretaceous pterosaurs. *Journal of Paleontology*, **63**, 669–677.

BENNETT, S. C. 1992. Sexual dimorphism of *Pteranodon* and other pterosaurs with comments on cranial crests. *Journal of Vertebrate Paleontology*, **12**, 422–434.

BENNETT, S. C. 1993. The ontogeny of *Pteranodon* and other pterosaurs. *Palaeobiology*, **19**, 92–106.

BENNETT, S. C. 1994. *Taxonomy and systematics of the Late Cretaceous pterosaur* Pteranodon *(Pterosauria, Pterodactyloidea)*. University of Kansas, Occasional Papers, **169**, 1–70.

BENNETT, S. C. 1995. A statistical study of *Rhamphorhynchus* from the Solnhofen Limestone of Germany – year-classes of a single large species. *Journal of Paleontology*, **69**, 569–580.

BENNETT, S. C. 1996a. Year-classes of pterosaurs from the Solnhofen Limestone of Germany: taxonomic and systematic implications. *Journal of Vertebrate Paleontology*, **16**, 432–444.

BENNETT, S. C. 1996b. The phylogenetic position of the Pterosauria within the Archosauromorpha. *Zoological Journal of the Linnean Society, London*, **118**, 261–308.

BENNETT, S. C. 1996c. On the taxonomic status of *Cycnorhamphus* and *Gallodactylus* (Pterosauria: Pterodactyloidea). *Journal of Paleontology*, **70**, 335–338.

BENNETT, S. C. 2000. New information on the skeleton of *Nyctosaurus*. *Journal of Vertebrate Paleontology*, **20**(3) (supplement), 27A.

BENNETT, S. C. 2001. The osteology and functional morphology of the Late Cretaceous pterosaur *Pteranodon*. *Palaeontographica A*, **260**, 1–153.

BENNETT, S. C. 2002. Soft tissue preservation of the cranial crest of the pterosaur *Germanodactylus* from Solnhofen. *Journal of Vertebrate Paleontology*, **22**, 43–48.

BENNETT, S. C. 2003. Morphological evolution of the pectoral girdle of pterosaurs: myology and function. *In*: BUFFETAUT, E. & MAZIN, J-M. (eds) *Evolution and Palaeobiology of Pterosaurs*. Geological Society, London, Special Publications, **217**, 191–215.

BENTON, M. J. 1982. The Diapsida: revolution in reptile relationships. *Nature*, **296**, 306–307.

BENTON, M. J. 1984. The relationships and early evolution of the Diapsida. *Symposia of the Zoological Society, London*, **52**, 575–596.

BENTON, M. J. 1985. Classification and phylogeny of the diapsid reptiles. *Zoological Journal of the Linnean Society, London*, **84**, 97–164.

BENTON, M. J. 1990. Origin and interrelationships of dinosaurs. *In*: WEISHAMPEL, D. B., DODSON, P. & OSMOLSKA, H. (eds) *The Dinosauria*. University of California Press, Berkeley, 11–30.

BENTON, M. J. 1995. Testing the time axis of phylogenies. *Philosophical Transactions of the Royal Society of London*, (B), **349**, 5–10.

BENTON, M. J. 1997. *Vertebrate Palaeontology*, 2nd Edition, Chapman and Hall, London, 452 pp.

BENTON, M. J. 1999. *Scleromochlus* and the origin of dinosaurs and pterosaurs. *Philosophical Transactions of the Royal Society, London*, (B), **354**, 1423–1446.

BININDA-EDMONDS, O. R. P., BRYANT, H. N. & RUSSELL, A. P. 1998. Supraspecific taxa as terminals in cladistic analysis: implicit assumptions of monophyly and a comparison of methods. *Biological Journal of the Linnean Society, London*, **64**, 101–133.

BONAPARTE, J. F. 1970. *Pterodaustro guinazui* gen. et sp. nov. pterosaurio de la Formacíon Lagarcito, Provincia de San Luis, Argentina y su significado en la geología regional (Pterodactylidae). *Acta Geologica Lilloana*, **10**, 207–226.

BONAPARTE, J. F. 1971. Descripcion del craneo y mandibulas de *Pterodaustro guinazui* (Pterodactyloidea – Pterodaustriidae nov.) de la Formacion Lagarcito, San Luis, Argentina. *Publicaciones del Museo Municipal de Ciências Naturales, Mar del Plata*, **1**, 263–272.

BONAPARTE, J. F. & SANCHEZ, T. M. 1975. Restos de un pterosaurio Puntanipterus globosus de la formación La Cruz, provincia de San Luis Argentina. *Actas 1 congresso Argentino Paleontologia y Biostratigrafía*, **2**, 105–113.

BOWERBANK, J. S. 1846. On a new species of pterodactyle found in the Upper Chalk of Kent. *Quarterly Journal of the Geological Society, London*, **2**, 7–8.

BOWERBANK, J. S. 1848. Microscopical observations on the structure of the bones of *Pterodactylus giganteus* and other fossil animals. *Quarterly Journal of the Geological Society, London*, **4**, 2–10.

BOWERBANK, J. S. 1851. On the pterodactyles of the Chalk formation. *Annals and Magazine of Natural History, Series 2*, **10**, 372–378.

BROCHU, C. A. 2002. Progress and future directions in archosaur phylogenies. *Journal of Paleontology*, **75**, 1185–1201.

BROILI, F. 1919. *Ctenochasma gracile* Oppel. *Geognostischen Jahresheften*, **29–30** (1916–1917), 299–306.

BROILI, F. 1924. *Ctenochasma* ist ein Flugsaurier. *Sitzungsberichte der Bayerischen Akademie der Wissenschaften, mathematisch-naturwissenschaftliche Abteilung*, 13–30.

BROILI, F. 1927. Ein *Rhamphorhynchus* mit Spuren von Haarbedeckung. *Sitzungsberichte der Bayerischen Akademie der Wissenschaften, mathematisch-naturwissenschaftliche Abteilung*, 49–67.

BROILI, F. 1936. Weitere Beobachtungen an *Ctenochasma*. *Sitzungsberichte der Bayerischen Akademie der Wissenschaften, mathematisch-naturwissenschaftliche Abteilung*, 137–156.

BROILI, F. 1938. Beobachtungen an *Pterodactylus*. *Sitzungsberichte der Bayerischen Akademie der Wissenschaften, mathematisch-naturwissenschaftliche Abteilung*, 139–154.

BROWN, G. W. 1986. Reassessment of *Nyctosaurus*: new wings for an old pterosaur. *Nebraska Academy of Sciences, Proceedings*, 47. [Abstract]

BUFFETAUT, E. 1999. Pterosauria from the Upper Cretaceous of Laño (Iberian Peninsula): a preliminary comparative study. *Estudios Museo Ciencias Naturales Alava*, **14**(1) (special), 289–294.

BUFFETAUT, E., GRIGORESCU, D. & CSIKI, Z. 2002. A new giant pterosaur with a robust skull from the latest Cretaceous of Romania. *Naturwissenschaften*, **89**, 180–184.

BUFFETAUT, E., LEPAGE, J-J. & LEPAGE, G. 1998. A new pterodactyloid pterosaur from the Kimmeridgian of the Cap de la Hève (Normandy, France). *Geological Magazine*, **135**, 719–722.

BUISONJÉ, P. H. DE. 1980. *Santandactylus brasilensis* nov. gen. nov. sp., a long necked, large pterosaurier from the Aptian of Brasil. II. *Proceedings of the Koninklijke Nederlandse Akademie van Wetenschappen, Series B*, **83**, 158–172.

BUISONJÉ, P. H. DE. 1981. *Ctenochasma porocristata* nov. sp. from the Solnhofen Limestone, with some remarks on other Ctenochasmatidae. *Proceedings of the Koninklijke Nederlandse Akademie van Wetenschappen, Series B*, **84**, 411–436.

CAI, Z. & WEI, F. 1994. On a new pterosaur (*Zhejiangopterus linhaiensis* gen. et sp. nov.) from Upper Cretaceous in Linhai, Zhejiang, China. *Vertebrata Palasiatica*, **32**, 181–194.

CAMPOS, D. A. & KELLNER, A. W. A. 1985. Um novo exemplar de *Anhanguera blittersdorffi* (Reptilia, Pterosauria) da formação Santana, Cretáceo Inferior do Nordeste do Brasil. *Boletim de Resumos, 9th Congresso Brasileiro de Paleontologia*, 13.

CAMPOS, D. A. & KELLNER, A. W. A. 1997. Short note on the first occurrence of Tapejaridae in the Crato Member (Aptian), Santana Formation, Araripe Basin, Northeast Brazil. *Anais da Academia Brasileira de Ciências*, **69**, 83–87.

CARPENTER, K., UNWIN, D. M., CLOWARD, K. & MILES, C. 2003. A new scaphognathine pterosaur from the Upper Jurassic Morrison Formation of Wyoming, USA. *In*: BUFFETAUT, E. & MAZIN, J-M. (eds) *Evolution and Palaeobiology of Pterosaurs*. Geological Society, London, Special Publications, **217**, 45–54.

CARROLL, R. L. 1987. *Vertebrate Paleontology and Evolution*. W. H. Freeman and Company, New York, 698 pp.

CARROLL, R. L. 1997. *Patterns and Processes of Vertebrate Evolution*. Cambridge Palaeobiology Series, Cambridge University Press, Cambridge, England, 448 pp.

CASAMIQUELA, R. M. 1975. *Herbstosaurus pigmaeus* (Coeluria, Compsognathidae) n. gen. n. sp. del Jurasico medio del Nequéri (Patagonia septentrionali). Uno de los mas pequenos dinosaurios conocidos. *Acta Primero Congreso Argentino Paleontologia et Biostratigrafia*, **2**, 87–102.

CHIAPPE, L. M., KELLNER, A. W. A., RIVAROLA, D., DAVILA, S. & FOX, M. 2000. Cranial morphology of *Pterodaustro guinazui* (Pterosauria: Pterodactyloidea) from the Lower Cretaceous of Argentina. *Natural History Museum of Los Angeles County, Contributions in Science*, **483**, 1–19.

CLARK, J. M., HOPSON, J. A. & HERNANDEZ, R. R. 1996. An uncrushed 'rhamphorhynchoid' pterosaur from the Middle Jurassic of Mexico. *Journal of Vertebrate Paleontology*, **16**(3) (supplement), 28A.

CLARK, J. M., HOPSON, J. A., HERNANDEZ, R. R., FASTOVSKY, D. E. & MONTELLANO, M. 1997. A primitive pterosaur from the Jurassic of Mexico and its implications for pterosaur foot posture. *Journal of Vertebrate Paleontology*, **17**(3) (supplement), 38A–39A.

CLARK, J. M., HOPSON, J. A., HERNANDEZ, R. R., FASTOVSKY, D. E. & MONTELLANO, M. 1998. Foot posture in a primitive pterosaur. *Nature*, **391**, 886–889.

CLARK, J. M., MONTELLANO, M., HOPSON, J. A., HERNANDEZ, R. & FASTOVSKY, D. E. 1994. An Early or Middle Jurassic tetrapod assemblage from the La Boca Formation, northeastern Mexico. *In*: FRASER, N. C. & SUES, H.-D. (eds) *In the Shadow of the Dinosaurs: Early Mesozoic Tetrapods*. Cambridge University Press, Cambridge, 295–302.

CLOWARD, K. & CARPENTER, K. 2001. A new scaphognathine (Rhamphorhynchidae) pterosaur from the Upper Jurassic Morrison Formation, USA. *In*: *Two Hundred Years of Pterosaurs. A Symposium on the Anatomy, Evolution, Palaeobiology and Environments of Mesozoic Flying Reptiles, Toulouse, France, September 5–8, 2001. Strata, Serie 1*, **11**, 32.

COLBERT, E. H. 1969. A Jurassic pterosaur from Cuba. *American Museum Novitates*, **2370**, 1–26.

COMPANY, J., RUIZ-OMEÑACA, J. I. & PEREDA-SUBERBIOLA,

X. 1999. A long-necked pterosaur (Pterodactyloidea, Azhdarchidae) from the Upper Cretaceous of Valencia, Spain. *Geologie en Mijnbouw*, **78**, 319–333.

CUVIER, G. 1809. Sur le squelette fossile d'un reptile volant des environs d'Aichstedt, que quelques naturalistes ont pris pour un oiseau, et dont nous formons un genre de sauriens, sous le nom de Petro-Dactyle. *Annales du Musee d'Histoire Naturelle, Paris*, **13**, 424–437.

DALLA VECCHIA, F. M. 1993. *Cearadactylus*? *ligabuei* nov. sp., a new early Cretaceous (Aptian) pterosaur from Chapada do Araripe (northeastern Brazil). *Bollettino della Società Paleontologica Italiana*, **32**, 401–409.

DALLA VECCHIA, F. W. 1995. A new pterosaur (Reptilia, Pterosauria) from the Norian (Late Triassic) of Friuli (Northeastern Italy). Preliminary note. *Gortania*, **16**, 59–66.

DALLA VECCHIA, F. W. 1996. The pterosaur *Eudimorphodon rosenfeldi* Dalla Vecchia, 1995 from the Late Triassic (Middle Norian) of Northeastern Italy and its habitat. *Journal of Vertebrate Paleontology*, **16**(3) (supplement), 30A.

DALLA VECCHIA, F. W. 1998. New observations on the osteology and taxonomic status of *Preondactylus buffarinii* Wild, 1984 (Reptilia, Pterosauria). *Bollettino della Società Paleontologica Italiana*, **36**, 355–366.

DALLA VECCHIA, F. W. 2001. Triassic pterosaurs: unravelling the puzzle. *In*: *Two Hundred Years of Pterosaurs. A Symposium on the Anatomy, Evolution, Palaeobiology and Environments of Mesozoic Flying Reptiles, Toulouse, France, September 5–8, 2001. Strata, Serie 1*, **11**, 33–35.

DALLA VECCHIA, F. M., MUSCIO, G. & WILD, R. 1989. Pterosaur remains in a gastric pellet from the Upper Triassic (Norian) of Rio Seazza Valley (Udine, Italy). *Gortania*, **10**, 121–132.

DALLA VECCHIA, F. M., WILD, R., HOPF, H. & REITNER, J. 2002. A crested rhamphorhynchoid pterosaur from the Late Triassic of Austria. *Journal of Vertebrate Paleontology*, **22**, 196–199.

DAVILA, S. & CHIAPPE, L. 2000. Adición a la anatomía postcraneana de *Pterodaustro guinazui* (Pterodactyloidea), San Luis, Argentina. *XV Jornadas Argentinas de Paleontología de Vertebrados Resúmenes, Ameghiniana*, **37** (4) (Suplemento), 23R.

DÖDERLEIN, L. 1923. *Anurognathus Ammoni* ein neuer Flugsaurier. *Sitzungsberichte der Bayerischen Akademie der Wissenschaften, mathematisch-naturwissenschaftliche Abteilung*, 117–164.

DÖDERLEIN, L. 1929. Über *Anurognathus Ammoni* Döderlein. *Sitzungsberichte der Bayerischen Akademie der Wissenschaften, mathematisch-naturwissenschaftliche Abteilung*, 47–63.

DONG, Z. 1982. On a new pterosaur (*Huanhepterus quingyangensis*, gen. et sp. nov.) from Ordos, China. *Vertebrata Palasiatica*, **20**, 115–121. [In Chinese]

EATON, C. F. 1910. *Osteology of* Pteranodon. Connecticut Academy of Arts and Sciences, *Memoirs*, **2**, 1–38.

ELVIDGE, D. J. & UNWIN, D. M. 2001. A morphometric analysis of the hind limbs of pterosaurs. *In*: *Two Hundred Years of Pterosaurs. A Symposium on the Anatomy, Evolution, Palaeobiology and Environments of Mesozoic Flying Reptiles, Toulouse, France, September 5–8, 2001. Strata, Serie 1*, **11**, 36.

FABRE, J. A. 1974. Un nouveau Pterodactylidae sur le Gisement Portlandien de Canjuers (Var): *Gallodactylus canjuerensis*, nov. gen., nov. sp. *Comptes rendus de l'Académie des Sciences, (D), Paris*, **279** (D), 2011–2014.

FABRE, J. A. 1976. Un nouveau Pterodactylidae du Gisement de Canjuers (Var) *Gallodactylus canjuersensis*, nov. gen., nov. sp. *Annals de Paléontologie (Vertébrés)*, **62**, 35–70.

FASTNACHT, M. 2001. First record of *Coloborhynchus* (Pterosauria) from the Santana Formation (Lower Cretaceous) of the Chapada do Araripe, Brazil. *Paläontologische Zeitschrift* **75**, 23–36.

FRAAS, O. 1878. Über *Pterodactylus suevicus*, Qu. von Nusplingen. *Palaeontographica*, **25**, 163–174.

FREY, E. & MARTILL, D. M. 1994. A new pterosaur from the Crato Formation (Lower Cretaceous, Aptian) of Brazil. *Neues Jahrbuch für Geologie und Paläontologie, Abhandlung*, **194**, 379–412.

FREY, E. & MARTILL, D. M. 1996. A reappraisal of *Arambourgiania*: the world's largest flying animal. *Neues Jahrbuch für Geologie und Paläontologie, Abhandlung*, **199**, 221–247.

FREY, E. & MARTILL, D. M. 1998. Soft tissue preservation in a specimen of *Pterodactylus kochi* (Wagner) from the Upper Jurassic of Germany. *Neues Jahrbuch für Geologie und Paläontologie, Abhandlung*, **210**, 421–441.

FREY, E. & TISCHLINGER, H. 2000. Weichteilanatomie der Flugsaurierfuße und Bau der Scheitelkämme: Neue Pterosaurierfunde aus den Solnhofener Schichten (Bayern) und der Crato-Formation (Brasilien). *Archaeopteryx*, **18**, 1–16.

FREY, E., MARTILL, D. M. & BUCHY, M.-C. 2003. A new crested ornithocheirid from the Lower Cretaceous of northeastern Brazil and the unusual death of an unusual pterosaur. *In*: BUFFETAUT, E. & MAZIN, J-M. (eds) *Evolution and Palaeobiology of Pterosaurs*. Geological Society, London, Special Publications, **217**, 55–63.

GALTON, P. M. 1980. Avian-like tibiotarsi of pterodactyloids (Reptilia: Pterosauria) from the Upper Jurassic of East Africa. *Paläontologische Zeitschrift*, **54**, 331–342.

GAUTHIER, J. A. 1986. Saurischian monophyly and the origin of birds. *In*: PADIAN, K. (ed.) *The origin of birds and the evolution of flight*. California Academy of Sciences, Memoirs, **8**, 1–55.

GAUTHIER, J. A. & PADIAN, K. 1985. Phylogenetic, functional, and aerodynamic analyses of the origin of birds and their flight. *In*: HECHT, M. K., OSTROM, J. H., VIOHL, G. & WELLNHOFER, P. (eds) *The Beginnings of Birds: Proceedings of the International Archaeopteryx Conference, 1984, Eichstätt, Germany*. Freunde des Jura-Museums, Eichstätt, 185–197.

GAUTHIER, J. A. & QUEIROZ, K. DE 2001. Feathered dinosaurs, flying dinosaurs, crown dinosaurs and the name "Aves". *In*: GAUTHIER, J. A. & GALL, L. F. (eds) *New Perspectives on the Origin and Early Evolution of Birds: Proceedings of the International Symposium in Honor of John H. Ostrom; 1999 Feb. 12–14; New Haven, CT*. Peabody Museum of Natural History, Yale University, New Haven, 7–41.

HARKSEN, J. C. 1966. *Pteranodon sternbergi*, a new fossil

pterodactyl from the Niobara Cretaceous of Kansas. *Proceedings of the South Dakota Academy of Sciences*, **45**, 74–77.

HARRIS, J. D. & CARPENTER, K. 1996. A large pterodactyloid from the Morrison Formation (Late Jurassic) of Garden Park, Colorado. *Neues Jahrbuch für Geologie und Paläontologie, Monatsheft*, **1996**(8), 473–484.

HAZLEHURST, G. & RAYNER, J. M. V. 1992. Flight characteristics of Jurassic and Triassic Pterosauria: an appraisal based on wing shape. *Paleobiology*, **18**, 447–463.

HE, X., YAN, D., & SU, C. 1983. A new pterosaur from the Middle Jurassic of Dashanpu, Zigong, Sichuan. *Journal of the Chengdo College of Geology (supplement)*, **1**, 27–32.

HOOLEY, R.W. 1913. On the skeleton of *Ornithodesmus latidens* from the Wealden of Atherfield (Isle of Wight). *Quarterly Journal of the Geological Society, London*, **69**, 372–422.

HOWSE, S. C. B. 1986. On the cervical vertebrae of the Pterodactyloidea (Reptilia: Archosauria). *Zoological Journal of the Linnean Society, London*, **88**, 307–328.

HOWSE, S. C. B. & MILNER, A. R. 1993. *Ornithodesmus* – a maniraptoran theropod dinosaur from the Lower Cretaceous of the Isle of Wight, England. *Palaeontology*, **36**, 425–437.

HOWSE, S. C. B. & MILNER, A. R. 1995. The pterodactyloids from the Purbeck Limestone Formation of Dorset. *Bulletin of the Natural History Museum, London (Geology)*, **51**, 73–88.

HOWSE, S. C. B., MILNER, A. R. & MARTILL, D. M. 2001. Pterosaurs. *In*: MARTILL, D. M. & NAISH, D. (eds), *Dinosaurs of the Isle of Wight*. Palaeontological Association, London, 324–355.

HUENE, F. VON. 1951. Zwei ausgezeichnet erhaltene Exemplare von *Pterodactylus* im Natur-Museum Senckenberg. *Senckenbergiana*, **32**, 1–7.

HUXLEY, T. H. 1859. On *Rhamphorhynchus Bucklandi*, a pterosaurian of the Stonesfield Slates. *Quarterly Journal of the Geological Society, London*, **15**, 658–670.

IVAKHNENKO, M. F. & KORABELNIKOV, V. A. 1987. [*Life of the Past World*]. Prosveshchenie Press, Moscow, 253 pp. [In Russian]

JENKINS, F. A. JR., SHUBIN, N. H., GATESY, S. M. & PADIAN, K. 1999. A primitive pterosaur of Late Triassic age from Greenland. *Journal of Vertebrate Paleontology*, **29**(3) (supplement), 56A.

JENKINS, F. A. JR., SHUBIN, N. H., GATESY, S. M. & PADIAN, K. 2001. A diminutive pterosaur (Pterosauria: Eudimorphodontidae) from the Greenlandic Triassic. *In*: JENKINS, F. A. JR, SHAPIRO, M. D. & OWERKOWICV, T. *Studies in Organismic and Evolutionary Biology in Honor of A. W. Crompton*. Bulletin of the Museum of Comparative Zoology, Harvard, **156**, 151–170.

JENSEN, J.A. & PADIAN, K. 1989. Small pterosaurs and dinosaurs from the Uncompahgre fauna (Brushy Basin Member, Morrison Formation, ?Tithonian), Late Jurassic, western Colorado. *Journal of Paleontology*, **63**, 363–374.

JI, S. & JI, Q. 1997. Discovery of a new pterosaur in western Liaoning, China. *Acta Geologica Sinica*, **71**, 1–6. [In Chinese]

JI, S. & JI, Q. 1998. A new fossil pterosaur (Rhamphorhynchoidea) from Liaoning. *Jiangsu Geology*, **22**, 199–206. [In Chinese]

JI, S., JI, Q. & PADIAN, K. 1999. Biostratigraphy of new pterosaurs from China. *Nature*, **398**, 573–574.

KAUP, J. J. 1834. Versuch einer Eintheilung der Saugethiere in 6 Stämme und der Amphibien in 6 Ordnungen. *Isis*, **3**, 311–315.

KELLNER, A. W. A. 1984. Ocorrência de uma mandíbula de pterosauria (*Brasileodactylus araripensis*, nov. gen.; nov. sp.) na formação Santana, Cretáceo da chapada do Araripe, Ceará, Brasil. *33 Congresso Brasileiro de Geologia, Anais*, **2**, 578–590.

KELLNER, A. W. A. 1989. A new edentate pterosaur of the Lower Cretaceous from the Araripe Basin, Northeast Brazil. *Anais da Academia Brasileira de Ciências*, **61**, 439–446.

KELLNER, A. W. A. 1991. The Santana Formation pterosaurs. Supplementary Notes and Comments. *In*: MAISEY, J. G. (ed.). *Santana Fossils: an Illustrated Atlas*. T.F.H. Publications Inc., Neptune City, New Jersey, 370–371.

KELLNER, A. W. A. 1995. The relationships of the Tapejaridae (Pterodactyloidea) with comments on pterosaur phylogeny. *Sixth Symposium on Mesozoic Terrestrial Ecosystems and Biota, Beijing, Short Papers*, 73–77.

KELLNER, A. W. A. 1996a. Pterosaur phylogeny. *Journal of Vertebrate Paleontology*, **16**(3) (supplement), 45A.

KELLNER, A. W. A. 1996b. *Description of new material of Tapejaridae and Anhangueridae (Pterosauria, Pterodactyloidea) and discussion of pterosaur phylogeny*. PhD thesis, Columbia University. [Published by University Microfilms International].

KELLNER, A. W. A. & CAMPOS, D. DE A. 1988. Sobre un novo pterossauro com crista sagital da Bacia do Araripe, Cretáceo Inferior do Nordeste do Brasil. (Pterosauria, *Tupuxuara*, Cretaceo, Brasil). *Anais da Academia Brasileira de Ciências*, **60**, 459–469.

KELLNER, A. W. A. & CAMPOS, D. DE A. 1994. A new species of *Tupuxuara* (Pterosauria, Tapejaridae) from the Early Cretaceous of Brazil. *Anais da Academia Brasileira de Ciências*, **66**, 467–473.

KELLNER, A. W. A. & HASEGAWA, Y. 1993. Postcranial skeleton of *Tupuxuara* (Pterosauria, Pterodactyloidea, Tapejaridae) from the Lower Cretaceous of Brazil. *Journal of Vertebrate Paleontology*, **13**, 44A.

KELLNER, A. W. A. & LANGSTON, W. 1996. Cranial remains of *Quetzalcoatlus* (Pterosauria, Azhdarchidae) from Late Cretaceous sediments of Big Bend National Park, Texas. *Journal of Vertebrate Paleontology*, **16**, 222–231.

KELLNER, A. W. A. & TOMIDA, Y. 2000. Description of a New Species of Anhangueridae (Pterodactyloidea) with Comments on the Pterosaur Fauna from the Santana Formation (Aptian–Albian), northeastern Brazil. *National Science Museum, Tokyo, Monographs*, **17**, 1–135.

KNOLL, F. 2000. Pterosaurs from the Lower Cretaceous (?Berriasian) of Anoual, Morocco. *Annales de Paléontologie*, **86**, 157–164.

KOH, T.-P. 1937. Untersuchungen über die Gattung *Rhamphorhynchus*. *Neues Jahrbuch für Geologie und Paläontologie, Abhandlung*, **77**, 455–506.

KUHN, O. 1967. *Die fossile Wirbeltierklasse Pterosauria*. Oeben, Krailling bei München, 52 pp.

LANGSTON, W. JR. 1981. Pterosaurs. *Scientific American*, February 1981, 122–136.

LAWSON, D. A. 1975a. Pterosaur from the latest Cretaceous of west Texas: discovery of the largest flying creature. *Science*, **187**, 947–948.

LAWSON D. A. 1975b. Could pterosaurs fly? *Science*, **188**, 676–678.

LEE, Y. N. 1994. The Early Cretaceous pterodactyloid pterosaur *Coloborhynchus* from North America. *Palaeontology*, **37**, 755–763.

LEONARDI, G. & BORGOMANERO, G. 1985. *Cearadactylus atrox* nov. gen., nov. sp.: novo Pterosauria (Pterodactyloidea) da Chapada do Araripe, Ceara, Brasil. *Coletânea de Trabalhos Paleontológicos, Brasilia, Série Geologia*, **27**, 75–80.

MADER, B. & KELLNER A. W. A. 1999. A new anhanguerid pterosaur from the Cretaceous of Morocco. *Boletim do Museu Nacional, Rio de Janeiro, Nova Série, Geologia*, **45**, 1–11.

MARSH, O. C. 1876. Notice of new sub-order of Pterosauria. *American Journal of Science, series 3*, **11**, 507–509.

MARTILL, D. M. & FREY, E. 1998. A new pterosaur lagerstätte in N.E. Brazil (Crato Formation; Aptian, Lower Cretaceous): preliminary observations. *Oryctos*, **1**, 79–85.

MARTILL, D. M., Frey, E., Diaz, G. C. & Bell, C. M. 2000. Reinterpretation of a Chilean pterosaur and the occurrence of Dsungaripteridae in South America. *Geological Magazine*, **137**, 19–25.

MARTILL, D. M., FREY, E., SADAQAH, R. M. & KHOURY, H. N. 1998. Discovery of the holotype of the giant pterosaur *Titanopteryx philadelphiae* Arambourg 1959, and the status of *Arambourgiania* and *Quetzalcoatlus*. *Neues Jahrbuch für Geologie und Paläontologie, Abhandlung*, **207**, 57–76.

MEYER, H. VON 1834. Beiträge zur Petrefactenkunde. [*Gnathosaurus subulatus*, ein Saurus aus dem lithographischen Schiefer von Solnhofen.] *Museum Senckenbergianum*, **1** (3), 1–26.

MEYER, H. VON 1847. *Homoeosaurus maximiliani* und *Rhamphorhynchus* (*Pterodactylus*) *longicaudus*, zwei fossile reptilien aus dem Kalkschiefer von Solenhofen im Naturalienkabinet seiner kaiserlich hoheit des Herzogs Maximilian von Leuchtenberg zu Eichstaedt. 3–11. Schmerber Verlag, Frankfurt am Main.

MEYER, H. VON 1851. *Ctenochasma römeri*. *Paläontographica*, **2**, 82–84.

MEYER, H. VON 1854. [Letter on various fossils.] *Neues Jahrbuch für Mineralogie, Geologie und Paläontologie*, (1854), 47–58.

MEYER, H. VON 1859. *Zur Fauna der Vorwelt. Vierte Abteilung: Reptilien aus dem lithographischen Schiefer des Jura in Deutschland und Frankreich*. H. Keller, Frankfurt am Main, 1–114.

MILLER, H. W. 1972. The taxonomy of the *Pteranodon* species from Kansas. *Transactions of the Kansas Academy of Sciences*, **74**, 1–19.

NESOV, L. A. 1984. [Upper Cretaceous pterosaurs and birds from Central Asia]. *Paleontologicheskii Zhurnal*, **1984** (1), 47–57. [In Russian]

NESOV, L. A. 1986. [The first finding of the Late Cretaceous bird *Ichthyornis* in the old world and some other bird bones from the Cretaceous and the Paleogene of Middle Asia]. *In*: POTAPOVA, R. L. (ed.) [*Ecological and Faunistic Investigation of Birds*]. *Proceedings of the Zoological Institute, Leningrad*, **147**, 31–38. [In Russian]

NESOV, L. A. 1991. [Giant flying reptiles of the family Azhdarchidae: I. Morphology and systematics]. *Vestnik Leningradskogo Universiteta, Series 7*, **2** (14), 14–23. [In Russian]

NESOV, L. A. & YARKOV, A. A. 1989. [New birds from the Cretaceous and Palaeogene of the USSR and some remarks on the origin and evolution of the Class Aves]. *In*: POTAPOV, R. L. (ed.) [*Faunistic and Ecological Studies of Eurasian Birds*]. *Proceedings of the Zoological Institute, Leningrad*, **197**, 78–97. [In Russian]

NEWTON, E. T. 1888. On the skull, brain and auditory organ of a new species of pterosaurian (*Scaphognathus purdoni*) from the Upper Lias near Whitby, Yorkshire. *Philosophical Transactions of the Royal Society, London*, **43**, 436–440.

NOPCSA, F. VON. 1928. The genera of reptiles. *Palaeobiologica*, **1**, 163–188.

OWEN, R. 1851. On a new species of pterodactyle (*Pterodactylus compressirostris*, Owen) from the Chalk; with some remarks on the nomenclature of the previously described species. *Proceedings of the Zoological Society, London*, 21–34.

OWEN, R. 1859. On remains of new and gigantic species of pterodactyle (*Pter. fittoni* and *Pter. sedgwickii*) from the Upper Greensand near Cambridge. *British Association for the Advancement of Science, Reports*, **28** (1858), 98–103.

OWEN, R. 1861. *Supplement (No. III) to the Monograph on the Fossil Reptilia of the Cretaceous Formations*. Monographs of the Palaeontographical Society, London, 1–19.

OWEN, R. 1870. *Monograph on the fossil Reptilia of the Liassic Formations*. Monographs of the Palaeontographical Society, London, **27**, 41–81.

OWEN, R. 1874. *Monograph on the Fossil Reptilia of the Mesozoic Formations*. Monographs of the Palaeontographical Society, London, 1–14.

PADIAN, K. 1983. Osteology and functional morphology of *Dimorphodon macronyx* (Buckland) (Pterosauria: Rhamphorhynchoidea) based on new material in the Yale Peabody Museum. *Postilla*, **189**, 1–44.

PADIAN, K. 1984a. A large pterodactyloid pterosaur from the Two Medicine Formation (Campanian) of Montana. *Journal of Vertebrate Paleontology*, **4**, 516–524.

PADIAN, K. 1984b. The origin of pterosaurs. *In*: REIF, W-E. & WESTPHAL, F. (eds) *Proceedings of the 3rd Symposium on Mesozoic Terrestrial Ecosystems and Biotu, Tübingen*, 163–168.

PADIAN, K. 1986. A taxonomic note on two pterodactyloid families. *Journal of Vertebrate Paleontology*, **6**, 289.

PADIAN, K. 1997. Pterosauromorpha. *In*: CURRIE, P. J. & PADIAN, K. (eds), *Encyclopedia of Dinosaurs*. Academic Press, 617–618.

PADIAN, K., RICQLÈS, A. J. DE & HORNER, J. R. 1995. Bone histology determines identification of a new fossil taxon of pterosaur (Reptilia: Archosauria). *Comptes rendus de l'Académie des Sciences, Paris, Séries II*, **320**, 77–84.

PADIAN, K. & SMITH, M. 1992. New light on Late Cretaceous pterosaur material from Santana. *Journal of Vertebrate Paleontology*, **12**, 87–92.

PADIAN, K. & WARHEIT, K. I. 1989. Morphometrics of the pterosaur wing: one sharp division, few trends. *Journal of Vertebrate Paleontology*, **9**(3) (supplement), 34A.

PADIAN, K. & WILD, R. 1992. Studies of Liassic Pterosauria: I. The holotype and referred specimens of the Liassic pterosaur *Dorygnathus banthensis* (Theodori) in the Petrefaktensammlung Banz, northern Bavaria. *Palaeontographica* A, **225**, 59–77.

PETERS, D. 1997. A new phylogeny for the Pterosauria. Journal of Vertebrate Paleontology, **17**(3) (supplement), 69A.

PETERS, D. 2000. A redescription of four prolacertiform genera and implications for pterosaur phylogenesis. *Rivista Italiana di Paleontologia e Stratigrafia*, **106**, 293–336.

PETERS, D. 2001. Sharov was right: *Sordes* is a dimorphodontid. *Journal of Vertebrate Paleontology*, **21**(3) (supplement), 88A.

PETRONIEVICS, B. 1928. Bemerkungen über *Anurognathus* Döderlein. *Anatomisches Anzeiger*, **65**, 214–222.

PLIENINGER, F. 1895. *Campylognathus Zitteli*, ein neuer Flugsaurier aus dem obersten Lias Schwabens. *Paläontographica*, **41**, 193–222.

PLIENINGER, F. 1901. Beiträge zur Kenntnis der Flugsaurier. *Paläontographica*, **48**, 65–90.

PLIENINGER, F. 1907. Die Pterosaurier der Juraformation Schwabens. *Paläontographica*, **53**, 209–313.

PLIENINGER, F. 1930. Pterosauria. *In*: POMPECKJ, J. F. (ed.) *Fossilium Catalogus 1. Animalia*. W. Junk, Berlin, Part 45, 1–84.

PRICE, L. I. 1953. A presenca de Pterosauria no Cretaceo Superior do Estada da Paraiba. *Divisao de Geologia e Mineralogia Notas Preliminaries e Estudos*, **71**, 1–10.

PRICE, L. I. 1971. A presença de Pterosauria no Cretáceo Inferior da Chapada do Araripe, Brasil. *Anais da Academia Brasileira de Ciências*, **43**, 451–461.

QUEIROZ, K. DE & GAUTHIER, J. A. 1992. Phylogenetic taxonomy. *Annual Review of Ecology and Systematics*, **23**, 449–480.

QUENSTEDT, F. A. 1855. Über *Pterodactylus suevicus* im lithographischen Schiefer Württembergs. 1–52. Universität Tübingen, Tübingen.

RYABININ, A. N. 1948. [Remarks on a flying reptile from the Jurassic of the Karatau]. *Transactions of the Palaeontological Institute, Academy of Science, USSR*, **15**, 86–93. [In Russian]

ROMER, A. S. 1956. *Osteology of the Reptiles*. University of Chicago Press, Chicago, 772 pp.

SALÉE, A. 1928. L'exemplaire de Louvain de *Dorygnathus banthensis* Theodori sp. *Mémoires de l'Institut de Géologique de l'Université de Louvain*, **4**, 289–344.

SANCHEZ, T. M. 1973. Redescription del craneo y mandibulas de *Pterodaustro guinazui* Bonaparte (Pterodactyloidea, Pterodaustriidae). *Ameghiniana*, **10**, 313–325.

SAYÃO, J. M. & KELLNER, A. W. A. 2000. Description of a pterosaur rostrum from the Crato Member, Santana Formation (Aptian-Albian) northeastern Brazil. *Boletim do Museu Nacional, Rio de Janeiro, Nova Serie, Geologia*, **54**, 1–8.

SCHALLER, D. 1985. Wing evolution. *In*: HECHT, M. K., OSTROM, J. H., VIOHL, G. & WELLNHOFER, P. (eds), *The Beginnings of Birds: Proceedings of the International Archaeopteryx conference, 1984, Eichstätt, Germany*. Freunde des Jura-Museums Eichstätt, 333–348.

SEELEY, H. G. 1869. *Index to the Fossil Remains of Aves, Ornithosauria and Reptilia, from the Secondary System of Strata arranged in the Woodwardian Museum of the University of Cambridge*. Deighton, Bell & Co., Cambridge, xxiii + 143 pp.

SEELEY, H. G. 1870. *The Ornithosauria: An Elementary Study of the Bones of Pterodactyles, made from Fossil Remains found in the Cambridge Upper Greensand, and Arranged in the Woodwardian Museum of the University of Cambridge*. Deighton, Bell & Co., Cambridge, xii + 135 pp.

SEELEY, H. G. 1871. Additional evidence of the structure of the head in ornithosaurs from the Cambridge Upper Greensand; being a supplement to 'The Ornithosauria'. *Annals and Magazine of Natural History*, **7** (37), 20–36.

SEELEY, H. G. 1880. On *Rhamphorhynchus Prestwichi*, Seeley, an ornithosaurian from the Stonesfield Slate of Kineton. *Quarterly Journal of the Geological Society, London*, **36**, 27–30.

SEELEY, H. G. 1891. On the shoulder girdle in Cretaceous Ornithosauria. *Annals and Magazine of Natural History*, **6** (7), 438–445.

SEELEY, H. G. 1901. *Dragons of The Air. An Account of Extinct Flying Reptiles*. Methuen, London, xiii + 239 pp.

SERENO, P. C. 1991. Basal archosaurs. Phylogenetic relationships and functional implications. Society of Vertebrate Paleontology, Memoir 2. *Journal of Vertebrate Paleontology*, **11**(4) (supplement), 1–53.

SERENO, P. C. 1996. The phylogenetic position of pterosaurs within Archosauria. *Journal of Vertebrate Paleontology*, **16**(3) (supplement), 65A.

SERENO, P. C. & ARCUCCI, A. 1990. The monophyly of crurotarsal archosaurs and the origin of bird and crocodile ankle joints. *Neues Jahrbuch für Geologie und Paläontologie, Abhandlung*, **180**, 21–52.

SERENO, P. C. & NOVAS, F. E. 1990. Dinosaur origins and the phylogenetic position of pterosaurs. *Journal of Vertebrate Paleontology*, **9**(3) (supplement), 42A.

SHAROV, A. G. 1971. [New flying reptiles from the Mesozoic of Kazakhstan and Kirghizia]. *Transactions of the Palaeontological Institute, Academy of Sciences, USSR*, **130**, 104–113. [In Russian]

SMITH, A. B. 1994. *Systematics and the Fossil Record*. Blackwell Scientific, Oxford, vii + 223 pp.

SMITH WOODWARD, A. 1902. On two skulls of ornithosaurian *Rhamphorhynchus*. *Annals and Magazine of Natural History, series 7*, **9**, 1–5.

STRAND, E. 1928. Miscellanea nomenclatorica Zoologica et Paleontologica. *Archiv für Naturgeschichte*, **92**, 30–75.

SWOFFORD, D. L. 1993. *PAUP: Phylogenetic Analysis Using Parsimony, Version 3.1.1*. Illinois Natural History Survey, Champaign, Illinois.

TAQUET, P. 1972 Un crane de *Ctenochasma* (Pterodactyloidae) du Portlandien inférieur de la Haute-

Marne, dans les collections du Musée de Saint-Dizier. *Comptes rendus de l'Académie des Sciences, Paris*, **274**, 362–364.

TISCHLINGER, H. 1994. Überlegungen zur Lebensweise der Pterosaurier anhand eines verheilten Oberschenkelbruches bei *Pterodactylus kochi* (Wagner). *Archaeopteryx*, **11**, 63–71.

UNWIN, D. M. 1988a. A new pterosaur from the Kimmeridge Clay of Kimmeridge, Dorset. *Proceedings of the Dorset Natural History and Archaeological Society*, **109**, 150–153.

UNWIN, D. M. 1988b. New remains of the pterosaur *Dimorphodon* (Pterosauria: Rhamphorhynchoidea) and the terrestrial ability of early pterosaurs. *Modern Geology*, **13**, 57–68.

UNWIN, D. M. 1991. *The morphology, systematics and evolutionary history of pterosaurs from the Cretaceous Cambridge Greensand of England*. PhD. thesis, Reading University.

UNWIN, D. M. 1992. The phylogeny of the Pterosauria. *Journal of Vertebrate Paleontology*, **12**(3) (supplement), 57A.

UNWIN, D. M. 1995a. Preliminary results of a phylogenetic analysis of the Pterosauria (Diapsida: Archosauria). *In*: SUN, A. & WANG, Y. (eds), *Sixth Symposium on Mesozoic Terrestrial Ecosystems and Biota, Beijing. Short Papers*, 69–72.

UNWIN, D. M. 1995b. A reassessment, using ontogenetic data, of the systematic status of pterosaurs (Reptilia: Diapsida) from the Upper Jurassic Lithographic Limestones of southern Germany. *II International Symposium on Lithographic Limestones, Lleida-Cuenca, Spain, Ediciones de la Universidad Autónoma de Madrid*, 151–154.

UNWIN, D. M. 1996. The fossil record of Middle Jurassic pterosaurs. *In*: MORALES, M. (ed.) *The Continental Jurassic. Museum of Northern Arizona Bulletin*, **60**, 291–304.

UNWIN, D. M. 1999. Pterosaurs: back to the traditional model? *Trends in Evolution and Ecology*, **14**, 263–268.

UNWIN, D. M. 2001. An overview of the pterosaur assemblage from the Cambridge Greensand (Cretaceous) of Eastern England. *Mitteilungen aus dem Museum für Naturkunde, Berlin, Geowissenschaftlichen Reihe*, **4**, 189–222.

UNWIN, D. M. 2002. On the systematic relationships of *Cearadactylus atrox* Leonardi and Borgomanero, 1985, an enigmatic Lower Cretaceous pterodactyloid pterosaur from the Santana Formation of Brazil. *Mitteilungen aus dem Museum für Naturkunde, Berlin, Geowissenschaftlichen Reihe*, **5**, 237–261

UNWIN, D. M. & BAKHURINA, N. N. 1994. *Sordes pilosus* and the nature of the pterosaur flight apparatus. *Nature*, **371**, 62–64.

UNWIN, D. M. & BAKHURINA, N. N. 2000. Pterosaurs from Russia, Middle Asia and Mongolia. *In*: BENTON, M. J., SHISKIN, M., UNWIN, D. M. & KUROCHKIN, E. (eds) *The Age of Dinosaurs in Russia and Mongolia*. Cambridge University Press, Cambridge, 420–433.

UNWIN, D. M., FREY, E., MARTILL, D. M., CLARKE, J. & RIESS, J. 1996. On the nature of the pteroid in pterosaurs. *Proceedings of the Royal Society, London: Biological Sciences*, **263**, 45–52.

UNWIN, D. M. & HEINRICH, W.-D. 1999. On a pterosaur jaw from the Upper Jurassic of Tendaguru (Tanzania). *Mitteilungen aus dem Museum für Naturkunde, Berlin, Geowissenschaftlichen Reihe*, **2**, 121–134.

UNWIN, D. M. & LÜ, J. 1997. On *Zhejiangopterus* and the relationships of pterodactyloid pterosaurs. *Historical Biology*, **12**, 199–210.

UNWIN, D. M., LÜ, J. & BAKHURINA, N. N. 2000. On the systematic and stratigraphic significance of pterosaurs from the Lower Cretaceous Yixian Formation (Jehol Group) of Liaoning, China. *Mitteilungen aus dem Museum für Naturkunde, Berlin, Geowissenschaftlichen Reihe*, **3**, 181–206.

VIOHL, G. 2000. *Drachen der Lüfte. Entwicklung und Leben der Flugsaurier. Führer zur gleichnamigen Ausstellung des Jura-Museums vom 9 Juni 2000 bis zum 7 Januar 2001*. Pfeil Verlag, München, 48 pp.

VISCARDI, P., DYKE, G. J., WILKINSON, M. & RAYNER, J. M. V. 1999. Missing data and the phylogeny of the Pterosauria. *Journal of Vertebrate Paleontology*, **19**(3) (supplement), 83A.

WAGNER, J. A. 1858. Neue Beiträge zur Kenntnis der urweltlichen Fauna des lithographischen Schiefers. 1. Abt. Saurier. *Abhandlung der Bayerischen Akademie der Wissenschaften Mathematisch-Physik Klasse*, **8**, 413–528.

WAGNER, J. A. 1860. Bemerkungen über die Arten von Fischen und Saurien welche im unteren wie im oberen Lias vorkommen sollen. *Sitzungsberichte der Konigliche Bayerischen Akademie der Wissenschaften Mathematisch-physik Klasse*, (1860), 36–52.

WAGNER, J. A. 1861. Übersicht über die fossilen Reptilen des lithographischen Schiefers in Bayern nach ihren Gattungen und Arten. *Sitzungsberichte der Konigliche Bayerischen Akademie der Wissenschaften Mathematisch-Physik Klasse* (1861), 467–535.

WANG, X. & LÜ, J. 2001. Discovery of a pterodactylid pterosaur from the Yixian Formation of western Liaoning, China. *Chinese Science Bulletin*, **46**, 1112–1117.

WANG, X., ZHOU, Z., ZHANG, F. & XU, X. 2002. A nearly completely articulated rhamphorhynchoid pterosaur with exceptionally well-preserved wing membranes and 'hairs' from Inner Mongolia, northeast China. *Chinese Science Bulletin*, **47**, 225–232.

WELLNHOFER, P. 1968. Über *Pterodactylus kochi* (Wagner 1837). *Neues Jahrbuch für Geologie und Paläontologie, Abhandlung*, **132**, 97–126.

WELLNHOFER, P. 1970. Die Pterodactyloidea (Pterosauria) der Oberjura Plattenkalke Süddeutschlands. *Bayerische Akademie der Wissenschaften, Mathematisch-naturwissentschaftliche Klasse, Abhandlungen*, **141**, 1–133.

WELLNHOFER, P. 1974. *Campylognathoides liasicus* (Quenstedt), an Upper Liassic pterosaur from Holzmaden. The Pittsburgh specimen. *Annals of the Carnegie Museum, Pittsburgh*, **45**, 5–34.

WELLNHOFER, P. 1975a. Die Rhamphorhynchoidea (Pterosauria) der Oberjura-Plattenkalke Süddeutschlands. III. Palökologie und Stammesgeschichte. *Palaeontographica A*, **149**, 1–30.

WELLNHOFER, P. 1975b. Die Rhamphorhynchoidea (Pterosauria) der Oberjura-Plattenkalke Süddeutschlands. II. Systematische Beschreibung. *Palaeontographica A*, **148**, 132–186.

WELLNHOFER, P. 1975c. Die Rhamphorhynchoidea (Pterosauria) der Oberjura-Plattenkalke Süddeutschlands. I. Allgemeine Skelettmorphologie. *Palaeontographica A*, **148**, 1–33.

WELLNHOFER, P. 1978. Pterosauria. *In*: WELLNHOFER, P. (ed.) *Handbuch der Paläoherpetologie*, Gustav Fischer, Stuttgart, Teil 19, 82 pp.

WELLNHOFER, P. 1985. Neue Pterosaurier aus der Santana Formation (Apt) der Chapada do Araripe, Brasilien. *Palaeontographica A*, **187**, 105–182.

WELLNHOFER, P. 1987. New crested pterosaurs from the Lower Cretaceous of Brazil. *Mitteilungen der Bayerischen Staatssammlung für Paläontologie und Historische Geologie*, **27**, 175–186.

WELLNHOFER, P. 1991a. *The Illustrated Encyclopedia of Pterosaurs*. Salamander Books, London, 192 pp.

WELLNHOFER, P. 1991b. Weitere Pterosaurierfunde aus der Santana-Formation (Apt) der Chapada do Araripe, Brasilien. *Palaeontographica A*, **215**, 43–101.

WELLNHOFER, P. 1991c. The Santana Formation pterosaurs. In Maisey, J. G. (ed.), *Santana fossils: An Illustrated Atlas*. T.F.H. Publications Inc., Neptune City, New Jersey, 351–370.

WELLNHOFER, P. 2001. A Late Triassic pterosaur from the Northern Calcareous Alps. *Two Hundred Years of Pterosaurs. A Symposium on the Anatomy, Evolution, Palaeobiology and Environments of Mesozoic Flying Reptiles, Toulouse, France, September 5–8, 2001. Strata, Serie 1*, **11**, 99–100.

WELLNOFER, P. 2003. A Late Triassic pterosaur from the Northern Calcareous Alps (Tyrol, Austria). *In*: BUFFETAUT, E. & MAZIN, J-M. (eds) *Evolution and Palaeobiology of Pterosaurs.* Geological Society, London, Special Publications, **217**, 5–22.

WELLNHOFER, P. & KELLNER, A. W. A. 1991. The skull of *Tapejara wellnhoferi* Kellner (Reptilia: Pterosauria) from the Lower Cretaceous Santana Formation of the Araripe Basin, Northeastern Brazil. *Mitteilungen der Bayerischen Staatssammlung für Paläontologie und Historische Geologie*, **31**, 89–106.

WILD, R. 1971. *Dorygnathus mistelgauensis* n. sp., ein neuer Flugsaurier aus dem Lias Epsilon von Mistelgau (Fränkischer Jura). *Geologische Blatter Nord Ost-Bayern*, **21**, 178–195.

WILD, R. 1975. Ein Flugsaurier-rest aus dem Lias Epsilon (Toarcium) von Erzingen (Schwäbischer Jura). *Stuttgarter Beiträge zur Naturkunde, serie B*, **17**, 1–16.

WILD, R. 1978. Die Flugsaurier (Reptilia, Pterosauria) aus der Oberen Trias von Cene bei Bergamo. *Bollettino della Società Paleontologica Italiana*, **17**, 176–256.

WILD, R. 1984a. A new pterosaur (Reptilia: Pterosauria) from the Upper Triassic (Norian) of Friuli, Italy. *Gortania*, **5**, 45–62.

WILD, R. 1984b. Flugsaurier aus der Ober Trias von Italien. *Naturwissenschaften*, **71**, 1–11.

WILD, R. 1994. A juvenile specimen of *Eudimorphodon ranzii* Zambelli (Reptilia, Pterosauria) from the Upper Triassic (Norian) of Bergamo. *Rivista del Museo Civico di Scienze Naturali 'E. Caffi' Bergamo*, **16**, 95–120.

WILKINSON, M. 1995. Coping with abundant missing entries in phylogenetic inference using parsimony. *Systematic Biology*, **44**, 501–514.

WILKINSON, M. & BENTON, M. J. 1995. Missing data and rhynchosaur phylogeny. *Historical Biology*, **10**, 137–150.

WILLISTON, S. W. 1902. On the skull of *Nyctodactylus*, an Upper Cretaceous pterodactyl. *Journal of Geology*, **10**, 520–531.

WILLISTON, S. W. 1903. On the osteology of *Nyctosaurus* (*Nyctodactylus*), with notes on American pterosaurs. *Field Columbian Museum, Geological Series, Publication Series* 2, 125–163.

WIMAN, C. 1923. Über *Dorygnathus* und andere Flugsaurier. *Bulletin of the Geological Institute of the University of Uppsala*, **19**, 23–54.

WINKLER, T. C. 1883. Note sur une espece de *Rhamphorhynchus* du Musée Teyler. *Archives du Musée Teyler, series 2*, **1**.

YOUNG, C. C. 1964. On a new pterosaurian from Sinkiang, China. *Vertebrata Palasiatica*, **8**, 221–255.

YOUNG, C. C. 1973. Pterosaurian fauna from Wuerho, Sinkiang. *In*: *Reports of Palaeontological Expeditions to Sinkiang, vol. III.* Institute of Vertebrate Palaeontology and Palaeoanthropology, Academia Sinica, Special Publications, **11**, 18–34. [In Chinese]

ZAMBELLI, R. 1973. *Eudimorphodon ranzii* gen. nov., sp. nov., uno pterosauro triassico (nota preliminare). *Rendiconti dell Istituto Lombardo di Scienze e Lettre (B)*, **107**, 27–32.

ZITTEL, K. A. 1882. Über Flugsaurier aus dem lithographischen Schiefer Bayerns. *Paläontographica*, **29**, 47–80.

Morphological evolution of the pectoral girdle of pterosaurs: myology and function

S. CHRISTOPHER BENNETT

College of Chiropractic, University of Bridgeport, Bridgeport, Connecticut, CT 06601–2449, USA (e-mail: cbennett@bridgeport.edu)

Abstract: The musculature of the pectoral region of representative rhamphorhynchoid (*Campylognathoides*) and large pterodactyloid (*Anhanguera*) pterosaurs was reconstructed in order to examine the function of various muscles and the functional consequences of the evolution of the advanced pectoral girdle of large pterodactyloids. The reconstructions suggest that *m. supracoracoideus* was not an elevator of the wing, but instead depressed and flexed the humerus. *m. latissimus dorsi*, *m. teres major*, *m. deltoides scapularis*, and *m. scapulohumeralis anterior* were wing elevators. Comparison of the origin, insertion and function of muscles in the rhamphorhynchoid and the large pterodactyloid suggests that the evolution of the advanced pectoral girdle: (1) straightened the pull of *m. pectoralis*, *m. deltoides scapularis* and *m. teres major*, improving their function in wing elevation; (2) allowed ligaments rather than muscles to resist the tendency of those muscles to move the scapula; and (3) braced the pectoral girdle against the vertebral column so that the tendency of *m. latissimus dorsi* and of aerodynamic lift on the wing to move the scapulocoracoid medially and dorsally, thereby compressing the thorax, could be resisted. The osteological and myological complexity of the advanced pectoral girdle, its uniqueness among tetrapods and its association with other complex osteological features argue that the advanced pectoral girdle is a synapomorphy complex of a single clade of large pterodactyloids, rather than a mere correlate of large size evolved convergently in various lineages.

The pectoral girdle of basal pterosaurs (e.g. *Eudimorphodon*, *Peteinosaurus*; Wild 1978, 1993) consisted of a long strap-like scapula that passed posteriorly over the ribcage; an elongate strut-like coracoid that articulated with the sternum; the scapula and coracoid fused at an acute angle and together forming a deep, saddle-shaped glenoid fossa for articulation with the humerus; and a broad sternum formed from the fusion of the interclavicle, paired clavicles and paired sternal plates. The pectoral girdle was superficially similar to that of extant birds (e.g. *Aquila*, *Corvus*; Fürbringer 1900; Seeley 1901; Padian 1983a, 1983b). Based on that similarity, Fürbringer (1900) suggested that *m. subcoracoideus* and *m. supracoracoideus* of pterosaurs might have originated in part from the sternum, as in birds, while others (Kripp 1943; Padian 1983b, 1985) interpreted the pectoral girdle of pterosaurs as functionally like that of birds. Thus, it was proposed that *m. supracoracoideus*, originating on the sternum and having the direction of its pull reversed by a pulley-like acrocoracoid process, was the primary elevator of the wing in pterosaurs. This interpretation has been widely accepted (e.g. Wellnhofer 1991a, 1991b; Monastersky 2001), yet has not been examined critically.

The essential morphology of the pterosaurian pectoral girdle did not change as the rhamphorhynchoids (i.e. non-pterodactyloid pterosaurs) radiated throughout the Jurassic or as the pterodactyloids evolved and eventually replaced the rhamphorhynchoids. However, in the large pterodactyloids (e.g. pteranodontoids, dsungaripterids and azhdarchoids) the pectoral girdle was modified from the condition in basal pterosaurs in that the posterior end of the scapula was rotated medially and articulated with the notarium formed of fused anterior dorsal vertebrae, and the coracoid was likewise rotated outward so that it was directed laterally rather than anterolaterally. Although this unusual condition was first described over 100 years ago (Marsh 1876; Seeley 1891), little has been done to understand the functional consequences of the change from the condition in basal pterosaurs, and it has been generally described as simply strengthening and stabilizing the pectoral girdle (e.g., Wellnhofer 1991b; Monastersky 2001).

This paper presents reconstructions of the pectoral musculature of a representative rhamphorhynchoid (*Campylognathoides*) and a representative large pterodactyloid (*Anhanguera*), in an attempt to determine the function of *m. supracoracoideus* and the functional consequences of the evolution of the advanced pectoral girdle. Reconstructions of the musculature of fossil vertebrates have been made for many years (e.g. Romer 1923, 1927; Galton 1969), and have been somewhat controversial (see Dilkes 2000 for a thorough review). Various authors have reconstructed a few muscles here and there on pterosaurs: Fürbringer (1900) discussed points of attachment of some pectoral muscles, Kripp (1943), Wellnhofer (1991b) and Monastersky (2001) illustrated a few reconstructed pectoral muscles, and

From: BUFFETAUT, E. & MAZIN, J-M. (eds) 2003. *Evolution and Palaeobiology of Pterosaurs*. Geological Society, London, Special Publications, **217**, 191–215. 0305–8719/03/$15

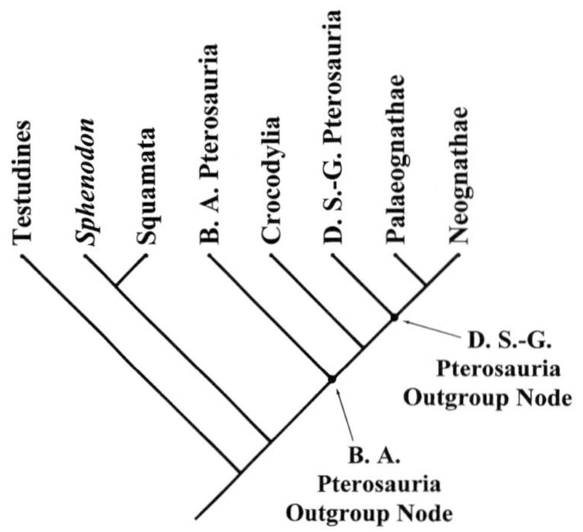

Fig. 1. Cladogram illustrating two possible positions of the Pterosauria relative to extant reptiles. If the Pterosauria was the major sister-group of the Dinosauria (i.e. D.S.-G. Pterosauria in the cladogram), then it is bracketed by extant crocodilians and birds, and the character states of muscles is inferred from the condition at the D.S.-G. outgroup node. Alternatively, if pterosaurs were basal archosauromorphs (i.e. B.A. Pterosauria in the cladogram), then they are bracketed by extant lepidosaurs and crocodilians, and the character states of muscles is inferred from the condition at the B.A. outgroup node.

Padian (1983b) reconstructed *m. supracoracoideus*, but no attempt has been made to reconstruct all the muscles of the pectoral region of pterosaurs.

Some authors (McGowan 1979, 1982; Nicholls & Russell 1985; but see Bryant & Seymour 1990) have argued that muscle scars do not accurately reflect the extent of muscle attachments in extant vertebrates and so are unreliable sources of information for the reconstruction of the musculature of fossil vertebrates. However, while there may be some truth in this, that is no reason to ignore completely the information available in the muscle scars that can be identified on the bones of fossil vertebrates, and Dilkes (2000) has shown that muscles can be reconstructed with some confidence from muscle scars and bony processes using the extant phylogenetic bracket method of Witmer (1995; see also Bryant & Russell 1992).

The use of the extant phylogenetic bracket method is complicated in this case by the fact that the relationship of the Pterosauria to other diapsids is not clear (Fig. 1). For 15 years it has been generally accepted that the Pterosauria was the major sister-group of the Dinosauria (Padian 1984a; Sereno 1991; Benton 1999). Recently, the author presented a phylogenetic analysis which suggested that the Pterosauria were basal archosauromorphs (Bennett 1996), while others have placed the Pterosauria within the Prolacertiformes (Peters 2000; Unwin 2000). There are problems with the latter cladistic analyses, but based on his analysis the author prefers the interpretation of the Pterosauria as basal archosauromorphs to that of the Pterosauria as the sister-group of the Dinosauria, and so will reconstruct muscles based on that relationship. However, where the different interpretations of relationship would affect the reconstruction they are noted and reference is made to 'basal archosauromorph pterosaurs' and 'dinosaur sister-group pterosaurs' in order to differentiate between the two different interpretations of pterosaur relationship.

Institution Abbreviations: AMNH, American Museum of Natural History, New York, USA; BSP, Bayerische Staatssammlung für Paläontologie und historische Geologie, Munich, Germany; BYU, Brigham Young University, Provo, Utah, USA; CM, Carnegie Museum of Natural History, Pittsburgh, USA; KUVP, Museum of Natural History, University of Kansas, USA; NSM, National Science Museum, Tokyo, Japan; PTH, Jura-Museum, Eichstätt, Germany; TMM, Texas Memorial Museum, University of Texas, Austin, USA; USNM, US National Museum, Washington, D.C., USA; and YPM, Peabody Museum of Natural History, Yale University, New Haven, USA.

Methodology

The extant phylogenetic bracket method of Witmer (1995) relies on the association between soft tissues (e.g. muscles) and their osteological correlates (e.g. processes and muscle scars) in the extant relatives of the fossil taxon of interest to permit the presence of soft tissues in the fossil taxon to be inferred from the presence of the osteological correlates. The character states of the closest extant relatives of the fossil taxon of interest are used to infer the character state of the most recent common ancestor of the extant taxa (at the bracket node), which is also an ancestor of the fossil taxon of interest. The character state at the bracket node is then used to infer the likely character state in the most recent common ancestor of the fossil taxon of interest and its closest extant relative (at the outgroup node), and that character state is applied to the fossil taxon of interest. Note that, depending on the character states in the extant taxa, the inference may be decisive or equivocal as to the character state at the outgroup node.

This paper firstly describes the processes and muscle scars on the bones of the pectoral region of pterosaurs. Secondly, there is a discussion of the character states of the muscles that can be inferred from the character states found in the extant relatives of pterosaurs, and lastly an attempt is made to match the inferred characters states of the muscles to the processes and muscle scars. As noted above, there is uncertainty as to the relationships of the Pterosauria to other diapsids. If the Pterosauria were the major sister-group of the Dinosauria, then the closest extant relatives of the Pterosauria would be crocodilians and birds, just as is the case for the dinosaur *Maiasaura*. Consequently, it has been possible to use the character states of muscles inferred by Dilkes (2000) for *Maiasaura* and to apply them directly to dinosaur sister-group pterosaurs. If the Pterosauria were basal archosauromorphs, then the closest extant relatives of the Pterosauria would be lepidosaurs and crocodilians. Dilkes (2000) provided sufficient information in his paper to infer the character states of muscles at the basal archosauromorph pterosaur outgroup node. Various papers (Fürbringer 1876, 1900; Gregory & Camp 1918; Jenkins & Goslow 1983) were consulted to review and confirm some of the information in Dilkes' paper, but no attempt was made to duplicate his thorough review of the literature on the myology of the pectoral region of extant reptiles. Note also that Dilkes (2000) has given a full discussion of his methods and, rather than duplicate his discussion here, the reader is referred to his paper.

Osteological correlates of muscles

Here those bony processes and muscle scars that may have provided origins and insertions for pectoral muscles are reviewed. Large pterodactyloids, probably simply because of the their size, show more muscle scars than smaller pterosaurs. In addition, *Anhanguera* is represented by three-dimensionally preserved material, much of which is completely freed of matrix, whereas the few specimens of *Campylognathoides* are crushed and preserved in slabs. Therefore, the evidence of processes and muscle scars on *Anhanguera* and other large pterodactyloids is first reviewed before turning to *Campylognathoides*.

The morphology of a bone often changes markedly during ontogeny as it is remodelled due to the external forces applied to it and as processes and tubercles ossify (Brinkman 1988). Likewise, the appearance of muscle scars changes through ontogeny. On the immature bones of most specimens of *Anhanguera* a muscle scar may be a groove or a low ovoid elevation, whereas the scar in the same location on an adult specimen of *Pteranodon* will be a rugose ridge. In addition, in an earlier study of a large sample of *Pteranodon* (Bennett 1993, 2001) it was noted that some muscle scars were present on almost all adult specimens, whereas other scars were present only on a small subset of adult specimens. The subset is interpreted as consisting of unusually old specimens, and the muscle scars, which consisted of large areas of light scarring on the shafts of the humerus, radius, and ulna in areas where one would expect there to be fleshy origins of muscles (e.g., M. brachialis, muscles of the carpus and digits), are interpreted as the scars of fleshy muscle attachments. Thus, the presence of these scars suggests that there may be other fleshy muscle attachments that have left no scars at all on the available bones.

Anhanguera

The reconstruction of the pectoral region and associated structures of *Anhanguera* is based on *Araripesaurus santanae* (AMNH 22555; Wellnhofer 1991a), with additional information on the humerus and sternum from *A. piscator* (NSM-PV 19892; Kellner & Tomida 2000) and *Santanadactylus araripensis* (BSP 1982 I 89; Wellnhofer 1985). Information on muscle scars was drawn from those specimens plus *S. brasilensis* (BSP 1987 I 65; Wellnhofer 1991b), *S. pricei* (BSP 1980 I 43, BSP 1980 I 122, AMNH 22552; Wellnhofer 1985, 1991a), *Pteranodon* (Bennett 2001), and the indeterminate pteranodontoids USNM 13804 and 20711. Note that I argued previously (Bennett 1993) that *S. araripensis* and *A. santanae* were separated entirely on the basis of ontogenetic characters, with *Anhanguera* consisting of immature specimens and *Santanadactylus* consisting of mature specimens. Examination of additional specimens and descriptions has not changed my opinion that specimens assigned to the two genera are congeneric and may be conspecific, the revision of Kellner & Tomida (2000) notwithstanding. Numbers 1–36 in the following text refer to the numbered processes, muscles scars and inferred areas of muscle attachment illustrated in Figures 2 and 3.

Vertebrae and Ribs

The cervical and dorsal vertebrae have prominent neural spines, which in extant tetrapods provide areas for the origin and insertion of various muscles. NSM-PV 19892 and specimens of *Pteranodon* have muscle scars on the neural spines of some cervical vertebrae that were probably associated with intervertebral muscles. Likewise some specimens of *Pteranodon* exhibit muscle scars on the ventral surface of the cervical centra that are probably for hypaxial muscles. In AMNH 22555 the neural spines of the notarial vertebrae are expanded anteroposteriorly such that they contact one another, and they presumably were fused together in mature specimens to form a supraneural plate, as in the specimen described by Wellnhofer *et al.* (1983). In *Pteranodon* the neural spines are narrow anteroposteriorly, but there is a supraneural plate formed of ossified interspinous ligaments. Both supraneural plates could provide large areas for the origin of muscles. No muscle scars have been noted on ribs except in *Pteranodon*, in which the first notarial rib exhibits a rugose muscle scar running down the entire posterior margin of the rib.

Sternum

AMNH 22555 does not preserve a sternum, but NSM-PV 19892 shows the morphology well. The cristospine is large and its anterior margin is crenellated, presumably for the attachment of muscles. *Pteranodon* does not exhibit such crenellations, but the margins of the cristospine are rugose, again presumably for the attachment of muscles. The posterior plate of the sternum of NSM-PV 19892 is broad and smooth, and does not exhibit muscle scars other than a median rugosity, 5–8 mm wide, which runs posteriorly from the posterior end of the inferior margin of the cristospine. Kellner & Tomida (2000) called the rugosity a 'keel', but while the sternum is keeled in that the two halves of the posterior plate meet at an angle at the mid-line, it is not keeled in the same way as the sternum of carinate birds. In some specimens of *Pteranodon* there is a low median ridge on the anterior half of the posterior plate, 2–3 mm wide and in places as much as 3–4 mm high, which may probably be called a keel. Such a keel might have been developed in mature specimens of *Anhanguera*. Various authors (e.g. Padian 1983b; Kellner & Tomida 2000; Bennett 2001) have suggested that the sternum of some pterosaurs may have borne a keel extended ventrally by cartilage; however, the median rugosity of NSM-PV 19892 is not what one would expect as the base of a cartilaginous keel. In the specimen of *Pteranodon* (YPM 2546) which suggested that a cartilaginous keel may have been present, the keel was flat-topped with squared-off margins, much like the base on the cartilaginous keel in an immature domestic chicken (*Gallus gallus*), whereas the rugose ridge of NSM-PV 19892 is quite unlike the condition in *Gallus* and instead seems to consist of muscle scars from the medial margins of the origin of pectoral muscles. In *Pteranodon* the posterior plate of the sternum has some small rugosities adjacent to the costal articulations that were interpreted as ligament scars, but the broad expanse of the posterior plate, like that of NSM-PV 19892, does not exhibit any other muscle scarring.

Scapulocoracoid

The scapulocoracoid of *Anhanguera* is illustrated in Figure 2. AMNH 22555 has a very slightly elevated area (#1) on the anterior surface of the proximal scapula. Various specimens of *Pteranodon* have a prominent rugose scar in the same area, and KUVP 968 is preserved with ossified ligaments or tendons contacting the scar as if they had been attached to it. There is a prominent rounded process (#3) superior and anterior to the glenoid fossa. In AMNH 22555 it has a textured surface similar to that of the glenoid fossa and the terminal expansion of the deltopectoral crest; however, it clearly was non-articular because the surface extends onto the anterior surface of the buttress. The biceps tubercle is an elongate process anterior to the glenoid fossa and separated from the body of the coracoid by a groove with a pneumatic foramen at the bottom. In AMNH 22555, its top (#4) is gently rounded and textured, with the texturing extending slightly down the lateral and posterior surfaces. NSM-PV 19892 has a somewhat lower, more laterally directed biceps tubercle, with a shallower groove than in AMNH 22555, and the same is true of BSP 1987 I 65. Scapulocoracoids of *Pteranodon* are crushed such that it is difficult to determine the shape of the biceps tubercle, but here, too, the tubercle is not as tall as in AMNH 22555. In *Pteranodon* there is a pneumatic foramen medial to the biceps tubercle, but there is no significant groove between the tubercle and the body of the coracoid. AMNH 22555 has a long shallow rugose groove (#5) extending along the anterolateral surface of the coracoid, and various specimens of *Pteranodon* also exhibit the scar. AMNH 22555 has a series of three muscle scars on the dorsal surface of the scapula, a short distance from the glenoid fossa. These include a low oval elevation (#7), a long, tapering rugose depression (#8) adjacent to it, and a curving rugose depression (#11) just above the posterior part of the glenoid fossa. Specimens of *Pteranodon* (e.g. YPM 1175 and 2373) seem to show the latter scar just above the glenoid fossa. There is a prominent posterior tubercle (#10) on the posterior margin of the scapula, which in AMNH 22555 has a distinctive texturing that wraps around onto the posterior and ventral surfaces of the tubercle. The tubercle is also prominent in NSM-PV 19892, BSP 1987 I 65, and specimens of *Pteranodon*. AMNH 22555 has a small tubercle (#12) on the lateral surface of the coracoid a short distance below the glenoid fossa. The tubercle is somewhat larger in NSM-PV 19892. In both specimens the tubercle lies at the dorsal end of a long shallow rugose depression (#13) on the posterolateral surface of the coracoid. In *Pteranodon* the same feature is represented by a long rugose scar, which is perhaps twice as wide as it is in AMNH 22555. In *Pteranodon* the lateral margin of the coracoid is somewhat expanded into a coracoid flange, and the long rugose scar (#13) is on the posterior aspect of the flange. The size of the flange in *Pteranodon* seems to be variable, although crushing prevents accurate comparisons of the size of specimens, and it is possible that the size of the flange increased during ontogeny. Some other pterosaurs also exhibit a coracoid flange, and it is extremely large in speci-

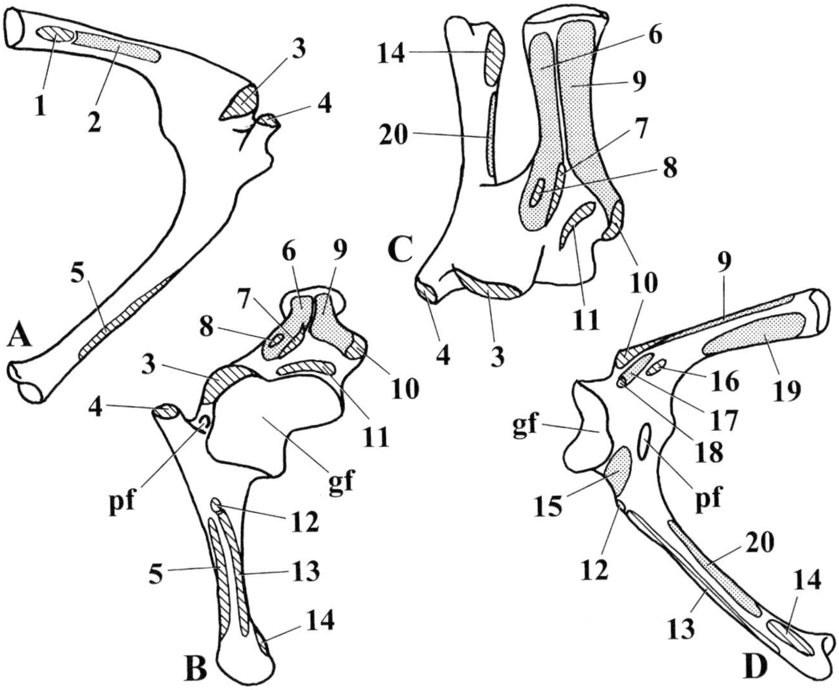

Fig. 2. Left scapulocoracoid of *Anhanguera* based on AMNH 22555 in (**a**) anterior, (**b**) lateral, (**c**) dorsal, and (**d**) posterior views, showing muscle scars and other features (hatched) and inferred areas of muscle origin and insertion (stippled). Identifications and/or inferred muscle attachments are as follows: 1, scapulonotarial ligament and/or *m. trapezius* and *m. levator scapulae* [I]; 2, *m. trapezius* and *m. levator scapulae* [I]; 3, *m. scapulohumeralis anterior* [O]; 4, biceps tubercle, *m. biceps* [O]; 5, *m. supracoracoideus* [O]; 6–8, *m. deltoideus scapularis* [O]; 9 & 10, *m. teres major* [O]; 11, scapular head of *M. triceps* [O]; 12 & 13, *m. coracobrachialis* [O]; 14, *m. sternocoracoideus* [I]; 15, coracoid head of *m. triceps* [O]; 16, *m. subscapularis* [O]; 17 & 18, *m. scapulohumeralis posterior* [O]; 19, *mm. serratus profundus et superficialis* [I]; 20, *m. costosternocoracoideus* [I]. See text for description of the numbered features and inferred muscle attachments. gf, glenoid fossa; pf, pneumatic foramen.

mens on *Quetzalcoatlus* (e.g. TMM 42138–1). AMNH 22555 has a low oval scar and a slight flange (#14) on the posterior aspect of the ventral end of the coracoid. In specimens of *Pteranodon* the scar is a low rugose ridge bent into the shape of an inverted J. AMNH 22555 has a low oval elevation and scar (#16) on the ventral aspect of the scapula near the glenoid, and slightly lateral to it is a shallow rounded pit (#18). One specimen of *Pteranodon* (YPM 2373) has a series of three small muscle scars on the underside of the scapula. It is not clear how they correspond to the scar (#16) and pit (#18) of AMNH 22555. BSP 1987 I 65, which seems to be from a fully mature individual, has a single long ridge (Wellnhofer 1991a, fig. 40, 'pof') in about the same position.

Humerus

The humerus of *Anhanguera* (Fig. 3) has the warped deltopectoral crest characteristic of pteranodontoid pterosaurs. In AMNH 22555 the terminal expansion (#24) at the end of the deltopectoral crest has a relatively flat surface, set off from the rest of the crest and with a distinctive texturing. There is a prominent ridge that runs from the end of the terminal expansion proximally toward the posterior tuberosity. This ridge might be a structural consequence of the warped deltopectoral crest, but it also provides additional area for muscle attachment. In AMNH 22555 and 22552 the area between that ridge and the terminal expansion has a number of irregular scars. Specimens of *Pteranodon* also have irregular scars is this area. AMNH 22555 has a weak ridge (#25) that runs roughly parallel to the long axis of the humerus, near the base of the deltopectoral crest. In AMNH 22552 it is more prominent and is rugose, and it is a narrow rugose ridge in USNM 13804 and *Pteranodon* (e.g. YPM 1164). AMNH 22552 has a long shallow rugose groove (#26) running down the medial side of the shaft of the humerus. The groove is also present in NSM-PV 19892. AMNH 22552 has a very faint area of scarring (#27) that is just

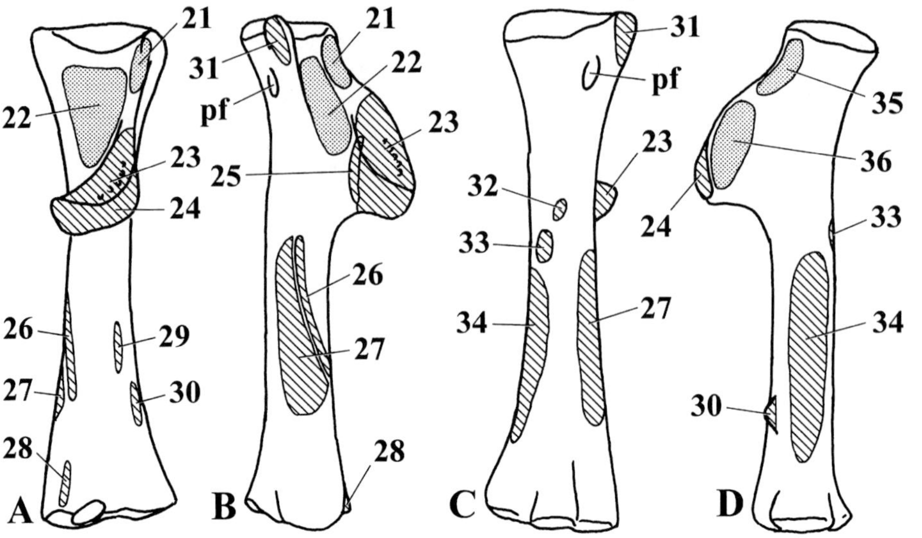

Fig. 3. Left humerus of *Anhanguera* based on AMNH 22555 and BSP 1982 I 89 (Wellnhofer 1985) in (**a**) ventral, (**b**) posterior, (**c**) dorsal, and (**d**) anterior views, showing muscle scars and other features (hatched) and inferred areas of muscle attachment (stippled). Identifications and/or inferred muscle attachments are as follows: 21, *m. supracoracoideus* [I]; 22, *M. coracobrachialis* [I]; 23–25, *m. pectoralis* [I]; 26 & 27, medial head of *m. triceps* [O]; 28, extensors of the carpus and digits [O]; 29, *M. brachialis*? [O]; 30, flexors of the carpus and digits [O]; 31, *m. subscapularis* and *m. scapulohumeralis posterior* [I]; 32, *m. teres major* [I]; 33, *m. latissimus dorsi* [I]; 34, lateral head of *m. triceps* [O]; 35, *m. scapulohumeralis anterior* [I]; 36, *m. deltoideus scapularis* [I]. See text for description of the numbered features and inferred muscle attachments. pf, pneumatic foramen.

dorsal to the long groove (#26). A few specimens of *Pteranodon* (e.g. YPM 1175) have a larger area of light scarring in the same position, which is interpreted as the rare scar of a fleshy muscle attachment in a very old specimen. AMNH 22552 has a small narrow rugose process (#28) on the ventral surface just proximal to the distal end of the humerus. AMNH 22552 has two scars on the lateral side of the ventral surface of the shaft. One (#29) is a low oval rugose flange. The other (#30) is a more prominent flange, the supracondylar process. It is also present in NSM-PV 19892, BSP 1980 I 43, BSP 1980 I 122 and in specimens of *Pteranodon*. AMNH 22555 and 22552 have a rugose area (#31) on the medial and dorsal surface of the posterior tuberosity of the humerus. It is also present in *Pteranodon*, and in YPM 1164 it extends somewhat further down the shaft of the humerus than is illustrated in Figure 3. AMNH 22552 and NSM-PV 19892 have an oval rugosity (#33) on the dorsal surface of the shaft, just distal to the deltopectoral crest. The same scar is present in specimens of *Pteranodon*, where it is a cluster of small circular to oval scars. In USNM 20711 it is a single oval scar, hardly depressed below the surrounding bone. Just proximal and medial to that scar (#33) is a small scar of varying shape (#32). In BSP 1982 I 92 it is a low oval scar, as illustrated in Figure 3. In *Pteranodon* it is a small cluster of scars, while in USNM 20711 it is an oval scar hardly depressed below the surrounding bone. However, in AMNH 22552, it is a narrow rugose ridge twice as long as the scar illustrated in Figure 3. The situation is further complicated by the fact that BSP 1982 I 89 (Wellnhofer 1985, fig. 8) and BSP 1980 I 43 (Wellnhofer 1985, fig. 22) are illustrated as having a single elongate ridge in the area of the two scars (#32 & 33 in Fig. 3). It was not possible to re-examine those specimens to confirm the illustrations and determine whether a second scar is present. A few specimens of *Pteranodon* (e.g. YPM 1175) have a large area of light scarring (#34) on the lateral surface of the shaft of the humerus, which is interpreted as the rare scar of a fleshy muscle attachment in a very old specimen. A few specimens of *Pteranodon* (e.g. YPM 1175, 2499) have a small distinct muscle scar on the dorsal surface of the deltopectoral crest. It is not illustrated in Figure 3, but it extends from the middle of inferred muscle attachment area #36 to the edge of the terminal expansion. Lastly, AMNH 22552 also may have had very weak texturing in the inferred muscle attachment area #21 on the ventral surface of the proximal deltopectoral crest; however, there was no convincing evidence to show that the texturing represented a soft-tissue attachment rather than a structural response to forces applied to the bone.

Radius and Ulna
The muscles of the brachium and antebrachium are beyond the scope of this paper, but it is necessary to note two prominent muscle scars on the ulna. There is a prominent biceps tubercle on the anterior face of the ulna, a short distance distal to the elbow in AMNH 22522 and BSP 1980 I 122. The tubercle is also present in *Pteranodon*, where it is divided into two parts by a transverse groove, and some specimens have weak muscle scarring around the biceps tubercle. Specimens of *Pteranodon* have a rugose crest on the posterior side of the proximal end of the ulna. The crest is also present in AMNH 22552 and other specimens from the Santana Formation, but in those specimens it is not noticeably rugose.

Campylognathoides

The reconstruction of the pectoral region and associated structures of *Campylognathoides* is based on *C. liasicus* CM 11424 (Wellnhofer 1974), with additional information from published descriptions of *C. liasicus* (Plieninger 1907; Wiman 1923) and *C. zitteli* (Plieninger 1895, 1907; Wiman 1923; Wellnhofer 1974). Information on muscle scars was drawn from those specimens plus BYU 9488 (Jensen & Padian 1989), assigned to *Mesadactylus ornithosphyos*, and a specimen of *Rhamphorhynchus muensteri* (PTH-49–4). Unless specifically referring to numbered features of *Anhanguera*, numbers 1–26 refer to numbered processes, muscles scars and inferred areas of muscle attachment illustrated in Figure 4.

Vertebrae and Rib
No specimen of *Campylognathoides* provides useful information about the vertebral column. CM 11424 is preserved with the ventral aspect exposed and the vertebral column obscured. The vertebral column is very poorly preserved in the specimens of Plieninger (1907) and Wiman (1923). No scars have been noted on any ribs of rhamphorhynchoids.

Sternum
The sternum consists of a narrow and deep cristospine bearing the coracoid articulations on its dorsolateral surfaces and a broad posterior plate. Although the surface and margins of the cristospine and the broad expanse of the posterior plate of the sternum seem well suited for the attachment of muscles, no muscle scars have been identified on the sternum.

Scapulocoracoid
The scapulocoracoid of *Campylognathoides* is illustrated in Figure 4. The scapulocoracoids of CM 11424 have a low ovoid tubercle (#3) on the anterolateral surface of the scapula just above the glenoid. It seems to correspond to scar #3 of *Anhanguera*. BYU 9488 has two prominent processes near the glenoid fossa. The first is a large process (#5) anterior to the ventral margin of the glenoid fossa, whereas the second (#7) is on the posterolateral surface of the coracoid just below the glenoid fossa. Padian (1983b; Jensen & Padian 1989) interpreted the first (#5) as an acrocoracoid process and the second (#7) as the biceps tubercle. Here the process anterior to the glenoid (#5) is interpreted as the biceps tubercle corresponding to #4 of *Anhanguera*, and the tubercle below the glenoid (#7) as the coracoid tubercle corresponding to #12 of *Anhanguera*. The biceps tubercle is present in CM 11424 and all well-preserved specimens of *Rhamphorhynchus* and other rhamphorhynchoid pterosaurs. The coracoid tubercle is not visible in CM 11424 because the scapulocoracoids are preserved with the anteromedial surface exposed. It was not noted in specimens of *Rhamphorhynchus*, and it is possible that the process does not develop significantly until later in ontogeny. There is a distinct flange (#14) on the posterior margin of the ventral coracoid, just above the sternal articulations. It is of moderate size in CM 11424, but is considerable larger in BYU 9488. It seems to correspond to #14 of *Anhanguera*.

Humerus
The humerus of *Campylognathoides* is illustrated in Figure 4. All available humeri are preserved with the ventral side exposed. The deltopectoral crest of *Campylognathoides* is large and is roughly the shape of a parallelogram. There is a small terminal expansion at its lateral end (#18), which may correspond to #24 of *Anhanguera*. The deltopectoral crest also has a low ridge parallel to the shaft of the humerus at the base of the crest. This feature is unlabelled in Figure 4, but the inferred muscle attachment area (#17) is between it and the end of the deltopectoral crest, so it may correspond to #25 of *Anhanguera*. One specimen of *Rhamphorhynchus* (PTH-49–4) has a large oval rugose elevation (#24) on the dorsal surface of the humerus, just distal to the deltopectoral crest. It presumably corresponds to #33 of *Anhanguera*. The same specimen also has a narrow rugose ridge (#26) extending along the posterolateral surface of the shaft of the humerus, which may correspond to the dorsal margin of #34 of *Anhanguera*.

Radius and ulna
There was no indication of any muscle scars on the ulna that correspond to the biceps tubercle of *Santanadactylus* and *Pteranodon*. However, various specimens of *Rhamphorhynchus* have a prominent crest on the posterior surface of the proximal ulna that corresponds to the rugose crest in the same position on *Pteranodon* ulnae.

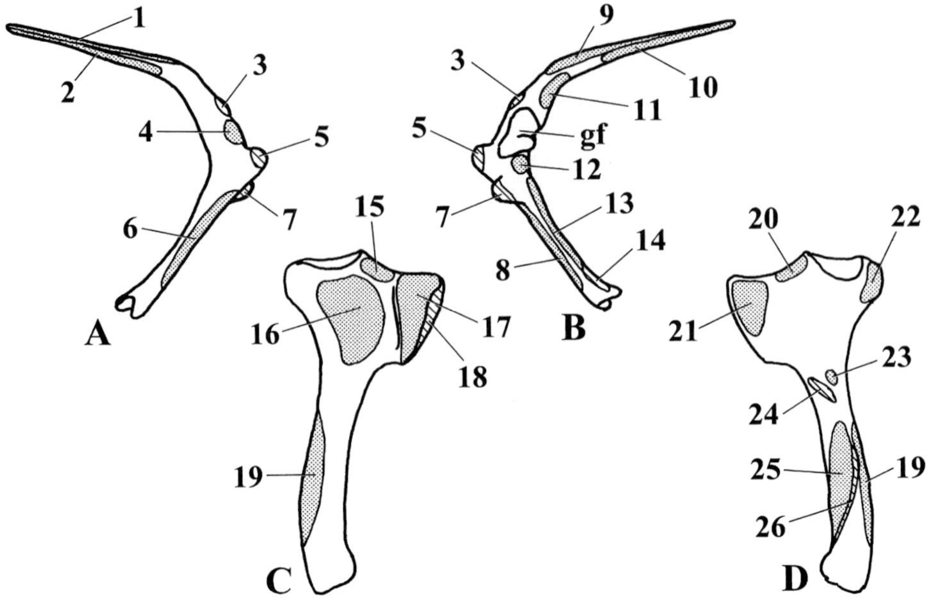

Fig. 4. (**a, b**) Left scapulocoracoid and (**c, d**) humerus of *Campylognathoides* based on CM 11424 (Wellnhofer 1974), in anteromedial (**a**), posterolateral (**b**), ventral (**c**), and dorsal views (**d**), showing muscle scars and other features (hatched) and inferred areas of muscle attachment (stippled). Identifications and/or inferred muscle attachments are as follows: 1, *m. deltoides scapularis* [O]; 2, *m. trapezius* and *m. levator scapulae* [I]; 3, scapular head of *m. triceps* [O]; 4, *m. scapulohumeralis anterior*; 5, *m. biceps* [O]; 6, *m. supracoracoideus* [O]; 7 & 8, *m. coracobrachialis* [O]; 9, *m. teres major* [O]; 10, *Mm. serratus superficialis et profundus* [I]; 11, *m. subscapularis* and *m. scapulohumeralis posterior* [O]; 12, coracoid head of *m. triceps* [O]; 13, *m. costosternocoracoideus* [I]; 14, *sternocoracoideus* [I]; 15, *m. supracoracoideus* [I]; 16, coracobrachialis [I]; 17 & 18, *m. pectoralis* [I]; 19, medial head of *m. triceps* [O]; 20, *m. scapulohumeralis anterior* [I]; 21, *m. subscapularis* and *m. scapulohumeralis posterior* [I]; 22, *m. deltoides scapularis* [I]; 23, *m. latissimus dorsi* [I]; 24, *m. teres major* [I]; 25 & 26, lateral head of *m. triceps* [O]. gf, glenoid fossa.

Inferred myology

Below, for each muscle of the pectoral region, the condition of the muscle in the extant relatives of pterosaurs is briefly reviewed and its probable character state in pterosaurs is inferred from character state distribution in the extant relatives. As noted above, the information on the character state distribution in the extant relatives of pterosaurs is drawn primarily from Dilkes (2000). Then the inferred character state of the muscle is matched with the processes and muscle scars described above. As in the preceding section, *Anhanguera* is discussed first because more muscle scars are known for it. Some muscles are reconstructed without there being any clear muscle scars in the fossils that can be matched to them. However, those muscles in extant reptiles typically have fleshy attachments to bones and are not associated with prominent muscle scars. Therefore, it is reasonable to suppose that the muscle in the fossil organisms also had a fleshy attachment that did not leave visible muscle scars.

Reconstructions of the pectoral region of *Campylognathoides* and *Anhanguera*, based on the interpretation of pterosaurs as basal archosauromorphs, are shown in Figures 5–13. Where the evidence as to the presence of a muscle or a division of a muscle is equivocal, the muscle or division is not reconstructed. Thus, *m. rhomboideus*, *m. subcoracoideus* and the second divisions of various muscles (see Table 1) are not reconstructed; also *m. deltoides clavicularis* is not reconstructed because there is uncertainty as to the site of its origin. With the exception of the fusiform muscles (i.e. *m. biceps* and long heads of *m. triceps*) and the curving *m. sternocleidomastoideus*, the muscles are drawn with straight or slightly curving lines extending between the areas of origin and insertion. The area of a muscle on a reconstruction should not be taken as necessarily indicative of the size and power of the muscle.

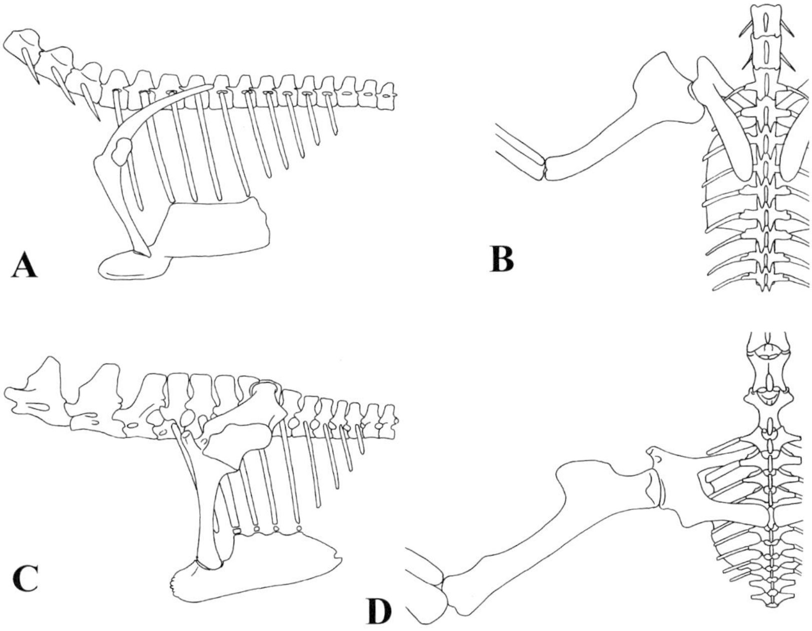

Fig. 5. Reconstructions of the left pectoral region of the skeleton of (**a, b**) *Campylognathoides*, and (**c, d**) *Anhanguera* in left lateral (a, c) and dorsal (b, d) views. The reconstructions include the sixth cervical to eleventh dorsal vertebrae and their ribs, sternum and scapulocoracoid. (**b**) and (**d**) also include the humerus and proximal radius and ulna. Not to scale.

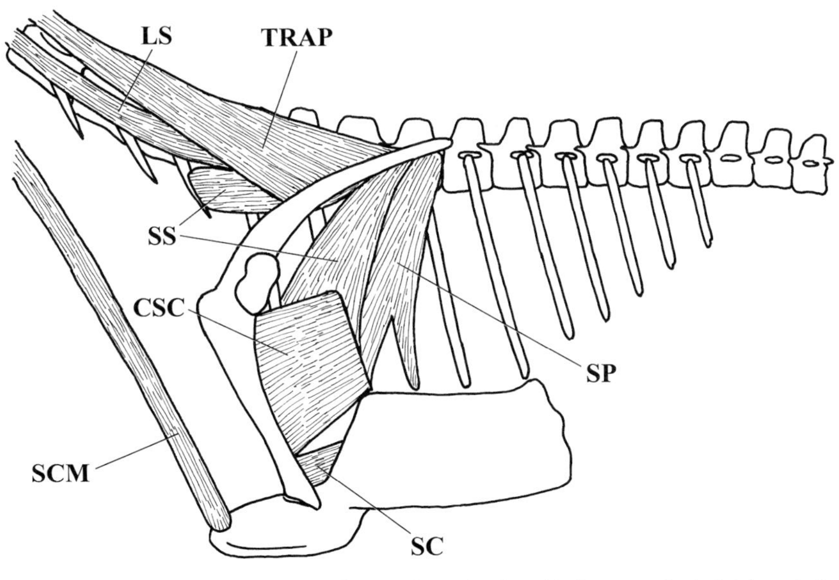

Fig. 6. Reconstruction of the left pectoral region of *Campylognathoides* with deep muscles in left lateral view. CSC, *m. costosternocoracoideus*; LS, *m. levator scapulae*; SC, *m. sternocoracoideus*; SCM, *m. sternocleidomastoideus*; SP, *m. serratus profundus*; SS, *m. serratus superficialis*; TRAP, *m. trapezius*.

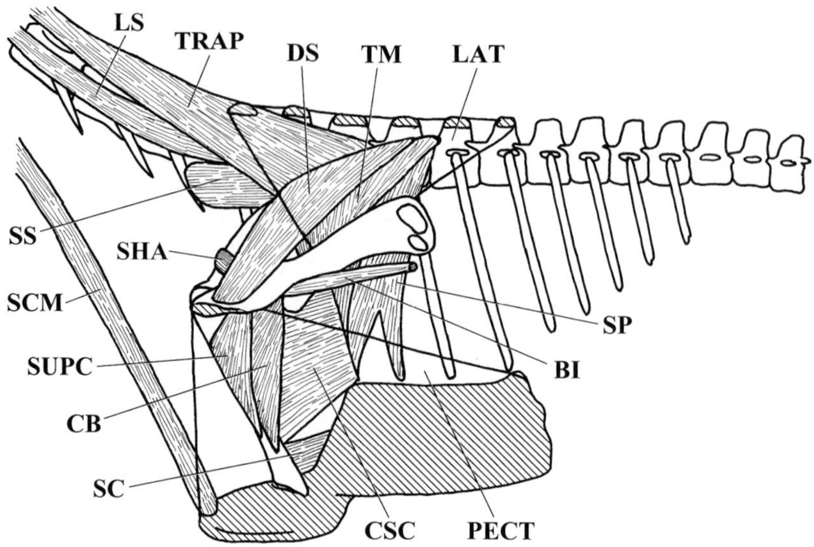

Fig. 7. Reconstruction of the left pectoral region of *Campylognathoides* with superficial and deep muscles in left lateral view. *M. pectoralis* is indicated only by the hatched origin on the sternum and insertion on the deltopectoral crest and its outlines between, and *m. latissimus dorsi* is indicated only by the hatched origin on the neural spines and its outlines directed toward the insertion on the posterior humerus, so as not to obscure deeper muscles. *M. triceps* is not included because it would be hidden behind the humerus. BI, *m. biceps*; CB, *m. coracobrachialis*; CSC, *m. costosternocoracoideus*; DS, *m. deltoides scapularis*; LAT, *m. latissimus dorsi*; LS, *m. levator scapulae*; PECT, *m. pectoralis*; SC, *m. sternocoracoideus*; SCM, *m. sternocleidomastoideus*; SP, *m. serratus profundus*; SS, *m. serratus superficialis*; SUPC, *m. supracoracoideus*; TM, *m. teres major*; TRAP, *m. trapezius*.

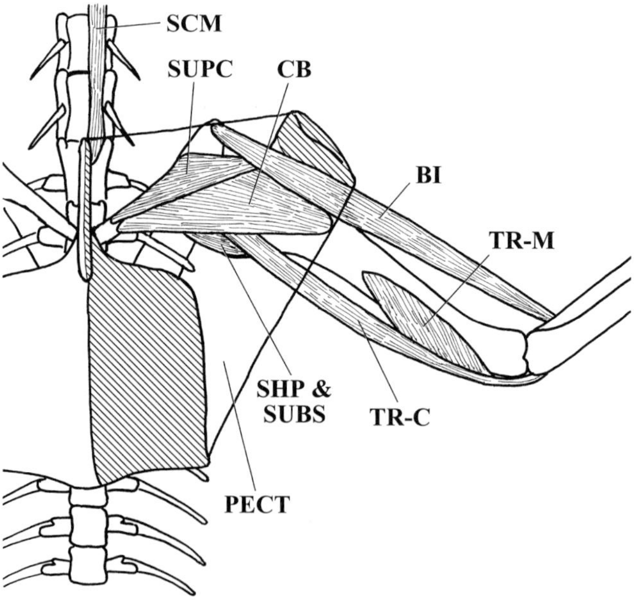

Fig. 8. Reconstruction of the left pectoral region of *Campylognathoides* in ventral view. *M. pectoralis* is indicated only by the hatched origin on the sternum and insertion on the deltopectoral crest and its outlines between, so as not to obscure deeper muscles. BI, *m. biceps*; CB, *m. coracobrachialis*; PECT, *m. pectoralis*; SCM, *m. sternocleidomastoideus*; SHP, *m. scapulohumeralis posterior*; SUBS, *m. subscapularis*; SUPC, *m. supracoracoideus*; TR-C, coracoid head of *m. triceps*; TR-M, medial head of *m. triceps*.

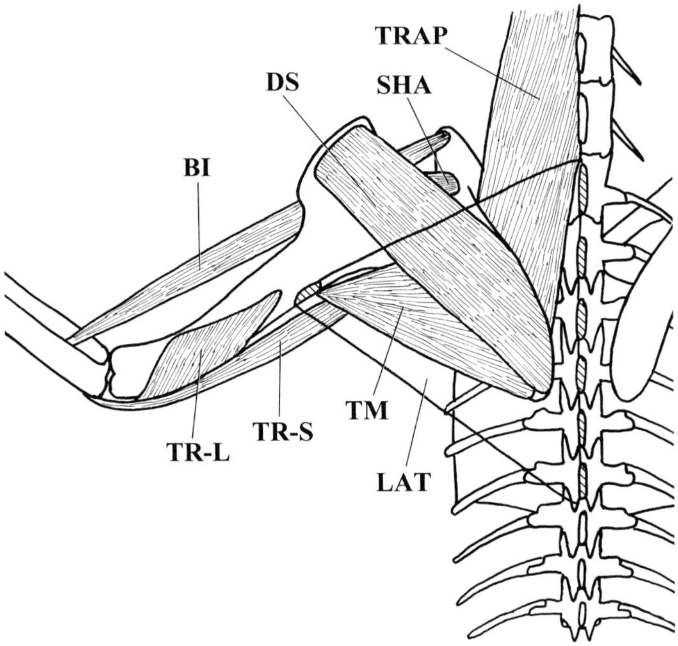

Fig. 9. Reconstruction of the left pectoral region of *Campylognathoides* in dorsal view. *M. latissimus dorsi* is indicated only by the hatched origin on the neural spines and insertion of the posterior humerus and its outlines between, so as not to obscure deeper muscles. BI, *m. biceps*; DS, *m. deltoides scapularis*; LAT, *m. latissimus dorsi*; SHA, *m. scapulohumeralis anterior*; TM, *m. teres major*; TR-L, lateral head of *m. triceps*; TR-S, scapular head of *m. triceps*; TRAP, *m. trapezius*.

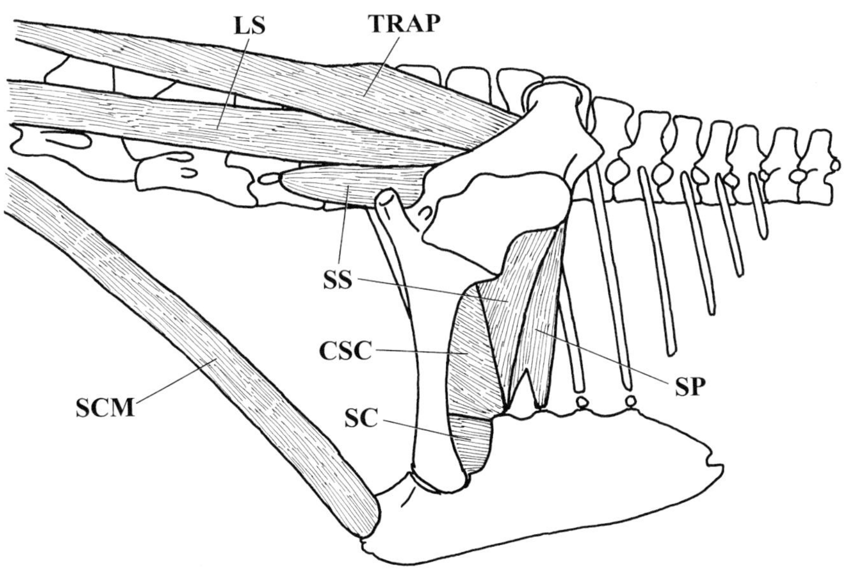

Fig. 10. Reconstruction of the left pectoral region of *Anhanguera* with deep pectoral muscles in left lateral view. CSC, *m. costosternocoracoideus*; LS, *m. levator scapulae*; SC, *m. sternocoracoideus*; SCM, *m. sternocleidomastoideus*; SP, *m. serratus profundus*; SS, *m. serratus superficialis*; TRAP, *m. trapezius*.

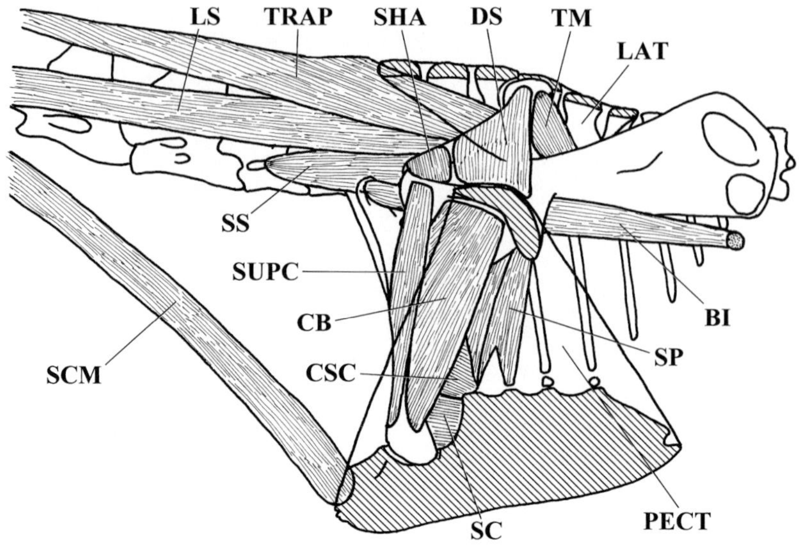

Fig. 11. Reconstruction of the left pectoral region of *Anhanguera* with superficial and deep pectoral muscles in left lateral view. *M. pectoralis* is indicated only by the hatched origin on the sternum and insertion on the deltopectoral crest and its outlines between, and *m. latissimus dorsi* is indicated only by the hatched origin on the neural spines and its outlines directed toward the insertion on the posterior humerus, so as not to obscure deeper muscles. M. triceps is not included because it would be hidden behind the humerus. BI, *m. biceps*; CB, *m. coracobrachialis*; CSC, *m. costosternocoracoideus*; DS, *m. deltoides scapularis*; LAT, *m. latissimus dorsi*; LS, *m. levator scapulae*; PECT, *m. pectoralis*; SC, *m. sternocoracoideus*; SCM, *m. sternocleidomastoideus*; SP, *m. serratus profundus*; SS, *m. serratus superficialis*; SUPC, *m. supracoracoideus*; TM, *m. teres major*; TRAP, *m. trapezius*.

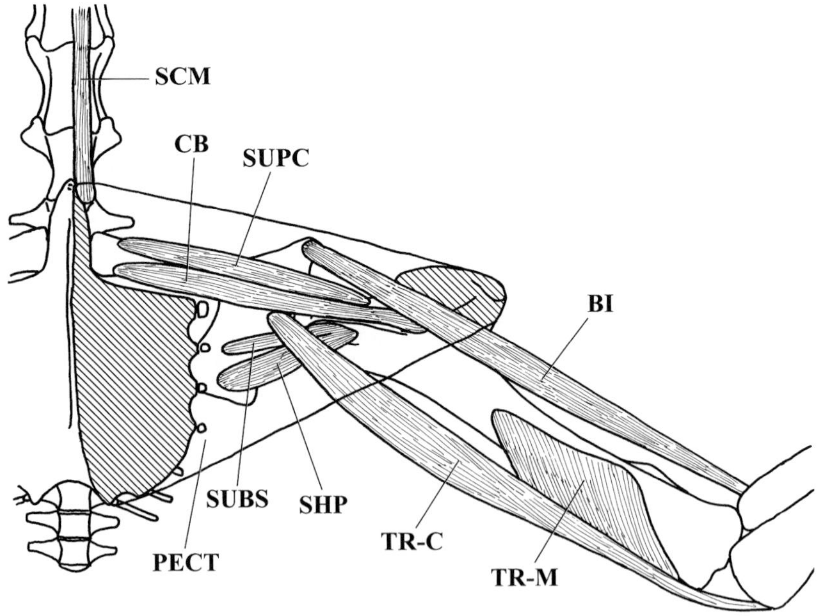

Fig. 12. Reconstruction of the left pectoral region of *Anhanguera* in ventral view. *M. pectoralis* is indicated only by the hatched origin on the sternum and insertion on the deltopectoral crest and its outlines between, so as not to obscure deeper muscles. BI, *m. biceps*; CB, *m. coracobrachialis*; PECT, *m. pectoralis*; SCM, *m. sternocleidomastoideus*; SHP, *m. scapulohumeralis posterior*; SUBS, *m. subscapularis*; SUPC, *m. supracoracoideus*; TR-C, coracoid head of *m. triceps*; TR-M, medial head of *m. triceps*.

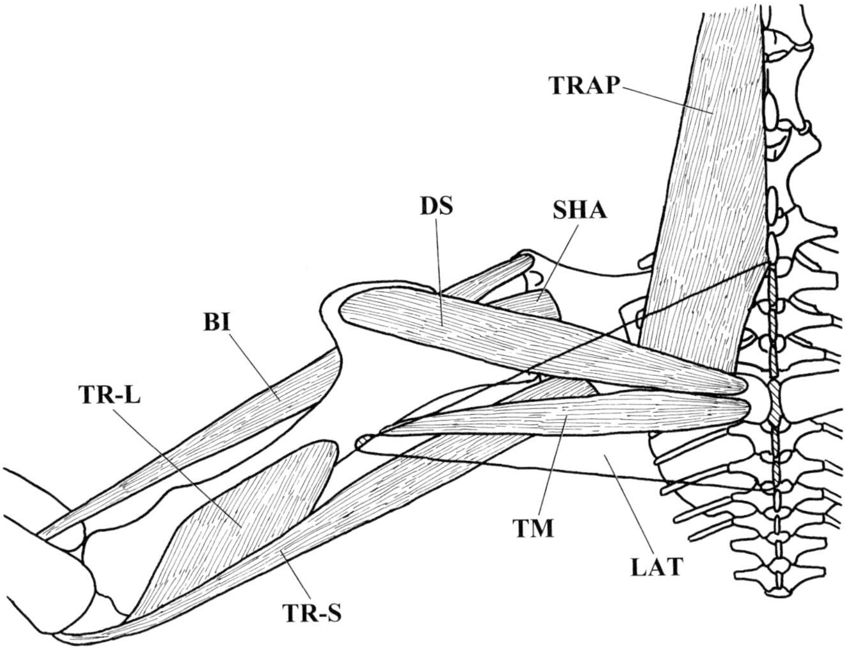

Fig. 13. Reconstruction of the left pectoral region of *Anhanguera* in dorsal view. M. latissimus dorsi is indicated only by the hatched origin on the neural spines and insertion of the posterior humerus and its outlines between, so as not to obscure deeper muscles. BI, *m. biceps*; DS, *m. deltoides scapularis*; LAT, *m. latissimus dorsi*; SHA, *m. scapulohumeralis anterior*; TM, *m. teres major*; TR-L, lateral head of *m. triceps*; TR-S, scapular head of *m. triceps*; and TRAP, *m. trapezius*.

Muscles

M. cucullaris

Lepidosaurs and crocodilians have a two-part *m. cucullaris* composed of *m. sternocleidomastoideus* and *m. trapezius*, and birds also have a two-part *m. cucullaris*, although the parts are not necessarily homologous with those of lepidosaurs and crocodilians. Based on Dilkes' (2000) analysis, the dinosaur sister-group pterosaur outgroup node would be decisively positive for the presence of *m. sternocleidomastoideus*, and the same is the case for basal archosauromorph pterosaurs. *M. sternocleidomastoideus* would be expected to originate from the clavicles and insert on the skull and the first cervical rib. In pterosaurs the clavicles have been incorporated into the sternum (Wild 1993), and it is reasonable to assume that *m. sternocleidomastoideus* originated on the sternum, as it does in crocodilians. Based on Dilkes' analysis, the dinosaur sister-group pterosaur outgroup node would be equivocal for the presence of *m. trapezius*, whereas the basal archosauromorph pterosaur outgroup node is decisive and positive for its presence. *M. trapezius* would have originated from the neural spines of the posterior cervical vertebrae and perhaps one or more of the anterior dorsals and would insert on the anterior margin of the scapula.

It seems likely that the crenellated and rugose anterior margin of the cristospine in *Anhanguera* represents the origin of *m. sternocleidomastoideus*, and the muscle is reconstructed as originating from the anterior end of the cristospine in *Anhanguera* and *Campylognathoides*. The insertion on the skull may have been on the tuberosity of the squamosal (Bennett 2001) in *Anhanguera* and on the posterior margin of the squamosal in *Campylognathoides*. There are no muscle scars on the cervical vertebrae of *Anhanguera* or *Campylognathoides* that can be associated with *m. trapezius*, but it is reconstructed as originating from the neural spines of all cervical vertebrae and inserting on the anterior border of the scapula in *Anhanguera* (#1 and/or #2) and *Campylognathoides* (#2).

M. levator scapulae

Based on Dilkes' (2000) analysis, the dinosaur sister-group pterosaur outgroup node is equivocal for the presence of *m. levator scapulae* and, if present, then equivocal as to whether it had one division as in crocodilians or two as in lepidosaurs. The two divisions (*superficialis* and *profundus*) insert on the medial and lateral surfaces of the scapula in lepidosaurs. The basal archosauromorph pterosaur outgroup node is decisively positive for the presence

Table 1. *Differences between inferred muscle character states*

Muscle	Dinosaur sister-group pterosaurs	Basal archosauromorph pterosaurs
M. trapezius	Presence equivocal	Present
M. levator scapulae	Presence equivocal	Present
M. serratus superficialis	Number of divisions (1 or 2) equivocal	One division
M. serratus profundus	One division	Number of divisions (1 or 2) equivocal
M. rhomboideus	Present	Presence equivocal
M. costosternocoracoideus	Presence equivocal	Present
M. teres major	Presence equivocal	Present
M. deltoides scapularis	Number of divisions (1 or 2) equivocal	One division
M. deltoides clavicularis	Number of divisions (1 or 2) equivocal	One division
M. triceps	Presence of coracoid and 2nd humeral heads equivocal	Coracoid and 2nd humeral heads present
M. supracoracoideus	Location of insertion equivocal	Insertion on flexor side of deltopectoral crest

of *m. levator scapulae*, but it is equivocal as to whether it would have one or two divisions. Here only one division is reconstructed, but note that Fürbringer (1876) mentioned a single specimen of *Alligator* with indications of a division of *m. levator scapulae* into two parts.

M. levator scapulae is reconstructed as originating from the ribs and transverse processes of the first four cervical vertebrae and inserting on the entire anterior border of the scapula in *Anhanguera* (#1 and/or #2) and *Campylognathoides* (#2).

M. serratus superficialis

Based on Dilkes' (2000) analysis, this muscle would have been present in dinosaur sister-group pterosaurs, but it is unclear whether it would have one division as in lepidosaurs or two divisions as in birds. The basal archosauromorph pterosaur outgroup node is decisive for the presence of a single division. *M. serratus superficialis* would have originated from the ventral parts of the last cervical rib and the anterior two or three dorsal ribs. The insertion would be along the posterior margin of the scapula.

The rugose muscle scar on the inferior part of the first notarial rib in *Pteranodon* may be part of the origin of this muscle. It is reconstructed in *Anhanguera* as originating from the last cervical and first three dorsal ribs and passing upwards to insert on the posteromedial scapula (#19). In *Campylognathoides* it is reconstructed as inserting on the upper medial scapula (#10).

M. serratus profundus

Based on Dilkes' (2000) analysis, dinosaur sister-group pterosaurs would have had an *m. serratus profundus* that originated from the inferior parts of the first three dorsal ribs and inserted on the superior part of the medial scapula. The basal archosauromorph pterosaur outgroup node is decisively positive for the presence of *m. serratus profundus*, but it is equivocal as to whether the muscle would have had two divisions as in lepidosaurs or just one as in crocodilians and birds. Where there are two divisions, the insertions are adjacent to one another, so the uncertainty is not likely to significantly affect the reconstruction.

In both *Anhanguera* and *Campylognathoides* this muscle is reconstructed as originating from the posterior cervical vertebrae and the lower parts of the first three dorsal ribs and passing up to insert on the scapula with *m. serratus superficialis*.

M. rhomboideus

This muscle is present only in crocodilians and birds, therefore it would have been present in dinosaur sister-group pterosaurs, but its presence is equivocal in basal archosauromorph pterosaurs. If present, it would have originated from the eighth and/or ninth vertebrae and passed posteriorly to insert on the medial surface of the superior end of the scapula; however, this muscle is not reconstructed because the basal archosauromorph pterosaur outgroup node is equivocal for its presence and there is no evidence on the fossils that it was present.

M. sternocoracoideus

Based on Dilkes' (2000) analysis, this muscle would be present in dinosaur sister-group pterosaurs and basal archosauromorph pterosaurs, but it is unclear whether it would have had two divisions as in lepidosaurs or only a single division as in birds (*m. sternocoracoideus* is absent in crocodilians). The point is not important because where there are two divisions, their origins and insertions are adjacent to one another. *M. sternocoracoideus* would originate on the sternum adjacent to the sternocoracoid joint and insert on the coracoid a short distance above the joint.

No scar or distinct area of origin for *M. sternocoracoideus* can be identified on the sternum of either *Anhanguera* or *Campylognathoides*, so it is reconstructed as originating from the anterior margin or anterior part of the dorsal surface of the posterior plate of the sternum. In *Anhanguera* the insertion can be associated with the muscle scar and flange (#14) on the posterior surface of the coracoid just above the sternocoracoid articulation, and in *Campylognathoides* the insertion can be associated with the prominent flange (#14) in the same place.

M. costosternocoracoideus
Based on Dilkes' (2000) analysis the dinosaur sister-group pterosaur outgroup node is equivocal for the presence of this muscle, but the basal archosauromorph pterosaur outgroup node is decisively positive for its presence. If present the muscle would probably have originated from the posterior border of the last cervical rib and the first sternal rib and inserted on the entire posterior margin of the coracoid.

In *Anhanguera* the origin of this muscle might be associated with the scarring on the posterior surface of the first notarial rib, although that scarring might also be from *mm. serratus superficialis et profundus* or intercostals. No scarring is present in *Anhanguera* on the posterior margin of the coracoid other than the flange and scar (#14) for *m. sternocoracoideus*; however, it is likely that this muscle would insert along the entire posterior border (#20). No scars that can be associated with this muscle are present in *Campylognathoides*, but it is reconstructed as originating from the last cervical rib and first sternal rib and inserting on the posterior border of the coracoid (#13).

M. latissimus dorsi
Turtles, lepidosaurs, crocodilians and birds all have *m. latissimus dorsi*, and the same would be expected of dinosaur sister-group and basal archosauromorph pterosaurs. *M. latissimus dorsi* would originate from the neural spines of the last cervical vertebra and the anterior dorsal vertebrae. It would insert on the proximal part of the posterior aspect of the humerus.

No scarring is present on any of the neural spines of *Anhanguera* or *Campylognathoides* to indicate how extensive the origin might have been. In *Anhanguera* it is reconstructed as originating from the first six dorsal vertebrae (here counting the first notarial as the first dorsal) because the notarium of large pterodactyloids usually consists of up to six fused vertebrae. In *Campylognathoides* it is also reconstructed as originating from the first six dorsal vertebrae. However, it is possible that the origin included one or more additional vertebrae cranially and/or caudally. The insertion can be associated with the prominent scar on the dorsal surface (#33 of *Anhanguera* and #24 in *Campylognathoides*) distal to the end of the deltopectoral crest.

M. teres major
This muscle is present in turtles, lepidosaurs and crocodilians, but is absent in birds. Thus, the dinosaur sister-group pterosaur outgroup node is equivocal for the presence of *M. teres major*, whereas the basal archosauromorph pterosaur outgroup node is decisively positive for its presence. If present, it would have originated from the posterolateral surface of the scapula and inserted with *m. latissimus dorsi* on the posterior surface of the humerus. Dilkes (2000) did not reconstruct the *m. teres major* of *Maiasaura*, noting that he could not find a separate muscle scar adjacent to that of *m. latissimus dorsi*; however, it is not clear from published accounts (Fürbringer 1876) that separate muscle scars would be present.

In *Anhanguera* the origin of this muscle may be associated with the prominent posterior process of the scapula (#10), and it is reconstructed as originating from that process and the posterior half of the dorsal surface of the scapula proximal to the process (#9). The insertion can be associated with the scar (#32) proximal and medial to the insertion of *m. latissimus dorsi* (#33). In *Campylognathoides* no muscle scars can be associated with the origin or insertion of *m. teres major*, and there does not seem to be a homologue of the posterior process on the scapula. Despite this, *m. teres major* is reconstructed as originating from the lateral half of the dorsal surface of the scapular blade (#9) and inserting just proximal and medial (#23) to *m. latissimus dorsi*.

M. deltoides scapularis
Based on Dilkes' (2000) analysis, the dinosaur sister-group pterosaur outgroup node would be decisively positive for the presence of *m. deltoides scapularis*, but equivocal as to whether it would have had one division as in turtles, lepidosaurs and crocodilians or two divisions as in birds. The basal archosauromorph pterosaur outgroup node would be decisive for the presence of a single division. For dinosaur sister-group pterosaurs it is unclear whether it would originate from the lateral scapula or just the acromion process, but Dilkes (2000) opted for an origin from the lateral scapula for *Maiasaura*. In basal archosauromorph pterosaurs the origin would be from the lateral to anterolateral surface of the scapula adjacent to the origin of *m. teres major*. *M. deltoides scapularis* would insert on the deltopectoral crest distal to the insertion of *m. scapulohumeralis anterior*.

In *Anhanguera* there are two muscle scars (#7 & 8) on the dorsal surface of the scapula that may be associated with the origin of this muscle; however, it is also reconstructed as originating from the anterior half of the dorsal surface of the scapula (#6). The insertion is reconstructed as on the dorsal surface of the deltopectoral crest (#36). In *Campylognathoides* no scars can be associated with this muscle and it is reconstructed as originating from the anteromedial half of the scapular blade (#1) and inserting on the dorsal surface of the deltopectoral crest (#21).

M. deltoides clavicularis
Based on Dilkes' (2000) analysis, this muscle would be present in dinosaur sister-group pterosaurs, but it is equivocal as to whether it would have had one division as in turtles, lepidosaurs and crocodilians or two as in birds. The basal archosauromorph pterosaur outgroup node is decisively positive for the presence of a single division of the muscle. It originates from the clavicle in lepidosaurs and birds, and from the scapula in crocodilians, which lack clavicles. Dilkes (2000) reconstructed *m. deltoides clavicularis* as originating from the scapula in *Maiasaura* because hadrosaurs lack clavicles. In pterosaurs the clavicle was incorporated into the sternum, and so it is possible that the origin shifted to the scapula, as it did in crocodilians, in which case it would have originated on the inferior part of the anterior border of the scapula, near the glenoid. However, it is also possible that the origin remained on the clavicles as they were incorporated into the sternum, in which case it would have originated from the anterior sternum. Although the basal archosauromorph pterosaur outgroup node is decisively positive for the presence of this muscle, it cannot be determined whether it originated from the scapula or the sternum, and therefore, the muscle is not reconstructed.

M. scapulohumeralis anterior
This muscle is absent in turtles, and in crocodilians it is either absent or homologous with the *m. deltoides clavicularis* of lepidosaurs or fused to the overlying *m. deltoides clavicularis*. Dilkes (2000) opted for the last interpretation and, based on his analysis, dinosaur sister-group pterosaurs would have had an *m. scapulohumeralis anterior*, and the same applies to basal archosauromorph pterosaurs. Dilkes (2000) suggested that it might have been fused to the *m. deltoides clavicularis*, as in crocodilians, and this is also possible for pterosaurs because they lack distinct clavicles and the insertion of *m. deltoides clavicularis* might have migrated to the scapula. *M. scapulohumeralis anterior* would originate from the scapula anterior to the glenoid and insert either with or adjacent to the *m. deltoides clavicularis* on the dorsal surface of the proximal end of the deltopectoral crest, proximal to the insertion of *m. deltoides scapularis*.

In *Anhanguera* the large scar on the process anterior to the glenoid fossa (#3) may be associated with this muscle, and so it is reconstructed as originating from it. No muscle scars can be associated with the insertion, but it is reconstructed as inserting on the dorsal surface of the proximal part of the deltopectoral crest. There are no scars in *Campylognathoides* that can be associated with this muscle, and it is reconstructed as originating immediately anterior to the glenoid fossa (#4) and inserting on dorsal surface of the proximal part of the deltopectoral crest (#20).

M. scapulohumeralis posterior
Based on Dilkes' (2000) analysis, this muscle would be present in dinosaur sister-group pterosaurs and also in basal archosauromorph pterosaurs. It would originate from the posterior margin of the scapula a short distance above the glenoid and insert on the dorsal surface of the posterior process distal to *m. subscapularis*.

In *Anhanguera* the origin of this muscle may be associated with the small pit (#18) on the posteroventral surface of the scapula, and it is also reconstructed originating from a small area proximal to the pit (#17). The origin can be associated with the muscle scar (#31) on the dorsal and posterior surface of the posterior tuberosity. In *Campylognathoides* no scars can be associated with the muscle, and it is reconstructed as originating on the posterolateral surface of the scapula above the glenoid (#11) and inserting on the dorsal surface of the posterior tuberosity (#22).

M. subcoracoscapularis
This muscle consists of two parts (*m. subcoracoideus* and *m. subscapularis*) in lepidosaurs and birds, whereas turtles and crocodilians have only a single *m. subscapularis*. Based on Dilkes' (2000) analysis dinosaur sister-group pterosaurs would have had an *m. subscapularis*, but it is uncertain whether they would have had an *m. subcoracoideus*. The same applies to basal archosauromorph pterosaurs. The *m. subscapularis* would have originated along the medial surface and/or posterior margin of the scapula and inserted on the dorsal surface of the posterior tuberosity. If *m. subcoracoideus* was present it would probably originate from the medial surface of the coracoid and insert on the dorsal surface of the posterior tuberosity.

In *Anhanguera* the small oval scar on the undersurface of the scapula (#16) may be associated with the origin of this muscle and the insertion can be associated with the muscle scar (#31) on the posterior tuberosity. In *Campylognathoides* no scars can be associated with the muscle and it is reconstructed originating with *m. scapulohumeralis posterior* on the posterolateral surface of the scapula above the glenoid (#11) and inserting with *m. scapulohumeralis posterior* on the dorsal surface of the posterior tuberosity (#22).

M. triceps
This muscle is present in all extant relatives of pterosaurs, but varies in the number of heads. Dilkes'

(2000) analysis suggests that the dinosaur sister-group pterosaur outgroup node is decisively positive for the presence of scapular and humeral heads, but equivocal as to whether they have a coracoid head, a posticum head, or a second humeral head. The basal archosauromorph pterosaur outgroup node is decisively positive for all but the posticum head, and equivocal for that. The scapular head would originate from the dorsal border of the glenoid fossa. The coracoid head would have originated from the posterior coracoid a short distance below the glenoid. The humeral head(s) would have originated from a large area on the extensor side of the shaft of the humerus. All heads would converge to insert by a tendon to the olecranon process of the ulna.

In *Anhanguera* the origins of three of the heads can be associated with scars: the scapular head with the curved scar (#11) above the glenoid; the medial head with the groove (#26) and the broad area of weak scarring (#27) on the medial surface of the shaft of the humerus; and the lateral head with the broad area of weak scarring (#34) on the lateral surface of the shaft of the humerus. No muscle scar can be associated with the origin of the coracoid head of *m. triceps*, but it is reconstructed as originating from the posterior surface of the coracoid below the glenoid. The insertion of *m. triceps* can be associated with the rugose crest on the posterior surface of the proximal ulna. In *Campylognathoides* the origin of the scapular head may be associated with the low tuberosity (#3) above the glenoid, and the origin of the medial head can be associated with narrow ridge on the dorsal humerus (#26). The medial head is reconstructed as originating from a large area (#25) on the medial surface of the shaft of the humerus in addition to the ridge (#26). The lateral head is reconstructed as originating from a large area (#19) on the lateral surface of the shaft of the humerus, and the coracoid head is reconstructed as originating on the posterior coracoid below the glenoid (#12). As in *Anhanguera*, the insertion can be associated with the crest on the posterior ulna.

M. pectoralis
This muscle is present in all extant relatives and therefore its presence can be inferred in pterosaurs as well. *M. pectoralis* would have originated from the sternum and perhaps the inferior ends of the ribs and inserted on the end and ventral surface of the deltopectoral crest.

In *Anhanguera m. pectoralis* can be associated with the median rugosity on the sternum and the muscle presumably originated from the ventral and lateral surfaces of the cristospine and the entire ventral surface of the posterior plate of the sternum. Its insertion can be associated with the terminal expansion (#24) of the deltopectoral crest and almost the entire ventral surface of the deltopectoral crest (#23 & #25). In *Campylognathoides* the origin would also have been the ventral and lateral surfaces of the cristospine and the entire ventral surface of the posterior plate of the sternum. The insertion can be associated with the terminal expansion of the deltopectoral crest (#18) and the muscle presumably also inserted on the entire ventral surface of the deltopectoral crest (#17).

M. supracoracoideus
Based on Dilkes' (2000) analysis, in dinosaur sister-group pterosaurs the *m. supracoracoideus* would have originated from the lateral surface of the coracoid inferior to the glenoid and might have had a second area of origin on the scapula, as in crocodilians and *Maiasaura* (Dilkes 2000). The insertion in turtles, lepidosaurs and crocodilians is on the flexor side of the proximal humerus, whereas in birds it is on the extensor side. Thus, the site of insertion would be uncertain in dinosaur sister-group pterosaurs, although Dilkes opted for an insertion on the flexor side of the deltopectoral crest in his reconstruction of *Maiasaura*. The basal archosauromorph pterosaur outgroup node is decisively positive for the presence of this muscle originating from the coracoid and inserting on the proximal part of the flexor side of the deltopectoral crest, but is equivocal as to whether there was a second head originating from the scapula, as in crocodilians.

In *Anhanguera* the origin of *m. supracoracoideus* can be associated with the long shallow groove (#5) on the anterolateral surface of the coracoid. The insertion is reconstructed as on the ventral surface of the proximal deltopectoral crest (#21). No scars can be associated with this muscle in *Campylognathoides*, but it is reconstructed originating from the anterolateral surface of the coracoid (#6) and inserting on the ventral surface of the proximal end of the deltopectoral crest (#15).

M. coracobrachialis
Turtles and lepidosaurs have two divisions (*longus* and *brevis*) of *m. coracobrachialis*, crocodilians have only one division (*brevis*), and birds have two divisions (*cranialis* and *caudalis*), which may or may not be homologous with the two divisions of turtles and lepidosaurs (Dilkes 2000). Therefore, it is equivocal as to whether dinosaur sister-group and basal archosauromorph pterosaurs had one or two divisions, and so only one division is reconstructed. *M. coracobrachialis* would originate from the external surface of the coracoid, and the *brevis* head would insert on the proximal part of the flexor side of the humerus. If a *longus* head was present it would insert on the medial supracondylar region of the humerus.

In *Anhanguera* the origin of this muscle can be associated with the coracoid tubercle (#12) and the long rugose groove (#13) extending down the posterolateral surface of the coracoid. No muscle scars can be associated with the insertion of the muscle, but it is reconstructed as inserting in the broad area (#22) on the proximal end of the ventral surface of the humerus. Note that in *Pteranodon* the coracoid flange bears a larger muscle scar than is present in the reconstructed specimen of *Anhanguera*. The larger muscle scar suggests that *m. coracobrachialis* may have been relatively larger than in *Anhanguera*, and the ventrolaterally directed flange may have improved its mechanical advantage. Likewise the very deep flange noted in *Quetzalcoatlus* suggests a large and powerful *m. coracobrachialis* and an altered mechanical advantage. In *Campylognathoides* this muscle can be associated with the prominent coracoid tubercle (#7) and it is also reconstructed as originating from the anterolateral surface (#8) of the coracoid and inserting on the broad area (#16) on the proximal end of the ventral surface of the humerus.

M. biceps
Extant lepidosaurs, crocodilians and birds have an *m. biceps* originating from the lateral surface of the coracoid anterior to *m. coracobrachialis* and inserting on the proximal ulna and radius. That would also be the case for both dinosaur sister-group and basal archosauromorph pterosaurs.

In *Anhanguera* the origin of this muscle can be associated with the prominent biceps tubercle (#4) of the coracoid, and the insertion can be associated with the biceps tubercle on the ulna a short distance distal to the elbow. In *Pteranodon* this tubercle is weakly divided in two by a groove running perpendicular to the long axis of the bone, and it is possible that this groove divides the insertion of *m. biceps* from that of *m. brachialis*, which probably would have inserted on the ulna near the insertion of *m. biceps*. No tubercle or scarred area has been noted on the radius and it is likely that the insertion on the radius was fleshy or absent. In *Campylognathoides* the origin can be associated with the prominent biceps tubercle (#5) anterior to the glenoid fossa and, although no muscle scars can be associated with its insertion, it probably inserted on the proximal radius and ulna.

Discussion

The reconstructions of the pectoral region of *Campylognathoides* and *Anhanguera* are based on the interpretation of pterosaurs as basal archosauromorphs, and although there would be some differences if the reconstructions were based on the interpretation of pterosaurs as a dinosaur sister-group (Table 1), these differences are minor and would not affect the conclusions of the following discussion. The most significant difference is in the evidence for the presence of *m. teres major*, which is decisively positive according to the interpretation of pterosaurs as basal archosauromorphs and equivocal according to the interpretation of pterosaurs as a dinosaur sister-group. *M. teres major* figures prominently in the following discussions and, if it were absent, its absence might affect part the conclusions. However, a number of specimens have muscle scars adjacent to the scar from the insertion of *m. latissimus dorsi* that suggest that *m. teres major* was present in pterosaurs regardless of their relationship to other diapsids.

Functional morphology

As noted above, Fürbringer (1900) and Padian (1983b) suggested that the origin of *m. supracoracoideus* might have been extended onto the large sternum as in birds. According to the reconstructions presented here (Figs 7, 8, 11 & 12), *m. supracoracoideus* originated from only the coracoid. There is no evidence, either in the form of muscle scars or from the origins of the muscle in extant relatives of pterosaurs, to suggest that the origin had been extended onto the sternum. Padian (1983b) suggested that *m. supracoracoideus* was the main elevator of the wing in pterosaurs, as it is in birds. He argued that the tendon of *m. supracoracoideus* passed up and over an acrocoracoid process before passing down to insert on the proximal humerus. The tendon was supposed to pass through a groove on the acrocoracoid process, and although the groove was labelled in one figure (Padian 1983b, fig. 7) and said to be clearly visible in other illustrations (e.g. Wellnhofer 1975a, fig. 9, according to Padian 1983b; Jensen and Padian 1989, fig. 2), It was not possible to identify a groove for a tendon on any rhamphorhynchoid scapulocoracoid. In birds, the groove for the tendon of *m. supracoracoideus* is prominent and readily identifiable, and its surface is smooth and distinct as it passes over the acrocoracoid process. No such prominent groove is present in rhamphorhynchoids.

There is a prominent groove between the biceps tubercle and the body of the coracoid in *Anhanguera*, and that groove has been interpreted as being for the tendon of *m. supracoracoideus* (Wellnhofer 1991a). However, the mere presence of a groove does not indicate that the groove was for a tendon. In the human scapula, the coracoid process is prolonged anteriorly so that the origins of *m. coracobrachialis* and the short head of *m. biceps brachii* are anterior to the center of the glenohumeral joint and can flex the humerus. At the base of the human coracoid process is the suprascapular notch, which appears well suited to act as a pulley for a tendon, as does the groove in *Anhanguera*, but which instead only provides a passage for the suprascapular nerve. Moreover, the bottom of the groove in AMNH 22555 is not smooth and flat, but instead has a ridge running across it between the anterior margin of the biceps tubercle and the body of the coracoid. Posterior to the ridge, a large pneumatic foramen opens on the floor of the groove. The ridge and the pneumatic foramen suggest that the groove was for a pneumatic duct and perhaps also vessels and nerves, not for a tendon. It is probable that the biceps tubercle of *Anhanguera* was prolonged dorsally so that the origin of *m. biceps* was at the level of the glenoid fossa and the muscle could flex the humerus without depressing it, and the groove may have been developed simply because of the pneumatic duct and other structures piercing the coracoid.

There is no evidence of the proposed groove for a tendon in any pterosaur but, despite that, let us consider Padian's (1983b) suggestion that Ostrom's (1976) explanation of how the action of *m. supracoracoideus* was reversed in birds could shed light upon the condition in pterosaurs. In extant birds *m. supracoracoideus* originates on the coracoid and sternum and inserts on the lateral process of the humerus. Its tendon passes upwards through the triosseal canal, over the acrocoracoid process and down to the humerus. Because it passes down to the humerus, *m. supracoracoideus* can elevate it, and *m. supracoracoideus* is the principal elevator of the humerus in flight (George & Berger 1966; Dial *et al.* 1988). This condition differs markedly from that of extant reptiles, in which the humerus is elevated by dorsal muscles (*m. latissimus dorsi*, *m. teres major*, *m. deltoides scapularis* and *m. scapulohumeralis anterior*; Jenkins & Goslow 1983) and depressed by ventral muscles (*m. pectoralis*, *m. coracobrachialis* and *m. supracoracoideus*). Ostrom (1976) presented a series of hypothetical steps in which the pull of *m. supracoracoideus* could be redirected so that its action would be changed from depression to elevation. In *Archaeopteryx*, *m. supracoracoideus* was reconstructed as originating on the coracoid and passing up and back before inserting on the proximal humerus; thus its action was depression and anterior flexion of the humerus. The muscle was dorsal to the biceps tubercle, but the tubercle did not alter the course of the muscle. Subsequently, the biceps tubercle was enlarged and shifted upwards in order to change the direction of pull of *m. biceps brachii* or for some other reason. *M. supracoracoideus* could not move ventrally past the biceps tubercle because of the curvature of the coracoid (and the adjacent clavicle), and therefore the course and pull of *m. supracoracoideus* was altered as the biceps tubercle moved upward. Eventually the course of *m. supracoracoideus* was altered so that it elevated the humerus, and thereafter its role in wing elevation could be directly selected for. The biceps tubercle thus became a pulley for the redirection of *m. supracoracoideus* in addition to the origin of *m. biceps brachii*.

There are two potential problems with applying Ostrom's scenario to pterosaurs: (1) the suggestion that the acrocoracoid process and the biceps tubercle were two separate structures, and (2) the lack of curvature of the coracoid (and an adjacent clavicle) that could prevent *m. supracoracoideus* from moving ventrally past an upward-migrating biceps tubercle. The first is only a problem if we accept the interpretation of a separate acrocoracoid process and biceps tubercle. If the tubercle on the coracoid (Figs 3, #12 & 4, #7) is the biceps tubercle, then the process anterior to the glenoid fossa, the acrocoracoid process of Padian (1983b), would not be associated with *m. biceps*. No explanation has been offered as to why a separate acrocoracoid process would have evolved or why it would have migrated upwards, diverting the course of *m. supracoracoideus*. If the coracoid tubercle was associated with the origin of *m. coracobrachialis* and the process anterior to the glenoid was the biceps tubercle, then it would be conceivable that the biceps tubercle migrated upwards in order to alter the pull of *m. biceps* or for some other reason; and if *m. supracoracoideus* was trapped above it, then that upward migration of the biceps tubercle would have diverted the course of *m. supracoracoideus*. However, there is no evidence that the coracoid in basal pterosaurs or their ancestors was curved or otherwise shaped so that *m. supracoracoideus* was trapped above the biceps tubercle, and no evidence that an upward migration of the biceps tubercle could in any way have altered the course of *m. supracoracoideus*.

It must be concluded that there was no groove over the biceps tubercle for the tendon of a muscle, that *m. supracoracoideus* did not pass over the biceps tubercle, and therefore that the course of *m. supracoracoideus* had not been altered by the biceps tubercle. In the reconstructions of *Campylognathoides* and *Anhanguera*, *m. supracoracoideus* originates on the anterolateral and anterior surfaces of the coracoid, respectively, and inserts on the proximal deltopectoral crest. In *Campylognathoides* it passes up and back to its insertion (Figs 7 & 8) and so could depress, flex and medially rotate the humerus. In *Anhanguera*, the muscle does not pass backwards as much as in *Campylognathoides* (Figs 11 & 12), but it still would depress, flex and medially rotate the humerus.

If *m. supracoracoideus* did not elevate the humerus, then elevation of the humerus must have been accomplished by dorsal muscles: *m. latissimus dorsi*, *m. teres major*, *m. deltoides scapularis* and *m. scapulohumeralis anterior*. In *Campylognathoides*, *m. latissimus dorsi* originated from the neural spines of six or more vertebrae, inserted on the posterior humerus, and was a large muscle (Figs 7 & 9). Its anterior part would elevate the humerus and also flex the humerus anteriorly somewhat, while its posterior parts would elevate and extend the humerus posteriorly. Note that it is possible that *m. latissimus dorsi* also originated from the posterior cervical vertebrae, in which case the anterior part, which is particularly well situated for humeral elevation, would have been more extensive. *M. teres major* originated from the length of the scapular blade, inserted on the posterior humerus, and was of moderate size. Like the posterior part of *m. latissimus dorsi*, it would elevate and extend the humerus posteriorly. *M. deltoides scapularis*, inserting on the deltopectoral crest, would elevate and flex the humerus anteriorly. *M. scapulohumeralis anterior* may not have been a particularly powerful elevator, but it would also be well situated to flex the humerus anteriorly. Thus, *m. scapulohumeralis anterior* and *m. deltoides scapularis* could oppose the tendency of the posterior part of *m. latissimus dorsi* and *m. teres major* to extend the humerus posteriorly while elevating it.

Depression of the humerus was accomplished by *m. pectoralis*, *m. coracobrachialis* and *m. supracoracoideus*. *M. pectoralis* was clearly the most powerful and important depressor, whereas *m. coracobrachialis* seems to have been of moderate size (Figs 7 & 8). However, *m. coracobrachialis* was probably relatively larger in those pterosaurs in which there is a coracoid flange (e.g. *Pteranodon*, *Quetzalcoatlus*). Anterior flexion of the humerus could be accomplished by *m. biceps*, *m. scapulohumeralis anterior*, *m. deltoides scapularis* and *m. supracoracoideus*. Posterior extension of the humerus could be accomplished by *m. teres major*, the scapular and coracoid heads of *m. triceps*, *m. scapulohumeralis posterior* and *m. subscapularis*. In addition, it is likely that the anterior and posterior parts of *m. latissimus dorsi* and *m. pectoralis* could assist in flexion and extension of the humerus.

M. supracoracoideus, *m. subscapularis* and *mm. scapulohumeralis anterior et posterior*, which inserted near the head of the humerus, probably collaborated to translate the head of the humerus through the saddle-shaped glenoid fossa and thereby to rotate the humerus about its long axis. In addition, they were probably important for stabilizing the glenohumeral joint, and it is likely that *m. biceps* and the scapular and coracoid heads of *m. triceps*, which originated near the glenoid fossa, also contributed to the stabilization of the glenohumeral joint.

Lastly, those muscles originating on the axial skeleton and inserting on the pectoral girdle were important for fixing the girdle so that the muscles originating on the girdle could move the wing. Thus, in *Campylognathoides*, *m. trapezius*, *m. levator scapulae* and *mm. serratus superficialis et profundus* (Fig. 6) were important for fixing and stabilizing the scapular blade so that it would not move laterally or ventrally when *m. deltoides scapularis* and *m. teres major* contracted.

Jenkins & Goslow (1983) noted that the functions of many pectoral muscles in the lizard *Varanus* were equivalent to those of the homologous muscles in the mammal *Didelphis* and hypothesised that this represented a retention of the primitive tetrapod functional pattern. The functions of many muscles interpreted from the muscle reconstruction of pterosaurs are similar to those of *Varanus* and *Didelphis* and suggest that the pattern of muscles and their function in basal pterosaurs was similar to that found in primitive tetrapods. Comparison of the reconstructed muscle pattern of *Campylognathoides* with that of *Crocodylus* and *Varanus* suggests that the musculature was little changed from the primitive pattern in archosauromorphs. Pterosaurs had evolved important specializations for flight (e.g. saddle-shaped glenohumeral joint, large plate-like sternum and large *m. pectoralis*), but the overall pattern of pectoral muscles does not seem to have been particularly specialized.

Functional consequences of the advanced pectoral girdle

The osteological consequences of the evolution of the advanced pectoral girdle of large pterodactyloids were: (1) rotation of the scapula and coracoid so that they were directed laterally rather than anterolaterally; (2) moving the glenoid fossa posteriorly relative to the sternum; and (3) an articulation between the scapula and the neural spines of the dorsal vertebrae so that the scapula mirrored the coracoid in resisting compressive forces between the glenoid and the axial skeleton. These osteological changes reflect corresponding changes in the associated muscles and ligaments. The change in the orientation of the scapula meant that the origins of *m. deltoides scapularis* and *m. teres major* were moved anteriorly relative to their insertions of the humerus and their direction of pull was changed from posteromedial (Fig. 9) to medial (Fig. 13). These changes probably improved the ability of *m. deltoides scapularis* to flex the humerus anteriorly and reduced the tendency of *m. teres major* to extend the humerus posteriorly during elevation, and so improved their function in wing elevation. The change in the orientation of the coracoid moved the glenoid posteriorly relative to the sternum and thus moved the insertion of *m. pectoralis* posteriorly relative to its origin on the sternum. This changed the direction of pull of *m. pectoralis* from posteromedial (Fig. 8) to medial (Fig. 12), and probably reduced the tendency of *m. pectoralis* to extend the humerus posteriorly during depression, improving its function in wing depression.

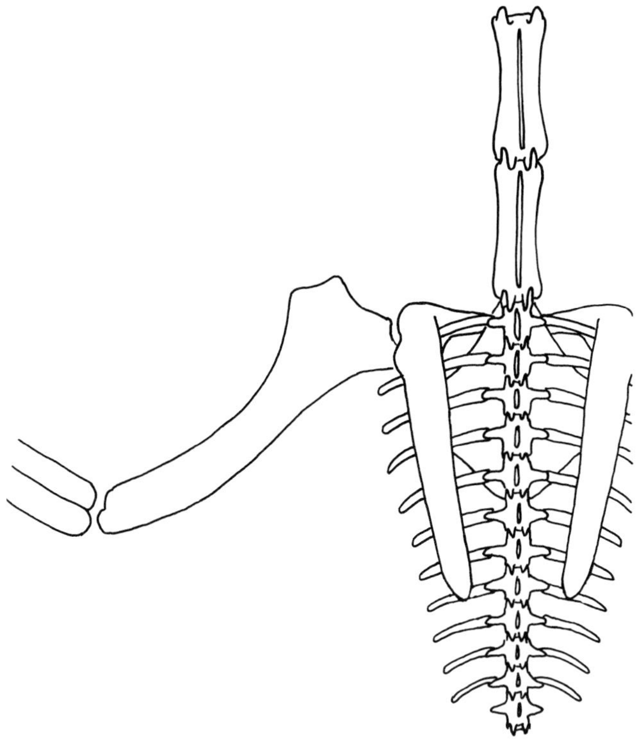

Fig. 14. Reconstruction of the left pectoral region of the skeleton of *Pterodactylus antiquus* in dorsal view (modified from Wellnhofer 1970, fig. 24). Note the posterior elongation of the scapula to the level of the eighth dorsal vertebra.

The articulation of the scapula with the neural spines of the dorsal vertebrae would have had three consequences. Firstly it would have braced the pectoral girdle against the vertebral column and the opposite pectoral girdle so that contraction of *m. latissimus dorsi* would not tend to pull the pectoral girdle medially, compressing or distorting the thorax. It is also possible that the bracing would prevent aerodynamic lift on the wing from pushing the girdle medially and compressing or distorting the thorax. This might have permitted *m. latissimus dorsi* to increase in size and power and to increase its importance in wing elevation without compressing or distorting the thorax. It is risky to predict the size or power of a muscle based on its projected area in a reconstruction, but comparison of the projected areas of *m. deltoides scapularis* in *Campylognathoides* (Fig. 9) and *Anhanguera* (Fig. 13) suggests that it was smaller and perhaps less important in wing elevation in *Anhanguera*.

The second consequence of the scapulonotarial articulation is that the role of *m. trapezius*, *m. levator scapulae* and *mm. serratus superficialis et profundus* in stabilizing the scapula would have been largely taken over by ligaments extending between the notarium and the medial scapula. This would mean that large pterodactyloids would not have to expend energy to fix the scapula while *m. deltoides scapularis* and *m. teres major* contracted. *M. trapezius*, *m. levator scapulae*, *m. sternocoracoideus* and *m. costosternocoracoideus* may have continued to control flexion and extension of the pectoral girdle on the sternum and notarium, and *m. trapezius* and *m. levator* scapulae may also have continued to move the neck and head, but it seems likely that there would be little need for *mm. serratus superficialis et profundus* to stabilize the scapula. *Mm. serratus superficialis et profundus* would also probably not be needed to move or stabilize the ribcage because the notarial ribs were fused to their vertebrae, and so it is probable that the muscles were reduced or lost.

The third consequence is that the dorsoventral depth of the thorax would have been essentially fixed by the pectoral girdle. That, plus the fusion of the notarial ribs to their vertebrae, would mean that the diameter of the thorax would be fixed as well, and so changing the thoracic volume in order to ventilate the lungs would require some unusual mechanism. Bennett (2001) suggested that rocking of the sternum on the scapulocoracoids combined with dorsoventral movements of the posteroventral body wall might have been sufficient for lung ventilation, but it is also possible that a hepatic piston, like that seen in extant crocodilians, was used to change the thoracic volume.

It is possible that the evolution of the advanced pectoral girdle began with the rotation of the scapulocoracoid, either to straighten the pull of *m. deltoides scapularis* and *m. teres major* relative to the wing or to move the glenoid fossa posteriorly relative to the sternum and straighten the pull of *m. pectoralis*. Whatever the cause, rotation of the scapulocoracoid would have brought the posteromedial end of the scapula in contact with the neural spine of the fourth dorsal vertebra. Selection for a bracing function of the scapula could only begin after contact between the scapula and the neural spines had been evolved for some other reason. Once an outwardly rotated scapulocoracoid in contact with the vertebral column had been achieved, selection for further improvements to the direction of pull of *m. deltoides scapularis*, *m. teres major* and *m. pectoralis* would lead to further rotation of the scapulocoracoid and development of a synovial joint between the scapula and vertebral column. It is possible that, after the scapula was braced against the vertebral column, selection for improved wing elevation led to increases in the size and power of *m. latissimus dorsi*, which in turn might have led to reduced selection on the size and power of *m. deltoides scapularis* and *m. teres major*, and a resultant decrease in their size and power. Fusion of the anterior dorsal vertebrae might have evolved in response to the stresses applied to the vertebral column by the scapular articulation, but it is also possible that the fusion of the vertebrae evolved before girdle rotation and the scapular articulation in response to the stresses placed on the vertebral column by a large and powerful *m. latissimus dorsi*.

The pectoral girdle of early pterodactyloids (e.g. *Pterodactylus antiquus*) is essentially unchanged from the condition found in basal pterosaurs. However, the scapula of *Pterodactylus* is considerably longer than in *Campylognathoides* and other rhamphorhynchoids, and has been reconstructed as extending posteriorly subparallel to the vertebral column (Fig. 14; Wellnhofer 1970). In this position the posterior end of the scapula is at the level of the eighth dorsal vertebra. The posterior elongation of the scapula would have extended the origins of *m. deltoides scapularis* and *m. teres major* posteriorly. This might have increased the size and power of the muscles, but it would also have increased the posterior component of their direction of pull, particularly in the case of *m. teres major*. The change might have improved the function of the muscles in wing elevation, but it also would have required increased power from *m. scapulohumeralis anterior* to counteract the tendency of *m. teres major* to extend the humerus posteriorly. Increased power from *m. teres major* would also have required increased power from *m. trapezius* and *m. levator scapulae* to fix the scapula as *m. teres major* contracted. These changes in *Pterodactylus* can be viewed as a response to a need for increased power or improved function in wing elevation, in the same way that the changes associated with the advanced pectoral girdle can be interpreted as a response to that need; however, the two responses are quite different.

Phylogenetic implications of the advanced pectoral girdle

A cladistic analysis of the Pterodactyloidea is beyond the scope of this paper, but it is necessary to discuss the phylogenetic implications of the advanced pectoral girdle. The significance of the advanced pectoral girdle was first noted by Young (1964), who erected a suborder Dsungaripteroidea to form a series of three grades with the Rhamphorhynchoidea and Pterodactyloidea. The Dsungaripteroidea was to include those pterosaurs with a notarium, and Young included all the then-known large pterodactyloids (the Dsungaripteridae, Ornithodesmidae, Ornithocheiridae, Pteranodontidae and Criorhynchidae). Each of the five families in the Dsungaripteroidea was viewed as descending from a different Jurassic taxon; thus the suborder was conceived as a paraphyletic grade taxon.

Bennett (1989, 1996) performed a cladistic analysis of the pterodactyloids and recognized a clade, Nyctosauridae + Dsungaripteridae + Pteranodontidae (*sensu* Padian 1984b, 1986 and Bennett 1989) + Azhdarchidae, which shared six synapomorphies, including notarium of fused dorsal vertebrae and dorsal ribs, and scapula rotated so that it forms a steep angle to the vertebral column. This clade was equivalent to Young's grade Dsungaripteroidea in composition. Within the clade, the Nyctosauridae were viewed as the sister-group to a clade Ornithocheiroidea (= Dsungaripteridae + Pteranodontidae + Azhdarchidae; tapejarids were inadequately known at the time), which shared the synapomorphies of neural spines of the notarium joined into a supraneural plate, and scapula articulating with a facet on the supraneural plate.

Kellner (1996; 2003) presented a cladistic analysis of the Pterosauria in which he recognized a clade Dsungaripteroidea (including the Nyctosauridae, Pteranodontoidea [= Pteranodontidae of Padian 1984b, 1986 and Bennett 1989] and Tapejaroidea [consisting of the Dsungaripteridae, Tapejaridae and Azhdarchidae]) sharing nine synapomorphies, including a notarium of at least three fused vertebrae. Kellner, like Bennett (1994) viewed the Nyctosauridae as the sister-group of the Ornithocheiroidea, which shared four synapomorphies.

Kellner (1996) only used a single character related to the advanced pectoral girdle (notarium of at least three fused vertebrae), but its distribution is fully consistent with a monophyletic Dsungaripteroidea and a single evolution of the advanced pectoral girdle among pterosaurs. Thus, Bennett (1989, 1994) and Kellner (1996) shared the view that there was a single clade of large pterodactyloids, including nyctosaurids, pteranodontoids, dsungaripterids and azhdarchids, with a single evolution of the advanced pectoral girdle. Both Bennett and Kellner also viewed the Nyctosauridae as the sister-group to the Ornithocheiroidea because *Nyctosaurus* seemed to lack a few ornithocheiroid characters. However, *Nyctosaurus* may have been coded incorrectly for some of those characters – the exact relationship between the scapula and the notarium in *Nyctosaurus* is still unclear. The systematic position of *Nyctosaurus* is beyond the scope of this paper, but the important point is that the Dsungaripteroidea and the Ornithocheiroidea may be synonymous.

Unwin (1995, 2003; Unwin & Lü 1997) presented an alternative view in which the advanced pectoral girdle was viewed as a correlate of large body size that was evolved convergently in several separate lineages of pterosaurs. In his cladogram the Ornithocheiroidea (= Pteranodontidae of Padian 1986 and Bennett 1989, and the Pteranodontoidea of Kellner 1996) is the sister-group to all other pterodactyloids, *Germanodactylus* is the sister-taxon to the Dsungaripteridae, *Germanodactylus* + Dsungaripteridae form the Dsungaripteroidea (not equivalent to that of Young 1964 or Kellner 1996), which in turn is the sister-group to the Azhdarchoidea (= Tapejaridae + Azhdarchidae). Thus, Unwin envisioned the advanced pectoral girdle evolving at least three times: in the ornithocheiroids (= pteranodontoids of Kellner 1996), dsungaripterids and azhdarchoids.

Bennett (1989, 1994) presented two characters related to the advanced pectoral girdle as synapomorphies of the Dsungaripteroidea and two additional pectoral characters as synapomorphies of the Ornithocheiroidea. Bennett (1989, 1994) and Kellner (1996) presented 11 other non-pectoral characters as synapomorphies of the Dsungaripteroidea and 5 other characters as synapomorphies of the Ornithocheiroidea. These 20 characters provide considerable support for the Dsungaripteroidea and the Ornithocheiroidea. However, the reconstruction of the pectoral myology presented in this paper demonstrates that previous analyses greatly underestimated the complexity of the advanced pectoral girdle complex. It is not merely the presence of a notarium of three or more fused dorsal vertebrae and their ribs, with a supraneural plate, and with the scapula rotated and articulating with that supraneural plate. It is also: the articulation for the scapula primarily on the neural spine of the fourth notarial vertebrae; rotation of the coracoid so that the coracoid is directed laterally rather than anterolaterally; the rearrangement of the course and direction of pull of *m. deltoides scapularis* and *m. teres major* originating on the scapula the rearrangement of the function of *m. levator scapulae*, *m. trapezius* and *mm. serratus superficialis et profundus* that had functioned primarily to fix the scapula; and the rearrangement of *m. coracobrachialis*, *m. supracoracoideus* and other muscles originating or inserting on the coracoid. It may also be a rearrangement of lung ventilation, because the pectoral girdle articulations and rib fusion would make normal rib and sternal movements impossible. I could continue, but the point is that the advanced pectoral girdle is merely the osteological expression of a very complex reorganization of the pterodactyloid thorax and forelimb. As such it is improbable that it could be evolved convergently in different lineages of pterodactyloids, and future cladistic analyses should include characters relating to the advanced pectoral girdle unless good cause can be brought to exclude them.

In conclusion, if one accepts that the advanced pectoral girdle is a synapomorphy of a monophyletic Dsungaripteroidea (*sensu* Kellner 1996), then one may consider which pterodactyloids might be close to the ancestor of the Dsungaripteroidea. Although the various small specimens of *Pterodactylus* have long been viewed as primitive and presumably ancestral to all other pterodactyloids (Bennett 1996, 2002), recent work has suggested that they only appear primitive because they are immature. The elongate scapula of *P. antiquus* supports that view because the pectoral girdle of *Pterodactylus* is too derived to have been ancestral to the advanced pectoral girdle of large pterodactyloids. Rotation of the scapulocoracoid in *Pterodactylus* would have brought the posterior end of the scapula into contact with the neural spine of the seventh or eighth dorsal vertebra, and it is unlikely that the scapula would have been shortened and the contact would have moved anteriorly to arrive at the advanced pectoral girdle condition. A short scapula with its posterior end adjacent to the neural spine of the fourth dorsal vertebra could be more easily modified to produce the advanced pectoral girdle of large pterodactyloids. Therefore, we should look for closest relatives of the Dsungaripteroidea in those pterodactyloids that still retained the primitive pectoral girdle of the rhamphorhynchoids.

Conclusions

This paper presents the first reconstructions of the myology of the pectoral regions of a representative rhamphorhynchoid exhibiting a primitive pectoral girdle and a large pterodactyloid exhibiting an advanced pectoral girdle. The reconstructions were based on inferences about the character states of the muscles drawn from the extant relatives of pterosaurs. The reconstructions show that *m. supracoracoideus* was not an elevator of the wing and that the pterosaurian pectoral girdle was not functionally like that of birds. Instead, the pectoral girdle, although specialized in many respects, retained much of the primitive tetrapod functional pattern. The reconstructions also show that the pectoral region of large pterodactyloids was highly specialized for flight. The complexity of the osteological and myological specialization of the advanced pectoral girdle makes it unlikely that it could have been evolved convergently in separate lineages, and supports a monophyletic Dsungaripteroidea.

I thank M. A. Norell (American Museum of Natural History) and K. L. Stadtman (Brigham Young University) for access to and assistance with specimens under their care. P. Christianssen and J.-M. Mazin kindly reviewed an earlier version of the manuscript.

References

BENNETT, S. C. 1989. A pteranodontid pterosaur from the Early Cretaceous of Peru, with comments on the relationships of Cretaceous pterosaurs. *Journal of Paleontology*, **63**, 669–677.

BENNETT, S. C. 1993. The ontogeny of *Pteranodon* and other pterosaurs. *Paleobiology*, **19**, 92–106.

BENNETT, S. C. 1994. *Taxonomy and systematics of the Late Cretaceous pterosaur* Pteranodon *(Pterosauria, Pterodactyloidea)*. Museum of Natural History, University of Kansas, Occasional Papers, **169**, 70 pp.

BENNETT, S. C. 1996. The phylogenetic position of the Pterosauria within the Archosauromorpha. *Zoological Journal of the Linnean Society, London*, **118**, 261–309.

BENNETT, S. C. 2000. New information on the skeleton of *Nyctosaurus*. *Journal of Vertebrate Paleontology*, **20**(3) (Supplement) 29A.

BENNETT, S. C. 2001. The osteology and functional morphology of the Late Cretaceous pterosaur *Pteranodon*. Part I. General description of osteology. Part II. Functional Morphology. *Palaeontographica, A*, **260**, 1–112, 113–153.

BENNETT, S. C. 2002. Soft tissue preservation of the cranial crest of the pterosaur *Germanodactylus* from Solnhofen. *Journal of Vertebrate Paleontology*, **22**, 43–48.

BENTON, M. J. 1999. *Scleromochlus taylori* and the origin of dinosaurs and pterosaurs. *Philosophical Transactions of the Royal Society, London (B)*, **354**, 1423–1446.

BRINKMAN, D. 1988. Size-independent criteria for estimating relative age in *Ophiacodon* and *Dimetrodon* (Reptilia, Pelycosauria) from the Admiral and Lower Belle Plains Formations of west-central Texas. *Journal of Vertebrate Paleontology*, **8**,172–180.

BRYANT, H. N. & RUSSELL, A. P. 1992. The role of phylogenetic analysis in the inference of unpreserved attributes of extinct taxa. *Philosophical Transactions of the Royal Society, London (B)*, **337**, 405–418.

BRYANT, H. N. & SEYMOUR, K. L. 1990. Observations and comments on the reliability of muscle reconstruction in fossil vertebrates. *Journal of Morphology*, **206**, 109–117.

DIAL, K. P., KAPLAN, S. R., GOSLOW, G. E. Jr. & JENKINS, F. A. Jr. 1988. A functional analysis of the primary upstroke and downstroke muscles in the domestic pigeon (*Columba livia*) during flight. *Journal of Experimental Biology*, **134**, 1–16.

DILKES, D. W. 2000. Appendicular myology of the hadrosaurian dinosaur *Maiasaura peeblesorum* from the Late Cretaceous (Campanian) of Montana. *Transactions of the Royal Society of Edinburgh, Earth Sciences*, **90**, 87–125.

FÜRBRINGER, M. 1876. Zur vergleichenden Anatomie der Schultermuskeln. *Morphologisches Jahrbuch*, **1**, 636–816.

FÜRBRINGER, M. 1900. Zur vergleichenden Anatomie des Brustschulterapparates und der Schultermuskeln. *Jenaische Zeitschrift für Naturwissenschaft*, **34**, 215–718.

GALTON, P. M. 1969. The pelvic musculature of the dinosaur *Hypsilophodon* (Reptilia: Ornithischia). *Postilla*, **131**, 1–64.

GEORGE, J. C. & BERGER, A. J. 1966. *Avian Myology*. Academic Press, New York.

GREGORY, W. K. & CAMP, C. L. 1918. Studies in comparative myology and osteology. No. III. *Bulletin of the American Museum of Natural History*, **38**, 447–563.

HOWSE, S. C. B. 1986. On the cervical vertebrae of the Pterodactyloidea (Reptilia: Archosauria). *Zoological Journal of the Linnean Society*, **88**, 307–328.

JENKINS, F. A., Jr & GOSLOW, D. E. Jr. 1983. The functional anatomy of the shoulder of the Savannah Monitor Lizard (*Varanus exanthematicus*). *Journal of Morphology*, **175**, 195–216.

JENSEN, J. A. & PADIAN, K. 1989. Small pterosaurs and dinosaurs from the Uncompahgre Fauna (Brushy Basin Member, Morrison Formation: ?Tithonian), Late Jurassic, western Colorado. *Journal of Paleontology*, **63**, 364–373.

KELLNER, A. W. A. 1996. *Description of new material of Tapejaridae and Anhangueridae (Pterosauria, Pterodactyloidea) and discussion of pterosaur phylogeny*. Unpublished PhD dissertation, Columbia University, New York.

KELLNER, A. W. A. 2003. Pterosaur phylogeny and comments on the evolutionary history of the group. *In*: BUFFETAUT, E. & MAZIN, J-M. (eds) *Evolution and Palaeobiology of Pterosaurs*. Geological Society, London, Special Publications, **217**, 105–137.

KELLNER, A. W. A. & TOMIDA, Y. 2000. *Description of a New Species of Anhangueridae (Pterodactyloidea) with Comments on the Pterosaur Fauna from the Santana Formation (Aptian–Albian), Northeastern Brazil*. National Science Museum, Tokyo, Monographs, **17**, 135 pp.

KRIPP, D. VON. 1943. Ein Lebensbild von *Pteranodon ingens* auf flugtechnisher Grundlage. *Nova Acta Leopoldina, Neue Folge*, **12** (83), 16–32.

MARSH, O. C. 1876. Principal characters of American pterodactyls. *American Journal of Science, Series 3*, **12**, 479–480.

McGowan, C. 1979. The hind limb musculature of the Brown Kiwi, *Apteryx australis mantelli*. *Journal of Morphology*, **160**, 33–73.

McGowan, C. 1982. The wing musculature of the Brown Kiwi *Apteryx australis mantelli* and its bearing on ratite affinities. *Journal of Zoology*, **197**, 173–219.

Monastersky, R. 2001. Pterosaurs. *National Geographic Magazine*, **199** (5), 86–105.

Nicholls, E. L. & Russell, A. P. 1985. Structure and function of the pectoral girdle and forelimb of *Struthiomimus altus* (Theropoda: Ornithomimidae). *Palaeontology*, **28**, 643–677.

Ostrom, J. H. 1976. Some hypothetical anatomical stages in the evolution of avian flight. *Smithsonian Contributions to Paleobiology*, **21**, 1–27.

Padian, K. 1983a. Osteology and functional morphology of *Dimorphodon macronyx* (Buckland) (Pterosauria: Rhamphorhynchoidea) based on new material in the Yale Peabody Museum. *Postilla*, **189**, 1–44.

Padian, K. 1983b. A functional analysis of flying and walking in pterosaurs. *Paleobiology*, **9**, 218–239.

Padian, K. 1984a. The origin of pterosaurs. *In*: Reif, W.-E. & Westphal, F. (eds) *Third Symposium on Mesozoic Terrestrial Ecosystems and Biota, Tubingen*. 163–168.

Padian, K. 1984b. A large pterodactyloid pterosaur from the Two Medicine Formation (Campanian) of Montana. *Journal of Vertebrate Paleontology*, **4**, 516–524.

Padian, K. 1985. The origins and aerodynamics of flight in extinct vertebrates. *Palaeontology*, **28**, 413–433.

Padian, K. 1986. A taxonomic note on two pterodactyloid families. *Journal of Vertebrate Paleontology*, **6**, 289.

Peters, D. 2000. A reexamination of four prolacertiforms with implications for pterosaur phylogenesis. *Rivista Italiana di Paleontologia e Stratigraphia*, **106**, 293–336.

Plieninger, F. 1895. *Campylognathus zitteli*. Ein neuer Flugsaurier aus dem Oberen Lias Schwabens. *Palaeontographica*, **41**, 193–222.

Plieninger, F. 1907. Die Pterosaurier der Juraformation Swabens. *Palaeontographica*, **53**, 209–313.

Romer, A. S. 1923. The pelvic musculature of saurischian dinosaurs. *Bulletin of the American Museum of Natural History*, **48**, 605–617.

Romer, A. S. 1927. The pelvic muscles of ornithischian dinosaurs. *Acta Zoologica*, **8**, 225–275.

Seeley, H. G. 1870. *The Ornithosauria: An Elementary Study of the Bones of Pterodactyles*. Deighton, Bell & Co. Cambridge, 135 pp.

Seeley, H. G. 1881. On evidence of two ornithosaurians referable to the genus *Ornithocheirus*, from the Upper Greensand of Cambridge, preserved in the collection of W. Reed, Esq., F.G.S. *Geological Magazine*, **8**, 13–20.

Seeley, H. G. 1891. On the shoulder-girdle in the Cretaceous Ornithosauria. *Annals and Magazine of Natural History, Series 6*, **7**, 438–445.

Seeley, H. G. 1901. *Dragons of the Air: An Account of Extinct Flying Reptiles*. Methuen and Co., London.

Sereno, P. C. 1991. Basal archosaurs: phylogenetic relationships and functional implications. *Journal of Vertebrate Paleontology*, **11** (4)(supplement), 1–53.

Unwin, D. M. 1995. Preliminary results of a phylogenetic analysis of the Pterosauria (Diapsida: Archosauria). *In*: Ailing, S. & Yuanqing, W. (eds) *Sixth Symposium on Mesozoic Terrestrial Ecosysytems and Biota, Beijing, Short Papers*, 69–72.

Unwin, D. M. 2000. *Sharovipteryx* and its significance for the origin of the pterosaurs flight apparatus. *Fifth European Workshop on Vertebrate Palaeontology, 27.06–01.07.2000, Karlsruhe*. [Abstract]

Unwin, D. M. & Lü, J. 1997. On *Zhejiangopterus* and the relationships of pterodactyloid pterosaurs. *Historical Biology*, **12**, 199–210.

Wellnhofer, P. 1970. Die Pterodactyloidea (Pterosauria) der Oberjura-Plattenkalke Süddeutschlands. *Bayerische Akademie der Wissenschaften, Mathematisch-Wissenschaftlichen Klasse, Abhandlungen*, **141**, 133 pp.

Wellnhofer, P. 1974. *Campylognathoides liasicus* (Quenstedt), an Upper Liassic pterosaur from Holzmaden—The Pittsburgh specimen. *Annals of the Carnegie Museum, Pittsburgh*, **45**, 5–34.

Wellnhofer, P. 1975a. Die Rhamphorhynchoidea (Pterosauria) der Oberjura-Plattenkalke Süddeutschlands. Teil I. Allgemeine Skelettmorphologie. *Palaeontographica A* ,**148**, 1–33.

Wellnhofer, P. 1975b. Die Rhamphorhynchoidea (Pterosauria) der Oberjura-Plattenkalke Süddeutschlands. Teil II. Systematische Beschreibung. *Palaeontographica A*, **148**,132–186.

Wellnhofer, P. 1975c. Die Rhamphorhynchoidea (Pterosauria) der Oberjura-Plattenkalke Süddeutschlands. Teil III. Palaökologie und Stammesgeschichte. *Palaeontographica A*, **149**,1–30.

Wellnhofer, P. 1978. *Pterosauria*. *In*: Wellnhofer, P. (ed.) *Handbuch der Palaeoherpetologie*. Gustav Fischer, Stuttgart, Teil 19, 82 pp.

Wellnhofer, P. 1985. Neue Pterosaurier aus der Santana-Formation der Chapada do Araripe, Brasilien. *Palaeontographica A*, **187**, 105–182.

Wellnhofer, P. 1991a. Weitere Pterosaurierfunde aus der Santana-Formation (Apt) der Chapada do Araripe, Brasilien. *Palaeontographica A*, **215**, 43–101.

Wellnhofer, P. 1991b. *The Illustrated Encyclopedia of Pterosaurs*. Salamander, London, 192 pp.

Wellnhofer, P., Buffetaut, E. & Gigase, P. 1983. A pterosaurian notarium from the Lower Cretaceous of Brazil. *Paläontologische Zeitschrift*, **57**, 147–157.

Wild, R. 1978. Die Flugsaurier (Reptilia, Pterosauria) aus der Oberen Trias von Cene bei Bergamo, Italien. *Bolletino della Societa Paleontologica Italiana*, **17**, 176–256.

Wild, R. 1993. A juvenile specimen of *Eudimorphodon ranzii* Zambelli (Reptilia, Pterosauria) from the Upper Triassic (Norian) of Bergamo. *Rivista del Museo Civico di Scienze Naturali 'Enrico Caffi', Bergamo*, **16**, 95–120.

Wiman, C. 1923. Über *Dorygnathus* und andere Flugsaurier. *Bulletin of the Geological Institution of the University of Uppsala*, **19**, 23–54.

Witmer, L. M. 1995. The extant phylogenetic bracket and the importance of reconstructing soft tissues in fossils. *In*: Thomason, J. (ed.), *Functional Morphology in Vertebrate Paleontology*. Cambridge University Press, New York, 19–33.

Young, C. C. 1964. On a new pterosaurian from Sinkiang, China. *Vertebrata Palasiatica*, **8**, 221–255.

The detailed anatomy of *Rhamphorhynchus*: axial pneumaticity and its implications

NIELS BONDE[1] & PER CHRISTIANSEN[2]

Geological Institute, Øster Voldgade 10, 1350 Copenhagen K, Denmark (e-mail: nielsb@geo.geol.ku.dk)
Department of Vertebrates, Zoological Museum, Universitetsparken 15, 2100 Copenhagen Ř, Denmark (e-mail: p2christiansen@zmuc.ku.dk)

Abstract: An acid- and transfer-prepared, juvenile *Rhamphorhynchus muensteri*, despite some fragmentation, is in an excellent state of three-dimensional preservation, exposing exquisite anatomical details hitherto unknown in other pterosaurs. Here we describe the axial pneumatizations of the cervical and anterior dorsal vertebrae and the sternum. The interior of the cervical centra is subdivided into a pair of large camerae, presumably by air sacs entering by large pleurocoels in the sides of the centra. This so-called 'camerate' type of pneumatization is hitherto unknown in pterosaurs. Another excavation enters from the ventral side into the base of the neural arch and stretches between the pre- and postzygapophyses. This type of cavity also penetrates from the ventral side into the base of the first few transverse processes of the dorsal vertebrae, although these lack central pleurocoels. The cristospine also has a complex pneumatic foramen.

Skeletal pneumaticity is most probably a result of a highly derived pulmonary system, as in extant birds. Morphologically similar pneumatic features are present in most saurischian dinosaurs and it is possible that they are the result of convergence. Because basal members of the various groups, including Triassic pterosaurs, appear to lack skeletal pneumaticity, convergence seems likely, although the stem-ornithodiran parsimoniously possessed a more bird-like than 'reptile'-like pulmonary system, albeit non-invasive. This points to possible tachymetabolism in these forms, which is in accord with the distribution of other factors such as integumentary structures and bone histology. It is concluded that evolution of this suite of advanced features, surprisingly, was among the earliest events in the ornithodiran lineage soon after it split off from its crocodilian sister-group.

For more than a century it has been recognized that pterosaurs possessed light skeletons, with thin and very compact bone walls and a meshwork of trabecular struts inside the hollow long bones for maintenance of mechanical strength (e.g. Wellnhofer 1991a; Ricqlès *et al.* 2000), as in extant birds (e.g. Rogers & LaBarbera 1993). Additionally, it has long been recognized that they also possessed distinct pneumatic fossae and foramina throughout the axial and appendicular skeleton (e.g. Seeley 1870, 1901; Marsh 1871, 1872; Eaton 1910; Wild 1971; Wellnhofer 1975a, b, 1978, 1980, 1991a, b; Kellner 1991; Bennett 1994, 2001a, b; Frey & Martill 1996; Viohl 2000). However, as Padian (1983a) pointed out, there is still no thorough systematic overview of the distribution of pneumaticity in pterosaurs and its possible systematic implications (but see Bennett 1994, Unwin 1995 and Unwin & Lü 1997 for discussion of pneumatic characters in pterodactyloid systematics). Despite the very common reference to pneumatopores when describing the osteology of pterosaurs, most references remain anecdotal, simply noting their presence with few if any comments on the detailed morphology and its systematic or soft-tissue implications. It appears, however, that the large pterodactyloids have more extensively pneumatized skeletons than the smaller 'rhamphorhynchoids', and the cervical vertebral column in particular is often extensively pneumatized (see e.g. Eaton 1910, Kellner 1991, Bennett 1994, 2001a).

Pterosaurs, like theropod dinosaurs and mammals, have hollow longbones, but hollow bones do not in themselves imply the presence of pneumaticity. Wild (1971), for instance, notes that *Dorygnathus* has pneumatic bones and subsequently goes into some detail about why the pneumatic bones almost certainly lacked external pneumatopores! Wellnhofer (1975a, 23) claims that the forearm bones of *Rhamphorhynchus* are pneumatic but state that they lack pneumatopores. Both statements are contradictory, because pneumatization is defined as the process by which air diverticulae from the lungs resorb bony tissue, thus invading the bone from the exterior (Baer 1896; Bellairs & Jenkin 1960; King 1966). Accordingly, a very important parameter in the evaluation of the presence of pneumaticity is examination of the bone exterior for properly sized external foramina that communicate with the bone interior (Hogg 1980; Witmer 1990; Britt *et al.* 1998). Such features are indeed common in many pterosaurs, particularly in the cervicals of pterodactyloids, but the internal structure of these chambers is virtually unknown in most cases, unless the bones are very well preserved but fractured.

From: BUFFETAUT, E. & MAZIN, J-M. (eds) 2003. *Evolution and Palaeobiology of Pterosaurs*. Geological Society, London, Special Publications, **217**, 217–232. 0305–8719/03/$15

The 'rhamphorhynchoid' pterosaur *Rhamphorhynchus* from the Late Jurassic (Tithonian) Solnhofen limestone of southern Germany (Barthel 1978; Viohl 1985, 1998) is known from a large number of skeletal specimens (over 100), many of which are nearly complete (Wellnhofer 1975a–c), and some of the Solnhofen specimens are known with soft-tissue preservation, most notably the chiropatagia (Wellnhofer 1975b, 1991a; Viohl 2000). Accordingly, with the possible exception of the edentulous and considerably larger North American *Pteranodon* from the Late Cretaceous, which is also known from very abundant fossil material (e.g. Eaton 1903, 1910; Bennett 1994, 2001a, b), *Rhamphorhynchus* must be one of the most well-known and well-studied pterosaurs in the world. The specimens vary considerably in size, by more than a factor of four when comparing the length of the wings, and differ in skeletal ossification as well. Initially much of this variation was considered species specific, and as many as five species were recognized (e.g. Wellnhofer 1975b, c, 1991a). Subsequently, this variation has been attributed to differences in ontogeny within the same species, *Rhamphorhynchus muensteri* (Bennett 1995, 1996b). A wide size range within presumably adult specimens is also recognized in other species of pterosaurs and is attributed to prolonged, albeit slower rates of growth (Bennett 1993; Unwin 2001) succeeding the very rapid initial growth phases characteristic of endothermic vertebrates.

Rhamphorhynchus is the type genus of the family Rhamphorhynchidae, which is considered the sister-taxon to the monophyletic Pterydactyloidea (Unwin 1992, 1995), implying paraphyly of the 'Rhamphorhynchoidea'. Although *Rhamphorhynchus* is known from several hundred individuals a large number of detailed anatomical characters have remained relatively obscure, due to the state of preservation and particularly the method of preparation of virtually all the known specimens. Even complete and well-preserved specimens have nearly always been mechanically prepared, thus only exposing one face of the bones and obscuring from view the potentially very delicate details of the vertebral and cranial anatomy. The present specimen, housed at the Geological Museum in Copenhagen as MGUH 1891.738, differs substantially from this pattern. It is a disarticulated, partial skeleton that is very well preserved. The specimen was originally assigned to the type species *R. muensteri* by Wellnhofer (1975b) and, accordingly, has not been synonymized following the recognition that characters previously considered species-specific are probably ontogenetic (Bennett 1995). The specimen was figured and a few details were described by Wellnhofer (1975a, b), who based reconstructions of several anatomical details on this specimen (his specimen 71); but apart from also noting the extraordinary nature of the specimen, he did not elaborate on many details of its anatomy. The mode of preparation, rather than the original state of preservation, appears, however, to be the primary agent responsible for the degree of details visible in the specimen. In this paper we consider the slab to be the part containing the major part of the skeleton, as opposed to Wellnhofer (1975a, b) who called this part the counterslab ('Gegenplatte').

Materials and methods

The specimen was recovered from the Solnhofen Lithographic Limestone (locality on the old label of 1891 states 'Solnhofen', without additional details). Initially not too much of the specimen was visible on the surface of the slabs, but it was apparent that soft tissues were probably not preserved, in which case the specimen would probably have had to be prepared mechanically. Accordingly, it was decided to attempt acid preparation. This was carried out by N.B. and other students during a student laboratory course in the early 1960s, under the supervision of E. Nielsen, at the time the only vertebrate palaeontologist in Denmark. The slabs were encased in a two-component artificial resin, which has subsequently turned yellow, although it retains much of its original transparency, and was prepared using the acetic acid (5–10%) transfer method of Toombs and Rixon (1959). Virtually all the original limestone matrix was dissolved; neither the slab nor counterslab sides have subsequently been filled with resin and, consequently, are fully open. Following acid preparation the specimen was carefully rinsed with water for an extended period.

Unexpectedly, the slabs proved to contain a completely uncrushed, disarticulated, partial skeleton (*contra* Wellnhofer [1975b, p. 155], who says that the wing phalanges appear crushed). The state of preservation and subsequent exposure by the acetic acid is such that the skeleton looks nearly extant and the apparent thinness and lightness of the bones belie the fact that they do not appear particularly fragile. In fact, several ribs were still somewhat elastic when the specimen was initially exposed, but most of this elasticity has subsequently diminished. The main slab has been on exhibit for almost three decades at the Geological Museum in Copenhagen, standing on one end and leaning against the back wall of a wooden showcase, without dampening tissue on the bottom edge. The specimen has deteriorated since Wellnhofer's (1975a–c) study. When the specimen was finally removed from the exhibit it was discovered to be slightly damaged, and that parts had come loose, probably as a result of slight, but nearly constant tremors from the subway that runs directly below the museum. The absence of humidity and temperature control in the showcase has resulted in the bones losing their initial elasticity, and has perhaps also contributed to the slight damage of the specimen. In Wellnhofer (1975b, pls 23 (fig. 6) & 24) several ribs can be seen on the slab. The posterior-most of these ribs is now missing, and part of one lower jaw and some teeth and a few fragments of other skull bones were loose and displaced (now attached with glue to the resin).

The skeleton lacks parts of the braincase and skull, one wing and the caudal vertebral column (Fig. 1a, b), but most of the skull and braincase are preserved in undistorted, three-dimensional detail unprecedented in a rhamphorhynchoid pterosaur. The postcranium is also three-dimensional and displays intricate details of the axial and appendicular skeleton. The entire cervical vertebral column, consisting of nine vertebrae, is present and articulated, albeit disarticulated from the dorsal vertebral column, and the dorsal vertebral column, consisting of 14 vertebrae, is also present and articulated. The sacrum consists of three articulated sacrals. In several vertebrae the lamella is absent or damaged, meaning that not only the external parts but also the internal parts of the vertebrae are exposed in great detail. Ironically, the excellent state of this specimen makes detailed comparisons with other *Rhamphorhynchus* specimens difficult. Compared, for instance, to the famous material in the collections of the Bayerische Staatssammlung in München, it is evident that many anatomical details of this specimen cannot be compared to those of most other specimens, even complete skeletons, as such details are simply not visible after mechanical preparation.

General description

The aim of this paper is to provide a detailed description of the pneumatic features, not the osteology of the specimen, but a few general remarks on other osteological characters seem appropriate. Wellnhofer's papers (1975a–c) remain the quintessential works on the anatomy of *Rhamphorhynchus*, but the descriptions of osteology (Wellnhofer 1975a) are often very brief, noting only a few details. The present specimen displays characters not featured in Wellnhofer (1975a–c) and, significantly, in some cases the observed characters in the present specimen differ from Wellnhofer's observations.

The slab (Fig. 1a) contains most of the preserved skeletal parts, and particularly the complete, articulated cervical vertebral column, exposed from the ventral side, is exquisitely preserved. The dorsal vertebral column is also present in articulation and is exposed from the dorsal side, although tilted to the left. In addition to the axial skeleton the slab contains the entire right wing, hindlimb, pelvic and sternal elements, the disarticulated skull, and the three-dimensionally exposed, undistorted anterior end of the upper jaws. The counterslab (Fig. 1b) contains much less bone, but the wing is also clearly visible and a lower jaw ramus is fully exposed. Significantly, the counterslab contains the 14 articulated dorsal centra, exposed ventrally and up to about the level of the transverse processes. It also contains a three-dimensional prepubis, the entire cristospine and parts of the anterior face of the sternum, exposed dorsally and showing the articulating facets for the coracoids. Also visible on the counterslab is the three-dimensionally preserved row of cervical neural spines.

The humerus (Fig. 2) is medially exposed and shows a distinct, elongate sinusoid muscle scar on the medial face of the diaphysis, presumably the fleshy insertion of the *m. latissimus dorsi*. Such a morphology of this muscle scar is hitherto unknown in *Rhamphorhynchus*. Despite its size the humerus appears unpneumatized (Fig. 2). In the jaws replacement teeth can be seen in a few places and the slab and counterslab combined provide excellent exposures of nearly every aspect of the complicated carpus. The pelvis is dorsally exposed and undistorted (Fig. 2). It clearly demonstrates that the acetabulum was laterodorsally exposed, thus potentially contributing to the debate on pterosaur terrestrial locomotion (see e.g. Padian 1983a, b; Wellnhofer 1988, 1991b; Unwin 1996; Bennett 1997; Henderson & Unwin 2001a, b) by seemingly making bipedal running very awkward, if possible at all.

As noted by Wellnhofer (1975a) there are nine procoelous cervicals and the atlas is a small, short, ring-shaped element, about 1 mm long. Anteriorly the atlas is concave and, thus, resembles a 'normal' procoelous centrum. The neurapophyses of the atlas are slender, 3-mm long bones, paired and pointing backwards from the lateral sides of the atlantal centrum. Caudodorsally the neurapophyses touch the tall neural arch of the axis at a small distinct articular area. The two atlantal neurapophyses do not meet at all in the mid-line. Wellnhofer (1975a) states that there is a single triangular proatlas that extends anteriorly over the atlas. However, on the counterslab, adjacent to the characteristically tall and triangular neural spine of the axis, are exposed the paired neural arches of the proatlas (Fig. 1b, pa), lying at about a 50° angle to each other, each triangular in shape and most closely resembling an arrowhead. The proximal articulating facet of a proatlas is complex, sinusoid and irregular, indicating some mobility. In Wellnhofer (1975a, fig. 6a), the axis intercentrum is shown as a wedge-shaped bone with the sharp end directed ventrally. The present specimen, however, indicates that this is incorrect. In *Rhamphorhynchus*, as in other vertebrates, the sharp end of the wedge pointed dorsally. The ventral length of the axis intercentrum

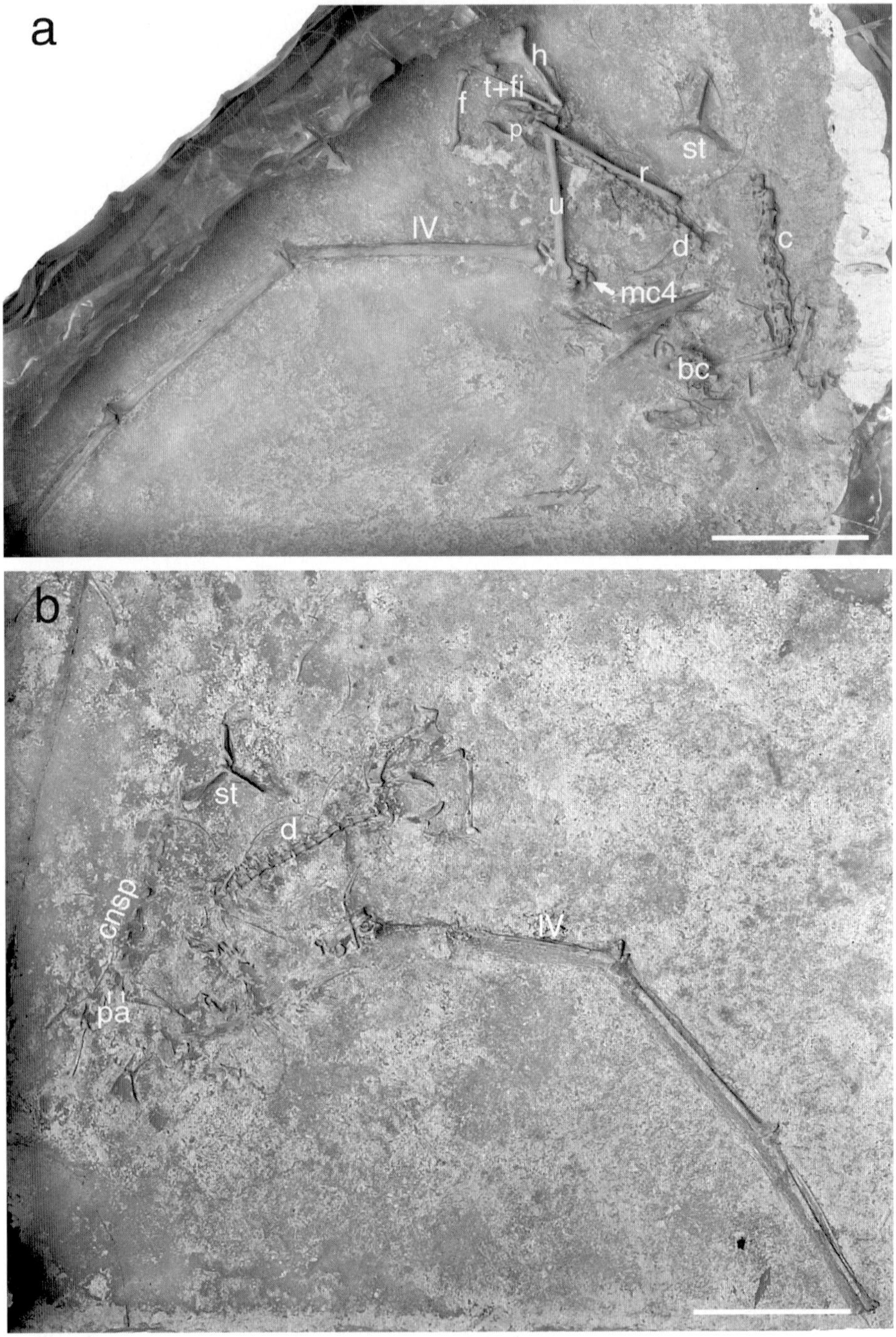

Fig. 1. *Rhamphorhynchus muensteri* MGUH 1891.738: (**a**) main slab and (**b**) counterslab. The specimen is encased in artificial resin and most bones are three-dimensionally exposed and virtually undistorted. bc, brain case; c, cervical vertebral column; cnsp, cervical neural spines; d, dorsal vertebral column; f, femur; h, humerus; IV, fourth digit; mc4, fourth metacarpal; p, pelvis; pa, proatlas (2); r, radius; st, sternum; t + fi, tibia and fibula; u, ulna. Scale bars 5 cm.

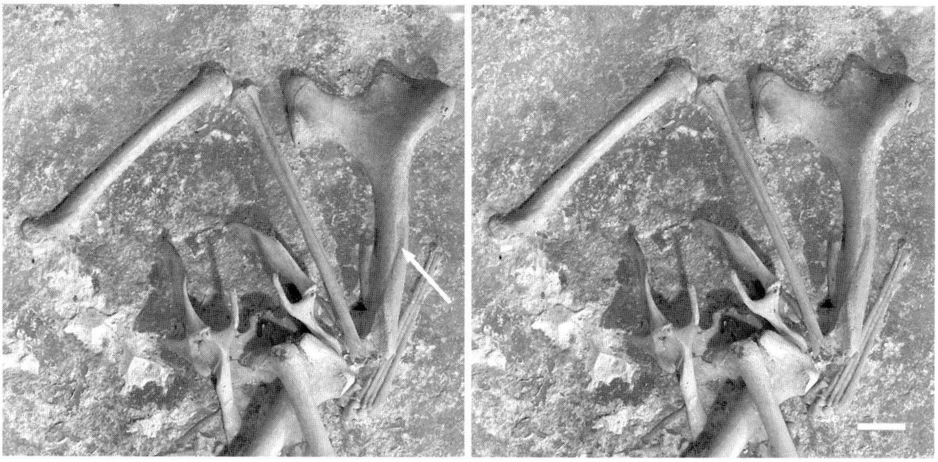

Fig. 2. Stereo close-up of right humerus, hindlimb and pelvis (viewed dorsally) from the main slab. Notice the elongate scar for m. latissimus dorsi on the humeral shaft (arrow), and the scar for *m. pectoralis* on the deltopectoral crest. The right metatarsals can be seen just below the humerus. Note also the position of the acetabula in the three-dimensionally preserved and undistorted pelvis. Scale bar 5 mm.

slightly exceeds that of the atlas. The intercentrum can be seen both anterior and posterior to the lower jaw on the slab (Fig. 7).

There is evidence of many hollow bones. The articulated dorsal centra on the counterslab have smaller pieces of lamella missing in several places around the distinctly procoelous articulating facets, revealing the trabeculated interior of the bones. Although the dorsal centra are fully uncrushed, very well preserved and expose their lateral sides on the counterslab, there is no evidence of pleurocoels on any of the centra (but see below). The appendicular bones were also hollow. The articulating facets of the humerus and radius have pieces of the lamella missing in several places, exposing the spongy interior of the elements. Adjacent to the prepubis on the counterslab there is a pedal phalanx, also with slight damage to the lamella. Even this small bone has a spongy interior. At the anterior end of the dorsal series is a poorly preserved rib, missing part of the exterior and the upper part. This bone is distinctly hollow, but the absence of the proximal part precludes determination of pneumaticity, because this would seem a likely place for the air diverticulae to enter, as in birds and some non-avian dinosaurs (e.g. Janensch 1947, 1950; Britt 1993). Where small pieces of other ribs are missing it is evident that these were also hollow. Thus, the entire skeleton appears to have been very lightly constructed.

Pneumatic features

The counterslab has the entire cristospine and anterior part of the sternum three-dimensionally exposed. Just posterior to the facets for the coracoids the sternum widens markedly and, in the transition from sternum to cristospine, there is a very large foramen, exposed laterally and posteriorly. The internal walls of the foramen are not smooth and level; rather, the internal structure is a complex meshwork of excavations and lamellar struts, probably for mechanical support (Fig. 3). This morphology is hitherto unknown for *Rhamphorhynchus*. Morphologically it bears a distinct resemblance to the interior of avian bones that have been excavated by air diverticulae. In extant birds the clavicular air sac is found in this area and pneumatizes the sternum, humerus and furcula among others (e.g. King 1957, 1966, 1979; Bellairs & Jenkin 1960). But it is uncertain, although not unlikely, whether a similar air diverticulum in pterosaurs produced this excavation. *Pteranodon* has a dorsal pneumatic foramen on the sternal plate penetrating into the cristospine (Bennett 2001a, pp. 65–66).

The proximal part of the left lower jaw is exposed from its ventral side and overlies the atlas and second intercentrum. The medial surface of the jaw ventral to the articulation shows a large excavation at the posteromedial face (see Figs 6 & 7), which represents the mandibular foramen. At the dorsomedial internal edge is a small foramen that extends into the mandibular foramen, and this foramen may represent a pneumatic foramen (illustrated in figure 7 as a lightened area inside the mandibular foramen).

The illustration by Wellnhofer (1975a, p. 14, fig. 6a) of the ventral view of the cervical vertebral column of *Rhamphorhynchus* is based on the present specimen. Between the zygapophysial rami and the centra are distinct spaces that Wellnhofer (1975a)

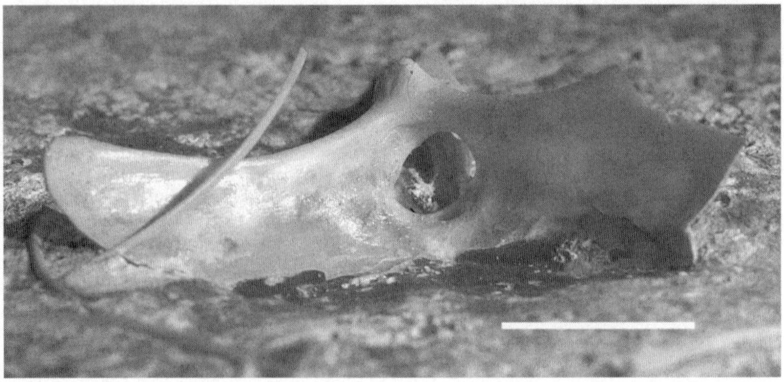

Fig. 3. Sternum (counterslab) viewed directly posteriorly. A large foramen is present on the anterior edge of the sternum and extends slightly into the posterior portion of the cristospine. The foramen opens laterally on both sides as well. The interior of the foramen is a complex meshwork of trabeculae which is partly visible on the right of the foramen in this figure. Scale bar 5 mm.

illustrated as cross-hatched. We can confirm that these spaces extend all the way into the interior of (but not through) the zygapophysial rami and thus are excavations not into the centra, but simply into the bases of the neural arches (cf. *Pteranodon*, Eaton 1910, pl. VI; Bennett 2001a, pp. 39–46, 52; further details below). Wellnhofer (1975a) also indicates spaces with cross-hatching on the ventral sides of the transverse processes of all the dorsal vertebrae (p. 14, fig. 6f, also based on this specimen). The counterslab exposes the articulated dorsal vertebral column in ventral aspect and, in a few places, also the transverse processes of the left side. On the slab the entire dorsal series is exposed dorsally, and here nearly all the transverse processes are present and rather well preserved, though most are broken. Wellnhofer does not describe these spaces but, because the cervical vertebrae have large distinct spaces, and because Wellnhofer illustrates the spaces on the cervical and dorsal vertebrae identically, this implies that all the dorsals also have holes/cavities, albeit up into the base of all the transverse processes. This implication does not correspond to the facts, as noted below.

On the slab the dorsals are exposed dorsally and laterally (Figs 4 & 5). Ventrally on the right transverse process of the first dorsal there is an elongate fossa in the anteromedial section of the process. Laterally the fossa is more narrow but as it approaches the centrum it expands anteroposteriorly and dorsoventrally into a distinct excavation. This bears resemblance to a pneumatic fossa. The following dorsal also displays a fossa, but it cannot be asserted whether or not they extend into the centrum. Evidence from the counterslab suggests that they do not. On dorsal 3 the parapophysis has migrated upwards and there is a slender pillar of bone that extends from the transverse process to the parapophysis, near the centrum. Just posterior to this pillar is a distinct excavation, probably similar to the preceding ones, but morphologically slightly different because of the position of the parapophysis. On dorsal 4 the parapophysis is situated at the anterior edge of the transverse process, and no fossa can be discerned on this vertebra on the slab.

On the counterslab, however, it is possible to view one of these foramina in detail, because the articulated row of dorsal centra have parts of the transverse processes preserved as well. Usually they are incomplete or nearly absent, but the most complete is from dorsal 4. The left transverse process of this vertebra is nearly fully preserved and displays a large foramen, although distinctly wider and more subcircular than indicated in Wellnhofer (1975a, p. 14, fig. 6f). These fossae are clearly not just holes, however, because none of the transverse processes of the slab and mainly the counterslab show any indication of a perforation on the dorsal faces of the transverse processes. Rather, close examination reveals that they are foramina that extend into the transverse processes from the ventral surface. The size and unusual location would indicate that they represent pneumatization of the dorsals, although not through pleurocoels, as is the case in saurischian dinosaurs (extant and extinct), but through the transverse processes. Several theropods and most birds pneumatize the vertebrae via the arch – not the centrum.

Examination of dorsal 5 and more posterior dorsals on the slab failed to indicate any fossae, and the preserved remains of the transverse processes on the counterslab corroborate this. Thus, the fossae probably did not extend posteriorly in the dorsal series, as indicated in Wellnhofer (1975a). On dorsal 5 the parapophysis has migrated to the anteroventral part of the transverse process, and it is possible that this simply precludes a fossa. Despite extensive

Fig. 4. Stereo close-up of dorsal vertebral column, viewed dorsally, from main slab. Scale bar 10 mm.

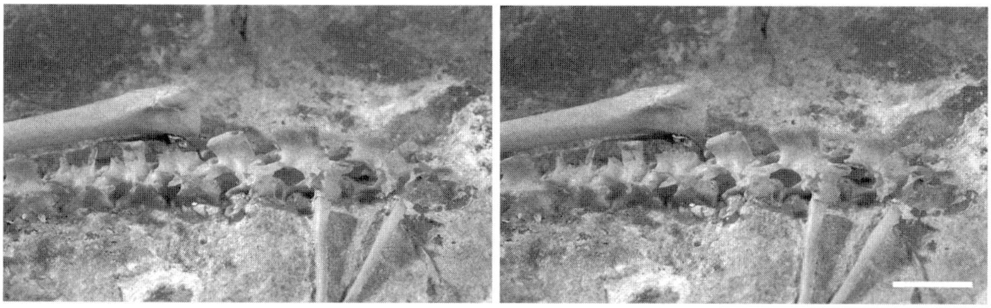

Fig. 5. Stereo close-up of anterior dorsals, viewed from right lateral side, from main slab. Notice the pneumatic foramina in the upper lateral part of the anterior two centra. Scale bar 5 mm.

pneumatization of the skeleton, even the carpals (Wellnhofer 1985), the posterior dorsals of pterodactyloids are frequently unpneumatized (e.g. Eaton 1910; Wellnhofer 1991a, b; Bennett 2001a, b), in the sense of lacking pleurocoels. They do, however, often have pneumatic notaria vertebrae (e.g. Bennett 2001a). However, if the dorsals without pleurocoels of this more basal pterosaur, a 'rhamphorhynchoid', in at least some instances appear to be pneumatized from the transverse processes, it would seem sensible to look for something similar in more advanced pterosaurs. Wellnhofer (1978) notes that *Rhamphorhynchus* has pneumatized dorsals, but it is uncertain whether he meant to implicate the centra. The counterslab offers ample opportunity to examine the entire ventral and most of the lateral faces of the dorsal centra, and none have pneumatic foramina or fossae. If the centra were pneumatic this would have to be via the transverse processes; the medial walls of these cavities, however, appear to be intact. In some instances bones are even pneumatized indirectly from other bones (particularly in the skull; Witmer 1990). At present it cannot be verified whether the centra of our fossil are pneumatized indirectly from the transverse processes.

The best evidence for pneumaticity stems, however, from the ventrally exposed cervical vertebral column of the slab (Figs 6–9). The tiny atlas described above does not show any evidence of pneumatization, but all of the succeeding eight cervicals do. The ventral surface lamella of the second centrum is broken away and exposes the interior, which shows a slightly asymmetrical thin medial wall with some complicated perforations between left and right sides. Because of the breakage the entrances to these cavities in the lateral face of the centrum cannot be seen (Figs 6 & 7). However, these pleurocoels are very evident in the third centrum, which is almost intact, apart from an opening in the rear, convex articular surface. Through this hole the interior medial wall, perforated by large irregular holes, can be seen. The excavations are clearly of the camerate type (cf. Britt 1993, 1997). The lateral sides of this centrum are perfectly preserved and show three large perforations into cavities in the centrum and the base of the neural arch (Fig. 7). The neural arch is perforated by a large foramen on the ventral surface of the strong ridge between the pre- and postzygapophyses. This hole is oval and quite large, about 2 mm long. The cavity inside probably extends the entire length between the two zygapophyses. There is an equally large pleurocoel into the mid-dorsal part of the centrum that leads into the two large lateral chambers divided by the incomplete medial wall. Immediately anterior to this pleurocoel is another, and slightly smaller, one that excavates the anteroventral part of the centrum. Paired chambers are thus formed, separated from the two main air chambers by a thin subhorizontal wall. Thus there are at least three pairs of large excavations into

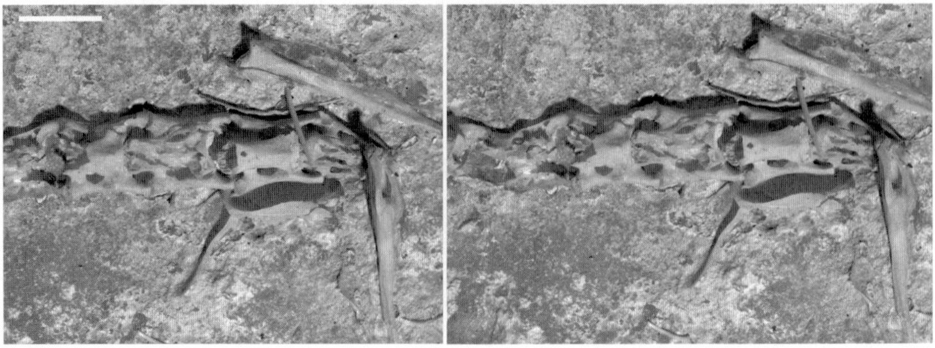

Fig. 6. Stereo close-up of the anterior cervicals and left posterior lower jaw ramus, viewed ventrally, from main slab. Notice the marked excavations along the central sides between the zygapophyses and the large mandibular fenestra in the posterior part of the lower jaw ramus. Scale bar 5 mm.

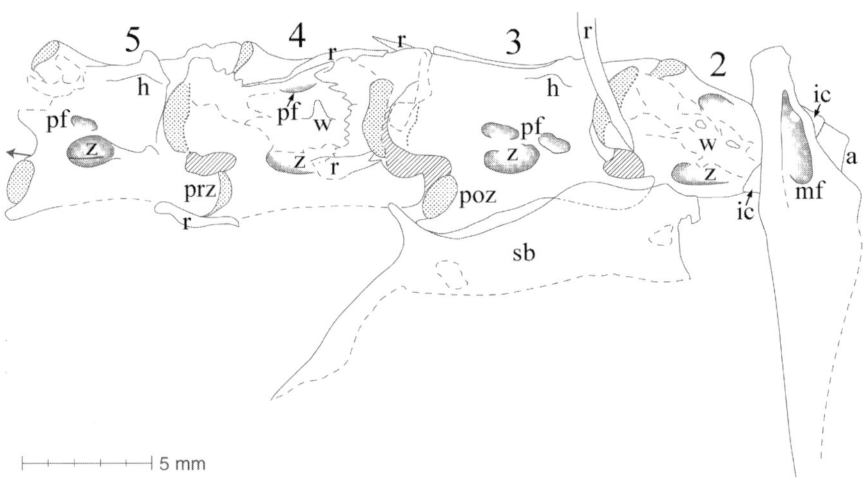

Fig. 7. The five anterior cervicals, including the atlas. - - - -, broken bone or limit to resin; · - · - · - ·, outline hidden behind bone; · · · · · · ·, boundary between centra and articular surfaces; dotted areas, articulating surfaces; cross-hatched areas, spaces between bones. 2–5, second to fifth cervical vertebra; a, atlas; h, hypapophysis; ic, axis intercentrum; mf, mandibular fenestra; pf, pneumatic foramen; poz, postzygapophysis; prz, prezygapophysis; r, rib; sb, unidentified skull bone; w, wall of bone; z, zygapophysial foramen; long arrow through posterior wall of neural arch. (Camera lucida drawing by N.B.)

each vertebra. Also the neural spines appear hollow, as seen from a few that are broken, but if there are pneumatopores into these spines they must penetrate via the neural arches, because there are no more external perforations. The separate antero-ventro-lateral cavities are actually also indicated, but less evident, in the second and fourth centrum, and so is the medial irregular wall, of which small remnants are seen in the broken fourth centrum (Figs 7–8).

On the right side of the fourth centrum a small, double-headed cervical rib, about 5 mm long, is articulated to the parapophysis, a small tubercle extending ventrolaterally near the anterior, concave articular surface of the centrum (Figs 7 & 8). Similar, but slightly broken tubercles are seen on the third centrum, and between the two there is a medial low hypapophysis, also weakly indicated on the fifth centrum. The anterior head of the rib cannot be seen articulating with the vertebra. The second to sixth centra are all a little over 6 mm long, while the seventh is approximately 5 mm, the eighth c. 4½ mm, and the ninth c. 4 mm long. The fifth centrum has lost its posterior end, so the internal medial wall is also visible here, and one parapophysis is broken to show its hollow interior. The fourth and the fifth vertebrae are the only ones in which the posterior wall of the neural arches is well exposed, and they show a small perforation probably penetrating into the zygapophyseal chamber (Fig. 7). Whether there are foramina holes in the anterior walls of the neural

Fig. 8. Close up of cervical 4 from main slab, viewed ventrally. The cervical is missing the entire lamella along its ventral side, exposing the interior in great detail. Notice the two large, elongate chambers running anteroposteriorly inside the vertebra (arrows). These communicate to the exterior via narrow canals that open up on the upper, anterior surface of the centrum as large foramina, approximately 20% of centrum length. Scale bar 5 mm.

arches cannot be observed. Such pneumatic foramina are seen in *Pteranodon* (Eaton 1910, pl. VI, figs 13 & 14; Bennett 2001a, figs 35, 39–42) and azdarchids (Britt 1993).

The entire centrum of the sixth vertebra is lost; only secondary calcite in the large neural canal and the neural arch is preserved. The seventh cervical vertebra is turned more ventrolaterally and exposes a 1 mm long pleurocoel rather low on the lateral surface of the centrum (Fig. 9). The eighth and ninth vertebrae lack pleurocoels, but have foramina into the ventral surface of the zygapophyseal region. A well-preserved double-headed cervical rib may belong to one of those two vertebrae. The seventh and eighth centra show small paired apophyses (traditionally called exapophyses, but see Bennett 2001a for discussion) lateroventrally near the intervertebral articular surfaces, which are in all vertebrae clearly procoelous. The eighth cervical and subsequent vertebrae seem to lack pleurocoels (Fig. 9). Only a few more dorsal vertebrae have lateral cavities excavated into the bases of the transverse processes (Fig. 5).

The neural spines of the third to eighth cervicals are triangular and slightly smaller and less pointed than that of the axis (visible on the counterslab), while the ninth and those further back are thinner and more laterally compressed; as mentioned some are broken and show a hollow interior. Their dorsolateral faces are mainly preserved as imprints in the resin and, as mentioned above, they show no indications of external perforations.

In summary, cervical vertebrae from the second to the seventh are extensively pneumatized with three pairs of large pores, two leading into paired camerae in the centrum, one into the base of the neural arch forming a cavity between pre- and postzygapophyses. These camerae and cavities take up most by far of the internal space of the vertebrae. Some or most of the cervicals also have pores through the posterior wall of the neural arch, perhaps into the zygapophysial cavity. Cervical 8 and backwards to dorsal 4 have large perforations into the ventral surface of the transverse processes, perhaps penetrating further into the centra, which are very hollow like the neural arches are. From the fifth dorsal and backwards no perforations or pleurocoels of the vertebrae can be observed, but they are very hollow anyway.

Physiological implications of pneumatic bones in pterosaurs

The presence of pneumatopores in much of the skeleton is strongly suggestive of a pulmonary system radically different from that of other 'reptiles', except most saurischian dinosaurs (including Neornithes). Pneumatopores indicate the presence of air diverticulae extending from the lungs proper into the bones and, presumably, into the body cavity as well, as in birds (Baer 1896; Salt & Zeuthen 1960; King 1966). The presence of external pneumatopores, communicating with internal chambers, and the internal resorption of bone tissue, results in an unusual and distinctive morphology that cannot be confused with any other process among extant vertebrates. This strongly indicates that one can extend such inferences to fossil taxa as well. Pneumatopores, and the inferred highly derived pulmonary system, have been taken as indicative of tachymetabolic endothermy (e.g. Seeley 1870; Viohl 2000), although this was doubted by Bakhurina & Unwin (1995a). However, it remains a fact that only animals that display many other distinct signs highly suggestive of tachymetabolic endothermy (extinct saurischian dinosaurs) or that are known to be tachymetabolic endotherms (extant saurischian dinosaurs) possess these features. Coupled with the evidence for pterosaur 'hair' (see Bakhurina & Unwin 1995a, b for discussion), along with the highly energy-demanding task of powered flight, which appears highly unlikely for non-tachymetabolic vertebrates (e.g. Maina 2000), this strongly points to tachymetabolic endothermy in pterosaurs, a conclusion also supported by their bone histology and inferred growth rates (e.g. Bennett 1993; Ricqlès *et al.* 2000; Unwin 2001).

It has been suggested that the function of the air diverticulae in pterosaurs was either to make the respiratory system more efficient, and perhaps also to cool the blood during powered flight (e.g. Viohl 2000; Bennett 2001b), or, to heat the inhaled air (Wellnhofer 1991a). However, the suggestion that pneumatic bones imply a greater surface area for gas

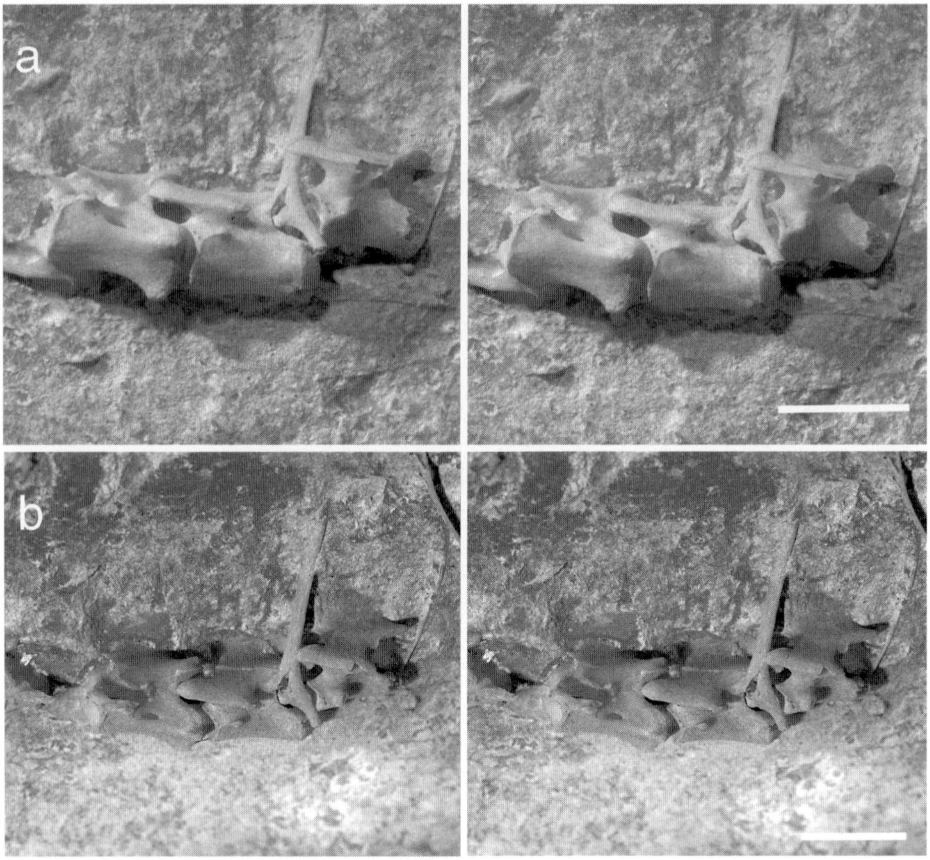

Fig. 9. Stereo close-up of three posterior cervicals (7–9) from main slab: (**a**) ventrolateral aspect, (**b**) lateral aspect. Notice the elongate pneumatic foramen on the centrum of cervical 7 and the short, blunt 'exapophyses'. Scale bar 5 mm.

exchange is highly dubious, not just because the amount of ventilation that could pass through these rather narrow spaces is limited (Bennett 2001b), but also because the air diverticulae in the only extant analogues, neornithine birds, are nearly avascular and serve the unidirectional flow of air though the parabronchial lumen, not the uptake of oxygen (Salt & Zeuthen 1960; Schmidt-Nielsen 1975; Scheid 1979). The cooling or heating hypotheses are possible, but have not been demonstrated in birds. Lacking sweat glands, extant birds pant to combat hyperthermia, and a large part of the evaporation from the respiratory system takes place in the air sacs (Salt & Zeuthen 1960). However, the primary function of air sacs appears to be as participators in an advanced oxygen-uptake system. Although heating or cooling purposes as a selective driving force for the evolution of air sacs cannot be ruled out, it would seem excessive to develop such an advanced respiratory system merely to enhance capabilities already possible with normal tetrapod lungs. Hypothermia in pteterosaurs could probably also be prevented from heat loss through the flight membrane, as in bats but unlike birds.

Among the two extant groups of flying vertebrates pterosaurs have more often been compared to bats than to birds, mainly because of their 'leathery wings' supported by a bony strut. However, it seems likely that the bat analogy is probably less justified with respect to the pulmonary and vascular systems. Birds and bats are both tachymetabolic endotherms, though many, particularly bats, are not homeothermic (e.g. Schmidt-Nielsen 1995), and both groups achieve high levels of oxygen consumption during powered flight. This, however, is facilitated by different means. Birds have smaller lungs than comparably sized mammals (Maina *et al.* 1989) and their lungs are nearly inexpansible (Jones *et al.* 1985), although some movement of the ribs does occur, thus contributing to ventilation (Salt & Zeuthen 1960). The very complex avian respiratory system, with its voluminous air sacs and associated, advanced venous blood flow perpendicular to the unidirectional air flow through the parabronchial lumen, makes the avian respiratory system very efficient in terms of oxygen uptake (Schmidt-Nielsen 1975; Schied 1979). This seems to be the primary function and advantage of the air sacs.

Bats, on the other hand, do not possess this very efficient uniflow respiratory system, and their lungs are basically typically mammalian, albeit proportionally much more capacious than in terrestrial mammals (Maina & King 1984). Additionally, they have proportionally much larger hearts and cardiac output (e.g. Snyder 1976) and substantially greater haematocrit values and blood–oxygen carrying capacity (Riedesel 1977). The net effect is that their effective oxygen uptake rivals that of birds (Thomas 1987). With respect to the pulmonary system, pterosaurs appear to have been more similar to their sauropsidan cousins than their synapsidan counterparts. Potentially, this could have influenced overall body size. It is well known that the giant azdarchids and pteranodontids particularly, but in effect all pterosaurs, possessed very compact and foreshortened bodies (see e.g. Viohl 2000, p. 25 for comparison between *Rhamphorhynchus* and a seagull). This would clearly not have been possible if they had mimicked the condition of bats with enlarged internal organs.

The compact bodies of giant pterosaurs compared to birds of similar wing span was perhaps one of the factors allowing certain forms to exploit giant size, simply allowing them to be lighter than a bird at any given wing span (see e.g. Paul 1990, 1991). However, the extent of the influence of pneumaticity on these factors is hard to evaluate, but the avian-style respiratory system would seem to be a prerequisite for reducing body volume to the extent attained in huge pterosaurs. Large birds also tend to have more compact bodies, but not to the same extent as in giant pterosaurs.

Homology of the air diverticulae in pterosaurs and birds

There are striking similarities between the excavations of the cervical and anterior dorsal vertebrae in *Rhamphorhynchus* and those that are correlated with pneumatization in modern birds. It is, however, not clear that the two are strictly homologous; they are much more likely to have been convergently derived in the two groups. The distribution of air sacs penetrating the axial skeleton in archosaurs is as follows (see Britt 1993). There are air sacs in the vertebrae, at least in the cervicals and some dorsals, in most modern birds, but there is a large variation in extent and distribution within the skeleton. There are pleurocoels in most theropods (also present in *Archaeopteryx* – see Britt *et al.* 1998 and Christiansen & Bonde 2000 – although we subsequently failed to find good evidence for pneumatic foramina in the third species [Bonde 1996], *A. bavarica* Wellnhofer 1993, in München) and advanced sauropods (see below), and similarly in many pterosaurs. But there is not a convincing overview of the distribution within this group (Padian 1983a), although Bennett (1994) and Unwin & Lü (1997) used pneumatic characters in discussions of pterodactyloid systematics. Crocodiles and more basal fossil archosaurs are generally supposed not to have had pneumatized skeletons (but see below). This also applies to the most primitive theropods (herrerasaurids) and prosauropods (Padian & Brit, pers. comm., report pneumaticity also in basal prosauropods) and probably to the more plesiomorphic sauropods(?), as well as all ornithischian dinosaurs. In fact, pneumatizations have not been described for basal pterosaurs either.

The Triassic pterosaurs of northern Italy were recently examined by one of us (N.B.) with equivocal results. In *Eudimorphodon ranzii* (Zambelli 1973) there may be a pneumatic foramen as a small slit or tiny hole in the middle of the broad proximal plate of the humerus in all three specimens – juveniles and an adult – figured by Wild (1978, fig. 13, indicated only in two of them as a rather shallow depression), but this is on the dorsal side of the humerus, and the foramina are too small to be convincing as pneumatic. The adult type further has a deep slit covered by a shelf in the distal end near the condyle, which might also be a pneumatic foramen. The juvenile No. 8950 in Bergamo (with skin and 'hairy' impressions, see Wild 1994) is also equivocal concerning the vertebrae. Although most vertebrae are exposed from the lateral face, the cervical transverse processes more or less obscure depressions in the centra which might be pleurocoels. Similar depressions lateral in the dorsal centra may also be pleurocoels, but they are not very convincing. Some holes in the 'right place' in a few of the crushed vertebrae of the 'juvenile' *Eudimorphodon* in Milano (MPUM 7309) seem artificial.

In the large *E. ranzii* holotype in Bergamo (No. 2888, Wild 1978) it is also not evident whether narrow depressions (filled with sediment) in the region between the centrum and the neural arch of the 'lumbar' vertebrae could be pneumatic or are simply an effect of the transverse process being pressed down towards the centrum during fossilization, because most of the vertebrae are exposed from the ventral side. Similarly there is no safe indication of pneumatics in the cervicals, mostly preserved in dorsal view (see Monastersky 2001, pp. 100–101). The wing bones are all hollow and crushed, but some of them show rather small but distinct depressions or pits, often near the ends of the bones. Some of these pits might be pneumatic, but are still not very convincing.

The holotype of *Peteinosaurus* has no vertebrae preserved, but the alleged *Peteinosaurus* (Bergamo no. 3359 without skull, but with a long 'rhamphorhynchoid' tail, see Wild 1978) shows the two posterior cervicals from the dorsal side rather than laterally (as indicated by Wild 1978), while the succeeding dorsals are exposed laterally but show no depressions to indicate pleurocoels. The last dorsals and one 'lumbar' are seen ventrolaterally and show only very weak depressions, if any at all, and one has a tiny foramen; however, there are no real indications for pneumaticity. Neither do the limb bones show signs of pneumatic foramina, unless a hole most proximal in the humerus (Wild 1978, fig. 35) is taken as such indication. (*Preondactylus* [Wild 1984] is preserved only as an imprint [see also Dalla Vecchia 1998], and it is unlikely to indicate anything about pneumatization).

In conclusion there appears to be no reliable evidence of skeletal pneumatization in the more well-preserved specimens of *Eudimorphodon* and *Peteinosaurus*, although it cannot be entirely excluded.

The problem therefore is that advanced pterosaurs, advanced (perhaps all) sauropodomorphs, and advanced theropods, including extant forms (birds), all appear to develop pneumatic air sacs, invading the axial and appendicular skeleton, but basal members of the three groups apparently did not possess such features. Accordingly, the pneumatic foramina in the vertebrae of these groups are presently best interpreted as the results of convergent (or parallel) evolution. We suggest that this raises some interesting questions about the soft-tissue morphology of the last common ancestor of the above groups.

Reconstructing primitive (morphotypic), ancestral dinosaur features from skeletal evidence alone might conceivably indicate a form without air sacs and with a crocodile-like respiratory system (implied e.g. by Ruben *et al.* 1997, 1999). But the evidence from the nearest sister-group, the Pterosauria, which has advanced air-sac systems in the axial skeleton of semi-advanced members such as *Rhamphorhynchus*, indicates an additional possibility: The last common ancestor of all ornithodirans (*sensu* Gauthier 1986 and Benton 1990) at least had air-sac systems, although they did not penetrate the skeleton. This again implies that, quite early after the split from the crocodilian sister-group, the ornithodiran line evolved an advanced respiratory system and presumably an advanced physiology (*contra* Ruben *et al.* 1997, 1999). Major, non-invasive air diverticulae are present in crown-clade Saurischia (Salt & Zeuthen 1960; King 1966, 1979).

Discussion

Other character distributions might support this non-traditional conclusion. Indeed there are other 'exotic' features with a similar distribution, namely integumental structures, feathers and 'protofeathers' (or cryptoptiles, Bonde & Christiansen 2002a, b; Christiansen & Bonde 2003), which are classically found in birds and avialans, including *Archaeopteryx*. With the recent discovery of several Lower Cretaceous (Swisher *et al.* 1999; Smith *et al.* 2001) small and medium-sized theropod dinosaurs with pennaceous or plumulaceous feathers and hair-like 'protofeathers' (Currie 1998; Ji *et al.* 1998; Xu *et al.* 1999a; Xu *et al.* 1999b) it is most likely that true pennaceous feathers had already evolved in basal Maniraptora, including the oviraptorosaurs (represented by feathered *Caudipteryx*) as well as dromaeosaurs (Norell *et al.* 2002) and avialans (Padian 1998; Holtz 2000; Padian *et al.* 2001). Additionally, the much more primitive compsognathid coelurosaurs have hair-like protofeathers covering most of the body and neck (Ackerman 1998; Chen et al. 1998; Currie & Chen 2001). Such cryptoptiles seem preserved in many more advanced theropods including early 'birds'. In primitive theropods the evidence is more indirect: the impressions of 'feather-like' structures at some footprints of the Early Jurassic (Gierlinski 1997). Ornithischian dinosaurs apparently do not have similar integumental structures. Some mummified hadrosaurs show rather 'scale-like' skin impressions, – but as-yet unverified reports of the little *'Psittacosaurus',* from the same Early Cretaceous deposits as the above feathered theropods, have found very odd, long, thin and curved appendices attached to the dorsal side of the tail, and some apparently have 'hair-like' structures on the body. If so, potentially, ornithischian dinosaurs also possessed some sort of proto-feathers.

Several pterosaurs show 'hair-like' integumentary structures, most notable the famous *Sordes pilosus* (Bakhurina & Unwin 1995a, b; Viohl 2000), but so does one of the Triassic forms, *Eudimorphodon* (Wild 1994, specimen studied by N.B.), and it is now generally agreed that pterosaurs were probably all covered in a 'hairy' coat (Wellnhofer 1991a; Cherkas & Ji 2002), most likely some sort of 'proto-feathers'. Thus, the possible distribution of air sacs and of an insulating coat of 'feathers' or proto-feathers is roughly similar to the distribution of skeletal pneumatization (a possible difference in the pattern of these features concerns ornithischians(?) and sauropods). This may indicate that the basal ornithodirans were already endothermic with isolating proto-feathers and an advanced respiratory 'flow-through' system with air sacs distributed between the internal organs (but not yet penetrating the skeleton). This is in accordance with the advanced physiology, probably tachymetabolic endothermy, that is also indicated by bone histology and inferred growth rates (Bennett 1993, Padian & Rayner 1993, Ricqlès *et al.* 2000).

We are well aware that the ornithodiran hypothesis is not necessarily correct. Some analyses indicate pterosaurs to be an earlier, stem-archosaurian lineage, split off from basal archosaurs before the crocodilian lineage (Wild 1978, Unwin 1995, 1999; Bennett 1996a; Peters 2000). The most radically different opinion is the hypothesis of Peters (2000), who concludes a cladistic study by indicating that pterosaurs belong with the Prolacertiformes, quite close to such former 'thecodonts' as *Longisquama* and *Sharovipteryx*, i.e. a very basal stem-archosaur lineage. We will not enter detailed discussion of this model here, but only note that, if correct, our conclusion that the last common ancestor of pterosaurs and birds already had this advanced physiology and air sacs becomes much less probable. This is because one would then have to assume that crocodiles secondarily lost these advanced features, which is not a very likely event. If Peters (2000) is correct, and the advanced features of pterosaurs are moved to a phylogenetic position far from the origin of dinosaurs, then it is also less likely that early dinosaurs, including the entire lineage of ornithischians (in the lack of some of the 'direct' evidence), possessed this advanced physiology. However, there are some dubious traces of possible pneumatics in some advanced 'thecodonts' with crocodylian affinities, such as rauisuchians, according to Gower (2001); although not particularly convincing, this might save the hypothesis of crocodiles reversing to secondarily primitive physiology.

Here we accept the traditional and well-discussed hypothesis of ornithodirans as an (advanced) sister-group of the crocodilian lineage. This prompts us to conclude, surprising as it may seem, that the evolution of advanced physiology and anatomy may have been one of the earliest events in the ornithodiran lineage, shortly after the split from the crocodilian lineage. In fact, both pterosaurs and dinosaurs were quite likely much more bird-like than is generally assumed. Cuvier after all was not right (1801, 1809): The famous 'Ptero-Dactyle' should not be called a 'reptile' (in the traditional sense), but was rather like a bird.

We thank the organisers of this excellent and very informative first symposium on pterosaurs for a very fine meeting in hospitable locations and a pleasant excursion to the pterosaur footprints at Crayssac, quite an extraordinary experience. We are most grateful to P. Wellnhofer for his interest and advice during our recent short visit to München, and to H. Mayr, curator of the Paläontologische Staatssammlung, as well as our colleagues in the Senckenberg Museum in Frankfurt and the Museum für Naturkunde in Berlin, and G. Viohl of the Jura-Museum, Eichstätt, for allowing our study of pterosaurs, *Archaeopteryx* and other birds in their care during the one-week visit, in June 2001, to those collections. K. Padian and B. Britt provided many helpful comments regarding improvements to the manuscript. We are indebted to G. Brovad, Zoological Museum, for work on the photos, and B. Munch, Geological Institute, for partly preparing Figure 7, and to the Geological Museum for lending us the specimen. N.B. thanks his institute for financial support for the symposium journey as well as the short study trips to Germany (mentioned above), Milano and Bergamo. N.B. heartily thanks A. Tintori and S. Renesto and their students for assistance and hospitality during the stay in Milano and also thanks A. Paganoni and her staff for their valuable help and hospitality during the visit to Bergamo, where he also had the good fortune of discussion with R. Zambelli, describer of the first *Eudimorphodon*. The late E. Nielsen is thanked for teaching N.B. important laboratory techniques as a student. This work was supported by a grant from the National Science Foundation to P.C.

References

ACKERMAN, J. 1998. Dinosaurs take wing. *National Geographic*, **194**, 74–99.

BAER, M. 1896. Beiträge zur kenntnis der Anatomie und Physiologie der Atemwerkzeuge bei den Vögel. *Zeitschrift für Wissenschaftliche Zoologie*, **61**, 1–82.

BAKHURINA, N. & UNWIN, D. M. 1995a. A survey of pterosaurs from the Jurassic and Cretaceous of the former Soviet Union and Mongolia. *Historical Biology*, **10**, 197–245.

BAKHURINA, N. & UNWIN, D. M. 1995b. A preliminary report on the evidence for 'hair' in *Sordes pilosus*, an Upper Jurassic pterosaur from Middle Asia. *Sixth Symposium on Mesozoic Terrestrial Ecosystems and Biota, Beijing*, Short Papers, 79–82.

BARTHEL, K. W. 1978. *Solnhofen – Ein Blick in die Erdgeschichte*. Otto Verlag, Thun, 393 pp.

BELLAIRS, A. D. A. & JENKIN, C. R. 1960. The skeleton of birds. *In*: MARSHALL, A. J. (ed.) *Biology and Comparative Physiology of Birds*. Academic Press, New York, 241–300.

BENNETT, S. C. 1993. The ontogeny of *Pteranodon* and other pterosaurs. *Paleobiology*, **19**, 92–106.

BENNETT, S. C. 1994. *Anatomy and systematics of the Late Cretaceous pterosaur* Pteranodon (Pterosauria, Pterodactyloidea). Natural History Museum, University of Kansas, Occasional Papers, **169**, 70 pp.

BENNETT, S. C. 1995. A statistical study of *Rhamphorhynchus* from the Solnhofen limestone of Germany: Year-classes of a single large species. *Journal of Palaeontology*, **69**, 369–380.

BENNETT, S. C. 1996a. The phylogenetic position of the Pterosauria within the Archosauromorpha. *Zoological Journal of the Linnean Society, London*, **118**, 261–308.

BENNETT, S. C. 1996b. Year-classes of pterosaurs from the Solnhofen limestone of Germany: Taxonomic and systematic implications. *Journal of Vertebrate Paleontology*, **16**, 432–444.

BENNETT, S. C. 1997. Terrestrial locomotion of pterosaurs: A reconstruction based on *Pteraichnus* trackways. *Journal of Vertebrate Paleontology*, **17**, 104–113.

BENNETT, S. C. 2001a. The osteology and functional morphology of the Late Cretaceous pterosaur *Pteranodon*. Part I: General description and osteology. *Palaeontographica A*, **260**, 1–112.

BENNETT, S. C. 2001b. The osteology and functional morphology of the Late Cretaceous pterosaur *Pteranodon*. Part II: Size and functional morphology. *Palaeontographica A*, **260**, 113–153.

BENTON, M. 1990. Origin and interrelationships of dinosaurs. *In*: WEISHAMPEL, D. B., DODSON, P. & OSMOLSKA, H. (eds) *The Dinosauria*. University of California Press, Berkeley, 11–30.

BONDE, N. 1996. The systematic and classificatory status of *Archaeopteryx. Museum of Northern Arizona, Bulletin*, **60**, 193–199.

BONDE, N. & CHRISTIANSEN, P. 2002a. The evolution of plumage – from dinosaur to early bird. *In*: BONDE, N. (compiler) *Bird Evolution*, Symposium, Palaeontology Club, Geological Insitute, Copenhagen; Abstr. (ext.), 12 pp.

BONDE, N. & CHRISTIANSEN, P. 2002b. Evolution of plumage from dinosaurs to birds. Abstracts of the 23rd International Ornithological Congress, Beijing, 2002, p. 333.

BONDE, N. & CHRISTIANSEN, P. (in prep). The problems of becoming feathered.

BRITT, B. B. 1993. *Pneumatic postcranial bones in dinosaurs and other archosaurs*. PhD thesis, University of Calgary, Canada.

BRITT, B. B. 1997. Postcranial pneumaticity. *In*: CURRIE, P. J. & PADIAN, K. (eds) *Encyclopedia of Dinosaurs*. Academic Press, San Diego, 590–593.

BRITT, B. B., MAKOVICKY, P. J., GAUTHIER, J. & BONDE, N. 1998. Postcranial pneumatization in *Archaeopteryx. Nature*, **395**, 374–376.

CHEN, P-J., DONG, Z-M. & ZHEN, S-N. 1998. An exceptionally well-preserved theropod dinosaur from the Yixian Formation of China. *Nature*, **391**, 147–152.

CHRISTIANSEN, P. & BONDE, N. 2000. Axial and appendicular pneumeticity in Archaeopteryx. *Proceedings of the Royal Society, London (Series B)*, **267**, 2501–2505.

CHRISTIANSEN, P. & BONDE, N. 2003. Body plumage *Archaeopteryx*: A review, and new evidence from the Berlin specimen. *Comptes Rendus Palevol* (submitted).

CURRIE, P. J. 1998. *Caudipteryx* revealed. *National Geographic*, **194**, 86–89.

CURRIE, P. J. & CHEN, P.-J. 2001. Anatomy of *Sinosauropteryx prima* from Liaoning, northeastern China. *Canadian Journal of Earth Science*, **38**, 1705–1727.

CUVIER, G. 1801. Reptile volant. *Magazin Encyclopédique, Paris*, **9**, 60–82.

CUVIER, G. 1809. Sur le squelette d'un reptile volant des environs d'Aichstedt, que quelques naturalistes ont pris pur un oiseau, et dont nous formons un genre de sauriens, sou le nom Ptero-Dactyle. *Annales de Musee d'Histoire Naturelle, Paris*, **13**, 424–437.

CZERKAS, S. A. & JI, Q. 2002. A new rhamphorhynchoid with a headcrest and complex integumentary structures. *In*: CZERKAS, S. J. (ed.) *Feathered Dinosaurs and the Origin of Flight. Dinosaur Museum Journal*, **1**, 15–41.

DALLA VECCHIA, F. 1998. New observations on the osteology and taxonomic status of *Preondactylus buffarinii*, Wild, 1984 (Reptilia, Pterosauria). *Bolletino della Societa Paleontologica Italiana*, **36**, 355–366.

EATON, G. F. 1903. The characters of *Pteranodon. American Journal of Science*, **16**, 82–86.

EATON, G. F. 1910. *Osteology of* Pteranodon. Connecticut Academy of Arts and Science, Memoirs, **2**, 1–38.

FREY, E. & MARTILL, D. M. 1996. A reappraisal of *Arambourgiania* (Pterosauria, Pterodactyloidea): One of the World's largest flying animals. *Neues Jahrbuch für Geologie und Paläontologie, Abhandlungen*, **199**, 221–247.

GIERLINSKI, G. 1997. What type of feathers could non-avian dinosaurs have, according to an Early Jurassic ichnological evidence from Massachusetts? *Przeglad Geologiczny*. **45**, 419–422.

GOWER, D. J. 2001. Possible postcranial pneumaticity in the last common ancestor of birds and crocodilians: Evidence from *Erythrosuchus* and other mesozoic archosaurs. *Naturwissenschaften*, **88**, 119–122.

HENDERSON, D. M. & UNWIN, D. M. 2001a. Investigating hypotheses of rhamphorhyncid pterosaur terrestrial locomotion with computer models. *Strata, Serie 1*, **11**, 47.

HENDERSON, D. M. & UNWIN, D. M. 2001b Mathematical and computational modelling of a walking pterodactyloid pterosaur. *Strata, Serie 1*, **11**, 47.

HOGG, D. A. 1980. A comparative evaluation of methods for identification of pneumatization in the avian skeleton. *Ibis*, **122**, 359–363.

HOLTZ, T. R. JR 2000. A new phylogeny of the carnivorous dinosaurs. *Gaia*, **15**, 5–61.

JANENSCH, W. 1947. Pneumazität bei Wirbeln von sauropoden und anderen Saurischiern. *Palaeontographica*, **3**, (supplement 7), 1–25.

JANENSCH, W. 1950. Die Wirbelsäule von *Brachiosaurus brancai. Palaeontographica*, **3**, (supplement 7), 27–92.

JI, Q., CURRIE, P. J., NORELL, M. A. & JI, S-A. 1998. Two feathered dinosaurs from northeastern China. *Nature*, **393**, 753–761.

JONES, J. H., EFFMANN, E. L., & SCHMIDT-NIELSEN, K. 1985. Lung volume changes during respiration in ducks. *Respiratory Physiology*, **59**, 15–25.

KELLNER, A. 1991. The Santana Formation pterasaurs. *In*: MAISEY, J. (ed) *Santana Fossils: An Illustrated Atlas*. T. F. H. Publications Inc., Neptune City, New Jersey, 370–371.

KING, A. S. 1957. The aerated bones of *Gallus domesticus. Acta Anatomica*, **31**, 220–230.

KING, A. S. 1966. Structural and functional aspects of the avian lungs and air sacs. *International Review of General and Experimental Zoology*, **2**, 171–267.

KING, A. S. 1979. Systema respiratorium. *In*: BAUMEL, J. J., KING, A. S., LUCAS, A. M., BREAZILE, J. E. & EVANS, H. E. (eds) *Nomina Anatomica Avium*. Academic Press, London, 227–265.

MAINA, J. N. 2000. What it takes to fly: The structural and functional respiratory refinements in birds and bats. *Journal of Experimental Biology*, **203**, 3045–3064.

MAINA, J. N. & KING, A. S. 1984. The structural functional correlation in the design of the bat lung: A morphometric study. *Journal of Experimental Biology*, **111**, 43–63.

MAINA, J. N., KING, A. S. & SETTLE, G. 1989. An allometric study of the pulmonary morphometric parameters in birds. *Philosophical Transactions of the Royal Society, London (Series B)*, **326**, 1–57.

MARSH, O. C. 1871. Note on a new and gigantic species of

pterodactyle. *American Journal of Science (Series 3)*, **1**, 472.

MARSH, O. C. 1872. Discovery of additional remains of Pterosauria with descriptions of two new species. *American Journal of Science (Series 3)*, **3**, 241–248.

MONASTERSKY, R. 2001. Pterosaurs – lords of the ancient Skies. *National Geographic*, **199**, 87–105.

NORELL, M. A., Ji, Q. et al. 2002. 'Modern' feathers of a non-avian dinosaur. *Nature*, **416**, 36–37.

PADIAN, K. 1983a. Osteology and functional morphology of *Dimorphodon macronyx* (Buckland) (Pterosauria: Rhamphorhyncoidea) based on new material in the Yale Peabody Museum. *Postilla*, **189**, 1–44.

PADIAN, K. 1983b. A functional analysis of flying and walking in pterosaurs. *Paleobiology*, **9**, 218–239.

PADIAN, K. 1998. When a bird is not a bird. *Nature*, **393**, 729–730.

PADIAN, K. & RAYNER, J. M. V. 1993. The wings of pterosaurs. *American Journal of Science*, **293-A**, 91–166.

PADIAN, K., JI, Q. & JI, S-A. 2001. Feathered dinosaurs and the origin of flight. *In*: TANKE, D. & CARPENTER, K. (eds) *Mesozoic Vertebrate Life*. Indiana University Press, Bloomington, 117–135.

PAUL, G. S. 1990. A re-evaluation of the mass and flight of giant pterosaurs. *Journal of Vertebrate Paleontology*, **10**, (Suppl. 3), 37A.

PAUL, G. S. 1991. The many myths , some old, some new, of dinosaurology. *Modern Geology*, **16**, 69–99.

PETERS, D. 2000. A reexamination of four prolacertiforms with implications for pterosaur phylogenesis. *Rivista Italiana di Paleontologica e Strateigrafia*, **106**, 293–336.

RICQLES, A. J. DE, PADIAN, K., HORNER, J. R. & FRANCILLON-VIEILLOT, H. 2000. Palaeohistology of the bones of pterosaurs (Reptilia: Archosauria): Anatomy, ontogeny, and biomechanical implications. *Zoological Journal of the Linnean Society, London*, **129** , 349–385.

RIEDESEL, M. L. 1977. Blood physiology. *In*: WIMSTATT, W. A. (ed.) *Biology of Bats*. Academic Press, New York, vol. II, 485–517.

ROGERS, R. R. & LABARBERA, M. 1993. Contribution of internal bony trabeculae to the mechanical properties of the humerus of the pigeon (*Columba livia*). *Journal of Zoology*, London, **230**, 433–441.

RUBEN, J. A., DAL SASSO, C., GEIST, N. R., HILLENIUS, W. J., JONES, T. D. & SIGNORE, M. 1999. Pulmonary function and metabolic physiology of theropod dinosaurs. *Science*, **283**, 514–516.

RUBEN, J. A., JONES, T. D., GEIST, N. R. & HILLENIUS, W. J. 1997. Lung structure and ventilation in theropod dinosaurs and early birds. *Science*, **278**, 1267–1270.

SALT, G. W. & ZEUTHEN, E. 1960. The respiratory system. *In*: MARSHALL, A. J. (Ed.) Biology and *Comparative Physiology of Birds*. Academic Press, New York, 363–409.

SCHIED, P. 1979. Mechanisms of gas exchange in bird lungs. *Reviews of Physiology, Biochemistry and Pharmacology*, **86**, 137–186.

SCHMIDT-NIELSEN, K. 1975. Recent advances in avian respiration. *In*: PEAKER, M. (ed.) *Avian Physiology*. Academic Press, London, 33–47.

SCHMIDT-NIELSEN, K. 1995. *Animal Physiology: Adaptation and Environment*, 4th edition. Cambridge University Press, Cambridge, 602 pp.

SEELEY, H. G. 1870. Remarks on Professor Owen's monograph on *Dimorphodon*. *Annals and Magazine of Natural History*, **4**, 129–152.

SEELEY, H. G. 1901. *Dragons of the Air: An Account of Extinct Flying Reptiles*. Methuen & Co. London, 239 pp.

SMITH, J.B., DODSON, P. *et al.* 2001. Biostratigraphy and avian origins in northeastern China. *In*: GAUTHIER, J. & GALL, L. F. (eds) *New Perspectives on the Origin and Early Evolution of Birds*. Museum of Natural History, Yale University, New Haven, Special Publications, 549–589.

SNYDER, G. K. 1976. Respiratory characteristics of whole blood and selected aspects of circulatory physiology in the common short-nosed bat *Cynopterus brachyotes*. *Respiratory Physiology*, **28**, 239–247.

SWISHER, C. C. III, WANG, Y.-Q., WANG, X.-L., XU, X. & WANG, Y. 1999. Cretaceous age for the feathered dinosaurs of Liaoning, China. *Nature*, **400**, 58–61.

THOMAS, S. P. 1987. The physiology of bat flight. *In*: FENTON, M. B., RACEY, P. & RAYNER, J. M. V. (eds) *Recent Advances in the Study of Bats*. Cambridge University Press, Cambridge, 75–99.

TOOMBS, H. A. & RIXON, A. E. 1959. The use of acids in the preparation of vertebrate fossils. *Curator*, **2**, 304–312.

UNWIN, D. M. 1992. The phylogeny of the Pterosauria. *Journal of Vertebrate Paleontology*, **12** (3) (supplement), 57A.

UNWIN, D. M. 1995. Preliminary results of a phylogenetic analysis of the Pterosauria (Diapsida: Archosauria). *Sixth Symposium on Mesozoi Terrestrial Ecosystems and Biota, Beijing*, Short Papers, 69–72.

UNWIN, D. M. 1996. Pterosaur tracks and the terrestrial ability of pterosaurs. *Lethaia*, **29**, 373–386.

UNWIN, D. M. 1999. Pterosaurs: back to the traditional model? *Trends in Evolutionary Ecology*, **14**, 263–268.

UNWIN, D. M. 2001. Patterns and processes of gigantism in pterosaurs. *Journal of Morphology*, **248**, 293.

UNWIN, D. M. & LÜ, J. 1997. On *Zhejiangopterus* and the relationships of pterydactyloid pterosaurs. *Historical Biology*, **12**, 199–210.

VIOHL, G. 1985. Geology of the Solnhofen Lithographic Limestone and the habitat of *Archaeopteryx*. *In*: HECHT, M., OSTROM, J., VIOHL, G. & WELLNHOFER, P. (eds) *The Beginning of Birds*. Proceedings of the International *Archaeopteryx* Conference, 1984. Freunde des Jura-Museums, Eichstätt, 31–44.

VIOHL, G. 1998. Die Solnhofener Plattenkalke – Entstehung und Lebensräume. *Archaeopteryx*, **16**, 37–68.

VIOHL, G. 2000. *Drachen der Lüfte. Entwicklung und Leben der Flugsaurier. Führer zur gleichnamigen Ausstellung des Jura-Museums vom 9, Juni 2000 bis zum 7 Januar 2001*. Friedrich Pfeil, München, 48 pp.

WELLNHOFER, P. 1975a. Die Rhamphorhyncoidea (Pterosauria) der Oberjura Plattenkalke Süddeutschlands. Teil I. Allgemeine Skelettmorphologie. *Palaeontographica A*, **148**, 1–33.

WELLNHOFER, P. 1975b. Die Rhamphorhyncoidea (Pterosauria) der Oberjura Plattenkalke Süddeutschlands. Teil II. Systematische Beschreibung. *Palaeontographica A*, **148**, 132–186.

WELLNHOFER, P. 1975c. Die Rhamphorhyncoidea (Pterosauria) der Oberjura Plattenkalke Süddeutschlands.

Teil III. Palökologie und Stammesgeschichte. *Palaeontographica A*, **149**, 1–30.

WELLNHOFER, P. 1978. Pterosauria. *In*: WELLNHOFER, P. (ed.) *Handbuch der Paläoherpetologie*. Gustav Fischer, Stuttgart, Teil 19, 82 pp.

WELLNHOFER, P. 1980. Flugsaurierreste aus der Gosau-Kreide von Muthmannsdorf (Niederösterreich) – ein Beitrag zur Kiefermechanik der Pterosaurier. *Mitteilungen der Bayerische Staatssammlung für Paläontologie & Historische Geologie*, **20**, 95–112.

WELLNHOFER, P. 1985. Neue pterosaurier aus der Santana Formation (Apt) der Chapada do Araripe, Brasilien. *Palaeontographica A*, **187**, 105–182.

WELLNHOFER, P. 1988. Terrestrial locomotion in pterosaurs. *Historical Biology*, **1**, 3–16.

WELLNHOFER, P. 1991a. *The Illustrated Encyclopedia of Pterosaurs*. Salamander, London, 192 pp.

WELLNHOFER, P. 1991b. Weitere Pterosaurierfunde aus der Santana-Formation (Apt) der Chapada do Araripe, Brasilien. *Palaeontographica A*, **215**, 43–101.

WELLNHOFER, P. 1993. Das siebte Exemplar von *Archaeopteryx* aus der Solnhofer Schichten. *Archaeopteryx*, **11**, 1–48.

WILD, R. 1971. *Dorygnathus mistelgauensis n.sp.*, ein neuer Flugsaurier aus dem Lias Epsilon von Mittelgau (Fränkischer Jura). *Geologische Blätter von Nord-Ost Bayern*, **21**, 178–195.

WILD, R. 1978. Die Flugsaurier (Reptilia, Pterosauria) aus der Oberen Trias von Cene bei Bergamo, Italien. *Bolletino della Societa Paleontologica Italiana*, **17**(2), 176–256.

WILD, R. 1984. A new pterosaur (Reptilia, Pterosauria) from the Upper Triassic (Norian) of Friuli, Italy. *Gortania*, **5**, 45–62.

WILD, R. 1993. A juvenile specimen of *Eudimorphodon ranzii* Zambelli (Reptilia: Pterosauria) from the Upper Triassic (Norian) of Bergamo. *Rivista Museo Civico Sciencias Naturales "E. Caffi" Bergamo*, **16**, 91–115.

WITMER, L. M. 1990. The craniofacial air sac system of Mesozoic birds. *Zoological Journal of the Linnean Society, London*, **100**, 327–378.

XU, X., TANG, Z-I. & WANG, X-L. 1999a. A therizinosauroid dinosaur with integumentary structures from China. *Nature*, **399**, 350–354.

XU, X., WANG, X-L. & WU, X-C. 1999b. A dromaeosaurid dinosaur with a filamentous integument from the Yixian Formation of China. *Nature*, **401**, 262–266.

ZAMBELLI, R. 1973. *Eudimorphodon ranzii*, gen. nov., sp. nov., un pterosauro Triassico. *Rendiconti dell Istituto Lombardo di Scienze & Lettere (B)*, **107**, 27–32.

New specimens of Pterosauria (Reptilia) with soft parts with implications for pterosaurian anatomy and locomotion

EBERHARD FREY[1], HELMUT TISCHLINGER[2], MARIE-CÉLINE BUCHY[3] & DAVID M. MARTILL[4]

[1]*Staatliches Museum für Naturkunde Karlsruhe, D-76133 Karlsruhe, Germany*
[2]*Tannenweg 16, D-85134 Stammham, Germany*
[3]*Universität Karlsruhe, Geologisches Institut, Postfach 6980, D-76128 Karlsruhe, Germany*
[4]*School of Earth and Environmental Sciences, University of Portsmouth, Portsmouth PO1 3QL, UK*

Abstract: New specimens of pterosaurs with soft-part preservation from the Solnhofen Lithographic Limestone (S Germany) and the Crato Formation (northeastern Brazil) yield hitherto unknown and unexpected details of pterosaur anatomy: the presence and internal anatomy of soft-tissue crests, the internal anatomy of the brachiopatagium, including a blood vessel system and structural details of foot and hand. Some consequences for pterosaurian flight, thermoregulation and aspects of evolution are discussed.

Introduction

Preservation of non-mineralized tissues in pterosaurs has been reported from the bituminous shales at Holzmaden (Early Jurassic, southern Germany), the Solnhofen Lithographic Limestone (Late Jurassic, southern Germany), the Late Jurassic limestone deposits near Karatau (Kazachstan) and the Early Cretaceous laminated limestone of the Crato Formation, as well as from the concretions in the marls of the Santana Formation from the Chapada do Araripe, northeastern Brazil. The latter formations are of Early Cretaceous age. Soft-part preservation in pterosaurs has been reviewed by Padian & Rayner 1993, Frey & Martill 1998 and Bennett 2000.

Institute abbreviations: BSP, Bayerische Staatssammlung für Paläontologie und Geologie, Munich, Germany; JME, Jura-Museum Eichstätt, Germany; NHMW, Naturhistorisches Museum, Wien, Germany; SMNK, Staatliches Museum für Naturkunde, Karlsruhe, Germany.

Solnhofen Lithographic Limestone specimens

Soft-part preservation: The earliest report on pterosaur soft-part preservation is based on a specimen of *Scaphognathus* from the Solnhofen Lithographic Limestone described by Goldfuss (1831). He reported imprints of wing membrane, a fur-like body coverage and a "mane" of long fibres along the neck. H. v. Meyer (1846, 1859) and other palaeontologists rejected his opinion and stated that all these structures were of anorganic origin. Recently a reinvestigation of the *Scaphognathus* specimen under ultraviolet light prooved that Goldfuss was right (Tischlinger 2002).

Webbed feet. Skin-like structures between the metatarsals and digits of the foot were first discovered on a specimen of *Rhamphorhynchus* (Broili 1927a) and shortly after were also proved for *Pterodactylus* (Döderlein 1929; Broili 1938). The webbing of the toes has been interpreted as a swimming device (Abel 1907, 1919, 1927). Broili (1927a, p. 39) explicitly mentions the smooth surface of the webbing without any structures resembling that of a brachiopatagium. Fibres in the webbing of both feet were described in a recently discovered specimen of *Pterodactylus* sp. (Frey & Tischlinger 2000; Tischlinger & Frey 2001). In the same specimen, a heel and sole pad with subcircular scales as well as the keratinous sheaths of the claws of hand and foot are preserved (Frey & Tischlinger 2000; Tischlinger & Frey 2001).

Throat pouches, rhamphothecae, soft-tissue crests and occipital cones. Other soft parts were discovered in *Pterodactylus* ventral to the mandible and were considered to represent a throat pouch (Broili 1927a, b). Recently a similar structure was found in the Edinburgh specimen of *Rhamphorhynchus* (Unwin in prep.). Soft-tissue crests or skin flaps in the occipital region are known in *Rhamphorhynchus* (Wanderer 1908; Broili 1927a, b), *Pterodactylus* (Broili 1927a, b; Frey & Tischlinger 2000; Tischlinger & Frey 2001) and *Germanodactylus* (Tischlinger 1998). The *Pterodactylus* specimen described by Frey & Tischlinger (2000) not only shows subparallel internal fibres in the crest but also spiral fibres inside an occipital soft-part cone. Furthermore, a wrinkled epidermis surface of the crest is reported (Tischlinger & Frey 2001).

From: BUFFETAUT, E. & MAZIN, J-M. (eds) 2003. *Evolution and Palaeobiology of Pterosaurs*. Geological Society, London, Special Publications, **217**, 233–266. 0305–8719/03/$15 © The Geological Society of London 2003.

Additionally, the specimen possesses a tiny keratinous beak which is the first evidence of such a structure for *Pterodactylus* (Tischlinger & Frey 2001).

Body hair. Hints of the presence of fur in the form of fine pits and fibres in the body area date from 1908 and were discovered on a specimen of *Rhamphorhynchus* (Wanderer 1908), but it was Broili (1927a) who suggested these structures as being hair papillae or hairs matted together. These pits and fibres are exceptionally well preserved in a *Rhamphorhynchus* specimen in the Leich collection, a former private collection which is now accessible to the public in the Fossilium Tierpark Bochum (Germany). A fur-like body coverage in *Pterodactylus* was first noted by Broili (1938), while Frey & Martill (1998) described a mat of bristles along the neck of a specimen of *Pterodactylus kochi*. The same specimen also preserves a uropatagium extending from the fifth metatarsal to the tip of the tail. Bristles along the neck are also seen in a specimen of *Germanodactylus* (Tischlinger 1998).

Patagia. Pterosaurian patagia were first reported from the Solnhofen Lithographic Limestone by Goldfuss (1831) in a specimen of *Scaphoguathus crassirostris* and by Winkler (1874) in a specimen of *Pterodactylus kochi*. A few years later, further patagia of *Rhamphorhynchus* were described (Marsh 1882; Zittel 1882). Meanwhile there are numerous reports on brachiopatagium preservation in *Rhamphorhynchus*, *Pterodactylus* (for a review see Padian & Rayner 1993 and Bennett 2000) and, recently also *Anurognathus* (Frey & Buchy pers obs.), whereby the record of the brachiopatagium preservation in *Rhamphorhynchus* is by far the best and most detailed. Two types of membrane preservation are until now reported from the Solnhofen Plattenkalk: external moulds, or substance preservation whereby the collagen elements are replaced by phosphate minerals.

The brachiopatagium consists of two parts: the aktinopatagium and the tenopatagium, the proximal part of patagium adjacent to the body (Schaller 1985). Bennett (2000) follows Schaller (1985) in the definition of the parts of the brachiopatagium, but uses the English transcription "actinopatagium". The aktinopatagium is characterised by straight support fibres which have been named "Aktinofibrillen" by Wellnhofer (1987b). In previous publications he called these fibres 'elastische Fasern [elastic fibres]' (Wellnhofer 1975c, 1980). As was shown by Bennett (2000) these fibres could consist of dense collagen, keratin or cartilage. We here propose to use the spelling "aktinofibrils" to emphasize that there is no evidence for the presence of actin.

Until now there has been no evidence from the Solnhofen Lithographic Limestone that the brachiopatagium attached to the ankle region. In one specimen of *Pterodactylus* (the 'Vienna specimen' NHMW 1975/1756; Wellnhofer 1987b, 1991; Bennett 2000) the brachiopatagium appears to insert at the upper leg close to the thigh. Wellnhofer (1987, 1991) assumed that 'the brachiopatagium was attached at the upper leg and extended to the side of the upper part of the lower leg' (Wellnhofer 1991, p. 149). Bennett (2000) postulated for *Rhamphorhynchus* an insertion of the brachiopatagium at the knee.

Remnants of a propatagium are known from the 'Zittel wing' (see above; Zittel 1882) and from some specimens of *Pterodactylus* (Wellnhofer 1970, 1975c, 1978, 1987, 1991).

Tail vane. In several *Rhamphorhynchus* specimens a tail vane of rhomboid outline is preserved (Wellnhofer 1975a, b, 1991). Some of these vanes contain fibres running perpendicular to the vertebral column. Until now it is unclear if this vane was held vertically (Wellnhofer 1975a, 1991) or horizontally (Holst 1957a, b).

Karatau Formation, Kazakhstan

Two species of pterosaurs are reported from the Karatau deposits at Michajlovka in Kazakhstan: *Sordes pilosus* Sharov 1971, named after the bristles that cover the body area, and *Batrachognathus volans* Rjabinin 1948, which has bristles around the mouth. Unwin & Bakhurina (1994) reinvestigated *Sordes* and found that most of the bristle-like structures represent brachiopatagium fibres released from the wing during decomposition, especially those close to the lateral margins of the body. However, Unwin & Bakhurina (1994) confirmed that there are fibre types, distinct from those of the brachiopatagium, which might be part of a body fur and are structurally equivalent to the bristles found in the neck of *Pterodactylus kochi* (Frey & Martill 1998). The fibres preserved in the brachiopatagium of *Sordes* appear to be coincident from their structure and orientation to those found in *Rhamphorhynchus*. The brachiopatagium in *Sordes* is attached at the ankle with an almost straight trailing edge and from there to the lateral terminus of the wing finger. The uropatagium is attached to the apex of the contralateral hooked fifth metatarsals and is marked by a reverse v-shaped recessus, symmetrical with the median line. Possible propatagia are preserved cranial to the wing skeleton of the holotype (Sharov 1971) and a narrow longitudinally oval tail vane is preserved around the distal end of the tail.

Santana Formation, Ceará, northeastern Brazil

Pterosaur remains are relatively abundant in the Romualdo Member of the Santana Formation, occurring more frequently than the remains of aquatic and semi-aquatic tetrapods, such as chelonians and crocodilians (Martill 1993). When enclosed in early diagenetic concretions their remains are often preserved three-dimensionally and may occasionally have soft-part preservation (Martill & Unwin 1989). The pterosaur fauna of the Romualdo Member is not only abundant, but is also diverse, with the genera *Tapejara*, *Tupuxuara*, *Coloborhynchus* and *Criorhynchus* known for certain (Kellner & Tomida 1996, 2000; Fastnacht 2001). The taxonomic validity of other genera (e.g. *Brasileodactylus*, *Santanadactylus*, *Araripesaurus* and possibly *Anhanguera*) is in doubt (Unwin 2001; Frey *et al.* 2003a).

The first pterosaur remains from the Romualdo Member nodules were reported by Price (1971), who described a partial skeleton for which he erected the taxon *Araripesaurus castilhoi*. Since that time many pterosaur specimens, often in a remarkable state of preservation, have been recovered from these deposits (e.g. Kellner, 1984, 1990; Campos & Kellner 1985; Wellnhofer 1985, 1987a, 1991; Kellner & Campos 1989; Kellner & Tomida 2000). The first evidence for soft-part preservation in Romualdo Member pterosaurs was reported by Campos *et al.* (1984) for a specimen of indeterminate pterodactyloid pterosaur in which a region of dark-coloured, possibly organic material exhibiting a fibrous texture is preserved adjacent to a wing bone. This material shows the fibres lying parallel to the long axis of the wing bone and strongly resembles similar features seen in mouldic soft tissues from Solnhofen Limestone examples of *Rhamphorhynchus*. A colour illustration of this specimen can be found in Wellnhofer (1991, p. 152).

The mode of preservation exhibited by this specimen is unusual for the Romualdo Member nodule beds. Martill (1988) and Martill & Unwin (1989) have shown that soft-tissue preservation in the nodules usually occurs as metasomatic replacements or bacterial autolithifications of soft tissues by calcium phosphate. This mode of preservation is not uncommon for soft tissue in both vertebrates and invertebrates in the Romualdo nodules, and has been found in ostracods (Smith 2000), elopomorph fishes (Martill 1988), elasmobranch fishes (Brito & Ferreira 1989), pterosaurs (Martill & Unwin 1989; Martill *et al.* 1990) and theropod dinosaurs (Kellner 1996).

Only one pterosaur has been reported with phosphatized soft tissues, though it has proved to be controversial. Martill & Unwin (1989) reported the three-dimensional, high-fidelity preservation of a portion of wing membrane associated with skeletal elements of an indeterminate pterodactyloid pterosaur. This specimen was particularly remarkable as it showed detail of several tissue types within the wing membrane, including an upper layer interpreted as a 10 μm-thick epidermis, underlain by a 20–40 μm-thick layer of spongy tissue, followed by a 180–200 μm-thick layer of structural fibres arranged in bundles and a basal layer (as preserved) of striated muscle fibres. This latter layer shows a thickness between 500 and 450 μm. Later, Kellner (1994, 1996) cast doubt on the wing membrane origin for the preserved soft tissues, suggesting instead that it represented a portion of the dermis and epidermis from the body. Kellner (1994, 1996) failed to explain the presence of the structural fibres that are structurally homologous with those fibres found in wing membranes of Solnhofen and other pterosaurs. Bennett (2000) agrees with Kellner (1994, 1996) in that the nature of the structural fibres as 'actinofibrils' cannot be confirmed.

Crato Formation, Ceará, northeastern Brazil

Most of the pterosaur material from the Crato Formation comprises fragmentary but articulated skeletons of wings and legs. Until now, only two virtually complete postcrania are known (Frey & Martill 1994; an unpublished ornithocheiroid SMNK PAL 3854), but several isolated crania, some lacking the mandible were described (Campos & Kellner 1997; Martill & Frey 1998). Those crania referable to the Tapejaridae and some wings and leg skeletons of possible azhdarchid pterosaurs preserve supreme soft parts, such as dermal cranial crests with internal fibres, patches of brachiopatagium, webbing and the keratinous claw sheaths of hand and foot, and heel and sole pads with subcircular scales (Frey & Tischlinger 2000; Tischlinger & Frey 2001).

Pterosaurs with soft-part preservation from the Solnhofen Lithographic Limestone

A comparison between the Solnhofen and the Crato material yields new insights into aspects of the external and internal soft-part anatomy of pterosaurs. Here, we describe a new *Rhamphorhynchus* specimen which preserves the brachiopatagium to a hitherto unknown detail. We also briefly describe those pterosaur soft-part specimens from Bavaria and northeastern Brazil which have been more extensively described in German by Frey & Tischlinger (2000) and Tischlinger & Frey (2001). On the basis of the new discoveries, we discuss aspects of the morphogenesis of cranial crests, the implications for aerial manoeuvrability, the impact on the reconstruction of the plantar surface of the foot and the interpretation of pterosaurian foot prints, and introduce ideas concerning thermoregulation in pterosaurs.

Methods: ultraviolet light photography

Essential details of the preserved soft parts can only be demonstrated by long-wave ultraviolet light photography. Best results are obtained with a wave length of 365–366 nanometers. According to the producer, the ultraviolet lamps guarantee an UV intensity between 4000 and more than 10000 microwatts per 10 mm^2, depending from the distance and the number of lamps. The application of different filters allows a selective visualization of peculiar fine structures. The filters were directly mounted on the lens of the stereo microscope or the camera lens. The optimum of filtering and exposure time was tested in a series of experiments (see Tischlinger 2002).

Pterodactylus sp.

Preservation. A new specimen of *Pterodactylus* sp. from the Upper Solnhofen Formation (Malm Žeta 2b, early Lower Tithonian) was discovered in the open-cast Schrandel quarry at Langenaltheim by the quarry-owner in the year 1998. Slab and counterslab are preserved and were mechanically prepared under ultraviolet light (by H. T.). The main slab to which most of the bones have been transferred is in the possession of the quarry owner but is accessible for scientific studies via the Jura-Museum in Eichstätt (Germany). One piece of the counterslab is housed in the Jura-Museum (accession number JME SOS 4784).

The skeleton is almost complete and articulated (Fig. 1a, b). The skull is detached from the vertebral column and now partially obscures the pelvic region. The mandible is fully depressed to bone lock. Its rostral part lies across the metatarsals of the right foot. The vertebral column of the neck is so strongly recurved that it forms a loop. Five cervical vertebrae are visible. The vertebral column of the trunk lies straight, in prolongation of the most caudally preserved cervical vertebra. Five thoracic neural spines are exposed alongside the right femur. The sacrum is slightly tilted ventrally against the vertebral column of the trunk. Both ischiopubis bones lie in the central region of the nasoantorbital fenestra with both praepubis bones being present. The vertebral column of the tail is angled 70° dorsally against the sacrum.

The extremities are all articulated, however, with most of the bones being bent beyond bone lock but still close to *in vivo* articulation. The right wing digit is bent medially with the most distal digital phalanx reaching the bottom-most part of the loop of the cervical column. The left one lies subparallel along the left side of the vertebral column of the trunk. The proximal digits of the left hand are straight and describe angles of 30° between the first and second digits and 45° between the second and third digits.

Both hindlimbs lie ventral to the trunk skeleton. The right femur parallels the vertebral column and is seen from medially. The right tibia has rotated medially, lies across the left tibia and, together with the fibula, exposes its facies cranialis. Femur and tibia of the right hindlimb are preserved in line. Both feet are flexed slightly dorsally with respect to the tibiae and seen from dorsally. The right foot has rotated medially, the left laterally.

Soft parts are preserved associated with the skull, the left hand and both feet.

Systematic palaeontology. With a skull length of 137 mm the specimen is a large pterodactyloid for the Solnhofen Lithographic Limestone. The biometry suggests that the new specimen is close to *Pterodactylus antiquus* but its tibia/femur ratio clearly differs from the values of *P. antiquus* (Wellnhofer 1970). Therefore we cautiously identify the specimen as *Pterodactylus* sp.
Order Pterosauria Kaup 1834
Superfamily Pterodactyloidea Plieninger 1901
Family Pterodactylidae Bonaparte 1838
Genus *Pterodactylus* Cuvier 1809
Species *Pterodactylus* sp.

Soft-part preservation
Skull (Fig. 1). The rostral terminus of both upper and lower jaws bears a keratinous hook which is of similar size to the rostral teeth, but is strongly phosphorescent under ultraviolet light (Figs 1b, e & 2a). This is the first evidence for a keratinous beak in *Pterodactylus*. Fibrous remnants of a medial crest are preserved extending from the caudal margin of the nasoantorbital fenetra to the occiput (Fig. 1a–c, f). The fibres are parallel and some reach a length of 14 mm (Fig. 1c, f). They are embedded in a whitish, slightly phosphorescent matrix which covers the caudal part of the skull, including the orbita (Fig. 1a–c). Skin-like structures, visible over a distance of 40 mm along the dorsal margin of the orbital and parietal part of the skull and are in direct contact with the fibres. These structures show a reticular surface (Fig. 1g) similar to the surface of the epidermis described by Martill & Unwin (1989, fig. 15 b). On the counterslab, the reticulated structures are better preserved (Fig. 1g). They must have formed the facies externa of the crest and most likely represent patches of epidermis.

From the parieto-occipital corner of the skull, a dorsally curved conical structure with a length of 22 mm arises (Fig. 1a, b). A similar cone was described for *Pterodactylus kochi* (Wellnhofer 1970). Inside the three dimensionally preserved occipital cone, fibres are preserved that form a multiple helix (Fig. 1d). Distally these fibres are phosphatized and

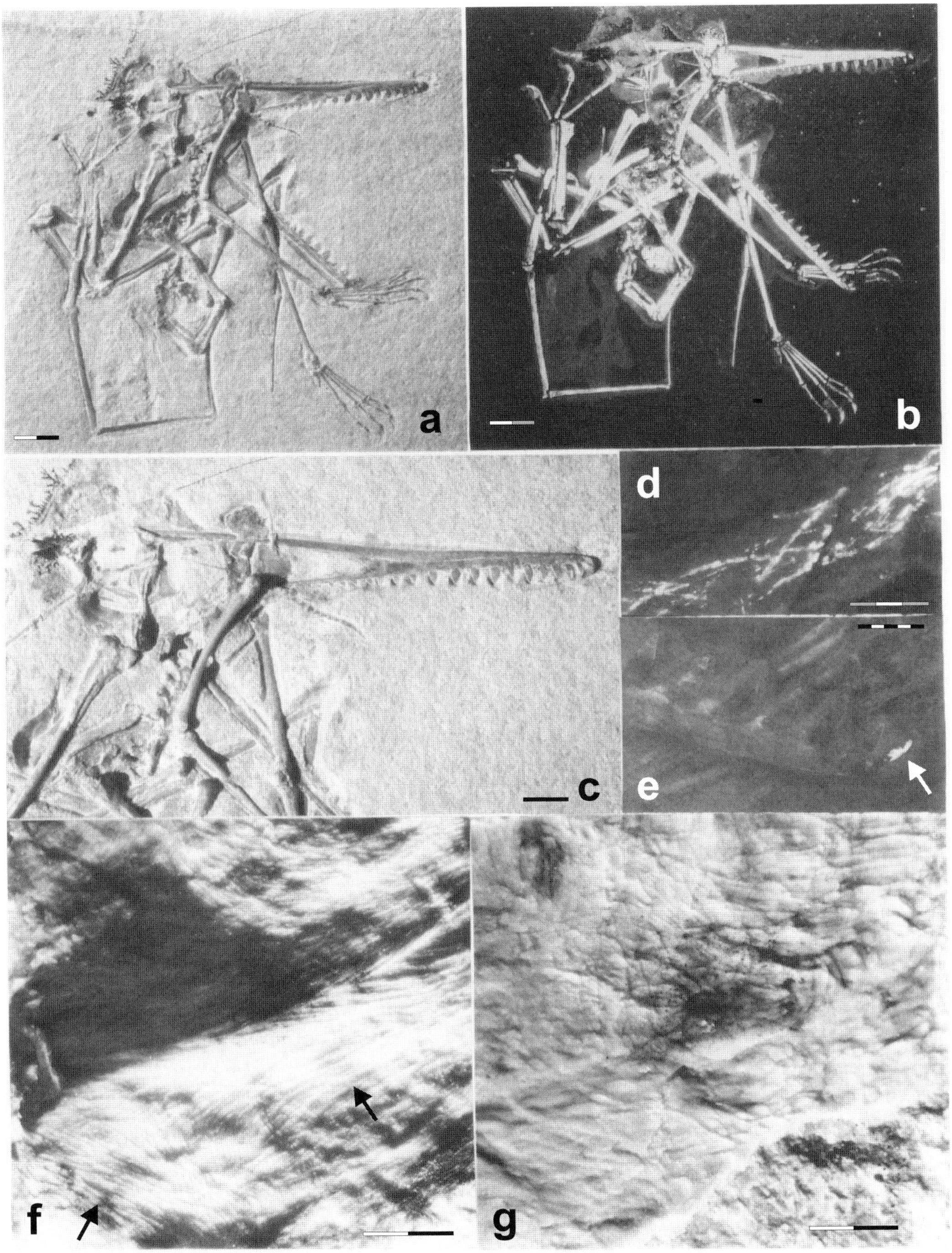

Fig. 1. *Pterodactylus* sp., soft-part preservation. (**a**) Specimen under normal light. Scale bar 20 mm. (**b**) Specimen under ultraviolet light. Scale bar 20 mm. (**c**) Close-up of skull with traces of a crest under normal light. Scale bar 10 mm. (**d**) Occipital cone with spiral fibres under ultraviolet light. Scale bar 3 mm. (**e**) Rostral terminus of mandible with a keratinous rhamphotheca (arrow). Scale bar 5 mm. (**f**) Close-up of the fibrous part of the crest overlapping the orbita under normal light (arrows point to fibres). Scale bar 2 mm. (**g**) Close-up of external mould of the epidermis surface of the crest under normal light (specimen JME SOS 4785, Jura-Museum, Eichstätt, Germany). Scale bar 2 mm.

phosphorescent under ultraviolet light; proximally they are blackened due to manganese oxide incrustation.

Feet (Fig. 2a–c). Soft parts are preserved between the metatarsals and the digits of both feet (Fig. 2a). In the area of the metatarsals, the soft-part preservation may represent remnants of the skin, muscles and tendons of the metatarsal, whereas between the digits a webbing reaches the base of the ungual phalanx (Fig. 2b, c). The distal margin of the webbing between two digits is concave (Fig. 2a–c). The webbing contains parallel fibres which are perpendicular to the distal margin of the webbing (Fig. 2b). The fibres are 0.2 mm apart, with 24 fibres in the 4.3 mm-wide webbing between digits III and IV of the left foot. Such fibres are preserved on both feet and, under daylight, they are best visible on the right one, where they are orange-brown due to a stain of iron oxide hydrate (Fig. 2c). The webbing is thickest between digits III and IV on both feet, where the fibres are strongly phosphorescent under ultraviolet light (Fig. 2a), indicating a phosphatic preservation.

All ungual phalanges bear keratinous claw sheaths that are approximately twice as long as the phalanx itself (Figs 1b & 2a). The pointed claw sheaths are strongly curved ventrally and phosphoresce under ultraviolet light. Such claw sheaths have not been previously reported for pterosaurs from the Solnhofen Lithographic Limestone.

A sole and heel pad is preserved that extends from the tarsus to the proximal end of digit IV of both feet (Fig. 2c). The metatarsal and phalanx V are enclosed therein but not covered in the specimen. The surface of these pads differs from all other surface structures in having fine squamation. The scales are subcircular, with a diameter of 0.2 mm, and are exposed on both sides of the tarsus. The squamation is also visible as a weak seam on both sides of the distal part of the tibiae over a distance of 20 mm. From the distal terminus of digit V there is a mediodistally running row of scales twice as large as the others. The scales are preserved as external moulds and are represented by shallow pits in the sediment, which indicates that the surface of each scale was slightly vaulted. Some of the scales at the proximal margin of metatarsal V are bordered by a stain of iron oxide hydrate.

Rhamphorhynchus muensteri

Preservation. The *Rhamphorhynchus muensteri* specimen (Fig. 3a, c; see also Frickhinger 1999, p. 149, figs 264 & 265; Viohl 2000, 2001) was discovered in a Plattenkalk quarry near Eichstätt, Germany. The skeleton is nearly fully articulated and all components are preserved three-dimensionally. Only the extremities of the tail and both wings were lost, together with the missing part of the slab during mining. The specimen is preserved on the bottom slab, as is indicated by a touchdown mark in the sediment adjacent to the dorsal margin of the skull, which is lying on its right side. This mode of preservation is unusual for the Solnhofen Lithographic Limestone: usually skeletons and soft parts occur in the top slab while the bottom slab displays an external mould of the fossil (Barthel 1964, 1972; Kemp 1999). The slabs normally are split in the boundary between top and bottom slab. Therefore, the specimen is seen from below with most of the skeleton sticking inside the top slab. In the preparation of the new *Rhamphorhynchus*, however, the integrity of the top and bottom slabs was maintained and the specimen was prepared by removing the matrix of the top slab from top to bottom, keeping the specimen on the bottom slab. Just one fragment of the top slab containing the soft parts of the ventral side of the left wing was split in the classical way between top and bottom slab. Under ultraviolet light this fragment shows extraordinarily preserved details of the brachiopatagium anatomy, including blood vessels (Fig. 4). The bottom slab (Fig. 3) remains with the private collector but will be transferred to the Jura-Museum Eichstätt in the future. The top slab with the brachiopatagium soft-part preservation (Fig. 4) is already housed in the Jura-Museum (collection number JME SOS 4785).

Systematic palaeontology. The pointed rostral terminus of the mandible with a circular cross-section as well as the short predental part of the upper jaw (shorter than the tooth # 3 in the upper jaw; Wellnhofer 1975b) allows us to refer the specimen to *Rhamphorhynchus muensteri* (Goldfuss 1831).

Order Pterosauria Kaup 1834
Superfamily Rhamphorhynchoidea Plieninger 1901
Family Rhamphorhynchidae Seeley 1870
Genus *Rhamphorhynchus* H. v Meyer 1846
Species *Rhamphorhynchus muensteri* (Goldfuss 1831)

Hard-parts preservation. The skull lies on its right side with the mandible preserved almost in full occlusion (fig. 3a, c). All teeth are in place. The vertebral column of the neck is strongly recurved. The skeleton of the trunk lies on the right side of its ventral surface, as does the vertebral column of the tail. The thorax, the left scapulocoracoid and the sternum are displaced 30° counterclockwise with respect to the median plane. In consequence, the last cervical vertebra and the first thoracic vertebra are disarticulated but lie close to their original positions. The caudal part of the presacral vertebral column is slightly kinked between thoracic vertebrae eight and nine (Fig. 3a, c). Five ribs are preserved in articulation on the right side of the body; on the left side there are only four. The ribs of thoracic vertebrae

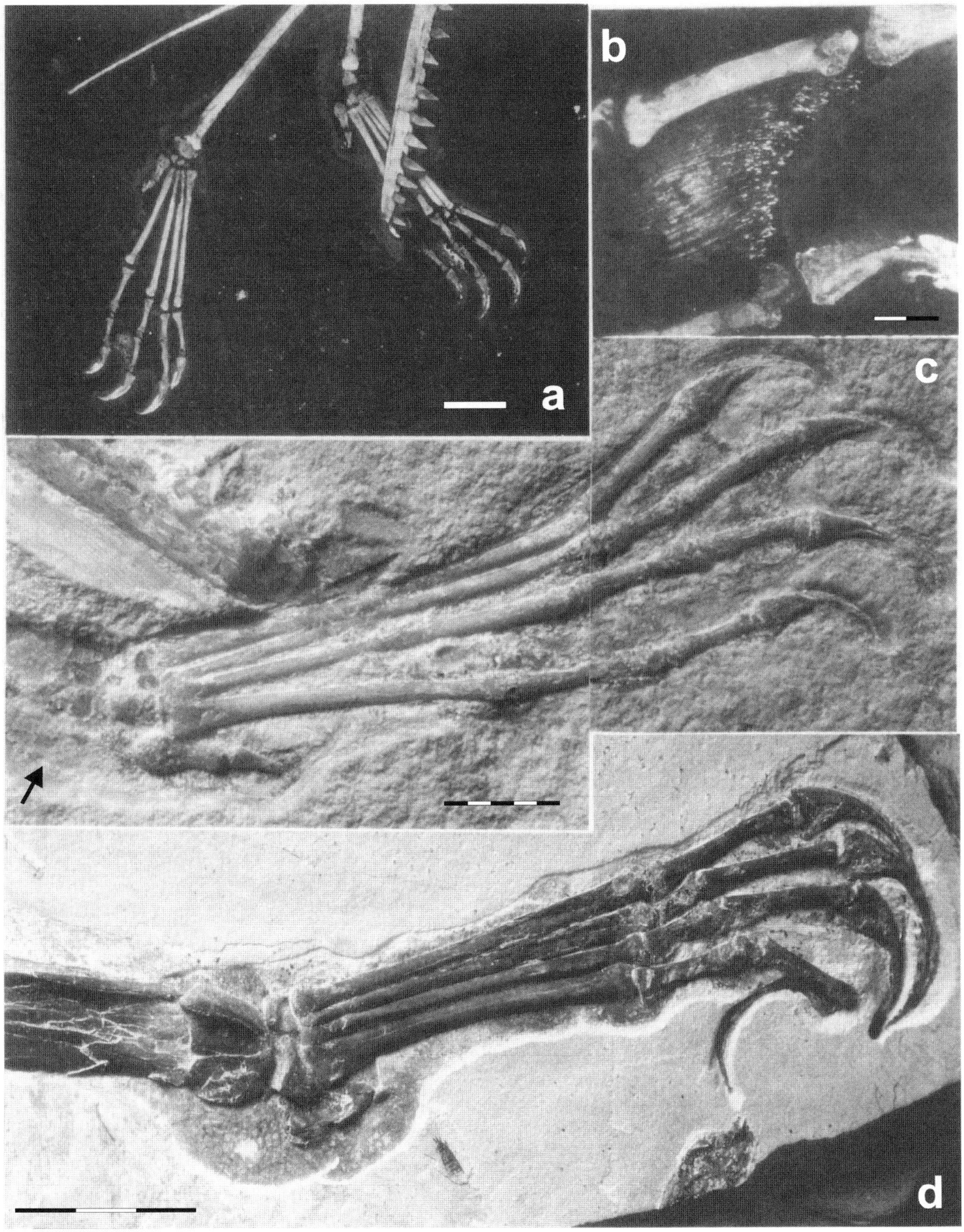

Fig. 2. *Pterodactylus* sp., soft-part preservation. (**a**) Close-up of both feet under ultraviolet light; note the huge claw sheaths and the webbing between the digits. Scale bar 10 mm. (**b**) Close-up of the webbing between digits III and IV on the left foot under ultraviolet light; note the subparallel-oriented fibres. Scale bar 2 mm. (**c**) Right foot under normal light; note the circular scales around the ankle joint (arrow). Scale bar 5 mm. (**d**) ?Azhdarchidae sp. (specimen SMNK PAL 3830, Staatliches Museum für Naturkunde Karlsruhe, Germany), soft-part preservation around the ankle region; note also the huge keratineous claw sheaths. The specimen was copied as a mirror image to clarify the close similarity to the foot of the Solnhofen *Pterodactylus*. Scale bar 30 mm.

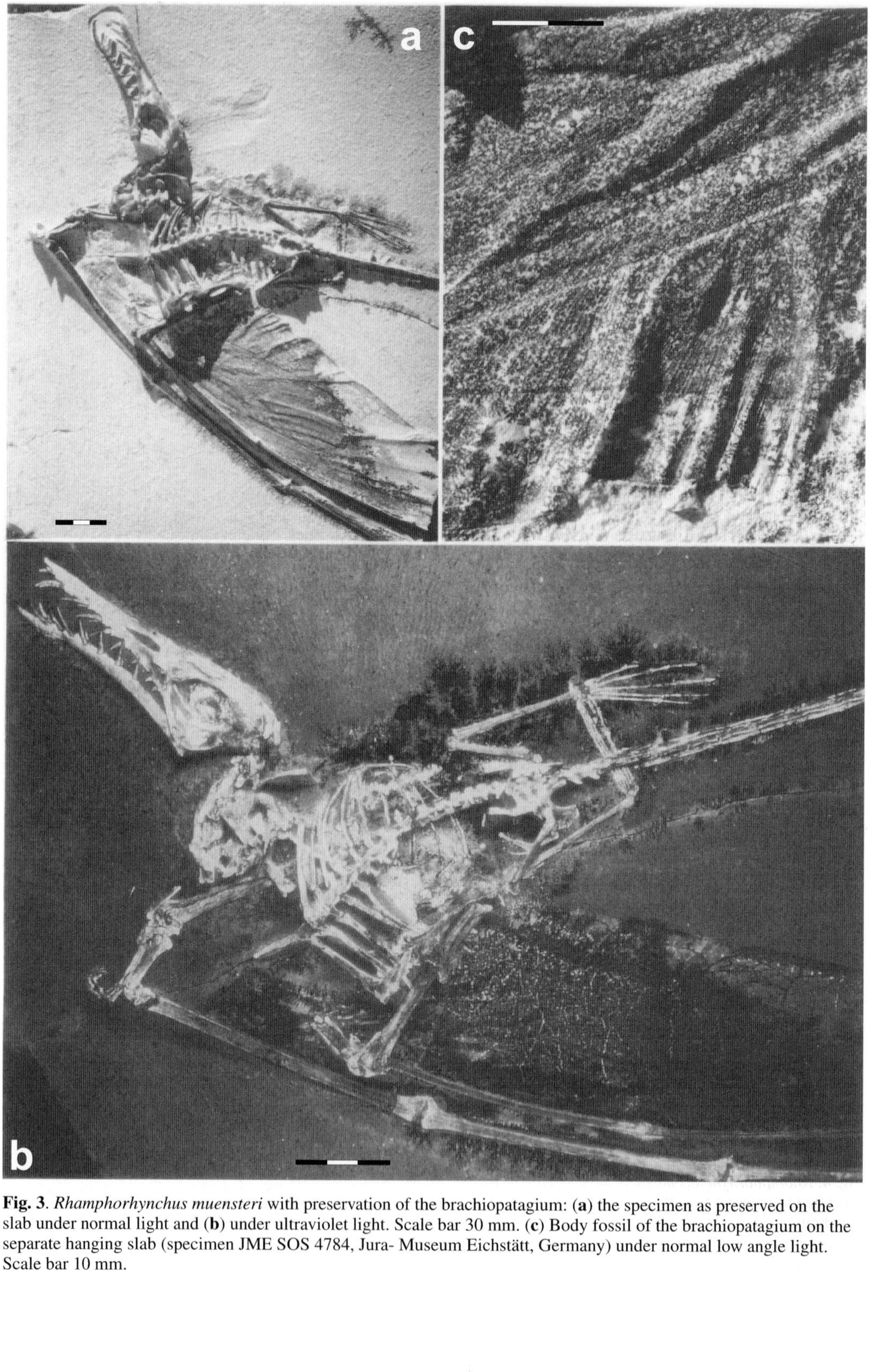

Fig. 3. *Rhamphorhynchus muensteri* with preservation of the brachiopatagium: (**a**) the specimen as preserved on the slab under normal light and (**b**) under ultraviolet light. Scale bar 30 mm. (**c**) Body fossil of the brachiopatagium on the separate hanging slab (specimen JME SOS 4784, Jura- Museum Eichstätt, Germany) under normal low angle light. Scale bar 10 mm.

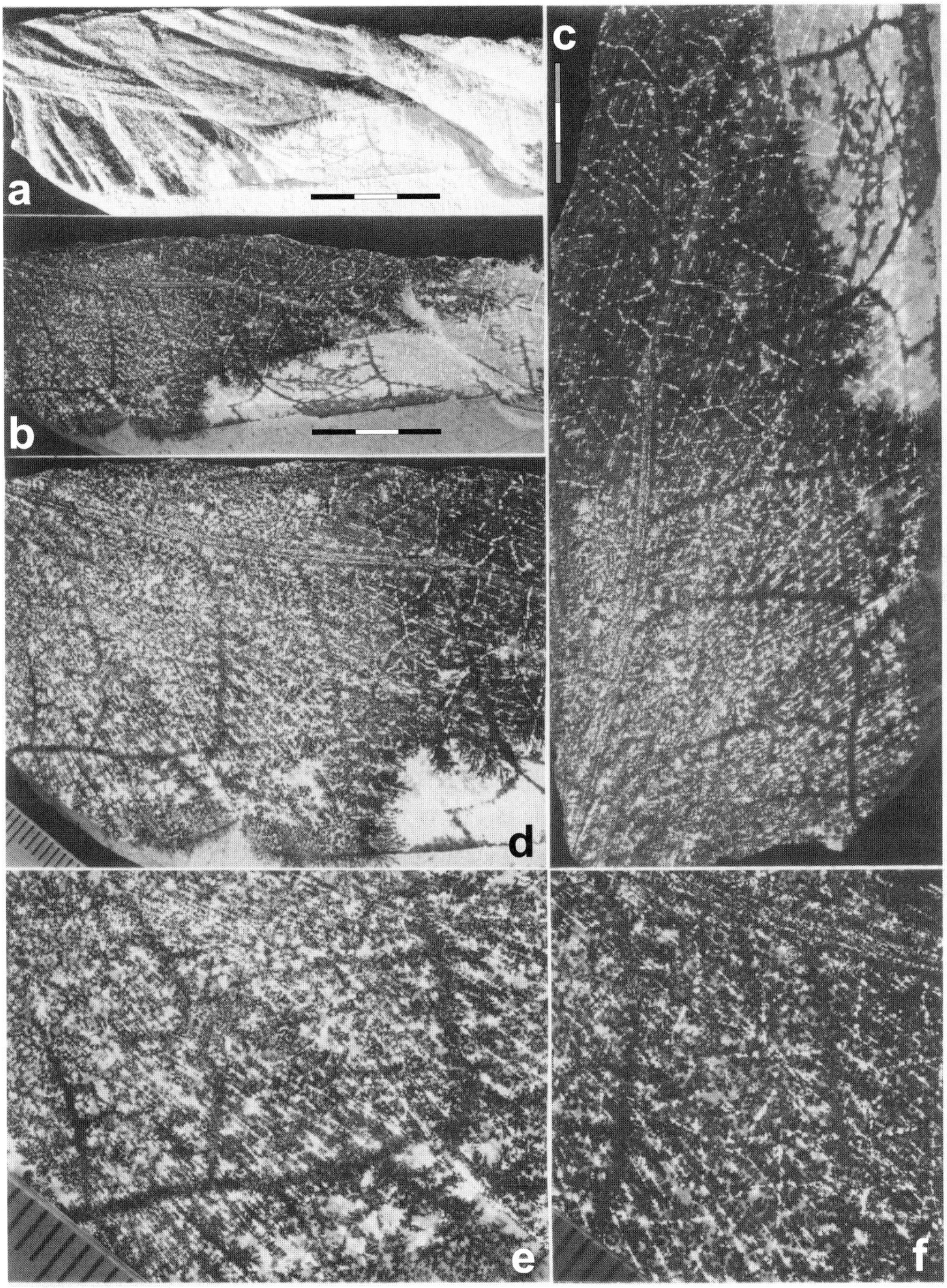

Fig. 4. *Rhamphorhynchus muensteri*, wing membrane (specimen JME SOS 4784, Jura-Museum Eichstätt, Germany): specimen as preserved (**a**) under normal light and (**b**) under ultraviolet light. Scale bar 30 mm. (**c**) Close-up of the membrane to show the blood vessel system, the ?fascia meshwork and the stabilizing fibres with a focus on the blood vessels. (**d**) Proximal part with a focus on the stabilizing fibres; the vessels are also visible. (**e**) Detail of the capillary system between the blood vessel loops. (**f**) Detail of two primary branchings from the main vessel; note the bright seam of tiny circular structures possibly representing remnants of muscle fibrils inside the wall of the main vessel. Scale segments 1 mm.

eight, nine and ten lie close to their original positions. Three ribs are scattered in the abdominal region, two to the left and two to the right of the vertebral column. Six gastralia are clustered caudal to the articulated ribs on the left side of the thorax (Fig. 3c). Due to the displacement of the thorax, the right scapulocoracoid is disarticulated (Fig. 3a, c). It lies on its caudal face with the glenoid fossa adjacent to the neural spines of cervical vertebrae three and four. The coracoid is directed medially and partly obscured by the disarticulated cervical vertebra eight and the thoracic vertebra one. The left scapulocoracoid is articulated with the sternum. The coracoid is oriented transversely, the scapula dorsoventrally forming an angle of approximately 100° with the coracoid. Of the sternum only the dorsal surface of the manubrium is visible (Fig. 3a, c). The sternal plate is obscured by the cranial part of the thorax which overlies it. The bones of both wings are near to their original articulation (Fig. 3a,c), although both elbow joints are slightly disarticulated. The metacarpal complex is a little distorted as a result of an obliquely directed compaction from the right side to the left. Of the right humerus only the ventral face of the proximal third is visible, with the deltoid process facing cranially. The distal two-thirds as well as the proximal third of the radius–ulna complex are obscured by the thorax. The radius–ulna complex is exposed in mediolateral view. All carpalia are in place, but the right wing finger metacarpal is seen from medially, indicating a slight torsion within the carpus. The right pteroid lies adjacent to the cranial margin of the carpus with its terminus pointing proximally (Fig. 3c). The metacarpals of the three small digits run parallel to the right wing finger metacarpal and are superimposed on it, as are the proximal elements of the digits. All digits are curved proximally, the most dorsally situated being more strongly curved than the others. The right wing finger articulates in an angle of 90° with its metacarpal.

The left humerus is seen in its dorsal aspect. The head of the humerus is directed cranially. The deltoid process points laterally and overlaps the radius and ulna. The humerus is displaced caudally from the left fossa glenoidea. The radius–ulna complex is visible from dorsally and points cranially. The elbow joint is flexed beyond its bone lock and therefore the radius and ulna are slightly shifted cranially with respect to the distal joint of the humerus. The carpus is dislocated medioventrally and lies adjacent to the distal articular surface of the radius–ulna complex. The proximal articular surface of the carpals is largely obscured by the radius and ulna. Due to the dislocation of the carpus, the metacarpal of the left wing finger points craniomedially. Only its cranial face is visible and the distal condyle is obscured by matrix. The metacarpals of the small digits are not visible. The three small digits lie subparallel to each other and point caudally, with their ungual phalanges curved medially. The smallest of the three digits lies medially, the largest laterally. The left wing finger is articulated with its metacarpal, points caudolaterally and is seen from its caudal face.

The pelvic girdle is attached to three sacral ribs (Fig. 3c). The left ilium and ischiopubis are fully exposed and are seen in dorsolateral view. On the right side the ilium is seen from dorsomedially, while the ischiopubis is obscured by the sacrum and matrix. All bones of the hindlegs are in articulation, with the head of both femora still sitting in the acetabular fossae.

The left femur (Fig. 3c) points craniolaterally at an angle of 45° to the long axis of the sacrum and, at the same time, is rotated ventrally at an angle of 20° with respect to the horizontal plane. The fourth trochanter is oriented caudolaterally. The proximal terminus of the tibia–fibula complex is obscured by the distal end of the femur and slightly displaced medially (Fig. 3a, c). The left tibia–fibula complex is directed caudally, forming an angle of 20° with the femur. The left foot is directed medially at an angle of 90° to the tibia, with all digits lying subparallel to each other (Fig. 3a, c). Digit I lies cranially, metatarsal IV caudally. All metatarsals and phalanges are seen from plantomedially, suggesting that the left foot was embedded in an overextended position, as is the right foot. The mid-part of the metatarsals is obscured by the vertebral column of the tail. The ungual phalanges and the distal element of the fifth metatarsal are obscured by the right hindlimb and the tail. The left tarsals show a slight displacement due to the unnatural medial position of the foot.

The right femur is pointing craniolaterally at an angle of 35° to the longitudinal axis of the sacrum. Its dorsolateral face is exposed, indicating a slightly pronated position. This is confirmed by the caudolateral orientation of the fourth trochanter and the almost vertical orientation of the articular condyle of the knee joint. The right tibia–fibula complex articulates with the femur at an angle of 30° and is directed caudally. The articular surface of the knee is obscured by the femur. The transverse axis of the distal articular condyle of the tibia extends from dorsomedially to ventrolaterally. The right foot is stretched out caudally in extension of the tibia–fibula complex, displaying its plantar surface (Fig. 3a, c). The tarsus lies in its natural position and all elements are articulated. The metatarsals of the right foot diverge distally and lie in line with their phalanges. Digit I is the most medial and digit II lies parallel to it. The ungual phalanges of both digits curve laterally. Digit II includes an angle of 5° with digit III and its ungual phalanx is directed caudally. The ungual phalanx of digit IV curves medially. The angle between digit IV and III is 10°. Metatarsal V lies subparallel to that of digit IV, but its distal part is obscured by matrix.

Soft-parts preservation. The brachiopatagium of the right wing is preserved continuously along the caudal margin of the visible elements of the arm. Its medial surface (ventral surface in flight position) is exposed (Fig. 3a, b). The right brachiopatagium is considerably obscured by the left brachiopatagium, which has flipped over medially so that the imprint of its medial surface (ventral in flight position) is visible (Fig. 3a). Level with the distal terminus of the left femur, the trailing edge of the left brachiopatagium curves caudomedially and attaches to the ankle of the left hindlimb (Fig. 3a). This contrasts with the assumption of Bennett (2000) that the brachiopatagium was attached level with the knee joint in *Rhamphorhynchus*. Both brachiopatagia show a complex pattern of folds and preserve the imprints of fibres in an arrangement as is seen in the 'Zittel wing' (Fig. 3b; Zittel 1882). These structures are not visible in the brachiopatagium area adjacent to the left hindleg (tenopatagium according to Schaller 1985 and Bennett 2000).

Most soft parts of the left wing are preserved on the top slab, which has been split during preparation (Fig, 4a–c). There, the aktinopatagium fragment is seen from its ventral surface and shows a three-dimensional preservation with high resolution of details (fig. 4).

Medial to the left tibia, and on both sides of the right tibia, soft parts are preserved that most probably represent remnants of the uropatagium (Figs 4a, c). On the left side the trailing edge of the uropatagium extends from the ankle to the caudal terminus of the left ischium. The right part of the uropatagium is better preserved. Its trailing edge extends from the right ankle to the acetabulum. Parallel to the lateral margin of the proximal part of the fibula, over a length of 21 mm, bundles of bristles are preserved in a blackish manganese stain (Fig. 5a). The bristles have a length of 2–4 mm. Distal to the fourth caudal vertebra the vertebral column is lined with a seam of soft parts, presumably skin.

Patches of soft part are seen in the trunk region but they cannot be assigned with confidence to the left wing or the integument of the body (Figs 4a, c). Faint traces of a uropatagium are preserved medial to both tibiae (Figs 4a, b & 5a).

Webbing is excellently preserved between metatarsals I-IV and digits I-V of the right pes (Fig. 5a, b). The webbing extends to the base of the claw phalanges. Between digits I and II, as well as between digits II and IV, phosphatized fibres are visible under ultraviolet light. These fibres run along side the digits and parallel each other (Fig. 5). Such phosphorescent fibres are also visible between digits III and IV of the left pes. Structure, position and dimension of these webbing fibres are coincident with those described for *Pterodactylus* (Frey & Tischlinger 2000, see above; Fig. 2a–c)

Tissue types in the brachiopatagium. With ultraviolet light and special filter techniques different phosphatized tissue types can be highlighted and made visible in great detail, especially in the piece of brachiopatagium which is preserved on the top slab (Fig. 4). The fibres in the brachiopatagium of the *Rhamphorhynchus* specimen JME SOS 4784 have an average diameter of 0.2 mm and show a length of between 30mm and more than 80 mm. They are densely packed but slightly diverging towards the trailing edge (Figs 4c–e & 5). Their mode of attachment to the wing bones cannot be established as the fibres terminate close to the bone surface without modification. Where the aktinopatagium is folded, the fibres lie above each other or cross over diagonally, but principally they are arranged in one single plane as was described by Zittel (1882) or Wellnhofer (1975a, b, 1991). Bifurcations of the fibres are also seen, especially close to the trailing edge (see also Wellnhofer 1991, p. 152). The fibres disappear in the brachiopatagium region lateral to the hindlimb. This area was therefore called 'tenopatagium' (Schaller 1985; Fig. 3a, c). The border between tenopatagium and aktinopatagium in the new specimen is confluent and we hypothesise that the fibres become gradually thinner towards the body and presumably had a reduced fossilization potential. The overall arrangement of the fibres in a radiating pattern resembles that in the 'Zittel wing'. Under ultraviolet light the fibres phosphoresce as homogenous strings.

Under ultraviolet light a second and hitherto unknown tissue type becomes visible. It is an irregular meshwork of longitudinal strings with a wavy appearance (Figs 4b–f & 6). These strings bifurcate and merge, and are found only in areas where aktinofibrils are present. The strings cross the fibres at angles between 30° and 90° and often form a rhombic lattice (Figs 4c, d & 6). Due to the restriction of these strings to the brachiopatagium and their arrangement, mineral, bacterial or fungal origins for the strings can be ruled out (Wuttke pers. comm.). All strings of this tissue type occur in one single layer ventral to the wing fibres.

A third soft-tissue type is seen as dark paired bands of parallel-sided structures with a slightly brighter band in between them (Figs 4b–d & 6). These bands represent tubular structures, as is indicated by tiny fibrils with a diameter of 2–3 microns which extend transversely across the bands and become gradually brighter in their opaque margins (Fig. 4d, f). The thickest of these tubular structures has a maximum diameter of 1.2 mm and lies

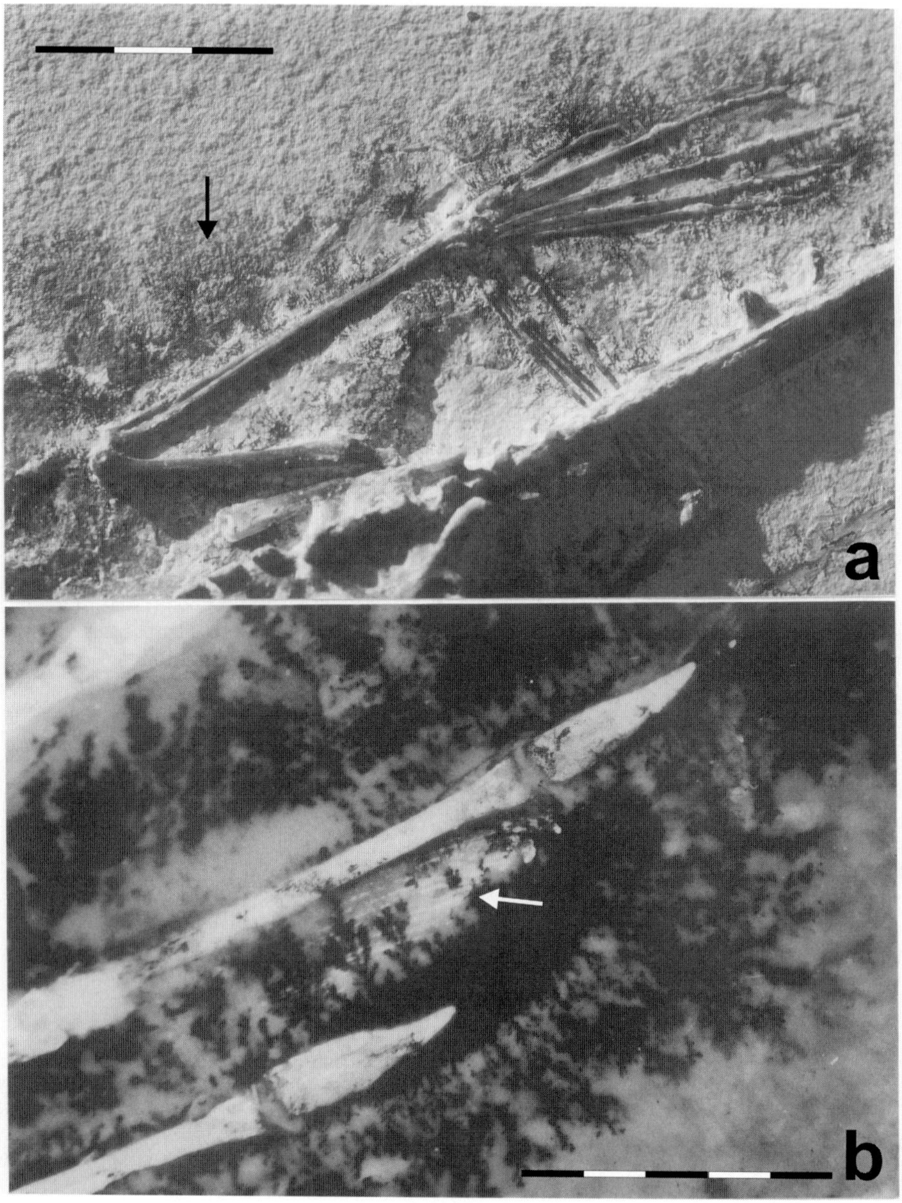

Fig. 5. *Rhamphorhynchus muensteri*. (**a**) Left hind leg showing the outlines of uropatagium and webbing; the arrow marks the bundle of bristles lateral to the tibia. Scale bar 30 mm. (**b**) Area of webbing between the digits I and II of the left pes; the arrow marks an area with preserved fibres. Scale bar 30 mm.

subparallel to the bones of the wing finger. In almost regular distances, tubes with a maximal diameter of 0.8 mm bifurcate towards the trailing edge of the brachiopatagium, where they are connected with arcade-like tubes. Continuously along the caudally directed tubes, thin tubules branch off medially and laterally where they bifurcate extensively, forming a dense meshwork (Figs 4b-d & 6).

The lumen of all these tubes is continuous and therefore we interpret this tube system as the remnants of a vessel system that formed a vascularized layer ventral to the wing fibres and the connective tissue meshwork. This vessel system theoretically could have transported air, lymph or blood. Pneumatic ducts resembling the vessel system in the *Rhamphorhynchus* wing in recent animals are

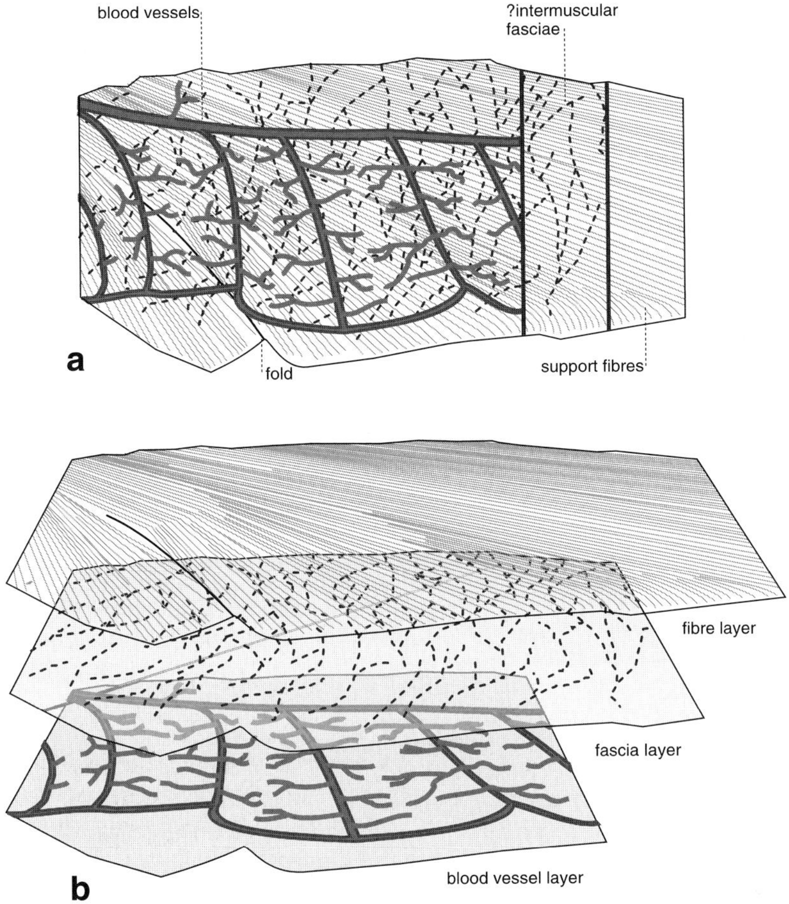

Fig. 6. *Rhamphorhynchus muensteri*, wing membrane (specimen JME SOS 4784, Jura-Museum Eichstätt, Germany). (**a**) Schematic line drawing of the different layers as preserved in the proximal two-thirds of the specimen. (**b**) The single layers drawn above each other.

only known from the Tracheata. In vertebrates pneumatic ducts never show such a pattern. The walls of pneumatic ducts and lymph vessels in recent vertebrates are extremely thin-walled structures (Leonhardt 1977; Kämpfe *et al.* 1993; Salomon 1993) whereas those in the pterosaurian brachiopatagium are thicker. Annular structures are unknown in the wall of pneumatic duct and lymph vessels. The wall thickness and the presence of annulae suggest a very high probability of a blood vessel origin for these structures. In another specimen of *Rhamphorhynchus* (NHNM, cf. Frickhinger 1999, p. 148, fig. 262 & p. 149, fig. 263) vessels of a similar type are seen in the brachiopatagium, but they are preserved as a string of iron oxide hydrates (Fig. 7c).

Two additional specimens of *Rhamphorhynchus* show such vessels under ultraviolet light: the 'Zittel wing' (Fig. 7a) and another *Rhamphorhynchus muensteri* housed in the Bayerische Staatssammlung (specimen number 1907 I 37; Fig. 7b). In the latter a tendon inside the trailing edge of the brachiopatagium becomes visible. In both specimens vessels are preserved but are not as distinct as those seen in the new specimen. They are, however, clearly visible and structurally identical.

Lastly, annular structures are reported from the walls of arteries and, to a lesser extent, in some veins, where muscle rings in the tunica media, together with rings of collagen fibres, guarantee pressure resistance and contractility (Kämpfe *et al.* 1993, pp. 176–177). In the wings of bats, veins are muscularized similar to arteries and are thus contractile as well (Kämpfe *et al.* 1993, pp. 176–177). We suggest that the vessels in the *Ramphorhynchus* wings transported blood, and that their preservation and distribution is consistent with arterial or veinal origin.

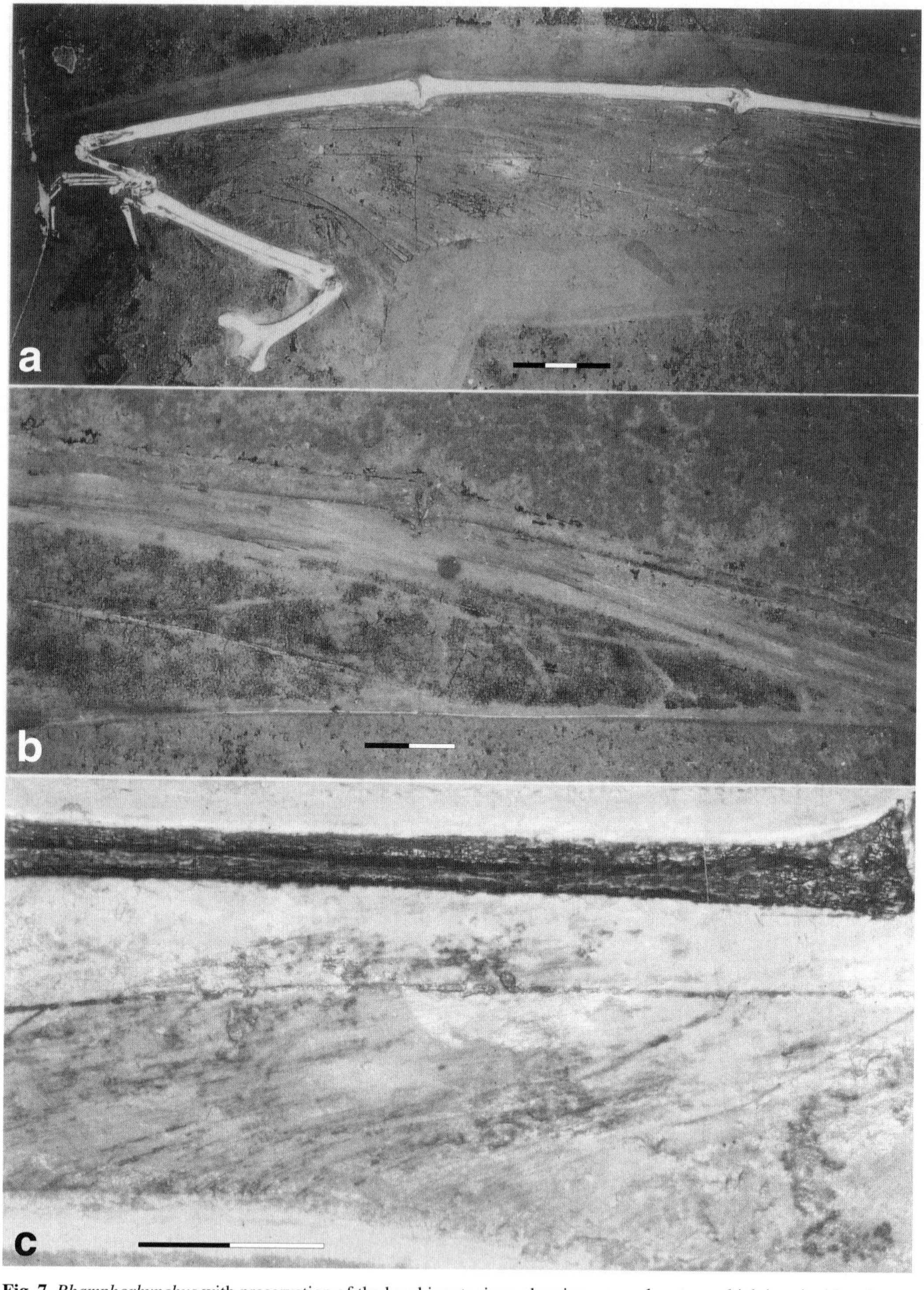

Fig. 7. *Rhamphorhynchus* with preservation of the brachiopatagium, showing a vessel system which is coincident in what is seen in *R. muensteri* (Figs 3 & 4). (**a**) 'Zittel wing' (specimen BSP AS I 771) Bayerische Staatssammlung Munich, Germany). Scale bar 30 mm. (**b**) BSP 1907 I 37. Scale bar 20 mm. (**c**) 'Vienna specimen'. Scale bar 20 mm.

Pterosaurs with soft-part preservation from the Crato Formation, northeastern Brazil

Several pterosaurs from the Nova Olinda Member of the Crato Formation have preserved soft-part structures that help to elucidate similar structures seen in the Solnhofen Lithographic Limestone pterosaur specimens described above. Three Crato specimens are particularly important: two isolated crania, both lacking the mandibles (Fig. 8a, b), and a wing with two hindlimbs (Fig. 9a). The Crato specimens represent animals with wing spans of approximately 3m and display excellent preservation of the bones and soft parts in three dimensions.

All three Crato specimens were obtained by the SMNK in an unprepared state. In SMNK 2344 PAL (Fig. 8a) the slab split along the right lateral surface of the head crest, while in SMNK 2343 PAL (Fig. 8b) the soft-part structure of the head crest split in the median plane. The split surface of the 'limb slab' (SMNK 3830 PAL, Fig. 9a) was located slightly above the bone surface, so that the bones were just visible as a series of smooth ridges.

Tapejaridae

The prominent premaxillary crest, and the shape and relative size of the nasoantorbital fenestra and of the orbit, indicate a referral to the Tapejaridae (Kellner 1990; Campos & Kellner 1997). However, according to the generic diagnosis given by Kellner (1990), neither specimen can be attributed to *Tapejara wellnhoferi* (Kellner 1989) from the Santana Formation, nor to *Tapejara imperator* (Kellner 1989) from the Crato Formation because both specimens lack an occipital spine (Campos & Kellner 1997). Therefore a new species was erected on the specimens discussed here: *Tapejara navigans* (Frey *et al.* 2003b).

Systematic palaeontology
Order Pterosauria Kaup 1834
Superfamily Pterodactyloidea Plieninger 1901
Family Tapejaridae Kellner, 1989
Genus *Tapejara* Kellner 1989
Species *Tapejara navigans* Frey *et al.* (sp. nov, 2003b)

Preservation. Both crania are likely to have been complete prior to excavation. Only the mandibles are lacking. Skull SMNK PAL 2343 (Fig. 8b, d) must have been found in the upper weathered layer of a quarry because recent roots were attached to the slab. Presumably the slab was broken due to weathering prior to discovery. Whereas the antorbital part of the skull is exceptionally well preserved, the orbital and occipital area are weathered. From the highest point of the premaxillary crest a mineralized vertical spur rises which is broken at the border of the slab. The crest extends over the entire dorsal margin of the skull. Soft parts are preserved between this spur and the dorsal margin of the skull, as well as in the rostral region of the beak and inside the nasoantorbital fenetra as a goethitic replacement probably after pyrite. The skull SMNK PAL 2344 (Fig. 8a) is complete and anatomically coincident with SMNK PAL 2343 in every detail. In addition the dimensions of the soft-part component of the crest is seen. It rises from the entire dorsal margin of the skull but its dorsal extremity was cut when the slab was trimmed by stoneyard workers.

Soft-part preservation
Beak sheath (Fig. 8b, c). The entire lateral surface of the beak is coated with patches of goethite in both specimens, indicating the presence of a former keratinous rhamphotheca. Towards the margins of the rostrum, the goethite layer increases in thickness and forms a reddish-brown seam around the rostral part of the beak. The keratinous tip of the beak appears to have extended 5 mm beyond the bony tip (Fig. 8b, c). Its termination is blunt and curved slightly ventrally. The keratinous lateroventral margins of the beak are compacted. Between them a low median blade is visible. It appears likely that the lateroventral margins of the rhamphotheca as well as the median blade were sharp. Craniodorsally the beak sheath forms a slightly convex blade which extends from the rostral terminus of the beak to the craniodorsal margin of the premaxillary crest (Fig. 8a, c). The goethite surface of this blade shows small scale wrinkles (Fig. 7c).

Internal cranial septa (Fig. 8a, b). Patches of goethite are preserved inside the nasoantorbital fenestra of both skulls and are especially intensive along the rostral margin and the caudal part of the dorsal margin. Due to their position in the median plane they probably represent remnants of a hitherto unknown median cranial septum inside the foramen nasoantorbitalis, possibly an internasal septum.

Crest (Fig. 8a, b, d). Dorsal to the bony crest of SMNK PAL 2343 are dense patches of goethite covering approximately 80% of the surface of the soft part of the crest (Fig. 8b). Between these patches, long subparallel fibres emerge from the dorsal margin of the bony part of the crest and gently curve rostrally. Where the fibres penetrate the bony crest, a striation is visible on the lateral surface of the bone (Fig. 8d). The striae have the same orientation as the fibres. Some of the fibres can be traced into the bony crest for up to 10 mm. There are approximately 15–20 fibres per 10 mm of crest. The most rostral fibres are fused by a phosphatic material and form a dorsally directed, slightly caudally curved spur

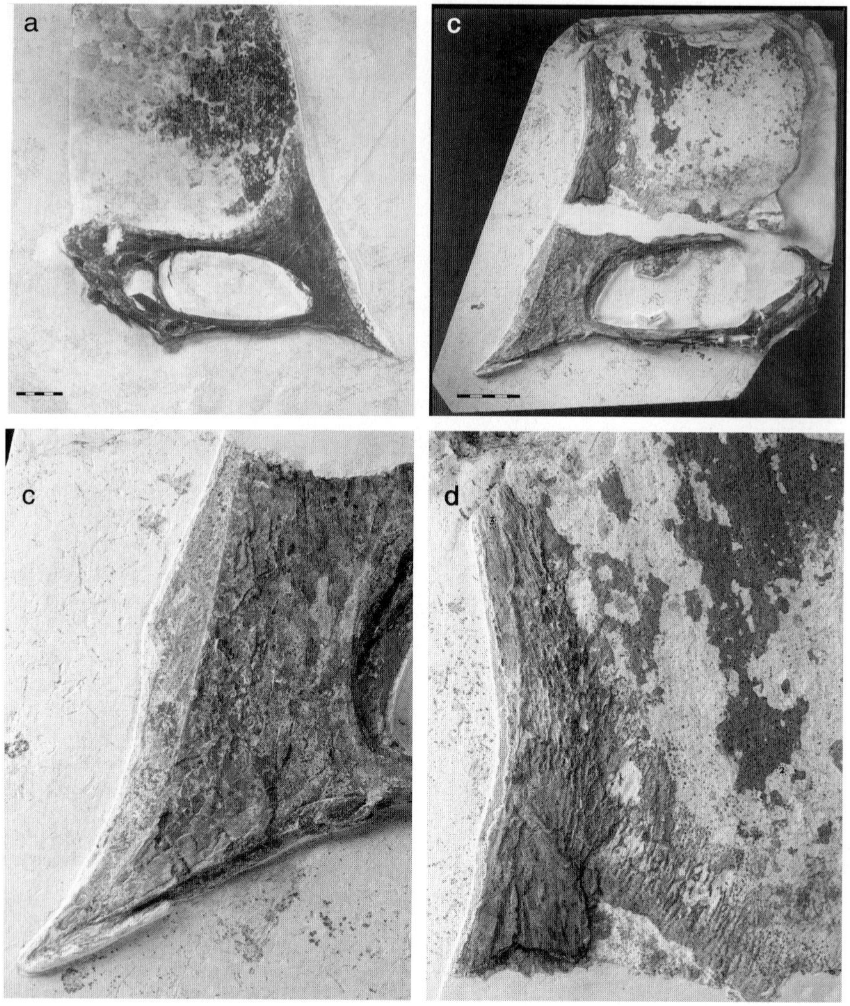

Fig. 8. *Tapejara navigans*. (**a**) holotype SMNK PAL 2344; note the enormous soft-part crest and the lack of an occipital process. Scale bar 50 mm. (**b**) Referred specimen SMNK PAL 2343. Scale bar 50 mm. (**c**) SMNK PAL 2343; close-up of the rostrum with the preserved rhamphotheca. Scale bar 30 mm. (**d**) SMNK PAL 2343; dorsal margin of the bony crest in the rostral area; note the bundles of mineralized fibres at the boundary between the bony and the soft tissue crest and the base of the supramaxillary spine. Scale bar 30 mm.

which tapers dorsally. This spur interdigitates with coarse striae on the highest point of the bony crest (Fig. 8d). Here the phosphatic material reaches into the bone. This is visible because the rostrolateral surface of the bony crest has broken off, exposing the basal anchoring of the suprapremaxillary spine. In SMNK PAL 2344 fibres are not visible because of a massive layer of goethite, which is thicker along the rostral margin of the crest (Fig. 8a). The original height of the crest can only be estimated due to trimming of the slab, but it must have been four to five times higher than the occipital height of the skull (Martill & Frey 1998). Whereas the leading edge of the crest is slightly convex, its trailing edge is almost vertical, straight and in places slightly wavy (Fig. 8a). A horizontal section through the soft-tissue crest shows a symmetrical airfoil with a rounded leading edge reinforced by the phosphatized suprapremaxillary spur as in SMNK PAL 2343. In contrast to SMNK PAL 2343, the base of the suprapremaxillary spur is laterally and rostrally overgrown by bone. With the exception of impressions of halite or probably pyrite pseudomorphs, no surface structures can be observed.

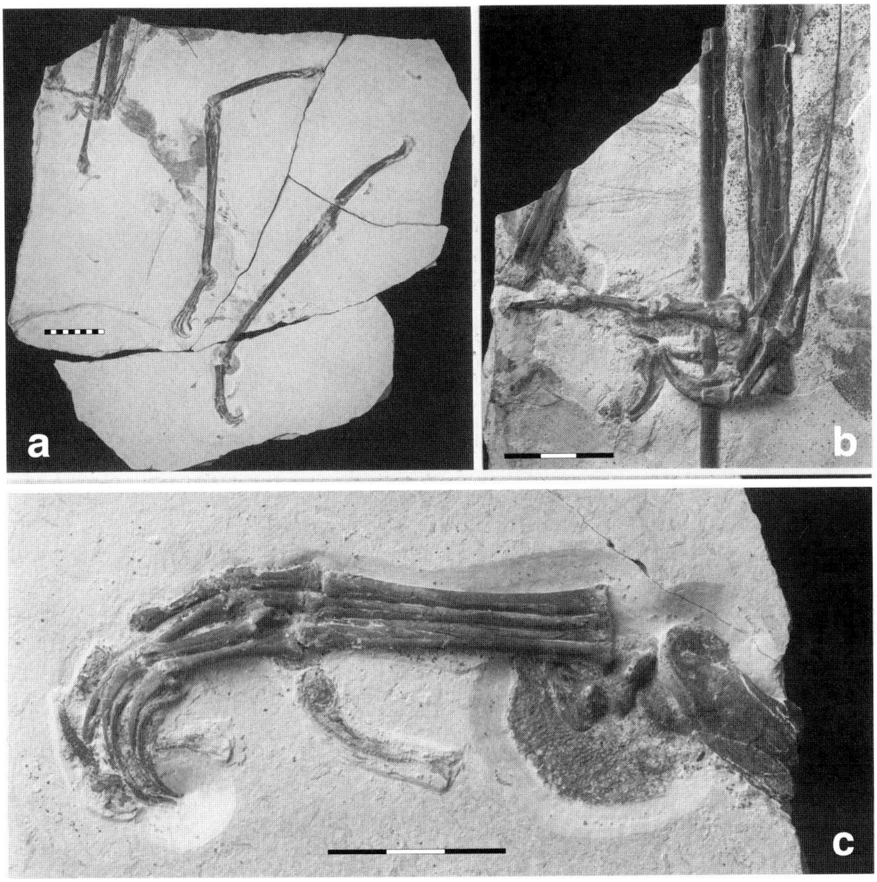

Fig. 9. Azhdarchidae indet. (SMNK PAL 3830). (**a**) Skeleton of both hindlimbs and a fragmentary left wing of one individual; note that the elements are held together by a trace of brachiopatagium, attaching to the ankle joint of the left foot. Scale bar 100 mm. (**b**) Close-up of the hand of the same specimen; note the keratineous sheaths covering the claw phalanges and the soft-part preservation between the digits. Scale bar 30 mm. (**c**) Close-up of the right foot of the same specimen, note the keratineous claw sheaths and the scaly sole pads. The left foot is depicted in Figure 2d (reverse). Scale bar 30 mm.

? Azhdarchidae indet.

Appendicular skeleton (Figs 9, 2d). Phalanx II of the wing finger is T-shaped in cross-section. A comparison with another more complete but hitherto undescribed pterosaur specimen (SMNK PAL 3855) from the Crato Formation suggests that the wing and hindlimbs come from an hitherto unknown azhdarchid pterosaur, probably a tapejarid.

Systematic palaeontology
Order Pterosauria Kaup, 1834
Superfamily Pterodactyloidea Plieninger, 1901
Family Azhdarchidae Nessov,1984, emend. Padian, 1986
Genus Azhdarchidae indet.

Preservation. The six parts of the slab (specimen # SMNK PAL 3830) fit perfectly together and contain an incomplete left wing and both complete hindlimbs of one pterosaur individual. All bones of the wing are articulated, but are bent beyond bone lock in a way that the elements almost parallel each other. The left hindlimb is visible in lateral view (Fig. 9a). The tibia is adducted at an angle of 60° against the femur and the facies dorsalis forms an angle of 120° with the tibia. The right hindlimb is straight and fully extended (Fig. 9a). Soft-part preservation comprises a portion of brachiopatagium connecting the left foot with the wing skeleton. It also consists of goethite and is present in areas of both feet as well as in the hand. Due to the incomplete status of preparation the full extent of the preserved soft parts cannot be determined.

Soft-part preservation in the foot

Webbing (Figs 2d, 9c). Webbing is preserved between all digits as a goethitic stain. Sadly, parts of the webbing have been destroyed in the left foot due to incompetent preparation, but in no place is any structural detail preserved inside the webbing. There is also a layer of goethite between and on the metatarsals which may represent the integument.

Claw sheaths (Figs 2d, 9c). Digits I–IV bear huge, sickle-shaped keratinous sheaths which are almost twice as long as the ungual phalanx. They are preserved on both feet but are better seen on the right one (Fig. 9c). The ungual phalanx of digit IV of the left foot has rotated caudally and now points towards the heel (Fig. 2d).

Sole and heel pads (Figs 2d, 9c). In the caudolateral area of the tarsus and, though expressed in its medial area, a seam of goethite is preserved in which a dense pavement of subcircular convex scales is visible. The scaled area encloses digit V and the adjacent metatarsal V entirely. Ventrolaterally to digit V the scales are slightly larger, with a diameter of approximately 2 mm, than in the other areas, where they reach a maximum diameter of 1 mm. At metatarsal IV only a thin seam of goethite with structures is visible. This structure is preserved in both feet and presumably represents the remnants of a cushioned heel and sole pad.

Soft-part preservation in the hand. The three short digits of the hand show patches of goethite on all ungual phalanges and remnants of the claw sheaths (Fig. 9b). In the most proximal of the small digits the claw sheath extends distally into the sediment and vanishes below the distal most digit, so that its length cannot be determined. Between proximal digits two and three a faint goethitic stain indicates the presence of webbing (Fig. 9b). The leading edge of the webbing almost reaches the proximal end of the ungual phalanges. A similar stain is also preserved on the medially directed side of digit III, extending from the base of the ungual phalanx caudally. It is still visible overlying the surface of wing finger phalanx I, in which it extends for a distance of 23 mm distally.

Brachiopatagium. A piece of brachiopatagium extends from the carpal area of the wing to the ankle region of the left foot (Fig. 9a). Although the membrane has several folds and displays traces of advanced decay, its attachment at the ankle is clearly visible (Fig. 9a). Some distal parts of the trailing edge are well preserved and show the typical striation produced by internal fibres. Bundles of the patagium fibres lie subparallel to the left foot. The fibres are almost circular in cross-section and approximately 0.1 mm thick and some of them are as long as 180 mm. Some of them branch. This specimen adds further evidence to the assumption above (see *Rhamphorhynchus*) that the tenopatagium is confluent with the aktinopatagium with respect to the existence of aktinofibrils. The fibrils apparently become thinner and more densely packed proximally but are still present.

Discussion

The specimens described above yield hitherto unknown details of pterosaurian soft part anatomy. The fact that the specimens are from a range of temporal and geographical settings and are referrable to a variety of taxa indicates that at least some of the structures were present in most pterosaurs. The soft parts described here have consequences for the reconstruction of the lifestyle and locomotion of pterosaurs. Furthermore the evolution of edentulous beaks and various head crests as well as aspects of the physiology can be re-evaluated.

Evolution of edentulous beaks

The new specimen of *Pterodactylus* shows a combination of teeth and keratinous beak which was hitherto unknown in this genus (Fig. 1e). The beak in *Pterodactylus* is tiny and covers only the rostralmost part of the jaws, which is too narrow for a pair of tooth sockets. This predental rostrum can either be enlarged by a reduction of the tooth rows from rostrally to caudally, or by an elongation of the edentulous part, or both. Among the pterodactyloid pterosaurs elongation of the rostrum is usually accompanied by a reduction of the tooth count, as in *Germanodactylus*, *Dsungaripterus*, *Phobetopter* and *Domeykodactylus* (Fig. 10). An elongation of the rostrum with retension of a high tooth count appears never to have occurred in pterodactyloid pterosaurs. This suggests that the evolution of an edentulous beak happened due to an increase of the keratinous rhamphotheca and is a consequence of a caudally directed step-by-step replacement of the teeth. This is highlighted by the mandible of *Domeykodactylus*, which retains a large tooth count for a dsungaripterid pterosaur, but whose rostralmost teeth are small compared with more caudally situated teeth. Without a reduction of tooth number and height, the rhamphotheca could not operate as a food-gathering beak with marginal blades. Only then can the teeth become obsolete as puncture and friction-generating devices (Figs 10 & 11). From the very beginning of the evolution of a keratinous rhamphotheca, the structure was subject to high acceleration during the closing movement of the jaws because of its distance from the jaw articulation. However, the short power arm of the adductor

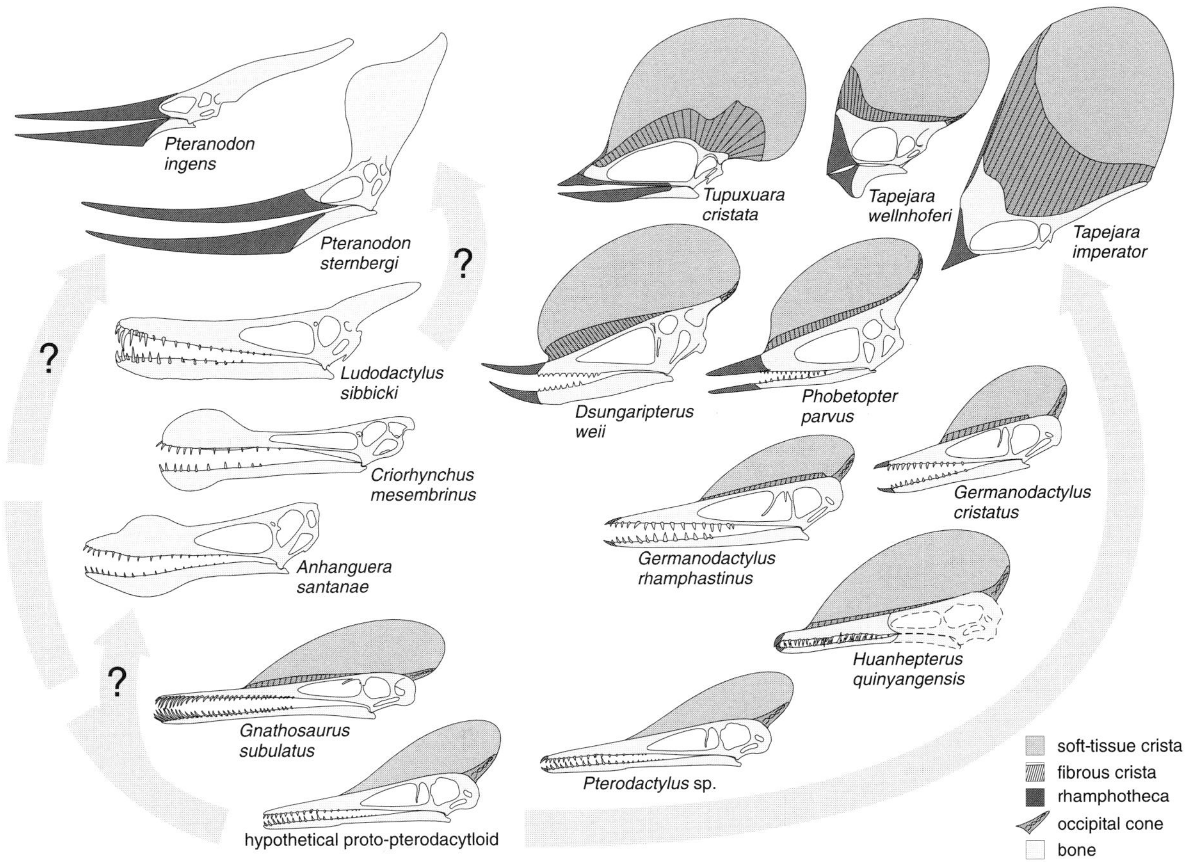

Fig. 10. Possible pathways for the evolution of edentulous jaws. While the evolutionary pathway to the right appears biomechanically conclusive, the evolution of the *Pteranodon*-type skull construction remains unclear.

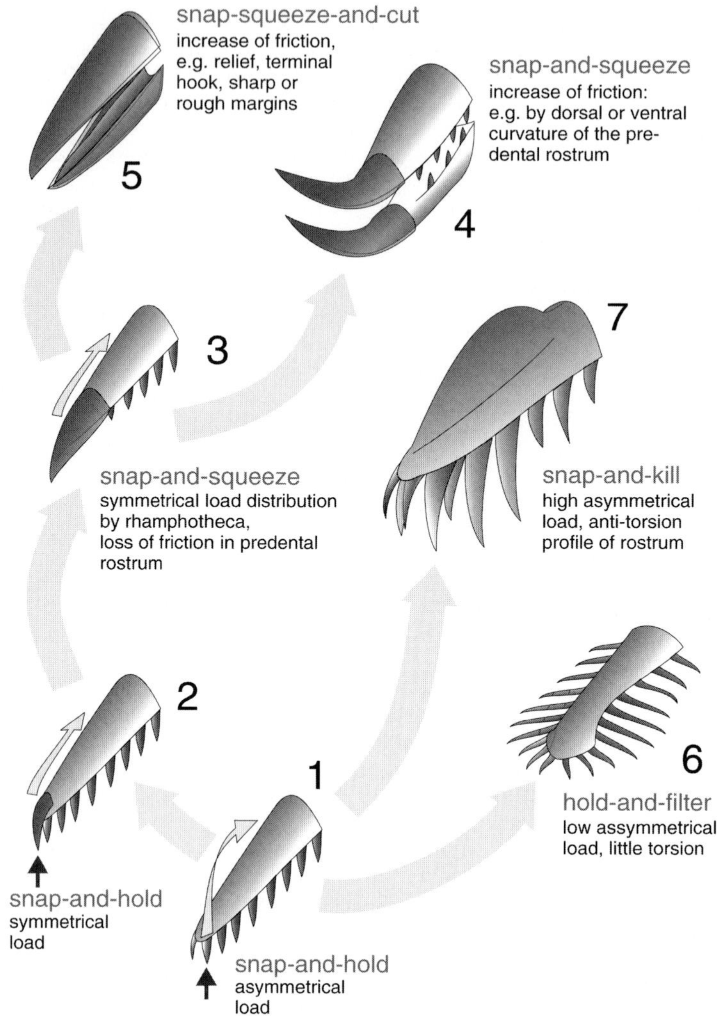

Fig. 11. Possible pathways for the evolution of the rostra of some pterodacyloid skull constructions from a hypothetical preconstruction (1). The evolution of a terminal keratineous hook (2, *Pterodactylus* type) changes the load conditions of the rostrum and opens the option for the evolution of different types of rhamphothecae: apical rhamphotheca (3, *Germanodactylus* type), apically upward-curved rhamphotheca with increased friction effect (4,*Dsungaripterus* type), rhamphotheca with structural friction device, edentulous (5,*Pteranodon* type). An elongation of the rostral tooth crowns leads either to filter-feeders (6,*Gnathosaurus* type) or piercing killers (7,*Coloborhynchus* type).

muscles would not have allowed for a powerful grip when the jaws were closed. Thus, the small rhamphotheca of *Pterodactylus*, from an engineering perspective, operates like a tooth, but, the torque load applied to the rostrum by such an uneven structure is reduced to the torque which teeth, being always even structures, would apply (see below; Fig 11.1). With an increase in the size of the predental rostrum (Fig. 11.2 & 3) the torsional forces become progressively smaller and, at the same time, the friction effect of the teeth is lost. This could be compensated for by a relief on the internal surface of the beak, either by longitudinal interdigitating ridges, as are seen in the tapejarids (Figs 8c & 11.5) or by transverse structures for which, as yet, no fossil evidence exists. Another possibility would be a dorsally directed curvature of the predental rostrum, which would prevent a prey item from slipping out again when the head is returned to a horizontal position. Such a construction is seen in *Dsungaripterus* (Fig. 11.4). A third option would be the shortening of the out-lever or a reinforcement of the in-lever. In *Germanodactylus*, *Dsungaripterus* and *Phobetopter*, the skull is short and high when compared with *Pterodactylus*

(Fig. 10). Such shortening results in an increase of the holding power, and possibly the bite power, combined with a decrease of bending moments in the rostrum of the aforementioned pterosaur skull constructions. The shortening of the skull is always combined with an increase in the size of the nasoantorbital fenestra and the development of occipital crests dorsal to the supratemporal fenestra (Fig. 10). These structural changes might have been accompanied by an increase of mandible adductor masses that could occupy additional bone and fascia surfaces. This is speculative and requires further data concerning the cranial anatomy and the mechanical loading capacity of the skull. A study of these parameters is in progress (Fastnacht, Hess, Frey & Weiser, DFG research project FR 1314/2–1). An extreme in the trend to shorten the skull is represented by the tapejarid pterosaurs, where the skull is almost as long as it is high (Figs 8a, b & 10).

The pteranodontid pterosaurs have a long-lever beak with a possible terminal hook which might have enhanced the effect of an impact bite (Fig. 10; Wellnhofer 1991, p. 139, reconstruction of the head of *Pteranodon ingens*). The nasoantorbital fenestra is small compared to that of the dsungaripterid or tapejarid pterosaurs and only on the proximal surface of the occipital crest was there an option for the attachment of additional muscle masses. That the rostrum is extremely long with respect to the nasoantorbital fenestra suggests an evolutionary pathway which took the option of increasing the predental part of the rostrum to greater extremes than in any other pterosaur constructions. For the moment it is impossible to tell from which constructional level the pteranodontid skull construction could evolve, but from a mechanical perspective the size and power of a prey item that *Pteranodon* could catch and hold was clearly limited, even with a pointed hook at the end of the beak. However, the aspect of prey manipulation could have been increased, which is difficult to prove.

A predental rostrum is also evident in many rhamphorhynchoid pterosaurs, e.g. in *Scaphognathus* and *Rhamphorhynchus* (e.g. Wellnhofer 1975a, 1991). However, in contrast to the pterodactyloid pterosaurs, the dentition remains an essential part of the prey-catching apparatus. In consequence, the teeth cannot be reduced and only a rostral elongation of the predental rostrum is possible, as in *Rhamphorhynchus*. Therefore, the evolution of keratinous friction devices as described above would make no sense. This is evidenced by the fact that rhamphorhynchoid beaks cannot be brought to tight occlusion at the lateral margins.

The evolution of keratinous beaks among pterosaurs should be discussed in the context of the forces applied to the tip of the rostrum and its option for food-gathering. Even the rostral-most teeth, as paired structures, can scarcely transmit symmetrical bending forces to the skull during a bite. Consequently the skull of toothed pterosaurs had to accommodate torque forces. The tip of a rhamphotheca, in contrast, always loads the skull symmetrically. Dorsoventral bending forces are distributed over a large attachment surface and inside one single bone unit, the premaxilla, instead of crossing sutures (Fig. 11.2 & 3). Whether or not this results in the option for reducing the teeth and thus weight in the skull remains to be demonstrated, but tooth reduction among some pterodactyloid pterosaurs probably saved some weight. With the evolution of beaks the snap-and-kill *modus operandi* is replaced by a snap-and-squeeze or snap-and-cut action, which in general might be coincident with a decrease of prey size and power for yet unknown reasons. However, a pointed beak is suitable for collecting and manipulating even tiny food items. Such a food resource is more difficult to handle for pterosaurs with large rostral fangs, such as the ornithocheirids *Coloborhynchus* and *Ornithocheirus*, which improved the snap-and-kill *modus operandi* (Fig. 11.7). The only pterosaurs that used the dentition for catching small prey items are *Gnathosaurus*, *Ctenochasma* and *Pterodaustro* (Figs 10 & 11.6) [*Gnathosaurus*]) but for these filter-feeders the prey spectrum is restricted to small items (Wellnhofer 1991). In conclusion, the pterodactyloid beak may have evolved in the context of extending the food range to small prey combined with a minimizing of the torque load during the handling of large prey. A starting point for the evolution of the pterodactyloid type of rhamphotheca might have been a jaw construction as is seen in *Pterodactylus* (Figs 1e & 11.2).

For an entire biomechanical analysis the reconstruction of the muscle masses and their effective vectors is essential but would surpass the scope of this paper, because that part is under work as a project on its own (Frey & Weiser, DFG research project FR 1314/2-1).

Evolution of cranial crests

Fibrous structures preserved in the cranial region of the new *Pterodactylus* specimen differ in length, diameter and their subparallel arrangement from the hair-like structures described for other pterosaurs (see Padian & Rayner 1993; Unwin & Bakhurina 1994; Frey & Martill 1998). Those fibres lying across the skull retained a subparallel arrangement (Fig. 1c, f) despite the deformation by folding, indicating that they must have been an integral part of a common structure. This is also suggested by the soft parts preserved dorsal to the skull, in which, especially rostrally, the site of attachment along the margo dorsalis of the skull is visible. The most

Fig. 12. Reconstruction of living *Pterodactylus antiquus* according to the lastest soft-part discoveries.

plausible interpretation of these soft parts is that they are remnants of a crest that consisted exclusively of soft tissue (Fig. 12). Such an interpretation is supported by the two tapejarid crania from the Crato Formation described above, in which the crest consists largely of skin sitting on a minor bony crest (Fig. 8a, b). The fibres inside the soft tissue of the crests of the latter are large compared with the fibres in the crest of *Pterodactylus*, but they show a similar subparallel arrangement and have a similar relative density (Tischlinger & Frey 2001). In the Crato tapejarids the fibres can be traced into the bony part of the crest, where they produce a characteristic striation on the lateral surfaces along the distal margin of the bony crest (Fig. 8d). Distally the fibres are mineralized individually, giving the bony crest a bristly outline, as is seen in a specimen of *Tapejara imperator* (SMNK 3829 PAL; Fig. 13). Thus the bony part of the crest of the tapejarid pterosaurs appears to represent a mineralization of fibres arising from their origin along the dorsal median line of the skull. Crests with similar striations are known from a variety of pterodactyloid pterosaurs, including *Dsungaripterus*, *Phobetopter*, *Domeykodactylus*, *Gallodactylus*, *Germanodactylus*, *Gnathosaurus* and *Normannognathus*, but also in *Tupuxuara* and *Quetzalcoatlus* sp. (Fig. 10; Young 1964, 1973; Bakhurina 1982; Martill *et al.* 2000; Wellnhofer 1970, 1991; Kellner & Campos 1989; Kellner 1990; Kellner & Langston 1996; Campos & Kellner 1997; Buffetaut *et al.* 1998). Such striations are strong evidence for a subsequent mineralization of a fibrous crest and indicate that there probably was a soft-tissue extension of the bony crest. The evolution of striated crests might have started with a soft-tissue keel along the dorsal median line of the skull supported by collagen fibres from the subdermal connective tissue, or the periosteum, or both. This ridge can only be enlarged with the development of vertical support fibres similar to the condition seen in the new *Pterodactylus* sp. where the crest reaches a height of 15–25 mm. The mineralization of the fibre attachment point would result in a stabilization of the anchoring of the soft-tissue crest and allow increased fibre heights. A low ridge of a bony striated crest would be seen along the dorsal mid-line of the skull. Such low striated crests are preserved, e.g. in *Germanodactylus* or *Gnathosaurus* (Fig. 10). Additional mineralization distally would result in increased stability, which perhaps explains the enormous crests seen in *Tupuxuara*. However, there is also the option to mineralize only parts of the crest, as is seen in the Tapejaridae, where a mineralized support in the leading edge of the crest appears sufficient for stabilization.

The fibrous cone arising from the occiput most likely stabilized the soft-tissue crest of *Pterodactylus* (Fig. 1d). The caudal-most fibres of the crest are anchored on this fibrous cone. In the referred specimen of *Tapejara imperator* (SMNK 3829 PAL) an occipital spine is preserved which is as long as the skull itself (cf. Campos & Kellner 1997). Isolated mineralized fibre bases are visible all along the entire dorsal margin of this spine (Fig. 13), suggesting that this spine is a secondary mineralization of an initially fibrous occipital cone. Possibly all occipital cones aligned with a striated crest, as seen in *Dsungaripterus* and *Phobetopter* might be secondary mineralisations (Fig. 10). Alternatively bones of the occipital area grew into the cone, but the proof of this scenario will only be resolved with histological studies.

The occipital crests of pteranodontid pterosaurs and the ornithocheirid pterosaur *Ludodactylus sibbicki* (Frey *et al.* 2003a; SMNK PAL 3828) from the Crato formation are characterised by smooth outlines. They are composed of the parietalia under participation of the supraoccipitale proximally (Eaton 1910; Bennett 2001) and are not associated with striated crests. Similar crests with smooth margins are seen in ornithocheirids, where the premaxillae, maxillae and the mandible may be extended vertically, e.g. *Criorhynchus* or *Coloborhynchus* (Fig. 10; Owen 1861, 1874; Wellnhofer 1987a; Lee 1994; Fastnacht 2001). These crests are neither fibrous nor do they show any striae nor thin and brittle margins. They are formed by the premaxillae or the dentalia, which are visible in tomography sections (Fastnacht 1997), and frequently show an external network of branching depressions which were probably occupied by blood vessels. In conclusion there are at least three different types of crests in pterosaurs which evolved under different biomechanical regimes. However, all of them change during ontogeny (Bennett 1993). Some taxa which have been diagnosed on the basis of the shape of the crests alone are probably ontogenetic stages or sexual dimorphs.

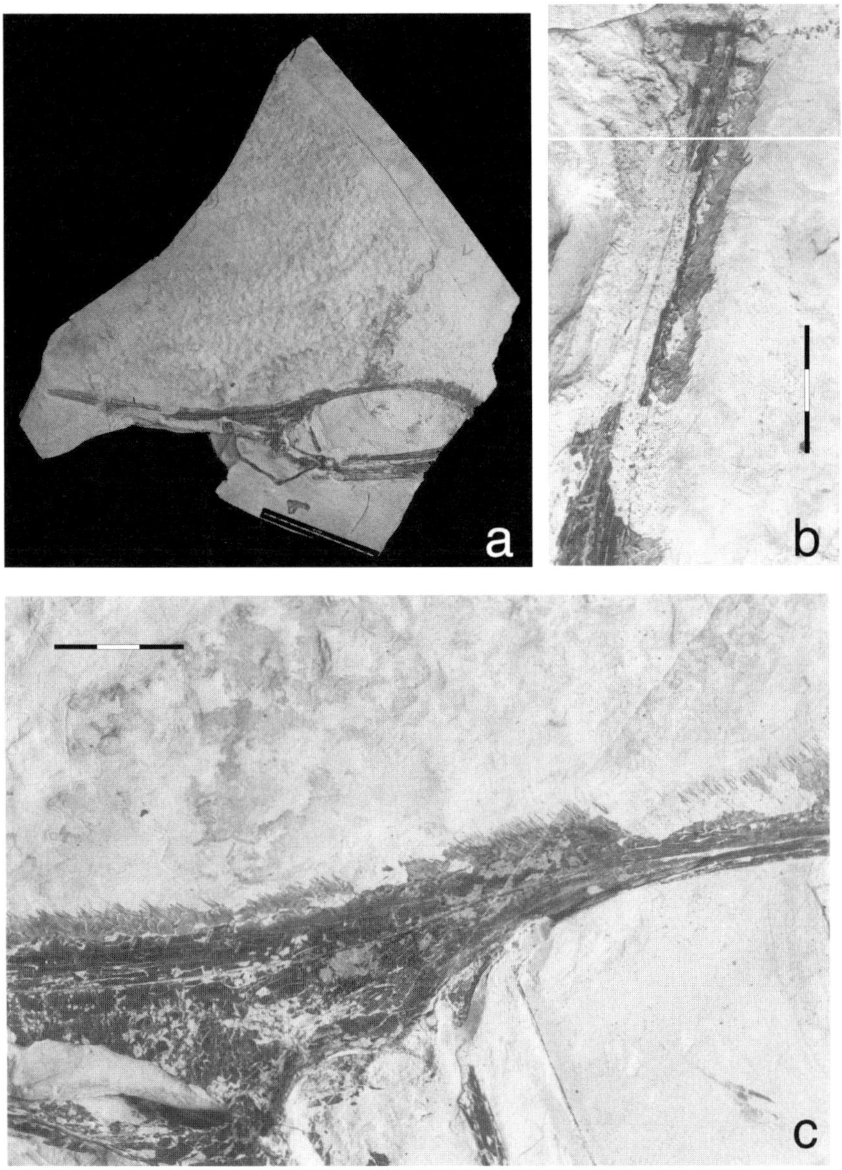

Fig. 13. *Tapejara imperator* (SMNK PAL 2839). (**a**) The specimen as preserved; note the occipital spine in which this species differs form *T. navigans*. Scale bar 200 mm. (**b**) Close-up of the occipital spine in vertical orientation (caudal is to the top) with the mineralized bases of the crest fibres. Scale bar 30 mm. (**c**) Bony crest dorsal to the orbita; note the seam of mineralized crest fibres along the dorsal margin of the bony crest. Scale bar 30 mm.

Soft-tissue crests and some aerodynamic implications

In the crests of the tapejarid pterosaurs from the Crato Formation, the fibres of the leading edge of the crest are co-mineralized, forming a vertical spine. This spine is the thickest part of the fibre layer. The fibres caudally adjacent to the spine are densely packed and, dorsal to the rostral part of the bony crest, form a multilayered mat. Further caudally the fibres rapidly become less densely arranged and form a single-layered mat. This condition is seen in SMNK 2343 PAL (Fig. 8b). In SMNK 2344 PAL, however, the lateral surface of the crest is preserved (Fig. 8a). At the cut of the slab dorsal to the skull the cross-section of the crest is clearly discernible. It resembles a symmetrical aerofoil with the thickest diameter approximately one-third caudal to the leading edge. Therefore, between the external surface of the epidermis and the fibre mat there must have been another tissue layer, probably connective or, more unlikely, adipose tissue. Evidence for such a layer comes from the *Pterodactylus* sp. from the Solnhofen Lithographic Limestone described above (JME SOS 4784). During preparation a layer of a white phosphatic powder was found between the fibre mat of the crest and the external mould of the crest on the counterslab. This suggests that the fibrous crests of at least some pterodactyloid pterosaurs were streamlined, with a stiff leading edge but a thin and probably flexible caudal part that in some cases was reinforced by a subvertical occipital cone, e.g. in *Pterodactylus*, *Tupuxuara*, *Dsungaripterus* and *Phobetopter* (Fig. 10). All these animals had a gradually caudodorsally rising leading edge. The crest probably reached its maximum height caudal to the orbita, most likely extending caudal to the occiput, and thus the occipital condyle. In both *Tapejara imperator* specimens known (Campos & Kellner 1997; SMNK 3829 PAL) the leading edge of the crest was oriented almost vertically and gradually curved caudally in its distal part. The vertically blade-like occipital cone is similar in length to that of the skull. The dorsal margin of the internal fibre mat preserved appears slightly concave (Campos & Kellner 1997), extending from almost the distal terminus of the leading edge to the caudal terminus of the occipital cone. However, in SMNK PAL 3829, there is extensive soft-part preservation dorsal to this margin, which indicates that the crest was highest in the postorbital region and dorsal to the occipital cone (Fig. 13a). This results in a crest which was possibly more than 3 times as large as the lateral surface of the skull. The question on the aerodynamic effect inevitably has to be examined. If the main surface of the crest is located caudally to the occipital condyle, each turn of the head would result in an automatic readjustment in the wind. The *modus operandi* would be like a weather vane that always turns into the wind direction (Fig. 14a, a´). In consequence, a weather-vane type of crest could have been used as a self re-adjusting rudder probably at very low flight velocities, an assumption that can be tested in a wind tunnel (Frey *et al.* in prep.). The pteranodontid type of crest may have operated in a similar fashion.

In *Tapejara navigans* from the Crato Formation (Figs 8 & 15), however, the weather-vane principle cannot have operated because the trailing edge of the crest stands vertical only a little caudally to the occipital condyle. Therefore the postoccipital surface of the crest is approximately one-fourth of the preoccipital one (Fig. 14b, b´). Such a crest cannot work as a self-adjusting rudder. The slightest

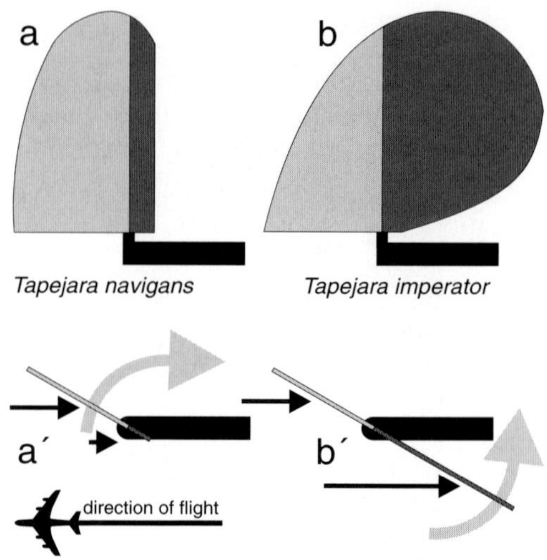

Fig. 14. Aerodynamical effects of the crests of (**a, a**′) *Tapejara navigans* and (**b, b**′) *T. imperator*. The area of the crest caudal to the occipital joint is light grey. If this area is larger than the one rostral to the occipital joint, lateral movements of the head are automatically compensated (b´). If it is smaller, the head would have been pushed to the side (a´).

lateral movement of the head of this tapejarid inevitably would force the animal in a curve that would increase the lateral pressure on the crest, and thus increase the turning forces; this would either simply reverse the animals' position until the head and crest formed a dragged and stable rudder or break its neck. Of course, biologically this makes no sense. In aircraft, the effect of any rudder surface increases with an increase in flying speed. Consequently, a very low flight velocity, mostly relying on wing beats, would have been a possibility for a controlled head-forward flight.

A seemingly bizarre option is that the crest could also have operated similar to a thrust-generating wind-surfing sail, provided that the pterosaur could maintain course by use of webbed feet in the water. The crest shares its shape and probably material properties with some low wind-surfing sails (Farke *et al.* 1994). This idea appears less bizarre when noting that the storm petrels (Hydrobathidae) sail over the sea with open wings, holding their feet into the sea for anchoring and steering (Burton 1990, p. 114–115).

Anatomy of the brachiopatagium with comments on thermoregulation in pterosaurs

The three-dimensionally preserved fragment of an aktinopatagium of the *Rhamphorhynchus muensteri*

Fig. 15. Reconstruction of the head of living *Tapejara navigans*.

specimen JME SOS 4785 from the Solnhofen Lithographic Limestone helps to interpret the pterosaur skin section from the Santana Formation published by Martill & Unwin (1989). In JME SOS 4785 three types of tissue layers are visible from dorsally to ventrally (Fig. 4):

(1) Straight subparallel to radial fibres ("aktinofibrils" of Wellnhofer 1982).
(2) A meshwork of fibres with an average angle of 45° to the latter.
(3) The vessel network layer.

The skin fragment from the Santana Formation pterosaur also shows a distinct layering (Fig. 16 a, c). The topmost layer consists of a sheet of epidermis 10 microns thick. An internally adjacent layer shows a spongy texture and might represent a hitherto unknown type of subdermis. Below this a layer of cylindrical support fibres, from their size most likely actinofibrils, is seen in transverse section. This is followed by a layer of striated muscle fibres. They are cut obliquely, which indicates that they ran at an angle of approximately 45° to the possible collagen fibres. Between the sheaths of muscle fibres some fascia-like structures are seen oriented in the same direction as the muscle fibres.

From their relative thickness and their subparallel arrangement, the fibres in the Santana specimen resemble the 'aktinofibrils'. The dense packing of the fibres might be due to diagenetic effects, such as shrinkage or their position inside the tenopatagial area of the brachiopatagium.

The meshwork of connective tissue ventral to the aktinofibrils in the Solnhofen *Rhamphorhynchus* could therefore represent the fascia-like structures seen in the muscle layer in the Santana brachiopatagium fragment. These two structures show a very similar angle to each other in both specimens. Provided these assumptions are correct, two conclusions are considered here:

(1) The brachiopatagium fragment from the Santana Formation comes from a wing membrane which was probably situated closer to the body than suggested by Martill & Unwin (1989). The sections published by Kellner (1994, 1996) show neither the parallel fibre bundles nor the muscle sheaths described by Martill & Unwin (1989) and thus might well come from the body wall or the arm adjacent to the brachiopatagium.
(2) The brachiopatagium of pterosaurs consisted of at least five layers (from dorsally to ventrally):
(a) A thin, naked epidermis, which in the Santana specimen is only 10 microns thick, with an irregular system of grooves on the external surface (Fig. 16b).
(b) A spongy subdermis, which in the Santana specimen is about 20–40 μm thick (Fig. 16 a, c).
(c) A layer of aktinofibrils, of 180–200 μm thickness in the Santana specimen (Figs 6 & 16a, c).
(d) A layer of dermal muscles which in the Santana specimen was approximately 500 – 450 microns thick. The sheaths were separated by bundles of fascia-like structures (Figs 6, 16 a, c).
(e) A vascular layer of unknown thickness (Fig. 6).

The construction of the brachiopatagium makes pterosaurs enigmatic among actively flying tetrapods. Bats also have multilayered wing membranes (Gupta 1967; Quay 1970; Holbrook & Odland 1978) but they are mostly nocturnal (Erkert 1982; Kunz 1982). Diurnal activities in Microchiroptera during the summer months are only reported from high latitudes (Nyholm 1965;

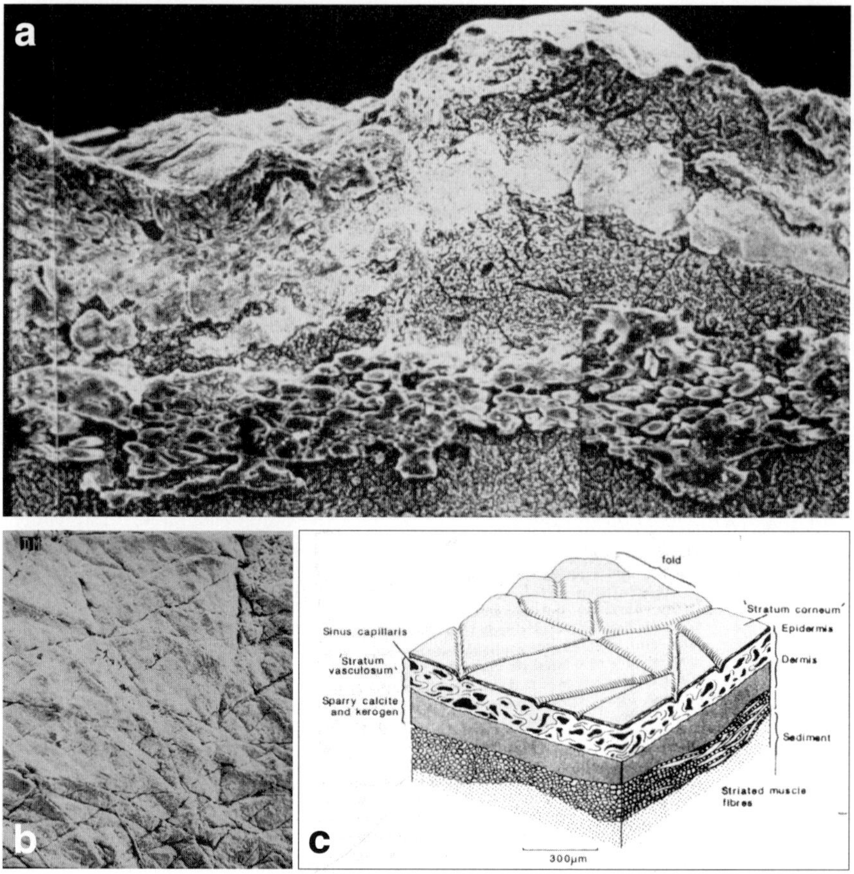

Fig. 16. (**a**) Cross-section showing the fibres below a spongy-looking layer of tissue and the sheath of muscle fibre below the fibril mat. (**b**) Surface of the dermis which looks very similar to the surface of the dermis of the crest found in *Pterodactylus* sp. (Fig. 1g). (**c**) Reconstruction of the pterosaurian wing membrane histology.

Speakman 1990, 1991) and a few island populations (Stoddart 1975; Bruner & Pratt 1979). As a consequence their wing membranes are never directly exposed to the sun for long periods. During the day bats roost in the shade. Large fruit bats that spent the daytime in trees and thus are exposed to sunlight, use their folded wing membranes for ventilation, but avoid flying over long distances under overhead sun (Nowak 1994, p. 13). Due to contractile arteries and veins in the flight membrane, bats can utilize their wings for conductive heat exchange. However, the thin membrane appears not to be heat-tolerant (Shipman 1998) and thus is mainly used as a cooling device to lose excess heat produced by the flight musculature (Hill & Smith 1984, p. 58). This might be one of the reasons why diurnal bats are exceedingly rare and probably excludes the evolution of gliding bats with large wing spans. The largest wing spans reported for bats are less than 1.5 m (Novak 1994).

Birds have small flight membranes – the propatagium and the plagiopatagium – compared with bats and pterosaurs (Salomon 1993). Both these patagia are covered by feathers which form a dense insulating layer on the arms. The major part of the wing surface consists of feathers composed of dead keratin and, thus, it is inert to heat loads. Flying in sunshine therefore causes no thermoregulatory problems for birds because they have no need to cope with the exposure of a large heat-exchange surface to the sun (King & King 1979: 18).

Pterosaurs had flight membranes and some achieved huge wing spans of more than 10 m (see Wellnhofer 1991). Thus, they were likely to be diurnal animals as is suggested by the size of the eye and the lack of any hint of the presence of an echolocation system. Consequently, in contrast to bats, the wing membrane of pterosaurs was probably exposed to the sun and thus must have been heat tolerant. In order to avoid overheating, cooling or insulation devices must have existed.

The spongy subepidermal tissue which appears to be unique for pterosaurs (Fig. 16a, c) might have been such a device. If this tissue could have been ventilated from the pneumatic system, it would have provided insulation to protect the tissues situated ventral to it from overheating. In addition a system of air capillaries could have been self-inflating when the wing was in the flight position and, like an insulation mattress, buffer temperature differences between the brachiopatagium´s dorsal and ventral surfaces. The thin epidermis could have been cooled down by the air flow. The system of grooves and humps seen in the Santana specimen indicates a large surface area, which would be advantageous for cooling in the same way as the ribs of an air-cooled motor. One can speculate that a light-reflecting colour, such as white or silver, could also have lowered the overheating hazard, at least in large pterosaurs. We might even speculate that the epidermis could have been translucent. In this case an air-filled structure adjacent to its internal face would result in a whitish-silvery colour of the dorsal wing surface, which would have had an optimal reflection effect.

The blood vessel system (Figs 4 & 6) close to the ventral surface of the brachiopatagium must have been effective for thermoregulation. Being situated on the ventral side of the wing, the shadow side, close to the surface, a pterosaur could cool down the patagia by dilating the blood vessels. At the same time the heat intake of the blood vessel system could have helped to maintain a high body temperature when gliding, especially at high altitudes or during the night. In phases of cool external temperatures, a constriction of the brachiopatagial blood vessels could have helped to maintain an elevated body temperature by retaining most of the blood in the body and neck. The coverage of body and neck with bristles would then be an effective insulation device, probably sufficient to guarantee a supply to the head of blood at an almost constant temperature. It is possible, hence speculative, that all membranous body parts in pterosaurs, such as crests and webbing, could have contributed to the thermoregulation of these animals. It is clear that the assumptions above may be difficult to prove but they are consistent with the anatomy of the animals.

Anatomy of foot and hand

The soft-part preservation of foot and hand is two dimensional but a correlation with numerous footprints helps to reconstruct the soft part anatomy of the foot and provides new clues to the anatomy of the soft parts of the hand. Furthermore the preserved soft parts help to confirm some footprints as those of pterosaurs and also help to clarify some problematic structures.

Presently only a single published footprint shows structures that coincide with the soft parts discovered on the feet of the new *Pterodactylus* and the Crato azhdarchid, thus allowing a detailed reconstruction of the plantar relief of this type of pterosaur (Mazin *et al.* 1995). The webbing extends to the bases of the keratinous claw sheaths. This suggests that the webbing has a maximum thickness of half a digit's diameter. An external relief of the webbing has not been identified. The pads ventral to the interdigital and the digitometatarsal joints have been described previously (Stokes 1957; Mazin *et al.* 1995, 1997). In most of the footprints the two medial claws dug into the substrate straight or slightly rotated laterally, while the two central ones were almost straight cranially. The claw of the lateral digit was mostly rotated medially. In the trackway described by Stokes (1957) the lateral digits of the foot show equally deep imprints of all pedal digits, whereas in the prints in the limestone of Crayssac (France) the medial and the lateral pedal digits are pressed deeper into the sediment than the two central ones (Mazin *et al.* 1995). This shows that the distal termini of the metatarsals in the living animal formed a shallow, ventrally concave arch.

Proximal to the digits there is a subtriangular plane plantar pad which sometimes shows faint imprints of the distal half of the metatarsals, indicating a very thin soft-tissue coverage of this part of the bones. Otherwise this pad is smooth, even in the best-preserved footprints (Stokes 1957). The distal margin of this triangular pad is sometimes interrupted centrally, which indicates that the plantar arch continued proximally towards the heel. The medial and lateral margins of the triangular pad line up with the adjacent digital prints and are straight. This is seen in the trackway described by Stokes (1957) and also in most of the trackways from Crayssac and other localities (Mazin *et al.* 1995, 1997).

An isolated single footprint from Crayssac (Mazin *et al.* 1995), as well as a pterosaur footprint from the Jurassic of Asturia, Spain (depicted on an unpublished tourist flyer by Ramos *et. al.* 2000), shows the imprint of a semicircular heel pad at the caudolateral margin of the triangular pad. This heel pad is the deepest part of the footprint and shows the same type of squamation seen in the *Pterodactylus* specimen from the Solnhofen Lithographic Limestone and the azhdarchid-type pterosaur from the Crato Formation (Frey & Tischlinger 2000). In these specimens the heel pad capped metatarsal V.

It is likely that pterosaurs normally walked semi-digitigrade, with the heel kept off the ground, as was suggested by Clark *et al.* (1998), probably because the brachiopatagium and the uropatagium were attached close to this point.

Handprints are more difficult to interpret, especially as there is rarely a clear differentiation

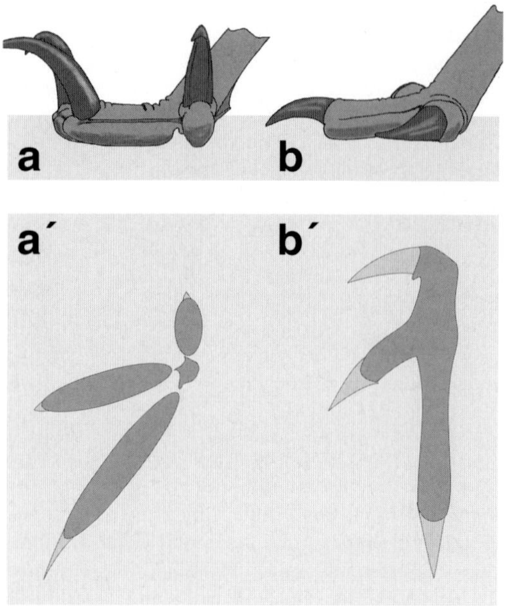

Fig. 17. Pterosaur hand anatomy and handprints. (**a**) Hand of an azhdarchid-type or *Pterodactylus*-type pterosaur with huge claws; in none of the known handprints did such claws leave a trace. Provided the claws were abducted during terrestrial locomotion, only the palmar pads would have left prints (**a**´) that would look similar to those described by Stokes (1957). (**b**) Hand of a short-clawed type of pterosaur; such a hand would have left prints as they are reported from Crayssac (**b**´; Mazin *et al.* 1995).

between the prints of the claws and the sole pad. In *Pterodactylus* and the azhdarchid type pterosaur from the Crato Formation the claws of the hand reached the size of an entire digit. As in the feet such claws should produce clear prints but presently there is no evidence for this. Possibly the claws were abducted slightly while walking and thus did not interact with the sediment (Fig. 17a). *Pteraichnus* (Stokes 1957) shows handprints with oval and stubby digit impressions which probably were produced only by sole pads. Regarding claws, only faint imprints of the base is seen (Fig. 17a). The handprints reported from Crayssac (Mazin *et al.* 1995) are rod-like, with two laterally pointing, hook-like impressions that might have been produced by claws which were tilted in such a way that the lateral surface contacted the mud (Fig. 17b). This suggests the existence of at least two different types of hand among pterosaurs.

Some of the handprints show imprints of webbing. The presence of webbing can be seen in the hand of the Crato azhdarchid but has not been detected in any pterosaur from the Solnhofen Lithographic Limestone. The palmar anatomy of the pterosaurian hand thus remains unclear for the moment.

The flight apparatus

The preserved flight membranes in pterosaurs allow us to comment on the efficiency of the pterosaurian flight apparatus. Because of the brachiopatagium attachment at the ankle and the presence of a uropatagium, the hindlimbs must have been held horizontally during flight and in approximately the same plane as the wings. It is likely that the femora in flight position pointed caudolaterally and the crures caudally and approximately parallel to each other. The hinge axis of the knee joint stood vertically, as did the ankle joint. Consequently, the feet were oriented vertically with the plantar surfaces facing medially (Fig. 18).

Due to their integration in the flight apparatus – combined brachiopatagium and uropatagium – the hindlimbs could influence both the tension and orientation of the brachiopatagia and contribute to flight control (Fig. 18). When the hindlimbs were synchronously rotated dorsally or ventrally, the pterosaur could control the pitch axis. When they were moved alternately dorsally or ventrally, movement in the roll axis would result. The yaw movement could be controlled by a parallel flexion and extension of the feet. If the feet were simultaneously extended they could have acted as an air brake, with the webbing of the digits providing the necessary surface. A strong ventral rotation of the hindlimbs would have resulted in an increase of the wing cambering at least in the proximal part of the wing (Bennett 2000), which would have allowed very low flight velocities but, at the same time, would have reduced velocity by acting as an additional air brake. Because the brachiopatagium in all pterosaurs in which the tenopatagial part of the brachiopatagium is fully preserved in attachment (see also the new *Rhamphorhynchus*), the cambering effect would have been much greater than that suggested by Bennett (2000), who assumes an attachment of the brachiopatagium level with the knee joint.

In such a context the shape of the trailing edge of the brachiopatagium had an influence on the manoeuvrability of a pterosaur. In *Sordes* the trailing edge of the brachiopatagium appears to have been straight or only slightly concave, resulting in a proximally very deep wing (fig. 17). As a consequence, each wing beat resulted in a follow-up beat of the hindlimbs similar to bats (Hill & Smith 1984, p. 46–47). Thus, the hindlimbs could not be moved independently to control the pitch and roll axes and control must have been provided by the entire wing as well as by the deep, shallowly forked uropatagium.

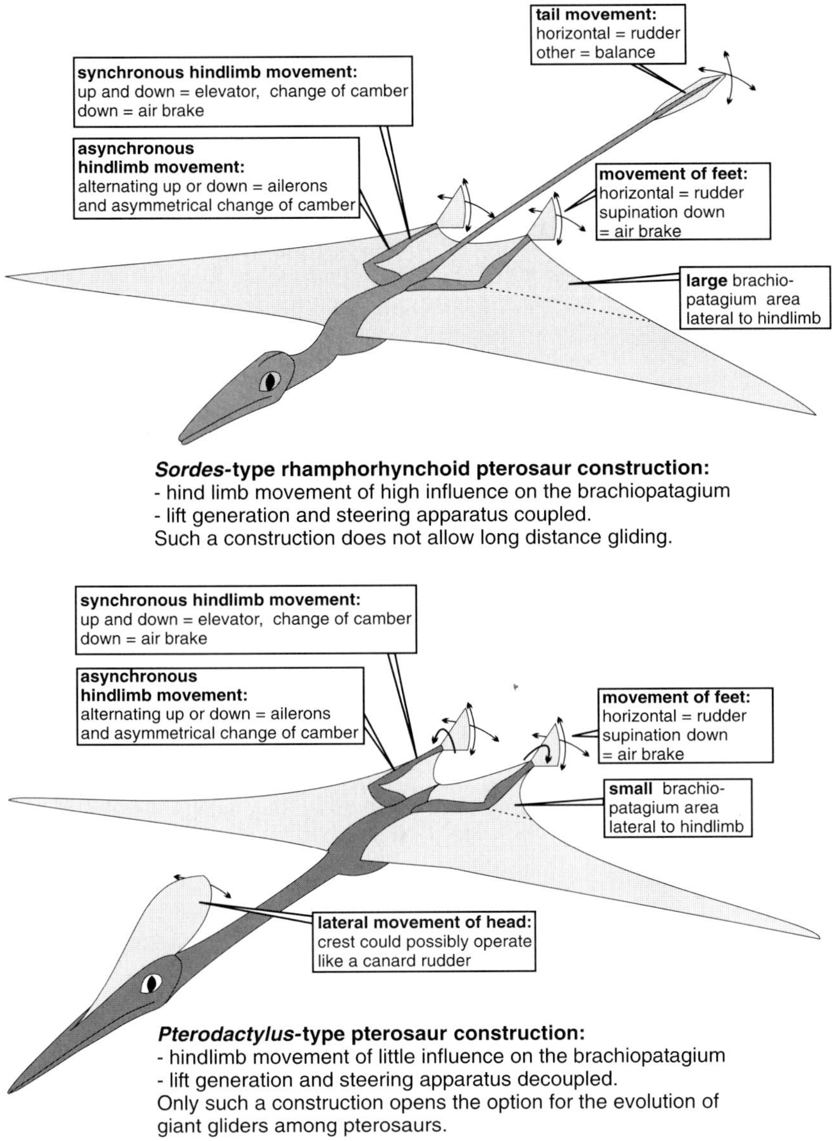

Fig. 18. Pterosaur flight control in a *Sordes*-type rhamphorhynchoid and a pterodactyloid pterosaur.

In the pterodactyloid type, and most likely also in *Rhamphorhynchus* (Figs 3a & 18, *Pterodactylus* type) the trailing edge of the brachiopatagium is deeply concave level with the crus. This results in a small membrane surface lateral to the hindlimbs which was aerodynamically less efficient during hindlimb movements. As a result of such a construction the hindlimbs were almost independent from the rest of the wing. Furthermore, the deeply forked uropatagium would have allowed independent mobility of the hindlimbs. Thus, the control elements are decoupled from the lift-and-thrust generating structures. Only this level of construction opens the option to an evolution of gigantic gliders among pterosaurs in which the steering movements of the hindlimbs had a minimal effect on the lift-generating wings.

The distribution of the propatagium is still under discussion. Bennett (2000, 2001) suggests that the pteroid bone articulated on the medial face of the lateral carpale and pointed medially, while Frey & Riess (1981) reconstructed a cranial orientation of the pteroid with a propatagium including the three digits. A problem with the reconstruction of Frey & Riess (1981) is that the joint facet that accommodates the pteroid sometimes contains a sesamoid (Bennett 2000, 2001), which would falsify the reconstruction of Frey & Riess (1981). However, there are hints of webbing between the digits of the hand from one specimen from the Crato Formation (Fig. 9b) that suggest the presence of a propatagium that could have included the digits. Theoretical constraints regarding the aerodynamic effect of a mobile propatagium to increase the wing camber for low flight velocities suggest a propatagium or similar structure to have been present. For the moment, for us, the anatomy in the pterosaurian carpus is unresolved and therefore the shape of the propatagium remains unclear.

However, an increase in the wing camber for low flight velocities during landing and take-off remains an aerodynamic necessity. Such a camber might have been brought about by the contraction of the muscle fibres within the brachiopatagium. We hypothesise that the aktinofibrils, together with the subdermal tissue, maintained the wing profile during flight. Muscle contraction resulted in a deformation of these tissues. Because the muscle fibres lay ventral to the aktinofibrils, the contraction could only have increased the wing camber. As soon as the muscles relax, the aktinofibrils returned the wing profile to a neutral position. Whether or not such a mechanism was aerodynamically effective is currently under investigation (Frey *et al.* in prep.).

Conclusions

The latest discoveries of pterosaur specimens with exceptionally well-preserved soft parts coming from different locations and belonging to different taxa provide new and important insights into several aspects of pterosaur anatomy.

The anatomy of the foot of rhamphorhynchoid-type pterosaurs cannot be reconstructed on the basis of the anatomy of the pterodactyloid foot because the skeletal structure is also different. The long fifth digit in the rhamphorhynchoid type of pterosaurs is not found in pterodactyloids and precludes the existence of a pterodactyloid-type heel pad in rhamphorhynchoid-type pterosaurs because the animals most probably controlled the tension of the uropatagium by means of this digit (see *Sordes pilosus*; Unwin & Bakhurina 1994).

The dermis of pterosaurs displays a variety of different surface structures, the distribution of which can be mapped on fossil evidence. The dermis of the body was coated with bristles. Neck bristles are only reported in pterodactyloid type pterosaurs from the Solnhofen Lithographic Limestone (Frey & Martill 1998; Tischlinger 1998). The epidermis of the crest of the *Pterodactylus* specimen described above shows remarkably similar structures to those of the wing membrane described by Martill & Unwin (1989). Because both specimens were fossilized under different geochemical regimes (Barthel 1964; Martill 1988), it appears unlikely that the surface structure in the form of irregular furrows represent a diagenetic artefact. The similarity of the soft-tissue crests and brachiopatagium is not restricted to the epidermal surface. Both structures share the presence of internal cylindrical support fibres of identical dimensions and preservation style. Furthermore, in both cases there are remnants of an unknown type of tissue between the epidermis and the fibrous layer. In the crest of *Pterodactylus* sp. it was preserved as a white powdery phosphate layer. In the Santana wing membrane there is a tubular type of tissue with a spongy appearance. Surface structures are not reported from the brachiopatagia of Solnhofen Lithographic Limestone pterodactyloids. In one case of a *Pterodactylus kochi*, the wing surface appears smooth (Frey & Martill 1998) but the material was not examined under SEM. Two further specimens of *Pterodactylus kochi* with a well-preserved brachiopatagium (Wellnhofer 1987b; Tischlinger 1993) also show an almost smooth surface.

Pterorhynchus wellnhoferi from China (Czerkas & Ji 2002) preserves a wing membrane with aktinofibrils and a fine pattern of rhombic fields on its surface which resemble the dermal surface structures described by Martill & Unwin (1989).

Cylindical support fibres also occur in the webbing of the feet and are of the same thickness and density as the support fibres in the brachiopatagium. Probably both structures evolved in the same context of surface enlargement, a precondition for the evolution of flight, which would mean that actinofibril-like structures were not restricted to the aktinopatagium only.

Further epidermal structures are the scales covering the plantar pad and the distal part of the tibia. These scales differ from the epidermis surface in the crest and brachiopatagium in being circular to slightly oval. In contrast to the irregular pattern of the reticulated epidermis, these scales form an almost regular pavement. Until now this type of squamation was known for pterodactyloids but not for rhamphorhynchoid types of pterosaurs.

Our study allows us to conclude that pterosaurs obviously were unique with respect to the soft-tissue morphology of their flight apparatus as well as the skeletal components, their epidermis differentiation and, most likely, also with respect to their physiology, especially their thermoregulation. These aspects have to be taken into account in any discussion on the origin of the Pterosauria and in any consideration of the taxonomy inside the group.

We cordially thank the reviewers of this paper, P. Christianssen (Copenhagen) and P. Griffith (Wolverhampton) for their thorough and helpful comments on the manuscript. For the supreme preparation work we thank R. Kastner, A. Anders and O. Dülfer (all Karlsruhe), as well as D. Kümpel (Wuppertal). The photography of the Brazil specimens was carried out by V. Griener (Karlsruhe), G. Bergmeier (Munich) gratefully put to our disposal the darkroom and processing equipment of the BSP. We also thank L. Zeilbeck (Kösching) for his assistance with stereomicroscope photography. Without the generous support of P. Wellnhofer and W. Werner (Munich), W. Langston Jr (Austin), G. Chong Diaz (Antofagasta), J. B. Filgueiras (Crato), G. Viohl (Eichstätt), R. Wild (Stuttgart) and M. Sander (Bonn) for accessing many of the specimens cited herein, this study would not have been possible. We are grateful to D. Kümpel (Wuppertal) and the owner of the Schrandel quarry at Langenaltheim for loaning us the main slabs of their specimens and donating the counter-slabs containing the crucial scientific information to the JME. For financial support for research trips to Brazil and Chile we wish to thank the Deutsche Forschungsgemeinschaft (DFG, Bonn) and the von Kettner-Stiftung, Karlsruhe. Part of the research, some results of which are presented here, was financed by DGF Project FR 1314/2-1. Special thanks go to E. Buffetaut (Paris) for his excellent editorial work.

References

ABEL, O. 1907. Bau und Lebensweise der Flugsaurier. *Verhandlungen der Kaiserlichen und Königlichen Zoologisch-Botanischen Gesellschaft Wien*, **1907**, 253–254, Vienna.

ABEL, O. 1919. Neue Rekonstruktion der Flugsauriergattungen *Pterodactylus* und *Rhamphorhynchus*. *Naturwissenschaften*, **7** (37), 661–665, Berlin.

ABEL, O. 1927. *Lebensbilder aus der Tierwelt der Vorzeit*, 2nd edition. Gustav Fischer, Jena, 714 pp.

BAKHURINA, N. N. 1982. Pterodactyl from the Lower Cretaceous of Mongolia. *Palaeontological Journal*, **4**, 104–108. [In Russian]

BARTHEL, K. W. 1964. Zur Entstehung der Solnhofener Plattenkalke (unteres Untertithon). *Mitteilungen der Bayerischen Staatssammlung für Paläontologie und Historische Geologie*, **4**, 37–69.

BARTHEL, K. W. 1972. The genesis of the Solnhofen lithographic limestone (Low. Tithonian): further data and comments. *Neues Jahrbuch für Geologie und Mineralogie, Monatshefte*, **1972** (3), 133–145.

BENNETT, S. C. 1993. The ontogeny of *Pteranodon* and other pterosaurs. *Paleobiology*, **19**, 92–106.

BENNETT, S. C. 2000. Pterosaur flight: The role of actinofibrils in wing function. *Historical Biology*, **14**, 255–284.

BENNETT, S. C. 2001. The osteology and functional morphology of the Late Cretaceous pterosaur *Pteranodon*. Part I. Size and functional morphology. *Palaeontographica A*, **260**, 113–153.

BRITO, P. M. & FERREIRA P. L. N. 1989. The first Hybodont shark, *Tribodus limae* n.g., n. sp., from the Lower Cretaceous of Chapada do Araripe (N-E Brazil). *Anais da Academia Brasileira de Ciênias*, **61**, 53– 57.

BROILI, F. 1927a. Ein *Rhamphorhynchus* mit Spuren von Haarbedeckung. *Sitzungberichte der Bayerischen Akademie der Wissenschaften, Mathematisch-Naturwissenschaftliche Abteilung*, **1927**, 49–67.

BROILI, F. 1927b. Ein Exemplar von *Rhamphorhynchus* mit Resten von Schwimmhaut. *Sitzungberichte der Bayerischen Akademie der Wissenschaften, Mathematisch-Naturwissenschaftliche Abteilung*, **1927**, 29–49.

BROILI, F. 1938. Beobachtungen an *Pterodactylus*. *Sitzungberichte der Bayerischen Akademie der Wissenschaften, Mathematisch-Naturwissenschaftliche Abteilung*, **1938**, 139–154.

BRUNER, L. P. & PRATT, H. D. 1979. Notes on the status and natural history of Micronesian bats. *Elepaio*, **40**, 1–4.

BUFFETAUT, E., LEPAGE, J-J. & LEPAGE, G. 1998. A new pterodactyloid pterosaur from the Kimmeridgian of the Cap de la Hève (Normandy, France). *Geological Magazine*, **135**, 719–722.

BURTON, R. 1990. *Bird Flight. An Illustrated Study of Birds´ Aerial Mastery*. Facts On File, New York, Oxford, Sydney, 160 pp.

CAMPOS, D.A. & KELLNER, A.W.A. 1985. Panorama of the flying reptiles of Brazil and South America. *Anais da Academia Brasileira de Ciências*, **57**, 453–466.

CAMPOS, D.A. & KELLNER, A.W.A. 1997. Short note on the first occurence of Tapejaridae in the Crato Member (Aptian) Santana Formation. Araripe Basin, northeast Brazil. *Anais da Academia Brasileira de Ciências*, **69** (1), 83–87.

CAMPOS, D.A., LIGABUE, D. & TAQUET, P. 1984. Wing membrane and wing supporting fibres of a flying reptile from the Lower Cretaceous of the Chapada do Araripe (Aptian; Ceará State, Brazil). *Third Symposium of Mesozoic Terrestrial Ecosystems & Biota, Tübingen*, 37–94.

CLARK, J. M., HOPSON, J. A., HERNANDEZ, R., FASTOVSKY, D. E. & MONTELLANO, M. 1998. Foot posture in a primitive pterosaur. *Nature*, **391**, 886–889.

CUVIER, G. 1809. Mémoire sur le squelette fossile d´un reptile volant des environs d´Aichstedt, que quelques naturalistes ont pris pour un oiseau, et dont nous formons un genre des sauriens, sous le nom de Pterodactyle. *Annales du Muséum D´Histoire Naturelle, Paris*, **13**, 424.

CZERKAS, S. A. & JI, Q. 2002. A new rhamphorhynchoid with headcrest and complex integumentary structures. *In*: CZERKAS, S. J. (ed.) *Feathered Dinosaurs and the Origin of Flight*, **1**, Blanding, Utah, 15–41.

DÖDERLEIN, L. 1929. Über *Rhamphorhynchus* und sein Schwanzsegel. *Sitzungberichte der Bayerischen Akademie der Wissenschaften, Mathematisch-Naturwissenschaftliche Abteilung*, **1929**, 117–164.

EATON, G. F. 1910. Osteology of *Pteranodon*. Yale University Press, Yale, 96 pp.

ERKERT, H. G. 1982. Ecological aspects of bat activity rhythms. *In:* KUNDZ, T. H. (ed.) *Ecology of Bats*. Plenum Press, New York, 201–242.

FARKE, U., MÖHLE, V. & SCHRÖDER, D. 1994. *Windsurf-Grundschein. Ich lerne surfen*. Delius Klasing Verlag, Bielefeld, 115 pp.

FASTNACHT, M. 1997. New light on tooth replacement of pterosaurs. *Journal of Morphology*, **232**, 253.

FASTNACHT, M. 2001. First record of *Coloborhynchus* (Pterosauria) from the Santana Formation (Lower

Cretaceous) of the Chapada do Araripe, Brazil. *Paläontologische Zeitschrift*, **75**, 23–36.

FASTNACHT, M., HESS, N., WEISER, P. & FREY, E. 2002 Finite element analysis in vertebrate palaeontology. *Senckenbergiana Lethaea*, **82**(1), 195–206.

FREY, E. & MARTILL, D. M. 1994. A new pterosaur from the Crato Formation (Lower Cretaceous, Aptian) of Brazil. *Neues Jahrbuch für Geologie und Paläontologie, Abhandlungen*, **194**, 379–412.

FREY, E. & MARTILL, D. M. 1998. Soft tissue preservation in a specimen of *Pterodactylus kochi* (WAGNER) from the Upper Jurassic of Germany. *Neues Jahrbuch für Geologie und Paläontologie, Abhandlungen*, **210**, 421–441.

FREY, E., MARTILL, D.M. & BUCHY, M.-C. 2003a. A new crested ornithocheirid from the Lower Cretaceous of northeastern Brazil and the unusual death of an unusual pterosaur. *In*: BUFFETAUT, E. & MAZIN, J.-M. (eds) *Evolution and Palaeobiology of Pterosaurs*. Geological Society, London, Special Publications, **217**, 55–63.

FREY, M. MARTILL, D. M. & BUCHY, M.-C. 2003b. A new species of tapejarid pterosaur with soft-tissue head crest. *In*: BUFFETAUT, E. & MAZIN, H.-M. (eds) *Evolution and Palaeobiology of Pterosaurs*. Geological Society, London, Special Publications, **217**, 65–72.

FREY, E. & RIESS, J. 1981. A new reconstruction of the pterosaur wing. *Neues Jhrbuch für Geologie and Paläontologie, Abhandlunge*, **161**, 1–27.

FREY, E. & TISCHLINGER, H. 2000. Weichteilanatomie der Flugsaurierfüße und Bau der Scheitelkämme: Neue Pterosaurierfunde aus den Solnhofener Schichten (Bayern) und der Crato-Formation (Brasilien). *Archaeopteryx*, **18**, 1–16.

FRICKHINGER, K.A. 1999. *Die Fossilien von Solnhofen –* Dokumentation der aus den Plattenkalken bekannten Tiere und Pflanzen, Band 2: Neue Funde, neue Details, neue Erkenntnisse; mit einer Einführung in die Geologie der Plattenkalke von H. Tischlinger. Goldschneck, Korb. 190 pp.

GOLDFUSS, A. 1831. Beiträge zur Kenntnis verschiedener Reptilien der Vorwelt. *Nova Acta Acadademia Leopoldina Carolinea*, **15**, 61–128.

GUPTA, B. B. 1967. The histology and musculature of plagiopatagium in bats. *Mammalia*, **31**, 313–321.

HILL, J. H. & SMITH, J. D. 1984. *Bats. A Natural History*. British Museum (Natural History), London, 243 pp.

HOLBROOK, K. A. & ODLAND, G. F. 1978. A collagen and elastic network in the wing of the bat. *Journal of Anatomy*, **126**, 21–36.

HOLST, E. von. 1957a. Wie flog *Rhamphorhynchus*? *Natur und Volk*, **87**, 81–87.

HOLST, E. von. 1957b. Der Saurierflug. *Palaeontologische Zeitschrift*, **33**, 15–22.

KÄMPFE, L., KITTEL, R. & KLAPPERSTÜCK, J. 1993. *Leitfaden der Anatomie der Wirbeltiere. 6. überarbeitete Auflage*. Gustav Fischer, Jena, Stuttgart, New York, 358 pp.

KAUP, J. 1834. *Isis*, **1834**, 315.

KELLNER, A.W.A. 1984. Ocorréncia de uma mandibula de Pterosauria (*Brasiliodactylus araripensis*, nov. gen., nov. spec.) na Formaçao Santana, Cretaceo da Chapada do Aararipe, Ceará – Brasil. *Annales de XXXIII Congreso Brasileiro Geologico de Rio de Janeiro*, 578–590.

KELLNER, A. W. A. 1989. A new edentate pterosaur of the Lower Cretaceous from the Araripe Basin, northeast Brazil. *Anais da Academia Brasileira de Ciências*, **61**, 439–446.

KELLNER, A. W. A. 1990. A new edentate pterosaur of the Lower Cretaceous from the Araripe Basin, northeast Brazil. *Anais da Academia Brasileira de Ciências*, **61**, 1–7.

KELLNER, A. W. A. 1994. Pterosaur soft tissue from Brazil: to be or not to be a wing membrane, that is the question. *Journal of Vertebrate Paleontology*, **14** (supplement), 31A–32A.

KELLNER, A. W. A. 1996. Reinterpretation of a remarkably well preserved pterosaur soft tissue from the Early Cretaceous of Brazil. *Journal of Vertebrate Paleontology*, **16**, 718–722.

KELLNER, A. W. A. 1996. Fossilised theropod soft tissues. *Nature*, **379**, 32.

KELLNER, A. W. A. & CAMPOS, D.A. 1989. Sobre um novo pterosauro com crista sagittal da Bacia do Araripe, Cretáceo Inferior do nordeste do Brasil. *Anais da Academia Brasileira de Ciências*, **60**, 459–469.

KELLNER, A. W. A. & LANGSTON, W. 1996. Cranial remains of *Quetzalcoatlus* (Pterosauria, Azhdarchidae) from Late Cretaceous sediments of Big Bend National Park, Texas. *Journal of Vertebrate Paleontology*, **16**, 222–231.

KELLNER, A. W. A. & TOMIDA, Y. 1996. New information on *Anhanguera* and the systematic position of the Anhangueridae (Pterosauria, Pterodactyloidea). *Journal of Vertebrate Paleontology*, **16**(3) (supplement), 45A.

KELLNER, A. W. A. & TOMIDA, Y. 2000. *Description of a New Species of Anhangueridae (Pterodactyloidea) with Comments on the Pterosaur Fauna from the Santana Formation (Aptian–Albian, Northeastern Brazil)*. National Science Museum Monographs, **17**, 135 pp.

KEMP, R. A. 1999. *Solnhofen tetrapod taphonomy*. PhD Thesis, University of Bristol.

KING, A. S. & KING, D. Z. 1979. Avian morphology: general principles. *In:* KING, A. S. & MCLELLAND, J. (eds) *Form and Function in Birds*. Academic Press, London, New York, Toronto, Sydney, San Francisco, vol. **1**, 1–38.

KUNZ, T. H. 1982. Roosting ecology in bats. *In:* KUNZ, T. H. (ed.) *Ecology of Bats*. Plenum Press, New York, 1–55.

LEE, Y. N. 1994. The Early Cretaceous pterodactyloid pterosaur *Coloborhynchus* from North America. *Palaeontology*, **37**, 755–763.

LEONHARDT, H. 1977. *Histologie, Zytologie und Mikroanatomie des Menschen für Ärzte und Studenten. 5. Neubearbeitete Auflage, Taschenlehrbuch der gesamten Anatomie*. Thieme Verlag, Stuttgart, vol. 3, 476 pp.

MARSH, O. C. 1882. The wings of pterodactyles. *American Journal of Science*, **23**, 251–256.

MARTILL, D. M. 1988. Preservation of fish in the Cretaceous of Brazil. *Palaeontology*, **31**, 1–18.

MARTILL, D. M. 1993. *Fossils of the Santana and Crato Formations*, Brazil. Field Guides to Fossils, Palaeontological Association, **5**, 159 pp.

MARTILL, D. M. & FREY, E. 1998. A new pterosaur Lagerstätte in N.E. Brazil (Crato Formation; Aptian, Lower Cretaceous): preliminary observations. *Oryctos*, **1**, 79–85.

MARTILL, D. M. & UNWIN, D. M. 1989. Exceptionally well preserved pterosaur wing membrane from the Cretaceous of Brazil. *Nature*, **340**, 138–140.

MARTILL, D.M., FREY, E., CHONG DIAZ, G. & BELL, C. M. 2000. Reinterpretation of a Chilean pterosaur and the occurrence of Dsungaripteridae in South America. *Geological Magazine*, **137**, 19–25.

MARTILL, D. M., UNWIN, D. M. & WILBY, P. 1990. Stripes on a pterosaur wing. *Nature*, **346**, 116.

MAZIN, J-M., HANZPERGUE, P., BASSOULLET, J-P., LAFAURIE, G. & VIGNAUD, P. 1997. Le gisement de Crayssac (Tithonien inférieur, Quercy, Lot, France): découverte de pistes de dinosaures en place et premier bilan ichnologique. *Comptes Rendus de l'Académie des Sciences, Paris, Serie IIa*, **325**, 733–739.

MAZIN, J-M., HANZPERGUE, P., LAFAURIE, G. & VIGNAUD, P. 1995. Des pistes de ptérosaures dans le Tithonien de Crayssac (Quercy, France). *Comptes Rendus de l'Académie des Sciences, Paris*, **321**, 417–424.

MEYER, H. VON. 1846. *Pterodactylus (Rhamphorhynchus) Gemmingi* aus dem Kalkschiefer von Solenhofen. *Palaeontographica*, **2**, 82.

MEYER, H. VON. 1859. Zur Fauna der Vorwelt, vierte Abt.: Reptilien aus dem lithographischen Schiefer des Jura in Deutschland und Frankreich. 1. Lief., 1–84, Frankfurt.

NESOV, L. A. 1984. [Pterosaurs and birds from the Upper Cretaceous of Middle Asia.] *Palaeontological Journal, Academy of Sciences, USSR*, **1984**, 47–57. [In Russian]

NOWAK, R. M. 1994. *Walker's Bats of the World*. The Johns Hopkins University Press, Baltimore and London, 287 pp.

NYHOLM, E. S. 1965. Zur Ökologie von *Myosotis mystacinus* (Leisl.) und *M. daubentoni* (Leisl.) (Chiroptera). *Annales Zoologici Fennici*, **2**, 77–123.

OWEN, R. 1861. *Monograph on the Fossil Reptilia of the Cretaceous Formations. Supplement III*. Palaeontographical Society, London, Monograph, **12**, 25 pp.

OWEN, R. 1874. *Monograph on the fossil Reptilia of the Mesozoic Formations. I. Pterosauria*. Palaeontographical Society Monograph, **27**, 14 pp.

PADIAN, K. 1986. A taxonomic note on two pterodactyloid families. *Journal of Vertebrate Paleontology* **4**, 516–524.

PADIAN, K. & RAYNER, J. M. V. 1993. The wings of pterosaurs. *American Journal of Science*, **293** (A), 91–116.

PLIENINGER, F. 1901. Beiträge zur Kenntnis der Flugsaurier. *Palaeontographica*, **53**, 209–213.

PRICE, I.L. 1971. A presença de Pterosauriado Cretaceo inferior da Chapada do Araripe, Brasil. *Anais da Academia Brasileira de Ciências*, **43** (supplement), 452–461.

QUAY, W. B. 1970. Integument and derivaties. *In* WIMSOTT, W. A. (ed.) *Biology of Bats*. Academic Press, New York, vol. 2, 1–56.

RAMOS, C. G., ARAMBURU, C., PIÑUELA, L. & LIRES, J. 2000. *La Costa de Los Dinosaurios. Rutas por el Jurásico de Asturias*. Gobierno del Principado de Asturias, Consejeria de Educacion y Cultura, 33 pp. [Tourist flyer].

RJABININ, A. N. 1948. [Remarks on a flying reptile from the Jurassic of the Kara-Tau.] *Transactions of the Palaeontological Institute, Academy of Sciences, USSR*, **15** (1), 86–93. [In Russian]

SALOMON, F. V. 1993. *Lehrbuch der Geflügelaantomie*. Gustav Fischer, Jena and Stuttgart, 479 pp.

SCHALLER, D. 1985. Wing evolution. *In*: HECHT, M. K., OSTROM, J. H., VIOHL, G. & WELLNHOFER, P. (eds). *The Beginnings of Birds. Proceedings of the International* Archaeopteryx *Conference, Eichstätt, 1985*. Freunde des Jura-Museums, Eichstätr, 333–348.

SEELEY, H. G. 1870. *The Ornithosauria: An Elementary Study of the Bones of Pterodactyles, Made from Fossil Remains Found in the Cambridge Greensand, and Arranged in the Woodwardian Museum of the University of Cambridge*. Deighton, Bell & Co., Cambridge, 137 pp.

SHAROV, A.G. 1971. New flying reptiles from the Mesozoic of Kazakhstan and Kirgizia., *Trudy Academia Nauk., Palaeontoloitscheki. Institut* **130**, 104–113. [In Russian]

SHIPMAN, P. 1998. *Taking Wing. Archaeopteryx and the Evolution of Bird Flight*, Simon & Schuster, New York, 336 pp.

Smith, R. 2000. Morphology and ontogeny of Cretaceous ostracods with preserved appendages from Brazil. *Palaeontology*, **43**, 63–98.

SPEAKMAN, J. R. 1990. Why do insectivorous bats not fly in daylight more frequently? *Functional Ecology*, **5**, 518–524.

SPEAKMAN, J. R. 1991. The impact of predation by birds on bat populations in the British Isles. *Mammalian Review*, **21**, 123–142.

STODDART, D. M. 1975. The diurnal flight of the azorean bat (*Nyctalus azoreum*) and the avifauna of the Azores. *Journal of the Zoological Society, London*, **177**, 483–506.

STOKES, W. L. 1957. Pterodactyl tracks from the Morrison Formation. *Journal of Paleontology*, **31**, 952–954.

TISCHLINGER, H. 1993. Überlegungen zur Lebensweise der Pterosaurier anhand eines verheilten Oberschenkelbruches bei *Pterodactylus kochi* (Wagner). *Archaeopteryx* **11**, 63–71.

TISCHLINGER, H. 1998. Die Flugkünstler der Jurazeit. Neue Erkenntnisse über das Leben der Flugsaurier. *Praxis der Naturwissenschaften. (Biologie)*, **47**, 29–34.

TISCHLINGER, H. 2002. Der Eichstätter *Archaeopteryx* im langwelligen UV-Licht. *Archaeopteryx*, **20**, 21–38.

TISCHLINGER, H. & FREY, E. 2001. "Mit Haut und Haar!" Flugsaurier-Neufunde mit spektakulärer Weichteilerhaltung. *Fossilien*, **3**, 151–158.

UNWIN, D. M., (in prep.), A new specimen of *Rhamphorhynchus* (Pterosauria: Rhamphorhynchidae) from the Upper Jurassic Solnhofen Limestone of Germany with evidence of a throat sac and other soft tissues.

UNWIN, D. M. (2001). An overview of the pterosaur assemblage from the Cambridge Greensand (Cretaceous) of eastern England. *Mitteilungen aus dem Museum für Naturkunde, Berlin, Geowissenschaftlich Reihe*, **4**, 189–217.

UNWIN, D. M. & BAKHURINA, N. N. 1994. *Sordes pilosus* and the nature of the pterosaur flight apparatus. *Nature*, **371**, 62–64.

VIOHL, G. 2000. *Drachen der Lüfte. Entwicklung und Leben der Flugsaurier. Führer zur gleichnamigen Ausstellung des Jura-Museums vom 9 Juni 2000 bis zum 7 Januar 2001*. Pfeil Verlag, München, 48 pp.

VIOHL, G. 2001. Jahresbericht 2000 des Jura-Museums. *Archaeopteryx*, **19**, 121–126.

WANDERER, K. 1908. *Rhamphorhynchus gemmingi* H. v., ein Exemplar mit teilweise erhaltener Flughaut aus dem Kgl. Mineralog. Geol. Museum zu Dresden. *Palaeontographica A*, **55**, 195–216.

WELLNHOFER, P. 1970. Die Pterodactyloidea (Pterosauria) der Oberjura-Plattenkalke Süddeutschlands. *Abhandlungen der Bayerischen Akademie der Wissenschaften, Neue Folge*, **141**, 1–133.

WELLNHOFER, P. 1975a. Die Rhamphorhynchoidea (Pterosauria) der Oberjura-Plattenkalke Süddeutschlands. I: Allgemeine Skelettmorphologie. *Palaeontographica*, **148**, 1–33.

WELLNHOFER, P. 1975b. Die Rhamphorhynchoidea (Pterosauria) der Oberjura-Plattenkalke Süddeutschlands. II: Systematische Beschreibung. *Palaeontographica*, **148**, 132–186.

WELLNHOFER, P. 1975c. Die Rhamphorhynchoidea (Pterosauria) der Oberjura-Plattenkalke Süddeutschlands. III: Palökologie und Stammesgeschichte. Palaeontographica, **149**, 1–30.

WELLNHOFER, P. 1978. Pterosauria. *In*: WELLNHOFER, P. (ed.) *Handbuch der Paläontologie*. Gustav Fisher, Stuttgart, Teil 19, 82 pp.

WELLNHOFER, P. 1980. Flugsaurier, Pterosauria. *Die Neue Brehm-Bücherei*, **534**, 1–135.

WELLNHOFER, P. 1982. Zur Biologie der Flugsaurier. *Natur und Museum*, **112**, 278–291.

WELLNOFER, P. 1985. Neue Pterosaurier aus der Santana-Formation (apt) der Chapada do Araripe, Brasilien, *Palaeontographica A*, **187**, 104–182.

WELLNHOFER, P. 1987a. New cristated pterosaurs from the Lower Cretaceous of Brazil. *Mitteilungen der Bayerischen Staatsammlung für Paläontologie und Historische Geologie*, **27**, 175–186.

WELLNHOFER, P. 1987b. Die Flughaut von *Pterodactylus* (Reptilia: Pterosauria) am Beispiel des Wiener Exemplares von *Pterodactylus kochi* (Wagner). *Annalen des Naturhistorischen Museums Wien*, **88** (A), 149–162.

WELLNHOFER, P. 1991. *The Illustrated Encyclopedia of Pterosaurs*. Salamander, London, 192 pp.

WELLNHOFER, P. 1993. *Die grosse Enzyklopädie der Flugsaurier*. Mosaik-Verlag, München, 192 pp.

WINKLER, T.C. 1874. Le *Pterodactylus* du Musée Teyler. *Archives du Musée Teyler*, **3** (4), 1–11.

YOUNG, C. C. 1964. On a new pterosaurian from Sinkiang, China. *Vertebrata Palasiatica*, **8**, 221–256.

YOUNG, C. C. 1973. Wuerho pterosaurs. *Institute of Vertebrate Palaeontology and Palaeoanthropology, Academica Sinica, Special Publications*, **11**, 18–34. [In Chinese]

ZITTEL, K.A. VON. 1882. Über Flugsaurier aus dem lithographischen Schiefer Bayerns. *Palaeontographica*, **29**, 47–80.

Middle- and bottom-decker Cretaceous pterosaurs: unique designs in active flying vertebrates

EBERHARD FREY[1], MARIE-CÉLINE BUCHY[2] & DAVID M. MARTILL[3]

[1]*Staatliches Museum für Naturkunde Karlsruhe, D-76133 Karlsruhe, Germany*
[2]*Universität Karlsruhe, Geologisches Institut, Postfach 6980, D-76128 Karlsruhe, Germany*
[3]*School of Earth and Environmental Sciences, University of Portsmouth, Portsmouth PO1 3QL, UK*

Abstract: Among Pterosauria there are three types of scapulocoracoid construction. In the ornithocheirid scapulocoracoid the scapula is oriented almost horizontally; it is shorter than the coracoid and the glenoid fossa is level with the ventral margin of the vertebral column. In the azhdarchid scapulocoracoid the scapula is curved ventrally and is as long as the coracoid. In this construction the glenoid fossa lies approximately in the mid-horizontal plane of the chest. In the tapejaroid construction, the scapula is about one-third longer than the coracoid, which is oriented subhorizontally, and the glenoid fossa is level with the dorsal rim of the sternal plate. Both latter conditions are hitherto unknown among flying vertebrates and result in an unstable, but manoeuvrable flight, probably powered with wing beats.

Introduction

In actively flying vertebrates the wings are modified pectoral fins or forelimbs, with the shoulder joint located level with the vertebral column or close to it (Gray 1953; Norberg 1981; Hildebrand 1982). Such dorsolateral articulation of the wings can only be achieved by an elongation of the shoulder elements ventral to the glenoid fossa, as is seen in birds and bats, while those elements dorsal to it either are reduced in relative size, as in bats (Vaughan 1959, 1966; Hill & Smith 1984; Norberg & Rayner 1987), or rotated into a horizontal position as in birds (Pennicuik 1972, 1975, 1989; Childress 1981; Burton 1990; Salomon 1993). The effect of such a construction is to bring the glenoid fossa above the centre of gravity. This helps to sustain an automatic gravity-triggered flight stabilization, as has been shown for Recent bats and birds (e.g. Vaughan 1959, 1966; Gray 1968; Pennicuik 1972, 1975, 1989; Childress 1981; Hildebrand 1982; Hill & Smith 1984; Nachtigall 1986; Norberg & Rayner 1987; Burton 1990). As in birds, the glenoid fossa in most pterosaurs is elevated by a dorsolaterally directed elongation of the coracoid and lies almost level with the vertebral column (Wellnhofer 1978, 1991a; Figs 1a, a' & 3a, a'). A gravity-triggered flight stabilization has also been postulated for such constructions (e.g. Scheffen 1926; Kripp 1943; Holst 1957; Heptonstall 1971; Bramwell & Whitfield 1974; Stein 1975; Frey & Riess 1981; Padian 1983; Pennicuik 1988).

Here we describe a unique type of wing articulation in some Cretaceous pterosaurs in which the scapula is inclined ventrolaterally while the shaft of the coracoid is almost horizontal. Consequently the glenoid fossa is level with the sternum, resulting in a ventrolateral articulation of the wings (Figs 2b, b', c, c', 3b, b', c, c').

The objective of this paper is a first approach to considering the consequences of such a wing construction on pterosaur flight dynamics and the constructional differences in the flight apparatus with respect to pterosaurs with a dorsal wing attachment.

Material

This study is based on the following specimens: an isolated scapulocoracoid with articulated humerus from the Crato Formation (Early Cretaceous, northeastern Brazil, SMNK PAL 3843); a fragmentary scapulocoracoid of a juvenile *Tapejara wellnhoferi* from the Santana Formation (Early Cretaceous, northeastern Brazil, SMNK PAL 1137); a scapulocoracoid and humerus of *Quetzalcoatlus* sp. from the Javelina Formation (Late Cretaceous, Texas, TMM 42138–1); and an entire shoulder girdle belonging to a partial skeleton of an undescribed *?Coloborhynchus* from the Santana Formation (Early Cretaceous, northeastern Brazil, SMNK PAL 1113). Other specimens include: the disarticulated shoulder girdle of a skeleton of an ornithocheirid, possibly a *Coloborhynchus* (SMNK PAL 1136); a partial wing skeleton of an undetermined ornithocheirid (SMNK PAL 1250); and isolated scapulocoracoids of ornithocheirid type (SMNK PAL 1135, 1269 and 2832). The latter all come from the Early Cretaceous Santana Formation of northeastern Brazil.

From: BUFFETAUT, E. & MAZIN, J-M. (eds) 2003. *Evolution and Palaeobiology of Pterosaurs.* Geological Society, London, Special Publications, **217**, 267–274. 0305–8719/03/$15 © The Geological Society of London 2003.

The pterosaurian shoulder girdle

The shoulder girdle in pterosaurs consists of a pair of scapulae and coracoids that during ontogeny, fuse to form a single element, the scapulocoracoid (Bennett 1993). The coracoids articulate with a sternal crest (cristospine) which emerges from the cranial margin of the sternal plate. In most Cretaceous pterosaurs the apex of the scapula is cylindrical, with a medially oriented, slightly concave articular surface that locates in a convex facet on the lateral face of the notarium. The notarium is formed by the co-ossification of the spinal processes of the cranial thoracic vertebrae. Depending on the orientations and relative sizes of the scapulae and coracoids, three different types of shoulder girdle construction can be recognized among Cretaceous pterosaurs.

In one shoulder girdle construction the shaft of the scapulae are oriented sub-horizontally, describing an angle of about 140° between each other (Fig. 1a, a', d, d'). The coracoids are longer than the scapulae and are marked by a caudally directed crest. The proximal part of the coracoids is oval in cross-section, with a diameter ratio of 2:3. The coracoids meet at their articulation with the cristospine at an angle of approximately 90°. Their medial terminus is characterized by an articulation surface with a V-shaped incision which is as deep as it is wide. This incision articulates on the dorsolateral surface of a median, subtriangular (in frontal view) articulation peg arising from the dorsal surface of the base of the cristospine. The incision is oriented transversely and its cranial shank fits into a pair of dorsally oriented sockets which lie cranially adjacent to the articulation peg, while the caudal shank articulates in a vertical groove caudal to the articulation peg (Fig. 1a, a', d, d'). The articulation between the coracoids and the sternum lies level with the dorsal rim of the sternal plate. The cristospine is directed cranially and is almost as long as the sternal plate.

The glenoid fossa is level with the ventral margin of the vertebral column at the notarium. As a consequence the wings articulate in the dorsal quarter of the thorax (Fig. 3a, a'). Additionally the scapulocoracoid shows processes and crests of low relief. This shoulder girdle construction is found in Pteranodontidae and Ornithocheiridae (Eaton 1910; Wellnhofer 1991a, b; Bennett 2001) and is here referred to as the 'ornithocheirid shoulder construction' without any phyletic implications.

In a second type of shoulder girdle the scapulae and coracoids are of almost equal length and the coracoids attach to the base of the cristospine at an angle of about 150° (Fig. 1b, b'). The proximal part of the shaft of the scapulae is oriented ventrolaterally at an angle of approximately 140°, as in the ornithocheirid shoulder girdle construction. At a point two-thirds along their entire lateral extension the scapulae curve ventrally into a vertical orientation at an angle of about 70°. The glenoid fossa lies almost exactly between the ventral margin of the vertebral column and the dorsal rim of the sternum. This position of the glenoid fossa indicates a wing attachment in the mid-horizontal plane of the body (Figs 1b, b' & 3b, b'). This type of shoulder girdle, until now, has only been reported for *Quetzalcoatlus* sp., which is referred to the Azhdarchidae. Therefore we call it the 'azhdarchid shoulder construction'.

In the third type of pterosaurian shoulder girdle construction the scapulae are about one-third longer than the coracoids. The angle that is included by the contralateral scapulae is 90° and the transition to the vertical part bearing the glenoid fossa is gradual (Fig. 1c, c'). The coracoids describe an angle of 170° between each other and are almost horizontal. The glenoid fossa occupies the ventral fourth of the scapulocoracoids in lateral aspect. The wings were attached level with the dorsal rim of the sternum. This type of scapulocoracoid is known from a specimen of unidentified pterosaur from the Crato Formation (Fig. 1c, c') and from a juvenile specimen identified as *Tapejara wellnhoferi* (SMNK PAL 1137). Because of the unclear taxonomy of the Crato scapulocoracoid we call this shoulder girdle type 'tapejaroid shoulder construction', again without any phyletic implications.

Both shoulder constructions with low wing articulation are characterized by a relief of expanded crests arising from the lateral surface of both the scapulae and coracoids (Fig. 1b, c). The crests extend from the glenoid fossa dorsomedially and ventromedially and are highest close to the fossa (Fig. 1b, c). The proximal part of the coracoids is dorsoventrally flat with a length/thickness ratio of about 5.5:1. The dorsal and ventral surfaces are slightly concave. In addition the recess on the medial articulation surface of the coracoids has a U shape which is 4× to 5× wider than deep (Fig. 1e, e'). This U-shaped recess of the coracoids articulates in a dorsoventral orientation with the convex lateral surface of the base of a cristospine, which has at best one-sixth of the length of the sternal plate and points cranioventrally (Figs 1e, e' & 3c, c'). The level of articulation of the coracoids with the sternum lies below the ventral surface of the sternal plate. There are no defined articulation sockets on the cristospine as there are in the ornithocheirid shoulder construction (Fig. 1d, c', e, e').

Wing attachments and the aerodynamic consequences in aircraft

Aircraft also show three basically different wing attachment constructions: the top-, middle- and bottom-decker attachment (Fig. 2a). In top-decker

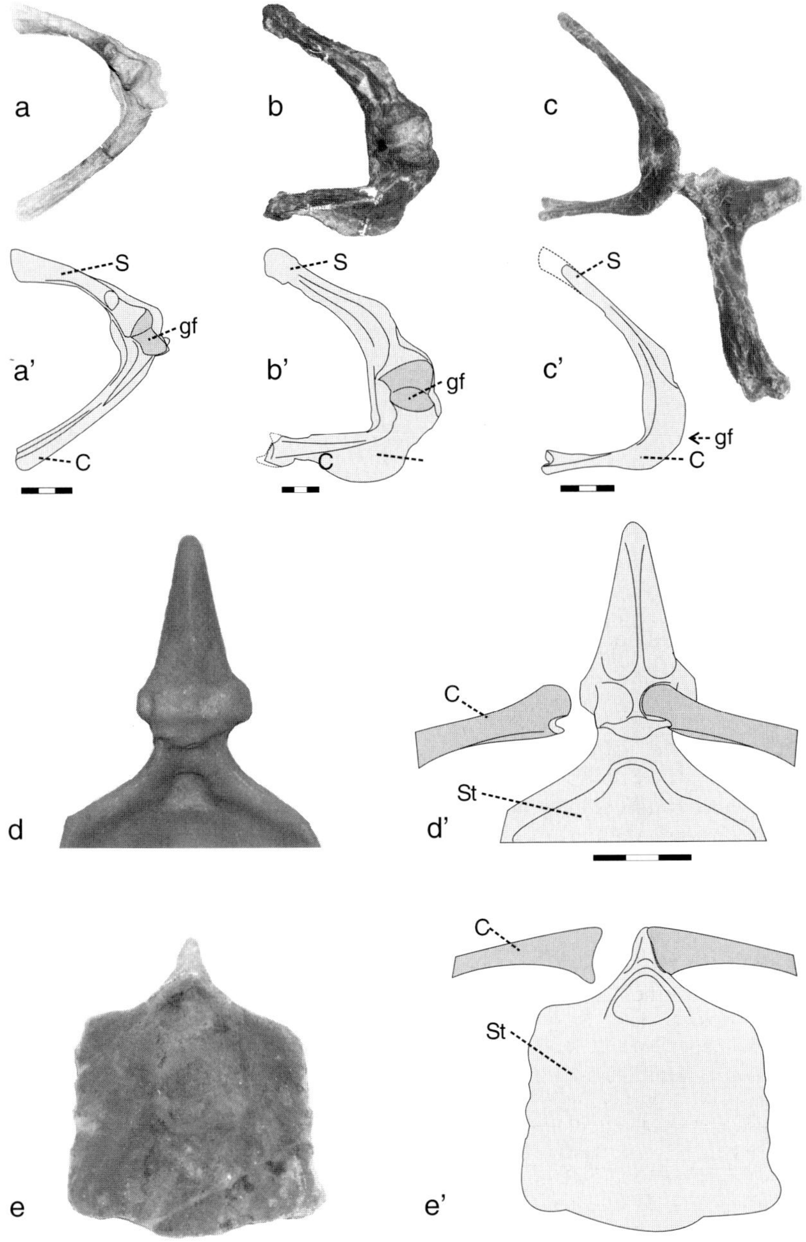

Fig. 1. Shoulder girdle constructions in pterosaurs. (**a**) Right scapulocoracoid of ?*Coloborhynchus* (SMNK PAL 1136) in caudal view. (**a′**) Line drawing of SMNK PAL 1136. (**b**) Right scapulocoracoid of *Quetzalcoatlus* sp. (TMM 42138–1) in caudal view; the dotted line marks the outline of the proximal terminus of the coracoid according to TMM 41544–25. (**b′**) Line drawing of TMM 42138–1 reconstructed. (**c**) Left scapulocoracoid of a possible tapejaroid pterosaur SMNK PAL 3843 from the Crato Formation (NE Brazil) in frontal view with the humerus articulated (the caudal surface is partly covered with matrix); the dotted line marks the outline of the proximal terminus of the scapula according to *Tapejara wellnhoferi* (SMNK PAL 1137). (**c′**) Line drawing of SMNK PAL 3843 reconstructed. (**d**) Cristospine of ?*Coloborhynchus* SMNK PAL 1136 in dorsal view. (**d′**) Line drawing of SMNK PAL 1136 with the right coracoid articulated; the left one shows the morphology of the proximal terminus. (**e**) Sternum of *Tapejara wellnhoferi* (SMNK PAL 1137) in dorsal view. (**e′**) Line drawing of SMNK PAL 1137 with the left coracoid articulated; the right one shows the morphology of its proximal terminus. C, coracoid; gf, glenoid fossa; S, scapula; St, sternum. Scale bar 50 mm.

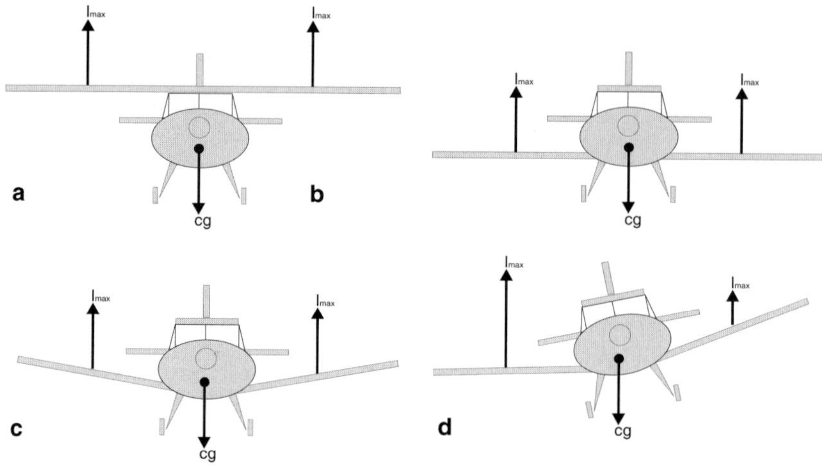

Fig. 2. (**a**) Top-decker. Relation between the effect of the maximum lift and the centre of gravity: (**b**) bottom-decker with horizontal wings, (**c**) bottom-decker with the wings in a dihedral angle, (**d**) aerodynamic effect of the lift of the dihedrally angled wings during curve flight.

aircraft the wing attachment is situated above the centre of gravity (Fig. 2a, c). This type of aircraft has automatic balance for roll movements due to the lift acting above the centre of gravity. This results in a stable flight, although these aircraft are less manoeuvrable, especially in the roll axis. The top-decker construction is seen, for example, in many training gliders (Brinkmann & Zacher 1999).

In bottom-deckers the wing attachment lies below the centre of gravity (Fig. 2a, b). If these aircraft are equipped with short straight and horizontal wings, they tend to roll easily, which requires greater control. Consequently manoeuvrability is increased at the cost of flight stability (Schlichting & Truckenbrodt 2001). Flight stability can be enhanced by increasing the wing span or by arranging the wings in a dihedral angle, as is seen, for example, in motor gliders (Fig. 2c; Brinkmann & Zacher 1999).

The stabilizing effect of dihedrally angled wings comes from the elevation of the centre of lift into (in bottom-deckers) or above the centre of gravity (in middle-deckers), but mainly because a rotation around the roll axis results in decrease of lift in that wing, which moves up (Fig. 2d, e; Schlichting & Tuckenbrodt 2001).

In middle-decker aircraft the wings are attached level with the centre of gravity. If the wings are oriented horizontally this type of aircraft is stable in all positions of the roll axis and thus is especially suitable for aerobatics. Again, as in a bottom-decker, a dihedral angling of the wings results in stable flight behaviour.

Pterosaurs as flying machines

Unlike aircraft, pterosaurs could move their wings in the horizontal and vertical plane. They could pronate and supinate them, alter the aspect ratio (square of the wing span divided by the wing area) by adduction and extension of the arm and the wing finger, and change the cambering of the wing with movements of the pteroid bone (and possibly the small fingers) in the area of the propatagium and with the hindlegs in the area of the brachiopatagium and uropatagium (Frey *et al.* 2003). They could also generate thrust with wing beats (Short 1914; Kripp 1943; Holst 1957; Heptonstall 1971; Bramwell & Whitfield 1974; Stein 1975; Frey & Riess 1981; Pennicuik 1988; Wellnhofer 1991b; Bennett 2000). Despite these differences, it is still possible to understand some of the aerodynamic effects of the three different shoulder constructions by comparing pterosaurs with flying machines.

If there is no or only little variation of gravity within an object, then the centre of mass (*cm*) is coincident with the centre of gravity (*cg*). In aircraft engineering *cg* is used to describe that point of flying machine in which its entire weight is balanced. Thus the mass has to be in buoyancy in *cg* too. In an atmospheric aircraft *cg* has to lie in line with the line of maximum lift (l_{max}), which is continuous with the maximum curvature of the top surface of the wing profile. The lift force is a vector directed into the opposite direction of gravity and has to exceed the maximum weight. If l_{max} lies in front of *cg* the aircraft is nose-heavy and would fly in a hyperbolic curve to the ground. If l_{max} lies behind *cg* an aircraft is tail-heavy and would fly in hopping movements. If *cg* moves away from l_{max}, e.g. when the load changes due to fuel consumption or differing passenger or baggage loads, an aircraft has to be trimmed. Trimming is a weight compensation which brings *cg* level with l_{max}. The main spar, with which the wing is attached to the fuselage in most gliders and other aircraft, lies level with l_{max}, because lift is then transmitted to the fuselage with minimum torque momentum.

For any flying machine two planes are defined to determine the vertical axis on which *cg* must lie; the cross-line between the median vertical plane and a transverse plane through l_{max} which is marked by the anchoring place of the main spar with the fuselage, where the force transmission occurs. The actual position of *cg* on this line is, by definition of *cg* dependent from the mass distribution.

Flying pterosaurs are subject to exactly the same physical principles concerning the position of *cg*. As in aircraft, the pterosaur must lie on the cross-line between the vertical median plane and a transverse plane where l_{max} is transmitted to the body: the vertical transverse plane through the anchoring areas of the wings, i.e. the contralateral glenoid fossae. As a consequence the masses of all organs and appendages necessarily have to be coincident on either side of both these planes during flight for adequate balance. During slight horizontal alterations of *cg*, trimming can be accomplished by minor adjustments of the appendages (arms and fingers, legs and toes, tail, neck and head; see Frey *et al.* 2003). However, the vertical position of *cg* on the cross-line of the two above-mentioned planes varied with the mass and the physiological state. Therefore, the position of *cg* can never be precisely determined by all three spatial coordinates, though its average level can be roughly estimated.

While the mass distribution of all internal organs and fluids in the body as well as the body muscles of pterosaurs is likely to be approximately equal, the distribution of the wing muscle masses appears to vary with the construction of the shoulder girdle (see below, Fig. 3). In any case, *cg* lies inside the chest, as in all Recent flying animals (e.g. Gray 1953; Pennicuik 1975; Norberg 1981; Hildebrand 1982; Nachtigall 1986; Norberg & Rayner 1987).

In pterosaurs with the ornithocheirid type of shoulder construction the flight muscles were dominated by the adductor masses of the arm, as can be concluded from their absolute length alongside the coracoideal pivot and their diameter as is indicated by the size of the attachment areas on the deltoid process of the humerus, the sternal plate and the cristospine. Compared to the adductor masses, the abductors appear to have been small, possibly not exceeding one-third the mass of the adductors. This brings most of the mass of the flight muscles ventral to the glenoid fossa. Therefore, the wings very probably articulated above *cg*, which makes the ornithocheirid construction a top-decker. With the wing articulation above the centre of gravity, roll movements are passively stabilized. The length of the wing was about 2.5× the length of the hindleg (Fig. 4a, a'). Because the trailing edge of the brachiopatagium was probably attached to the ankle in most pterosaurs (Unwin & Bakhurina 1994; Frey *et al.* 2003, Figs 3a,b, Fig. 9a, Fig. 18), the length of the hindlegs determines the aspect ratio of the wing, which in ornithocheirid-type pterosaurs was high: about 30 (Fig. 4a, a'). As in gliders, high aspect ratios enhance flight stability at the cost of manoeuvrability. Therefore, ornithocheirid-type pterosaurs show key features of stable gliders and were well able to control their flight with small adjustments of the wings made by the arms and legs (Brinkmann & Zacher 1999). Thermal, slope and dynamic gliding was certainly possible, as was postulated for example by Hankin & Watson (1914) and Bramwell & Whitfield (1974). The suggestion that pterosaurs of an ornithocheirid construction relied mostly on gliding is supported by the position of the deltoid process on the proximal third of the humerus. This and the vertically oriented coracoid allows the reconstruction of long-fibred wing adductor musculature extending from the deltoid crest to the ventral face of the sternum and to the lateral face of the cristospine (Fig. 3a, a'). Long-fibred wing adductor muscles would have produced a deep power stroke for the wing (Wellnhofer 1991a). However, the smaller (as reconstructed) adductor muscle diameters relative to the azhdarchid type (see below) did not permit a strong power upstroke. According to the low bone relief of the scapulocoracoid, the upstroke and the wing adjustment muscles must have been small and weak (Hankin & Watson 1914; Bennett 2003). Thus, much of the performance for flight in pterosaurs with the ornithocheirid construction was probably achieved by gliding and dynamic soaring, rather than flapping.

In pterosaurs of the azhdarchid and tapejaroid types prominent crests on the shoulder girdle and an enormous deltoid process on the humerus indicate a powerful flight musculature (Fig. 3b, c; Bennett 2003). However, due to the small cristospine, the sternal attachment surface for the adductor muscles is smaller than in an ornithocheirid construction of equal size but has a wider diameter at the insertion area of the cranioventrally expanded deltoid crest. Probably the overall mass of the wing adductors was equivalent or even slightly less than that of an ornithocheirid construction of equal body size. The wing abductors of the azhdarchid and tapejaroid constructions were likely to have been longer, independent of the length of the scapulae, and more powerful due to the crest relief of the scapula in comparison with an ornithocheirid construction. This construction allowed an increased muscle mass dorsal to the wing attachment relative to the ventrally situated adductor

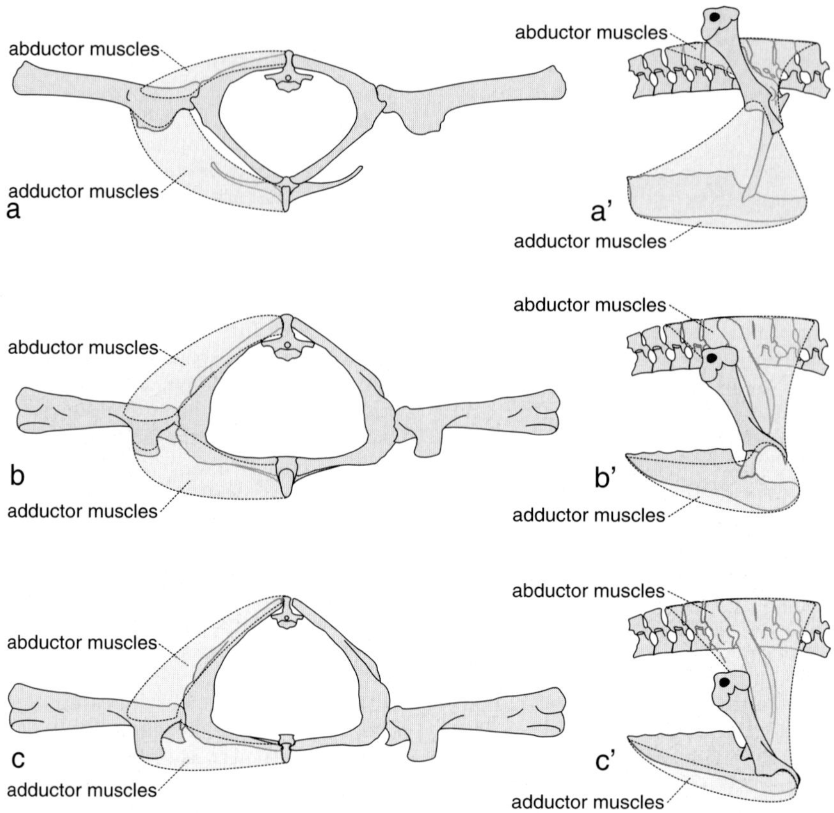

Fig. 3. Schematic reconstructions of three different types of wing attachment in pterosaurs. The upstroke and downstroke muscle masses are reconstructed according to Wellnhofer (1991a) and Bennett (2003). (**a**) Top-decker construction (ornithocheirid type) in frontal view, (**a′**) lateral view of the same. (**b**) Middle-decker construction (azhdarchid type) in frontal view, (**b′**) lateral view of the same. (**c**) Bottom-decker construction (tapejaroid type) in frontal view, (**c′**) lateral view of the same.

masses. This arrangement results in *cg* being higher than in an ornithocheirid construction. The *cg* lay at least level with the glenoid fossa in the azhdarchid and definitely above it in the tapejaroid construction. The azhdarchid construction therefore would represent a middle-decker and the tapejaroid a bottom-decker construction.

The wing articulation in or ventral to *cg* resulted in unstable, but more manoeuvrable flight when the wings were held horizontally or angled ventrally. Stabilization in the roll axis could have been achieved by bringing the wings into a dihedral angle (Fig. 2d, e). L_{max} could then be brought above *cg* and roll movements compensated for by an increase of lift at the wing facing in the direction of the roll (Fig. 2e). Pterosaurs of the azhdarchid and tapejaroid type could glide under stable conditions with their wings held in this V-position. They also could actively destabilize their flight by lowering their wings, e.g., for quick changes of direction.

The wing span of azhdarchid and tapejaroid pterosaur construction is smaller with respect to the body length than in an ornithocheirid construction. In contrast to the latter, the hindlimbs of the former almost reached half the length of a wing, with the femur being always longer than the humerus (Fig. 4b, b'). The consequence would be a low aspect ratio of approximately 20, indicating poor gliding performances compared with the ornithocheirid construction (Brinkmann & Zacher 1999). This, and the powerful upstroke muscles, indicate that the pterosaurs characterized by a ventrally located wing attachment were probably able to perform prolonged flapping flight using a powerful down and upstroke.

Conclusions

The 'giant wings' of the Cretaceous comprised at least three distinct anatomical types with differing

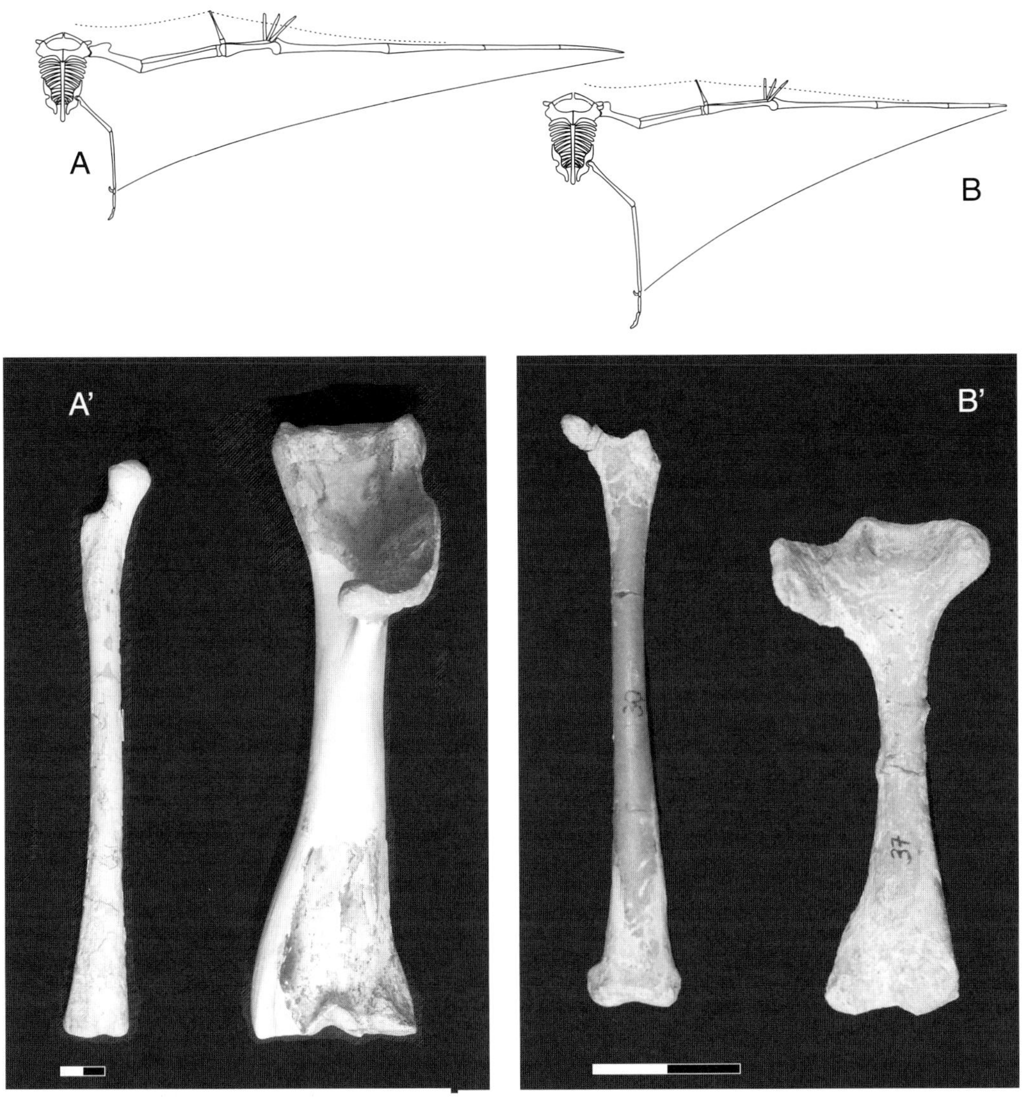

Fig 4. Body proportions in an ornithocheirid (**a**) and (**b**) an azhdarchid or tapejaroid construction (redrawn from Wellnhofer 1991a). The proportional difference of both constructions is also seen in the ratio of the length of the humerus versus the length of the femur. (**a′**) humerus and femur of the ornithocheirid pterosaur (?*Coloborhynchus*) SMNK PAL 1136; (**b′**) humerus and femur of *Tapejara wellnhoferi* (SMNK PAL 1137).

flight capabilities: the ornithocheirid-, azhdarchid- and tapejaroid-type pterosaurs. These differing flight capabilities are consistent with differences in the wing musculature which should now be subject to detailed studies building on the work of Bennett (2003) and examining the aspect ratio of the various pterosaur wing types. The ecological implications of this discovery require investigation, as does the origin of both types of flight apparatus. The question arises whether or not the azhdarchid type of wing evolved from a primary tetrapod condition, in which the glenoid fossa is level with the sternum, or from an ornithocheirid-type shoulder girdle. The latter hypothesis would require explanations of why, how and under what conditions such a constructional change could have occurred. Possibly both types of shoulder girdles observed in Cretaceous pterosaurs evolved from an intermediate condition. In order to solve this question the investigation of Triassic and Jurassic pterosaurs with respect to the type of wing attachment is essential.

We thank R. Kastner, D. Bauer and A. Anders (all Karlsruhe), who did fantastic preparation work on the specimens. Special thanks go to W. Langston Jr (Austin) for his hospitality during the investigation of the Big Bend pterosaur material. F. Buchy (Villepreux), R. Benischke (Karlsruhe) and T. Hubel (Darmstadt), as physicians with engineering competence, patiently discussed with us all kinds of centres (cm, cg). D. M. Henderson's (Calgary) critical comments helped enormously to improve the manuscript. V. Griener (Karlsruhe) did the photography.

References

BENNETT, S. C. 1993. The ontogeny of *Pteranodon* and other pterosaurs. *Palaeobiology*, **19**, 92–106.

BENNETT, S. C. 2000. Pterosaur flight: The role of actinofibrils in wing function. *Historical Biology*, **14**, 255–284.

BENNETT, S. C. 2001. The osteology and functional morphology of the Late Cretaceous pterosaur *Pteranodon*. Part I. Size and functional morphology. *Palaeontographica A*, **260**, 113–153.

BENNETT, S. C. 2003. Morphological evolution of the pectoral girdle of pterosaurs. *In*: BUFFETAUT E. & MAZIN, J-M. (eds) *Evolution and Palaeobiology of Pterosaurs*. Geological Society, London, Special Publications, **217**, 119–215.

BRAMWELL, C. D. & WHITFIELD, G.R. 1974. Biomechanics of *Pteranodon*. *Philosopical Transactions of the Royal Society, London (B)*, **267**, 503–581.

BRINKMANN, G. & ZACHER, H. 1999. *Die Deutsche Luftfahrt. Die Evolution der Segelflugzeuge*. Bernard & Graefe Verlag, Bonn, 290 pp.

BURTON, R. 1990. *Bird Flight. An Illustrated Study of Birds' Aerial Mastery*. Facts On File, New York, Oxford, Sydney, 160 pp.

CHILDRESS, S. 1981. *Mechanics of Swimming and Flying*. Cambridge University Press, Cambridge.

EATON, G. F. 1910. *Osteology of* Pteranodon. Connecticut Academy of Arts and Sciences, New Haven, Memoirs, 38 pp.

FREY, E. & RIESS, J. 1981. A new reconstruction of the pterosaur wing. *Neues Jahrbuch für Geologie und Paläontologie, Abhandlungen* **161**, 1–27, Stuttgart.

FREY, E., TISCHLINGER, H., BUCHY, M.-C. & MARTILL, D. M. 2003. New specimens of Pterosauria (Reptilia) with soft parts with implications for pterosaurian anatomy and locomotion. *In*: BUFFETAUT E. & MAZIN, J-M. (eds) *Evolution and Palaeobiology of Pterosaurs*. Geological Society, London, Special Publications, **217**, 233–266.

GRAY, J. 1953. *How Animals Move*. Cambridge University Press, Cambridge, 114 pp.

GRAY, J. 1968. *Animal Locomotion*. Weidenfeld & Nicholson, London, 479 pp.

HANKIN, E. H. & WATSON, D. M. S. 1914. On the flight of pterodactyles. *Aeronautical Journal*, **72**, 1–12.

HILDEBRAND, M. 1982. *Analysis of Vertebrate Structure*, 2nd edition. John Wiley & Sons, New York, Chichester, Brisbane, Toronto, Singapore, 654 pp.

HILL, J. H. & SMITH, J. D. 1984. *Bats. A Natural History*. British Museum (Natural History), London, 243 pp.

HEPTONSTALL, W. B. 1971. An analysis of the flight of the Cretaceous pterosaur *Pteranodon ingnes* (Marsh). *Scottish Journal of Geology*, **7**, 61–78.

HOLST, E. 1957. Der Saurierflug. *Paläontologische Zeitschrift*, **31**, 15–22.

KRIPP, D. 1943. Ein Lebensbild von *Pteranodon ingens*. *Nova Acta Leopoldina*, **12**, 217–286.

NACHTIGALL, W. 1986. *Bat Flight – Fledermausflug*, Biona Report 5. Gustav Fischer Verlag, Stuttgart.

NORBERG, U. M. 1981. Flight, morphology and the ecological niche in some birds and bats. *Symposia of the Zoological Society of London*, **48**, 173–197.

NORBERG, U. M. & RAYNER, J. M. V. 1987. Ecological morphology in bats (Mammalia; Chiroptera): wing adaptations, flight performance, foraging strategy and ecolocation. *Philosophical Transactions of the Royal Society, London (B)*, **316**, 335–427.

PADIAN, K. 1983. A functional analysis of flying and walking pterosaurs. *Paleobiology*, **9**, 218–239.

PENNICUIK, C. J. 1972. Soaring behaviour and performances of some East African birds from a motor glider. *Ibis*, **114**, 178–218.

PENNICUIK, C. J. 1975. Mechanics of flight. *In*: FARNER, D. S. & KING, J. R. (eds) *Avian Biology*. Academic Press, London, New York, Toronto, Sydney, San Francisco, vol. 5, 1–73.

PENNICUIK, C. J. 1988. On the reconstruction of pterosaurs and their manner of flight, with notes on vortex wakes. *Biological Review*, **63**, 299–331.

PENNICUIK, C. J. 1989. *Bird Flight Performance. A Practical Calculation Manual*. Oxford University Press, Oxford. 153 pp.

SALOMON, F-V. 1993. *Lehrbuch der Geflügelantomie*. Gustav Fischer, Jena, Stuttgart, 479 pp.

SCHEFFEN, W. 1926. Flugsaurier und Segelflug. *Natur und Museum*, **56**, 198–207.

SCHLICHTING, H. & TRUCKENBRODT, E. 2001. Aerodynamik des Tragflügels (Teil III), des Rumpfes, der Flügel-Rumpf-Anordnung und der Leitwerke. *In*: *Aerodynamik des Flugzeuges* **2**, 3rd edition, Springer-Verlag, Berlin, 514 pp.

SHORT, G. H. 1914. Wing adjustments of pterodactyls. *Aeronautical Journal*, **72**, 13–20.

STEIN, R. S. 1975. Dynamic analysis of *Pteranodon ingens*, a reptilian adaption to flight. *Journal of Paleontology*, **49**, 534–548.

UNWIN, D. M. & BAKHURINA, N. N. 1994. *Sordes pilosus* and the nature of the pterosaur flight apparatus. *Nature*, **371**, 62–64.

VAUGHAN, T. A. 1959. *Functional Morphology of Tree Bats*: Eumops, Myotis, Macrotus. Natural History Museum of Kansas, Publications, **12**, 153 pp.

VAUGHAN, T. A. 1966. Morphology and flight characteristics of molossid bats. *Journal of Mammology*, **47** (2), 249–260.

WELLNHOFER, P. 1978. Pterosauria. *In*: WELLNHOFER, P. (ed.) *Handbuch der Paläoherpetologie*. Gustav Fischer, Stuttgart, Teil 19, 82 pp.

WELLNHOFER, P. 1991a. *The Illustrated Encyclopedia of Pterosaurs*. Salamander, London. 192 pp.

WELLNHOFER, P. 1991b. Weitere Pterosaurierfunde aus der Santana-Formation (Apt) der Chapada do Araripe, Brasilien. *Palaeontographica A*, **215**, 43–101.

Pterosaur tracks from the latest Campanian Cerro del Pueblo Formation of southeastern Coahuila, Mexico

RUBEN A. RODRIGUEZ-DE LA ROSA

Laboratorio de Paleontología, Museo del Desierto, and Coordinación de Paleontología, SEPC, AP 307, C.P. 25000, Saltillo, Coahuila, México (e-mail: ruben_raptor@hotmail.com)

Abstract: A new vertebrate tracksite in southeastern Coahuila, northern México, known as the El Pelillal tracksite (Latest Campanian, Cerro del Pueblo Formation) has yielded an important ichnofauna that includes the tracks of turtles, crocodilians, a small theropod dinosaur, a mammal-like organism and the tracks of pterosaurs. The pterosaurian manus impressions share an extraordinary similarity, in size and morphology, with the Jurassic ichnogenus *Pteraichnus* Stokes 1957 and are referred to *Pteraichnus* sp.; however, the pes impressions differ in that they are rather very elongated. According to the geological features and ichnofauna the El Pelillal tracksite represents a shallow, freshwater or lacustrine deposit, which agrees with our knowledge about the habitat preference of Cretaceous pterosaurs. This newly discovered tracksite in southeastern Coahuila, Mexico, offers great potential for palaeoichnological research and is thus becoming important in the understanding of the palaeoecosystems during the Late Cretaceous in southern North America.

Introduction

The Mesozoic vertebrate track record in México is mainly dominated by dinosaurian tracks (Ferrusquía-Villafranca *et al.* 1978a, 1978b, 1993, 1995, 1996a,b; Aguillón-Martínez *et al.* 1998; Rodríguez-de la Rosa 1998; Ortíz-Mendieta *et al.* 2000). Recently new Late Cretaceous vertebrate tracksites have been discovered in the southeastern part of the State of Coahuila; these have yielded important ichnofaunas that include some of the first non-dinosaurian vertebrate tracks in Mexico (Rodríguez-de la Rosa *et al.* 2002).

One of these tracksites is now known as the El Pelillal Tracksite (Fig. 1). It was discovered by an amateur archaeologist in 1997 close to the town of Reata, in the Municipio de Ramos Arizpe; up to that time an isolated turtle track was the only specimen known from this site (Rodríguez-de la Rosa & Rodríguez-Garza 1998). In a subsequent visit, in the year 2000, tracks of invertebrates, turtles, crocodiles, a small theropod dinosaur, a couple of tracks similar to those of the ichnogenus *Brasilichnium* and those of pterosaurs were discovered.

The outcrop is composed of a reddish to light-brown, fine-grained, intensely bioturbated and ripple-marked sandstone that in some areas bears root traces. It is located stratigraphically high in the sequence of the Cerro del Pueblo Formation of the Late Campanian. This age is based on the stratigraphic position, lithology and correlation, as well as the fossil content of this formation (Murray *et al.* 1962; McBride *et al.* 1974; Vega-Vera *et al.* 1990).

The main purpose of this paper is to describe the pterosaurian tracks from the El Pelillal tracksite. They

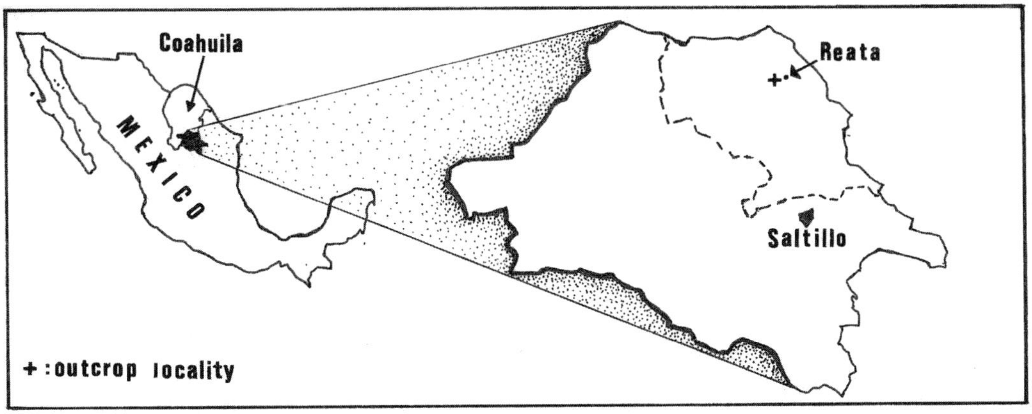

Fig. 1. Geographic location of the El Pelillal Tracksite (+) in southeastern Coahuila, Mexico. Dotted lines outline the area of the Municipio de Ramos Arizpe.

From: BUFFETAUT, E. & MAZIN, J-M. (eds) 2003. *Evolution and Palaeobiology of Pterosaurs.* Geological Society, London, Special Publications, **217**, 275–282. 0305–8719/03/$15

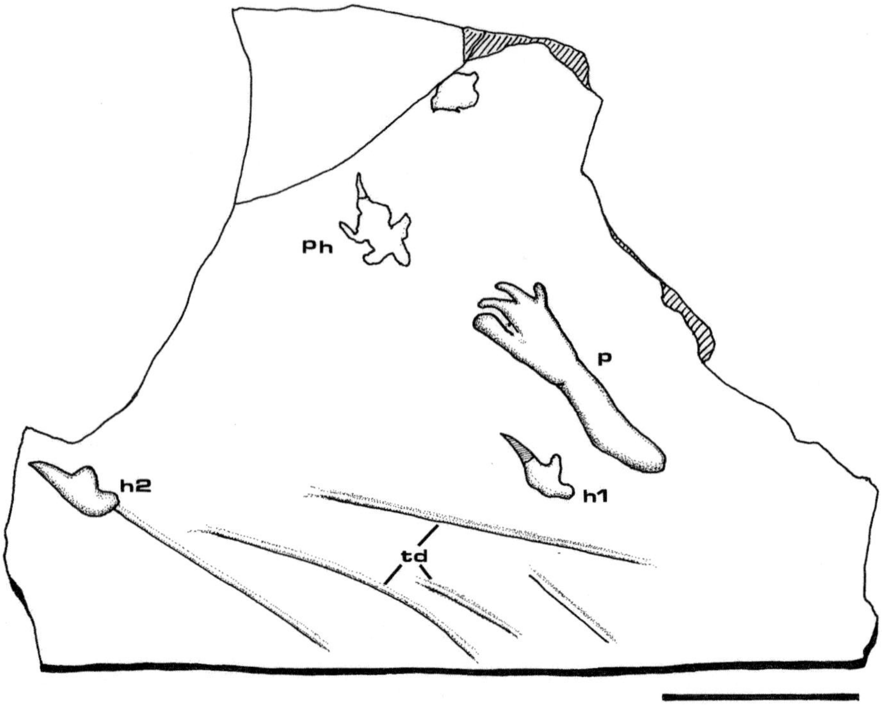

Fig. 2. SEPCP-48/250. Slab containing *Pteraichnus* sp. manus prints (h1, h2), a single pes print (p) and marks that are interpreted as tail drags (td) of unknown origin. Ph, possible overimpressed manus impressions. Scale bar 10 cm.

represent the first record of pterosaur tracks from México and became part of the increasing knowledge about Late Cretaceous pterosaur palaeoichnology.

Materials and method

Most of the drawings of the tracks were made on a plastic sheet with waterproof ink markers. The precise location of the El Pelillal tracksite is available directly from the files of the Coordinación de Paleontología, SEPC, and the Laboratory of Paleontology of the Museo del Desierto-Coahuila, in order to protect the site as well as the interests of the landowners. All the specimens are housed in the Paleontological Collection of the Coordinación de Paleontología de la Secretaría de Educación Pública de Coahuila (SEPCP) under the catalogue numbers SEPCP-48/250 to SEPCP-48/267 and SEPCP-1/268, SEPCP-1/269, SEPCP-51/270 to 274, SEPCP-1/275, and SEPCP-48/276.

Systematic description

Ichnofamily Pteraichnidae Lockley *et al.* 1995
Ichnogenus *Pteraichnus* Stokes 1957
Species *Pteraichnus* sp. (Figs 2–5)

Material. SEPCP-48/250 to SEPCP-48/267 and SEPCP-1/268, SEPCP-1/269, SEPCP-51/270 to 274, SEPCP-1/275, and SEPCP-48/276.

Description

Manus (Figs 2, 3). With the exception of a manus preserved on SEPCP-48/251-A (Fig. 5) most of the manus prints are found unassociated with pes prints, thus forming a manus-pes set or a trackway. The manus prints are tridactyl, with a strong asymmetry, and show the typical pterosaurian configuration (Fig. 3). These are deeply impressed in the sediment in comparison with the known pes prints (cf. Lockley *et al.* 1995).

The width of the manus prints ranges from 2.0 to 4.7 cm, mean of 3.4 cm; while the length ranges from 4.1 to 9.0 cm, mean of 6.9 cm.

The manus shows an anterior digit, or digit I, that is usually short, with a distal outline that varies from suboval to slightly subtriangular, sometimes blunt.

The middle digit, or digit II, is short to slightly elongated, the distal outline is generally suboval, some times with a blunt distal end.

The posterior digit, or digit III, is elongated with a distal end that ranges from rounded to fusiform in shape.

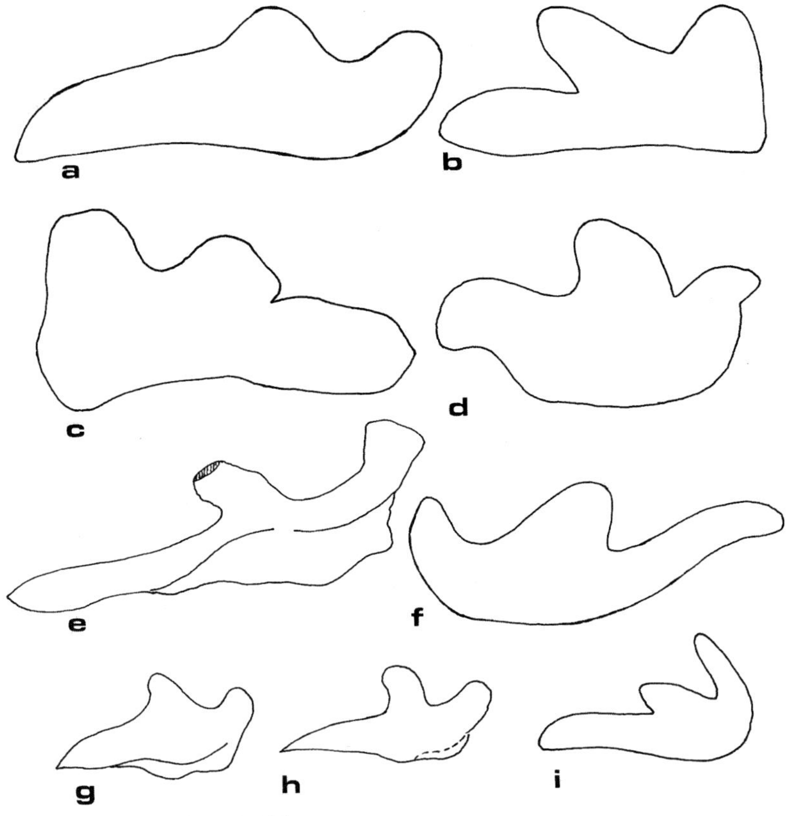

Fig. 3. *Pteraichnus* sp. manus impressions from the El Pelillal Tracksite. (**a**) SEPCP-48/251-A; (**b**) SEPCP-48/252; (**c**) SEPCP-48/264; (**d**) SEPCP-48/263; (**e**) SEPCP-48/255; (**f**) SEPCP-48/256; (**g**) SEPCP-48/250; (**h**) SEPCP-48/250; (**i**) SEPCP-48/257-A. Scale bar 5 cm.

Some manus prints bear a shallow concave area at the metacarpal–phalangeal joint (Fig. 3e).

The mean angle between digits I and II is 39.1° and between digits II and III is 63.9°.

Some manus prints are slender or thinner; however, this represents extramorphological variation due to collapse of the sediment. In these later prints the interdigital angulation is greater being of 89.4° between digits I and II and 44.3° between digits II and III.

Pes (Fig. 4)

At least three pes prints are preserved, the best pes print (Figs 2, 4) is preserved on SEPCP-48/250. This slab bears a supposed left print in convex hyporelief, with four digits, a rounded and well-defined plantar pad and a posterior projection (Fig. 4).

The total length of the footprint is 13.2 cm. The whole pes print is 4.6 cm in its maximum width (across the digits) and 2.0 cm posteriorly.

The distal portion of the pes print is subtriangular in shape, with a slightly rounded, well-defined plantar pad, and digits II and III slightly longer than digits I and IV (Fig. 4).

Digit I is the most conspicuous (Fig. 4), it is massive in comparison to digits II, III and IV. The first digit has a straight medial side and rounded anterior and lateral sides.

Digits II and III are more or less of equal dimensions and nearly straight; however, they curve anteriorly to the medial side and both have anterior rounded ends.

Digit IV is short, with a rounded anterior end and it is directed anterolaterally.

The points of divergence of the four digits are not the same. There is a large, subrectangular indentation between digits I and II, giving a thumb-like appearance to digit I. Digits II and III diverge more anteriorly, and it is the same situation for digit IV, although with a slightly more posterior point of divergence. The overall angulation between the impressions of digits I and IV is 35°.

The pes print bears a well-defined plantar pad, but a more or less straight posterior metatarsal impression

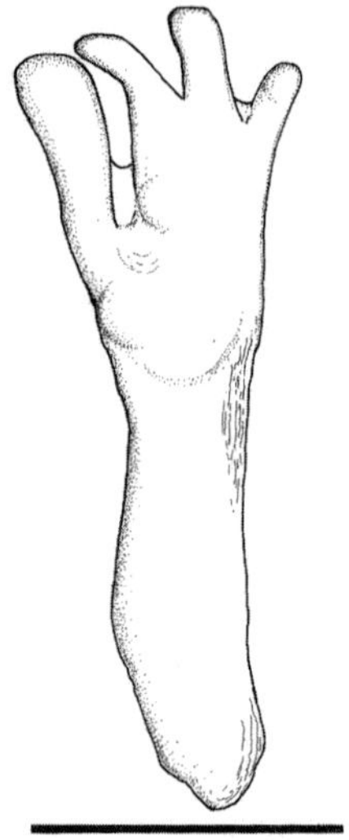

Fig. 4. *Pteraichnus* sp. pes impression preserved on SEPCP-48/250. Note the elongate pes morphology. Scale bar 5 cm.

is preserved, which is c.2.0 cm wide and c.7.0 cm long (Fig. 4). A second impression of the metatarsi is preserved on SEPCP-48/251-A (Fig. 5); it is 1.38 cm wide and 8.41 cm long, nearly straight, and opens into a small, fan-like area. This second metatarsal impression is associated with a manus print forming the typical pterosaurian manus-pes set (Fig. 5).

A third pes print was recently found in a small slab that became part of SEPCP-48/250. This pes print is composed of two areas. The distal part is subtriangular in shape and bears the impression of four digits; digits I and II are bulbous and of more or less equal dimensions, digits III and IV are of smaller size and of equal dimensions, which gives a human-like appearance to this pes print. The proximal part of the pes print represents the elongate metatarsi, but, in this area, the sediment became collapsed. This pes print is 12.9 cm long and 6.5 cm wide across the digits.

Discussion

Preliminary geological observations at the El Pelillal Tracksite suggest a shallow freshwater or lacustrine environment (Fig. 6); this is also ichnologically supported by the presence of invertebrate bioturbation, root traces and tracks of turtles, crocodilians, a small theropod dinosaur and a mammal-like organism (e.g. a *Brasilichnium*-like track). This ichnofaunal association also supports the notion of cosmopolitan pterosaurs in regard to their habitat (Calvo & Lockley 2001).

The presence of pterosaur tracks in the Late Cretaceous of southeastern Coahuila also agree with the observation of several authors that Cretaceous pterosaurs, and particularly pterosaurian tracks, are more frequently associated with deposits that represent freshwater and/or lacustrine environments (Kellner 1994; Lehman 1996; Lockley *et al.* 1995; 1997, 2001; Lockley 1998, 1999; Calvo & Lockley 2001).

The El Pelillal Tracksite is a manus-dominated pterosaurian assemblage that bears several manus prints against three pes prints. Other manus-dominated pterosaurian assemblages similar to that of the El Pelillal Tracksite are present in localities around the world, as in North America (e.g. Summerville and North Horn formations of Utah), Europe (e.g. Spain; Lockley *et al.* 1995) and Asia (Lockley *et al.* 1997).

This situation is explained by the fact that the centre of gravity of pterosaurs during quadrupedal walking was located anteriorly (Lockley *et al.* 1995, 2001; Bennett 1997); this resulted in having more deeply impressed manus prints and/or manus-dominated assemblages.

Lockley *et al.* (1995) suggested that the three preserved digit impressions on *Pteraichnus* manus correspond to digits II, III and IV. However, a more conservative proposal was that of Bennett 1997, based on anterior, middle and posterior digit inpressions. In fact, Bennett 1997 noted claw impressions on the manus posterior digit on a trackway of *Pteraichnus stokesi*. In this way, the correlation anterior digit I, middle digit II and posterior digit III seems to have some support. In fact, some findings from France and Utah show *Pteraichnus*-like manus prints with the impression of digit IV (Mazin *et al.* 1995; Lockley *et al.* 2001).

The difference in size of the manus prints from southern Coahuila surely corresponds to pterosaurs of different ages and the overall manus morphology strongly recalls that of *Pteraichnus stokesi* Lockley *et al.* 1995. However, the only pes print, or at least the elongate metatarsi impression, associated with a manus print (SEPCP-48/251-A, Fig. 5), shows a positive rotation similar to that observed in *P. saltwashensis* (Stokes 1957; Lockley *et al.* 1995); this condition clearly differs from the strong positive rotation observed in trackways of *P. stokesi*; however, this could be due to different modes of walking by individuals of related species.

Some manus prints show a great angulation between digits I and II, forming an obtuse angle (see Fig. 3e). This angle differs from the square and/or acute angle present in other pterosaurian ichnospecies. This interdigital angulation could have some taxonomical value for naming a new ichnospecies; however, for the time being, it is possible to assign these tracks to *Pteraichnus* sp.

An important aspect of the El Pelillal Tracksite is

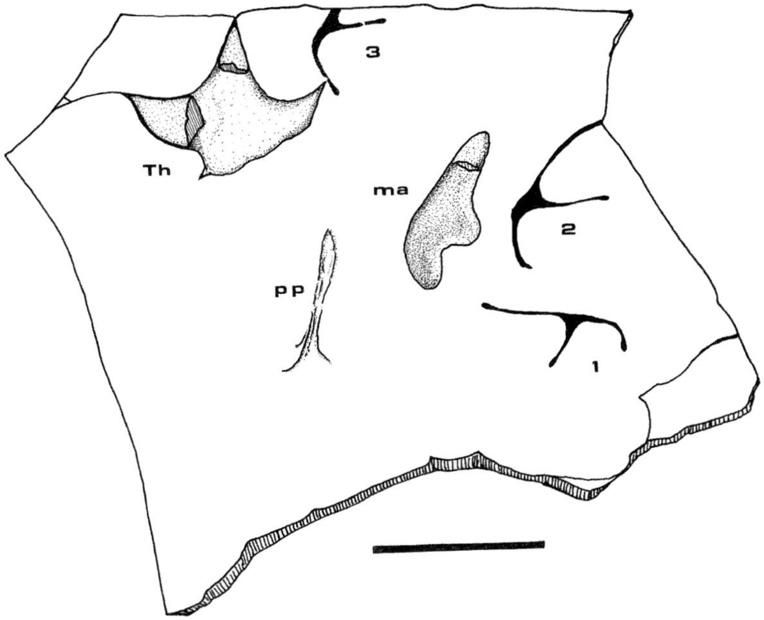

Fig. 5. SEPCP-48/251-A. Slab containing the only known manus-pes set (ma + pp) of *Pteraichnus* sp. from the El Pelillal Tracksite; however, the pes (pp) is represented just by the metatarsi impression. ma, manus impressions; th, theropod track; 1–3, pterosaurian manus impressions that show sediment collapse. Scale bar 10 cm.

that it adds a fifth record of Cretaceous pterosaur pes tracks, which adds to our understanding of pterosaurian pes morphology. Other Cretaceous tracksites containing pterosaur pes tracks are located in the Lower Cretaceous of Santa Cruz de las Yanguas, Spain (cf. Lockley *et al.* 1995); the Lower Cretaceous of Dorset, United Kingdom (Wright *et al.* 1997); the Lower Cretaceous of Neuquen Province, Argentina (Calvo & Lockley, 2001); the Late Cretaceous of Chollanam Province, South Korea (Lockley *et al.* 1997), and the Late Cretaceous of Emery County in Utah (Lockley 1999).

A remarkable feature of the pes tracks from southeastern Coahuila (SEPCP-48/250, SEPCP-48/251-A) is that they are very elongated. In regard to this, it is interesting to note that most of the Cretaceous pterosaur pes prints known to date are also very elongated (e.g. pes prints from Korea and Argentina, Lockley *et al.* 1997; Calvo & Lockley 2001).

Pterosaurs have been portrayed as habitual bipeds, facultative quadrupeds, facultative bipeds and habitual quadrupeds; either plantigrade and/or digitigrade (Padian 1983, 1996; Wellnhofer 1991). However, one of the best analyses of pterosaur locomotion has been given by Bennett 1997, and there are some authors that support his interpretations of a quadrupedal-plantigrade organism (Lockley *et al.* 1995, 2001; Clark *et al.* 1998).

Unwin 1989 proposed some pes morphologies based on osteological characters; some of his drawings (e.g. Unwin 1989; fig. 27.6b–c) show an elongate pes print similar to that observed in the El Pelillal samples. If we add a well-defined plantar pad, at the metatarsal–phalangeal joint, to some of the Unwin´s interpretations the similarity between this and the pes prints from El Pelillal Tracksite is striking.

In spite of the elongate metatarsal impression preserved in the El Pelillal pes prints, no trace of the conspicuous metatarsal and/or digit V is found. However, they would be expected for some pterodactyloid tracks and, in fact, their presence has been suggested (Lockley *et al.* 1997). However, the situation of the El Pelillal tracks can be explained by the fact that some authors have suggested that this structure (metatarsal and/or digit V) was embedded in the soft tissue (e.g. Wellnhofer 1978, 1991; Frey and Tischlinger 2000).

The elongate pes morphology observed in the Cretaceous of the El Pelillal Tracksite could be of some taxonomical and/or ecological significance in future studies of pterosaur palaeoichnology. In regard to this idea, it is interesting to note that the elongate pes prints from Neuquen (Calvo & Lockley 2001) could be related to pterosaurs such as *Pterodaustro*. In this way, the filter-feeding strategies of this bizarre pterosaur could be related to an elongated pes morphology capable of providing a stable stance while feeding. This could have been the situation for other Late Cretaceous pterosaurs that,

Fig. 6. Palaeoecological reconstruction of the El Pelillal Tracksite (Latest Campanian, Cerro del Pueblo Formation) in southeastern Coahuila, México. The evidence suggests the coexistence of a pterosaurian taxon (represented here by an ornithocheirid pterosaur and those seen on distance), a small theropod dinosaur plus turtles, crocodilians and a ?mammalian organism. The plants are reconstructed based upon the palaeobotanical evidence found in the Cerro del Pueblo Formation, e.g. fruits of *Tricostatocarpon silvapinedae* (Zingiberales, with the banana-like leaves) and those belonging to Phytolaccaceae. (Artwork by author.)

although of large size, have been regarded as feeders on invertebrates (e.g. gastropods) (Lehman 1996; Lockley *et al.* 1997).

Lockley 1999 noticed the possibility that some pterosaurs congregated in some areas in high densities. In addition to this, the El Pelillal Tracksite evidence suggests that this site was frequented by pterosaurs, at least, in repeated seasons. This idea has some support by slabs bearing true or positive tracks and additional pterosaur natural casts of manus prints on their underfaces (e.g. SEPCP-48/255).

Most of the El Pelillal Tracksite ichnotaxa agree with the osteological record of vertebrates found in the Cerro del Pueblo Formation; thus it is possible to speculate about the possible track-makers; in this way, crocodilians (such as goniopholidids and eusuchians) have been reported, nine taxa of turtles (including trionychids, chelydrids, pleurodirans such as *Bothremys*, etc.) are well known, and small theropod dinosaurs (troodontids and dromaeosaurids) are also known, although evidence of mammals is as yet non-existent (Rodríguez-de la Rosa & Cevallos-Ferriz 1998). Although the pterosaur remains are fragmentary, some of them suggest that the Family Ornithocheiridae is perhaps present in the Cerro del Pueblo Formation.

For the moment the study of the Late Campanian tracks from the El Pelillal Tracksite allowed us to add to our understanding of Late Cretaceous pterosaur palaeoichnology. However, there is great potential for palaeoichnological research in southeastern Coahuila, Mexico. Prior to this paper the only palaeoichnological studies dealt with a theropod trackway and the biological and palaeoecological implications of Campanian vertebrate coprolites; both were studied using material from the Cerro del Pueblo Formation (Aguillón-Martínez *et al.* 1998; Rodríguez-de la Rosa *et al.* 1998). However, the finding of new tracksites and their study will shed more light on the vertebrate palaeontology and palaeoichnology of this sedimentary sequence of southern North America.

I wish to thank to R. Rodríguez-Garza, J. Flores-Ventura, José López-Espinoza, I. Vallejo-González and C. R. Delgado-de Jesús for their help during fieldwork. My special gratitude goes to Magdalena S. Cárdenas-García, A. H. González, the staff at the Museo del Desierto-Coahuila, and to R. Gómez-Núñez (Coordinación de Paleontología, SEPC), for their support for the palaeontological research in southeastern Coahuila. Thanks are given to M. G. Lockley and C. A. Meyer for reviewing the manuscript. M. G. Lockley provided useful information regarding pterosaur paleoichnology. This paper is part of the results of the 'Parras Basin Dinosaur Project' carried out in southeastern Coahuila by a team composed of the Museo del Desierto/SEPC – Royal Tyrrell Museum of Palaeontology – Museum of Natural History of the University of Utah; thanks are also given to the National Geographic Society Grant No. 7171–01. Finally, very special thanks are given to the French Embassy in Mexico (J. Montemayor, A. Marchegay and J-P. Lecertua) for their kind support that permitted me to attend the 'Two Hundred Years of Pterosaurs' symposium.

References

AGUILLÓN-MARTÍNEZ, M. C., VALLEJO-GONZÁLEZ, I, HERNÁNDEZ, R. R. & KIRKLAND, J. I. 1998. Dinosaur trackway from the Cerro del Pueblo Formation, Difunta Group (Latest Campanian, Cretaceous), Coahuila, México. *Journal of Vertebrate Paleontology*, **18** (3) (supplement), 23A.

BENNETT, S. C. 1997. Terrestrial locomotion in pterosaurs: a reconstruction based on *Pteraichnus* trackways. *Journal of Vertebrate Paleontology*, **17** (1), 104–113.

CALVO, J. O. & LOCKLEY, M. G. 2001 The first pterosaur tracks from Gondwana. *Cretaceous Research*, **22**, 585–590.

CLARK, J. M., HOPSON, J. A., HERNANDEZ, R. R., FASTOVSKY, D. E. & MONTELLANO, M. 1998. Foot posture in a primitive pterosaur. *Nature*, **391**, 886–889.

FERRUSQUÍA-VILLAFRANCA, I., APPLEGATE, S. P. & ESPINOSA-ARRUBARRENA, L. 1978a. Las huellas más australes de dinosaurios en Norte América y su significación geobiológica. *Actas del Congreso de Paleontología y Bioestratigrafía y Congreso Latinoamericano de Paleontología, Buenos Aires, Argentina*, 249–263.

FERRUSQUÍA-VILLAFRANCA, I., APPLEGATE, S. P. & ESPINOSA-ARRUBARRENA, L. 1978b. Rocas volcanosedimentarias mesozoicas y huellas de dinosaurios en la región suroccidental pacífica de México. *Universidad Nacional Autónoma de México, Instituto de Geología, Revista*, **2**, 150–162.

FERRUSQUÍA-VILLAFRANCA, I., JIMÉNEZ-HIDALGO, E. & BRAVO-CUEVAS, V.M. 1995. Jurassic and Cretaceous dinosaur footprints from México: additions and revisions. *Journal of Vertebrate Paleontology*, **15** (3) (supplement), 28A.

FERRUSQUÍA-VILLAFRANCA, I., JIMÉNEZ-HIDALGO, E. & BRAVO-CUEVAS, V. M. 1996a. The Xochixtlapilco dinosaur ichnofauna from the Middle Jurassic of southeastern México. *Journal of Vertebrate Paleontology*, **16** (3) (supplement), 33A.

FERRUSQUÍA-VILLAFRANCA, I., JIMÉNEZ-HIDALGO, E. & BRAVO-CUEVAS, V. M. 1996b. Footprints of small sauropods from the Middle Jurassic of Oaxaca, southeastern México. *In*: MORALES, M. (ed.) *The Continental Jurassic*. Museum of Northern Arizona Bulletin, **60**, 119–126.

FERRUSQUÍA-VILLAFRANCA, I., TILTON, T. L., LANG, H. R., PITTMAN, J. G. & LOCKLEY, M. G., 1993. Dinosauricnitas tardicretácicas en Puebla suroccidental y su significación geológico-paleontológica. *Memorias del IV Congreso Nacional de Paleontología*, 33–34.

FREY, E. & TISCHLINGER, H., 2000. Weichteilanatomie der Flugsaurierfüße und Bau der Scheitelkämme: Neue Pterosaurierfunde aus den Solnhofener Schichten (Bayern) und der Crato-Formation (Brasilien). *Archaeopteryx*, **18**, 1–16.

KELLNER, A. W. A. 1994. Remarks on pterosaur taphonomy

and paleoecology. *Acta Geologica Leopoldensia*, **17**, 175–189.

LEHMAN, T. M. 1996. Habitat and behavior of *Quetzalcoatlus*: Paleoenvironmental reconstruction of the Javelina Formation (Upper Cretaceous), Big Bend National Park, Texas. *Journal of Vertebrate Paleontology*, **16** (3) (supplement), 48A.

LOCKLEY, M. G. 1998. The vertebrate track record. *Nature*, **396**, 429–432.

LOCKLEY, M. G. 1999. Pterosaur and bird tracks from a new Late Cretaceous locality in Utah. *In*: GILLETTE, D.D. (ed.) *Vertebrate Paleontology in Utah*. Utah Geological Survey, Miscellaneous Publications, **99** (1), 355–359.

LOCKLEY, M. G., HUH, M., LIM, S-K., YANG, S. Y., CHUN, S. S. & UNWIN, D.-M. 1997. First report of pterosaur tracks from Asia, Chollanam Province, Korea. *Journal of the Paleontological Society, Korea, Special Publications*, **2**, 17–32.

LOCKLEY, M. G., LOGUE, T. J., MORATALLA, J. J., HUNT, A. P., SCHULTZ, R. J. & ROBINSON, J. W. 1995. The fossil trackway *Pteraichnus* is pterosaurian, not crocodilian: implications for the global distribution of pterosaur tracks. *Ichnos*, **4**, 7–20.

LOCKLEY, M. G., WRIGHT, J. L., LANGSTON, W. & WEST, E. S. (2001) New pterosaur track specimens and tracksites in the Late Jurassic of Oklahoma and Colorado: their paleobiological significance and regional ichnological context. *Modern Geology*, **20**, 179–203.

MAZIN, J-M., HANTZPERGUE, P., LAFAURIE, G. & VIGNAUD, P. 1995. Des pistes de ptérosaures dans le Tithonien de Crayssac (Quercy, France). *Comptes Rendus de l'Académie des Sciences, Paris, Série IIa*, **321**, 417–424.

MCBRIDE, E. F., WEIDIE, A. E., WOLLEBEN, J. A. & LAUDON, R. C. 1974. Stratigraphy and structure of the Parras and La Popa basins, northeastern México. *Geological Society of America Bulletin*, **85**, 1603–1622.

MURRAY, G. E., WEIDIE, A.E. Jr, BOYD, D. R., FORDE, R. H. & LEWIS, P. D. Jr. 1962 – Formational divisions of Difunta Group, Parras Basin, Coahuila and Nuevo León, México. *Bulletin of the American Association of Petroleum Geologists*, **46**, 374–383.

ORTÍZ-MENDIETA, J.A., GARCÍA-OROZCO, J., FERRUSQUÍA-VILLAFRANCA, I., JIMÉNEZ-HIDALGO, E. & BRAVO-CUEVAS, V. M., 2000. Primer hallazgo de huellas de dinosaurio en Michoacán suroccidental: Reporte preliminar. *VII Congreso Nacional de Paleontología y I Simposio Geológico en el Noreste de México, Libro de Resúmenes*, 109–110.

PADIAN, K., 1983. A functional analysis of flying and walking in pterosaurs. *Paleobiology*, **9**, 218–239.

PADIAN, K. 1996. Terrestrial locomotion in pterosaurs. *Journal of Vertebrate Paleontology*, **16** (3) (supplement), 57A.

RODRÍGUEZ-DE LA ROSA, R. A., 1998. ¿Aves o dinosaurios en Durango, México?. *VI Congreso Nacional de Paleontología, Libro de Resúmenes*, 58–59.

RODRÍGUEZ-DE LA ROSA, R. A. & RODRÍGUEZ-GARZA, R., 1998. Una icnita de quelonio del Cretácico Superior de Coahuila, México. *VI Congreso Nacional de Paleontología, Libro de Resúmenes*, 62–63.

RODRÍGUEZ-DE LA ROSA, R. A. & CEVALLOS-FERRIZ, S. R. S. 1998. Vertebrates of the El Pelillal Locality (Campanian, Cerro del Pueblo Formation), southeastern Coahuila, Mexico. *Journal of Vertebrate Paleontology*, **18**, 751–764.

RODRÍGUEZ-DE LA ROSA, R. A., LÓPEZ-ESPINOZA, J. & AGUILLÓN-MARTÍNEZ, M.C. 2002. El registro fósil de huellas de vertebrados en México. Primer Congreso Latinoamericano de Paleontologia de Vertebradas, Chile, Resúmenes, **47**.

RODRÍGUEZ-DE LA ROSA, R. A., CEVALLOS-FERRIZ, S. R. S. & SILVA-PINEDA, A. 1998. Paleobiological implications of Campanian coprolites. *Palaeogeography, Palaeoclimatology, Palaeoecology*, **142**, 231–254.

STOKES, W. L. 1957. Pterodactyl tracks from the Morrison Formation. *Journal of Paleontology*, **31**, 952–954.

UNWIN, D. M. 1989. A predictive method for the identification of vertebrate ichnites and its application to pterosaur tracks. *In*: GILLETTE, D. D. & LOCKLEY, M. G. (eds) *Dinosaur Tracks and Traces*, Cambridge University Press, Cambridge, 259–274.

VEGA-VERA, F. J., MITRE-SALAZAR, L. M. & MARTÍNEZ-HERNÁNDEZ, E. 1990. Contribución al conocimiento de la estratigrafía del Grupo Difunta (Cretácico Superior-Terciario) en el noreste de México. *Universidad Nacional Autónoma de México, Instituto de Geología, Revista*, **8**, 179–187.

WELLNHOFER, P. 1978. Pterosauria. *In*: WELLNHOFER, P. (ed.) *Handbuch der Paläoherpetologie*. Gustav Fisher, Stuttgart, Teil 19, 82 pp.

WELLNHOFER, P. 1991. *The Illustrated Encyclopedia of Prehistoric Flying Reptiles*. Barnes and Noble Books, New York, 192 pp.

WRIGHT, J. L., UNWIN, D. M., LOCKLEY, M. G. & RAINFORTH, E. C. 1997. Pterosaur tracks from the Purbeck Limestone Formation of Dorset, England. *Proceedings of the Geologists' Association, London*, **108**, 39–48.

Ichnological evidence for quadrupedal locomotion in pterodactyloid pterosaurs: trackways from the Late Jurassic of Crayssac (southwestern France)

JEAN-MICHEL MAZIN[1], JEAN-PAUL BILLON-BRUYAT[1], PIERRE HANTZPERGUE[2] & GÉRARD LAFAURIE[3]

[1]*Laboratoire de Géobiologie, Biochronologie et Paléontologie humaine, Université de Poitiers, 40 avenue du recteur Pineau, 86022 Poitiers Cedex, France (e-mail: jmmazin@univ-poitiers.fr)*
[2]*Centre des Sciences de la Terre, Université Claude Bernard Lyon I, 43 boulevard du 11 Novembre, 69622 Villeurbanne Cedex, France*
[3]*10, rue Léon Baraillé, 46100 Figeac, France*

Abstract: More than 30 well-preserved pterosaurian trackways have been excavated from the site of Crayssac (Lower Tithonian, southwestern France), along with hundreds of isolated imprints. The general pattern of pterosaurian trackways is described and corresponds to the morphology of the ichnogenus *Pteraichnus* Stokes 1957. A crocodilian origin for *Pteraichnus* is rejected. The morphology of the pes and manus prints is described and discussed, and exhibits a great dynamic variability. The three manus digit prints are confirmed as the marks of digits I–III, but the high angle of divarication may reflect strong digit abduction as well as a backward folding of the digit III. The manus gait-width is as much as 3 × the pes gait-width. The narrowest manus trackways corresponds to low-velocity walking in a semi-erect stance, whereas the widest manus trackways corresponds to higher velocity walking (perhaps even running), with a subhorizontal body.

Since the first description of a pterosaur (Collini 1784), and its identification as a flying reptile (Cuvier 1801), the question of terrestrial locomotion in these animals has been debated. The bipedal hypothesis directly opposed the quadrupedal hypothesis and, as early as the beginning of the nineteenth century, Cuvier (1809) proposed a bipedal mode of terrestrial locomotion for his reptilian 'ptérodactyles' whereas Soemmerring (1812, 1817) argued for quadrupedal locomotion for his 'mammalian' *Onithocephalus*. This conflict remained alive for a long time, leading workers to consider pterosaurs as either unable to walk (Abel 1925) or bad walkers (i.e. Wellnhofer 1978). In all cases, it was assumed that pterosaurian forelimbs were unsuited for quadrupedal locomotion. This assumption has been emphasised by Padian (1983, 1984, 1985), who defended the bipedal hypothesis with osteological and phylogenetic arguments, considering that pterosaurs and dinosaurs shared a common bipedal ancestor. However, during the following decade, several authors defended the traditional quadrupedal interpretation with osteological arguments (i.e. Wellnhofer & Vahldiek 1986; Wellnhofer 1988; Unwin 1987, 1988, 1989; Unwin & Bakhurina 1994), or a quadrupedal mode of locomotion for the small pterodactyloids and non-pterodactyloids while the large pterodactyloids could be bipedal as suggested by Bennett (1990, 1997), with osteological arguments, but questioned by Lockley *et al.* (1997) from new ichnological data from Korea.

In 1957, an alternative method for resolving the problem was proposed by Stokes, who described and named the ichnotaxon *Pteraichnus saltwashensis* based on a short trackway from the Morrison Formation (Late Jurassic) of Apache County, northeastern Arizona, United States. From this trackway, composed of tetradactyl pes prints associated with tridactyl manus prints, Stokes assumed that the track-maker was a quadrupedal pterodactyloid pterosaur. Subsequently, several authors reported occurrences supporting Stokes' conclusion (Logue 1977; Stokes 1978; Stokes & Madsen 1979; see also Wellnhofer 1978). However Stokes' conclusion was strongly contested by Padian and Olsen (1984), who argued that the track-maker was, in fact, crocodilian. Consequently, several pterosaur specialists questioned the pterosaurian origin of Stokes' trackway and *Pteraichnus* was no longer regarded as an ichnotaxon of pterosaurian origin (Unwin 1986, 1989; Prince & Lockley 1989; Lockley 1991; Wellnhofer 1991; Bennett 1992). Thus, by the beginning of the 1990s, the problem of pterosaurian terrestrial locomotion was unresolved, neither from osteological and phylogenetical argument, nor from ichnological data.

New ichnological occurrences arose by the mid-1990s. The Stokes' specimen has been reinvestigated, while new data have been found from new sites (Logue 1994; Hunt et al. 1995; Lockley & Hunt 1995; Lockley *et al.* 1995; Mazin *et al.* 1995). Mazin *et al.* (1995) and Lockley *et al.* (1995) independently

From: BUFFETAUT, E. & MAZIN, J-M. (eds) 2003. *Evolution and Palaeobiology of Pterosaurs*. Geological Society, London, Special Publications, **217**, 283–296. 0305–8719/03/$15

demonstrated that the ichnotaxon *Pteraichnus* is actually pterosaurian and not crocodilian. In the following years, the pterodactyloid origin of *Pteraichnus* and of the so-called '*Pteraichnus*-like' trackways became widely accepted (Lockley *et al.* 1997; Mazin *et al.* 1997; Bennett 1997; Unwin 1997a).

Since that time, new specimens have been referred to *Pteraichnus* or *Pteraichnus*-like trackways and some problematic specimens have also been assigned to pterosaurian trackways (see Unwin 1997b for a list; Wright *et al.* 1997; Meijide Calvo & Fuentes Vidarte 1999; Rodriguez de la Rosa 2001). However, some of these specimens cannot be assigned with certainty to pterosaurs. Subsequently, several new ichnospecies, all referred to the ichnogenus *Pteraichnus*, have been recently named: *P. stokesi* from the Late Jurassic of Wyoming (Lockley *et al.* 1995); *P. palaciei-saenzi* from the Early Cretaceous of Soria, Spain (Pascual Arribas & Sanz Pérez 2000); *P. manueli*, *P. vetustior* and *P. cidacoi*, also from the Early Cretaceous of Soria, Spain (Fuentes Vidarte 2001; Meijide Calvo 2001; Meijide Fuentes 2001).

The discovery of a Late Jurassic Lagerstätte in Crayssac (Lower Tithonian, southwestern France), which yields particularly well-preserved vertebrate and invertebrate prints and trackways (Mazin *et al.* 1995, 1997), provides an excellent opportunity to resolve the question of pterosaurian terrestrial locomotion, at least for the Jurassic pterodactyloid pterosaurs. Among the hundreds of trackways excavated in Crayssac, more than 30 are referable to pterosaurs, ranging from sparrow to seagull size, well preserved in an ancient littoral calcareous mud.

The aim of this paper is to address the main questions that confuse pterosaurian ichnology and, subsequently, the analysis of the pterosaurian terrestrial locomotion.

General description

Even though the numerous tracks and trackways from Crayssac can obviously be assigned to different pterosaurian track-makers and probably correspond to several ichnospecies (which is beyond the scope of this paper), all the excavated trackways referred to pterodactyloids show the same general morphology (Fig. 1). Without exception all are quadrupedal trackways, consisting of an alternation of pes and manus prints without a tail trail. The pes prints are located near the mid-line of the trackway and are oriented parallel to the axis or with an outward rotation of up to 30°. This outward rotation is not constant and can vary along the same trackway. The pes print is anterior to or at the level of the manus print, which is located either on the same line as the pes print or further from the mid-line. The manus trackway can be 3 × wider than the pes trackway.

The pes prints are triangular, elongate and fully plantigrade (Fig. 2). They are tetradactyl with anterior claw marks. The four digit imprints are subequal in length and correspond to digits I–IV. On all the pes prints observed, the metatarsals diverge moderately, giving the trace a triangular shape ranging from 25° to 35°. The sole is flat but can show impressions of the long metatarsals, with the deepest area of the prints located at the metatarso-phalangeal joint, and sometimes in the heel area. When the posterior part of the pes is deeply marked, a posterolateral depression can be seen, which corresponds to the reduced fifth metacarpal, but this is uncommon. On some well-preserved pes prints, impression of the interdigital webbing occurs (Fig. 2d), but the digital pads are rarely preserved. When they are present, the phalangeal formula can be evaluated as 2, 3, >3, >4, ?, and the penultimate phalanx appears to be the longest on digits I and II (see Mazin *et al.* 1995).

The manus prints are asymmetric, with three digit imprints (Fig. 3). The first digit imprint is anteriorly to anterolaterally oriented and is the shortest. It is sometimes straight or laterally curved or folded and commonly bears a claw mark. The second digit imprint is laterally or posterolaterally orientated. It is commonly straight, rarely divided into two successive pads and sometimes has the impression of an ungual. The third digit imprint is posteriorly oriented and the longest. It is often bent, giving the digit mark a sinusoidal shape. A few of the hundreds of manus prints observed show a claw mark on digit III. In the area where the digit marks converge, a deeper central impression is present.

Variability of print morphology and substrate competence

Except in rare cases of exceptional preservation (i.e. a fine natural mould of a pes or of a manus with little dynamic disturbance), the features described above are rarely seen on a single print. As noted by Baird (1980), footprints record the dynamic interaction of the pes and the substrate. A print cannot be considered as a static cast of the pes. Hence, it is necessary to accurately discriminate between 'ichnoanatomic' characters and 'ichnodynamic' features. Such discrimination is helped by the observation of a great number of specimens, which is possible with the Crayssac pterosaurian trackways. Among the ichnoanatomic characters there is the number of digits, as well as the number of pads or the claw marks, when they are present and when they can be attested by observations on several specimens. Nevertheless,

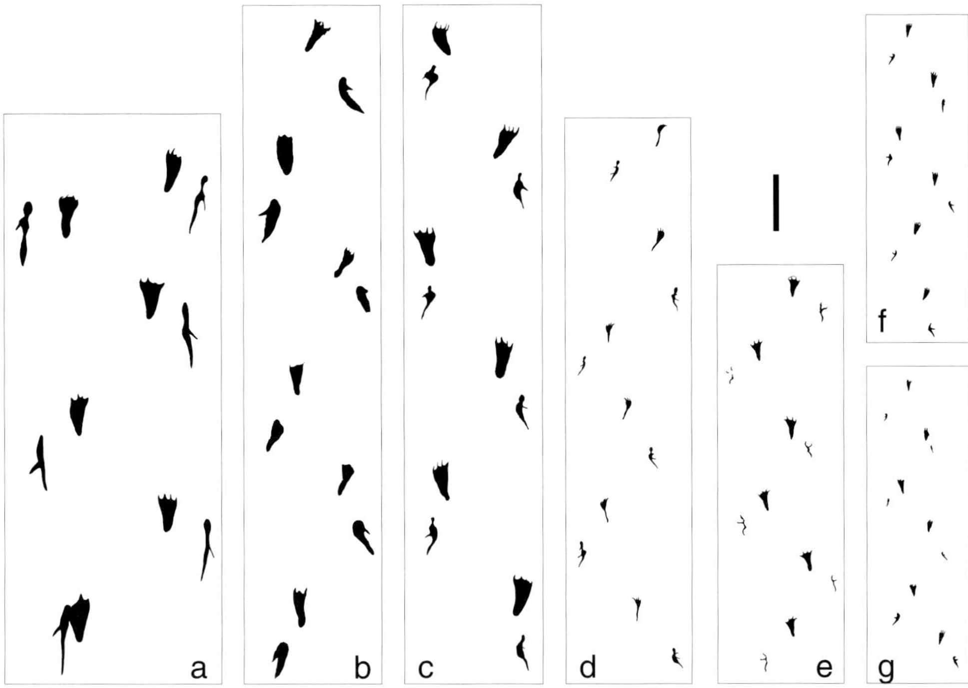

Fig. 1. Parts of some pterosaurian trackways from Crayssac (Lower Tithonian, southwestern France). For each example, an area of six sets of adjacent manus and pes prints is illustrated. (**a**) CR96.22, *Pteraichnus* isp., a 22-pace trackway. (**b**) CR98.29, *Pteraichnus* isp., a 17-pace trackway. (**c**) CR00.50, *Pteraichnus* isp., an 18-pace trackway. (**d**) CR01.10, *Pteraichnus* isp., a 12-pace trackway. (**e**) CR99.43, *Pteraichnus* isp., a 42-pace trackway. (**f**) CR97.32, *Pteraichnus* isp., a 28-pace trackway. (**g**) CR98.26, *Pteraichnus* isp., a 31-pace trackway. Scale bar 10 cm.

they can be modified by dynamics and all measurements on a print can be strongly affected. Among the ichnodynamic features, there are the bulges, drag and slide marks and, in general, all the relief and deformation due to various factors, such as stance, gait and velocity of the track-maker, and the competence of the substrate.

Most of these kinds of ichnodynamic features can be observed on the prints and trackways from the Crayssac laminated limestone. From the great number of potential surfaces from the 1.5 m-thick laminated limestone, 28 have been excavated over a large area (from 2–100 m^2). Several dozens of other surfaces are potentially available. All the excavated surfaces yield numerous vertebrate and invertebrate prints, of which the great variability in morphology reflects various substrate competence, from very fluid to dry, crusted mud. In the same way, all the excavated surfaces bear mud cracks, which attest long emersions. Statistical analysis of the sequences of the Crayssac tidalite reflects a diurnal tide cycle, which means only one immersion a day. Consequently, the emersion stage could have been very long between two high tides (Hantzpergue *et al.* in prep.). Experimental work with vertebrates (crocodiles, turtles, birds) and invertebrates (extant isopods) walking on a micritic mud reconstituted from pulverized Crayssac limestone confirms the particular mechanics of the original carbonated mud. It remains liquid and fluid for a long time and solidifies quite rapidly by superficial encrusting when drying, after a short stage during which the mud is plastic. From these observations, it appears that the original substrate of Crayssac remained mainly incompetent during the emersion period between two high tides, with a short period of competence before crusting.

Field observation clearly confirms that the prints and trackways were made at different stages of the tide cycle. Concerning the pterosaurs, most of the prints seem to have been made on a non-competent substrate, leading to prints which reflect the dynamic interaction of the feet and the incompetent substrate much more than the detailed anatomy of the feet (e.g. Figs 2f–g & 3e–g). In some rare cases, prints correspond to claw marks only, which reflects pterosaurs walking on a dry crusted mud only marked by the claws (e.g. Fig. 2h). Prints made on a competent substrate are less common, but can be found, e.g. the manus print of Figure 3c or the pes print of Figure

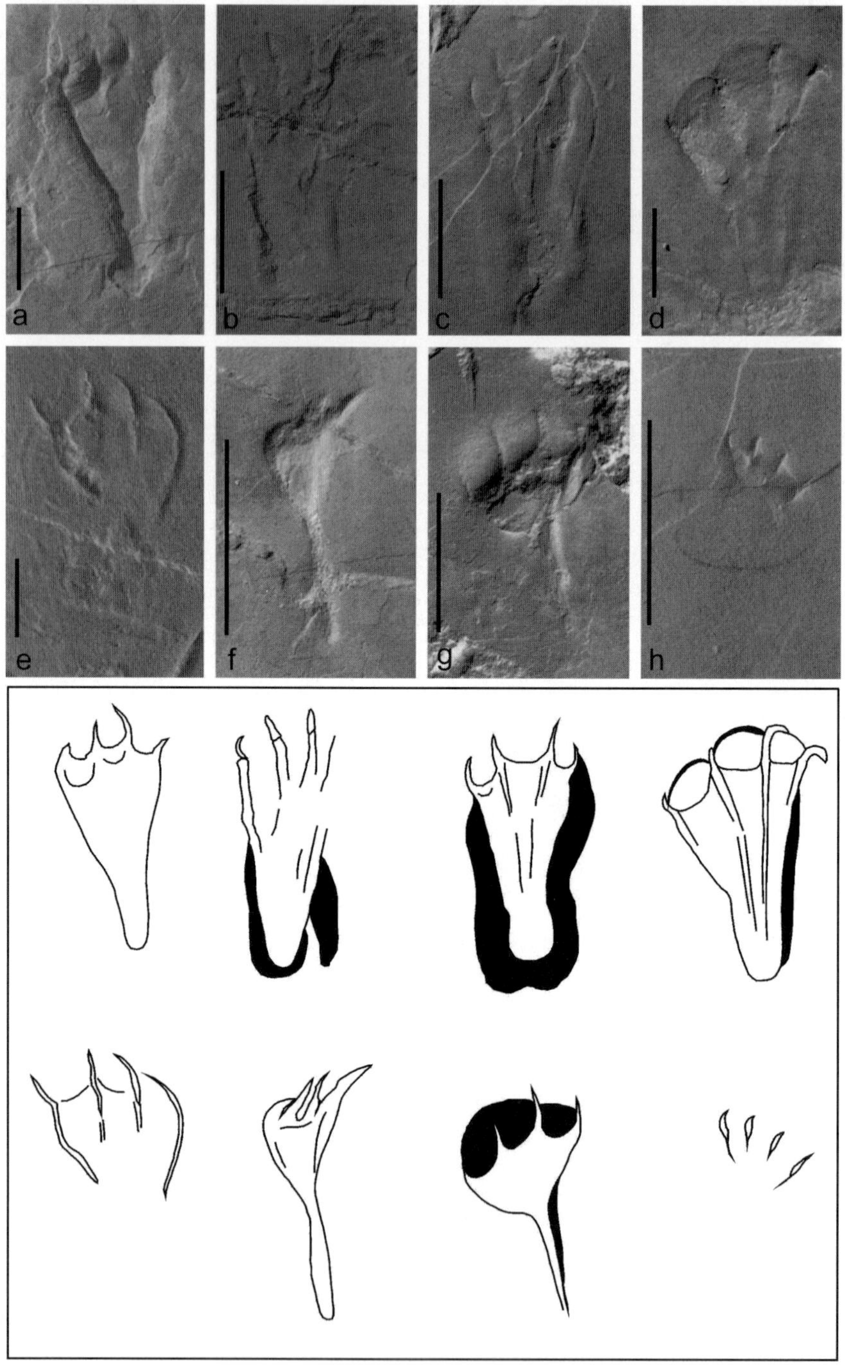

Fig. 2. *Pteraichnus* isp. (Crayssac, Lower Tithonian, southwestern France). Morphology of diverse pes prints. Photographs and interpretative drawings (white, imprint; black, bulges and crests). Such variability is partly due to 'ichnoanatomic' characters relevant to specific discrimination, but which are masked and modified by 'ichnodynamic' features consequent to the stance, gait and speed of the track-maker, as well as to the competence of the substrate. (**a**) Right pes print from the trackway CR01.03. (**b**) Right pes print from the trackway CR98.33. (**c**) Right pes print from the trackway CR98.33. (**d**) Left pes print from the trackway CR99.43. (**e**) Isolated right pes print. (**f**) Left pes print from the trackway CR01.10. (**g**) Isolated pes print. (**h**) Right pes print from the trackway CR95.64. Scale bars 2 cm.

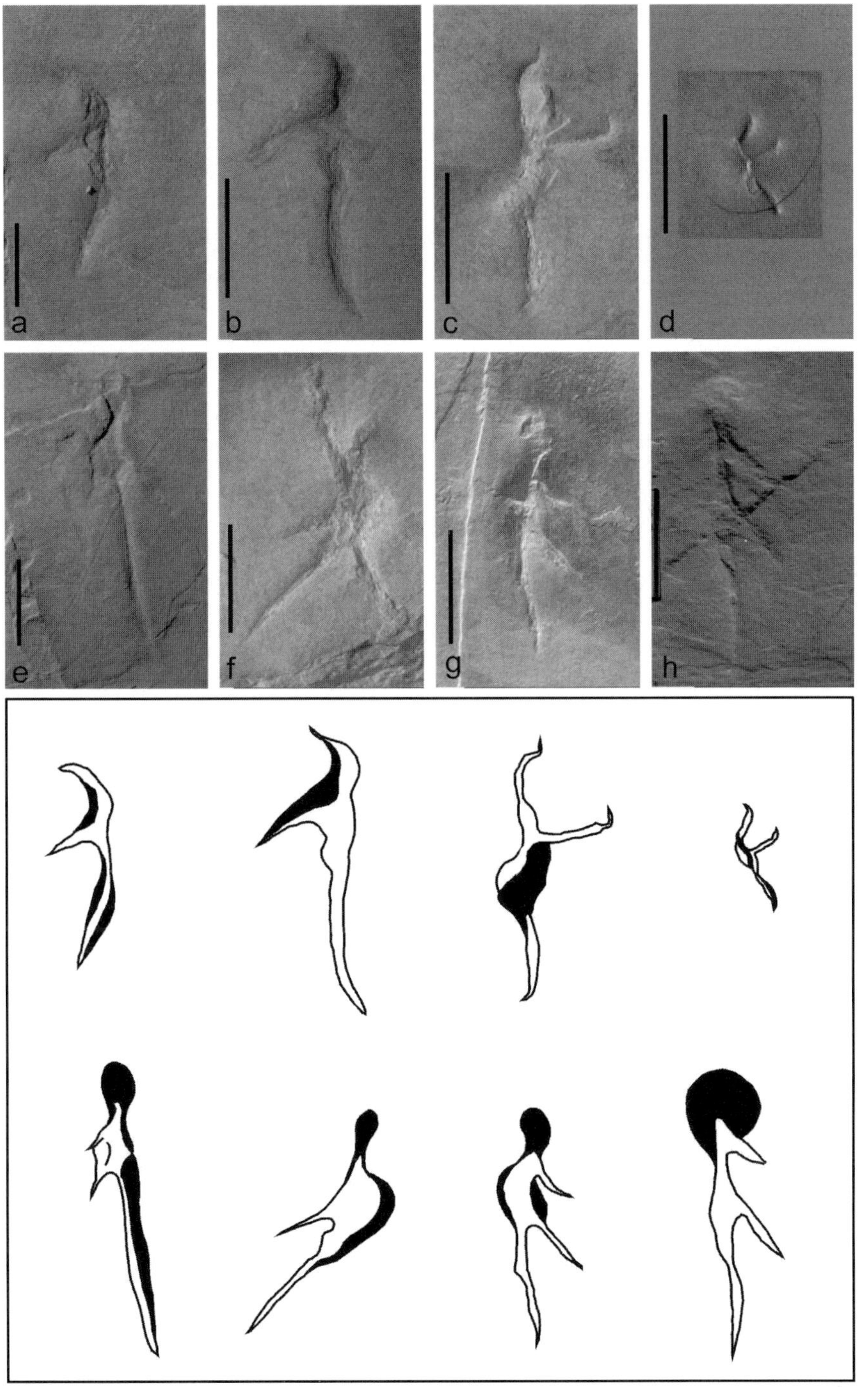

Fig. 3. *Pteraichnus* isp. (Crayssac, Lower Tithonian, southwestern France). Morphology of diverse manus prints. Photographs and interpretative drawings (white, imprint; black, bulges and crests). See the remark in Figure 2 about variability. (**a**) Left manus print from the trackway CR01.04. (**b**) Left isolated manus print. (**c**) Right manus print from the trackway CR99.43. (**d**) Right manus print from the trackway CR95.64. (**e**) Left manus print from the trackway CR96.22. (**f**) Left manus print from the trackway CR00.50. (**g**) Right manus print from the trackway CR00.50. (**h**) Right manus print from the trackway CR01.10. Scale bars 2 cm (except d: 1 cm).

2d, on which digit and metapod impressions are observable as well as webbing marks. In the same way, the pes print illustrated in Mazin *et al.* 1995 (p. 421m Fig. 2a; specimen CR94.02), originally belonging to a short trackway accidentally destroyed by a quarryman, was made on a quite competent substrate and shows phalangeal joint pads. Finally, it is noteworthy that, on almost all excavated surfaces, several generations of trackways and prints have been made at different stages of the evolution of the substrate competence during the emersion period.

From a biometrical point of view, there is a great variability in print morphology, as well as in trackway pattern. Some examples can be given. On Crayssac specimens, the pes print length can vary by 15% along the same trackway, and the manus print length variability can reach 100% due to bulges and drag marks. The digit spread of the pes can vary by 25° to 35° for the same pes along a single trackway. The angle of divarication (outward rotation) of the pes print can vary from 0° to 30° in the same trackway (e.g. CR99.43), and can sometimes be negative (inward rotation). Similarly, the rotation of the manus print can be negative (inward rotation), parallel to the axis, or positive (outward rotation). Thus, it is necessary to be cautious with pterosaurian ichnotaxonomy because many of the ichnoanatomic characters are masked by ichnodynamic features (see Billon-Bruyat & Mazin, 2003). The case of *Pteraichnus stokesi*, erected by Lockley *et al.* (1995) from a short trackway from the Sundance Formation of Wyoming, is very interesting on this point. This specimen is undeniably pterosaurian, and the ichnospecies seems to differ consistently from the type species *P. saltwashensis* by the outward rotation of the pes prints, the smaller pace angulation and the larger manus digit prints. If, according to the typological taxonomy commonly used in ichnology, this specimen clearly differs from the type specimen and can be considered as valid, these criteria of discrimination can be questioned. As noted above, on long trackways from Crayssac, the angle of divarication of the pes print as well as the pace angulation or the length of digit prints are dynamic features which vary in a same trackway. Thus, if one of the long trackways from Crayssac had been found in several isolated slabs on the field, it might have been possible to erect different new ichnospecies on the basis of the angle of divarication of the pes print, the pace angulation or the length of digit prints.

Crocodilian versus pterosaurian trackways

Besides the taxonomic discrimination of pterosaur prints, an important debate concerns the pterosaur origin for these tracks. The foremost opposition to the pterosaurian status of the Stokes' original trackway has been the possible misinterpretation suggested by Padian & Olsen (1984). From experimental trackways of a young *Caiman*, the authors argued that the Stokes' *Pteraichnus* could have been made not by a pterosaur, but by a crocodilian.

Confusion between these two kinds of trackmakers could be possible when tracks are badly preserved, and so it is not surprising that crocodilian tracks have been erroneously assigned to pterosaurs (i.e. Gillette & Thomas 1989; Bennett 1992). However, new investigations led Mazin *et al.* (1995), Lockley *et al.* (1995) and Bennett (1997) to conclude that *Pteraichnus* is pterosaurian. It is not necessary to reproduce the demonstration here, but new data reinforce the pterosaurian interpretation. Firstly, new experimental trackways have been made with *Crocodylus niloticus* by two of us (J-M. M. & J-P. B-B.), in order to study the variation of trackway pattern and print morphology on various substrates (wet controlled clay mud, sand and Crayssac mud reconstituted from pulverized limestone). None of the experimental trackways can be confused with *Pteraichnus*, whatever the substrate consistency or the animal velocity (Fig. 4). Subsequently, crocodilian trackways have been excavated at Crayssac, alongside those attributed to pterosaurs (Fig. 4). From the comparison of crocodilian trackways (experimental and fossil) with pterosaurian trackways, there is no longer any doubt or any confusion. A crocodilian trackway always bears a tail trail, whereas a pterosaurian trackway (at least the pterodactyloid trails described in this paper) never show a tail trail. The pterosaurian manus print is tridactyl whereas the crocodilian print is pentadactyl (anatomically and functionally). In crocodilians the manus gait-width is as wide as the pes gait-width, whereas in pterosaurs the manus gait-width can be up to 3 times the pes gait-width. Note that it is perhaps astonishing to use the expression 'manus or pes gait-width' in place of the expression 'manus or pes trackway width' in reference to the distance between a manus (or a pes) print and the virtual line between the two opposite manus (or pes) prints. According to a constructive discussion with C. Bennett, one of the reviewers, it appears that the expression 'manus or pes trackway width', which is commonly used, could incorrectly reflect this biometrical measurement, which actually corresponds to the height of the triangle constituted by the lines joining three successive manus (or pes) prints. Haubold (1971) uses the German word *Gangbreite*, which means 'width of the pace angulation pattern' or, more strictly, 'manus gait-width'. This latter expression is preferred in this work when referring to this biometrical measurement, which is particularly significant in ichnology.

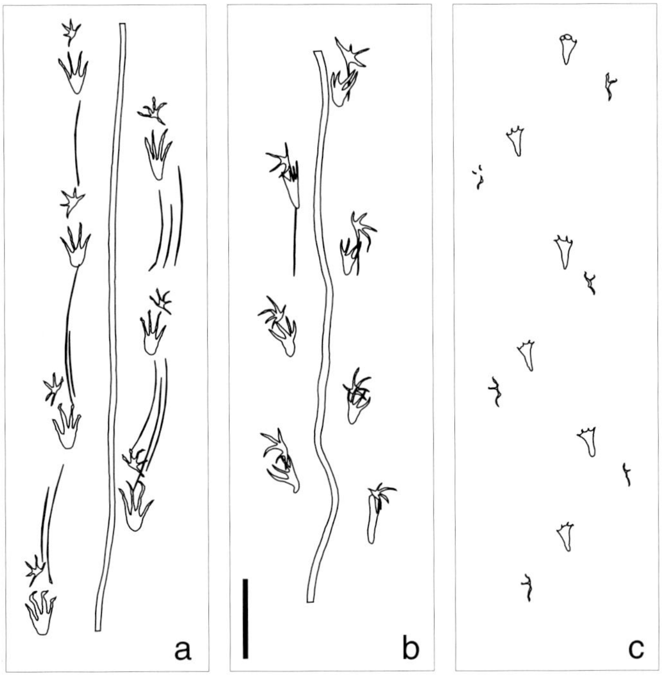

Fig. 4. Comparison of crocodilian and pterosaurian trackways. (**a**) *Crocodylus niloticus*, experimental trackway (snout-vent length 65 cm). (**b**) CR97.34, Crocodylia indet., part of a crocodilian trackway from Crayssac. (**c**) CR99.43, *Pteraichnus* isp. part of a pterosaurian trackway from Crayssac. Scale bar 10 cm.

Identification and position of manus digit imprints

As predicted by Unwin (1989), the pterodactyloid manus print is tridactyl. However, while the anatomy of the pterosaur manus predicted a three-digit imprint oriented anteriorly, the actual manus print shows three-digit imprints laterally and posteriorly directed. This unexpected print morphology generated a debate regarding the identification of the digits and the rotation of the manus during terrestrial locomotion. The excellent preservation of Crayssac traces and trackways allows us to suggest answers to these two questions.

In his original description of *Pteraichnus saltwashensis*, Stokes (1957) interpreted the three-digit impressions as those of digits II–IV. This assumption is supported and reinforced by Lockley *et al.* (1995), with several arguments. Firstly, they considered that such an interpretation does not require a strong outward rotation of the manus, which is in accordance with the supposed rotational ability of the pterosaurian wrist and elbow. Secondly, the shape of the pit in the central area of the print would accommodate the metacarpo-phalangeal joint imprint in a non-rotated position rather than in a rotated position. Thirdly, the posterior digit imprint is much too long in comparison with other digit imprints to be anything other than the proximal part of the first phalanx of digit IV.

Conversely, Mazin *et al.* (1995) interpreted these three-digit impressions as digits I–III, which was later confirmed by Bennett (1997) and Unwin (1997b). This second interpretation is the most satisfactory for several reasons. Firstly, there is no objective reason to explain why digit I should not leave an impression when the others deeply mark the substrate. Secondly, as Padian & Olsen (1984) noted, the long posterior digit imprint is often curved, which would not possible if this impression was due to the straight first phalanx of the wing digit. Bennett (1997) also argued that the Wyoming trackway described by Lockley *et al.* (1995) shows a manus print with a posterior digit that bears a large, strongly curved ungual. Finally, Mazin *et al.* (1995) noted that, on some rare manus prints from Crayssac, a shallow mesioposterior imprint could be interpreted as the mark of the proximal end of digit IV.

It is clear that the three-digit imprints of the pterodactyloid manus print correspond to digits I–III, but such an unexpected print morphology requires explanation. However, to avoid an oversimplified interpretation, it is necessary to consider print variability.

From the observation of hundreds of pterodactyloid manus prints from Crayssac, the manus prints show a range of morphologies (Fig. 3). Do they represent different track-makers and different taxa? Probably yes, but this is beyond the scope of this work. In the most frequent morphology (Fig. 3a, b), the digit I imprint is oriented anteriorly, but distally curved and hooked. It is clear that this straight anterior mark corresponds to the long phalanx of this digit while the lateral hook corresponds to an ungual, often well marked and sometimes extended by a drag mark. The digit II imprint is straight, oriented postero-aterally, always longer than the digit I imprint and without a claw mark. The digit III imprint, the longest, is oriented posteriorly and often slightly curved, but without an ungual mark. The central pit (interpreted as the metacarpo–phalangeal joint IV impression) may be present or absent. When present, it is oval, shallow and located at the intersection of the axes of digits II–III. In a second, less common morphology, the general pattern is quite similar (Fig. 3c-d). However, the digit II imprint is oriented laterally to slightly anterolaterally and always has a claw. The imprint of digit III is orientated posteriorly, separated into two segments, and ends with a sharp mark that could correspond to a claw. The metacarpo-phalangeal joint IV depression is very shallow and is laterally bordered by an oblique curved crest that masks the proximal part of the digit III imprint. A third type of morphology is characterized by a rounded bulge anteriorly (fig. 3e–h). The straight anterior digit I imprint is very short or absent and seems to be prolonged by this bulge. In fact, as observable on some manus prints (Fig. 3g–h), digit I is hooked and the bulge is a dynamic feature due to the insertion and extraction of digit I from the mud when the manus leaves the ground during walking. The digit II imprint is thin and strongly oriented posterolaterally, forming a low angle with the imprint of digit III. The digit III imprint is directed posteriorly and is notably longer in comparison with the two other digits. It is straight, oblong in its median part and lacks an ungual impression. The metacarpo-phalangeal IV joint imprint is deeply marked and can be large.

From these descriptions, it is possible to discuss the position of each digit during terrestrial locomotion. When the forelimb is fully extended in flight, digit IV is located posteriorly to the other three, but during folding of the wing, the forearm and the metarcapal are brought in to a subvertical plane while the wing digit is folded close to the body. For these reasons, digits I–III were thought to be anteriorly oriented, as Wellnhofer (1988) and Unwin (1989) logically concluded. In fact, as the tracks show, digits I–III are laterally and posteriorly curved. Mazin *et al.* (1995, 1997) and Unwin (1997b) consider that the orientation of the manus digits is a consequence of a spreading of the digits as well as a lateral rotation of the manus. Wellnhofer (1985) noted that the wrist region was able to rotate up to 45° in some pterosaurs. Bennett (1997) considers that the manus was placed on the substrate without any rotation of the metacarpus at the wrist, and that digits I–III were hyper-extended laterally under the fourth metacarpophalangeal joint. He also notes a large abductor tubercle on the proximal phalanx of digit III of *Pteranodon* and *Pterodactylus*, which could have been used to spread out digit III and, consequently, digits I and II, towards a lateral and posterior position. Unwin (1997b) noted that this joint allows a maximum digit III abduction up to and beyond 90° to digit I in several taxa. Thus the metacarpophalangeal joint was not strongly restrictive and 'permitted a considerable degree of lateral as well as vertical movement' (Unwin, 1997b, p. 378).

The lateral rotation of the manus as well as the considerable abduction of the digits can explain the manus print pattern observed in *Pteraichnus*. Such a situation is well illustrated by Unwin (1997b). However, some morphological characteristics of manus prints from Crayssac lead to the questioning of such an interpretation. Firstly, imprints of digit III exceptionally show an ungual mark. Secondly, the manus print orientation (angle of divarication) is not constant. The longitudinal axis of the manus print (understood here as the digit III print axis or the digit I–digit III print axis when they are opposed) can be parallel to the trackway axis, as well as with an outward or an inward divarication (Fig. 5). When the manus is parallel or inwardly rotated, which is the most frequent situation, as illustrated by Unwin's interpretative drawing (1997b, Fig. 4), such a spreading of digits and lateral rotation of the manus are allowable because of the rotation permissible at the wrist. However, in some cases, the angle of divarication is positive for the manus (Fig. 5) and the twisting of the wrist and of the metacarpo-phalangeal joint should become very strong. Thirdly, the divarication of the digit impressions themselves is inconsistent with the supposed movement of the metacarpo-phalangeal joints. In most of the manus prints from Crayssac, as well as those from Soria (Meijide Calvo & Fuentes Vidarte 1999; Pascual Arribas & Sanz Pérez 2000; Fuentes Vidarte 2001; Meijide Calvo 2001: Meijide Fuentes 2001) digit abduction is higher than 90° and can reach 180° or more (Fig. 5; see also Fig. 3). It cannot be excluded that the movement of the manus digits through the surface may exaggerate the apparent angle of divarication when the substrate is incompetent.

However, when digit IV is folded, digit III could be pulled backwards in the same motion, and digit III could lay on its mesiodorsal side when the manus is on the ground. This posteriorly directed folding of digit III, as a consequence of the folding of digit IV,

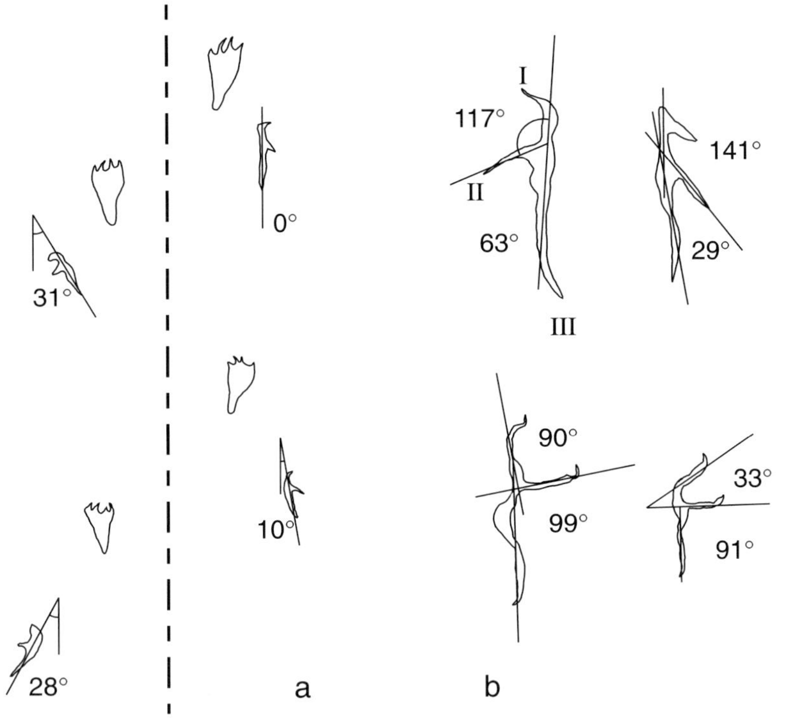

Fig. 5. *Pteraichnus* isp. (Crayssac, Lower Tithonian, southwestern France). (**a**) Four sets of manus and pes prints from the trackway CR98.26 showing various angles of divarication of the manus, from inward rotation to outward rotation. (**b**) Four manus prints with angles of abduction of the digits.

could explain the absence of digit III claw marks on the majority of prints, as well the very posterolaterally orientation of digit II imprint, closely abducted to the digit III, observable on several manus prints (Fig. 3e–h).

Another problem concerning the pterodactyloid manus tracks is the relative length of the digit imprints. As noted by all the authors who have described such manus prints, the length of the three-digit imprint increases from anterior to posterior, which is consistent with the skeletal observations. However, the digit III imprint appears relatively long in comparison to the other two. Unwin (1997b) noted that it can reach twice the length of digit II imprint, whereas digit III is never more than 1.4 times the length of digit II in pterosaurs. Measurements on manus imprints from Crayssac confirm this and show that the relative length of digit III imprints can even reach more than 3 times the digit II imprint length. Conversely, Unwin (1997b) noted that, the length of digit III imprint is 1.3 to 1.6 times the pes print length, which is consistent with skeletal data and leads us to consider that, in the manus prints, digit I and II imprints are relatively shortened. Unwin (1997b) noted that, in some specimens, notably in the holotype of *Pteraichnus saltwashensis*, the manus print is as long or longer than the pes print, which does not correspond to the situation in pterosaurs. He suggested that the relative shortening of the pes print could be due to a semi-plantigrade contact of the foot. Such remarks concerning biometry apply to the great number of pterodactyloid prints and trackways from Crayssac. An explanation for the differences between ichnological and osteologic measurements can be found in the fact that ichnodynamic features often mask or modify ichnoanatomic characters. For example, the elongate digit III imprint relative to other digits is due to a distal drag mark that extends the digit III imprint. Similarly, all the pes prints of these trackways are fully plantigrade with a well-marked heel, and the high manus print to pes print length ratio is due to a stretching of the manus print by the posterior drag mark and the anterior bulge, rather than to a shortening of the pes print.

The position of the manus prints on the trackways

On all the quadrupedal pterodactyloid trackways the pes print is located anterior to the manus print (Fig.

1), which is classically described as an 'overstepping'. However, Bennett (1997) noted that the relative position of the manus and pes prints does not reflect a true overstepping of the pes on the manus during the cycle of terrestrial locomotion: the pes does not go beyond the manus. When the pes touches the substrate, the ipsilateral manus has already left the ground (Bennett, 1997; Mazin *et al.* 2001a, c). Thus, there is no overstepping in terms of functional morphology.

Another problem of pterodactyloid trackways is the relative position of the pes and manus prints. Theoretically, as previously noted by several authors (e.g. Padian & Olsen 1984; Wellnhofer 1988; Unwin 1989), the manus prints must be further from the mid-line of the trackway than the pes prints because of the greater length of the forelimbs. The type specimen of *Pteraichnus* described by Stokes (1957) does not show such morphology. This has been one of the arguments used by Padian & Olsen (1984) to reject a pterosaurian origin for this trackway. The second *Pteraichnus* trackway described from the Jurassic of North America also shows a trackway with manus prints and pes prints at almost the same distance from the body (Lockley *et al.* 1995; Bennett 1997). This is also the case for the trackways from the Early Cretaceous of Spain (Pascual Arribas & Sanz Pérez 2000) and for some trackways from Crayssac (e.g. Fig. 1b, c). The trackways from Crayssac show great variability, ranging from specimens where the manus and pes gait-width are almost equal to specimens where the manus gait-width reaches 3 times that of the pes gait-width. Such variability reflects different stances and speeds of the track-makers, as well as taxonomic diversity. As noted by Bennett (1997, p. 111), 'the lateral placement of the manus would have increased the stability of the walking pterosaur' and it can be logically considered that the widest manus trackways correspond to slow-velocity walking with short paces, while the narrowest manus trackways correspond to faster velocity walking with longer paces. From the first biometric measurements and computer reconstructions on the collection of trackways from Crayssac (Mazin *et al.* 2001c), it appears that there is effectively a correlation between the velocity (expressed as the stride ratio, i.e. stride/pes print length) and the stance (expressed as the gait-width ratio, i.e. manus gait-width/pes gait-width). However, this correlation unexpectedly appears to be reversed: the shortest is the stride; the narrowest is the manus gait-width (Fig. 6). From the calculated sample, specimen CR00.50 (Fig. 1c) has the shortest strides and the narrowest manus gait-width, while specimen CR01.10 (Fig. 1d) has the longest strides and the widest manus gait-width. This unexpected conclusion could be related to the stance of the track-maker. At low-velocity walking, the body was held semi-erect, while at higher velocity walking (or even running) the body was held less erect, and even subhorizontal (fig. 7). A subhorizontal stance would allow the long forelimbs to reach the most distant anterior point when running. Conversely, when the body was held semi-erect, the forelimbs were held more vertically in an almost parasagittal plane and the manus could have been brought close to the axis of the trackway, but could not reach a distant anterior point.

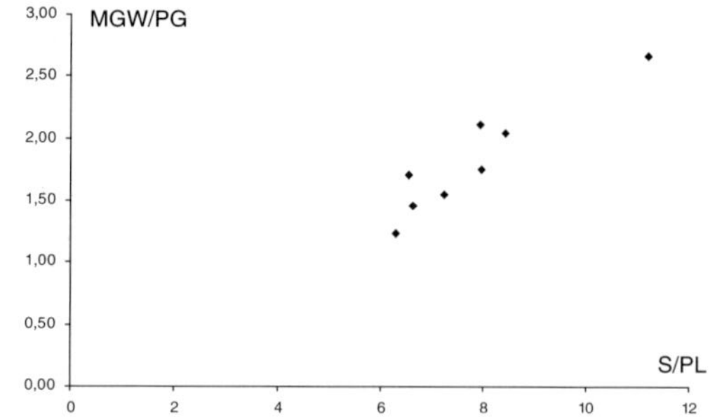

Fig. 6. Pterodactyloid stances according to speed. Stance is expressed as manus gait-width/pes gait-width (MGW/PGW); velocity is expressed as stride/pes print length (S/PL). Ratios have been calculated from six successive sets of manus and pes prints from straight and regular parts of the trackways. Eight specimens have been plotted, from left to right: CR00.50 (6.28; 1.24); CR96.22 (6.54; 1.71); CR98.27 (6.62; 1.47); CR98.29 (7.22; 1.56); CR99.43 (7.91; 2.12); CR97.32 (7.94; 1.76); CR98.26 (8.41; 2.05); CR01.10 (11.17; 2.67).

The problem of 'manus-only' trackways

An unusual feature in pterosaurian ichnology is the occurrence of so-called 'manus-only' trackways or trampling areas bearing exclusively manus prints. Parker & Basley (1989) reported such ichnites from the Late Cretaceous of Utah, United States. Pascual Arribas & Sanz Pérez (2000) reported several horizons bearing only, or mainly, manus prints in the Early Cretaceous of Soria, Spain. The same feature has been discussed by Lockley *et al.* (1995), who suggested that the manus prints were made by buoyant and feeding animals. No 'manus-only' trackways have been found at Crayssac, but on almost all the excavated surfaces, isolated manus prints or groups of isolated manus prints are very common. Bennett (1997), Unwin (1997b) and Mazin *et al.* (1997) also noted that the manus prints are significantly more deeply impressed than the pes prints when associated in trackways.

As noted by Unwin (1997b), it is unlikely that pterosaurs proceeded on forelimbs alone. Several authors (e.g. Wellnhofer 1991; Lockley*et al.* 1995;

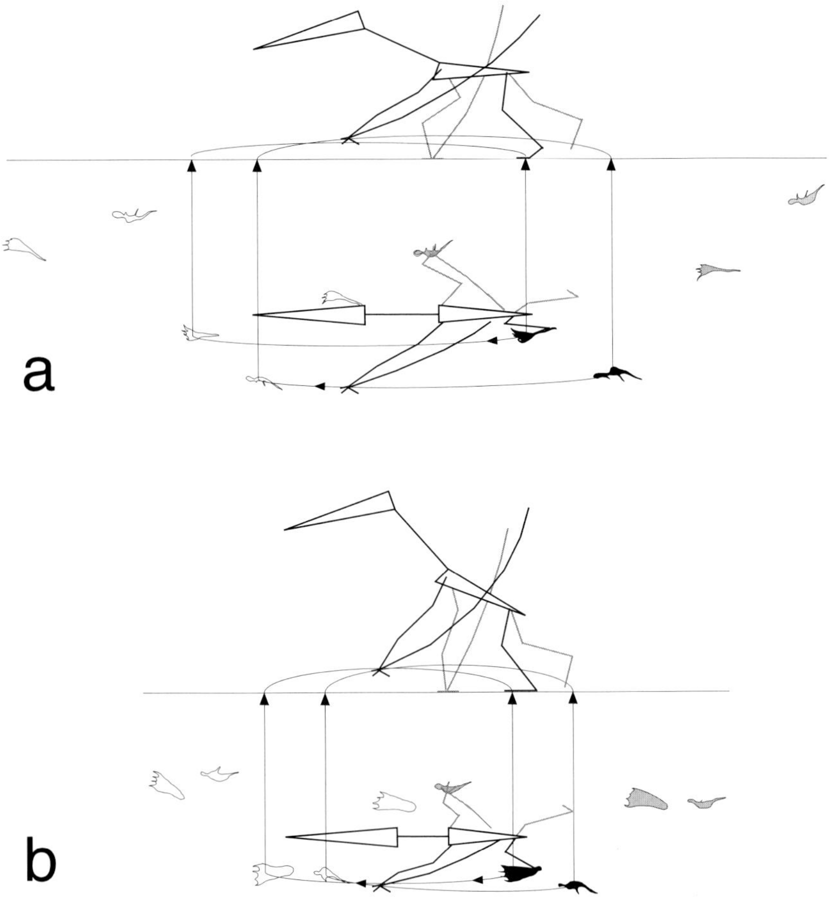

Fig. 7. Pterodactyloid stances according to velocity. (**a**) High-velocity walking or running. The body is subhorizontal, the strides are long and the manus trackway is wide. (**b**) Low-velocity walking. The body is semi-erect, the strides are relatively short and the manus trackway is narrow.

Bennett 1997; Unwin 1997b; Mazin *et al.* 2001a, c) pointed out that, in the particular body plan of pterodactyloid pterosaurs, the centre of mass is anteriorly displaced. Thus, during terrestrial locomotion, a larger part of the weight would have been transmitted through the forelimbs, resulting in deeper impression of the manus. Furthermore, a biometrical study of one particularly well-preserved trackway from Crayssac (specimen CR99.43) shows that the pes print surface area reaches 390 mm^2 (pes sole: 310 mm; webbing: 80 mm) while the manus print surface area is only 50 mm^2. Thus, if the fore- and hindlimbs had supported the same mass, the pressure would have been around 8 times higher for the manus than for the foot. Moreover, the anterior displacement of the centre of mass significantly increased the weight transmitted through the forelimbs and consequently increased the pressure of the manus. This higher pressure on the substrate is consistent with the deeper manus prints, but may be not sufficient to explain the 'manus-only' trackways, since they should correspond to trackways in which the manus penetrates the substrate while the pes does not mark at all. From field observations at Crayssac, several hypotheses can be proposed to explain such unexpected trackways. For example, it could be suggested that, on a quite dry and hard substrate, the pressure on the manus leads it to penetrate the substrate whereas the pes does not leave an impression because of the combination of a large area and a low weight loading. This suggestion can be questioned. The anterior part of the pes, at the metatarso-phalangeal joint is always the deepest part of the pes prints, probably due to the pes thrust when the pes is leaving the ground. This area, or the pes claws at least, should have impressed the substrate (see Fig. 2h). Furthermore, most, if not all of the isolated manus prints or the 'manus-only' trample areas observed at

Crayssac have been made on wet, non-competent substrate. Conversely, it could be suggested that, on a fluid and non-competent mud, manus and pes normally impress the substrate, as in all the common trackways, but the shallower pes prints are more quickly closed up or collapsed because of the mud fluidity. Such partially closed pes prints exist in some trackways at Crayssac (e.g. CR01.10, Figs 1d & 2g) but the closure concerns only the metatarsal area, probably because it is the shallowest part of the print. In the same way, the shallow pes prints could have been washed away by the high tide whereas the deeper manus prints were not erased. The condition of deposition of the micritic substrate in Crayssac, under very low energy, is not consistent with this suggestion (Mazin *et al.* 1995, 1997). Another explanation, linked to the non-competent substrate itself, could be suggested. If the animal walked on a superficial fluid substrate, the manus and pes penetrated this mud and could have left prints on the underlying level, leading to 'underprints', while the superficial fluid mud closed up and did not preserve traces. In such a situation, the manus could have reached the underlying level through the fluid mud and left an underprint, while the largest pes with a low weight loading did not reach the underlying level through the fluid mud. Thus, the enigmatic 'manus-only' pterodactyloid trackways could find explanation either in the combination of dynamics and substrate competence, or in pterosaur behaviour, as suggested by Lockley *et al.* (1995). The hundreds of pterodactyloid tracks available on the ichnite-bearing surfaces of Crayssac will raise arguments, and several works are in preparation on surfaces bearing sets of associated pterosaur simple prints and underprints (Mazin *et al.* in prep.) and sets of associated dinosaur and pterosaur prints (Griffiths *et al.* in prep.). Finally, it is noteworthy that such an ichnologic particularity is inconsistent with crocodilian trackways and provides a further argument against the crocodilian hypothesis (see above).

Conclusion: terrestrial abilities of pterodactyloids

After several decades of controversy, pterodactyloid trackways can now be recognized with certainty. However, the terrestrial abilities of pterosaurs is still quite controversial. There is little doubt that our knowledge of pterosaurian terrestrial locomotion will increase in the future, as more ichnological data become available. Presently, pterosaurian trackways from Crayssac allows us to conclude that:

(1) Almost all the pterosaurian trackways excavated in Crayssac can be referred to the ichnogenus *Pteraichnus* Stokes 1957, which is assigned to pterodactyloid pterosaurs. Concerning the very rare non-pterodactyloid pterosaurian trackways, manus and pes print morphology is clearly different (Mazin *et al.* 2001b).

(2) All the known pterodactyloid trackways show the same general pattern, which cannot be confused with any other known Mesozoic tetrapods, notably crocodilian or non-pterodactyloid trackways. In its general morphology (e.g. pes print overstepping, manus to pes gait-width ratio), as well as in the manus print morphology (strict tridactyly), *Pteraichnus* are strongly different from crocodilian trackways.

(3) All the pterodactyloid trackways from Crayssac, as well as pterodactyloid trackways from elsewhere, are quadrupedal.

(4) Regarding the pterosaurian trackways from Crayssac, the terrestrial capabilities of pterodactyloids were much greater than previously thought. Pterodactyloids were able to walk without apparent difficulties and some trackways demonstrate their unexpected high terrestrial velocity.

Nevertheless, it would be an oversimplification to conclude that one ichnological specimen (or even 30 specimens, as at Crayssac) reflects the diversity of the Pterosauria throughout the Mesozoic. From the diversity of pterosaurian morphology, it is clear that various modes of terrestrial locomotion existed in pterosaurs and, presently, the mechanisms of taking off and landing remain unknown.

The authors are grateful to the public councils and companies that have supported the field excavation of Crayssac. They also thank the two referees, C. Bennett and D. Martill, who have made very good and useful remarks to improve the manuscript.

References

ABEL, O. 1925. *Geschichte und Methode der Rekonstruktion Vorzeitlicher Wirbeltiere*. Gustav Fischer, Stuttgart.

BAIRD, D. 1980. A prosauropod dinosaur trackway from the Navajo Sandstone (Lower Jurassic). *In*: JACOBS, L. L. (ed.) *Aspects of Vertebrate History*, Museum of Northern Arizona Press, Flagstaff, 219–230.

BENNETT, S. C. 1990. A pterodactyloid pterosaur pelvis from the Santana Formation of Brazil: implications for terrestrial locomotion. *Journal of Vertebrate Paleontology*, **10**, 80–85.

BENNETT, S. C. 1992. Reinterpretation of problematical tracks at Clayton Lake State Park, New Mexico: not one pterosaur, but several crocodiles. *Ichnos*, **2**, 37–42.

BENNETT, S. C. 1997. Terrestrial locomotion of pterosaurs: a reconstruction based on *Pteraichnus* trackways. *Journal of Vertebrate Paleontology*, **17**, 104–113.

BILLON-BRUYAT, J-P. AND MAZIN, J-M., 2003. The systematic problem of Tetrapod ichnotaxa: the case study of

Pteraichnus Stokes, 1957 (Pterosauria, Pterodactyloidea. *In*: BUFFETAUT, E. & MAZIN, J.-M. (eds) *Evolution and Palaeobiology of Pterosaurs*. Geological Society, London, Special Publications, **217**, 315–324,

COLLINI, C. A. 1784. Sur quelques Zoolithes du Cabinet d'Histoire naturelle de S.A.S.E. Palatine et de Bavière, à Mannheim. *Acta Academiae Theodoro-Palatinae Mannheim*, pars physica, **5**, 58–103.

CUVIER, G. 1801. Reptile volant. *Magasin Encyclopédique, Paris*, **9**, 60–82, Paris.

CUVIER, G. 1809. Mémoire sur le squelette fossile d'un reptile volant des environs d'Aichstedt, que quelques naturalistes ont pris pour un oiseau, et dont nous formons un genre de Sauriens, sous le nom de Ptéro-Dactyle. *Annales du Muséum d'Histoire Naturelle, Paris*, **13**, 424–437.

FUENTES VIDARTE, C. 2001. A new species of *Pteraichnus* for the Spanish Lower Cretaceous: *Pteraichnus cidacoi*. *Strata, Serie 1*, **11**, 44–46.

GILLETTE, D. D & THOMAS, D. A. 1989. Problematical tracks and traces of Late Albian (Early Cretaceous) Age. Clayton Lake State Park, New Mexico, USA. *In*: GILLETTE D.D. & LOCKLEY M.G. (eds), *Dinosaur Tracks and Traces*. Cambridge University Press, Cambridge, 337–342.

HAUBOLD, H. 1971. Ichnia amphibiorum et reptiliorum fossilium. *In*: WELLNHOFER, P. (ed.) *Handbuch der Paläoherpetologie*. Gustav Fisher, Stuttgart, Teil 18, Verlag, 184 pp.

HUNT, A. P., LOCKLEY, M. G., HUPS, K. & SCHULTZ, R. 1995. Jurassic vertebrate paleontology of Cactus Park, west-central Colorado. *Geological Society of America, Abstracts with Program, Rocky Mountains Section*, **27**, 4, 15.

LOCKLEY, M. G. 1991. *Tracking Dinosaurs*. Cambridge University Press, Cambridge, 238 pp.

LOCKLEY, M. G. & HUNT, A.P. 1995. *Dinosaur Tracks and Other Fossil Footprints of the Western United States*. Columbia University Press, New York, 338 pp.

LOCKLEY, M. G., LIM, M., HUH, S-K., YANG, S-Y., CHUN, S.S. & UNWIN, D. M. 1997. First report of pterosaur tracks from Asia, Chollanam Province, Korea. *Journal of the Paleontological Society, Korea*, Special Publications, **2**, 17–32.

LOCKLEY, M. G., LOGUE, T. J., MORATALLA, J. J., HUNT, A. P., SCHULTZ, R.J. & ROBINSON, J. W. 1995. The fossil trackway *Pteraichnus* is pterosaurian, not crocodilian: implication for the global distribution of pterosaur tracks. *Ichnos*, **4**, 7–20.

LOGUE, T. J. 1977. Preliminary investigations of pterodactyl tracks at Alcova, Wyoming. Wyoming Geological Association. *Earth Science Bulletin*, **10**, 29–30.

LOGUE, T. J. 1994. Alcova, Wyoming tracks of *Pteraichnus saltwashensis* made by pterosaurs. *Geological Society of America, Abstracts with Program, South Central Region*, **26**, 10.

MAZIN, J-M., BILLON-BRUYAT, J-P., HANTZPERGUE, P., & LAFAURIE, G. 2001a. The pterosaurian trackways of Crayssac (southwestern France). *Strata, Serie 1*, **11**, 57–59.

MAZIN, J-M., BILLON-BRUYAT, J-P., HANTZPERGUE, P., & LAFAURIE, G. 2001b. Could they be the first rhamphorhynchid tracks? Yes! *Strata, Serie 1*, **11**, 64–65.

MAZIN, J-M., BILLON-BRUYAT, J-P. & ROLLER, P. 2001c. If they did it, they were able to do it! A computational reconstruction of pterodactyloid terrestrial locomotion from trackways. *Strata, Serie 1*, **11**, 60–63.

MAZIN, J-M., HANTZPERGUE, P., BASSOULLET, J-P., LAFAURIE, G. & VIGNAUD, P. 1997. Le gisement de Crayssac (Tithonien inférieur, Quercy, Lot, France): découverte de pistes de dinosaures en place et premier bilan ichnologique. *Comptes Rendus de l'Académie des Sciences, Paris*, **325**, 733–739.

MAZIN, J-M., HANTZPERGUE, P., LAFAURIE, G. & VIGNAUD, P. 1995. Des pistes de ptérosaures dans le Tithonien de Crayssac (Quercy, Lot). *Comptes Rendus de l'Académie des Sciences, Paris*, **321**, 417–424.

MEIJIDE FUENTES, F. 2001. Pterosaur tracks in Oncala Moutain Range (Soria, Spain). A new ichnospecies: *Pteraichnus vetustior*. *Strata, Serie 1*, **11**, 70–71.

MEIJIDE CALVO, M. 2001. Pterosaur trace in Oncala Berriasian (Soria, Spain). New ichnospecies: *Pteraichnus manueli*. *Strata, Serie 1*, **11**, 72–74.

MEIJIDE CALVO, M. & FUENTES VIDARTE, C. 1999. Huellas de pterosaurios en el Weald de Soria (Espana). *Actas I Jornadas Internacionales sobre Paleontologia de Dinosaurios y su Entorno*, Salas de los Infantes, Burgos, 397–406.

PADIAN, K. 1983. A functional analysis of flying and walking in pterosaurs. *Paleobiology*, **9**, 218–239.

PADIAN, K. 1984. The origin of pterosaurs. *In*: REIF, W. E. & WESTPHAL, F. (eds). *Third Symposium on Mesozoic Terrestrial Ecosystems and Biola, Tübingen*, 163–168.

PADIAN, K. 1985. The origin and aerodynamics of flight in extinct vertebrates. *Paleontology*, **28**, 413–433.

PADIAN, K. & OLSEN, P.E. 1984. The fossil trackway *Pteraichnus*: not pterosaurian, but crocodilian. *Journal of Paleontology*, **58**, 178–184.

PARKER, L. & BASLEY, J. 1989. Coal mines as localities for studying trace fossils. *In*: GILLETTE, D. D. AND LOCKLEY, M. G. (eds) *Dinosaur Tracks and Traces*. Cambridge University Press, Cambridge, 353–359.

PASCUAL ARRIBAS, C. & SANZ PÉREZ, E. 2000. Huellas de pterosaurios en el grupo Oncala (Soria, Espana). *Pteraichnus palaciei-saenzi*, nov. icnosp. *Estudios Geologicos*, **56**, 73–100.

PRINCE, N. K. & LOCKLEY, M. G. 1989. The sedimentology of the Purgatoire Tracksite Region, Morrison Formation of South-eastern Colorado. *In*: GILLETTE D. D. AND LOCKLEY, M. G. (eds) *Dinosaur Tracks and Traces*, Cambridge University Press, Cambridge, 155–163.

RODRIGUEZ DE LA ROSA, R. A. 2001. Pterosaur tracks from the Late Cretaceous of Northern Mexico: paleoecological and anatomical implications. *Strata, Serie 1*, **11**, 70–71.

SOEMMERRING, S. T. VON. 1812. Über einen *Ornithocephalus*. *Denkschriften der Akademie der Wissenschaften München, Mathematisch-Physik Klasse*, **3**, 89–158.

SOEMMERRING, S. T. VON. 1817. Über einen *Ornithocephalus brevirostris* der Vorwelt. *Denkschriften der Akademie der Wissenschaften München, Mathematisch-Physik Klasse*, **6**, 89–104.

STOKES, W. L. 1957. Pterodactyl tracks from the Morrison Formation. *Journal of Paleontology*, **31**, 952–954.

STOKES, W. L. 1978. Animal tracks in the Navajo-Nugget Sandstone. *University of Wyoming, Contributions to Geology*, **16**, 103–107.

STOKES, W. L. & MADSEN, J. H. JR. 1979. Environmental significance of pterosaur tracks in the Navaro Sandstone (Jurassic) Grand County, Utah. *Brigham Young University, Geological Studies*, **26**, 21–26.

UNWIN, D. M. 1986. Tracking the Dinosaurs. *Geology Today*, **2**, 168–169.

UNWIN, D. M. 1987. Pterosaur locomotion. Joggers or waddlers? *Nature*, **327**, 13–14.

UNWIN, D. M. 1988. New remains of the pterosaur *Dimorphodon* (Pterosauria: Rhamphorhynchoidea) and the terrestrial locomotion of early pterosaurs. *Modern Geology*, **13**, 57–68.

UNWIN, D. M. 1989. A predictive method for the identification of vertebrate ichnites and its application to pterosaur tracks. *In*: GILLETTE, D. D. & LOCKLEY, M. G. (eds), *Dinosaur Tracks and Traces*. Cambridge University Press, Cambridge, 259–274.

UNWIN, D. M. 1990. Pterosaurs: back to the traditional model? *Trends in Ecology and Evolution*, **14**, 263–268.

UNWIN, D. M. 1997a. Locomotory roles of the hind limbs in pterosaurs. *Journal of Vertebrate Paleontology*, **17**, 82A.

UNWIN, D. M. 1997b. Pterosaur tracks and the terrestrial ability of pterosaurs. *Lethaia*, **29**, 373–386.

UNWIN, D. M. & BAKHURINA, N. N. 1994. *Sordes pilosus* and the nature of the pterosaur flight apparatus. *Nature*, **371**, 62–64.

WELLNHOFER, P. 1978. Pterosauria. *In*: WELLNHOFER, P. (ed.) *Handbuch der Paläoherpetologie*. Gustav Fischer, Stuttgart, Teil 19, 82 pp.

WELLNHOFER, P. 1985. Neue Pterosaurier aus der Santana Formation (Apt) der Chapada do Araripe, Brasilien. *Palaeontographica A*, **187**, 105–182.

WELLNHOFER, P. 1988. Terrestrial locomotion in pterosaurs. *Historical Biology*, **1**, 3–16.

WELLNHOFER, P. 1991. *The Illustrated Encyclopedia of Pterosaurs*, Salamander, London, 192 pp.

WELLNHOFER, P. & VAHLDIEK, B-W. 1986. Ein Flugsaurier-Rest aus dem Posidonienschiefer (Unter-Toarcium) von Scmanuselah bei Braunschweig. *Paläontologische Zeitschrift*, **60**, 329–340.

WRIGHT, J. L., UNWIN, D. M., LOCKLEY, M. G. & RAINFORTH, E. 1997. Pterosaur tracks from the Purbeck Limestone Formation of Dorset, England. *Proceedings of the Geologists' Association, London*, **108**, 39–48.

Pterosaur swim tracks and other ichnological evidence of behaviour and ecology

MARTIN G. LOCKLEY & JOANNA L. WRIGHT

Department of Geology, University of Colorado at Denver, PO Box 173364, Denver, Colorado, CO 80217, USA (e-mail: mlockley@carbon.cudenver.edu)

Abstract: New finds of pterosaur tracks (cf. *Pteraichnus*) from the Summerville and Sundance formations (Late Jurassic) of western North America associated with elongate scrape marks and small circular paired depressions indicate that pterosaurs could swim, or at least float in water, and that they may have fed on small animals living at the sediment surface in shallow water. In this respect their behaviour resembled certain seabirds. The ichnological evidence for pterosaur 'swim' and 'feeding' traces consists of scrape marks interpreted as traces left when the paddling limb of a pterosaur registered on the substrate, and small circular paired depressions as traces left when the beak of a pterosaur was probing for food. Such traces resulting from aquatic activity are consistent with the nearshore environments in which pterosaur bones and trackways have been found.

There is growing evidence for an extensive and complex *Pteraichnus* ichnofacies in the Late Jurassic of western North America; over 80 specimens have been collected with many more remaining in the field. In particular we draw attention to the large quantitative data base that is available for morphometric, size-frequency studies and the potential for behavioural studies of individual locomotion and flocking. In addition the sites hold considerable promise for understanding what appears to be the world's largest pterosaur ichnofacies in the context of ancient depositional environments and regional sequence stratigraphy.

Introduction

Since the discovery of pterosaurs almost 150 years ago their morphology and lifestyle have been the subject of much discussion and speculation. The earliest known pterosaur tracks are from the Late Jurassic. This coincides with the appearance of pterodactyloid pterosaurs in the fossil record and a significant increase in global pterosaur diversity. Nearly all known pterosaur tracks appear to have been made by pterodactyloids.

The western United States is the cradle of pterosaur tracking. It was in this area that William Stokes found the first purported pterosaur tracks: ichnogenus *Pteraichnus* (Stokes 1957). Subsequent discoveries in Wyoming (Logue 1977) and Oklahoma (West 1978) suggested the widespread distribution of *Pteraichus*, as has subsequently been confirmed. Unfortunately Stokes' original find was disputed by Padian & Olsen (1984), who favoured a crocodilian track-maker. This led to a decade in which the existence of pterosaur tracks was either denied or ignored. Despite a vertebrate track renaissance in recent years, which has lead to the discovery and preliminary description of many new pterosaur tracksites (Lockley *et al.* 1995, 1996, 1997, 2001; Mazin *et al.* 1995, 1997; Wright *et al.* 1997; Calvo & Moratalla 1998; Meijide Calvo & Fuentes Vidarte 1999; Pascual Arribas & Sanz Pérez, 2000; Garcia-Ramos *et al.* 2000, 2001; Calvo & Lockley 2001) the ichnofaunas that are now available have not received the detailed attention they deserve. This paper is a contribution to current efforts to document this extensive ichnological record more fully.

Of interest in the present study are the pterosaur tracks of the Late Jurassic found in Arizona, Utah, Colorado, Wyoming and Oklahoma. They form part of a western North American regional complex of ichnofaunas best understood in their palaeobiogeographical and facies context. Stokes (1957) first showed clearly many of the characteristic features we now associate with pterosaur tracks: i.e. elongate four-toed pes imprints with subparallel toes and a narrow heel, and highly distinctive and asymmetric three-fingered elongate manus imprints. However Stokes' discovery of a single trackway from Arizona consisted of footprints that were not well enough preserved to show detailed anatomical structures of the track-maker's foot. It was not until a great deal more evidence accumulated, especially from North America (Lockley *et al.* 1995, 1996, 2001), but also from Europe (Mazin *et al.* 1995, 1997; Wright *et al.* 1997; Meijide Calvo & Fuentes Vidarte 1999; Meijide Calvo *et al.* in press; Pascual Arribas & Sanz Pérez 2000; Garcia Ramos *et al.* 2000, 2001), that the conclusion that prints of this morphology were indeed made by pterosaurs was inescapable. Further lively discussion has ensued and a number of pterosaur experts have come down strongly in favor of the interpretation that *Pteraichnus* is pterosaurian (Bennett 1997; Unwin 1997), rather than crocodilian as maintained by Padian & Olsen (1984) and Padian (1998). For earlier iterations of such discussions, that predate most *Pteraichnus* discoveries see

From: BUFFETAUT, E. & MAZIN, J-M. (eds) 2003. *Evolution and Palaeobiology of Pterosaurs.* Geological Society, London, Special Publications, **217**, 297–313. 0305–8719/03/$15 © The Geological Society of London 2003.

Padian (1983), Wellnhofer (1988) and Unwin (1987, 1989).

The Late Jurassic is an especially interesting time in pterosaur evolution. It was during this time period that the two main groups of pterosaurs, the rhamphorhynchoids and the pterodactyloids, were experiencing a major evolutionary transition or 'turnover.' The rhamphorhynchoids, the original pterosaurs which had had the skies to themselves, as far as vertebrates were concerned, since the Late Triassic, some 70 Ma previously, were being displaced by the pterodactyloids which were to replace completely them by the Cretaceous. Some of these derived Cretaceous forms reached sizes far in excess of any previous land vertebrate and, in fact, still hold the record for the all-time biggest flying vertebrate; the largest, e.g. *Quetzalcoatlus*, are estimated to have had wing spans of more than 12 m, with corresponding footprints about 30 cm in length (Lockley *et al.* 1997; Hwang *et al.* 2002).

Interestingly, pterosaur tracks are not known until the Late Jurassic and most pterosaur tracks so far described in the scientific literature have been attributed to pterodactyloid pterosaurs. The possible existence of one or two rhamphorhynchoid tracks hinges on the interpretation of faint traces of the fifth pes digit and is at best a tentative conclusion.

Walking v. swimming: the ichnological evidence

The vast majority of footprints and trackways attributed to fossil vertebrates appear to have been made by terrestrial animals progressing over relatively soft, emergent, subaerially exposed shoreline or wetland habitat substrates covered by very shallow water. In the majority of cases these trackways represent walking progression, though a few trackways have been interpreted as evidence of running. The same general conclusion has been reached in the study of pterosaur tracks: i.e. most represent walking progression on emergent shoreline substrates, presumably very close to the water's edge. However, some tracks and traces associated with the North American Late Jurassic pterosaur-dominated assemblages appear to represent 'swim' and/or 'feeding' traces. These form the subject of the descriptions and discussions presented below.

First, however, it is helpful to review what is known of vertebrate swim traces in general. Leaving aside fish, which may produce sinuous tail traces, or burrows, we focus on traces attributed to aquatic tetrapods, such as crocodiles and turtles, or terrestrial vertebrates that may have periodically taken to the water. Common sense dictates that such animals, and here we can include so-called 'waterbirds' and 'seabirds', which only leave traces when their feet or other appendages contact the substrate. In the case of diving animals this may happen in water of any depth greater than the animal's minimum body dimensions but, in the case of animals that only float or dabble, we are able to observe (in the case of modern animals) or estimate the shallow water depths involved, as in the case of gulls and ducks (Cadée, 1990).

There is a modest literature on the tracks attributed to fossil vertebrates involved in creating purported swim tracks. Unfortunately, for historical reasons, this literature began with ambiguous reports of swimming sauropods inferred from the presence of manus-dominated trackways (Bird 1944; Ishigaki 1989). These have subsequently been shown to be underprints transmitted during normal walking progression (Lockley & Rice 1990; Lockley 1991). Thus the focus of attention in the somewhat limited field of 'swim track studies' has shifted somewhat to an analysis of the tracks of habitually aquatic vertebrates, such as crocodiles and turtles.

In theory, traces made by vertebrates propelling themselves along in water would be expected to grade into tracks of normal terrestrial locomotion in one direction as the water became shallower and to disappear in the other direction as the water became deeper and the animal was free-floating (McAllister 1989a, b). Unfortunately exposures of track-bearing surfaces are not continuous enough to permit such observations, nor is animal behaviour or sedimentary environment preservation sufficiently predictable to allow formation and preservation of such linear emergent-submergent, or submergent-emergent trace fossil sequences, except perhaps in rare, as yet unreported cases. The reality of the situation is that aquatic or swimming vertebrates can engage in a variety of behaviours that are largely unconstrained by the effects of gravity. In such a buoyant state their appendages may touch the bottom substrate in an irregular or random fashion, with highly variable degrees of pressure and duration. Thus the resultant traces may be highly irregular, incomplete and often isolated.

The best we can hope for is that, in some cases, there will be clear evidence of progression in one direction while the appendages are engaged in regular paddling motion that makes contact with the bottom substrate, either inadvertently or deliberately to gain purchase during locomotion. Examples of such patterns are know for the famous turtle tracks of the Cerin lithographic limestones of France (Thulborn 1989; Lockley & Meyer 2000), which have also, in our opinion, been misinterpreted as hopping dinosaur tracks (Bernier *et al.* 1984). Another 'trackway' of turtle footprints, evidently made subaqueously, was reported from channel deposits in the Salt Wash Member of the Morrison Formation, where an animal was 'evidently walking' or 'shuffling' along the river bottom (Foster *et al.* 1999). We also, herein, present an example of pterosaur tracks inferred to show such regular swimming or paddling traces.

However, there are also other ichnological clues that help us to recognize and identify 'swim' tracks. Most notable among these are the preponderance and appearance of toe impressions. We already know that, in walking progression, the foot is in a forwards and downwards motion when it registers on the substrate. This inherent locomotor characteristic is accentuated when a swimming or buoyant animal touches the substrate with its toe tips. The result is to produce footprints that are 'incomplete', with such features as 'posterior overhangs and reflectures' and 'striations and claw marks' aligned in the direction of foot motion (McAllister 1989a, b). Using a large assemblage of such traces from the Cretaceous Dakota Group of Kansas as a forum for discussion of such 'swim traces', McAllister (1989a, b) inferred a series of swimming ornithischian traces. Such interpretations may have been influenced by Coombs (1980), who attributed somewhat similar parallel striations to swimming theropod dinosaurs. However, the Kansas tracks and several other similar assemblages from the Dakota Group of Colorado and New Mexico have also been interpreted as the tracks of crocodilians (Lockley & Rice 1990; Bennett 1993; Lockley & Hunt 1994, 1995). Thus, trace fossil evidence of swimming dinosaurs remains very dubious.

A final feature of purported tetrapods swim traces such as those reported from Kansas is that they often occur in significant numbers in strong parallel alignment. This suggests that current activity, prevailing winds or even the progression of gregarious animals in a single direction may also be a factor in producing such distinctive patterns. Examples from the *Pteraichnus* assemblages are given below.

In shoreline settings, where most tracks occur, it is also possible to find tracks suggestive of feeding behaviour. In the case of modern waterbirds or seabirds, we know that they dabble in shallow water and may leave traces made by the beak. The same was presumably true of ancient waterbirds and pterosaurs, and it has already been suggested that a few such traces, consisting of paired beak-tip impressions are found in the fossil record (Lockley *et al.* 1995; Wright *et al.* 1997). A few examples of such traces are illustrated herein. The probability of such feeding (and swim) traces being attributed to pterosaurs is increased by the fact that most of these assemblages are otherwise essentially monospecific, consisting only of pterosaur (*Pteraichnus*) walking traces, and were found in habitats that appear to have been unsuitable for other vertebrates.

Geological and palaeobiogeographical setting of pterosaur trace-bearing assemblages

The Late Jurassic (Oxfordian–Kimmeridgian) pterosaur tracks from the western United States are found in the upper units of the Sundance/Summerville Formations and in the lower/basal units of the Morrison (Salt Wash Member or its undifferentiated equivalents). The track-bearing portion of the Sundance Formation in Wyoming was deposited in shallow marine or periodically emergent tidal-flat environments and consists of carbonate deposits (Fig. 1). The roughly equivalent Summerville Formation of the more southerly states was also deposited under marine/coastal environments and consists of mainly terrigenous siliciclastic sediments. The Salt Wash Member and undifferentiated basal sequences of the overlying Morrison Formation are interpreted as fluvial and lacustrine deposits. To date tracks have been found in the following locations and stratigraphic units:

(1) Salt Wash Member of the Carrizo Mountains, northeastern Arizona (Stokes 1957). Type specimen is in the University of Utah collections.
(2) The top of the undifferentiated Summerville-Bluff Formation, Carrizo Mountains, northeastern Arizona (Lockley & Mickelson 1997).
(3) Upper Summerville of the Del Monte Mines area, near Bullfrog, eastern Utah (Lockley *et al.* 1995). Specimens include CU-MWC 188.1–9, 11–12, 15–16, 46–50, 52–58 and 64–73.
(4) Upper Summerville of the Ferron area east-central Utah (J. Bishop pers. comm.). Specimens include CU-MWC 188.74–99.
(5) Summerville Formation of the Cactus Park area, near Grand Junction, western Colorado. Specimens include CU-MWC 188.20 and 59–63.
(6) Summerville Formation Furnish Canyon area of Baca County, southeastern Colorado (Lockley *et al.* 1996, 2001). Specimens include CU-MWC 188.36–44.
(7) Morrison Formation of the Oklahoma panhandle (Lockley *et al.* 2001). Specimens include CU-MWC 188.45.
(8) Sundance Formation of the Alcova Lake area, central Wyoming (Logue 1977; Lockley *et al.* 1995). Specimens include University of Wyoming specimens (UW 12360–12372) and Tate Museum specimens (Casper, Wyoming) 0049.1–49.12.

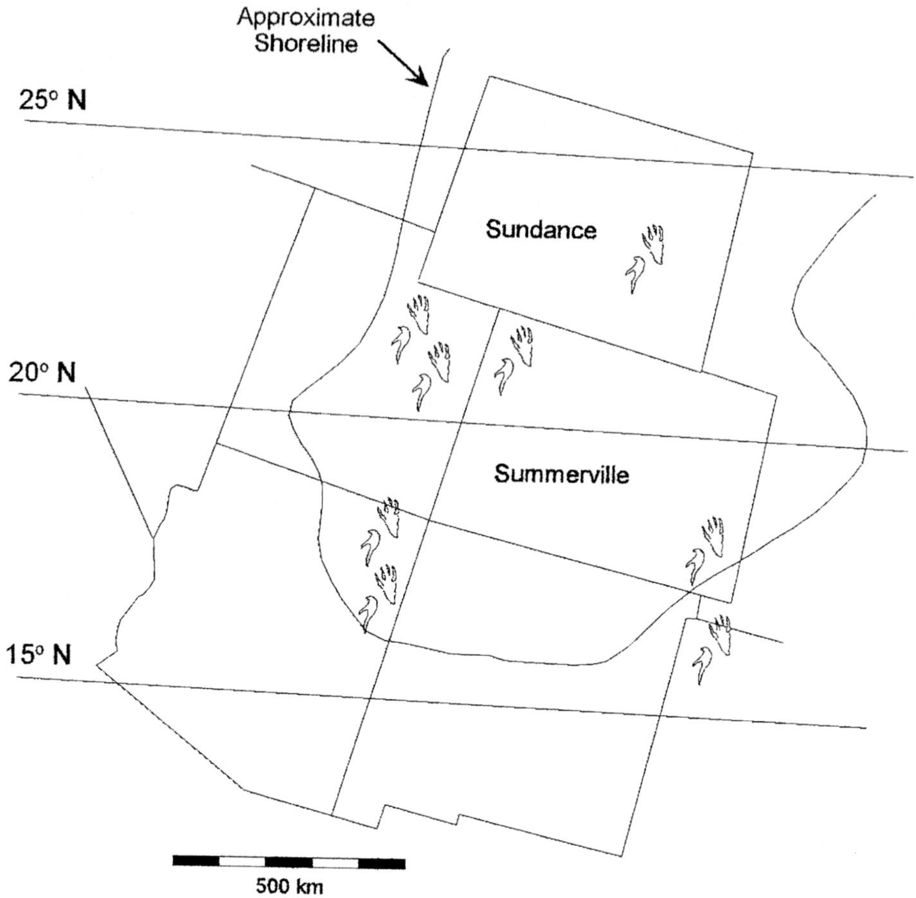

Fig. 1. Locality map showing important Late Jurassic sites mentioned in the text. States with pterosaur tracksites, clockwise from north, are Wyoming, Colorado, Oklahoma, Arizona and Utah.

Significant material from most of these units has been collected, especially in recent years, and is reposited in the University of Colorado at Denver–Museum of western Colorado joint collections (prefix CU-MWC) and in the University of Wyoming (UW) collections. Among the catalogued material listed above there are at least 80 specimens.

Thus the three main lithological units in which pterosaur tracks are found in the western United States are the Summerville and Sundance Formations and the overlying Morrison Formation, although, as noted below, there is some debate over the boundaries between the Summmerville–Sundance and the overlying Morrison Formation; these units are to some extent time equivalent and, where sequential track-bearing units occur in stratigraphic sequence, as for example in the southeastern Colorado and Oklahoma sites, the litho- and chronostratigraphic separation is slight. The Morrison Formation is universally considered to be Late Jurassic, while the Summerville Formation, although considered Mid-Jurassic in its basal and basinal part (Breihaupt *et al.* 2001; Kvale *et al.* 2001), is probably Late Jurassic in its upper part, which consists of the nearshore horizons where the pterosaur tracks are found.

For most, if not all of the Mid-Jurassic, Wyoming is thought to have been covered by a shallow sea, the Sundance Sea, while Colorado and Utah were also covered by this sea or were partially emergent as tidal flats (Anderson & Lucas 1996). The recent discovery of dinosaur tracks in the upper part of the Sundance Formation near Shell, northern Wyoming (Breihaupt *et al.* 2001; Kvale *et al.* 2001) upset this notion somewhat by implying that there was a regressive phase (or phases) during the Mid-Jurassic. This evidence also suggests either that the sea was quite shallow or that sea level fluctuated more than previously thought.

During the Late Jurassic the sea retreated northwards, and marine sediments were still being deposited in Wyoming while the fluvial sediments of the Salt Wash Member (lower part of the Morrison Formation) were being laid down in western Colorado and Utah (Anderson & Lucas 1996). In eastern Colorado, the Morrison Formation is undifferentiated, resembling the Upper Brushy Basin Member of western regions, and in one locality in Oklahoma it contains pterosaur tracks stratigraphically above Summerville deposits with similar pterosaur tracks (Lockley *et al.* 2001).

The climate in this area in the Mid-Jurassic is thought to have been arid but this aridity was progressively reduced by the Late Jurassic, with the movement of the continent into more temperate latitudes. The horizons within the Sundance, Summerville and Morrison Formations are all associated to some degree with relatively high sea level (and base level) in the Sundance–Summerville marine embayment or epeiric sea. Indeed the early Late Jurassic was a time of global rise in sea level, leading to widespread aggradation of platform carbonate and other shelf and continental margin sedimentary deposits that were particularly suited to the formation and preservation of extensive track-bearing deposits. However, this does not mean that the pterosaur track deposits were uniform in composition. The Sundance deposits consist predominantly of shallow-water carbonates, locally containing normal marine faunas such as brachiopods and belemnites. By contrast the Upper Summerville track-bearing deposits consist of siliciclastic red bed deposits that lack body fossils but are locally associated with gypsum (e.g. controversial Tidwell Member of Utah, placed by some in the top of the Summerville and others at the base of the Morrison).

The only hitherto described *Pteraichnus* tracksite from the Morrison is associated with a mixed carbonate-siliciclastic lacustrine sequence found in south-eastern Colorado and the Oklahoma panhandle (Lockley *et al.* 2001). In various sections in the southern part of this embayment the transition from marginal marine water-lain units of the Upper Summerville is marked by the presence of palaeosols, calcareous pedogenesis and localized mixed carbonate and clastic pond and lake deposits. Thus we have a picture of a shallow marine embayment in which there were more open marine conditions to the north, allowing for the development of normal marine carbonates, and more restricted conditions to the south, leading to the development of local evaporative basins. Landwards and to the south of these marginal marine deposits, as the Sundance Summerville seas retreated northwards we see the development of fluviolacustrine deposits and the influx of the famous terrestrial, dinosaur-rich Morrison fauna. In many ways the pterosaurian ichnofauna represents a forerunner or precursor of the famous Morrison faunas, although the presence of pterosaurian tracks and skeletal remains in the Morrison assemblages proves that ptersoaurs remained in the region as the facies and palaeoenvironment changed. Presumably the pterosaurian capacity for flight and affinity for marginal marine habitats, especially in the Jurassic, helps to explain their abundance in pre-Morrison deposits that lack fossil evidence of any other higher vertebrates.

Description of the traces

Walking traces

Pterosaur tracks are morphologically conservative. Even though we now think pterosaurs to have been reasonably competent quadrupedal animals on the ground, the tracks do not in general give a picture of particularly agile or fleet terrestrial animals. The pace length of walking pterosaur tracks is 1 or 2 times the foot length (FL) while the stride length is around 3 or 4 FL (Fig. 2). Moreover the internal width of the pes trackway is between 1 and 2 FL, indicating an animal with a semi-erect rather than erect stance (Lockley *et al.* 1995). Since it is our contention that Jurassic swim tracks are dominated by partial *Pteraichnus* pes traces, it is important to be familiar with the typical configuration of complete pes tracks as found in walking trackways, which are found at most of the eight sites listed above.

Jurassic pterosaur pes tracks are approximately triangular in shape; they have four toes, subequal in length and terminating in narrow curving claws, with a narrow heel which is often faintly impressed. The pterosaur tracks from these units fall within a fairly narrow size range. The four-toed pes impressions range in length from 25 to 115mm (although most are around 70–80 mm long), representing animals with an estimated wing span of 0.2–1.2m (average c. 0.75 m). The length of the digits comprises approximately half the footprint length. Manus impressions are usually about the same length as their corresponding pes, although they are usually slightly narrower. The tridactyl manus impressions are interpreted as having been made by digits I, II and III.

Footprint length to stride length ratios are approximately 1:2.7 – 1:5 and outer trackway widths are 1–3 times footprint length. Manus impressions are either in line with or outside the pes impressions and usually fall behind the pes impressions in the trackway. The angulation pattern of the trackway is between 80° and 135°. Pes impressions are angled slightly outwards from the trackway mid-line at angles of up to 30°, although this is extreme and it is usually less than 10°. These data indicate that the

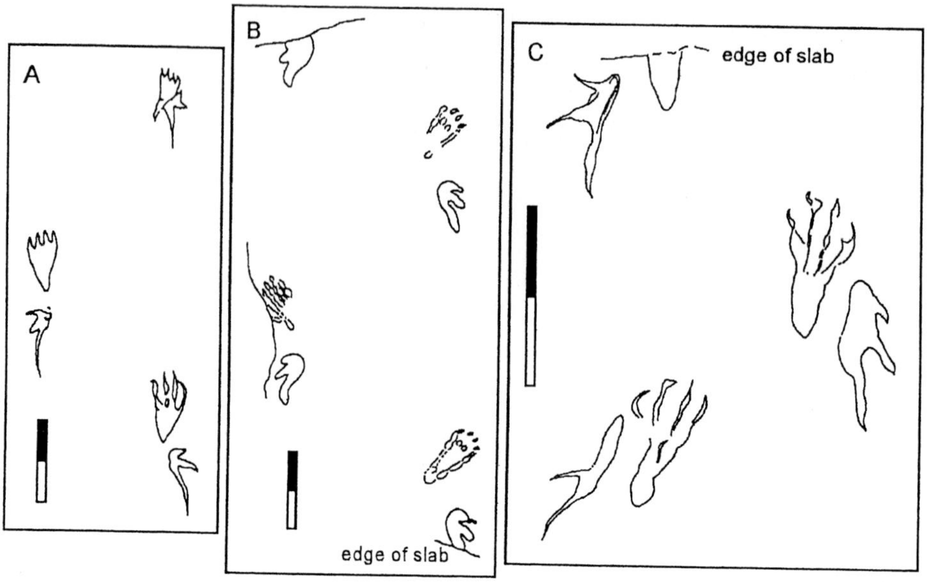

Fig. 2. Typical pterosaur walking trackways: (**a**) type specimen of *Pteraichnus saltwashensis* (after Stokes 1957). (**b**) type specimen of *P. stokesi* (after Lockley *et al.* 1995); (**c**) CU-MWC 188.1 (after Lockley *et al.* 1995). Scale bars 10 cm.

track-makers were not particularly speedy or graceful animals on land but the common occurrence of these tracks suggests that terrestrial forays were not rare.

Thus far only one possible Late Jurassic rhamphorhynchoid pterosaur track has been reported, implying that others may be attributed to this group. This track reveals a purported, faint impression of the fifth digit while the rest of the footprint is identical to other pterosaur tracks, usually interpreted as having been made by pterodactyloids. Pterodactyloid pterosaurs, however, do not have a fifth digit, so such tracks, if correctly interpreted, could not be attributed to this group. The impression of the fifth digit could have been made when the animal put down the extra digit for balance, or for some other behavioural reason, but because the hindlegs and the fifth digit of rhamphorhynchoids were so integral to the flight apparatus the rhamphorhynchoid inference is dubious. On balance it seems likely that rhamphorhynchoids had a much more limited terrestrial ability than pterodactyloids. Differences in the distribution of these two groups in the fossil record may reflect preservational biases and/or facies preferences. Since the track record of pterosaurs only starts in the Late Jurassic one could predict that it would be dominated by pterodactyloids, as the evidence seems to show. Thus the track record has the potential to indicate the relative abundance of these two groups.

Swim traces

Convincing examples of swim traces only occur at two of the eight tracksites listed above, i.e. at the Del Monte Mines and Alcova lakes localities. At least one of the tracks from the small sample at the Cactus Park locality may also be interpreted as swim tracks, though the evidence is less clear cut. The examples illustrated herein, however, are particularly well-preserved and instructive in arguing the case for swimming behaviour.

Specimen CU-MWC 186.46. This very well-preserved isolated pes print, is preserved as a natural cast in grey, indurated fine-grained sandstone (Fig. 3). The phalangeal pad and claw impressions are very clear. Two pads can be seen on digit I, three on digits II and III and possibly four on digit IV, although on this digit they all run into one another. One ?heel pad can be seen but, apart from this, the impression of the heel and the proximal part of the footprint is very shallow. Based on the association of this track with typical incomplete 'swim traces' we infer that the track may have been made by a buoyant or partially buoyant animal impressing most of the distal part of the foot on the substrate while progressing in/on shallow water. The track is much deeper and clearer than any known walking pes tracks. This suggests that the foot registered on a very soft, ductile but cohesive substrate, as might be found in a shallow subaqueous setting.

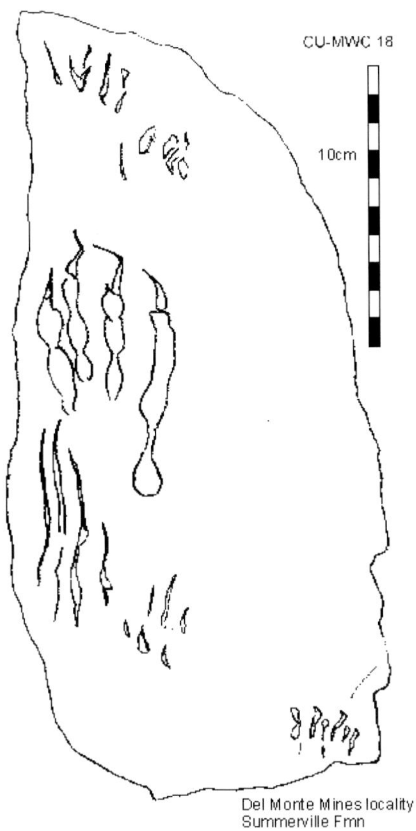

Fig. 3. Swim track specimen CU-MWC 186.46: one of the best examples of clear pes track definition with associated toe-tip traces/scratch marks.

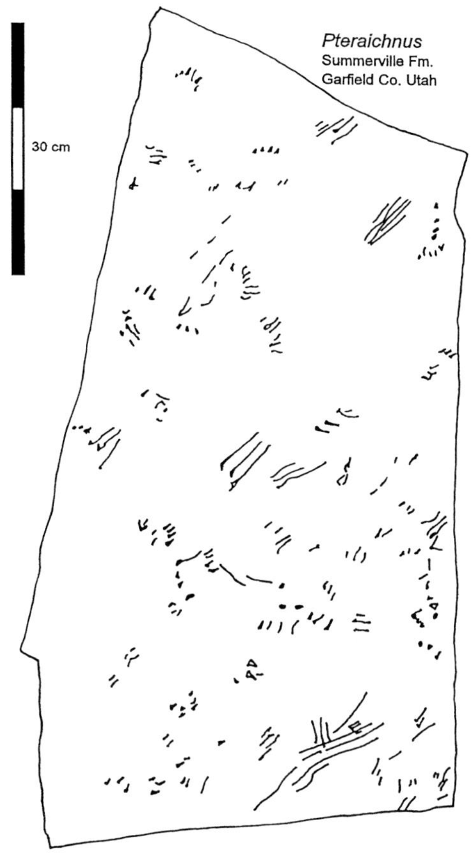

Fig. 4. Swim track specimen 188.58. For clarity only the best-defined swim tracks and striations marks are shown

In addition to the single fairly large footprint preserved here there are a number of other marks probably made by pterosaurs, perhaps even the same individual. The most obvious of these are some very fine (i.e. narrow) scrape marks which occur in groups of up to four on the same surface. Their association with the pterosaur footprint casts is further emphasised by the fact that they have the same orientation as the pterosaur claw traces. One can predict that, if a pterosaur scraped or touched its foot on the substrate, one would expect to see narrow toe impressions and elongate striations (cf. McAllister 1989a, b). Pterosaur claws are noted for being particularly sharp and narrow. The inner three claw impressions curve and are inclined sharply outwards and the outer one curves and is inclined slightly inwards. The outer toe diverges slightly more than the other three toes. This is matched by the scrape marks on the slab and so suggests a genetic relationship between the two. The two middle toes of a pterosaur foot are longer than the two outer ones and thus would be expected to make deeper impressions; again this is seen in the scrape marks, in this case and in certain other examples from this same assemblage. Although the scrape marks are rather closer together, one would expect the toes to be drawn closer together when not weight-bearing, e.g. at different phases in the paddling stroke. Otherwise one would have to postulate that the scrape marks were made by a different and smaller individual.

Specimens CU MWC 188.3, 188.9, 188.46, 188.49–50, 188.52, 188.54–56, 188.58, 188.66–67, 188.72–73. All these specimens (Figs 4–6) reveal patterns of parallel and subparallel scrape marks or striations. These typically consist of clusters of two, three, four or more short toe impressions or elongate scrape marks. If all the traces consisted of four discrete toe impressions or scrape marks, as is clearly the case in some examples, especially on slabs 188.9, 188.50, 188.56 and 188.58 (Figs 4 & 5), there would be little ambiguity in attributing them to pterosaur pes toe impressions. However, as indicated above, the paddling, scraping or foraging activities

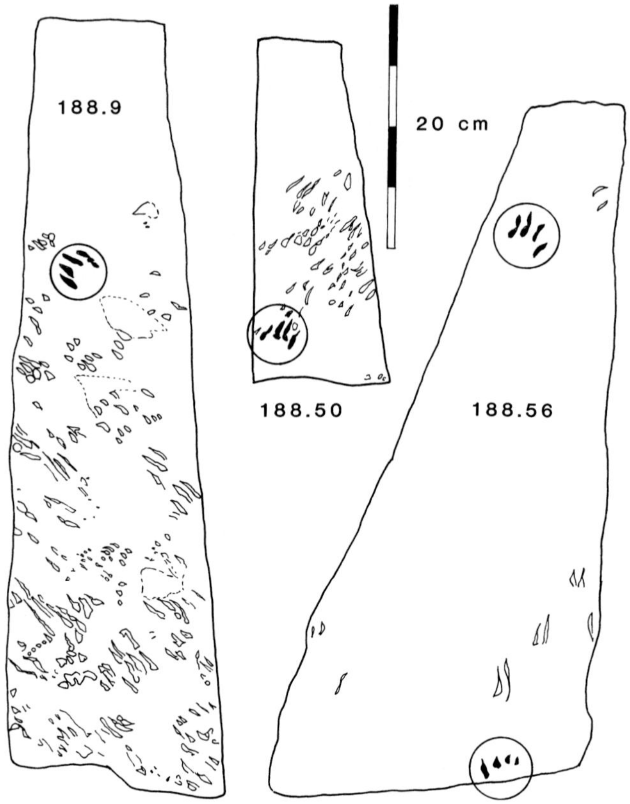

Fig. 5. Swim track specimens CU-MWC 188.9, 188. 50 and 188.56. Circles highlight toe-tip impressions with typical quadripartite pterosaurian pes characteristics.

of a buoyant, seabird-like animal are bound to lead to variety and irregularity in the resulting traces. This is exactly what we see.

The large slabs (188.9, 188.50, 188.56 and 188.58; Figs 4 & 5), were not found *in situ*, so the precise geological orientation is not known, even though the horizon of origin seems certain. The slabs nonetheless give the overall impression of several animals progressing in the same direction (from bottom right to top left for 188.9, from bottom to top for 188.56, and from bottom left to top right for 188. 50 and 188.58) without ever touching or scraping the substrate with more than their toe tips, i.e. there are no complete pes impressions as in specimen 188.46. If this interpretation is accepted one might infer that the water was just deep enough for a buoyant pterosaur of average size (say pes length 7.5 cm) to reach the bottom with its toe tips. If the hip socket was near the waterline then water depth would be about equivalent to the combined length of leg and foot. The strong parallel alignment is suggestive of an external control, such as water or wind current, that directed animals in this preferred direction. An alternative, biologically, rather than physically controlled, explanation is that a gregarious group (flock) of pterosaurs was intentionally paddling or 'tip-toeing' its way in a preferred direction.

The most remarkable slab so far discovered at the Del Monte Mines locality consists of about 1 m^2 of surface (Fig. 7) on which there are literally hundreds of parallel scratch marks or 'swim tracks' many consisting of four parallel striations of the type described and illustrated above. The abundance of such evidence, as recorded on at least a dozen smaller slabs, suggests that a substantial area of substrate was impacted by pterosaurian trace-makers, either while foraging for food or while moving in response to wind and/or water currents in shallow-water settings.

Although the Del Monte Mines locality is particularly rich in such evidence of parallel alignments of swim traces, it may not be the only site with such evidence. At least one slab (188.62), from the Cactus Park locality in western Colorado, can be interpreted as toe-tip traces and may indicate swimming rather than walking progression. Then, to use a metaphor from the introduction, we may consider the Del Monte Mines locality as the 'cradle' or type locality

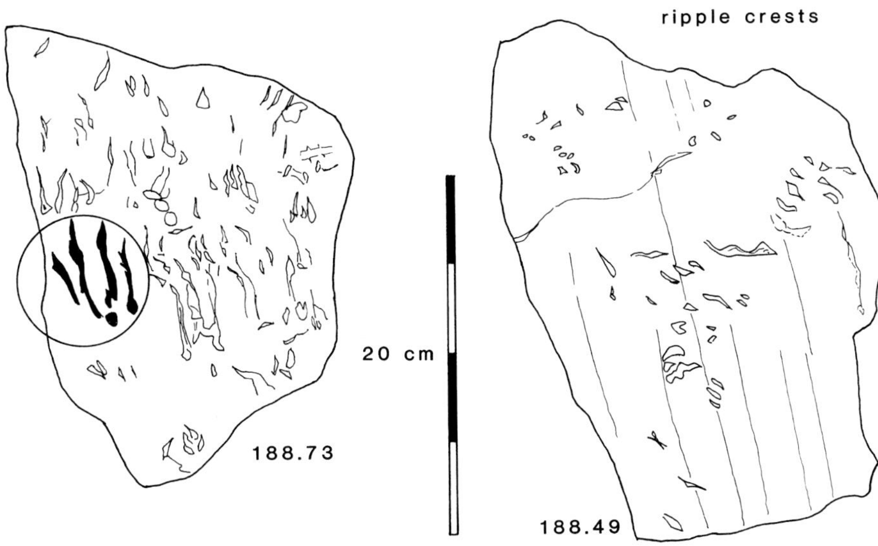

Fig. 6. Specimens 188.49 and 188.73. Circle highlights typical quadripartite pterosaurian pes configuration.

for pterosaur swim tracks, and use it as a guide to help identify similar traces from other regions.

A Sundance Formation trackway specimen. This specimen, from the Windy Hill Member, at Alcova Lake Wyoming, has been recorded by direct field-tracing of tracks and sedimentary structures (ripple marks) onto clear acetate film (tracing T 230 in our library; Fig. 8). The specimen is instructive because it shows five tracks arranged in a more or less regular configuration that appear to suggest paddling or swimming behaviour. The tracks consist of at least three parallel scrape marks that cut across a series of symmetric wave ripples (mean wavelength about 3.5 cm) with truncated crests. Such truncation is associated with wave action in very shallow water. Moreover, the tracks, which occur in a monospecific assemblage of *Pteraichnus* tracks, are situated 15–20 cm apart, thus comparing closely with the typical width of *Pteraichnus* trackways. However, there is no sign of manus tracks, and the tracks, rather than alternating, are more nearly arranged in a paired configuration. Such a configuration, seen in the Cerin turtle tracks, can be interpreted as synchronous strokes of appendages while swimming (Fig. 9). Walking *Pteraichnus* trackways often reveal manus tracks that are deeper than the pes tracks and, indeed, walking trackways with pes-only configurations are unknown. Thus an alternative explanation is required that takes into account the regular pattern indicative of some form of locomotion. Given that the hopping track-maker interpretation proved dubious in the case of the Cerin tracks, it would be speculative and unconvincing to introduce it here, especially in light of the overwhelming evidence that pterosaurs progressed quadrupedally on land. Thus the only reasonable explanation that is consistent with the wave-ripple marks on the substrate and the abundance of pes scrape marks recorded at other sites is that this represents a swimming pterosaur, presumably progressing buoyantly in shallow water.

Webbed feet for swimming

According to Wellnhofer (1991, pp. 158–9), the foot of pterodactyloid pterosaurs such as *Pterodactylus* was fully webbed, a fact which provides 'proof' that they could swim. This inference is probably correct since all living bird species with fully webbed feet (notably ducks, geese, gulls, auks, pelicans, cormorants, gannets and other seabirds) habitually float on water and dive or skim the surface for food. To date, the best evidence of web impressions associated with pterosaur tracks is probably that cited by Mazin *et al.* (1997, p. 735) who recognize 'un trace de palmure interdigitale' (a trace of interdigital webbing) in one case and 'présence d'une membrane interdigitale' (presence of an interdigital membrane) in another. Similarly Pascual Arribas & Sanz Perez (2000, fig. 23a) also illustrated what appears to be interdigital webbing for the ichnospecies *Pteraichnus palaciei saenzi* from the Late Jurassic/Early Cretaceous of Spain.

Such traces help with several lines of interpretation. First, they demonstrate that the trace fossil record confirms findings (web impressions) associated with well-preserved body fossils from the Solnhofen Limestone (Wellnhofer 1991). Second, this in turn further supports the argument that the tracks are pterosaurian in origin, i.e. this interpretation cannot be challenged because of a lack of

Fig. 7. Photo of large slab in field location at Del Monte Mines site.

evidence for webbing. Third, webbing evidence (from body and trace fossils) strongly supports the interpretation that pterodactyloid pterosaurs displayed aquatic adaptations.

In the interest of caution, it should be stressed that web traces are not often preserved in fossil footprints. Simple observation of duck tracks along modern lake and pond shorelines will confirm that the web trace is much more shallow and indistinct than that of the digits, and may be missing altogether. Sediment bulges may sometimes develop between digit impressions in large fossil footprints attributed to dinosaurs. These in turn may erode to produce curved exfoliation features between the digit traces. These have sometimes been mistaken for web traces (Lockley & Hunt 1995). To the best of our knowledge there are no convincing examples of web traces associated with well-documented dinosaur tracks.

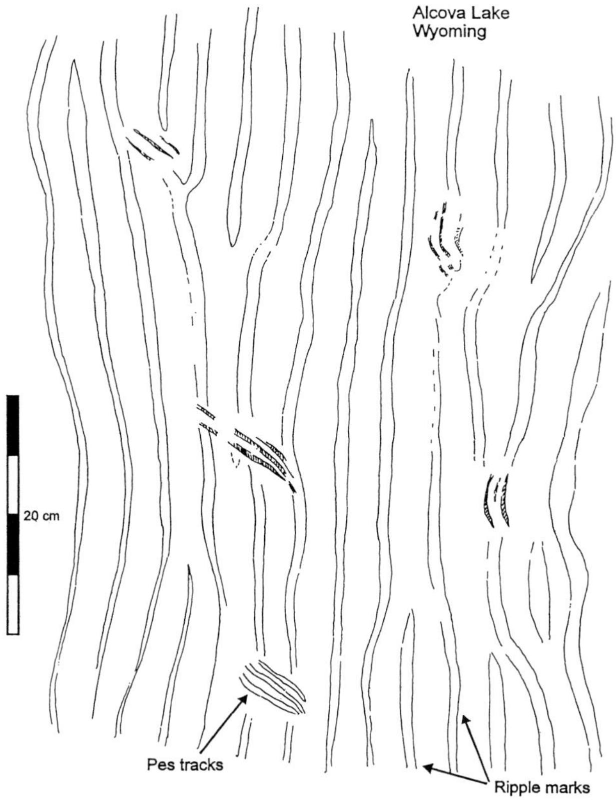

Fig. 8. Alcova Lake specimen showing pes traces interfering with ripple marks. Closely spaced lines are truncated crests, between wider troughs. Traced directly from field outcrop.

Fig. 9. Reconstruction of floating pterosaur. (Artwork courtesy of Judy Peterson.)

Pterosaur feeding traces

Paired rounded depressions (Fig. 10) associated with *Pteraichnus* walking and swim traces have been interpreted as beakprod marks (cf. Wright *et al.* 1997). These are frequently found on surfaces with invertebrate trails and may also be associated with partial swim tracks. They are evidently not the surface traces of U-shaped burrows. At the Del Monte localities in eastern Utah, they mostly occur in association with very thinly bedded units, in the same lithofacies and in the same narrowly constrained stratigraphic levels as the other pterosaur track assemblages. The depressions range in size from 1 to 5 mm in diameter and they occur 2–5 mm apart. Some are joined by a shallow elongate depression.

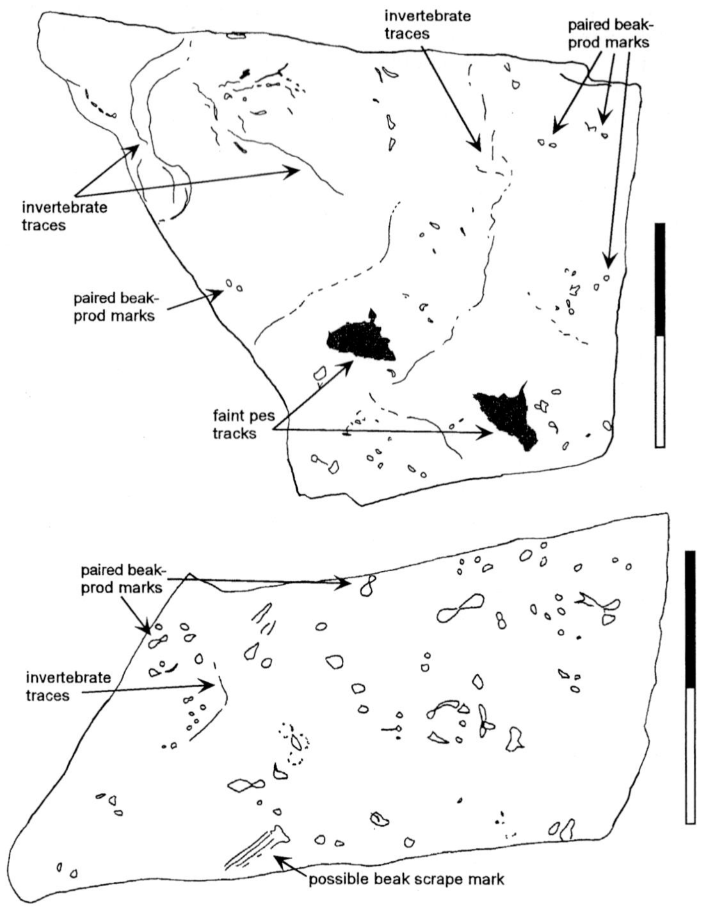

Fig. 10. Pterosaur feeding traces. Scale bars 10 cm.

Again there is no direct evidence that these were made by pterosaurs; however, their consistent size and spacing indicates that they are a genuine feature. In addition their association with invertebrate traces might provide a clue to their purpose: perhaps the pterosaurs were feeding on the invertebrates that produced the traces.

Discussion

Pterosaur tracks are being discovered faster than they can be described, not just in the Late Jurassic of western North America, but worldwide in the Late Jurassic and Cretaceous (as noted elsewhere in this volume). Thus the potential for detailed studies of pterosaur paleobiology, including foot and hand morphology, walking and swimming locomotion, size frequency, feeding behaviour, palaeoecology and palaeobiogeography-facies association, and biotic or ichnofacies association with other trace-makers is considerable. We should bear in mind that the examples given herein deal primarily with the swimming and feeding behaviour inferred from a single ichnogenus from a restricted stratigraphic interval, mostly at a single locality.

The pterosaur tracks from all three units (Summerville-Sundance and Morrison) are similar in morphology and occur in essentially monospecific assemblages. The best examples of 'swim tracks' occur at localities in eastern Utah and Wyoming and include the examples illustrated herein. No 'swimming' pterosaur tracks have been yet found in the Morrison Formation but this could be because of the smaller sample size, difference in facies association or a combination of such factors. The pterosaur tracks from eastern Utah and north-eastern Arizona are associated with the theropod dinosaur tracks *Megalosauripus* and *Therangospodus* (Lockley *et al.* 1996, 2000) and a few sauropod footprints, which are taken to be an indication of the increase in dinosaur activity that accompanied the stratigraphic, facies and palaeogeographic transition into the dinosaur-rich Morrison facies.

Pteraichnus tracks evidently characterize a *Pteraichnus* ichnofacies (defined as a repeat occurrences of *Pteraichnus*-dominated assemblages or ichnocoenoses) associated with a shallow-water shoreline environment favourable to invertebrate life in marginal marine or lagoonal settings. In the case of the Morrison assemblage the tracks evidently represent a carbonate lake system. It is interesting to note that, in the eastern Utah Summerville Formation assemblages, the tracks are confined to a restricted stratigraphic interval in which there are abundant invertebrate traces and a few dinosaur tracks. Most other intervals lack any sign of invertebrate bioturbation. Such an association suggests that pterosaurs were drawn to settings where there was significant biotic activity. In the case of the eastern Utah assemblage, the dinosaur tracks may also indicate the attraction of a favourable environment.

Most known pterosaur tracks are thought, on morphological grounds, to have been made by pterodactyloid pterosaurs and the appearance of pterosaur tracks in the fossil record coincides with the rise of this group in the Late Jurassic. The abundance of pterosaur walking tracks and swim traces indicates that pterosaurs may have been well adapted to terrestrial locomotion and shallow-water foraging along Late Jurassic marine shorelines, where we also find evidence for invertebrate faunas and saurischian dinosaurs. They may at this time also have begun to frequent lake basins away from marine shorelines. Indeed Cretaceous track evidence suggests that they became so well-adapted to lake systems that this appears to have been their preferred habitat. Indeed several authors have remarked that they exhibit a pronounced ecological shift in this direction (Unwin *et al.* 1997).

Until recently we really knew very little about the behaviour of pterosaurs, apart from the fact that they flew. Unlike the case in dinosaurs, their eggs or faeces have yet to be found. Until the mid-1990s most palaeontologists were even in denial about the existence of their tracks. Thus we were stuck in a conceptual dark age where pterosaurs hung like bats from cliffs and trees as the only perches from which they could launch themselves into flight. Padian (1987) tried to change this conception, but went too far in insisting that pterosaurs were bird-like bipeds (Padian 1983). Track discoveries have in many ways revolutionized our view of these creatures, proving that they were able quadrupeds that habitually congregated on flat shoreline substrates from which they presumably could take off without clambering up nearby trees or cliffs (a singularly poor adaptation for creatures that dominated the Mesozoic skies for more than 100 Ma). This view has generated resistance among those favouring a bipedal dinosaur-like or bird-like model for pterosaur locomotion. But it seems that like birds they could probably take to the air easily, without a long bipedal 'run up' or 'take off'. Moreover it seems they were swimmers or floaters and foragers on the substrate in shallow settings, much like modern seabirds and waterbirds (Cadée 1990). Thus we may also assume that it is likely that they could take off easily from the water. If this was not the case we would have to make the unparsimonious assumption that they swam to shore to take off.

Future directions

It is worth stressing that this paper, like many others cited herein, deals with only one relatively limited, descriptive aspect of pterosaur track research. For this reason we hope, in conclusion, to draw attention to the abundance of ichnological data still available for analysis. In less than a decade the available sample of Late Jurassic material curated in museum collections has increased to nearly 100 specimens, in addition to other specimens that remain at field locations. The potential for further studies falls into several discrete categories as follows:

There is a large volume of size-frequency data available for morphometric analysis. Tracks in the North American Late Jurassic sample range in size from 2.5 to c. 11 cm in length, comparable with the French Late Jurassic sample (Mazin *et al.* 1995, 1997), though larger tracks are known from the Late Jurassic of Spain (Garcia Ramos 2001). Those in the global sample, including many Cretaceous sites, range up to 33 cm in length (Fig. 11). Such a large sample holds promise beyond size-frequency analysis and should allow for allometric shape analyses and morphological studies which will help determine characters that may be diagnostic from a taxonomic/ichnotaxonomic point of view. As a convenient reference we summarize known sites in Table 1.

From the viewpoint of swim tracks, the Del Monte Mines sample discussed herein is perhaps the largest sample currently available, but the Wyoming sample is also large (more than 25 specimens), as is the new site from the Ferron area of Utah. The potential exists to compare all samples from a morphological point of view in order to assess the morphological variation in size and shape. For example the presence of small tracks in the Del Monte Mines and Wyoming samples has considerable palaeobiological significance, implying the presence of either small species or precocious juveniles. From finds in the Solnhofen limestone, it

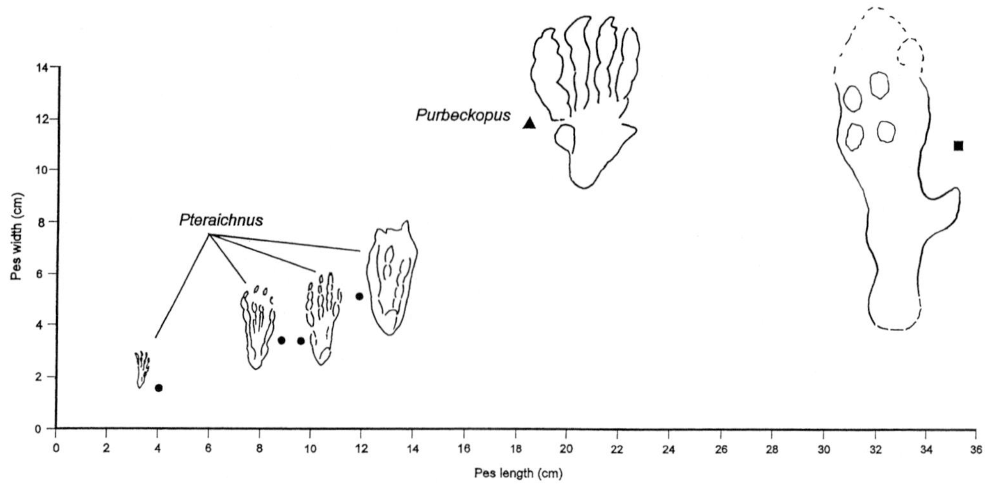

Fig. 11. Known size range for Jurassic and Cretaceous pterosaurian tracks. Note that most Late Jurassic tracks fall in the size range 2.5–11.0 cms. Large track (right) is from the Cretaceous of Korea (after Lockley *et al.* 1997; Hwang *et al.* 2002).

Table 1. *Well-documented pterosaur track occurrences*

Track type/Ichnotaxon	Locality/Age	Representative citations
Pteraichnus saltwashensis	Arizona / Late Jurassic	Stokes 1957
Pteraichnus stokesi	Wyoming / Late Jurassic	Lockley *et al.* 1995
Pteraichus ichnosp. indet.	Colorado-Utah, Oklahoma / multiple Late Jurassic sites	Lockley *et al.* 2001 and references therein
cf. Pteraichnus	SW France / Late Jurassic	Mazin *et al.* 1995, 1997
cf. Pteraichnus (large)	NW Spain/ Late Jurassic	Garcia Ramos *et al.* 2000
Pteraichnus palaciei-saenzi	Spain / multiple Late Jurassic or Early Cretaceous sites	Pascual Aribbas & Sanz-Perez 2000
Pteraichnus parvus	Spain / Early Cretaceous	Meijide Calvo *et al.* in press
Purbeckopus pentadactylus	England / basal Cretaceous	Wright *et al.* 1997
cf. Pteraichnus	Argentina / Early Cretaceous (Albian)	Calvo & Lockley 1991 Calvo & Moratala 1998
Haenamichnus	Korea / Late Cretaceous	Hwang *et al.* 2002
cf Pteraichnus	Mexico / Late Cretaceous	Rodriguez-de la Rosa 2003
cf Pteraichnus	Utah / Late Cretaceous	Lockley 1998

appears probable that small forms are juveniles, not small species (Bennett 1996). This suggests that many pterodactyloids were precocial. A detailed study of the available track sample would provide quantitative insight into the size-frequency structure of the Late Jurassic *Pteraichnus* ichnofaunas of western North America.

Analysis of swim tracks, especially at the large Del Monte Mines site, would allow an understanding of preferred orientation trends in the context of the local sedimentological succession. For example it should be possible to measure orientations of striations *in situ* at several sites where loose material has been recovered and to determine if there is one large preferred orientation trend at one stratigraphic level or whether there are several such trends at many levels. Such local ichnofacies analysis can be extended to the larger regional picture. One of the important questions that need to be addressed is to what extent can we correlate known *Pteraichnus* ichofaunas from site to site? It appears at present that there is a regional *Pteraichnus* ichnofacies that is likely to become known in more detail as we 'fill in' with the discovery of further sites at predictable horizons in the Upper Summerville and equivalent stratigraphic intervals. Indeed the very fact that we can predict further occurrences at such stratigraphic levels confirms the internal consistency of the regional *Pteraichnus* ichnofacies model.

While the regional distribution of pterosaur tracks

in the Upper Summerville, Upper Sundance and Lower Morrison may be considered a biostratigraphic phenomenon associated with an evolutionary radiation among pterodactyloids, it is also a palaeoecological and palaeogeographical phenomenon. The rise in sea level associated with the Summerville transgression is associated with a global rise in sea level during the early part of the Late Jurassic. Such fluctuations in sea level are enough to cause extensive aggradation of coastal plain and shelf sediments and create conditions suitable for the preservation of track-rich facies, as has been discussed in relation to dinosaur freeways or 'megatracksites' (Lockley 1997). Thus the regional *Pteraichnus* ichnofacies is, in part, a sequence stratigraphic phenomenon. This in turn means that it probably has some relationship to the Moab dinosaur megatracksite associated with the upper tongue of the Summerville, between the Moab Tongue Member of the Entrada Formation and the overlying Morrison Formation.

Although it is outside the scope of this paper to discuss this relationship in detail, it is evident that the Moab megatracksite, which extends over more than 2000 km^2, is situated within the geographical area of the *Pteraichnus* ichnofacies and also cannot be separated from it in time. If we assume that the *Pteraichnus* ichnofacies may prove to be a single continuous facies, genetically related to the Sundance Summerville embayment, at least during the latter part of its history, we come to the impressive conclusion that it is an ichnofacies that may be up to 1 000 000 km^2 in extent. Among the challenges facing ichnologists, and sedimentary geologists in general, are: (1) to determine whether this is a single complex ichnofacies or a mosaic of separate ichnofacies separated by barren intervals, and (2) to assess the spatial, temporal and biotic relationships of dinosaur-dominated ichnofacies and those of pterosaurian origin.

Much of the work conducted in western Colorado and Utah was carried out under the jurisdiction of Bureau of Land Management Permits. We also thank J. Ferguson, Kenton, Oklahoma, for help in the field and for access to information concerning sites in southeastern Colorado and Oklahoma.

References

ANDERSON, O. J. & LUCAS, S. G. 1996. The base of the Morrison Formation (Upper Jurassic) of northwestern New Mexico and adjacent areas. *In*: MORALES, M. (ed.) *The Continental Jurassic. Museum of Northern Arizona Bulletin*, **60**, 443–456.

BENNETT, S. C. 1993. Reinterpretation of problematic tracks at Clayton Lake State Park, New Mexico. Not one pterosaur but several crocodiles. *Ichnos*, **2**, 37–42.

BENNETT, S. C. 1996. Year-classes of pterosaurs from the Solnhofen Limestone of Germany: taxonomic and systematic implications. *Journal of Vertebrate Paleontology*, **16**, 432–444.

BENNETT, S. C. 1997. Terrestrial locomotion of pterosaurs: a reconstruction based on Pteraichnus trackways. *Journal of Vertebrate Paleontology*, **17**, 104–173.

BERNIER, P., BARALE, G. ET AL. 1984. Découverte de pistes de dinosaures sauteurs dans les calcaires lithographiques de Cerin (Kimméridgien supérieur, Ain France) – implications paléoécologiques. *Géobios Memoire Special*, **8**, 177–85.

BIRD. R. T. 1944. Did *Brontosaurus* ever walk on land? *Natural History*, **53**, 61–67.

BREIHAUPT, B. H., SOUTHWELL, E. H. ADAMS, T. L. & MATHEWS, N. A. 2001. Innovative documentation methodologies in the study of the most extensive dinosaur tracksite in Wyoming. *In*: SANTUCCI, V. & MCCLELLAND, L. (eds) *Proceedings of the 6th Fossil Resource Conference, Grand Junction Colorado, Sept. 2001*. Geological Resources Division Technical report NPS/NRGD/GRDTR-01/01, 113–122.

CADÉE, G. C. 1990. Feeding traces and bioturbation by birds on a tidal flat, Dutch Wadden Sea. *Ichnos*, **1**, 23–30.

CALVO, J, O. & LOCKLEY, M. G. 2001. The first pterosaur track record in Gondwanaland. *Cretaceous Research*, **22**, 585–590.

CALVO, J. O. & MORATALLA, J. J. 1998. First record of pterosaur tracks in southern continents. *III Encuentro Argentino de Icnología y Primera Reunión de Icnología del Mercosur*. Resúmenes Mar del Plata, 7–8.

COOMBS, W. P. JR. 1980. Swimming ability of carnivorous dinosaurs. *Science*, **207**, 1198–1200.

FOSTER, J. R., LOCKLEY, M. G. & BROCKETT, J. 1999. Problematic vertebrate (?turtle) tracks from the Morrison Formation of eastern Utah. *In*: GILLETTE, D. D. (ed). *Vertebrate Paleontology in Utah*. Utah Geological Survey. Miscellaneous Publications, **99–1**, 185–191.

GARCIA-RAMOS, J. C. ARAMBURU, C., PINUELA, L. & LIRES, J. 2001. *The Dinosaur Coast: Colunga-Ribadesella-Villaviciosa: Field Guide to the Jurassic of Asturias*. Consejeria de Education y Cultura del Principado de Asturias, Spain, 33 pp.

GARCIA-RAMOS, J. C., PINUELA, L., LIRES. J. & FERNANDEZ, L. A. 2000. Icnitas de reptiles voladores (pterosaurios) con impressiones de la piel en el Jurassic Superior de Asturias (N. de España). *In*: DIEZ, J. B. & BALBINO, A. C. (eds) *Primero Congresso Iberico de Paleontologia XVI Journadas de la Socieded Española de Paleontologia Evora (Portugal) 12–14 Octobre, 2000*, 87.

HWANG, K. G., HUH, M. LOCKLEY, M. G. UNWIN, D. M. & WRIGHT, J. L. 2002. New pterosaur tracks (Pteraichnidae) from the Late Cretaceous Uhangri Formation, S. W. Korea. *Geological Magazine*, **139**, 421–435.

ISHIGAKI, S. 1989. Footprints of swimming sauropods from Morocco. *In*: GILLETTE, D. D & LOCKLEY, M. G. (eds.) *Dinosaur Tracks and Traces*. Cambridge University Press, Cambridge, 83–86.

KVALE, E. P., JOHNSON, G. D., MICKELSON, D. L., KELLER, K. FURER, L. C. & ARCHER, A. W. 2001. Middle Jurassic (Bajocian and Bathonian) Dinosaur megatracksites, Bighorn Basin, Wyoming, U.S.A., *Palaois*, **16**, 233–254.

LOCKLEY, M. G. 1991. *Tracking Dinosaurs: A New Look at an Ancient World*. Cambridge University Press, Cambridge, 238p.

LOCKLEY, M. G. 1997. The paleoecological and paleoenvironmental importance of dinosaur footprints. *In*: FARLOW, J. O & BRETT-SURNAM, M. K. (eds) *The Complete Dinosaur*. Indiana University Press. Bloomington and Indianapolis, 554–578.

LOCKLEY. M. G. & HUNT, A. P. 1994. *Fossil Footprints of the Dinosaur Ridge Area*. Friends of Dinosaur Ridge and the University of Colorado at Denver Dinosaur Trackers Research Group, with the Morrison Museum of Natural History, 53 pp.

LOCKLEY, M. G. & HUNT, A. P. 1995. *Dinosaur Tracks and Other Fossil Footprints of the Western United States*. Columbia University Press, Berkeley 338 pp.

LOCKLEY, M. G. & MEYER, C. A. 2000. *Dinosaur Tracks and Other Fossil Footprints of Europe*. Columbia University Press, Berkeley. 323 pp.

LOCKLEY, M. G. & MICKELSON, D. 1997. Dinosaur and pterosaur tracks in the Summerville and Bluff (Jurassic) beds of eastern Utah and northeastern Arizona. *New Mexico Geological Society Guidebook, 48th Field Conference Four Corners Region*, 133–138.

LOCKLEY M. G. & RICE, A. 1990. Did brontosaurus ever swim out to sea? Evidence from brontosaur and other dinosaur footprints. *Ichnos*, **1**, 81–90.

LOCKLEY, M. G., HUH, M. LIM, S-K., YANG, S-Y. CHUN, S. S. & UNWIN, D. 1997. First report of pterosaur tracks from Asia, Chollanam Province Korea. *Journal of the Paleontological Society, Korea*, Special Publications, **2**, 17–32.

LOCKLEY, M. G., HUNT, A. P. & LUCAS, S. G. 1996. Vertebrate track assemblages from the Jurassic Summerville Formation and correlative deposits. *In*: MORALES, M. (ed.) *The Continental Jurassic. Museum of Northern Arizona Bulletin*, **60**, 249–254.

LOCKLEY, M. G., LOGUE, T. J., MORATELLA, J. J., HUNT, A. P., SCHULTZ, R. J. & ROBINSON, J. W. 1995. The fossil trackway *Pteraichnus* is pterosaurian, not crocodilian: implications for the global distribution of pterosaur tracks. *Ichnos*, **4**, 7–20.

LOCKLEY, M. G., MEYER, C. A. & SANTOS, V. F. DOS 2000. *Megalosauripus*, and the problematic concept of Megalosaur footprints. *Gaia: Revista de Geociências, Museu Nacional de Historia Natural, Lisbon, Portugal*, **15**, 313–337.

LOCKLEY, M. G., WRIGHT, J. L., LANGSTON, W. JR. & WEST, E.S. 2001. New pterosaur track specimens from the Late Jurassic of Oklahoma and Colorado: their paleobiological significance and regional ichnological context. *Modern Geology*, **20**, 179–203.

LOGUE, T. J. 1977. Preliminary investigation of pterosaur tracks at Alcova, Wyoming. *Wyoming Geological Association, Earth Science Bulletin*, **10**, 29–30.

MAZIN, J-M., HANTZPERGUE, P., BASSOLLET, J-P. LAFAURIE, G. & VIGNAUD, P. 1997. The Crayssac site (Lower Tithonian, Quercy, Lot, France): discovery of dinosaur tracks *in situ* and first ichnological results. *Comptes Rendus de l'Académie des Sciences, Paris, Série IIa*, **321**, 417–424.

MAZIN, J-M., HANTZPERGUE, P., LAFAURIE, G. & VIGNAUD, P. 1995. Des pistes de ptérosaures dans le Tithonien de Crayssac. *Paléontologie*, **321**, 417–424.

MCALLISTER, J. 1989a. Dakota Formation tracks from Kansas: implications for the recognition of subaqueous traces. *In*: GILLETTE, D. D & LOCKLEY, M. G. (eds) *Dinosaur Tracks and Traces*. Cambridge University Press, Cambridge, 343–348.

MCALLISTER, J. 1989b. *Subaqueous Vertebrate Footmarks from the Upper Dakota Formation (Cretaceous) of Kansas, U.S.A*. Natural History Museum, University of Kansas, Occasional Papers, **127**, 1–22.

MEIJIDE CALVO, M. & FUENTES VIDARTE, C. 1999. Huellas de pterosaurios en el Weald de Soria (España). *Actas I, Journadas Internacionales sobre Paleontologia de Dinosaurios y su Entorno*, Salas de los Infantes, Burgos, 397–406.

MEIJIDE CALVO, M. MEIJIEDE FUENTES, F. FUENTES VIDARTE, C. & MEIJIDE FUENTES, M. 2001. Huellas de pterosaurios en la Sierra de Oncola (Soria, España) Nueva icnospecies Pteracihnus parvas. *Actas II, Journadas Internacionales sobre Paleontologia de Dinosaurios y su Entorno*. Salas de los Infantes.

PADIAN, K. 1983. A functional analysis of flying and walking in pterosaurs. *Paleobiology*, **9**, 218–239.

PADIAN, K. 1987. The case of the bat-winged pterosaur, *In*: CZERKAS, S. J & OLSEN, E. C, (eds) *Dinosaurs Past and Present*. Natural History Museum of Los Angeles and University of Washington Press, vol. 2, 64–81.

PADIAN, K. 1998. Pterosaurs and ?avians from the Morrison Formation (Upper Jurassic, Western U.S.). *Modern Geology*, **23**, 57–68.

PADIAN, K. & OLSEN, P.E. 1984. The fossil trackway *Pteraichnus*: not pterosaurian but crocodilian. *Journal of Paleontology*, **58**, 178–184.

PASCUAL ARRIBAS, C. & SANZ PEREZ E. 2000. Huellas de pterosaurios en el groupo Oncala (Soria España). *Pteraichnus palaciei-saenzi*, nov. ichnosp. *Estudios Geologicos*, **56**, 73–100.

RODRIGUEZ DE LA ROSA, R. A. 2003. Pterosaur tracks from the latest Campanian Cerro del Pueblo Formation of southeastern Coahuila, Mexico. *In*: BUFFETAUT, E. & MAZIN, J-M. *Evolution and Palaeobiology of Pterosaurs*. Geological Society, London, Special Publications, **217**, 275–282.

STOKES, W.L. 1957. Pterodactyl tracks from the Morrison Formation. *Journal of Paleontology*, **31**, 952–954.

THULBORN, R. A. 1989. The gaits of dinosaurs. *In*: GILLETTE, D. D & LOCKLEY, M. G. (eds) *Dinosaur Tracks and Traces*, Cambridge University Press, Cambridge, 39–49.

UNWIN, D. M. 1987. Pterosaur locomotion: joggers or waddlers? *Nature*, **237**, 13–14.

UNWIN, D.M. 1989. A predictive method for the identification of vertebrate ichnites and its application to pterosaur tracks. *In*: GILLETTE, D. D & LOCKLEY, M. G. (eds) *Dinosaur Tracks and Traces*. Cambridge University Press, Cambridge, 259–274.

UNWIN, D.M. 1997. Pterosaur tracks and the terrestrial ability of pterosaurs. *Lethaia*, **29**, 373–386.

UNWIN, D. M. BAKHURINA, N. N., LOCKLEY, M. G., MANABE, M. & LU, J. 1997. *Pterosaurs from Asia. Paleontological Society, Korea, Special Publications*, **2**, 43–65.

WELLNHOFER, P. 1988. Terrestrial locomotion in pterosaurs. *Historical Biology*, **1**, 3–16.

WELLNHOFER, P. 1991. *The Illustrated Encyclopedia of Pterosaurs*. Crescent Books, New York, 192 pp.

WEST, E. S. 1978. *Biostratigraphy and paleoecology of the Lower Morrison Formation of Cimarron County, Oklahoma*. Masters dissertation, University of Kansas.

WRIGHT, J. L., UNWIN, D. M., LOCKLEY, M. G. & RAINFORTH, E. C. 1997. Pterosaur tracks from the Purbeck Limestone Formation of Dorset, England. *Proceedings of the Geologists' Association*, **108**, 39–48.

The systematic problem of tetrapod ichnotaxa: the case study of *Pteraichnus* Stokes, 1957 (Pterosauria, Pterodactyloidea)

JEAN-PAUL BILLON-BRUYAT & JEAN-MICHEL MAZIN

Université de Poitiers, Laboratoire de Géobiologie, Biochronologie et Paléontologie humaine, CNRS UMR 6046, 40 avenue du recteur Pineau, 86022 Poitiers, France
(e-mail: jp.billon@univ-poitiers.fr)

Abstract: This paper deals with the systematics of tetrapod ichnotaxa based on footprints. Beyond the consideration of the nomenclatural rules for ichnotaxa in the ICZN, this paper tries to determine how to establish an ichnotaxonomy that reflects the identity of the track-maker (the organism that has made the track) and how to include this ichnotaxonomy in the skeleton-based taxonomy. This leads to the establishment of several criteria, e.g. ichnospecies should be defined on the print morphology and the relative position of the prints (including the variability due to the track-maker's dynamics), two ichnospecies should represent different species, the ichnospecies and ichnogenus levels are sufficient to discriminate the ichnotaxa and link them to the skeleton-based taxonomy. These ichnotaxonomical criteria are applied to a revision of the ichnogenus *Pteraichnus* Stokes 1957 (Pterosauria, Pterodactyloidea). Only the type species *Pteraichnus saltwashensis* is considered as valid, the pterosaurian origin of *Purbeckopus* is questioned and *Agadirichnus* is highlighted because it could be a senior synonym of *Pteraichnus*. The result of this drastic revision underlines the importance of the proposed ichnotaxonomical principles to avoid the unfounded proliferation of tetrapod ichnotaxa.

Ichnotaxa are regulated by the fourth edition of the International Code of Zoological Nomenclature (ICZN 1999, Article 1.2.1). An ichnotaxon is defined as:

> 'a taxon based on the fossilized work of an organism, including fossilized trails, tracks or burrows (trace fossils) made by an animal',

the work being:

> 'the result of the activity of an animal (e.g. burrows, borings, galls, nests, worm tubes, cocoons, tracks), but not part of the animal. The term applies to trace fossils . . . but does not apply to such fossil evidence as internal moulds, external impressions, and replacements'.

This paper focuses on how to classify tetrapod ichnotaxa based on foot prints, a problem correlated to the difficulties in defining, discriminating ichnotaxa and to linking them to taxa (to identify the maker of the tracks). To deal with this issue, debated since the nineteenth century (see Sarjeant 1990), the paper consists of two parts:

(1) A discussion on the systematics of tetrapod ichnotaxa; that is,
(a) an explanation of how the ICZN treats ichnotaxa;
(b) a determination of how to define ichnospecies so that the ichnotaxonomy reflects the identity of the track-maker and how to include the ichnotaxonomy in the skeleton-based taxonomy.

(2) An application of these ichnotaxonomical principles to a revision of the ichnogenus *Pteraichnus* (Pterosauria, Pterodactyloidea), in order to demonstrate their importance to avoid an unfounded proliferation of tetrapod ichnotaxa.

Nomenclature

The ichnotaxa are principally regulated in the ICZN by the following provisions.

Principle of Priority

An ichnotaxon cannot compete with a taxon under the Principle of Priority (ICZN 1999, Article 23.7.3):

> 'A name established for an ichnotaxon does not compete in priority with a name established for an animal (even for the animal that formed, or may have formed, the trace fossil)'.

Type species fixation

In the fourth edition of the ICZN, and contrary to the third edition (ICZN 1985, Article 13b), a name

From: Buffetaut, E. & Mazin, J-M. (eds) 2003. *Evolution and Palaeobiology of Pterosaurs.* Geological Society, London, Special Publications, **217**, 315–324. 0305–8719/03/$15

established at the genus-group level after 1999 for an ichnotaxon needs a type species fixation (Article 66.1):

> 'An ichnotaxon at the genus-group level proposed after 1999 must have a type species fixed for its name to be available. If established before 2000 it does not require a type species; however, one may have been, or may be, fixed in accordance with Article 69' [see also Article 42.2.1 & Article 42.3.2].

Nevertheless, if a name published for an ichnotaxon

> 'is replaced after 1999 by a new replacement name (*nomen novum*) a type species must then be designated, if one has not already been fixed' [Article 13.3.3].

Eligibility as name-bearing types

According to Article 72.5.1: 'an example of the fossilized work of an animal' is eligible to be a name-bearing type of a nominal species-group taxon, but what is meant by 'an example'? According to the ICZN, the holotype is not intended to display the diagnostic features of the species; it is merely a name-bearing specimen. Hence, a partial print, a single print or a set of prints (*sensu* Leonardi 1987, prints of the manus and the pes of the same side, impressed in the same cycle of movement) can be a holotype. Nevertheless, it is useful when the holotype bears diagnostic features, notably if the ichnotaxonomy reflects the identity of the track-maker (see below). In this way, several successive pes prints (in bipeds) or sets of prints (in quadrupeds) are consensually recognized as the best possible basis to define an ichnospecies (see Sarjeant 1989). They give the relative position of the prints along the track which can be used as a diagnostic feature, as illustrated by the discrimination between pterosaur and crocodilian tracks (Padian & Olsen 1984, Lockley *et al.* 1995, Mazin *et al.* 1995, Mazin *et al.* 2003). As in the definition of the term 'trackway' (Haubold 1971, Leonardi 1987), a minimum type material should comprise at least three successive pes prints (in bipeds) or sets of prints (in quadrupeds), in order to exhibit all the characteristics of a complete locomotor cycle (Fig. 1).

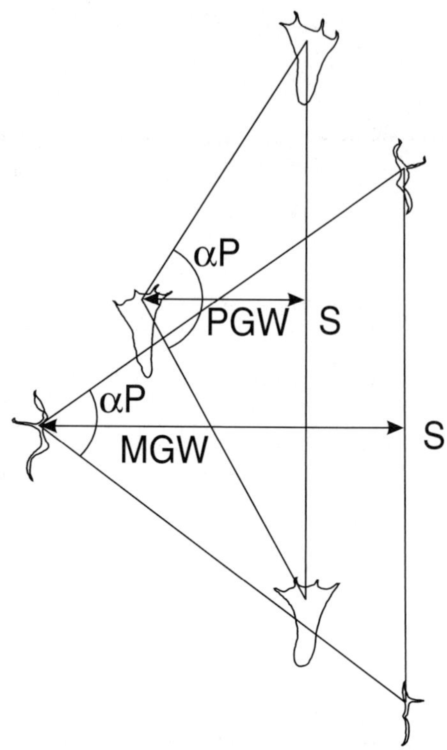

Fig. 1. A complete locomotor cycle as holotype (here in a pterodactyloid pterosaur track). Although the holotype is not intended to display the diagnostic features of the ichnospecies, the use of at least three successive pes prints (in bipeds) or three successive sets of prints (in quadrupeds) as type material is useful. It indicates all the characteristics of a complete locomotor cycle (stance, gait), which can be used as diagnostic features. αP: pace angle; MGW: manus gait-width; PGW: pes gait-width; S: stride.

Classification

As for species, several ichnospecies can be clustered in ichnogenera and higher taxa, i.e. in a classification. However, the aim of the classification and the way of establishing it must be defined.

Different types of classification

As mentioned by Thulborn (1990) concerning dinosaur ichnotaxa, there are three different approaches to classifying fossil tetrapod tracks: (1) to include ichnospecies and ichnogenera in the existing Linnean classification; (2) to create a Linnean-like system for ichnotaxa; and (3) to create an entirely independent system of classification for tracks. The third approach was strongly supported by Sarjeant (1990), who required changing behaviours to be pointed out by a nomenclatural differentiation, i.e. by defining the activities of the track-maker (even in a single track between walking and running phases). Such a classification leads to a proliferation of ichnotaxa and to a difficult correlation with the skeleton-based taxonomy. The second approach consists of establishing higher ichnotaxonomic levels, such

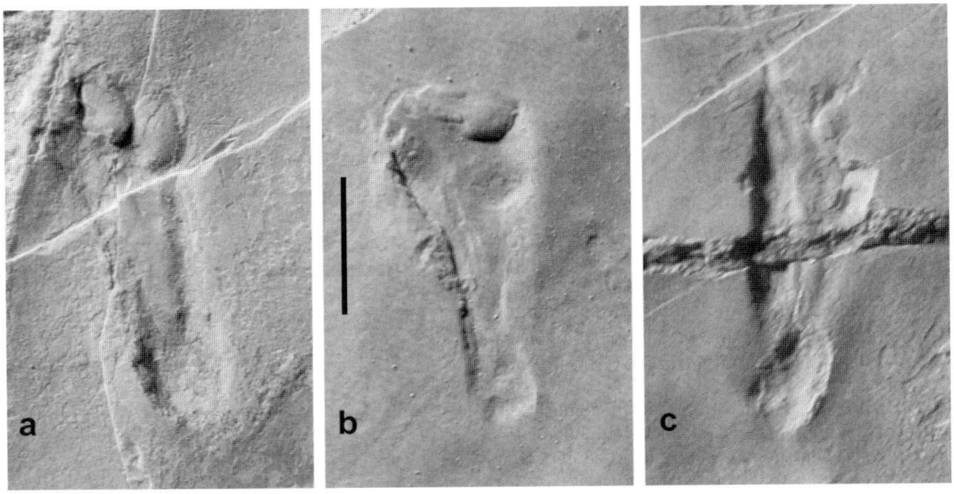

Fig. 2. Variation of the print morphology according to the substrate. The quality of impression of the autopod anatomy is closely linked to the nature and competence of the substrate, as shown here by three *Pteraichnus* isp. right pes prints (**a**, **b** & **c**) along a same track (CR98.28, Crayssac, Lower Tithonian, southwestern France). The pes prints present respectively a decrease in the record of pes anatomy, notably of digit impressions, owing to a wetter and wetter substrate. Scale bar 2 cm.

as ichnofamilies, ichno-orders, ichnoclasses, etc., and names are based on the identity of the track-maker (all tracks made by the same animal should be given the same name). Such a classification is more or less parallel to the skeleton-based taxonomy and permits only the separation of the ichnotaxonomy from the skeleton-based taxonomy. The first approach is also based on the systematic affinity of the track-makers but it uses only ichnospecies and ichnogenera and links them to the skeleton-based taxonomy as precisely as possible. Such a classification could be an ideal system but lacks, until now, principles to define ichnospecies and include them appropriately in the skeleton-based taxonomy.

Identification of the track-maker

To include ichnospecies in the skeleton-based taxonomy implies prior identification of the track-maker. A print is not a simple record of the autopod anatomy, because the nature and competence of the substrate, the soft tissues (which surround bones) and the stance, gait and velocity of the track-maker influence the print morphology, so a print is the result of the dynamic interaction between the autopod and the substrate (Baird 1980, Fig. 2). That is why it is difficult precisely to link a print to a species, because a single species makes different types of prints and, conversely, different species can produce more or less similar prints.

The usual method of identifying the track-maker consists of reconstructing the autopod anatomy from the prints, notably by the creation of a composite print showing all available skeletal characters of the track-maker's autopod (number and shape of digits, phalanges, etc.), and by determining the stance and gait from the track pattern. These characters are compared to those of the potential track-makers in the fossil record.

Conversely, Unwin (1989) suggested a useful 'predictive' method to identify fossil tetrapod prints with the construction of potential prints and tracks through the analysis of the autopod anatomy and functional morphology of fossil tetrapods. Unwin (1989) applied this method to pterosaurs but the variability observed on pterosaur tracks (see Mazin *et al.* 2003) is larger than that of the theoretical ones, because of unexpected stances and gaits.

Definition of ichnospecies

As for a species, diagnostic features must be used to establish an ichnospecies (Fig. 3). Three types of characters should be used to establish the diagnosis of an ichnospecies:

(1) The print morphology, reflecting to a certain extent the autopod anatomy, i.e. features which can be directly diagnostic and self-supporting but are generally distorted by the interaction between the track-maker's dynamics (gait and velocity) and the substrate (nature and competence).
(2) The track pattern (relative position of the prints

		Items	Ichnotaxonomy (Prints)		Taxonomy (Skeletal elements)
Ichnospecies	*Definition*	**Diagnostic features**	print morphology	← - - -	autopod anatomy
			track pattern (including variability due to the track-maker's dynamics)		functional morphology
		Ichnospecies vs species	1 ichnospecies	→	1 or several species
			2 ichnospecies	→	at least 2 different species
	Insertion in skeleton-based taxonomy	**Species**	ichnogenus and ichnospecies	→	genus and species
		Higher taxa	not useful, meaningless	→	family
				→	order
				→	class

Fig. 3. Definition of ichnospecies reflecting the identity of the track-maker and their insertion in the skeleton-based taxonomy. As for skeletal elements, prints display diagnostic features linked to a certain extent to the autopod anatomy and the functional morphology of the track-maker. The dotted arrow indicates that sometimes skeletal diagnostic features of the track-maker's autopod are observed on prints and can be used in the diagnosis of ichnospecies, as the elongate penultimate phalanges in manus and pes prints of pterosaurs. One ichnospecies can represent different osteological species but the diagnostic features should clearly show that two ichnospecies represent different osteological species. Considering only diagnostic features linked to the track-maker's autopod anatomy and functional morphology (including the variability due to the track-maker's dynamics), i.e. features related to some extent to the identity of the track-maker, it is possible to insert ichnospecies in skeleton-based taxonomy. The ichnospecies and ichnogenus levels appear sufficient to differentiate ichnotaxa and link them to taxa as precisely as possible according to the degree of identification.

along the track), reflecting the functional morphology (stance and gait, in other words the locomotor cycle), i.e. features bearing dynamic information which can be diagnostic but can be altered by the intra-ichnospecific variability due to the track-maker's dynamics (gait and velocity). This variability is well illustrated by the numerous *Pteraichnus* tracks (made by pterodactyloid pterosaurs) from Crayssac (Lower Tithonian, southwestern France), with modifications of the track pattern along a same track (Mazin *et al.* 2003).

(3) The dynamic characters corresponding to the interaction between the track-maker's dynamics (gait and velocity) and the substrate (nature and competence), i.e. potentially informative features concerning the interaction between the track-maker and the substrate; nevertheless these generally modify the two foregoing types of characters (Fig. 2), so that great caution should be exercised when using this third type of character as diagnostic.

Conversely, the discrimination between ichnospecies should exclude the intra-ichnospecific variability due to the track-maker's dynamics, as well as the variability resulting from to the interaction between the track-maker's dynamics and the substrate. It is difficult in practice to estimate fully their inference on the print morphology and track pattern because it implies having different types of tracks of the same ichnotaxon (with different track-maker's dynamics and substrates). Moreover, considering that one ichnospecies can represent several species (different species can produce more or less similar prints), the diagnostic features should clearly show that two ichnospecies represent different species.

Using only such diagnostic features correlated with the track-maker's autopod anatomy and functional morphology, i.e. features related to some extent to the identity of the track-maker, it is possible to link the ichnospecies to the skeleton-based taxonomy. The same criteria should be applied to the ichnogenus level.

Relationships between the ichnospecies and the skeleton-based taxonomy (Fig. 3)

Lull (1915) was the first to include ichnotaxa in the skeleton-based taxonomy. He defined ichnotaxa up to the ichnofamily level and placed the ichnofamilies alongside the skeleton-based families into orders. Sarjeant (1990) recognized that this approach was systematically informative but criticized this system because 'it gives an equal systematic status to the animal itself and to a random effect of its actions'. This is an extreme interpretation of Lull's method, which clearly differentiates ichnofamilies from skeleton-based families (which include the probable track-makers). Nevertheless, the use of the ichnofamily level is arguable because, as for ichnospecies compared to species, one ichnofamily can represent several families; this only serves to move the problem of the representativeness of ichnotaxa compared to taxa to the family level and to complicate the insertion of ichnotaxa in the skeleton-based taxonomy. The ichnospecies and ichnogenus levels appear sufficient and necessary to differentiate ichnotaxa and link them to taxa. According to the degree of identification of the track-maker, an ichnospecies should be ascribed to a taxon of a taxonomic level as low as possible: family, order or class group level and, in exceptional situations, to genus or species group level. Of course, ichnotaxa are clearly earmarked and delimited as non-organisms within the classification. Including ichnotaxa in skeleton-based taxonomy can also be useful in terms of cladistics, because ichnotaxa are potentially character-bearing and may help in adding some morpho-functional and soft-tissue characters in a data matrix of actual characters based on skeletons. In this respect, they can be taxonomically informative in the framework of a biological classification. Some palaeontologists argue that ichnotaxa should not be included in the skeleton-based taxonomy because footprints are inorganic objects and do not reproduce or evolve (e.g. Thulborn 1990). In fact, these inorganic objects are the reflection of the autopod anatomy and the functional morphology of track-makers, and thus the morphology of footprints and their position along the tracks follow the evolution of tetrapod locomotion.

Remark on the stratigraphical range of ichnotaxa

The large stratigraphical range of ichnotaxa can seem astonishing. The ichnogenus *Pteraichnus* for example, initially defined as pterodactyloid pterosaur prints from the Late Jurassic (Stokes 1957), is encountered up to the Late Cretaceous (Lockley *et al.* 1995, 1997). In fact, such a long time interval is unlikely to reflect the existence of a single genus; in other words this ichnogenus includes prints made by different species, genera and even families.

Revision of *Pteraichnus*

In 1957, Stokes described an unusual vertebrate track from the Late Jurassic of the Morrison Formation (Arizona, United States) and assigned it to a pterosaur track-maker, more precisely to a pterodactyloid pterosaur, fixing a new ichnogenus and a new ichnospecies: *Pteraichnus saltwashensis*. This interpretation was strongly challenged by Padian & Olsen (1984) who argued, by means of an experiment using an extant *Caiman*, that this track was not made by a pterosaur but by a crocodilian. Following this analysis all tracks similar to *Pteraichnus* were ascribed to crocodilians up to 1995. At this time, and independently, Lockley *et al.* (1995) and Mazin *et al.* (1995), without a doubt, referrred *Pteraichnus* to pterodactyloid pterosaur tracks, a reassessment driven by the discovery of new specimens of *Pteraichnus* from the United States and Europe. Subsequently and quickly, several new ichnospecies have been referred to *Pteraichnus*; their validity needs to be discussed, as well as the validity of other ichnospecies which could be referred to pterodactyloid pterosaurs.

Emended diagnosis of *Pteraichnus*

Stokes (1957) diagnosed *Pteraichnus* as follows:

> 'Aerial quadruped. Manus with 3 digits, 2 of which are small and bear short, blunt claws; the third, very much larger, supports the wing membrane and is capable of being bent backward and upward so as to leave only a short trailing impression. The pes has 4 digits. The heel is very narrow, the sole V-shaped and the toes slender with curving claws. The inferred length of the body exclusive of neck, head and tail, about 30 inches. In walking, the hindfoot oversteps the forefoot indicating the possibility of occasional bipedal stance. The apparent reduction of digits in both manus and pes is distinctive and is the chief reason for placing the animal in the Pterodactyloidea.'

In 1995, Lockley *et al.* emended this diagnosis as follows:

> 'Wide trackway of a quadrupedal animal with elongate, symmetrical, functional-tetradactyl, plantigrade pes impressions and asymmetrical tridactyl manus. Impressions of manus digit IV, elongate, curved and posteriorly directed, parallel to the trackway axis. Manus impressions often more deeply impressed than pes impressions.'

According to the new discoveries and interpretations of *Pteraichnus* tracks this last diagnosis must be updated. It is noteworthy that, in pterosaur prints, as generally in ichnology, some diagnostic characters can be observed only on exceptionally well-preserved prints. The diagnosis of *Pteraichnus* is emended as follows, based on personal observations on the site of Crayssac (Lower Tithonian, southwestern France) and several papers (Stokes 1957; Unwin 1989, 1997; Lockley *et al.* 1995; Mazin *et al.* 1995, 1997, 2003; Bennett 1996, 1997).

Emended diagnosis. Quadruped track; manus prints on the same axis or more laterally to the pes prints; pes print anterior to the ipsilateral manus print; manus prints as or more deeply impressed than pes prints; elongate, subtriangular, plantigrade and tetradactyl pes print; toe II and III prints are slightly longer than I and IV, toe I–IV prints are clawed;

elongate, asymmetrical, digitigrade and tridactyl manus print; digit I print anterior or anterolateral (generally with claw mark), digit II print anterolateral to posterolateral (rarely with claw mark), digit III print posterior (exceptionally with claw mark), digit prints increasing respectively in length; digit IV is rarely marked and limited to the impression of the proximal part of the fourth manus digit (oriented posteromedially); rounded impression in the medial margin of the manus print (impression of the fourth metacarpo-phalangeal joint); elongate penultimate phalanges can be observed in some exceptionally well-preserved manus and pes prints (inferred from digital pads when available).

Revision of the ichnospecies ascribed to pterodactyloid pterosaurs (Table 1)

Type ichnospecies: *Pteraichnus saltwashensis* Stokes 1957. In 1957, Stokes designated the ichnospecies *Pteraichnus saltwashensis* (Morrison Formation, Late Jurassic, Arizona, United States) as the type species of *Pteraichnus* (the type specimen is a track consisting of nine sets of successive prints). According to the ICZN, a type species for an ichnotaxon may have been fixed before 2000 (Article 66.1), although it was not required previously (see above), in which case *P. saltwashensis* is valid. Furthermore, *P. saltwashensis* is consistent with the features of the emended diagnosis of *Pteraichnus*.

Pteraichnus stokesi Lockley *et al.* 1995. In 1995, Lockley *et al.* fixed the new ichnospecies *Pteraichnus stokesi*, from the upper part of the (Mid- to) Late Jurassic Sundance Formation, Wyoming, United States. *P. stokesi* (type specimen is a track consisting of four partial sets of successive prints) shares the diagnostic features of *Pteraichnus*, notably the elongate pes penultimate phalanges. Lockley *et al.* (1995) erected this new ichnospecies on differences in the track width (with corresponding smaller pace angulation of about 20°), the footprint rotation (outwards to a greater degree) and the digit III print (shorter and wider) compared to *P. saltwashensis*. According to Lockley *et al.* (1995)

> 'Although some of these characteristics, such as manus digit impression length and pace angulation, could be attributed to differential preservation and variation in speed of progression, such considerations can not easily account for all the observed differences. It is therefore, on the basis of the combination of these three differences that we erect the new ichnospecies.'

We agree with the remark of Lockley *et al.* (1995) that the differential preservation and the variation in speed of progression cannot easily account for all the observed differences between *P. stokesi* and *P. saltwashensis*. However, the features considered by Lockley *et al.* (1995) when erecting this new ichnospecies are unsuitable if we consider the foregoing types of characters used to define an ichnospecies. The second type of characters (track pattern) show that the differences between *P. stokesi* and *P. saltwashensis* in terms of track width and footprint rotation can be linked to the intra-ichnospecific variability (due to the track-maker's dynamics). The third type of characters (dynamic characters linked to the interaction between the track-maker's dynamics and the substrate) has little influence because, as noted by Lockley *et al.* (1995), the prints are well preserved (notably with the presence of the pad impressions), i.e. made on a competent substrate. The absence of the generally observed distal drag mark extending the digit III print (due to uncompetent substrates, see Mazin *et al.* 2003), has probably led Lockley *et al.* (1995) to interpret the digit III prints of *P. stokesi* shorter and wider than *P. saltwashensis*. In fact, the digit III prints of *P. stokesi* merely reflect a good preservation of manus print on a competent substrate, and hence are close to the autopod anatomy. Furthermore, we agree with Bennett (1997) who underlines that 'there is insufficient evidence to indicate that the Arizona and Wyoming trackways were necessarily made by different species' (*sensu* different osteological species). At present, *P. stokesi* has no actual diagnostic feature, so it is not valid (it remains available) and must be considered as a junior synonym of *P. saltwashensis*.

Pteraichnus palaciei-saenzi Pascual Arribas & Sanz Pérez 2000 (renamed here *Pteraichnus palacieisaenzi* according to ICZN 1999, Article 32.5.2.3), *Pteraichnus cidacoi* Fuentes Vidarte 2001, *Pteraichnus manueli* Mejide Calvo 2001 and *Pteraichnus vetustior* Mejide Fuentes 2001. These four ichnospecies, from the Lower Cretaceous of Spain, are unavailable according to the ICZN because they lack a clearly mentioned diagnosis, an indication of the holotype deposition or, indeed, of a holotype. Furthermore, although these ichnospecies share some diagnostic features of *Pteraichnus* they have been erected with characters that could easily be attributed to intra-ichnospecific variability due to the track-maker's dynamics (see Mazin *et al.* 2003 for data), such as minor differences in the pace angle, the stride, the pes print orientation and the interdigital divarication. Consequently, these ichnospecies should be considered at the moment as *nomina nuda* and referred to *Pteraichnus* isp.

Purbeckopus pentadactylus Delair 1963. In 1997, Wright *et al.* re-examined the ichnospecies *Purbeckopus pentadactylus* Delair 1963, from the

Table 1. *Status of* Pteraichnus *ichnospecies and ichnogenera discussed in this paper*

Ichnotaxon	Locality	Formation	Age	Prints		Diagnostic features linked to the intra-ichnospecific variability due to the track-maker's dynamics	Name available (*sensu* ICZN)	Status	References
				Isolated	Organized in track				
Agadirichnus elegans	Agadir, Morroco	Tagrara anticline	Late Cretaceous	x			x	Possible senior synonym of *Pteraichnus*	Ambroggi & Lapparent 1954
Pteraichnus cidacoi	Valloria, Soria, Spain	Cameros Basin, Oncala Formation	Early Cretaceous	x		x		*Nomen nudum; Pteraichnus* isp.	Fuentes Vidarte 2001
Pteraichnus manueli	Villar del Río, Soria, Spain	Cameros Basin, Oncala Formation	Early Cretaceous		x	x		*Nomen nudum; Pteraichnus* isp.	Mejide Calvo 2001
Pteraichnus palacieisaenzi	Santa Cruz de Yanguas, Soria, Spain	Oncala Formation	Early Cretaceous		x	x		*Nomen nudum; Pteraichnus* isp.	Pascual Arribas & Sanz Pérez 2000
Pteraichnus saltwashensis	Apache County, Arizona, USA	Morrison Formation	Late Jurassic		x		x	Type ichnospecies of *Pteraichnus*	Stokes 1957
Pteraichnus stokesi	Alcova Lake, Wyoming, USA	Sundance Formation	(Mid-) to Late Jurassic		x	x	x	Junior synonym of *Pteraichnus saltwashensis*	Lockley *et al.* 1995
Pteraichnus vetustior	San Pedro Manrique-Fuentes de Magaña, Soria, Spain	Cameros Basin, Oncala Formation	Early Cretaceous	x		x		*Nomen nudum; Pteraichnus* isp.	Mejide Fuentes 2001
Purbeckopus pentadactylus	Dorset, UK	Purbeck Limestone Formation	Early Cretaceous	x			x	Valid ichnospecies; non pterosaurian	Delair 1963, Wright *et al.* 1997

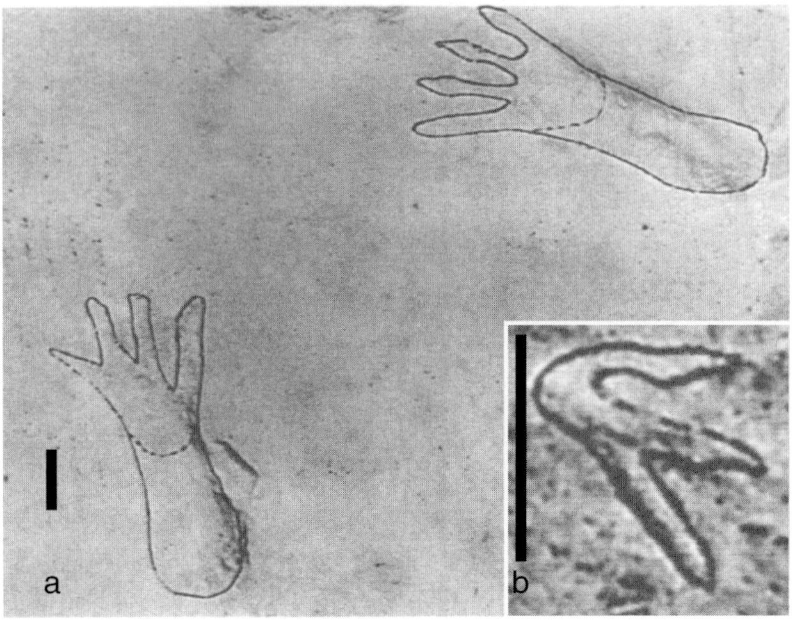

Fig. 4. *Agadirichnus* Ambroggi & Lapparent 1954 (Late Cretaceous, Morocco), a possible senior synonym of *Pteraichnus* Stokes 1957. (**a**) Holotype of *Agadirichnus elegans*: two elongate, plantigrade and tetradactyl pes prints with an important size of the digital pad impression between the penultimate and the ungual phalanges. (**b**) Elongate, asymmetrical, digitigrade and tridactyl print supposed to be a pterodactyloid pterosaur right manus print. Photographs of counterprints (from casts) and outlines from Ambroggi & Lapparent 1954.) Scale bar 2 cm.

Purbeck Limestone Formation (Early Cretaceous) of Dorset (southern United Kingdom). They reinterpreted *P. pentadactylus* as pterosaurian in origin because 'the pes tracks show indications of elongate penultimate phalanges' and 'the impression of the manus lies well outside that of the pes'. Nevertheless these prints are first of all poorly preserved and, as noted by Wright *et al.* (1997), the presence of elongate penultimate phalanges 'is difficult to distinguish in the *Purbeckopus* prints'; personal observation of the holotype leads to the same conclusion. Secondly, the type specimen includes only some unorganized prints, hence there is no evidence for the association of manus and pes prints (Wright *et al.* 1997, fig. 3b; pers. obs. of holotype). The prints of *Purbeckopus* share some diagnostic features of *Pteraichnus*; nevertheless the difficulty of being sure of the presence of elongate penultimate phalanges (because of the poorly preserved pes prints), and the doubtful association of manus and pes prints, questions the pterosaurian origin of these prints. *Purbeckopus pentadactylus* is considered as valid but its pterosaurian origin is dubious.

Agadirichnus elegans Ambroggi & Lapparent 1954. In 1954, Ambroggi & Lapparent reported prints from the Late Cretaceous (Maastrichtian) of Morocco. Among these prints, two unusual pes prints (Ambroggi & Lapparent 1954, p. 53; type C; Fig. 4a) led Ambroggi & Lapparent to erect a new ichnogenus and a new ichnospecies, *Agadirichnus elegans*, tentatively referred to a lacertilian. In the light of the present knowledge on pterosaur prints, these two isolated pes prints could be reinterpreted as pterodactyloid pterosaur in origin. According to Ambroggi & Lapparent the pes prints are elongated (up to 11.5 cm in length), plantigrade, constituted of four slender digits, subequal in length, with a low divarication, and large digital pad impressions, corresponding to the joint between the penultimate and the ungual phalanges. In addition, the pes print is subtriangular in shape with a rounded heel and the inner digits (II and III, and not III and IV *sensu* Ambroggi & Lapparent 1954) are slightly longer than the outer ones (I and IV). Furthermore the authors tentatively referred other prints from the same locality to an indeterminate small reptile (?lizard, ?pterosaur; Ambroggi & Lapparent 1954, p. 55, type D) and birds (Ambroggi & Lapparent 1954, p. 56, type E). Some of these prints could be referred to *Pteraichnus* manus prints, because these elongate, asymmetrical digitigrade prints have, as noted by Ambroggi & Lapparent (1954), three digit prints, two of them being shorter and appressed (Fig. 4b), like digits I and II of the *Pteraichnus* manus print (see Mazin *et al.* 2003, fig. 3g, h). On the grounds of the data reported by Ambroggi & Lapparent (1954) it is interesting to underline that *Agadirichnus*, never mentioned in any work on pterosaur tracks, could be a senior synonym of *Pteraichnus*. It is not possible at present to certify this because we have not observed the original material, so we prefer to consider *Agadirichnus* merely as a possible senior synonym of *Pteraichnus*.

Redefinition of *Pteraichnus*

Until now, six ichnospecies belonging to the ichnogenus *Pteraichnus* have been defined. Only the type ichnospecies *Pteraichnus saltwashensis* is valid, others are junior synonyms (*P. stokesi*) or *nomina nuda* (*P. cidacoi*, *P. manueli*, *P. palacieisaenzi*, *P. vetustior*). In addition, numerous prints (isolated or organized in tracks) have been interpreted as pterosaurian in origin and could be referred to *Pteraichnus* isp. from Argentina (see Calvo & Lockley 2001 for the most complete description and illustration of the material), Mexico (Rodriguez de la Rosa 2001), North America (see Lockley *et al.* 1995 for a previous list and Calvo & Lockley 2001 for references after 1995), Korea (Lockley *et al.* 1997), France (Mazin *et al.* 1995, 1997, 2001) and Spain (see Lockley *et al.* 1995 for a previous list; Meijide Calvo & Fuentes Vidarte 1999; Pascual Arribas & Sanz Pérez 2000; Fuentes Vidarte 2001; Mejide Calvo 2001; Mejide Fuentes 2001). According to the re-interpretation of the ichnotaxonomy of *Pteraichnus* and these numerous discoveries of *Pteraichnus* prints all over the world (leading to more than 30 Jurassic and Cretaceous pterosaur tracksites), the ichnogenus *Pteraichnus* Stokes 1957 can be redefined as follows:

Diagnosis. See the emended diagnosis.

Type species. *Pteraichnus saltwashensis* Stokes 1957.

Stratigraphical range. Kimmeridgian (?Oxfordian) to Maastrichtian. Remark: *Pteraichnus* should be encountered maybe from the Callovian (Unwin 1996), undoubtedly from the Oxfordian (Buffetaut & Guibert 2001) to the Maastrichian (Wellnhofer 1991), which corresponds to the known stratigraphical range of the Pterodactyloidea.

Etymology. The generic name is derived from Latin, *pteron* = wing and *ichnos* = print or track. Stokes (1957) ascribed *Pteraichnus* to Pterosauria and Pterodactyloidea, so even if the etymology of *Pteraichnus* is not taxonomically informative, *Pteraichnus* is related to pterodactyloid pterosaur prints.

Geographical range: America (Argentina, ?Canada, Mexico, United States), Asia (Korea), Europe (France, Spain).

In addition, it is noteworthy that the creation of the ichnofamily Pteraichnidae Lockley *et al.* 1995 is unnecessary. The ichnogenus *Pteraichnus* is sufficient to cluster this type of pterosaur prints and ascribe them to Pterodactyloidea. Consequently, the systematic palaeontology of *Pteraichnus* and its type species *Pteraichnus saltwashensis* is as follows:

Order Pterosauria Kaup 1834
Superfamily Pterodactyloidea Plieninger 1901
Ichnogenus *Pteraichnus Stokes 1957*
Ichnospecies *Pteraichnus saltwashensis* Stokes 1957

Conclusion

The present systematic work on pterodactyloid pterosaur ichnospecies, following an ichnotaxonomy reflecting the identity of the track-maker, demonstrates that they all belong to the ichnogenus *Pteraichnus* and that the only valid ichnospecies is the type species *Pteraichnus saltwashensis*. Nevertheless, it is obvious that *Pteraichnus* should include several ichnospecies; the difficulty lies in their discrimination independently of the intra-ichnospecific variability due to the track-maker's dynamics and considering the effects of the interaction between the track-maker's dynamics and the substrate.

Beyond its application to pterosaur tracks, the advocated classification system should be applied to all tetrapod ichnotaxa, in which the tangled systematics needs to be rationalized in order to avoid the unfounded proliferation of ichnospecies, a necessary prerequisite before using them, for example, in palaeoecology.

This classification system can be difficult to put into practice but is the most appropriate for defining ichnospecies and linking them to skeleton-based taxonomy. Otherwise, it is preferable not to name and classify doubtful footprints rather than of naming and classifying them to the utmost, the essential being to publish them to ensure their existence in the scientific records.

We thank P. Janvier and T. Thulborn for their reviews of the manuscript. We are grateful to the councils and private companies which have provided logistical and financial support during the excavations of The 'Pterosaur Beach' of Crayssac. We thank G. Lafaurie, who discovered the palaeontological site of Crayssac.

References

AMBROGGI, R. & LAPPARENT, A. F. DE 1954. Les empreintes de pas fossiles du Maestrichtien d'Agadir. *Notes du Service Géologique du Maroc*, **10** (122), 43–66.

BAIRD, D. 1980. A prosauropod dinosaur trackway from the Navajo sandstone (Lower Jurassic) of Arizona. *In*: JACOBS, L. L. (ed.) *Aspects of Vertebrate History*.

Museum of Northern Arizona Press, Flagstaff, 219–230.

BENNETT, S. C. 1996. The phylogenetic position of the Pterosauria within the Archosauromorpha. *Zoological Journal of the Linnean Society, London*, **118**, 261–308.

BENNETT, S. C. 1997. Terrestrial locomotion of pterosaurs: a reconstruction based on *Pteraichnus* trackways. *Journal of Vertebrate Paleontology*, **17**, 104–113.

BUFFETAUT, E. & GUIBERT, J-P. 2001. An early pterodactyloid pterosaur from the Oxfordian of Normandy (northwestern France). *Comptes Rendus de l'Académie des Sciences, Paris, Serie IIa*, **333**, 405–409.

CALVO, J. O. & LOCKLEY, M. G. 2001. The first pterosaur tracks from Gondwana. *Cretaceous Research*, **22**, 585–590.

DELAIR, J. B. 1963. Notes on Purbeck fossil footprints, with descriptions of two hitherto unknown forms from Dorset. *Proceedings of the Dorset Natural History and Archaeological Society*, **84**, 92–100.

FUENTES VIDARTE, C. 2001. A new species of *Pteraichnus* for the Spanish Lower Cretaceous *Pteraichnus cidacoi*. *Strata, Série 1*, **11**, 44–46.

HAUBOLD, H. 1971. Ichnia amphibiorum et reptiliorum fossilium. *In*: KUHN, O. (ed.) *Handbuch der Paläoherpetologie*. Gustav Fisher, Stuttgart, Teil 18, 124 pp.

International Commission on Zoological Nomenclature 1985. *International Code of Zoological Nomenclature*, 3rd edition. International Trust for Zoological Nomenclature, London, 338 pp.

International Commission on Zoological Nomenclature 1999. *International Code of Zoological Nomenclature*, 4th edition. International Trust for Zoological Nomenclature, London, 306 pp.

LEONARDI, G. (ed.) 1987. *Glossary and Manual of Tetrapod Footprint Palaeoichnology*. Departamento Nacional da Produçaõ Mineral, Brasilia, 75 pp.

LOCKLEY, M. G., HUH, M., LIM, S-K., YANG, S-Y., CHUN, S-S. & UNWIN, D. M. 1997. First report of pterosaur tracks from Asia, Chullanam Province, Korea. *Journal of the Paleontological Society, Korea*, **2**, 52–67.

LOCKLEY, M. G., LOGUE, T. J., MORATALLA, J. J., HUNT, A. P., SCHULTZ R. J. & ROBINSON, J. W. 1995. The fossil trackway *Pteraichnus* is pterosaurian, not crocodilian: implications for the global distribution of pterosaur tracks. *Ichnos*, **4**, 7–20.

LULL, R. S. 1915. Triassic life of the Connecticut Valley. *Connecticut Geological and Natural History Survey Bulletin*, **81**, 1–285.

MAZIN, J-M., BILLON-BRUYAT J-P., HANTZPERGUE P. & LAFAURIE, G. 2001. The pterosaurian trackways of Crayssac (south-western France). *Strata, Série 1*, **11**, 57–59.

MAZIN, J-M., BILLON-BRUYAT J-P., HANTZPERGUE P. & LAFAURIE, G. 2003. Ichnological evidence for quadrupedal locomotion in pterodactyloid pterosaurs: trackways from the Late Jurassic of Crayssac (southwestern France). *In*: BUFFETAUT, E. & MAZIN, J-M. *Evolution and Palaeobiology of Pterosaurs*. Geological Society, London, Special Publications, **217**, 283–296.

MAZIN, J.-M., HANTZPERGUE, P., BASSOULLET, J-P., LAFAURIE G. & VIGNAUD, P. 1997. Le gisement de Crayssac (Tithonien inférieur, Querçy, Lot, France): découverte de pistes de dinosaures en place et premier bilan ichnologique. *Comptes Rendus de l'Académie des Sciences, Paris, Série IIa*, **325**, 733–739.

MAZIN, J-M., HANTZPERGUE, P., LAFAURIE G. & VIGNAUD, P. 1995. Des pistes de ptérosaures dans le Tithonien de Crayssac (Querçy, Lot). *Comptes Rendus de l'Académie des Sciences, Paris, Série IIa*, **321**, 417–424.

MEIJIDE CALVO, M. 2001. Pterosaur trace in Oncala Berriasian (Soria, Spain) New ichnospecies *Pteraichnus manueli*. *Strata, Série 1*, **11**, 72–74.

MEIJIDE CALVO, M. & FUENTES VIDARTE, C. 1999. Huellas de pterosaurios en el Weald de Soria (España). *Actas de las I Jornadas Internacionales sobre Paleontología de Dinosaurios y su Entorno*, Salas de los Infantes, Burgos, 397–406.

MEIJIDE FUENTES, F. 2001. Pterosaur tracks in Oncala Mountain range (Soria, Spain) A new ichnospecies *Pteraichnus vetustior*. *Strata, Série 1*, **11**, 70–71.

PADIAN, K. & OLSEN, P. E. 1984. The fossil trackway *Pteraichnus*: not pterosaurian, but crocodilian. *Journal of Paleontology*, **58**, 178–184.

PASCUAL ARRIBAS, C. & SANZ PÉREZ, E. 2000. Huellas de pterosaurios en el grupo Oncala (Soria, España). *Pteraichnus palaciei-saenzi*, nov. icnosp. *Estudios Geológicos*, **56**, 73–100.

RODRIGUEZ DE LA ROSA, R. A. 2001. Pterosaur tracks from the Late Cretaceous of northern Mexico: paleoecological and anatomical implications. *Strata, Série 1*, **11**, 85–86.

SARJEANT, W. A. S. 1989. 'Ten paleoichnological commandments': a standardized procedure for the description of fossil vertebrate footprints. *In*: GILLETTE, D. D. & LOCKLEY, M. G. (eds) *Dinosaur Tracks and Traces*. Cambridge University Press, Cambridge, 369–370.

SARJEANT, W. A. S. 1990. A name for the trace of an act: approaches to the nomenclature and the classification of fossil vertebrate footprints. *In*: CARPENTER, K. & CURRIE, P. J. (eds) *Dinosaur Systematics: Approaches and Perspectives*. Cambridge University Press, Cambridge, 299–307.

STOKES, W. L. 1957. Pterodactyl tracks from the Morrisson Formation. *Journal of Paleontology*, **31**, 952–954.

THULBORN, R. A. 1990. *Dinosaur Tracks*. Chapman & Hall, London, 410 pp.

UNWIN, D. M. 1989. A predictive method for the identification of vertebrate ichnites and its application to pterosaur tracks. *In*: GILLETTE, D. D. & LOCKLEY, M. G. (eds) *Dinosaur Tracks and Traces*. Cambridge University Press, Cambridge, 259–274.

UNWIN, D. M. 1996. The fossil record of Middle Jurassic pterosaurs. *Museum of Northern Arizona Bulletin*, **60**, 291–304.

UNWIN, D. M. 1997. Pterosaur tracks and the terrestrial ability of pterosaurs. *Lethaia*, **29**, 373–386.

WELLNHOFER, P. 1991. *The Illustrated Encyclopedia of Pterosaurs*. Salamander, London, 192 pp.

WRIGHT, J. L., UNWIN, D. M., LOCKLEY, M. G. & RAINFORTH, E. 1997. Pterosaur tracks from the Purbeck Limestone Formation of Dorset, England. *Proceedings of the Geologists' Association*, **108**, 39–48.

The John Quekett sections and the earliest pterosaur histological studies

LORNA STEEL

School of Earth and Environmental Sciences, University of Portsmouth, Burnaby Building, Burnaby Road, Portsmouth PO1 3QL, UK
Correspondence address:
Dinosaur Isle, Culver Parade, Sandown, Isle of Wight PO36 8QA, UK
(e-mail: Lorna.steel@tiscali.co.uk)

Abstract: In 1855, John Quekett, assistant curator to Richard Owen of the Hunterian Museum in London, published a catalogue of the museum's thin-section specimens. Among them were 15 sections of pterosaur bone. These sections were cut from eight bones found in the Lower Chalk of Kent, the Wealden of Sussex and the Stonesfield Slate of Oxfordshire, and are believed to be the first pterosaur bone thin sections to have been produced and described. Six of these thin sections survived the bombing of London during the Second World War, although the bones from which they were cut were destroyed. The sections were re-examined in early 2001 and photographed in plane and crossed polarized light for the first time. The histology of one thin section is inconsistent with a pterosaurian origin. This case highlights the value of bone palaeohistology as an aid in confirming the identity of incomplete remains.

The earliest microscopic study of pterosaur bone was performed in the middle of the nineteenth century (Bowerbank 1848) by placing small fragments of bone in a drop of water under a microscope. The first thin-section studies of pterosaur bone followed shortly afterwards (Quekett 1849a, 1849b, 1855). Since those pioneering studies, there have been few investigations on pterosaur bone palaeohistology until recently. Many of these older studies described pterosaurian bone histology as part of broader palaeohistological reviews (e.g. Quekett 1849a, 1849b, 1855; Seitz 1907; Gross 1934; Enlow & Brown 1956, 1957, 1958). Recent work in particular has begun to interpret pterosaur hard tissues in terms of the biology of the living animal, and includes taxonomic investigations (Padian *et al.* 1995), and growth rate analysis (Ricqlès *et al.* 2000).

The Hunterian pterosaur bone thin sections

The earliest pterosaur bone thin sections were probably produced during the mid-1800s by John Quekett of the Hunterian Museum in London, and were described in a privately published museum catalogue (Quekett 1855). The histological thin-section collection of the Hunterian Museum at the time comprised mainly modern soft tissues and bones, but also contained a few fossil bones, including elements from ichthyosaurs, crocodiles and dinosaurs. A total of 15 thin sections of pterosaur bone are described in the catalogue. These transverse and longitudinal sections had been cut from eight different bones, most of which had been obtained by John Hunter, the founder of the Hunterian collection, but some had been donated by John Bowerbank. Three of the eight bones were found in the Late Cretaceous Lower Chalk of Kent, and two of them were associated. Two of the bones came from the Early Cretaceous Wealden Group of Sussex, and three were from the Middle Jurassic Stonesfield Slate Formation of Oxfordshire, United Kingdom. These sections were numbered Bb62, Bb63, Bb64, Bb65, Bb66, Bb67, Bb68, Bb69, Bb70, Bb71, Bb72, Bb73, Bb74, Bb75 and Bb76 (see Table 1). Quekett's catalogue provided a written histological description for each thin section, often with a detailed drawing to illustrate important features, such as the shape of the osteocyte lacunae and the orientation of canals.

Unfortunately, many important specimens in the Hunterian Museum were lost to enemy bombing raids on London during the Second World War. Of the 15 so-called pterosaur thin sections, only six survived (Fig. 1): Bb63, Bb65, Bb66, Bb67, Bb68 and Bb69. The original bones from which all the sections were cut were unfortunately destroyed, and it is no longer possible to identify the source skeletal element for the sections.

The importance of Quekett's work was recognized by Bramwell (1972), who re-examined the six surviving thin sections and reported that they were in poor condition. However, a re-examination in early 2001, using a petrological microscope, indicates that they are in excellent condition and suitable for histological analysis. The sections are set in a

From: BUFFETAUT, E. & MAZIN, J-M. (eds) 2003. *Evolution and Palaeobiology of Pterosaurs.* Geological Society, London, Special Publications, **217**, 325–334. 0305–8719/03/$15

Table 1. Pterosaur bone thin sections described by Quekett (1855)

Number	Direction	Specimen	Provenance
Bb62	Transverse	Wing bone	Lower Chalk, Kent
Bb63	Transverse	Same bone	As above
Bb64	Transverse	Another bone of above skeleton	As above
Bb65	Vertical	Same bone	As above
Bb66	Vertical	Same bone	As above
Bb67	Not given	Thin fragment of bone	As above
Bb68	Not given	Chippings from same bone	As above
Bb69	Transverse	Small tapering bone	Stonesfield Slate
Bb70	Vertical	As above	As above
Bb71	Transverse	Scapulocoracoid	As above
Bb72	Vertical	As above	As above
Bb73	Transverse	Flattened bone	As above
Bb74	Vertical	As above	As above
Bb75	Transverse and longitudinal	Wing bone	Wealden of Sussex
Bb76	Transverse	Long bone	Wealden

The descriptions given in this table are quoted from Quekett (1855, pp. 123–128). 'Vertical' probably means longitudinal.

resin (probably Canada balsam), which is still transparent, on glass slides of 25 $\times$60 mm, with glass coverslips. There are occasional scratch marks across the sections from the polishing process that are only seen at high magnifications. The sections have now been photographed in both plane polarized light (PPL) and crossed polarized light (XPL) for the first time. This study has shed doubts on the pterosaurian identity of one of the sections.

Procedure

The six thin sections were examined under a petrological microscope in PPL and XPL, with and without a tint plate, at high and low magnifications. The inclusion of the tint plate under crossed polars produces a brightly coloured image, which allows the orientation of the bone fibres to be determined. Without the tint plate, the orientation of the bone fibres can still be seen, but the black-and-white images produced in this way obscure histological detail and are also more difficult to photograph. All the sections were photographed with the exception of Bb68. This particular slide consisted of several minute chips of bone, which close examination proved to be uninformative and not worth photographing.

Pterosaur bone histology

It is generally agreed that pterosaur bone usually consists of a thin cortex surrounding a large lumen, occasionally crossed by trabeculae (Wellnhofer 1991), though the tissue is often cancellous at epiphyses. A thin endosteal layer often lines the lumen. The histology of the cortex is usually described as fibro-lamellar, i.e. composed of woven-fibred tissue with primary osteons (see Francillon-Vieillot *et al.* 1990). This type of tissue is generally accepted to be an indicator of rapid growth (Amprino 1947; Ricqlès *et al.* 2000). The cortex may be thickened in corners (Bramwell & Whitfield 1974).

Histological description of the Hunterian specimens

Remarkably good preservation in most of these fossil bones has provided the opportunity to examine and obtain images of their histology.

Five of the specimens (Bb63, Bb65, Bb66, Bb67, Bb68) are typically pterosaurian in their bone histology. They are composed of fibro-lamellar cellular bone (fl) of periosteal origin and contain a rather sparse reticular arrangement of primary osteons (po). A 'plywood' structure was observed in Bb63 (Fig. 2), a histology which occasionally occurs in pterosaur bones (pers. obs.; Ricqlès *et al.* 2000). A well-developed endosteal layer (el) is present in the same specimen. Secondary osteons (so) are absent in these five sections and have not been unequivocally observed by this author in any pterosaur bone. In most of the sections, typically elongate osteocyte lacunae are clearly visible, complete with radiating canaliculi (Figs 3 & 4). They are less clear in Bb67 (Fig. 5).

However, the sixth thin section (Bb69) is histologically distinct (Figs 6, 7, 8 & 9). This specimen is oval in cross-section and measures 3 mm along its longest axis and 2 mm along its shortest axis. It contains a central lumen that measures 1 mm along its longest axis and 0.5 mm along its shortest axis (Fig. 6). A thin layer of avascular parallel-fibred endosteal bone (el) lines the lumen. The periosteal bone contains two distinct tissues that are separated by a reversal line (r). The innermost periosteal tissue is dominated by large secondary osteons (so). Many of these have become truncated by periosteal resorption prior to the deposition of the overlying periosteal tissue, which is highly vascularized by numerous longitudinal primary osteons (po). These primary osteons become more abundant as the periosteal surface of the bone is approached. Within its tissues, this bone records a complex sequence of deposition, reworking, resorption and further deposition, which was probably rapid.

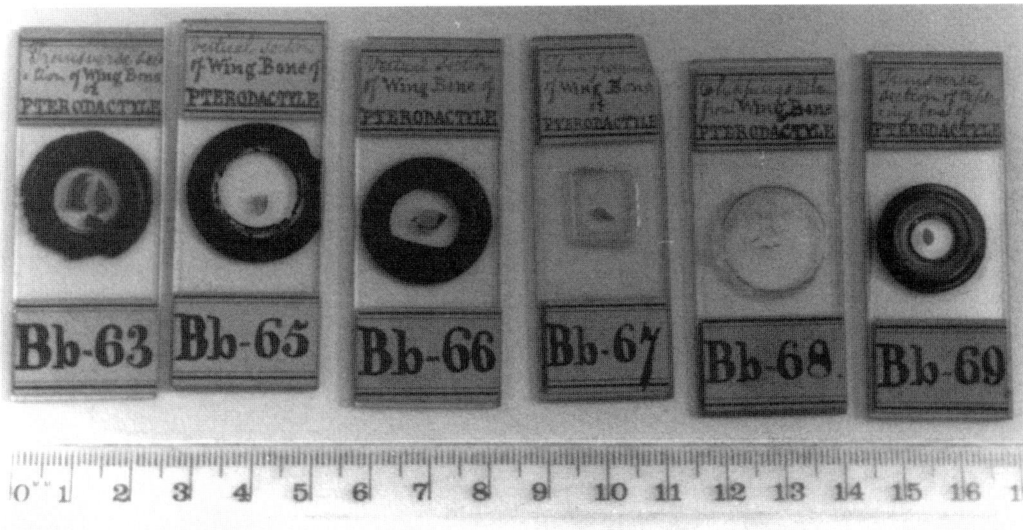

Fig. 1. The six surviving pterosaur bone thin sections in the Quekett microscopical collection at the Hunterian Museum.

What is Bb69?

The original specimen from which Bb69 was made was described as 'a small, tapering bone, probably of Pterodactyle, from the Stonesfield Slate' (Quekett 1855, p. 125). Although its slender appearance led Quekett to identify it as a small pterosaur bone, its histology is atypical for a pterosaur origin.

The Stonesfield Slate is well known for its Bathonian (Middle Jurassic) vertebrates, including chelonians (e.g. Gillham 1994), dinosaurs (e.g. Buckland 1824; Galton 1975), early mammals (e.g. Evans & Milner 1991), fish, crocodilians and rhamphorhynchoid pterosaurs (e.g. Huxley 1859; Phillips 1871; Seeley 1880). Only one pterosaur bone from this area has been studied in thin section (Seitz 1907). It has been referred to *Rhamphocephalus bucklandi* (Seitz 1907, figs 39, 40, 41 & 42). Its histology is consistent with a pterosaurian assignment and is quite unlike that of Bb69.

Further investigation of the fossil vertebrates of the Stonesfield Slate may indicate to which group Bb69 belongs. The histology of the bone and the presence of a small lumen is suggestive of a small dinosaur.

Conclusion

Quekett (1855) identified Bb69 as pterosaurian, partly on the basis of its thin walls. However, other vertebrates occasionally have a similar bone structure. Birds are a well-known example, but Martill *et al.* (2000) reported similarly thin cortical bone in a small dinosaur femur from the Santana Formation of Brazil. Thus, the identification of incomplete bones as pterosaurian purely on the basis of their slenderness and/or thin walls should be treated with caution.

The distinctive histology of pterosaurian long bones provides a useful tool for distinguishing pterosaur bones from other vertebrates in mixed or fragmentary assemblages. However, further work is required as bone histology appears to be somewhat convergent among pterosaurs and birds, as indeed many authors have remarked (e.g. Bowerbank 1848; Phillips 1871, pp. 225–226; Seitz 1907; Gross 1934; Enlow & Brown 1957; Bennett 1993; Ricqlès *et al.* 2000). Due to histological convergence between various vertebrate groups, perhaps Quekett's aim to use palaeohistology for 'determining the affinities of minute fragments of organic remains' (Quekett 1849) is still some way off.

Special thanks are due to M. Cooke and S. Chaplin of the Royal College of Surgeons of London, UK, for their help and cooperation in this study. R. Pulley and R. Loveridge assisted with photomicrography. Thanks to D. Martill for reading an early draft of the manuscript. This work was supported by the School of Earth and Environmental Sciences, University of Portsmouth.

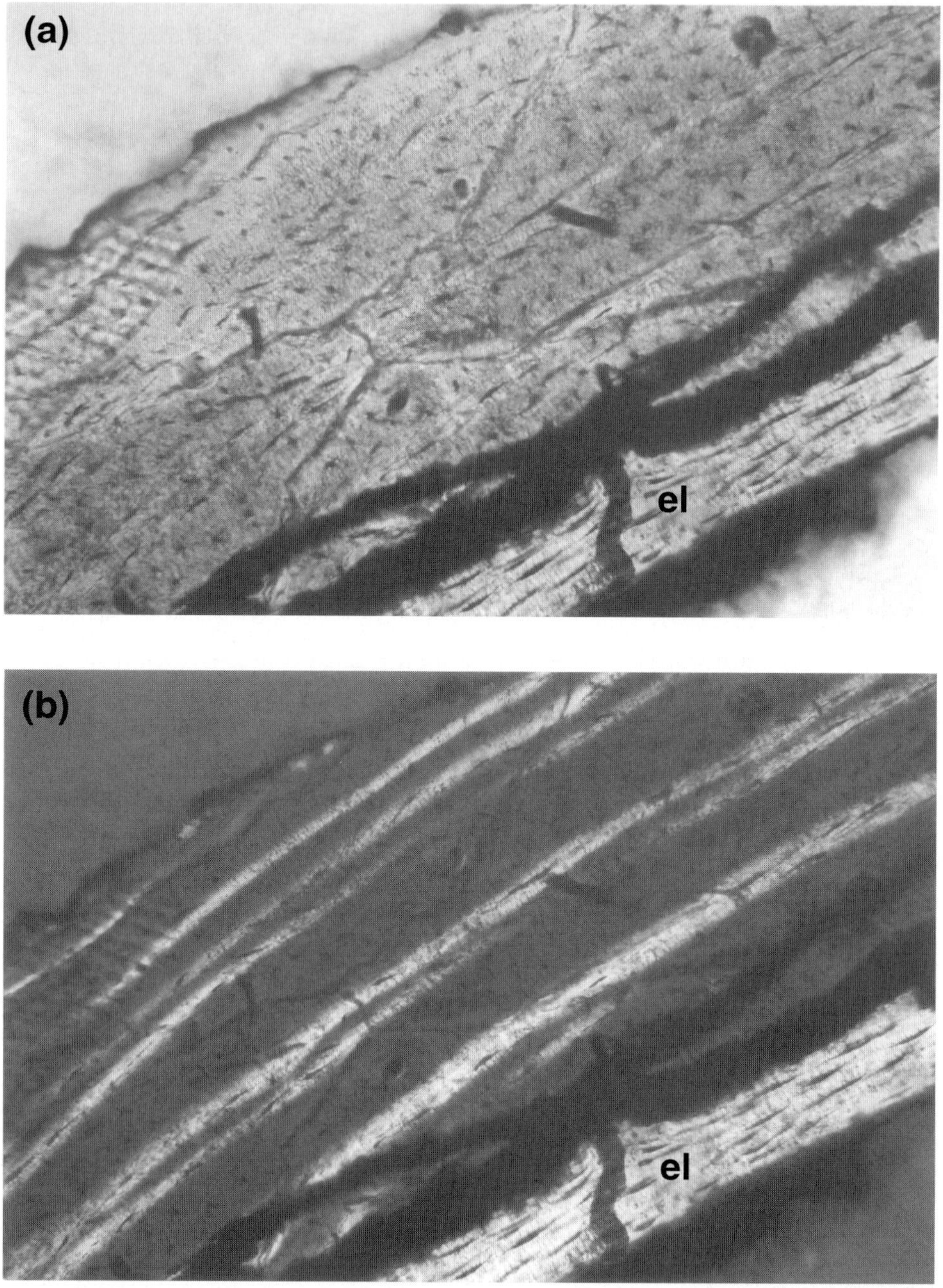

Fig. 2. Bb63 in (**a**) PPL, and (**b**) XPL with tint plate. The periosteal tissue contains several small vascular canals and is lined by an avascular endosteal layer (el). The periosteal surface appears to have suffered abrasion damage. The 'plywood' structure is much clearer in XPL. Field of view for both figures: 0.8 mm × 0.54 mm.

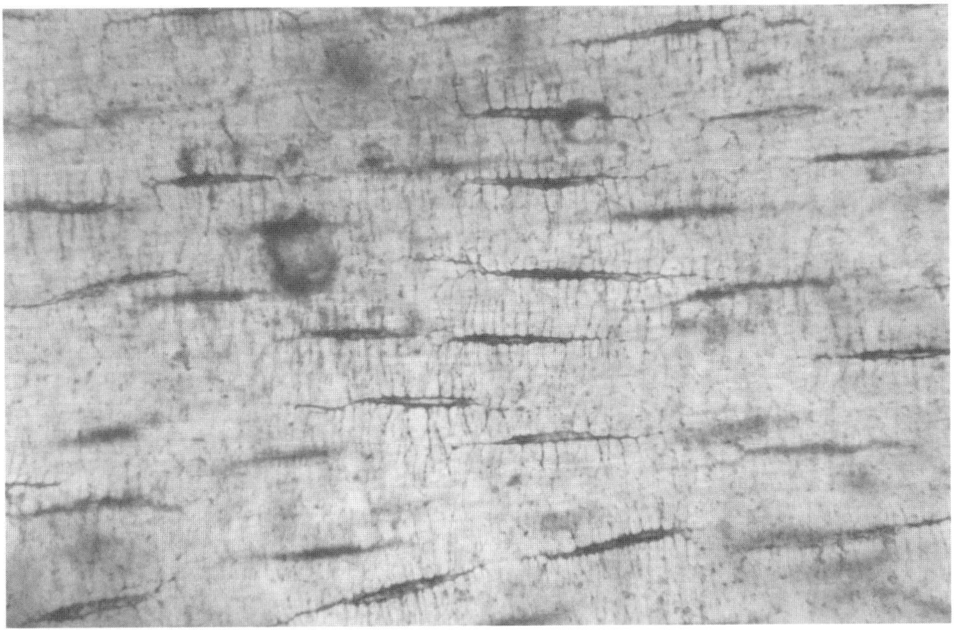

Fig. 3. Bb65 in PPL, exhibiting well-preserved osteocyte lacunae. Field of view 0.3 mm × 0.2 mm.

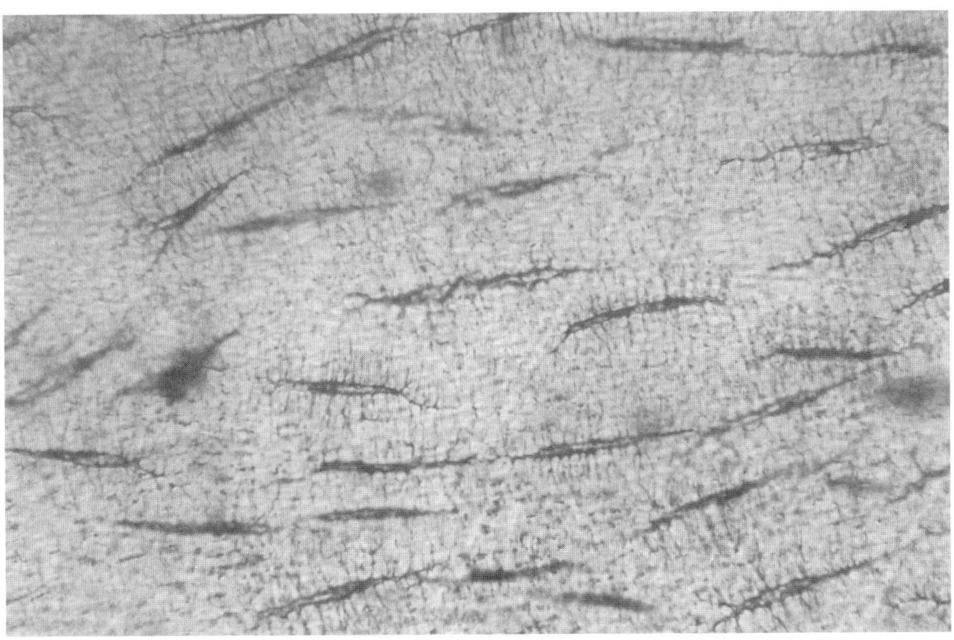

Fig. 4. Bb66 in PPL, exhibiting well preserved osteocyte lacunae. Field of view 0.3 mm × 0.2 mm.

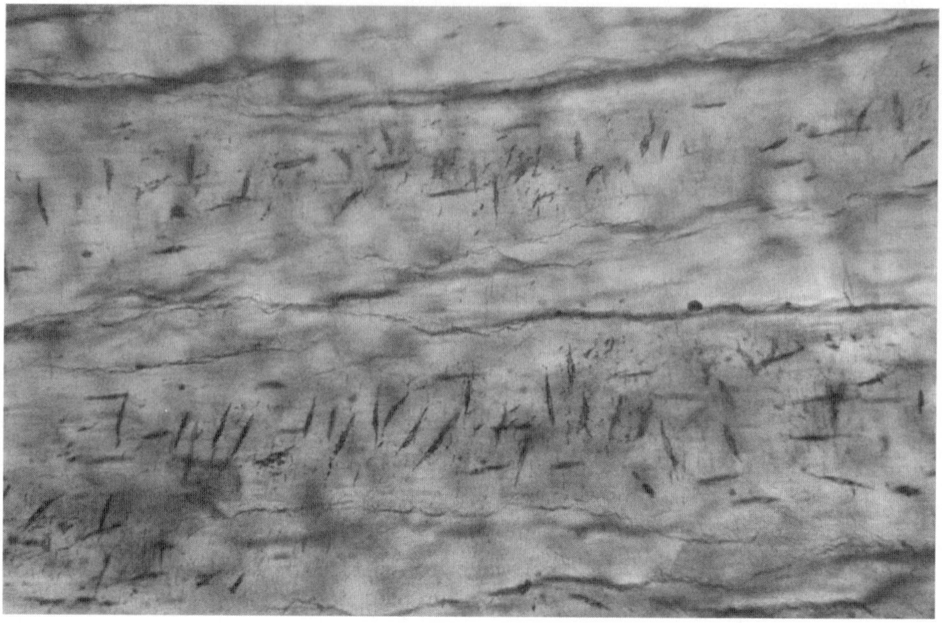

Fig. 5. Bb67 in PPL, exhibiting rather poor preservation of histological detail. The osteocyte lacunae are indistinct in comparison with the other sections (compare with Figs 3 & 4). Field of view: 0.8 mm × 0.54 mm.

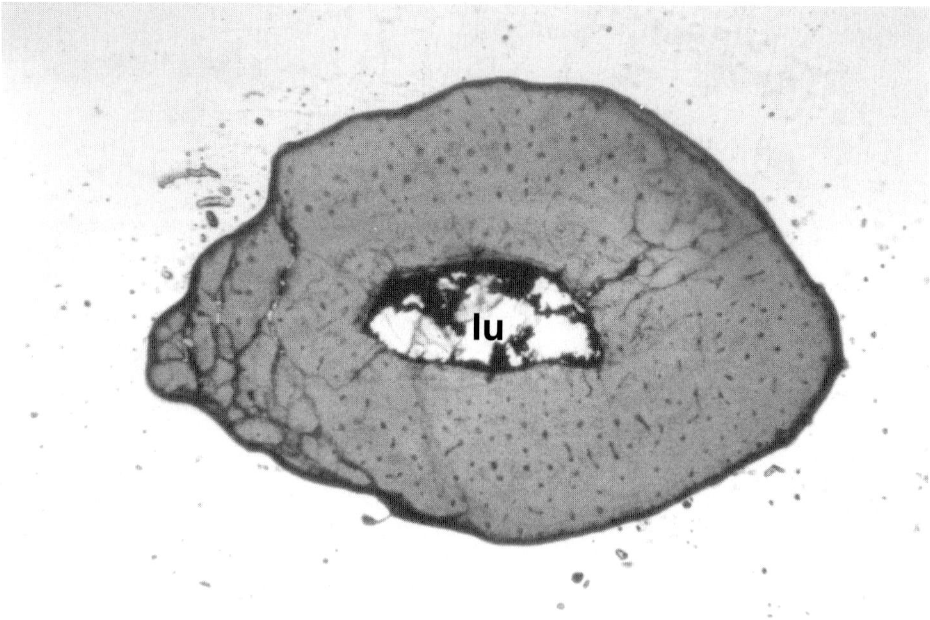

Fig. 6. Bb69 in PPL, showing the complete section. This specimen has a small lumen (lu) and the cortex is highly vascular. Field of view: 3.9 mm × 2.9 mm.

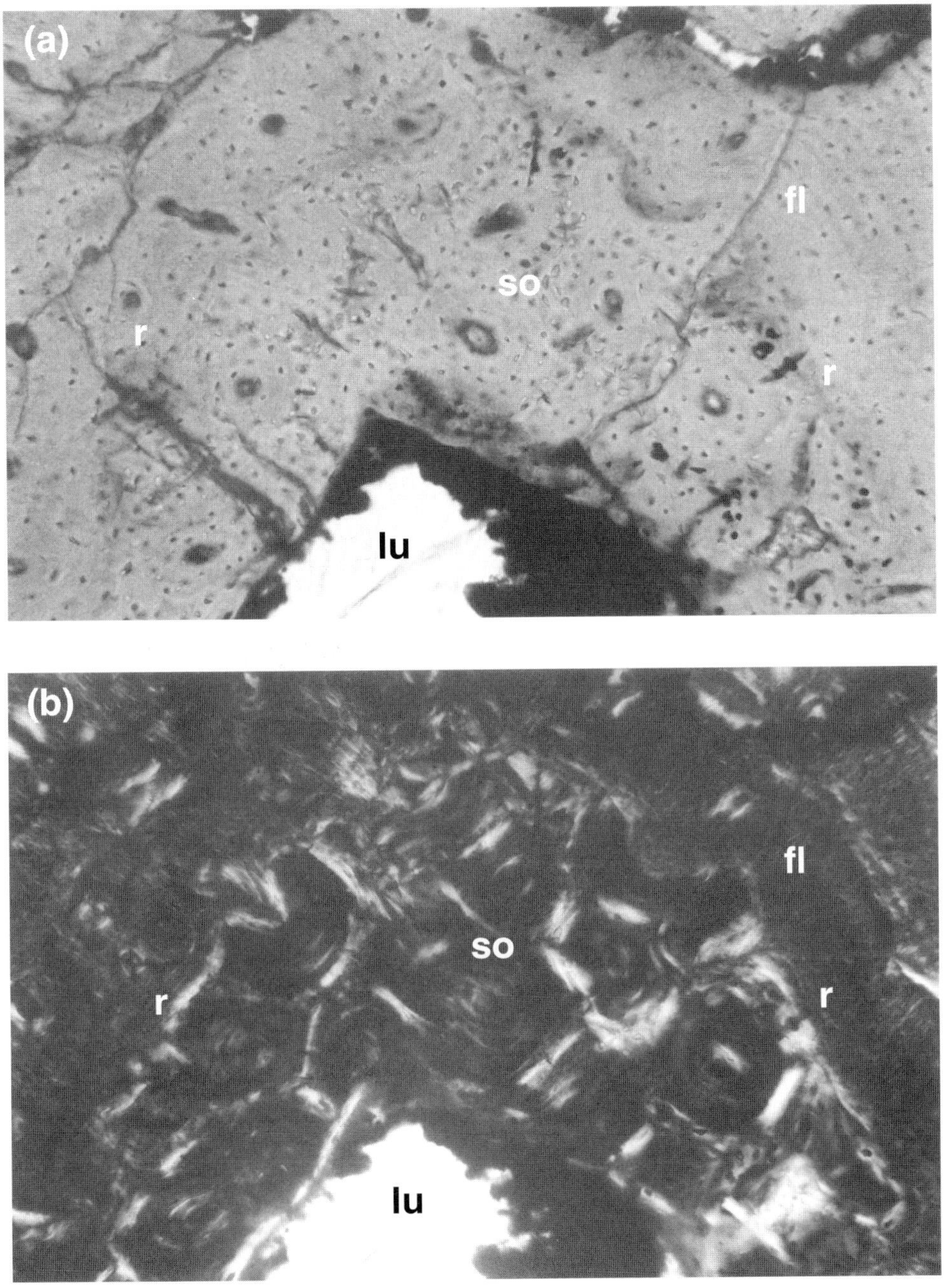

Fig. 7. Bb69 in (**a**) PPL, and (**b**) XPL showing part of the lumen (lu), with secondary osteons (so), reversal line (r) and primary fibro-lamellar bone (fl) with numerous longitudinal primary osteons. Field of view: 0.8 mm $\times$ 0.54 mm.

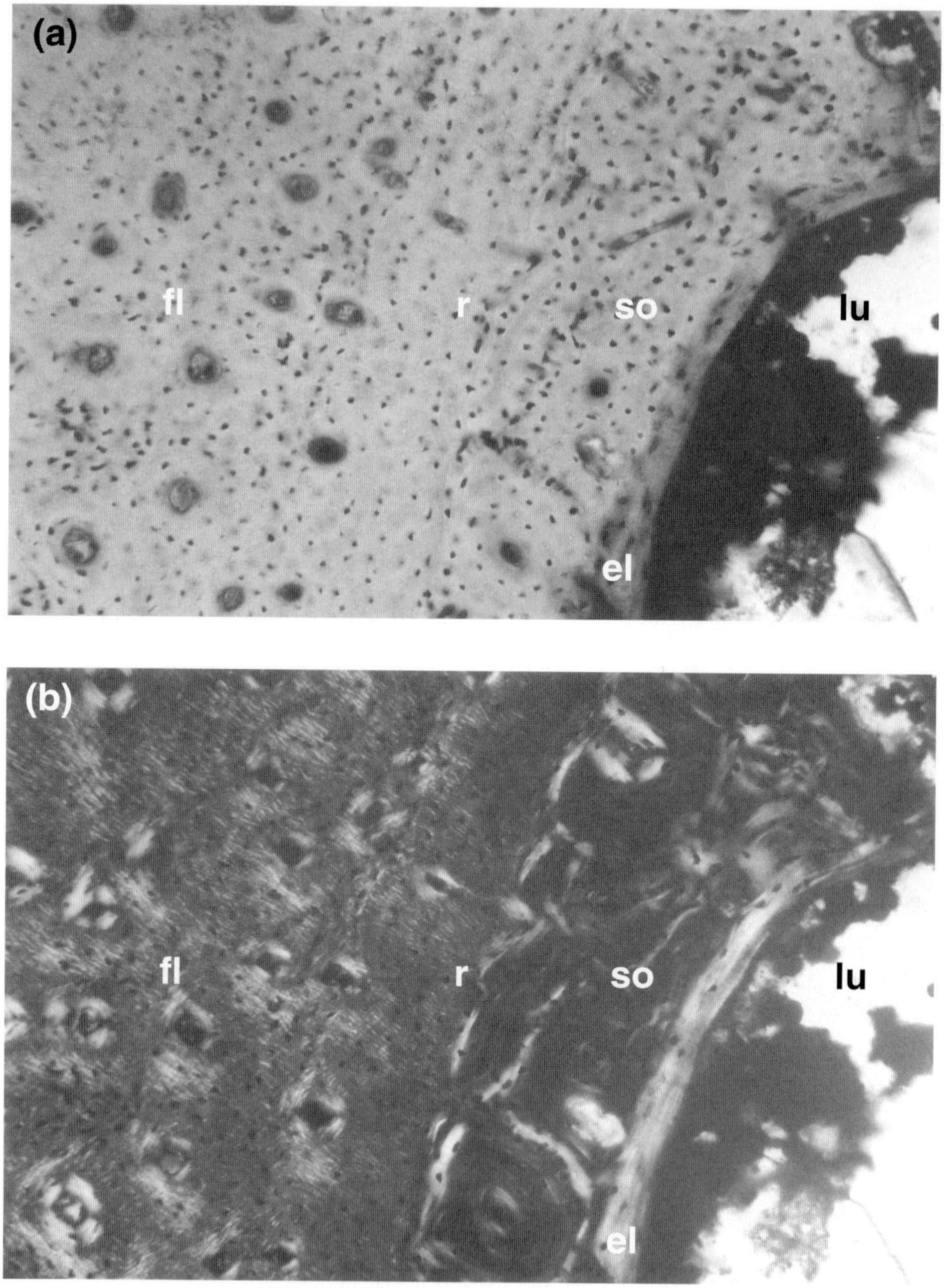

Fig. 8. Bb69 in (**a**) PPL, and (**b**) XPL showing another part of the lumen (lu), lined by an endosteal layer (el). The remaining bone tissues comprise a region of secondary osteons (so), a reversal line (r) and highly vascular fibro-lamellar bone (fl). Field of view: 0.8 mm × 0.54 mm.

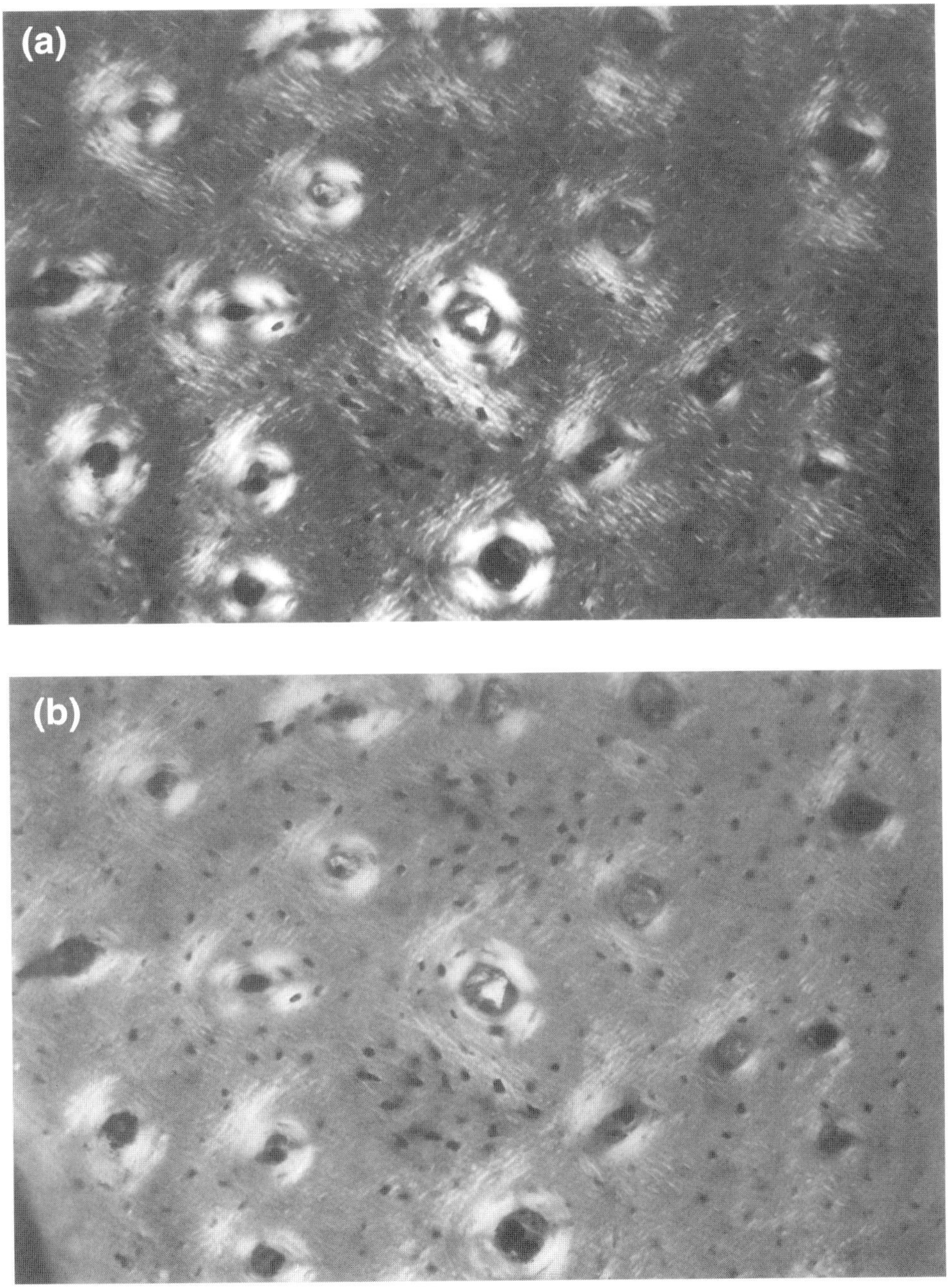

Fig. 9. Fibro-lamellar tissue forming the majority of the cortex in Bb69: (**a**) in XPL, and (**b**) XPL with tint plate. The lamellae of the primary osteons are clearly seen in contrast to their woven matrix. Field of view: 0.5 mm × 0.36 mm.

References

AMPRINO, R. 1947. La structure du tissu osseux envisagée comme expression de différences dans la vitesse de l'accroissement. *Archives de Biologie*, **58**, 315–330.

BENNETT, S. C. 1993. The ontogeny of *Pteranodon* and other pterosaurs. *Paleobiology*, **19**, 92–106.

BOWERBANK, J. S. 1848. Microscopical observations on the structure of the bones of *Pterodactylus giganteus* and other fossil animals. *Quarterly Journal of the Geological Society, London*, **4**, 2–10.

BRAMWELL, C. D. 1972. The contribution of John Thomas Quekett to the study of pterodactyls. *Microscopy*, **32**, 189–195.

BRAMWELL, C. D. & WHITFIELD, G. R. 1974. Biomechanics of *Pteranodon*. *Philosophical Transactions of the Royal Society, London, Series B*, **267**, 503–581.

BUCKLAND, W. 1824. Notice on the *Megalosaurus* or great fossil lizard of Stonesfield. *Transactions of the Geological Society, London, Series 2*, **1**, 390–396.

ENLOW, D. H. & BROWN, S. O. 1956. A comparative histological study of fossil and recent bone tissues. Part 1. *Texas Journal of Science*, **8**, 405–443.

ENLOW, D. H. & BROWN, S. O. 1957. A comparative histological study of fossil and recent bone tissues. Part 2. *Texas Journal of Science*, **9**, 186–214.

ENLOW, D. H. & BROWN, S. O. 1958. A comparative histological study of fossil and recent bone tissues. Part 3. *Texas Journal of Science*, **10**, 187–230.

EVANS, S. E. & MILNER, A. R. 1991. Middle Jurassic microvertebrate faunas from the British Isles. *Fifth Symposium on Mesozoic Terrestrial Ecosystems and Biota*, University of Oslo, 21–22.

FRANCILLON-VIEILLOT, H., DE BUFFRENIL, V. *ET AL.* 1990. Microstructure and mineralisation of vertebrate skeletal tissues. *In:* CARTER, J. G. (ed.) *Skeletal Biomineralisation: Patterns, Processes and Evolutionary Trends.* Van Nostrand Reinhold, New York, vol. 1, 471–530.

GALTON, P. M. 1975. English hypsilophodontid dinosaurs (Reptilia: Ornithischia). *Palaeontology*, **18**, 741–752.

GILLHAM, C. 1994. A fossil turtle (Reptilia, Chelonia) from the Middle Jurassic of Oxfordshire, England. *Neues Jahrbuch für Geologie und Paläontologie, Monatshefte*, **10**, 581–596.

GROSS, W. 1934. Die Typen des mikroskopischen Knochenbaues bei fossilen Stegocephalen und Reptilien. *Zeitschrift für Anatomie*, **103**, 731–764.

HUXLEY, T. H. 1859. On *Rhamphorhynchus bucklandi*, a pterosaurian from the Stonesfield Slate. *Quarterly Journal of the Geological Society, London*, **15**, 658.

MARTILL, D. M., FREY, E., SUES, H-D. & CRUICKSHANK, A. R. 2000. Skeletal remains of a small theropod dinosaur with associated soft structures from the Lower Cretaceous Santana Formation of northeastern Brazil. *Canadian Journal of Earth Sciences*, **37**, 891–900.

PADIAN, K., RICQLÈS, A. J. DE & HORNER, J. R. 1995. Bone histology determines identification of a new fossil taxon of pterosaur (Reptilia: Archosauria). *Comptes Rendus de l'Academie des Sciences, Paris, Serie IIa*, **320**, 77–84.

PHILLIPS, J. 1871. *Geology of Oxford and the Valley of the Thames*. Clarendon Press, Oxford, UK.

QUEKETT, J. T. 1849a. On the intimate structure of bone, as composing the skeleton in the four great classes of animals, viz., mammals, birds, reptiles and fishes, with some remarks on the great value of the knowledge of such structure in determining the affinities of minute fragments of organic remains. *Transactions of the Microscopical Society, London*, **2**, 46–58.

QUEKETT, J. T. 1849b. Additional observations on the intimate structure of bone. *Transactions of the Microscopical Society, London*, **2**, 59–64.

QUEKETT, J. T. 1855. *Descriptive and Illustrated Catalogue of the Histological Series Contained in the Museum of the Royal College of Surgeons of England, Vol. 2.* London, UK.

RICQLÈS, A. J. DE, PADIAN, K., HORNER, J. R. & FRANCILLON-VIEILLOT, H. 2000. Palaeohistology of the bones of pterosaurs (Reptilia: Archosauria): anatomy, ontogeny, and biomechanical implications. *Zoological Journal of the Linnean Society, London*, **129**, 349–385.

SEELEY, H. G. 1880. On *Rhamphorhynchus prestwichi. Quarterly Journal of the Geological Society, London*, **36**, 27.

SEITZ, A. L. 1907. Vergleichende Studien über den mikroskopischen Knochenbau fossiler und rezenter Reptilien und dessen Bedeutung für das Wachstum und umbildung des Knochengewebes im allgemeinen. *Nova Acta Abhandlungen der Kaiserlichen Leopold-Carolignischen Deutschen Akademie der Naturforscher*, **87**, 229–370.

WELLNHOFER, P. 1991. *The Illustrated Encyclopedia of Pterosaurs*. Salamander Books, London, 192 pp.

Histovariability in bones of two pterodactyloid pterosaurs from the Santana Formation, Araripe Basin, Brazil: preliminary results

JULIANA M. SAYÃO

Setor de Paleovertebrados, Museu Nacional/UFRJ, Quinta da Boa Vista, s/n São Cristóvão CEP 20940–040, Rio de Janeiro, Brazil (e-mail: jsayao@mn.ufrj.br)

Abstract: Thin sections of pterosaur bones have not been extensively studied so far. Until now mainly isolated bones were the subject of this type of analysis. Here we present preliminary results of the histological analysis of two pterodactyloid pterosaurs from the Romualdo and Crato Members of the Santana Formation. The first specimen (Museu Nacional, MN 4809-V) comprises both wings (from humerus to the third phalanx of the fourth digit) and one hindlimb (tibia, fibula and pes). The second (MN 6527-V) consists of a partial skeleton with the incomplete left wing (humerus, radius, ulna, carpals, pteroid, wing metacarpal, two phalanges of the fourth digit) and fragments of one hindlimb (complete tibia and proximal articulation of the femur). Thin sections of the diaphyseal portion of each bone were ground in transverse, longitudinal and oblique orientation and the histological structures were compared. The following features can vary in the first observed specimen (MN 4809-V): presence of primary and secondary tissues in the cortex; absence or presence and position of lines of arrested growth; resorption of primary structures. These variations in different bones of the same individual represent differential growth rates. The second specimen (MN 6527-V) is well vascularized and has fibro-lamellar bone indicative of rapid growth in the thin section of the radius, ulna and first phalanx of the fourth digit. This last feature is very similar to the condition observed in most living birds.

Introduction

Palaeohistological studies on fossil bones are a very important tool since they could help to solve several biological problems regarding extinct animals, such as skeletal maturity, growth rates and physiology, and could lead to broad ecological considerations (e.g. Bennett 1993; Chinsamy *et al.* 1995; Horner *et al.* 1999; Horner *et al.* 2001). Up to now this methodology has not been extensively used in pterosaurs, mainly because of the lack of suitable material.

The first published thin section of this group of extinct flying reptiles was from a humeral shaft presented by Quekett (1849). In 1855 Quekett published a catalogue of the thin sections housed in the Hunterian Museum (London), which includes 15 sections of pterosaur bones. Some of these were lost in the bombing of London during the Second World War (Steel 2001). Sixty years later Seitz (1907) presented numerous thin sections of different extinct reptiles, including the pterosaur *Rhamphocephalus bucklandi* (Meyer 1832; Jurassic of Germany) and *Pteranodon* (Cretaceous of Kansas). More palaeohistological analyses of pterosaurs were made in subsequent decades, mainly based on isolated elements (e.g. Gross 1934; Enlow & Brown 1957; Bennett 1993). Among the most interesting contributions in this field was the identification of a new pterosaur taxon from the Late Cretaceous of Montana (Padian *et al.* 1995) and the identification of the pteroid as a true bone and not a secondarily calcified structure (Unwin *et al.* 1996). The most comprehensive work on pterosaur histology was made by Ricqlès *et al.* (2000) who described the histological features of similar bones from individuals of different sizes from the same taxon in detail and discussed bone structure, ontogeny and biomechanical implications.

Despite being one of the most important pterosaur deposits in the world, the specimens from the Santana Formation have not until now been the subject of detailed palaeohistological studies. With the exception of the pteroid (Unwin *et al.* 1996) and the initial results from some thin sections of long bones briefly mentioned before (Sayão *et al.* 2000) no palaeohistological study of pterosaur material from this stratigraphic unit has been made. Here is presented a multi-skeletal element analysis of the bone histology of two pterosaur specimens from the Santana Formation (Crato and Romualdo Members). Partial results of this study were mentioned previously (Sayão 2001) and a more comprehensive study is presented here.

Geological setting

The Araripe Basin is located between the States of Ceará, Pernambuco and Piauí, in northeastern Brazil. From the palaeontological point of view the foremost lithostratigraphic unit is the Santana Formation (Maisey 1991), which is divided into three members: Crato, Ipubi and Romualdo (Beurlen 1971). Palynological data suggest that

From: BUFFETAUT, E. & MAZIN, J-M. (eds) 2003. *Evolution and Palaeobiology of Pterosaurs.* Geological Society, London, Special Publications, **217**, 335–342. 0305–8719/03/$15

these deposits are Aptian to Albian (Pons *et al.* 1990).

The Santana Formation embodies two different *Lagerstätten*, corresponding to the Crato and Romualdo Members, distinguished by the depositional environment and the type of fossilization (e.g. Kellner 1998; Sayão & Kellner 2000). The base of the sequence is the Crato Member, composed essentially of freshwater laminated limestone. This unit is very rich in fossils, predominantly insects (Martins Neto 1991) and fishes, and, more seldom, pterosaur remains (Frey & Martill 1994; Kellner 1994; Sayão & Kellner 2000). The specimens of flying reptiles from this unit, such as MN 6527-V, are very compacted and tend to be more fragmented than the material from the Romualdo Member. The latter is located on top of the Santana Formation and is composed of shales and marls. Pterosaurs are found in limestone concretions (e.g. Kellner & Tomida 2000), including specimen MN 4809-V. All tend to be very well preserved, with the bones showing almost no distortion.

Material and methods

The material used for the palaeohistological analyses conducted here consists of two partial skeletons housed in the palaeovertebrate collection of the Museu Nacional/UFRJ, under the numbers MN 4809-V and MN 6527-V. MN 4809-V was preserved in a limestone concretion from the Romualdo Member. It comprises both pterosaur wings from humerus to the third phalanx, right tibia and femur (Fig. 1a, b). It shares one synapomorphy of the Pteranodontoidea (*sensu* Kellner 1996): the deltopectoral crest of the humerus is warped. Furthermore the distal portion of deltopectoral crest is expanded with a markedly concave surface, which is currently regarded as a potential synapomorphy of the Anhangueridae (Kellner, in press.).

The second specimen studied here (MN 6527-V, Fig. 2), preserved in laminated limestone from the Crato Member, comprises an incomplete and articulated left wing, with humerus, radius, ulna, proximal, distal and lateral-distal carpals, wing metacarpal, two indeterminate phalanges, tibia and proximal articulation of femur. It represents an azhdarchoid pterosaur, on the basis of the proportions of the distal articulation of the wing metacarpal relative to its total length (Sayão & Kellner 2001).

All bones were measured and pictures of specimens MN 4809-V and MN 6527-V were taken. Before sectioning, all bones were cast. Each bone shaft was sectioned and the sectioned portion was embedded in epoxy resin, ground into a thin section and polished and examined under ordinary and polarized light. Sections in transverse, longitudinal and oblique orientations were processed. Similar bones from the different specimens and bones of the same specimen were compared. Standard thin sectioning, grinding and polishing techniques, such as those described by Chinsamy and Raath (1992), were used.

Palaeohistology

Pterosaurs have hollow bones, with extremely thin bone walls, which implies a large medullar cavity (Ricqlès *et al.* 2000). As pointed out by many authors, their bones are sometimes even thinner than those of birds (e.g. Owen 1870; Seeley 1870; Padian 1985; Wellnhofer 1991). Those conditions were also verified in the material studied here (Fig. 3), although the medullar cavity in MN 6527-V could not be measured since the specimen is very compacted. A detailed description of each one is presented below.

Specimen MN 4809-V

In the transversal thin section of the ulna (diameter 21.98 mm, bone wall 1.03 mm) there is a thin layer of avascular periosteal bone covering the surface of the bone. It corresponds to the so-called 'external fundamental system' (EFS) (*sensu* Ricqlès *et al.* 2000) that was first described by Cormack (1987) in adult mammals. These are part of several structural changes in the internal cortex which occur while the animal is reaching the adult ontogenetic stage. It further consists of the deposition of a thin layer of lamellar, poorly vascularized bone at the periphery of the cortex (Ricqlès *et al.* 2000). The compacta is composed of many round and longitudinally oriented primary osteons embedded within the woven matrix (Fig. 4a). A thin coat of endosteal bone surrounds the medullar cavity. On the anterior part of the bone (where the bone wall is thicker) a large erosion room in the primary bone can be seen (Fig 4b). In this area part of the primary tissue is still present.

The pteroid has the thinner bone wall of this specimen (7.78 mm diameter, 0.57 mm bone wall). The periosteal bone is lamellar and is formed by a few primary osteons within a woven matrix. They are displayed in the cortex side by side, arranged in just one line.

In the thin section of the wing metacarpal (diameter 24.50 mm, bone wall 2.41 mm) a distinct compact bone region surrounds a large central medullar cavity. The compacta consists of both primary and secondary bone tissue. The peripheral region of the bone is composed of lamellar periosteal bone. The primary bone occupies the mid-cortical region and comprises primary osteons embedded within the woven bone matrix (Fig. 5c). Deep in the

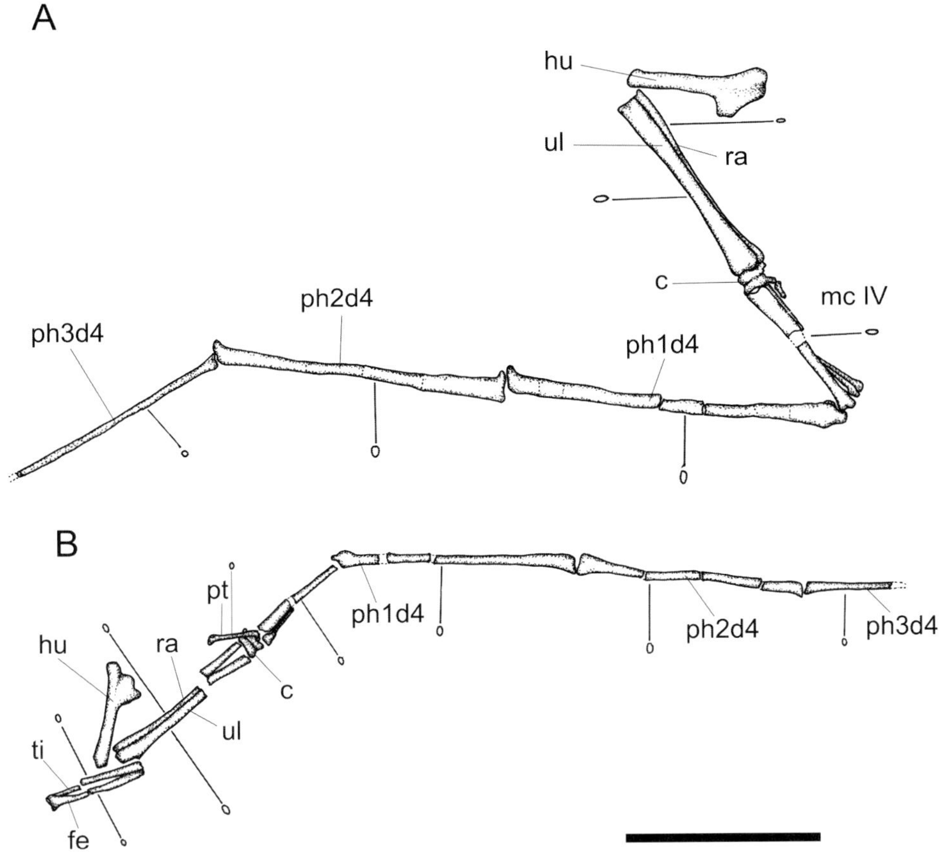

Fig. 1. Specimen MN 4809-V. Sketch of (**a**) the right and (**b**) the left wings. The circle shows sectioned bones. hu, humerus; ul, ulna; c, carpals; pt, pteroid; mc IV, wing metacarpal; ph1d4, first wing phalanx; ph2d4, second wing phalanx; ph3d4, third wing phalanx; fe, femur; ti, tibia. Scale bar 300 mm (**a**) and 380 mm (**b**).

cortex, the vascular canals are larger in size than those formed later. Internal to the primary tissue lies a region of secondary reconstruction. Here a few secondary osteons occur. The latter are distinguished by the cement line, which marks the furthest extent of bone resorption. In this case the circumferentially orientated osteocytes are located around the vascular canals. One line of arrested growth (LAG) can be observed closer to the medullar cavity. A comparison of the left and right metacarpals, as expected, showed that the LAGs were in exactly the same position. A thin coat of endosteal bone, with its fibres running circularly, surrounds the marrow cavity.

On the first phalanx of the fourth digit (diameter 22.66 mm, bone wall 0.91 mm) the transverse thin section presents a cortex basically composed of secondary tissue. A thin, avascular lamellar periosteal layer, the so-called EFS, is observed in the upper region of the cortex (Fig. 5d). The layer situated below the EFS shows many secondary osteons (Fig. 5e). In anterior view, where the bone is thicker, two big irregular erosion rooms are formed in the perimedullary region, eroding the original primary tissue (Fig. 5f). These are big and could be seen non-microscopically. Another erosion room, less developed, is located ventrally.

Under polarized light, using a quartz wedge, the oblique section of the second phalanx of the fourth digit (ph2d4, diameter 18.66 mm, bone wall 0.79 mm) shows a reticular vascularization of 'plywood-like' layers. Each ply is thick and presents an homogeneous, dense, parallel-fibred bone. The cells are spindle-shaped and oriented according to the fibres in the ply. Each ply is oriented perpendicularly relative to the previous one, and the bone cells follow this pattern (see Ricqlès *et al.* 2000 for a review).

The general features observed in the second phalanx could be observed also in the section of the third phalanx of the fourth digit (ph3d4). Otherwise ph3d4 presents two LAGs, one closer to the lamellar periosteal layer and the second closer to the endosteal bone.

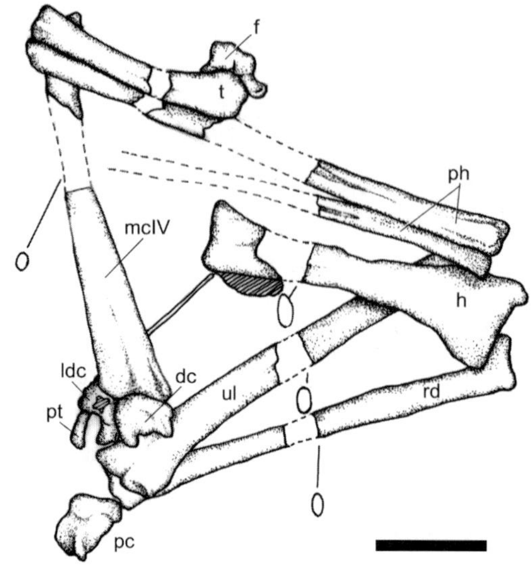

Fig. 2. Sketch of pterosaur MN 6527-V. The circle shows the sectioned parts of the bones. hu, humerus; ra, radius; ul, ulna; pc, proximal carpal; dc, distal carpal; ldc, lateral distal carpal; pt, pteroid; mc IV, wing metacarpal; ph, wing phalanges; fe, femur; ti, tibia. Scale bar 150 mm.

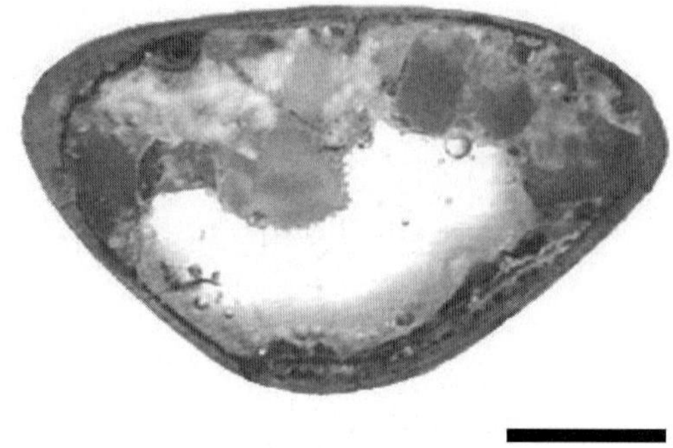

Fig. 3. Mid-diaphyseal transverse section of the first wing phalange of MN 4890-V, showing the extremely thin bone walls of the referred specimen. Scale bar 90 mm.

The thin section of the femur (diameter 10.81 mm, bone wall 1.00 mm) shows the compacta consisting mostly of secondary bone tissue (Fig. 5a, b). The peripheral region of the bone is composed of lamellar periosteal bone. In the cortex many secondary osteons are observed in a woven matrix. No LAG can be seen. The most interesting feature of this bone is the presence of pneumatic canals (Fig. 5a). They consist of two cavities displayed close to the internal surface of the bone, surrounded by endosteal tissue, as usual in bones that possess this kind of structure (Ricqlès *et al.* 2000). A few secondary osteons are located between the two cavities.

To summarize, the examined histological sections of MN 4809-V show a thin, nearly avascular periosteal layer covering the outer surface of the bone (the 'external fundamental system' *sensu* Ricqlès *et. al.* 2000). In a thin and compact cortex made of lamellar periosteal bone, the primary vascular canals have reached their ultimate diameters by centripetal deposition of the primary osteons. In a few sections (ph1d4, femur and mcIV) several generations of secondary osteons are developed in the most internal part of the cortex. In most bones, the laminar cortex presents one or more LAGS.

Several histological variations were observed in different bones. The main difference is in the cortex. The ulna, pteroid and second and third wing phalanges have a cortex composed basically of primary osteons. In the wing metacarpal both primary and secondary osteons are present. The femur shows many secondary osteons. Other bones, such as ph1d4 and ulna, show big erosion rooms in the primary tissue for the formation of the medullar cavity. Another difference between the bones is the presence or absence and number of LAGs.

Specimen MN 6527-V

The cortex of the humerus is composed of primary tissue (Fig. 6a). Most of the structures, up to its subperiosteal surface, are made of reticular fibro-lamellar bone. The vascular canals are numerous and small in diameter and their orientation is longitudinal with irregular anastomoses. Evidence of an erosion room is located close to the medullar cavity (Fig. 6b). Since the degree of compaction of this bone is high, the shape and extension of this erosion room cannot be observed properly. This histological pattern is commonly reported in primary fast-growing tissue, involved in the formation of the cortex.

The radius, ulna and wing metacarpal have histological features similar to those observed in the humerus. A transversal thin section of the diaphysis presents a dense reticular tissue type. The periosteal bone is fibro-lamellar. There is no evidence of bone erosion and reconstruction, nor of deposition of an avascular covering of the periosteal layer. The main histological difference between these bones and the humerus is the lack of erosion rooms observed in the latter.

There is no evidence of LAGs nor of secondary osteons in this specimen.

Discussion

Overall, pterosaurs share the main categories of bone tissues generally observed in other tetrapods. As in other thin sections of these flying reptiles, the 'plywood-like' tissue was observed and seems to be

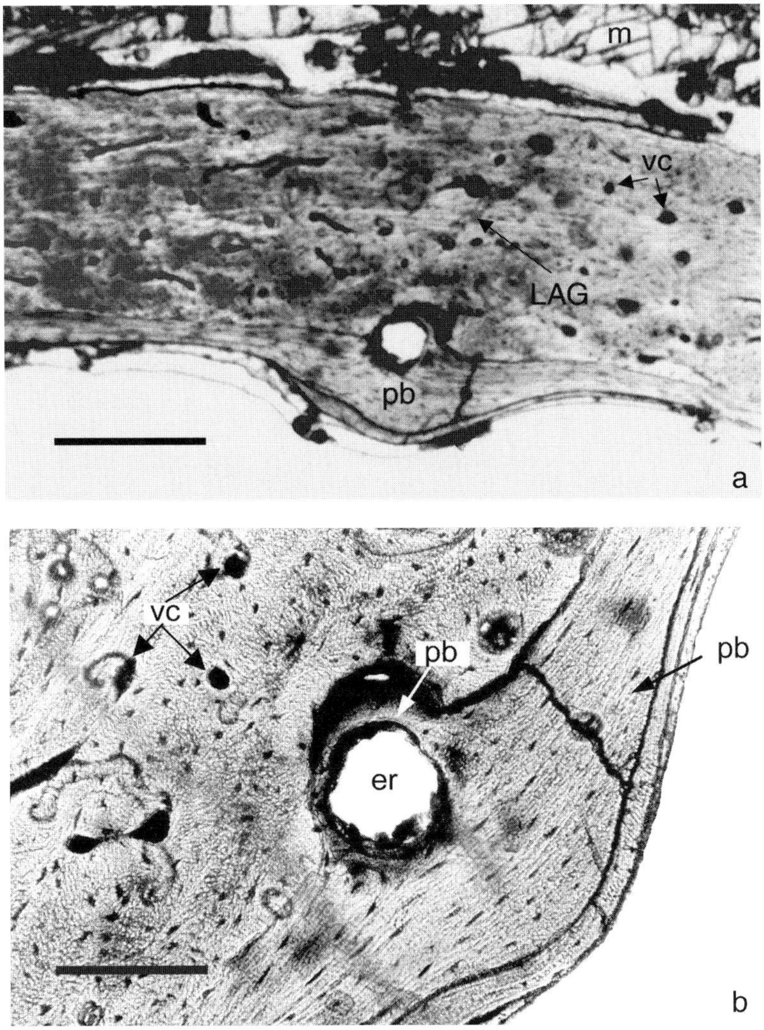

Fig. 4. Specimen MN 4809-V. (**a**) Dorsal region of the transverse section of the ulna, showing primary osteons and the location of the large erosion of the primary bone. (**b**) Detail of the erosion room, within the primary bone. pb, primary bone; LAG, line of arrested growth; vc, vascular canals; m, matrix; er, erosion room. Scale bar 500 μm in (**a**) and 300 μm in (**b**).

unique to this group, as proposed before (Ricqlès *et al.* 2000).

The present study also allowed the integration of gross anatomical and histological observations to determine the ontogenetic stage of pterosaur specimens. Authors have used the fusion and ossification of some bones as an indicator of ontogenetic maturity (Wellnhofer 1975; Bennett 1993; Kellner & Tomida 2000), which is followed here. In MN 4809-V several bones are fused, such as the carpals and the extensor tendon process of the first wing phalanx, which is consistent with the interpretation of this specimen as representing an adult individual. The histological sections have shown the presence of a thin coat of lamellar, poorly vascularized bone in the external cortex in all bones studied and secondary osteons in some elements (mcIV, ph1d4). These histological features were also reported in the adult pterosaur *Montanazhdarcho* (Padian *et al.* 1995). Therefore it can be concluded that MN 4809-V represents an adult individual.

The second specimen (MN 6527-V) shows unfused carpals (no ph1d4 is preserved), a feature typical of non-adult individuals. The histological sections lack secondary osteons and show fibrolamellar reticular bone. Based on these features it can be concluded that MN 6527-V represents a young animal that is still in the process of growing. It

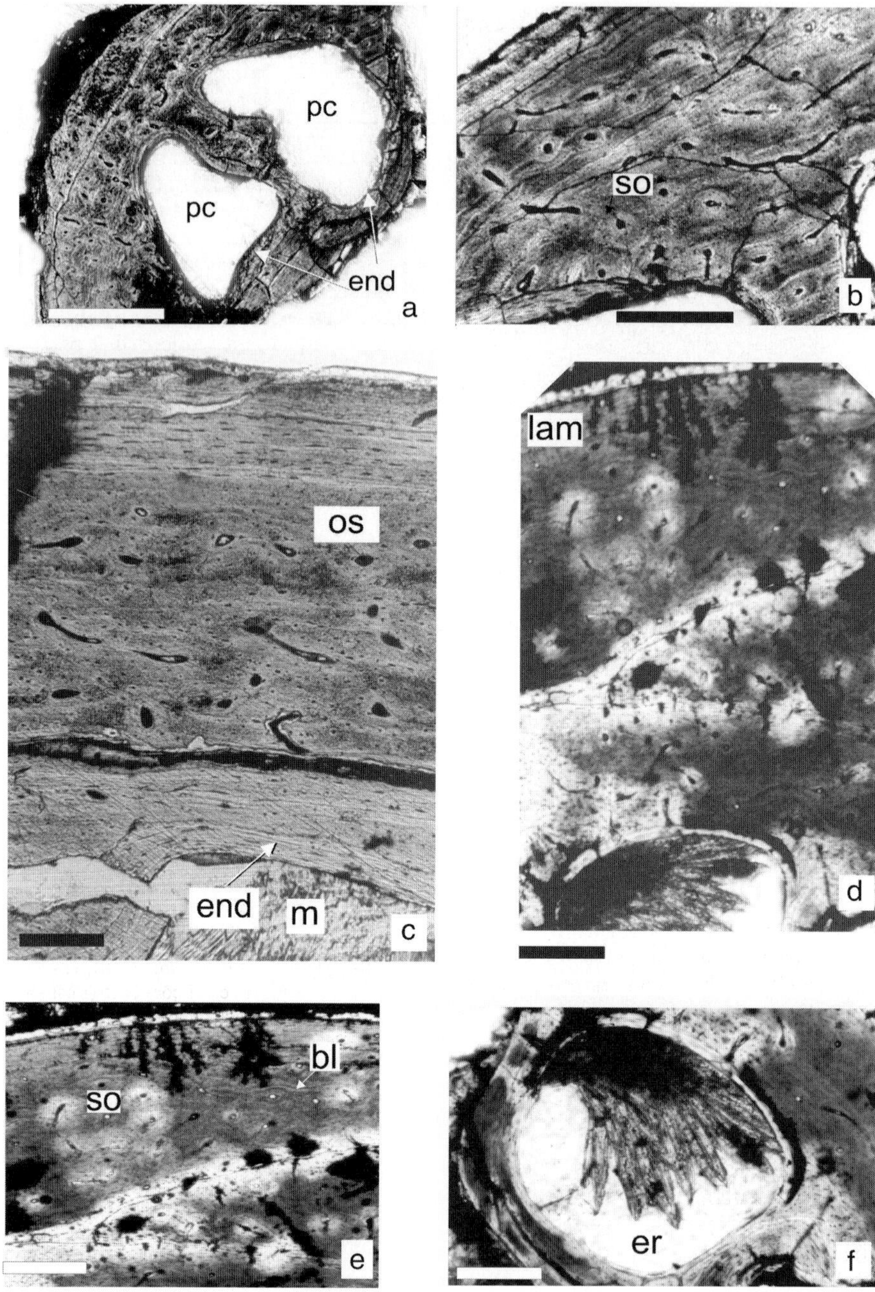

Fig. 5. Specimen MN 4809-V. (**a**) Dorsal part of the transverse section of the femur with two large pneumatic canals surrounded by the endosteal layer. (**b**) Detail of the anterior section, showing the secondary osteons of the femur, closer to the cavity. (**c**) Anterior part of the transverse section of the wing metacarpal; many primary osteons can be observed in this part of the section. Avascular lamellar periosteal bone is deposited in the outer region of the cortex. (**d**) Anterior view of the transverse section of the first wing phalanx. In the outer region of the cortex a thin layer of lamellar periosteal bone is deposited. In the layer below it, some secondary osteons are present, and a large erosion room is formed endosteally. (**e**) Detail of the secondary osteons of the first wing phalanx. (**f**) Detail of the large erosion room of the same bone; inside the cavity large idiomorphous calcite crystals are formed. (pc, pneumatic cavity; vc, vascular canals; m, matrix; er, erosion room; os, primary osteon; end, endosteal bone; lam, lamellar periosteal bone; so, secondary osteons; bl, bright line. Scale bar 200 μm (**a**), 400 μm (**b**) and 170 μm (**c–f**).

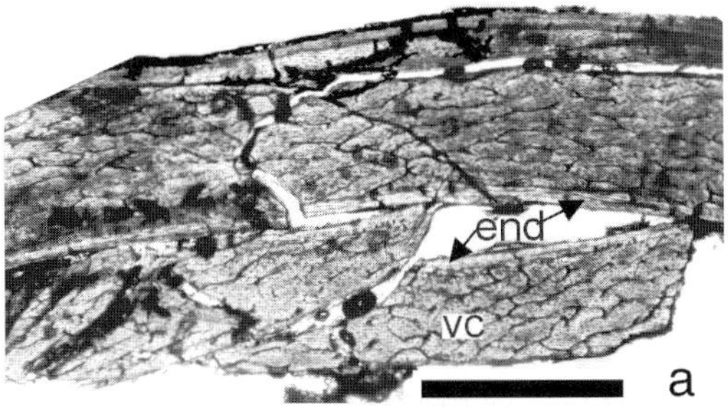

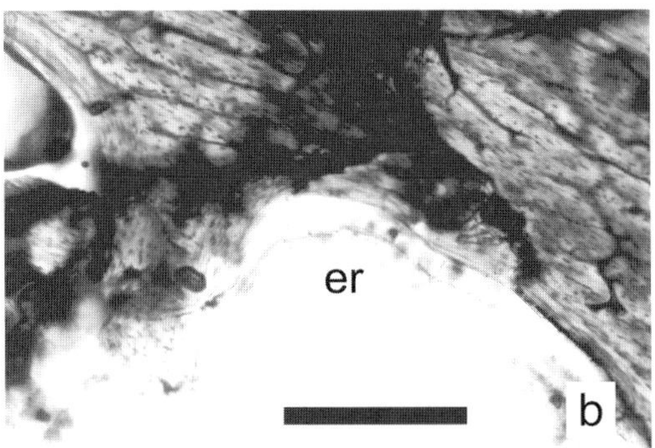

Fig. 6. Specimen MN 6527-V. (**a**) Part of the transverse section of the humerus composed of reticular fibro-lamellar tissue, indicative of rapid growth. Note the degree of compaction of this bone which presents parts of the section displaced within the matrix. (**b**) Part of the erosion room present in the humerus section. For abbreviations see Figure 5. Scale bar 200 μm (**a**) and 400 μm (**b**).

should be noted that the well-vascularized, reticular type of bone observed in this specimen is very similar to the condition observed in birds (Enlow & Brown 1957; Ricqlès 1978; Chinsamy *et al.* 1995) and is commonly associated with rapid growth.

Among the most interesting results obtained from this study is the histovariability observed in specimen MN 4809-V (from the Romualdo Member). This variation indicates that bones have differential growth rates. For example, ph1d4 already has secondary osteons and its external cortex is poorly vascularized, indicating that growth has ceased. Other bones, e.g. the ulna, show only primary osteons, despite the deposition of poorly vascularized tissue, suggesting that the growth of this bone was ceasing and that it had almost reached its maximum size. The presence of LAGs also varies and they were only observed in the mcIV, ulna and ph3d4. The absence of LAGs in most bones could be explained by the high resorption rates of pterosaur bones. Also, LAGs may have been obliterated by secondary bone. Contrary to that, in MN 6527-V, there is no histological variation at all. Representing a young animal, all observed histological features indicate that the animal was still growing and the resorption rates were very high. The comparisons between the histological features of both specimens corroborates the hypothesis that only the most recent histological components are recorded (Ricqlès *et al.* 2000).

Another interesting feature is the presence of pneumatic canals in the femur of MN 4809-V, reported here for the first time in pterosaurs. This indicates that the femur was pneumatized like many other bones of the pterosaur skeleton.

To conclude, the present study shows that histological features can vary among the bones of the same individual and suggests that pterosaur bones present different growth rates in the same individual.

The author thanks L. Leal and L. Carvalho (Museu Nacional/UFRJ, Rio de Janeiro) for their help with the preparations of the bone sections and M. S. de Oliveira (MN/UFRJ) for preparing Figures 1 and 2; H. de Paula Silva (MN/UFRJ) prepared specimen MN 4809-V and A. W. A. Kellner (MN/UFRJ) provided comments on an earlier version of the manuscript. M. Sander (Universität Bonn, Germany) and A. Chinsamy-Turan (University of Witwatersrand, South Africa) are thanked for their valuable suggestions, which greatly improved this paper. I would also like to express my gratitude to E. Buffetaut (CNRS, Paris) for the invitation to contribute to this volume and for support during the elaboration of this article. This project was partially funded by Conselho Nacional de Desenvolvimento científico e Tecnológico (CNPq – fellowship to J. Sayão) and Fundação Carlos Chagas Filho de Amparo a Pesquisa do Estado do Rio de Janeiro (FAPERJ – fellowship to A. Kellner and J. Sayão).

References

BENNETT, S. C. 1993. The ontogeny of *Pteranodon* and other pterosaurs. *Paleobiology*, **19**, 92–106.

BEURLEN, K. 1971. As condições ecológicas e faciológicas da Formação Santana na Chapada do Araripe (Nordeste do Brasil). *Anais da Academia Brasileira de Ciências*, **43**, 411–415.

CHINSAMY, A. & RAATH, M. A. 1992. Preparation of fossil bone for histological examination. *Palaeontologia Africana*, **29**, 39–44.

CHINSAMY, A. CHIAPPE, L. M. & DODSON, P. 1995. Mesozoic avian bone microstructure: physiological implications. *Paleobiology*, **21**, 561–574.

CORMACK, D. 1987. *Ham's Histology*. Lippincott,New York, 732 pp.

ENLOW, D. H. & BROWN, S. O. 1957. A comparative histological study of fossil and recent bone tissues. Part II. *Texas Journal of Sciences*, **9**, 185–214.

FREY, E. & MARTILL, D. 1994. A new pterosaur from the Crato Formation (Lower Cretaceous, Aptian) of Brazil. *Neues Jahrbuch für Geologie und Paläontologie, Abhandlungen*, **194**, 379–412.

GROSS, W. 1934. *Die* Typen des mikroskopien Knochenbaues bei fossilen Stegocephalen und Reptilien. *Zeitschrift für Anatomie*, **103**, 731–764.

HORNER, J. R., PADIAN, K. & RICQLÈS, A. J. DE 2001. Comparative osteohistology of some embryonic and perinatal archosaurs: developmental and behavioral implications for dinosaurs. *Paleobiology*, **27**, 39–58.

HORNER, J. R., RICQLÈS, A. J. DE & PADIAN, K. 1999. Variation in dinosaur skeletochronology indicators: implications for age assessment and physiology. *Paleobiology*, **25**, 295–304.

KELLNER, A. W. A. 1994. Remarks on pterosaur taphonomy and paleoecology. *Acta Geologica Leopoldensia*, **39**, 175–189.

KELLNER, A. W. A. 1996. *Description of new material of Tapejaridae and Anhangueridae (Pterosauria, Pterodactyloidea) and discussion of pterosaur phylogeny*. PhD thesis, Columbia University. [Published by University Microfilms International]

KELLNER, A. W. A. 1998. Panorama e perspectivas do estudo de répteis fósseis no Brasil. *Anais da Academia Brasileira de Ciências*, **70**, 647–676.

KELLNER, A. W. A. & TOMIDA, Y. 2000. *Description of a New Species of Anhangueridae (Pterodactyloidea) with Comments on the Pterosaur Fauna from Santana Formation (Aptian–Albian), Northeastern Brazil.* National Science Museum Monographs, **17**, 135 pp.

MARTINS NETO, R. G. 1991. Sistemática dos Ensifera (Insecta, Orthopteroidea) da Formação Satana, Cretáceo inferior do nordeste do Brasil. *Acta Geologica Leopoldensia*, **32**, 3–162.

MAISEY, J. G. 1991. *Santana Fossils: An Illustrated Atlas*. T.F.H. Publications Inc, Neptune, New Jersey, 459 pp.

MEYER, H. V. 1832. Palaeologica zur Geschichte der Erde. S. Schmerber, Frankfurt.

OWEN, R. 1870. Monograph on the fossil reptilia of Liassic formations. Palaentographical Society, London, Monographs, **27**, 41–86.

PADIAN, K. 1985. The origins and aerodynamics of flight in extinct vertebrates. *Palaeontology*, **28**, 413–433.

PADIAN, K., RICQLÈS, A. J. DE & HORNER, J. R. 1995. Bone histology determinates identification of a new fossil taxon of pterosaur (Reptilia: Archosauria). *Comptes Rendus de l'Academie des Sciences, Paris, Serie IIa*, **320**, 77–84.

PONS, D. BERTHOU, P. Y. & CAMPOS, D. A. 1990. Quelques observations sur la palynologie del' Aptien supérieur et l' Albien du bassin d'Araripe (N.E. du Brésil). *In*: CAMPOS, D. A., VIANA, M. S. S., BRITO, P. M. & BEURLEN, G. (eds) *Atas do 1º Simpósio sobre a Bacia do Araripe e Bacias Interiores do Nordeste*, 241–252.

QUEKETT, J. T. 1849. Additional observations on the intimate structure of bone. *Transactions of the Microscopical Society, London*, **2**, 59–64.

QUEKETT, J. T., 1855. *Descriptive and Illustrated Catalogue of the Histological Series Contained in the Museum of the Royal College of Surgeons of England*, vol. 2. London, UK.

RICQLÈS, A. J. DE 1975. Recherches paléohistologiques sur les os longs des Tétrapodes, VII, sur la classification fonctionnelle et l'histoire des tissue osseux des Térapodes: Première partie, Structures. *Annales de Paléontologie (Vertebrés)*, **61**, 49–129.

RICQLÈS, A. J. DE 1978. Recherches paléohistologiques sur les os longs des Tétrapodes, VII, sur la classification fonctionelle et l'histoire des tissue osseux des Tétrapodes: Troisième partie, fin. *Annales de Paléontologie (Vertebrés)*, **64**, 153–176.

RICQLÈS, A. J., DE PADIAN, K., HORNER, J. R. & FRANCILLON-VIEILLOT, H., 2000. Paleohistology of the bones of pterosaurs (Reptilia: Archosauria): anatomy, ontogeny, and biomechanical implications. *Zoological Journal of the Linnean Society, London*, **129** (3), 349–385.

SAYÃO, J. M. 2001. Comments on pterosaur bone histology. *Strata, Serie I*, **11**, 89.

SAYÃO, J. M. & KELLNER, A. W. A. 2000. Description of a pterosaur rostrum from the Crato member, Santana Formation (Aptian–Albian) northeastern Brazil. *Boletim do Museu Nacional, Rio de Janeiro, Nova Série, Geologia*, **54**, 1–8.

SAYÃO, J. M. & KELLNER, A. W. A. 2001. Comments on the pterosaur fauna from Tendaguru, Upper Jurassic of Africa, with the identification of a possible azhdarchid. *XVIII Congresso Brasileiro de Paleontologia, Universidade Federal do Acre*, 145.

SAYÃO, J. M., CARVALHO, L. B. & KELLNER, A. W. A. 2000. Preliminary histological analysis of a pterosaur (Pterodactyloidea, Anhangueridae) from the Araripe Basin (Aptian–Albian), Brazil. *Journal of Vertebrate Paleontology*, **20** (3), 67A.

SEELEY, H. G. 1870. Remarks on Prof. Owen's monograph on *Dimorphodon. Annals and Magazine of Natural History*, **4** (6), 129–152.

SEITZ, A. L. 1907. Vergleichende Studien uber den mikroskopischen Knochenbau fossiler und rezenter Reptilien. *Nova Acta Leopoldina*, **37**, 230–370.

STEEL, L. 2001. Pterosaur bone histology: A re-interpretation of some of the earliest histological sections. *Strata, Serie 1*, **11**, 90–91.

UNWIN, D. M., FREY, E. MARTILL, D. M., CLARKE, J. B. & RIESS, J. 1996. On the nature of the pteroid in pterosaurs. *Proceedings of the Royal Society, London (B)*, **263**, 45–52.

WELLNHOFER, P. 1975. Die Rhamphorhynchoidea (Pterosauria) der Oberjura-plattenkalke Süddeutschlands. *Palaeontographica A*, **148** (1–3), 1–33.

WELLNHOFER, P. 1991. *The Illustrated Encyclopedia of Pterosaurs*. Salamander, London, 192 pp.

Index

Page numbers in italic, e.g. *153*, refer to figures. Page numbers in **bold**, e.g. **321**, signify entries in tables.